2nd International Conference on Data Mining, Communications and Information Technology (DMCIT 2018)

Journal of Physics: Conference Series
Volume 1060

Shanghai, China
25 – 27 May 2018

ISBN: 978-1-5108-6827-4
ISSN: 1742-6588

Printed from e-media with permission by:

Curran Associates, Inc.
57 Morehouse Lane
Red Hook, NY 12571

Some format issues inherent in the e-media version may also appear in this print version.

Printed by Curran Associates, Inc. (2018)

For permission requests, please contact the Institute of Physics
at the address below.

Institute of Physics
Dirac House, Temple Back
Bristol BS1 6BE UK

Phone: 44 1 17 929 7481
Fax: 44 1 17 920 0979

techtracking@iop.org

Additional copies of this publication are available from:

Curran Associates, Inc.
57 Morehouse Lane
Red Hook, NY 12571 USA
Phone: 845-758-0400
Fax: 845-758-2633
Email: curran@proceedings.com
Web: www.proceedings.com

Table of contents

Volume 1060

2018 2nd International Conference on Data Mining, Communications and Information Technology (DMCIT 2018)

25–27 May 2018, Shanghai, China

Accepted papers received: 6 July 2018
Published online: 23 July 2018

Preface

Papers

Data Mininng

Communication Engineering

Simulation, Modeling and Optimization

IOP Conf. Series: Journal of Physics: Conf. Series **1060** (2018) 011001 doi:10.1088/1742-6596/1060/1/011001

PREFACE

DMCIT 2018 was held in Shanghai, China during May.25-27, 2018. DMCIT 2018 was organized by Asia Pacific Institute of Science and Engineering (APISE). The conference provides a useful and wide platform both for display the latest research and for exchange of research results and thoughts in enviromental and energy engineering. The participants of the conference were from almost every part of the world, with background of either academia or industry, even well-known enterprise. The success and prosperity of the conference is reflected high level of the papers received.

The proceedings are a compilation of the accepted papers and represent an interesting outcome of the conference. This book covers 3 chapters: Data Mining, Communication Engineering and Simulation, Modeling and Optimization.

We would like to acknowledge all of those who supported DMCIT 2018. Each individual and institutional help were very important for the success of this conference. Especially we would like to thank the organizing committee for their valuable advices in the organization and helpful peer review of the papers.

We sincerely hope that DMCIT 2018 will be a forum for excellent discussions that will put forward new ideas and promote collaborative researches. We are sure that the proceedings will serve as an important research source of references and the knowledge, which will lead to not only scientific and engineering progress but also other new products and processes.

Anders Lindquist, Shanghai Jiao Tong University, China
Norman C. Beaulieu Beijing University of Posts and Telecommunications, China

IOP Conf. Series: Journal of Physics: Conf. Series **1060** (2018) 011002 doi:10.1088/1742-6596/1060/1/011002

COMMITTEES

Conference Chair

Anders Lindquist, Shanghai Jiao Tong University, China

Norman C. Beaulieu, Beijing University of Posts and Telecommunications, China

Program Committee Chair

Arbee L.P. Chen Vice President at Asia University, Taiwan

Publicity Chair

Wei Jia, University of Chinese Academy of Science, China

Publication Chair

Rung-Ching Chen, Chaoyang University of Technology, Taiwan

International Technical Committee

J. Amudhavel, Department of Computer Science & Engineering, KL University, India

Dong HUANG, South China Agricultural University, China

M. A. JABBAR, Vardhaman college of Engineering Hyderabad, Telangana, India

Hongchun Qu, Chongqing University of Posts and Telecommunications, China

RajaniKanth Aluvalu, Vardhaman college of Engineering Hyderabad, Telangana, India

Ka-Chun Wong, City University of Hong Kong, Hong kong

Mustansar Ali Ghazanfar, University of Engineering and Technology, Taxila, Pakistan

Guan-Ling Lee, National Dong Hwa University, Taiwan

M. Awais Azam, University of Engineering and Technology, Taxila, Pakistan

Peer review statement

All papers published in this volume of *Journal of Physics: Conference Series* have been peer reviewed through processes administered by the proceedings Editors. Reviews were conducted by expert referees to the professional and scientific standards expected of a proceedings journal published by IOP Publishing.

Deep Matrix Factorization for Recommender Systems with Missing Data not at Random

Fei Zhang[1a], **Jiaxing Song**[1b] **and Shiyu Peng**[1c]

[1]Department of Computer Science and Technology, Tsinghua University, Beijing, China

[a]zhangfei15@mails.tsinghua.edu.cn, [b]jxsong@tsinghua.edu.cn, [c]psy16@mails.tsinghua.edu.cn

Abstract Matrix Factorization is a widely used collaborative filtering method in recommender systems. However, most of them are under the assumption that the rating data is *missing at random* (MAR), which may not be very common. For some users, they may only rate those movies they like, so the inferences will be biased in previous models. In this paper, we proposed a deep matrix factorization method based on *missing not at random* (MNAR) assumption. As far as we know, this model firstly uses deep learning method to address MNAR issue. The model consists of a complete data model (CDM) and a missing data model (MDM), which are both learned by neural networks. The CDM is nonlinearly determined by two factors, the user latent features and item latent features like other matrix factorization methods. And the MDM also use these two factors but taking the rating value as extra information while training. We used variational Bayesian inference to generate the posterior distribution of our proposed model. Through extensive experiments on different kind of datasets, our proposed model produce gains in some widely used metrics, comparing with several state-of-the-art models. We also explore the performance of our model within different experimental settings.

1. Introduction

In the era of information explosion, information overload makes it difficult for users to obtain their interested items and for service providers to process accurate item delivery. To overcome this problem, recommender systems are widely used, since they can help service providers to offer pertinent items to individual consumers. Nowadays, recommender systems are fundamental in various commerce sites, such as recommendation of movies at Netflix, commodities at Amazon and photos at Instagram.

Collaborative filtering (CF) recommender systems are the most representative and effective approaches in industry. Among various CF techniques, matrix factorization (MF) is the most popular one since Netflix Prize [1] [2]. MF aims at decomposing a user-item matrix into a low-dimensional space, and predicting unknown value by using the inner product of latent vectors. MF has been enhanced by many researchers in early works [3] [4]. And some methods are focus on modeling implicit feedback [3] [5] [6], since users can't explicitly rate items in many scenarios. Recently, deep learning methods have been shown to improve performances of recommender systems [7]-[9]. MF methods are also adapted to take advantage of learning abilities of neural networks. The simplest way is concatenating latent vectors and their inner product as the input layer [10]. And neural collaborative filtering (NCF) use a multi-layer feedforward network to model user-item interactions [11].

Most of these MF methods are under the assumption that user-item interaction matrix is MAR [12]. That is, the process of selecting observed data is independent of the value of unobserved data. When data is MAR, the model trained by observed data can be applied to the whole data set. However, this assumption may be incorrect while considering following scenarios in a movie rating application: i) user may only watch movies that they like, and may only rate movies that they have watched; ii) user may rate movies with extreme bias. In these scenarios, the dependence between the observation process and the missing data can't be ignored, and consequently the data is not MAR. An online survey organized by Yahoo Music also shows that the distribution of ratings for random songs was quite different with the distribution of voluntary ratings [13]. When the MAR assumption fails, the prediction accuracy of those inferences will be suffered. To address the lack of robustness of MAR, several adaptive traditional CF methods based on MNAR were proposed. The most common way to deal with MNAR data is to learn jointly a complete data model (CDM) and a missing data model (MDM) [12]. Some early works based on MNAR used traditional CF methods to train CDM and MDM [13]-[15]. However, these models are lacking of flexibility and they are out of date.

In this paper, we originally use deep learning techniques to train the CDM and MDM based on MNAR assumption. Firstly, we construct two deep matrix factorization models to obtain CDM and MDM with user-item rating matrix as input. By combining the MDM with the CDM, we build a joint model for prediction. A new loss function was also proposed by considering the accuracy of CDM and MDM at the same time. In summary, the contributions of our work are outlined as follows:

1) We employ novel deep matrix factorization to model MNAR data, which is premier to address MNAR issue by deep learning.
2) We design a new neural network to combine the CDM model and MDM model with elaborate inference. And a new loss function is applied while training the network.
3) We perform extensive experiments on open datasets, which indicates our model produce gains in various evaluation metrics.

2. Related Work

In this section, we first review the theory of missing data and discuss the common mechanisms under different assumption. Then, we will describe how to process deep matrix factorization by using Multilayer Perceptron (MLP).

2.1. *Principles in Missing Data Model*

The main principles to handle missing data are developed by Litter & Rubin [12]. We assume that our input data is the rating matrix $\mathbf{R} \in \mathbb{R}^{M \times N}$. Each element r_{ui}, where $u = 1,\ldots,M$ and $i = 1,\ldots,N$, represents the rating value of user u on item i. We can divide the matrix into two parts, $\mathbf{R} = \{\mathbf{R}^+, \mathbf{R}^-\}$, where $\mathbf{R}^+$ and $\mathbf{R}^-$ denote the set of *observed* and *missing* entries in $\mathbf{R}$, respectively. And we also introduce a companion matrix $\mathbf{X}$, where x_{ui} in $\mathbf{X}$ is a Bernoulli random variable. The value of x_{ui} indicates whether r_{ui} is observed ($x_{ui} = 1$) or not ($x_{ui} = 0$). To obtain the matrix $\mathbf{R}$ and $\mathbf{X}$, we assume that $\mathbf{R}$ is generated by the CDM with parameters Ψ and $\mathbf{X}$ is generated by the MDM with parameters Θ. And both models may share a set of latent variables $\mathbf{Z}$. As such, the joint distribution for $\mathbf{R}$, $\mathbf{X}$ and $\mathbf{Z}$ determined by Ψ and Θ is

$$p(\mathbf{X},\mathbf{R},\mathbf{Z} \mid \Psi,\Theta) = p(\mathbf{X} \mid \mathbf{R},\mathbf{Z},\Theta)\, p(\mathbf{R},\mathbf{Z} \mid \Psi) \quad (1)$$

There are three types of methods to process missing data. The most restrictive assumption is *completely missing at random* (CMAR), which refers the probability of observing a rating is independent of any ratings and latent variables, that is, $p(\mathbf{X} \mid \mathbf{R},\mathbf{Z},\Theta) = p(\mathbf{X} \mid \Theta)$. The second assumption is MAR, which refers that the observation probability depends only upon the observed data and Θ, that is, $p(\mathbf{X} \mid \mathbf{R},\mathbf{Z},\Theta) = p(\mathbf{X} \mid \mathbf{R}^+,\Theta)$. In this sense, the MDM can be ignored when learning Θ, since $p(\mathbf{X} \mid \mathbf{R}^+,\Theta)$ is constant with respect to Θ [13]. However, the inferences will be biased if the MAR assumption is used when data is MNAR. To correct the observational bias, we must jointly learn CDM and MDM and predict unknown values by combining them together.

2.2. Deep Matrix Factorization

Matrix Factorization is a popular CF method which associates each user and item with a real-valued latent vector representing features. Formally, we let $\mathbf{p}_u$ denotes the latent vector of user u, and $.\mathbf{q}_i.$ denotes the latent vector of item i. Then we collect all the $\mathbf{p}_u$ to matrix $\mathbf{P} \in ¡^{M\times K}$, which denotes the user latent matrix. Similarly, $\mathbf{Q} \in ¡^{N\times K}$ is the item latent matrix. K is the dimension of the latent space, and usually K is far less than M and N. In the traditional MF methods, the dot production of $\mathbf{p}_u$ and $\mathbf{q}_i$ is simply considered as the prediction of $\hat{r}_{ui}$, as shown by the following equation,

$$f(u,i \mid \mathbf{p}_u, \mathbf{q}_i) = \mathbf{p}_u^T \mathbf{q}_q = \sum_{k=1}^{K} p_{uk} q_{ik} \tag{2}$$

The inner production linearly combines each dimension of latent vector with same weight, which may limit the expression of MF [11]. To resolve the limitation, we can use MLP to capture the non-liner relationship between latent vectors.

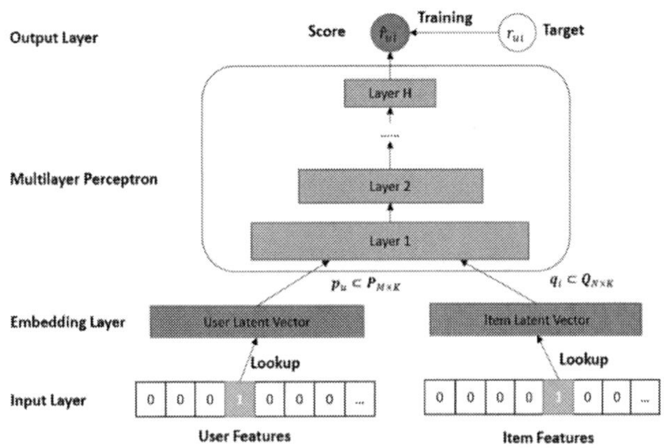

Figure 1. Framework of Deep Matrix Factorization

Figure 1 illustrates the general framework of a MF methods using MLP. The bottom layer takes the features of user and item as input. These features can be generated by various models, such as content-based [16], neighbor-based and social-based. Since MF method is a CF method on rating matrix, it only takes the identities of user and item as input. Following the input layer, the embedding layer converts the spare input features into dense latent vectors, same as the vector $\mathbf{p}_u$ and $\mathbf{q}_i$ mentioned above. After that, the latent vectors are fed into H hidden layers, where each layer will capture the latent knowledge between user-item interactions. At last, the output layer will generate the predicted value $\hat{r}_{ui}$. To train the neural network, a loss function should be defined between predicted value $\hat{r}_{ui}$ and target value r_{ij}, usually a pointwise loss function will be applied as following equation,

$$L = \sum_{u,i} (r_{ui} - \hat{r}_{ui})^2 + \lambda \left(\sum_{u} \| \mathbf{p}_u \|^2 + \sum_{i} \| \mathbf{q}_i \|^2\right), \tag{3}$$

where λ is a regularization hyperparameter. Several model extend this general framework, and train model on observed explicit or implicit feedback data. In our work, we adopt this general framework to train the CDM and MDM respectively.

3. DMF-MNAR

In this section, a probabilistic model based on DMF will be firstly introduced as our CDM. Then we describe MDM, which will generate the $\mathbf{X}$ based on $\mathbf{R}$. Finally, we will construct a joint model combining two models together.

3.1. The Complete Data Model

To build a complete data model with MF method, we assume that $\mathbf{R}$ is generated by two low dimension matrix $\mathbf{P}$ and $\mathbf{Q}$. And each discrete rating r_{ui} is determined by the output of a MLP $y_{ui} = f_{cdm}(\mathbf{p}_u, \mathbf{q}_i)$, where $\mathbf{p}_u$ and $\mathbf{q}_i$ are served as input. More precisely, the interaction in MLP can be defined as the following equations,

$$
\begin{aligned}
f_{cdm}(\mathbf{p}_u, \mathbf{q}_i) &= \phi_{out}(\phi_H(\phi_{H-1}(\ldots(\phi_1(\mathbf{p}_u, \mathbf{q}_i))))), \\
\mathbf{t}_1 = \phi_1(\mathbf{p}_u, \mathbf{q}_i) &= \mathbf{W}_1 \begin{bmatrix} \mathbf{p}_u \\ \mathbf{q}_i \end{bmatrix} + \mathbf{b}_1, \\
\phi_h(\mathbf{t}_{h-1}) &= f_a(\mathbf{W}_h \mathbf{t}_{h-1} + \mathbf{b}_i), h = 2,3,\ldots,H, \\
\phi_{out} &= \sigma(\phi_H(\mathbf{t}_{H-1})),
\end{aligned} \tag{4}
$$

where $\mathbf{W}_h$, $\mathbf{b}_h$ and $\mathbf{t}_h$ denote the weight matrix, bias vector and output of the corresponding h-th layer of MLP, respectively. For the activation function f_a, it can be chosen among some basic functions, such as sigmod, hyperbolic tangent (tanh) and Rectifier (ReLU). In this work, we opt ReLU as the activation function because it is more biologically plausible and proved to be non-saturated [17]; moreover, it is more suitable for spare data. To avoid the predictions outside of valid ratings, we can use logistic function $\sigma(x) = 1/(1+exp(-x))$ to bound the output into $(0,1)$ .At same time, we map the rantings 1,2,…,V to the same interval by using the function $g(x) = (x-1)/(V-1)$, where V is the maximum rating value.

When y_{ui} is given, to generate the r_{ui}, we assume that the ratings are distributed under a Gaussian distribution with y_{ui} as mean value. And we also place zero-mean fully-factorized Gaussian priors $\mathbf{P}$ and $\mathbf{Q}$. So that, the distributions of $\mathbf{R}$, $\mathbf{P}$ and $\mathbf{Q}$ are

$$p(\mathbf{R} \mid \mathbf{P}, \mathbf{Q}, \sigma_0^2) = \prod_{u=1}^{M} \prod_{i=1}^{N} \mathrm{N}(r_{ui} \mid y_{ui}, \sigma_0^2) \tag{5}$$

$$p(\mathbf{P} \mid \sigma_P^2) = \prod_{u=1}^{M} \mathrm{N}(\mathbf{p}_u \mid 0, \sigma_P^2), \qquad p(\mathbf{Q} \mid \sigma_Q^2) = \prod_{i=1}^{N} \mathrm{N}(\mathbf{q}_i \mid 0, \sigma_Q^2), \tag{6}$$

where the σ_0^2, σ_P^2 and σ_Q^2 are the variances of these distributions. Let Ψ be the set of parameters , so that the joint distribution for $\mathbf{R}$ and Ψ is

$$p(\mathbf{R}, \Psi) = p(\mathbf{R} \mid \mathbf{P}, \mathbf{Q}) p(\mathbf{P}) p(\mathbf{Q}), \tag{7}$$

which denotes the CDM. Now we will describe the MDM that can explain which entry in $\mathbf{R}$ will be missing, that is specified by the matrix $\mathbf{X}$.

3.2. The Missing Data Model

We also obtain the MDM by using a DMF method, but it has some differences with the CDM. Since the MDM generates $\mathbf{X}$ as a function of $\mathbf{R}$, the rating values also determine whether the data can be observed. As such, we assume that $\mathbf{X}$ is obtained by three factors $\mathbf{E} \in \mathbb{R}^{M \times K}$, $\mathbf{F} \in \mathbb{R}^{N \times K}$ and $\mathbf{G} \in \mathbb{R}^{V \times K}$, each of them is relative to the latent matrix of users, items and rating values. In particularly, the v-th row of G represents the latent vector of rating value v. Similarly, each discrete entry x_{ui} in $\mathbf{X}$ is determined by the output of another MLP network $z_{uiv} = f_{mdm}(\mathbf{e}_u, \mathbf{f}_i, \mathbf{g}_v)$, where $\mathbf{e}_u$, $\mathbf{f}_i$ and $\mathbf{g}_v$ are the inputs of the networks. The interactions in MDM network are similar with equation (4), except the first layer is modified by

$$\mathbf{s}_1 = \pi_1(\mathbf{e}_u, \mathbf{f}_i, \mathbf{g}_v) = \mathbf{V}_1 \begin{bmatrix} \mathbf{e}_u \\ \mathbf{f}_i \\ \mathbf{g}_v \end{bmatrix} + \mathbf{d}_1, \tag{8}$$

where we use π_h, $\mathbf{s}_h$, $\mathbf{V}_h$ and $\mathbf{d}_h$ to represent the transfer function, output, weight matrix and bias vector of each layer, respectively.

We also use a sigmod function at last layer to bound the output into (0,1). Since x_{ui} is a binary value, we can directly use z_{uiv} to represent the probability of $x_{ui}=1$ when $r_{ui}=v$, that is, $p(x_{ui}=1\,|\,r_{ui}=v)=z_{uiv}$. And we also place zero-mean Gaussian priors on $\mathbf{E}$, $\mathbf{F}$ and $\mathbf{G}$. So that, the distributions of $\mathbf{X}$ is,

$$p(\mathbf{X}\,|\,\mathbf{E},\mathbf{F},\mathbf{G},\mathbf{R})=\prod_{u=1}^{M}\prod_{i=1}^{N} z_{uiv}^{x_{ui}}(1-z_{uiv})^{(1-x_{ui})}, \tag{9}$$

As the same way, we let the Θ be the set of parameters, $\Theta=\{\mathbf{E},\mathbf{F},\mathbf{G}\}$. So our MDM can be described by the following equation,

$$p(\mathbf{X},\Theta\,|\,\mathbf{R})=p(\mathbf{X}\,|\,\mathbf{E},\mathbf{F},\mathbf{G},\mathbf{R})p(\mathbf{E})p(\mathbf{F})p(\mathbf{G}). \tag{10}$$

3.3. *The Joint Model*

As we have the CDM and MDM, we can combine two model together to obtain the joint model *Deep Matrix Factorization with Missing data Not At Random* (DMF-MNAR). This model generates the probability that each entry r_{ui} takes a specific rating value v conditioned by the indicator x_{ui}. When x_{ui} is closest to 1, it means that the entry can be observed standing a good chance. On the contrary, if x_{ui} is closest to 0, we can make a discount on the rating values. For instance, the probability of an observed entry r_{ui} taking values of v is,

$$p^{DMF-MNAR}(r_{ui}=v,x_{ui}=1)=p^{CDM}(r_{ui})p^{MDM}(x_{ui}=1\,|\,r_{ui}=v), \tag{11}$$

where $p^{CDM}(r_{ui})$ and $p^{MDM}(x_{ui}=1\,|\,r_{ui}=v)$ are the individual predictions of the CDM and MDM. In this way, the joint model improves the predictions of the CDM by using the information available in MDM.

4. Training Method of DMF-MNAR

As mentioned in section 2.2, a DMF method can be trained by a simple MLP network. However, our proposed model is quite different from the simple DMF method. In the first place, we can't simply apply equation (3) as our loss function, because the CDM and MDM are probabilistic models. Moreover, our model DMF-MNAR contains two individual DMF networks to train CDM and MDM respectively. So in this section, we firstly introduce an approximation method considering the accuracy of CDM and MDM at the same time. And we proposed a pairwise neural networks for DMF-MNAR.

4.1. *Approximate Interface*

We assume that the observed data is $\mathbf{R}^+$ and indicator matrix $\mathbf{X}$ are given, so the posterior distribution of the parameters Ψ, Θ and missing data $\mathbf{R}^-$ is

$$p(\Psi,\Theta,\mathbf{R}^-\,|\,\mathbf{R}^+,\mathbf{X})=\frac{p(\mathbf{X},\Theta\,|\,\mathbf{R})p(\mathbf{R},\Psi)}{p(\mathbf{R}^+,\mathbf{X})}, \tag{12}$$

where $p(\mathbf{R}^+,\mathbf{X})$ is a normalization constant. As most Bayesian models, the posterior is intractable so that we use a variational Bayesian (VB) inference to approximate [18].

Since the $\mathbf{R}$ and $\mathbf{X}$ are generated by some latent matrixes $\Psi=\{\mathbf{P},\mathbf{Q}\},\Theta=\{\mathbf{E},\mathbf{F},\mathbf{G}\}$, we assume that $q(\Psi)=\{q(\mathbf{P}),q(\mathbf{Q})\},q(\Theta)=\{q(\mathbf{E}),q(\mathbf{F}),q(\mathbf{G})\}$ are the variational approximations to the true posteriors of these latent matrixes respectively. As we use a stochastic version of VB called SVI [15], so that our model maximizes the ELBO as following equation,

$$\begin{aligned} &\mathrm{L} = \mathrm{L}_{CDM} + \mathrm{L}_{MDM}, \\ &\mathrm{L}_{CDM} = \mathrm{E}_{q(\Psi)}[log \quad p(\mathbf{R} \mid \Psi)] - KL[q(\Psi) \parallel p(\Psi)], \\ &\mathrm{L}_{MDM} = \mathrm{E}_{q(\Theta)}[log \quad p(\mathbf{X} \mid \Theta, \mathbf{R})] - KL[q(\Theta) \parallel p(\Theta)], \end{aligned} \tag{13}$$

where the L_{CDM} denotes the accuracy of CDM and L_{MDM} is for MDM. If we approximate the log likelihood with a Monte Carlo method meaning for each entry we use a single sample from posterior to estimate in every iteration, we can further decompose the equation. Because the probabilistic distribution of CDM and MDM are quite different, the expressions of ELBO are not same.

For the CDM, the ratings and latent vectors are distributed under Gaussian distributions. So that the L_{CDM} can be rewritten as,

$$\mathrm{L}_{CDM} = -\frac{1}{2\sigma_0^2}\sum_{u=1}^{M}\sum_{i=1}^{N}(r_{ui} - y_{ui})^2 - KL(q(\mathbf{P}) \parallel p(\mathbf{P})) - KL(q(\mathbf{Q} \parallel p(\mathbf{Q})) \tag{14}$$

And for the MDM, the indicator x_{ui} is under a Bernoulli distribution. So that the L_{MDM} can be rewritten as,

$$\begin{aligned} \mathrm{L}_{MDM} = &\sum_{u=1}^{M}\sum_{i=1}^{N} x_{ui} z_{uiv} + (1 - x_{ui})(1 - z_{uiv}) \\ &-KL(q(\mathbf{E} \parallel p(\mathbf{E}))) - KL(q(\mathbf{F} \parallel p(\mathbf{F}))) - KL(q(\mathbf{G} \parallel p(\mathbf{G}))) \end{aligned} \tag{15}$$

4.2. Neural Networks for DMF-MNAR

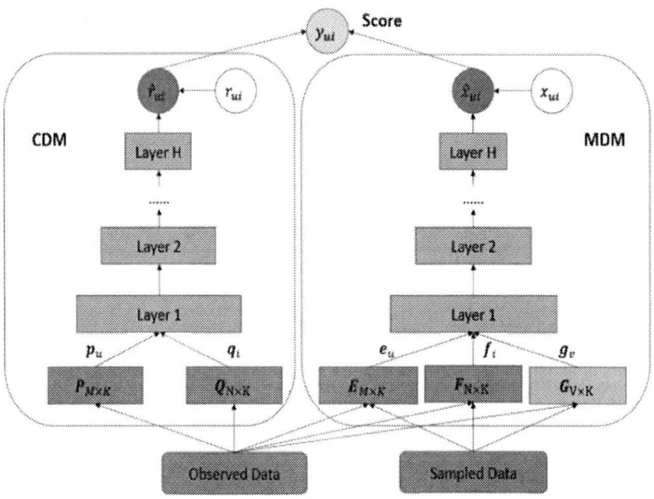

Figure2. The framework of DMF-MNAR

One import aspect of deep learning is how to organize the layers in neural networks. Figure 2 illustrates the framework of our proposed model DMF-MNAR, which has two individual parts representing the CDM and MDM, respectively.As shown in the figure, the CDM and MDM both take rating matrix as the input. However, the formats of input data have small difference between two models. The CDM only take two factors represented by user and item, so that we can use a set of (u,i,r_{ui}) as the train data. But in the MDM, we also use the information contained by rating values, so the format of input data is (u,i,r_{ui},x_{ui}). The value of x_{ui} indicates whether the entry can be observed. But in most actual situations, the data we collected are all observed, which makes training MDM intractable if we only have positive samples. So we sample items randomly from the list that user never rated before as the negative samples, where we set the rating value $r_{ui} = 1$ and indicator $x_{ui} = 0$. This is a widely used method for implicit feedback, but we think it's also suitable for explicit feedback.

Because the user's preference inferred from the discrepancy of rating values is not sufficient, according our MNAR assumption. So the negative samples will help in training the MDM, but they are redundant to the CDM.

To avoid the influences of negative samples on the CDM, we also adjust the embedding layer to suit our model. There are five latent vectors in the embedding layer, two of them belong to the CDM and the others belong to the MDM. A straightforward way is to let CDM and MDM share the same user and item latent vectors. But in that way, the negative samples will bias the predicted values. So we allow CDM and MDM to learn separate embedding layers from different input data. What's more, this strategy can improve the flexibility of our proposed model.

Since two models are generated from different input data, it's also hard to maximize the objective function in equation (14) at the same time. So we trained the two models alternately by their individual object functions represented in equation (15) and (16). We adopt the Adaptive Moment Estimation (Adam) in our networks. For each iteration, we firstly randomly select a small batch D^+ from observed data, and we train the CDM on D^+. After that we generated another batch D^- containing the negative samples for each user in D^+, and we trained the MDM on $D^+ \cup D^-$.

5. Experiments

In this section, we conduct experiments to demonstrate the performance achieved by our proposed model on various open datasets.

5.1. Experimental Settings

Datasets. We conduct experiments on three publicly accessible datasets: *Yahoo! Music ratings for User Selected and Randomly Selected songs, version 1.0, MovieLens 100k* and *Amazon Music*.

1) **Yahoo! Music** This dataset provides a unique opportunity to test collaborative filtering methods that incorporate missing data model [13]. This dataset consists of ratings collected during normal user interaction with Yahoo's LaunchCast internet radio service, as well as ratings for items collected using an online survey. There are 15400 users, 1000 songs, and 311704 ratings in observed dataset. And for 5400 of these users, exactly 10 songs are randomly selected and presented to each user during the online survey. The user can listen the songs repeatedly and are enforced to rate on them. As that the dataset for randomly selected items contains 54000 ratings. The distribution of ratings for two parts are significantly different because normal interactions have more high ratings than online survey ratings. This is a strong evidence that Yahoo! dataset is under MNAR condition.
2) **MovieLens** This movie rating dataset has been widely used to evaluate collaborative filtering methods. We used the version containing one million ratings, where there are 6040 users, 3706 items and at least 20 ratings for each user. Although this dataset only consists of observed ratings, we randomly sample user-item interactions as the random selected ratings to train our MDM.
3) **Amazon Music** Amazon Music is a large dataset contains of 4.6 million reviews. As for collaborative filtering methods, we only use the rating information in the reviews. To make the dataset more appropriate for training, we filtered the dataset that only users with at least 20 interactions and items with at least 5 interactions are retained. Similar with MovieLens dataset, we also randomly sample some interactions as the random selected data.

Experimental Protocol. The experimental protocol we adopted is different from typical methods, since we deal with two rating datasets (observed dataset and random selected dataset). But our idea is very simple: we train model on observed dataset and random selected dataset and test on random selected items (*Yahoo! Music*) or user-selected items (*MovieLens*, *Amazon Music*). For *Yahoo! Music* dataset, we collect all online survey ratings as our test items. But for *MovieLens* and Amazon Music, the ten latest interactions for each user are hang out for test and remaining data is considered as the observed data. For each dataset, our CDM model was trained on observed data and our MDM was trained on the combination of observed data and random selected data. The final prediction value is generated by the joint model, which mentioned in section 3.3.

Evaluation Metrics. We evaluate the precision of predicted rating by root-mean-square error (RMSE), which is widely used in various recommendation systems. What's more, the output of a recommendation system usually is a ranking list where front items are more likely to be chosen. So we also evaluate the ranking performance using standard metric in information retrieval ranking applications, the normalized discounted cumulative gain (nDCG@L), where the L is the length of ranked list.

5.2. Performance Comparison

In this section, we compare our proposed model DMF-MNAR with some novel methods. As our model aims to model data under MNAR assumption by using neural networks, we mainly compare with similar methods.

- **MF-MNAR** It is a matrix factorization method learning from MNAR data [15]. The major difference between our model is that we employed neural networks based MANR assumption, which breaks the liner limitation of traditional MF methods. We adjust model according its original idea so that it can be measured by the metrics mentioned before.
- **NNMF** It is a state-of-the-art method using neural networks to train MF methods [10]. This method replaces the inner product by an arbitrary function which combines the latent vectors of user and item. However, this model is based on MAR assumption as most neural collaborative filtering methods. We implemented this method and tuned the parameters as the author mentioned.
- **DMF-MNAR** This is our proposed deep matrix factorization model based MNAR assumption. As far as we know, our model is the first to build a neural collaborative filtering method based on MNAR assumption. We implement our mode on Tensorflow, which is convenient to build neural networks. By default, we set the *neg_ratio* to 1, the depth of hidden layers to 3, the batch size to 25000 and the learning rate to 0.01.

The results of the comparison are summarized in Table 1. It shows that our model performed well in RMSE or nDCG in different datasets. For the *Yahoo! Music* dataset, our model achieves the best performance in both metrics, which can conclude that our model is more suitable for MNAR data. But for the *MovieLens* and *Amazon Music*, our model only get gains in RMSE with a close nDCG. Although our model has limitation in two datasets, we think the model is more realistic. Because the aim of recommender system is rating missing data rather than past observed data.

Table 1.RMSE and nDCG@10 comparisons of different methods

Datasets	Metrics	MF-MNAR	NNMF	DMF-MNAR
Yahoo Music!	RMSE	1.4875	1.0853	**1.0536**
	nDCG@10	0.8209	0.8167	**0.8224**
MovieLens	RMSE	1.043	1.0465	**1.0313**
	nDCG@10	0.889	**0.8901**	0.8611
Amazon Music	RMSE	1.2101	1.0839	**1.0405**
	nDCG@10	0.8826	**0.9136**	0.8965

5.3. Sensitivity to Hyper-Parameters

We also explore the performance of our proposed model under different parameters settings.

Depth of Hidden Layers In the neural network of our proposed model, we used MLP to map the users and items to latent vectors. The depth of hidden layers H determines the model's capability, and we conduct some elaborate experiments to find effective depth. Figure 3 illustrates the performance of different layers while training on *Yahoo! Music*. As shown in the results, a single layer gets the worst performance because of lesser learning ability. And the 2-layer and 3-layer have similar results, but it seems that 3-layer converge to stable state in less iterations. We also tried deeper layers, but it's not useful to produce additional gains. And the higher depth will make it more time consuming to train the model. So the best depth of layers we suggested is 2 or 3 according the experiments.

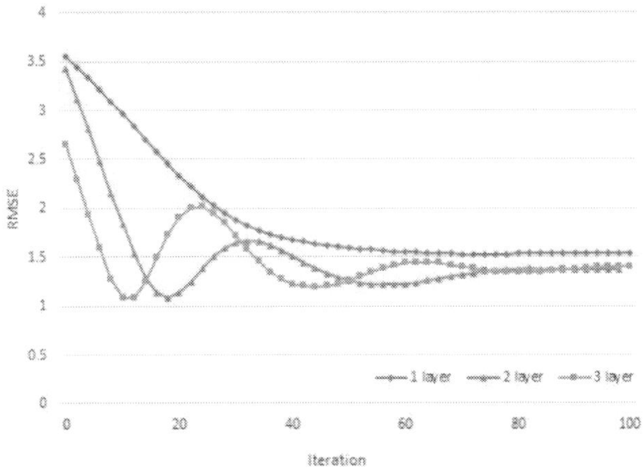

Figure 3. The performances of DMF-MNAR with different layers

Size of Latent Vectors Same as the other MF methods, the size of latent vectors in embedding layer is possibly a sensitive parameter in our model. Usually a small vector will limit the generalization ability of model, but a large vector may reduce the accuracy of model. So we also conduct some experiments with different sizes of latent vectors on *Yahoo! Music* to explore suitable parameters setting. As shown in Table 2, the model with 16-size latent vectors performs best compared with other parameters. So our experimental result also proves that small or large vectors are ineffective to our model, which similar with other MF methods.

Table 2. Results of different K on Yahoo! Music

Metrics	8	16	32	64
RMSE	1.0535	**1.0553**	1.0678	1.2898
nDCG@10	0.8060	**0.8128**	0.8108	0.8001

6. Conclusion and Future Work

In this work, we study the problem of learning collaborative filtering method based MNAR assumption. Our proposed model is the first practical implementation to use deep matrix factorization for ordinal matrix data with entries missing not at random. In our model, there are two parts representing the CDM model and the MDM model, respectively. And we combined two models together as the joint model to describe the whole MNAR data. We also proposed a method using variational bayesian inference to approximate the distribution of our model. In the other hand, we designed a pairwise neural networks to train the model, which used generation ability of deep learning. The experiments on several datasets demonstrate that our proposed model produces gains in various metrics.

Up to now, we only use the rating matrix as our input data, and the latent features are generated by identity of user and item, which may limit the expression of those features. In the future, we will use extra data, such as social relation, review text and history data, to model features of user and item. What's more, to make the recommender system more intelligent, we will consider toemploy reinforcement learning in our model. Therefore, we can evaluate a long-term reward while users are interacting with recommender system. We believe that the mechanism of deep learning can improve the recommender system.

References

[1] Yehuda Koren, Robert Bell, and Chris Volinsky. Matrix factorization techniques for recommender systems. *Computer*, 42(8), 2009.

[2] Yunhong Zhou, Dennis Wilkinson, Robert Schreiber, and Rong Pan. Largescale parallel collaborative filtering for the netflix prize. *Lecture Notes in Computer Science*, 5034:337–348, 2008.

[3] Yehuda Koren. Factorization meets the neighborhood: a multifaceted collaborative filtering model. In *Proceedings of the 14th ACM SIGKDD international conference on Knowledge discovery and data mining*, pages 426– 434. ACM, 2008.

[4] Andriy Mnih and Ruslan R Salakhutdinov. Probabilistic matrix factorization. In *Advances in neural information processing systems*, pages 1257– 1264, 2008.

[5] Xiangnan He, Hanwang Zhang, Min-Yen Kan, and Tat-Seng Chua. Fast matrix factorization for online recommendation with implicit feedback. In *Proceedings of the 39th International ACM SIGIR conference on Research and Development in Information Retrieval*, pages 549–558. ACM, 2016.

[6] Robin Devooght, Nicolas Kourtellis, and Amin Mantrach. Dynamic matrix factorization with priors on unknown values. In *Proceedings of the 21th ACM SIGKDD international conference on knowledge discovery and data mining*, pages 189–198. ACM, 2015.

[7] Heng-Tze Cheng, Levent Koc, Jeremiah Harmsen, Tal Shaked, Tushar Chandra, Hrishi Aradhye, Glen Anderson, Greg Corrado, Wei Chai, Mustafa Ispir, et al. Wide & deep learning for recommender systems. In *Proceedings of the 1st Workshop on Deep Learning for Recommender Systems*, pages 7–10. ACM, 2016.

[8] Sheng Li, Jaya Kawale, and Yun Fu. Deep collaborative filtering via marginalized denoising auto-encoder. In *Proceedings of the 24th ACM International on Conference on Information and Knowledge Management*, pages 811–820. ACM, 2015.

[9] Florian Strub and J´er´emie Mary. Collaborative filtering with stacked denoising autoencoders and sparse inputs. In *NIPS workshop on machine learning for eCommerce*, 2015.

[10] Gintare Karolina Dziugaite and Daniel M Roy. Neural network matrix factorization. *arXiv preprint arXiv:1511.06443*, 2015.

[11] Xiangnan He, Lizi Liao, Hanwang Zhang, Liqiang Nie, Xia Hu, and TatSeng Chua. Neural collaborative filtering. In *Proceedings of the 26th International Conference on World Wide Web*, pages 173–182. International World Wide Web Conferences Steering Committee, 2017.

[12] Roderick JA Little and Donald B Rubin. *Statistical analysis with missing data*. John Wiley & Sons, 2014.

[13] Benjamin M Marlin and Richard S Zemel. Collaborative prediction and ranking with non-random missing data. In *Proceedings of the third ACM conference on Recommender systems*, pages 5–12. ACM, 2009.

[14] Harald Steck. Training and testing of recommender systems on data missing not at random. In *Proceedings of the 16th ACM SIGKDD international conference on Knowledge discovery and data mining*, pages 713–722. ACM, 2010.

[15] Jos´e Miguel Hern´andez-Lobato, Neil Houlsby, and Zoubin Ghahramani. Probabilistic matrix factorization with non-random missing data. In *International Conference on Machine Learning*, pages 1512–1520, 2014.

[16] Tao Chen, Xiangnan He, and Min-Yen Kan. Context-aware image tweet modelling and recommendation. In *Proceedings of the 2016 ACM on Multimedia Conference*, pages 1018–1027. ACM, 2016.

[17] Xavier Glorot, Antoine Bordes, and Yoshua Bengio. Deep sparse rectifier neural networks. In *Proceedings of the Fourteenth International Conference on Artificial Intelligence and Statistics*, pages 315–323, 2011.

[18] Manfred Opper and David Saad. *Advanced mean field methods: Theory and practice*. MIT press, 2001.

Passive Intermodulation Interference Suppression through Sparse Discrete Fractional Fourier Transform

JING OU[1], LU TIAN[2], JIANCHUN LIU[1] and CELUN LIU[1]

[1] Beijing Institute of Technology, 5 Zhongguancun South Street, Haidian District, Beijing, China
[2] China Electronics Technology Group Coorporation, Academy of Electronics and Information Technology, 11 Shuangyuan Road, Shijingshan Dsitrcit, Beijing, China

tianlu218@gmail.com

Abstract. With the wide application of duplex mode antenna, passive intermodulation (PIM) interference has become a severe problem. And the traditional methods have shortcomings that can not be overcome. In order to solve this issue, this paper proposes an algorithm to suppress passive intermodulation (PIM) interference based on digital signal processing (DSP). We choose power series model to express PIM signal, which is also called PIM interference model. In the proposed scheme, satellite transmit pilot signal during the pilot period. With the transmitted pilot signal and the received signal, the real-time parameters of PIM interference model can be obtained. Then during the data transmission period, data is transmitted in uplink and downlink separately. The PIM interference can be rebuilt through the downlink transmitted data and the PIM interference model. In this way, the PIM interference is suppressed. We chose chirp signal to be pilot signal for its wide bandwidth, which can cover all frequency in modulated system. In order to obtain the PIM interference model parameters, we use Sparse Discrete Fractional Fourier Transform (SDFrFT) through Pei method due to its low computational complexity. The simulation results show that this algorithm can reconstruct the original signal with a relatively low complexity.

1. Introduction

Augment of satellite systems' transmit power, number of carriers and bandwidth makes PIM become a severe problem. Once satellite system is affected by PIM, the current state-of-the-art in electronic circuits impels it to turn on spare antenna and disable duplex mode. The generally way to handle this problem is promoting hardware performance in order to reduce nonlinearity, whose cost is extremely expensive.

This motivates researches on PIM suppression based on DSP. [1] presents a general framework for the adaptive feedforward PIM cancellation, but it costs extra hardware, which is impossible due to limitation in satellites resource. Works in [2] propose a method to cancel PIM signal based on DSP. However, this method needs to interrupt communication to estimate the parameters. [3] provides a novel digital PIM suppression algorithm.

In order to reduce algorithm's computational complexity, we turn to Discrete Fractional Fourier Transform (DFrFT) for it is an efficient way to solve many challenging engineering problems. Among all DFrFT algorithms, Pei's algorithm has the lowest complexity [4] and its complexity can be further reduced through sparse Fourier transform [5-6]. [7] gives a modified Cauchy distribution model of high-order PIM signal. However, there is no complete model expression of PIM interference [8], we select power series model to express PIM signal.

In this paper, we propose a digital way to estimate and cancel PIM interference through power series model. We chose chirp signal to be pilot signal for its wide bandwidth, which can cover all frequency in modulated system. Because of the sparse characteristic of chirp signal in fractional Fourier domain, SDFrFT is adopted to estimate parameters of PIM signal.

The rest of this paper is organized as follows. Section 2 details PIM signal reconstruct algorithm. Simulation results are presented in section 3. Finally, conclusions are drawn in Section 4.

2. PIM signal reconstruct algorithm

As shown in figure 1, pilot signal c_0 and data c are transmitted in t_1 and t_2 respectively. During t_1, received signal d_0 is used to estimate parameters of PIM signal and reconstruct PIM signal $\widehat{p_0}$. During t_2, estimated signal $\hat{c}$ can be obtained by $\mathrm{d} - \hat{p}$. figure 2 demonstrates this process in detail.

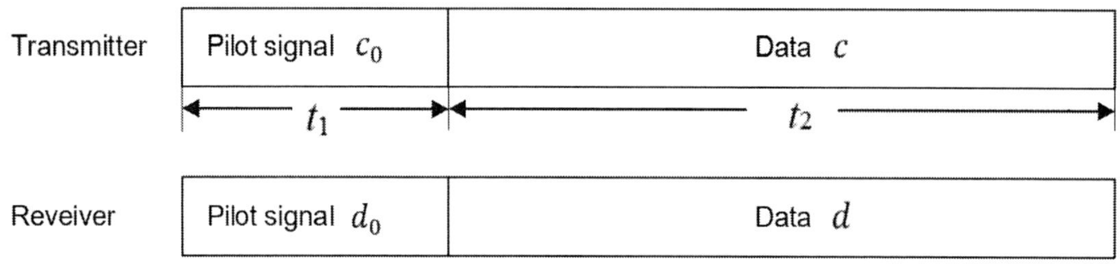

Figure 1. Architecture of transmitted and received signal.

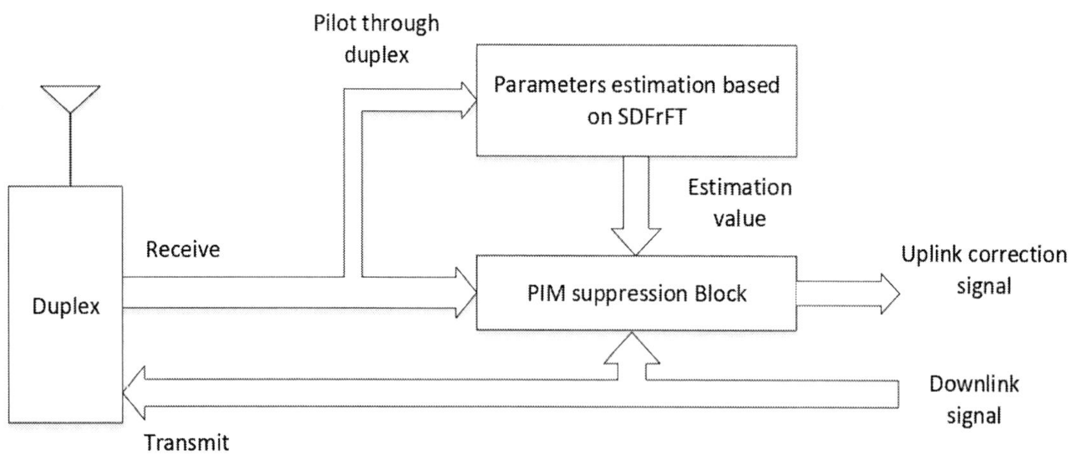

Figure 2. Architecture of overall PIM suppression algorithm.

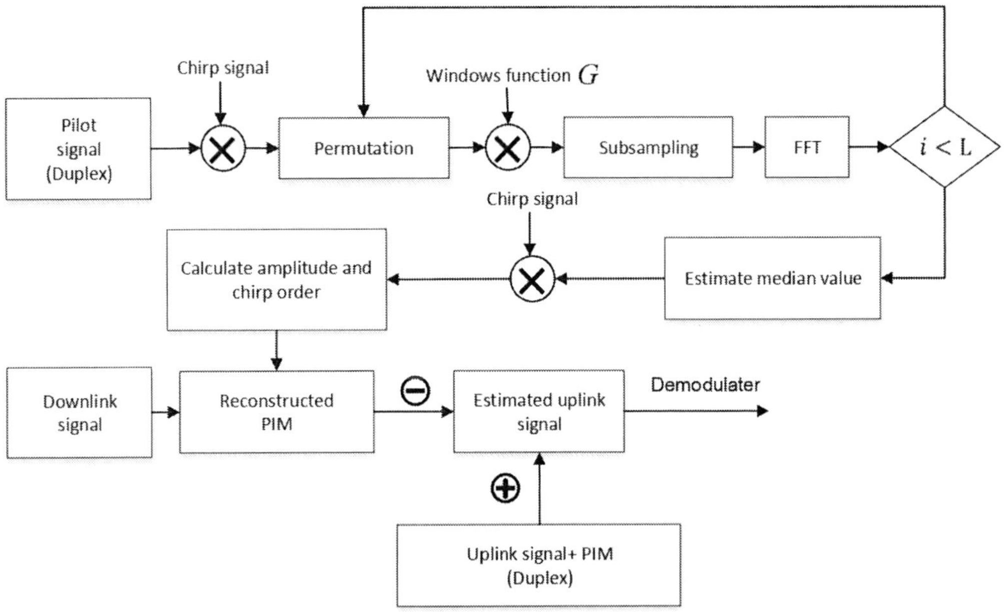

Figure 3. Flow chart of overall PIM suppression algorithm.

Flow chart of the overall algorithm is showed in figure 3, where L is the time of repetitions.

2.1. *PIM parameters' estimation*

The PIM signal can be expressed in power series model [8]

$$y(t) = \sum_{i=1}^{N} a_i x(t)^i \quad (1)$$

where a_i represents the amplitudes of different order $i \in \{1,2,\cdots,N\}$, N is the number of sampling points.

Chirp signal is given by

$$x(t) = \exp[\mathrm{j}(2\pi f_0 t + \pi k t^2)] \quad (2)$$

where f_0 is start frequency of chirp signal and k represents frequency modulation rate.

Using equation (1) and (2), PIM signal becomes

$$y(t) = \sum_{i=1}^{N} a_i \exp[\mathrm{j}(2\pi f_0 t + \pi k t^2)] = \sum_{i=1}^{N} y_i(t) \quad (3)$$

Using Pei sampling method, we obtain

$$y(n) = y(n\Delta t) \quad (4)$$

$$X_\alpha(m) = F(m\Delta u) \quad (5)$$

$$X_\alpha(m) = \begin{cases} \left(\frac{sin\alpha - jcos\alpha}{2M+1}\right)^{1/2} \exp\left(\mathrm{j}\frac{1}{2}cot\alpha m^2\Delta u^2\right) \sum_{n=-N}^{N} \exp\left(\mathrm{j}\frac{1}{2}cot\alpha n^2\Delta t^2 - \mathrm{j}\frac{2\pi mn}{2M+1}\right) y(n) & \text{if } \sin\alpha > 0 \\ \left(\frac{jcos\alpha - sin\alpha}{2M+1}\right)^{1/2} \exp\left(\mathrm{j}\frac{1}{2}cot\alpha m^2\Delta u^2\right) \sum_{n=-N}^{N} \exp\left(\mathrm{j}\frac{1}{2}cot\alpha n^2\Delta t^2 - \mathrm{j}\frac{2\pi mn}{2M+1}\right) y(n) & \text{if } \sin\alpha < 0 \end{cases} \quad (6)$$

where Δt is the reverse of sampling interval, α is a real number that represents rotation angle of fractional Fourier transform (FrFT) and $\Delta u = \frac{2\pi|sin\alpha|}{\Delta t(2M+1)}$. Besides, $M \geq N$ should be satisfied.

Taking linear superposition of FrFT into consideration, we process one argument of $y_0(n)$ at a time. Assuming the sampling rate is f_s, which means $\Delta t = 1/f_s$, then $X_{0\alpha}(m)$ and $x(n)$ become

$$x(n) = \exp[\mathrm{j}(2\pi f_0 n/f_s + \pi k(n/f_s)^2)] \quad (7)$$

$$X_\alpha(m) = \left(\frac{sin\alpha - jcos\alpha}{2M+1}\right)^{1/2} \exp\left(\mathrm{j}\frac{1}{2}cot\alpha m^2\Delta u^2\right) \sum_{n=-N}^{N} \exp\left(\mathrm{j}\frac{1}{2}cot\alpha n^2(n/f_s)^2 - \mathrm{j}\frac{2\pi mn}{2M+1}\right) y_i(n)$$

$$= a_i A_\alpha \exp\left(\mathrm{j}\frac{1}{2}cot\alpha m^2\Delta u^2\right) \sum_{n=-N}^{N} \exp\left(\mathrm{j}\frac{1}{2}cot\alpha n^2(n/f_s)^2 - \mathrm{j}\frac{2\pi mn}{2M+1}\right) \exp[\mathrm{j}i(2\pi f_0 n/f_s + \pi k(n/f_s)^2)] \quad (8)$$

It is obvious that the maximum value of $|X_\alpha(m)|$ can be achieved when [9]

$$\left(\frac{1}{2}cot\alpha + \pi ik\right)(n/f_s)^2 = 0 \tag{9}$$

$$\max|X_\alpha(m)| = (2\mathrm{N}+1)|a_i A_\alpha| \tag{10}$$

Then i and a_i is

$$i = -\frac{cot\alpha}{2\pi k} \tag{11}$$

$$a_i = \frac{\max|X_\alpha(m)|}{(2\mathrm{N}+1)|A_\alpha|} \tag{12}$$

2.2. *Normalization*

Sampling time is $T = \mathrm{N}/f_s$, and range of time and frequency domain is $[-T/2, T/2] \times [-f_s/2, f_s/2]$ [10]. Adopt time dimension factor $S = (T/f_s)^{1/2}$. New coordinates are defined as $t \to t/S$ and $f \to f \times S$, then range of time and frequency domain become $[-(Tf_s)^{1/2}/2, (Tf_s)^{1/2}/2] \times [-(Tf_s)^{1/2}/2, (Tf_s)^{1/2}/2]$. Let $\Delta x = (Tf_s)^{1/2}$, then time-frequency distribution of signal after normalization is a circle centered on origin with a radius of Δx. Therefore, the sampling interval becomes $1/\Delta x$, and we have $\mathrm{N} = (\Delta x)^2$.

After normalization, sampling data changes from $x\left(\frac{n}{f_s}\right)$ into $x\left(\frac{n}{\Delta x}\right)$. The DFrFT of sampling data becomes

$$X_\alpha(m) = \left(\frac{sin\alpha - jcos\alpha}{2\mathrm{M}+1}\right)^{1/2} \exp\left(\mathrm{j}\frac{1}{2}cot\alpha m^2 \Delta u^2\right) \sum_{n=-\mathrm{N}}^{\mathrm{N}} \exp\left[\frac{\mathrm{j}cot\alpha n^2}{2(\Delta x)^2} - \mathrm{j}\frac{2\pi mn}{2\mathrm{M}+1}\right] y_i(n) \tag{13}$$

Equation (9) and (11) turn into

$$\frac{cot\alpha n^2}{2(\Delta x)^2} + \pi ik(n/f_s)^2 = 0 \tag{14}$$

$$i = -\frac{cot\alpha}{2\pi k(\Delta x/f_s)^2} \tag{15}$$

2.3. *SDFrFT*

Steps of SDFrFT with complexity $O\left((nklogn)^{1/2} logn\right)$ [5] detail in figure 4.

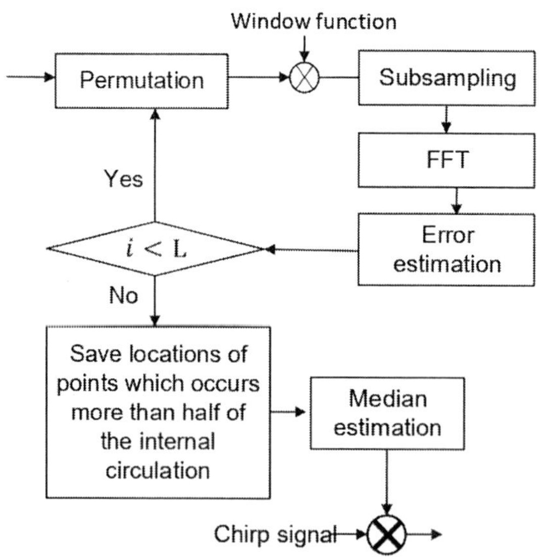

Figure 4. Flow chart of SDFrFT algorithm.

The first step, permutation, enables the algorithm to distinguish signals whose spectrums are close to each other. And window function reduces the spectrum leakage. After subsampling and fast Fourier transform (FFT), error estimation takes part through Hash function and penalty function. After L repetitions, save locations of points that occur more than half of the internal circulation and estimate the median. Multiple the result by chirp signal to get its FrFT.

2.4. Multi-component signal process and PIM signal reconstruction
Setting chirp signal as pilot signal makes PIM signal feature several peaks in fractional domain. Set a threshold value $P_{threshold}$. In order to get all $|\hat{F}_\alpha(m)|$ and α, we first search the maximum magnitude value $\max|\hat{F}_\alpha(m)|$ in two-dimension fractional domain. Next, obtain i and a_i from (15) and (12) respectively if $\max|\hat{F}_\alpha(m)| > P_{threshold}$. Then, we apply a narrow band filter $G(m)$ to weaken the maximum peak, where $\hat{F}'_\alpha(m) = \hat{F}_\alpha(m)G(m)$, and find the next maximum value by steps above until $\max|\hat{F}_\alpha(m)| < P_{threshold}$.

3. Simulation

Set f_s as 112MHz/s, N as 8192, f_0 as 0.5MHz and k as 17.2GHz. The order of FrFT ranges from 0 to 2 with step of 0.001.The number of sparse spectrum line is set to 1, and the number of loop is 9. The order of two components are 7.3 and 8.9 respectively.

Figure 5 shows the original signal, the signal with PIM interference and the signal after cancellation in time domain. As can be seen from figure 5, although the power of the PIM interference is larger than that of the signal, the signal after cancellation and the original signal are almost superposition.

Figure 6 shows the power spectral density (PSD) of original signal, PIM interference, signal with interference and signal after cancellation separately. Figure 6(c) shows that the original signal is submerged in PIM interference. However, as can be seen from figure 6(d), although there is several sub-peak in PSD of the signal after cancellation, the PSD of the original signal can be identified easily.

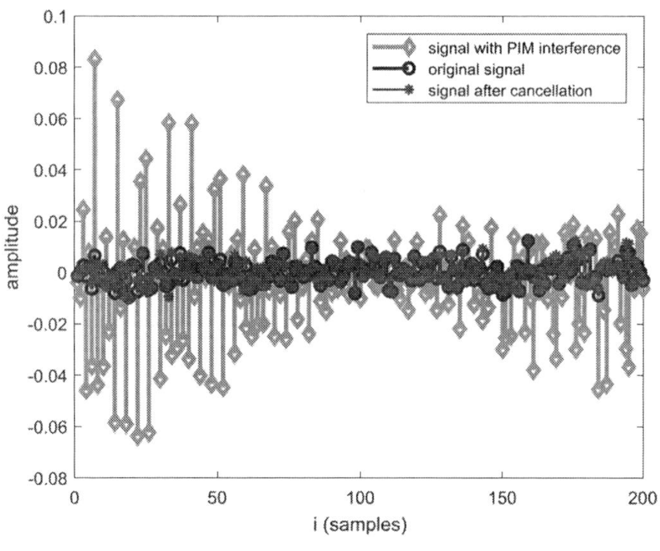

Figure 5. Cancellation results in time domain.

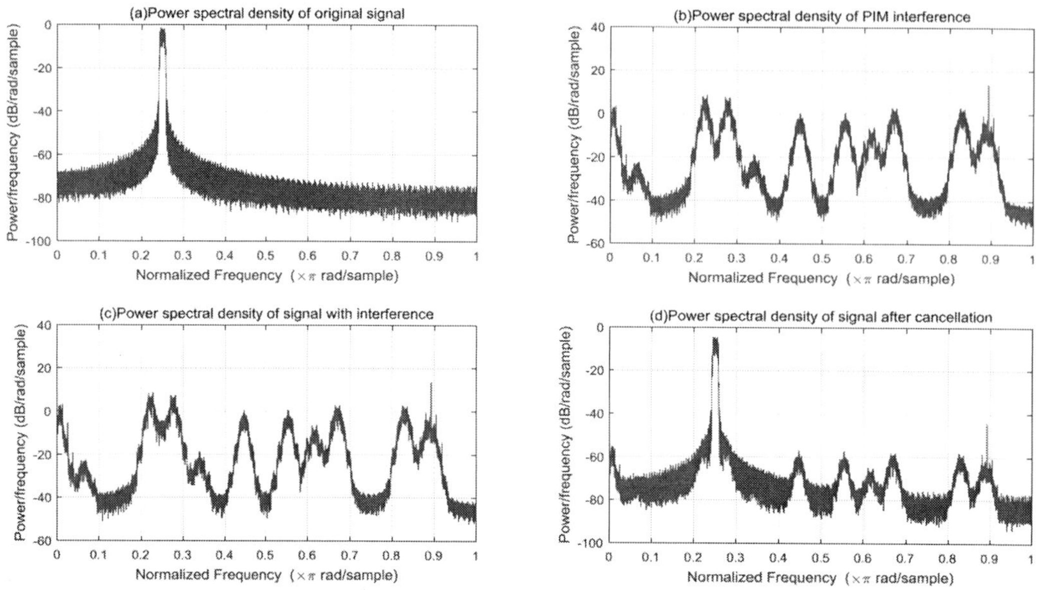

Figure 6. PSD of original signal, PIM interference, signal with interference and signal after cancellation.

4. Conclusion

In this paper, we proposes a PIM suppression algorithm with low computational complexity. We use SDFrFT to reconstruct PIM signal in power series model. Simulation results show that this algorithm performs well in PIM cancellation.

Acknowledgments

This work was supported by the National Natural Science Foundation of China under Grants 61601027

References

[1] Keehr E. A. and Hajimiri A. 2011 Successive regeneration and adaptive cancellation of higher order intermodulation products in rf receivers *J IEEE Transactions on Microwave Theory and Techniques*. **59** pp 1379–96

[2] Wang M, Li S, Baldemair R, et al 2012 Dynamic cancellation of passive intermodulation interference *P US, US 20120295558* A1

[3] Lu T., Han H. C., et al 2017 Adaptive suppression of passive intermodulation in digital satellite transceivers *J Chinese Journal of Aeronautics* **30** pp 1154-60

[4] Pei S. C., Ding J. J. 2000 Closed-form discrete fractional and affine Fourier transforms *J Transactions on Signal Processing* **48** pp 1338-53

[5] Hassanieh H., Indyk P., et al 2012 Simple and practical algorithm for sparse Fourier transform *C Acm-Siam Symposium on Discrete Algorithms Society for Industrial and Applied Mathematics* pp 1183-94

[6] Gilbert A. C., Indyk P., et al 2014 Recent developments in the sparse Fourier transform: A compressed Fourier transform for big data *J IEEE Signal Process* **31** pp 91–100

[7] Lu T., Yi W., et al 2016 Modified Cauchy distribution model of high-order passive intermodulation *C Int. Symp. on Antennas and Propagation* pp 586-7

[8] Al-Mudhafar A. and Hartnagel H. 2012 Bit Error Probability in the Presence of Passive Intermodulation *J IEEE Communication Letters* **16** pp 1145-48

[9] Bing D., Junbao L. and Shiqi C. 2014 Analysis of parameter estimation using the sampling-type algorithm of discrete fractional Fourier transform *J Defense technology* **10** pp 321-7

[10] Tanli C., Lianghao G. and Zaixiao G. 2017 The concise Fractional Fourier transform and its application in detection and parameter estimation of the linear frequency-modulated signal *J Chinese Journal of Acustica* **1** pp 70-86

IOP Conf. Series: Journal of Physics: Conf. Series **1060** (2018) 012003 doi:10.1088/1742-6596/1060/1/012003

BDT Performance and Time Offset Analysis

WANG Xiang-lei and DU Yan

Beijing Satellite Navigation Center

chxywxl@163.com

Abstract. The nature of satellite navigation can be seen as a time measurement. Time measurement is the basic for satellite navigation. First, the time scale of satellite navigation system is the basic for the system operation, and the accuracy of time measurement determines the accuracy of satellite orbit. It can be said that the accuracy of the time determines the accuracy of satellite navigation to a certain extent. BDT is the system time of the beidou satellite navigatin, It is established and maintained by the ground operation control system, BDT is the reference time of the entire ground operation control system and is also the reference time of satellite, thus ensuring the consistency of the satellite time and the ground time. The definition and realization of the BDT are introduced. The time and frequency system to maintain BDT and its performance are briefly described. The measurement methods of BDT/GPST time offset are introducted, the measurement results are analysied, the prediction modelling time offset is studited, the 1st, 2nd and 3rd order ploynomial model are given, 1~2ns precision could be got in BDT-GPST prediction, if the system time stability or the receiver observation precision are improved, the higher precision could be reached in future.

1. Introduction

All we know that time system plays a significant role in Beidou system, Accurate and reliable time is the base of PNT service. How is about the performance of Beidou system time? It is a very concered qustion for many GNSS users, so it is beneficial and indispensable to evaluate the BDT Performance[1-2]. The information of time offset is importance for the user, satellite is invisual some time during the day, so the time offset information is depond on the prediction when the satellite is not in the visual field of user. In the following sections the definition and realization of BDT is introduced and its performance is described, then the BDT evalution is analyzed which from GPS CV, BD CV and TWSTFT, and then the measurement and prediction of time offset is studied. Finally the results and conclusions are detailed.

2. Definition and realization of BDT

Analogous to GPS system time — (GPST) and different from UTC, the BDT is an internal and continuous uniform navigation time scale without any leap second. The basic unit of BDT is the SI second, and the largest unit is one week defined as 604,800 seconds. BDT is counted with the week number (WN) and the second of week (SoW), from 0 to 604,799. The zero point of BDT is January 1 , 2006（Sunday）UTC 00h 00m 00s, which means all the numerical values of WN and SoW are equal to zero at that time.

BDT is realized in the "composite clocks" approach and maintained by TFS located at MCS. According to the functions, TFS is mainly composed of 5 parts: the clock ensemble (CE), the inter measurement element (IME), the outer comparison element (OCE), the data processing element (DPE) and the signal generation element (SGE). The structure of TFS is illustrated in Figure.1.

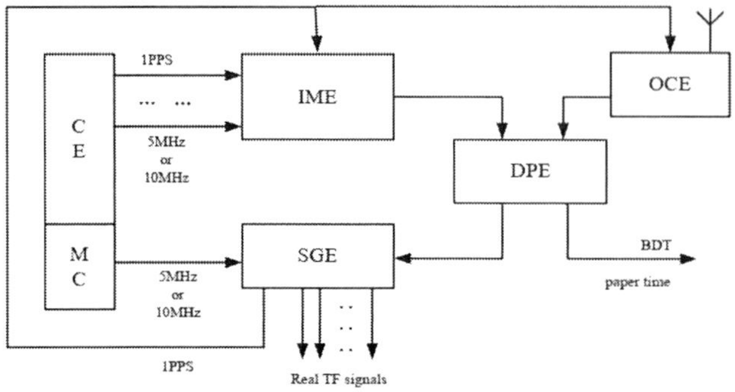

Figure 1. The Structure of the BDT and Frequency System

The clocks used in time keeping are more than ten hydrogen masers, which form a clock ensemble to provide time and frequency signals. The inter measurement element provides measurements to the original time and frequency signals from CE, and gives out the clock differences both in time and frequency in a circular pattern. The deviations of BDT with respect to other time scales, especially to UTC, are obtained by the outer comparison element. The data processing element completes the calculation with the given algorithm to give a relative uniform time scale, based on all the information from IME and OCE, which is called BDT which works as the time reference for the whole navigation system. In line with BDT, the signal generation element exerts a frequency adjustment to the frequency signal from the Master Clock (MC), and generates all the real physical time and frequency signals required in the Master Control Station[3-5].

The algorithm used in BDT calculation is well designed to form a good composite clock. The frequency offset, drift and instability of the free clocks have been taken into account. The weight of clock is determined by its Allan variance (the frequency drift is taken off) and a limited weight is also used based on the robust estimation principle. In order to be as consistent as possible with UTC, BDT may be steered with a interposed frequency adjustment after a period of time (more than 30 days) according to the situation, but the quantity of the interposed frequency adjustment is not allowed to be more than 5×10-15. At present, the performance of BDT is as follows[5]:

Time (frequency) accuracy : < 2.0E-14

Time (frequency) stability : < 2.0E-14 /1 day

< 1.4E-14 /7days

Time deviation: |BDT-UTC| < 100ns (modulo one second).

3. BDT performance evaluation

The BDT time comparison links are illustrated in figure 2.

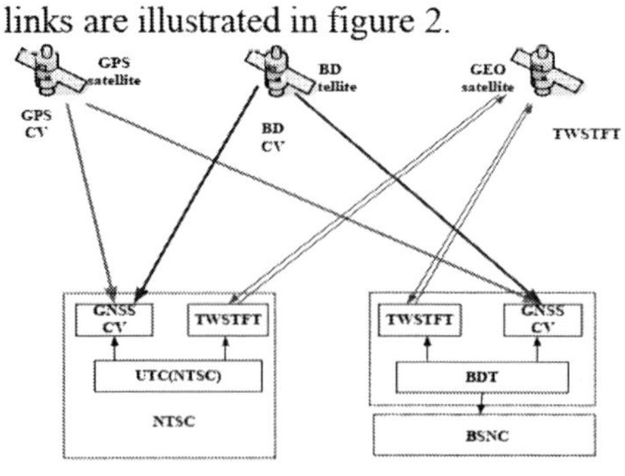

Figure2. BDT time comparison links

IOP Conf. Series: Journal of Physics: Conf. Series **1060** (2018) 012003 doi:10.1088/1742-6596/1060/1/012003

Time offset can be attained via TWSTFT, GPS CV, and BDT CV, and the precision of TWSTFT is the best, the precision of GPS CV and BDT CV are comparative. the principle of TWSTFT, GPS CV, and BDT CV would be intrduced in the following section.

The basic principle of two way time transfer is as follows: the satellites and stations generate and transmit pseudo range signals to be controlled by their local clocks, then the uplink pseudo range ρ_u and downlink pseudo range ρ_d are measured by the satelites and the stations respectively, the clock offsets between satellites and stations are computed in the master station by using the uplink pseudo ranges and the downlink pseudo ranges.

The basic principle of GPS CV is as follows: two time and frequency laboratory, which the distance is far, observe the same satellite at the same time, and the GPS satellite clock time is their common reference, the time offset between GPS satellite clock time and the laboratory is measured, then, the time offset of two laboratories is received via make difference for the measurement result of two laboratories.

Analogous to GPS CV, the time offset of two laboratories can be received, the only different is the reference, the reference time is Beidou satellite clock time.

The result of time offset between BDT and UTC(NTSC), which using GPS CV, BD CV and TWSTFT, are given in Figure 3.

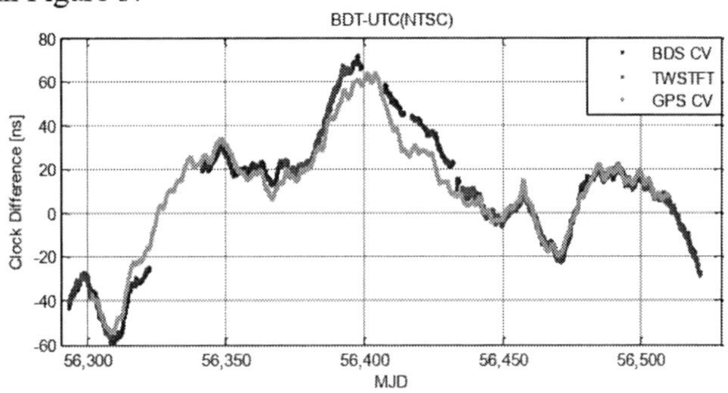

Figure 3. The time offset between BDT and UTC(NTSC)

The change direction of BDT relative to UTC(NTSC) is conformably, and BD CV and TWSTFT have the higher precision. The result of GPS CV is bad, and it is confirm that the reason is the receiver of GPS CV is single frequency, and the correct error of ionosphere is larger. The time offset between UTC and UTC(NTSC) is given in Figure4.

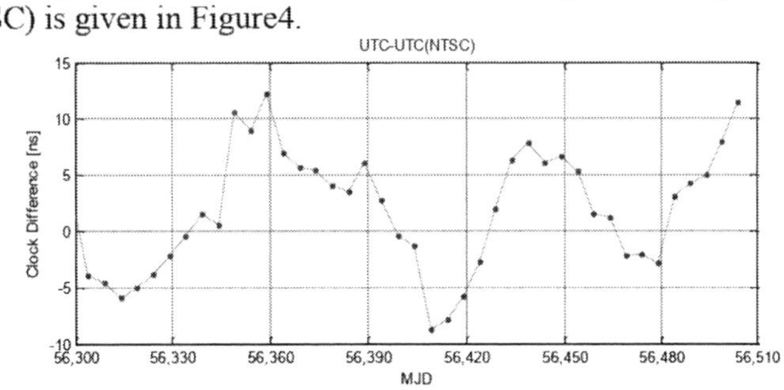

Figure4. The time offset between UTC and UTC(NTSC)

So, the time offset between BDT and UTC can be received via make plus for BDT-UTC(NTSC) andUTC(NTSC)-UTC , and the result is given in Figure5.

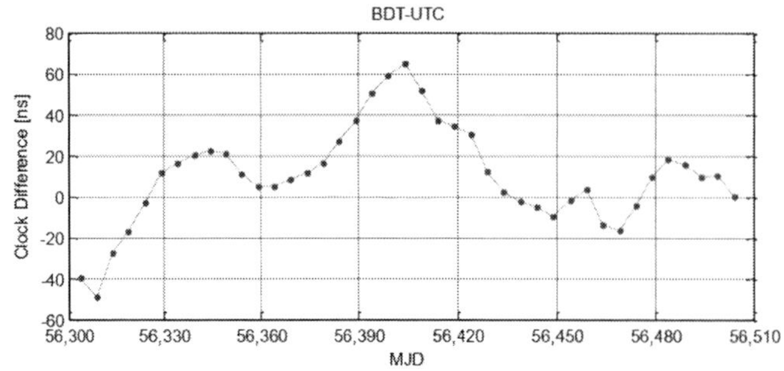

Figure5. The time offset between BDT and UTC

The max of time offset between BDT and UTC is 64.8ns. The performance evaluation results of BDT relative to UTC(NTSC) is given in Table1, and the frequency stability is given in Figure6.

Table1. Performance evaluation results of BDT relative to UTC(NTSC)

Measurement method	Frequency accuracy	Frequency stability		Time
GPS CV	8.89E-14	2.58E-14/d	1.20E-14/7d	Aug.-Sept.
BDS CV	7.67E-14	1.75E-14/d	1.08E-14/7d	Aug.-Sept.
TWSTFT	6.90 E-14	2.16 E-14/d	6.33E-15/7d	Aug.-Sept.

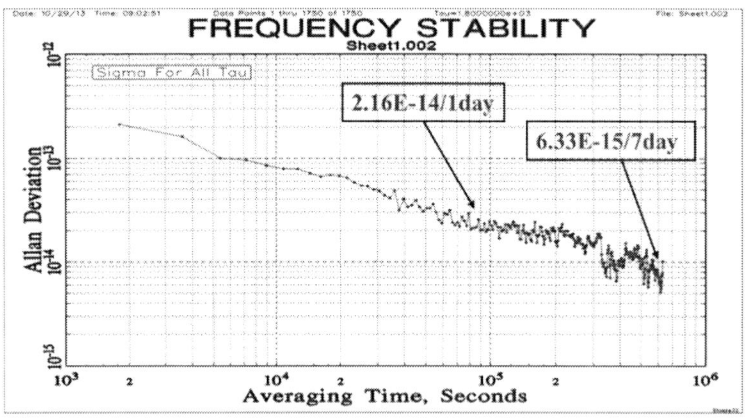

Figure6. The frequency stability of BDT relative to UTC(NTSC)

So, the day frequency stability is 2.16E-14/day, and the seven days frequency stability is 6.33E-15/7days.

4. BDT/GPST time offset measurement

The measurement principle of BDT/GPST time offset is given in Figure7, BDT is reference, and GPST is measurement, the result is BDT-GPST.

The data come from the Septentrio 4 GPS CV receiver. The measurement result is given in Figure8.

IOP Conf. Series: Journal of Physics: Conf. Series **1060** (2018) 012003 doi:10.1088/1742-6596/1060/1/012003

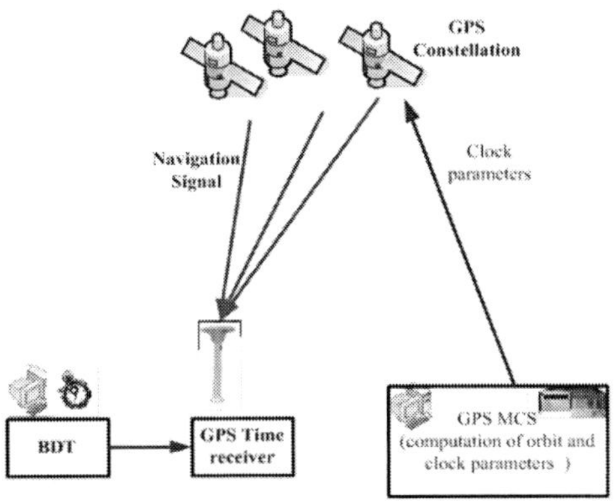

Figure7. The measurement principle of BDT/GPST time offset

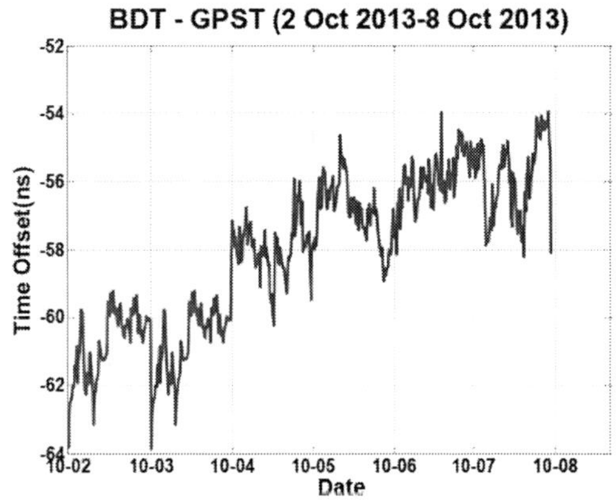

Figure8. The time offset between BDT and GPST

The measurement data is from 2 Oct. 2013 to 10 Oct. 2013. The time offset is not measured when the satellite is not in visual, so the time offset will be predicted during the satellite is invisual. The different prediction model will give the different prediction result, and the different result will be given using different fitting data. In this section, time offset prediction algorithms based different ploynomial models: 1st model, 2nd model, 3rd model are studied, and the different fitting data will be analyzed. The different case are given in Table2.

Table2. Time offset data modelling and prediction

Model	1st order model	2nd order model	3rd order model
Data usage	Prediction: 1 day Based on: 3 days	Prediction: 1 day Based on: 3 days	Prediction: 1 day Based on: 3 days
	Prediction: 1 day Based on : 6 days	Prediction: 1 day Based on : 6 days	Prediction: 1 day Based on : 6 days

4.1. Three days→one day

Prediction error comparation of three models are given in Figure9 and Figure10.

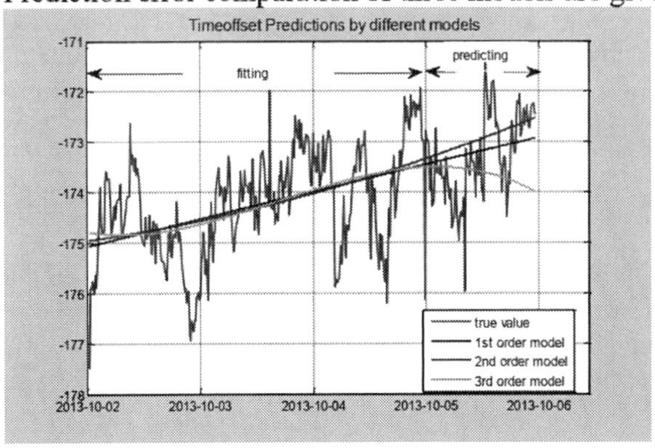

Figure9. the fitting results and the prediction results

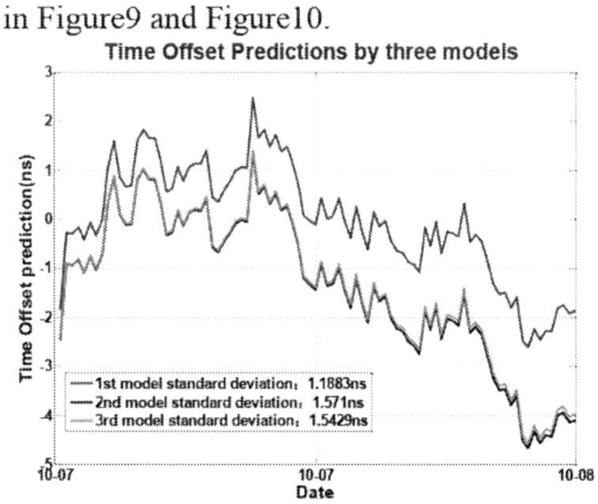

Figure10. Prediction error comparation of three models

4.2. Six days→one day

Prediction error comparation of three models are given in Figure11 and Figure12, and the comparative results are given in the Table3.

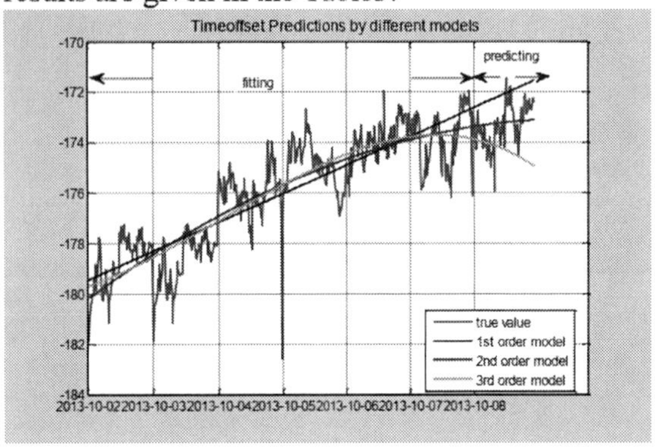

Figure11. the fitting results and the prediction results

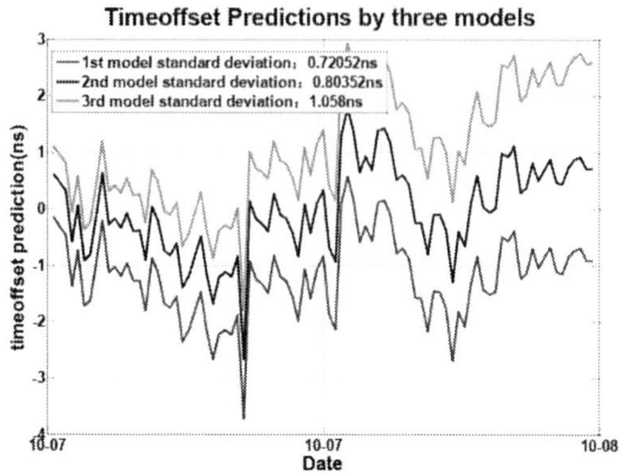

Figure12. Prediction error comparation of three models

Table3. Result comparation and analysis

Prediction model	Std Dev(3 days' data => 1day)	Std Dev(6 days' data => 1day)
1st order poly	1.19	0.72
2nd order poly	1.57	0.80
3rd order poly	1.54	1.05

The followed conclusions can be given from the above experiments:

(1)The longer the data used, the better the prediction precision. 6days' data is enough for the prediction in the above experiments.

(2)Low order(1-2order) polynomial could get good precision for modelling and prediction. Higher order(over 3 order) polynomial is not necessary .

(3)1 ~ 2 ns precision could be got in BDT-GPST prediciton. If the system time stability or the receiver observation precision are improved, higher precision could be reached in future.

5. Conclusion

Beidou system have been supplied service for users, BDT is the basic of Beidou system. BDT is stated by WN and SoW from Jan.1, 2006 UTC 00h 00m 00s. BDT is maintained by the Time and Frequency System of the master station. The frequency accuracy of BDT is superior to 2.0E-14 and its stability is better than 6.0E-15/30 days. The time offset between BDT and UTC was kept within 100ns and the frequency stability of BDT is about 3.0E-14.

Reference

[1] Huang Qiaohua. "Development of BeiDou Navigation Satellite System," presented at 5th Meeting of International Committee on GNSS (ICG-5),Beijing, China,2012.

[2] Ran C, The development plan of BeiDou Navigation Satellite System (in Chinese). China Satellite Navigation Conference 2010 (CSNC2010), Beijing, China.

[3] Cai Z W etc."BDT Performance and Time Offset Analysis", presented at ICG-8 Conference.2013.

[4] Han C H. "Time Measurement Within the Frame of Relativity", Progress in Astronomy, Vol.20, No.2, P107-113,2002.

[5] Liu L etc. The Method and Experiment Analysis for Two-Way Satellite Common-View Time Transfer, 2011, China Satellite Navigation Conference 2011 (CSNC 2011), 2011, Shanghai, China.

An Implementation of a XML Documents Secure Circulation System

Miao Liu and Ya Li

School of Computer Science and Educational Software, Guangzhou University, Guangzhou, 510006, China

E-mail: liumiao@gzhu.edu.cn

Abstract. XML is already one standard of data exchanging among applications. How to assure XML data security is an important problem. The paper designs and implements a flexible XML documents secure circulation system by using S2SH frameworks. The system assures XML documents confidentiality by providing XML documents encryption and decryption, assures XML documents integrity by implementing digital signature and verification. It allows owners of documents to define access control policies by XACML, and assures strict and flexible access control for XML documents. It also uses digital certificates to authenticate users, to create digital signatures using corresponding private keys, and assures non-repudiation for updating XML documents.

1. Introduction

The XML is the de facto standard for storing and exchanging information now. When developers agree with the XML schemas in advance, the applications which they developed could then exchange XML data automatically. The XML is also the basis of the semantic web, and so more and more XML data will be exchanged in the Internet. A lot of XML data deals with personal privacy and some other secrets, thus, it is critical to ensure XML data security[1-3].

There are many applications which process and exchange XML documents among them, such as electronic medical records, electronic official documents and transaction lists. These applications must provide mechanisms to prevent those XML documents from unauthorized accessing and modifying, to ensure non-repudiation and to allow owners to define flexible access control rules. We design and implement a flexible XML documents secure circulation system which is suggested as a solution to these goals. The system is a general XML documents secure management system, no matter how different XML documents schemas are.

This paper is organized as follows. In section 2, we show the composition of the system, and the design of four core modules. In section 3, we explain the implementation architecture of the system, the implementation of the access control function and the documents circulation function. Conclusions are discussed in section 4.

2. System design

2.1. The composition of the system

The system consists of four first class modules, which are IssuingDocuments, Outbox, Inbox, and PAP. The module IssuingDocuments is divided into four second class modules; they are the Create, Modify,

Delete and Designate. The Designate module designates receivers who will audit the issuing documents. The Outbox module is divided into two modules which are Unfinished and Finished. The Unfinished module manages documents which are not finished auditing yet, and the Finished module manages audited documents. The Inbox module is divided into three modules which are Auditing, AuditedUnfini and AuditedFini modules. The Auditing module manages documents which are being audited, the AuditedUnfini module manages those documents which is audited by the user, but not finished auditing by the other receivers, and the Auditedfini module manages those documents which are finished auditing by all receivers. The PAP module is the policy administration point, which is divided into two modules, and they are the UserAdmin and Policy modules. The UserAdmin module manages users and their certificates, and the Policy module edits and checks access control policies. When users log in the system, they could use the above modules to operate on their XML documents. The module structure diagram of the system is illustrated as Figure 1.

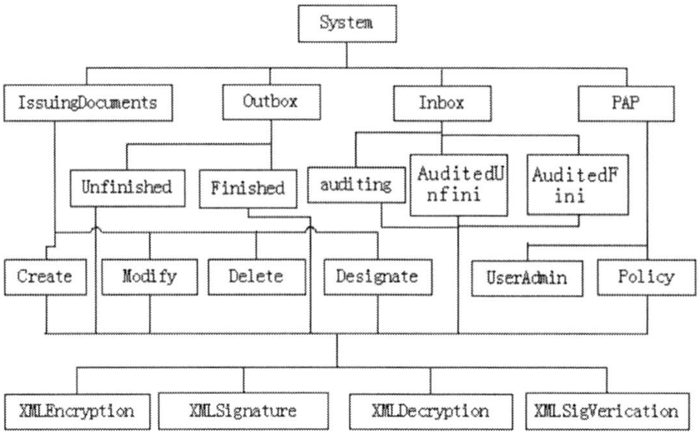

Figure 1. The module structure diagram of the system

As illustrated in the Figure 1, the four bottom mouldes are XMLEncryption, XMLSignature, XMLDecryption and XMLSigVerication moudles. They are the core modules of the system, because they implement XML Encryption and XML Signature standards, and many functional modules need to invoke them.

2.2. Design of XMLEncryption and XMLDecryption modules

XML Encryption[4] is not a new encryption algorithm, it defines how to use XML to represent encrypted data and cipher data, and it also specifies the standard encryption and decryption processes. In the XML encryption, the encrypted data may be all of a XML document, a XML element or its content, and non-XML data. The encryption data may be placed in the encrypted document, or placed in another file, and be referenced in the encrypted document. If the XML documents have many receivers, XML Encryption allows encrypting different document parts by using different keys, and thus we may send the same XML document to different receivers. XML Encryption also defines the standard decryption process according to the cipher data contained or referenced in the encrypted XML document.

The flowcharts of the XMLEncryption and XMLDecryption modules are illustrated in the Figure 2.

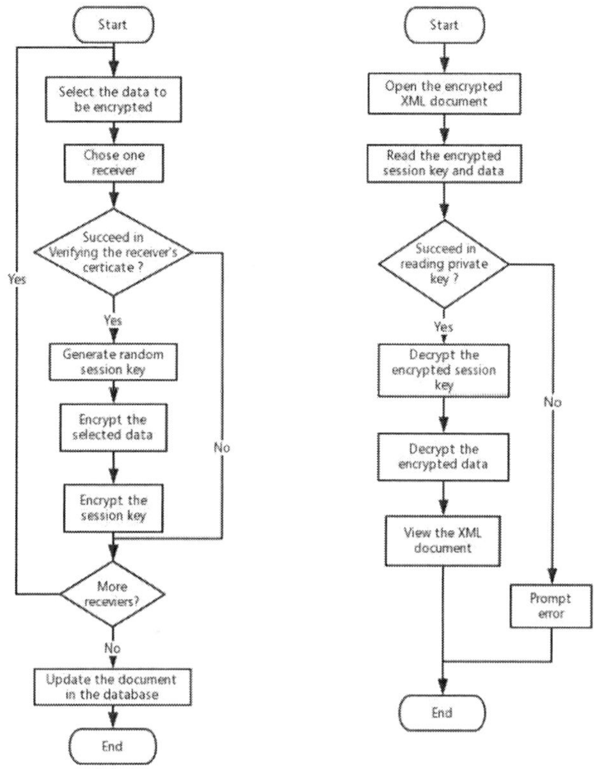

Figure 2. Flowcharts of XMLEncryption and XMLDecryption modules

In the XMLEncryption module, the sender may select one or more receivers of his document to be sent. The session key is generated randomly and encrypted by using the receiver's public key. The data to be encrypted is encrypted by using the session key. Each receiver's encrypted data will be encrypted with different session key.

In the XMLDecryption module, the receiver must provide his private key to decrypt the session key first, if succeed, and then decrypt the encrypted data which sent to him.

2.3. Design of XMLSignature and XMLSigVerification modules

XML Signature[5] is not a new signature algorithm, it defines how to use XML to represent signature data and cipher data. In the XML Signature, the data to be signed may be all of a XML document, a XML element or its content, and non-XML data. The signature may be placed in the signed document, or placed in another file, and also could be detached from the signed data.

XML Signature also specifies the standard signing and verification processes. The process of signing includes two phases: generating Reference element and generating Signature element. One or more Reference elements identify the data that is digested via a URI. The Reference element contains: an option Transforms element that in turn contains a list of one or more Transform elements describing a transformation algorithm used to transform the data before they are digested; the DigestMethod element specifies the method used to generate the digest value reported in the DigestValue element. Thus, the process of verifying also includes two steps: calculating the digest values of the referenced data and compares them with the values in the Reference element, if same, then proceed to verify the corresponding digital signature values. The flowcharts of the XMLSignature and XMLSigVerification modules are illustrated in the Figure 3.

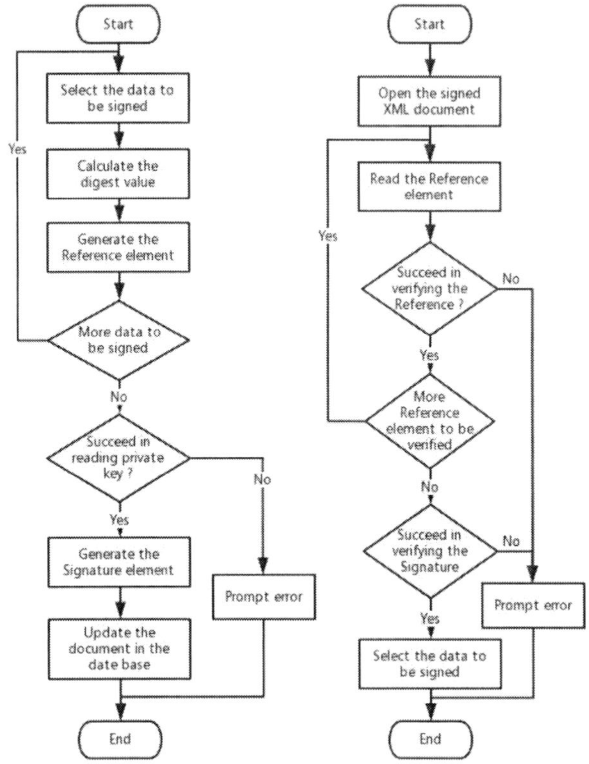

Figure 3. Flowcharts of XMLSignature and XMLSigVerification modules

3. The implementation of the system

3.1. The system architecture

The system adopts ExtJs + S2SH (Struts2 + Spring + Hibernate) frameworks to construct the system, it includes four tiers: the presentation tier, the controller tier, the business logic tier and the database tier. The system architecture is illustrated as the Figure 4.

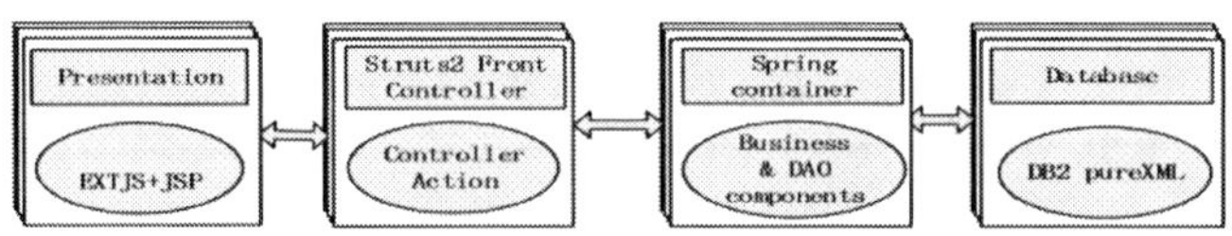

Figure 4. The system architecture diagram

3.2. The implementation of access control

XACML (XML access control markup language)[6-7] defines some XML elements to represent access control policies for XML documents, and also specifies the XML access control model. The model includes PEP (Policy Enforce Point), PDP (Policy Decision Point) and PAP (Policy Administration Point). The PEP intercepts access requests, and queries the PDP whether allows them or not. The PDP makes decisions according to the subject information, the access control policies and context parameters. The PAP is used to create access control policies and checks whether access control policies are conflict with each other.

The PEP is a struts2 filter which acts as AOP access proxyer; it intercepts all requests, and sends user's information and the request to the PDP. The PDP decides whether allow or disallow the user to access the data. The owner of XML documents uses PAP to write access control policies for his

IOP Conf. Series: Journal of Physics: Conf. Series **1060** (2018) 012004 doi:10.1088/1742-6596/1060/1/012004

documents, and check whether they are consistent or not. The access control flow chart is illustrated in the Figure 5.

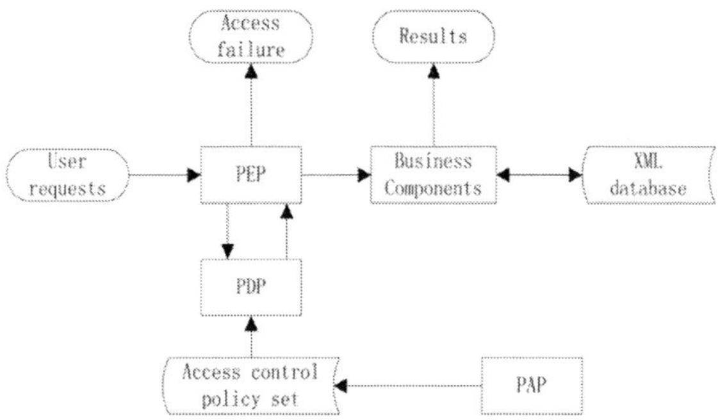

Figure 5. The access control flow chart

3.3. The implementation of documents circulation

The designate module in the Figure 1 is used to assign the documents receivers and specify the order of them. The XML documents will circulate among the receivers in the predefined order. Each receiver reads or updates the documents according to access control policies which the owner specified before.

The XML documents circulate according to predefined workflow. We design two tables in the database to help push the documents circulation: one is node table and the other is workflow table. Each row in the node table represents a receiver, and it contains userID, status,flowstep and active fields. The status field has four values to chose: 0 represents starting, 1 represents finished, 2 represents to be audited and 3 represents audited. The flowstep field records the receiving order, the default value is 0. The active field records if the document circulates to the current node. The value 0 represents not received yet, and 1 represents already received. One workflow record associates with many records in the node table.

When sending one XML document, the system sets the flowstep values of the receiver nodes according to the receiving order. The first node's status value is set to 0, the last node's status value is set to 1, and all immediate nodes' status values are set to 2. In the meantime, the first node's active value is set to 1, and the other nodes' status values are set to 0.

When the first receiver logs in the system, the auditing in his Inbox has the XML document which is associated with the first node's wfid field value. When finished auditing, the system sets the node's status value to 3, then judges if the current node is the ending node or not. If yes, set the corresponding workflow's status value to 1, meaning the workflow is finished. If no, find the next node, which has the same wfid field value ,and the flowstep field value is current node's flowstep field value plus 1, then sets the next node's active value to 1. The flow chart of documents Circulation is illustrated in Figure 6.

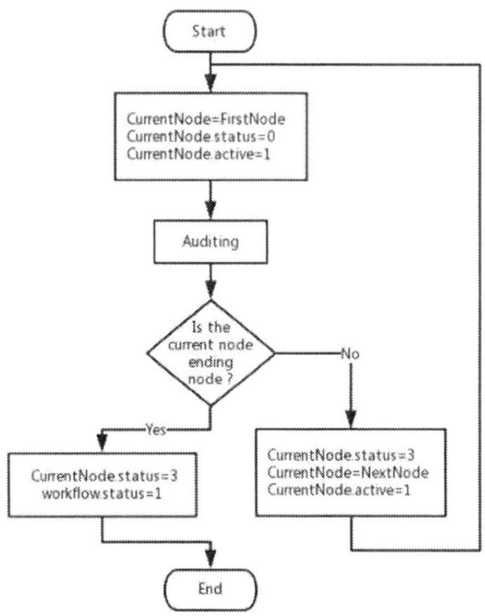

Figure 6. The flow chart of documents circulation

4. Conclusions

In this paper, we design and implement a flexible XML documents secure circulation system. The system deals with all aspects of XML security. The system is consistent with the XML Encryption, XML Signature and XACML standards, thus, it is a general framework which circulates XML documents, and could be used in the circulation of any XML documents. The system allows owners of XML documents define and modify access control policies at any time, and implements dynamic access control for all XML documents.

Acknowledgments

The paper's work is supported by Guangzhou Municipal Universities project 1201620342.

References

[1] Sookhak M, Gani A, Khan MK, Buyya R, Dynamic remote data auditing for securing big data storage in cloud computing. Inf Sci 380:101---116, 2017.
[2] Y. Li, K. Gai, Z. Ming, H. Zhao, and M. Qiu. Intercrossed access control for secure financial services on multimedia big data in cloud systems. ACM Transactions on Multimedia Computing Communications and Applications, 12(4s):67, 2016.
[3] Azhar Rauf, Waqas Ali, Maher Ahmed, Shah Khusro, Shaukat Ali, Efficient XQuery over Encrypted XML Documents, The 10th International Conference on Computer Science & Education (ICCSE 2015), July 22–24, 2015.
[4] Luigi Coppolino, Salvatore D'Antonio, Giovanni Mazzeo, and Luigi Romano, Cloud security: Emerging threats and current solutions. Computers and Electrical Engineering 59, Supplement C (2017), 126--140, 2017.
[5] C. Mainka, M. Jensen, L. Lo Iacono, J. Schwenk, I. Ivanov, M. Sinderen, F. Leymann, T. Shan, "Making XML Signatures Immune to XML Signature Wrapping Attacks", Cloud Computing and Services Science Springer International Publishing, vol. 367, pp. 151-167, 2013.
[6] S. Jan, C. D. Nguyen, L. Briand,Automated and Effective Testing of Web Services for XML Injection Attacks, Proceedings of the 2016 International Symposium on Software Testing and Analysis (ISSTA), Jul 2016.
[7] Panagiotis Ilia , Barbara Carminati , Elena Ferrari , Paraskevi Fragopoulou , SAMPAC: Socially-Aware collaborative Multi-Party Access Control, Proceedings of the Seventh ACM on Conference on Data and Application Security and Privacy, March 22-24, 2017.

The Solution of Disaster Recovery System on Cloud Computing Environment

Chang Yup Choo [1] and Kwang Sik Chung [2,*]

[1] KT Corp., Enterprise Business Performance Unit, BunDang-gu, GyeongGi-Do, seongnam, Republic of Korea
[2] Dept. Of Computer Science, Korea National Open University, Jongno-gu, Dongsung-Dong, Seoul, Republic of Korea

*E-mail : kchung0825@knou.ac.kr

Abstract. As the number of cases of cloud environment development in the financial industry has increased, cases of disaster recovery system in the financial industry are increasing. However, the establishment of the cloud environment for the main tasks in the financial industry is being restricted or limited in the cloud environment due to the security restriction regulation on the personal information and the financial business information. Unlike the establishment of a disaster recovery system for a general enterprise cloud environment, the financial industry of the cloud environment has a wide range of personal information security and protection regulations, and there are many limitations according to the respective country. Therefore, it is necessary to consider the existing one in constructing the dis-aster recovery system of the cloud environment in the financial industry. This research analyzes the solution and technology trends for existing cloud disaster recovery system, and extracts the problems of existing cloud disaster recovery system. In addition, the financial business classification criteria based on the services to be provided in the cloud environment, and determine the scope of introduction of the legacy system and the cloud system according to the cloud business classification standard are defined in this paper. In the standard of financial service to be provided in the cloud environment, the classification standard of the financial business is newly defined based on the range of influence of the personal information and the service continuity of the financial service from the core/context theory. Lastly disaster recovery system in cloud environment for financial business services is defined.

1. Introduction

In order to minimize the loss of financial services due to natural or man-made disasters, a disaster recovery system has been established and operated in financial industry. The current disaster recovery system is mostly built and operated with the same computer architecture of the main data center. However, the change management of the disaster recovery system against the change of the computerized environment is not performed smoothly. In order to compensate for this situation, there is a growing interest in building a disaster recovery system in the cloud environment and the number of cases of disaster recovery system in the cloud environment has been gradually increasing. However, there is a limitation in constructing a disaster recovery system for the cloud environment. In this paper, cloud deployment environment is defined based on financial business classification in order to build a disaster recovery system in a cloud environment, and a disaster recovery system operation scheme is proposed.

This paper is organized as follows. First previous related cloud disaster recovery technique and venders are analyzed. Then, the financial tacks classification metrics for cloud system adaptation and operations of cloud disaster recovery system are proposed. Finally, research conclusion in chapter 4 is presented.

2. Related Works

The cloud market has not reached the maturity stage, and the public cloud service is mainly provided by the telecom companies. But it does not include the disaster recovery service in the cloud service. In terms of hardware of disaster recovery sys-tem, main components are divided into server, storage, and network. Servers and networks are cloud environments that can be configured to provide a virtual environment by the cloud services while maintaining fail-over or active-active state during the data center disaster. However, in the case of storage, since the data is continuously generated and changed through the server providing the main service, remote data replication should be performed through the data replication function provided by the cloud service.

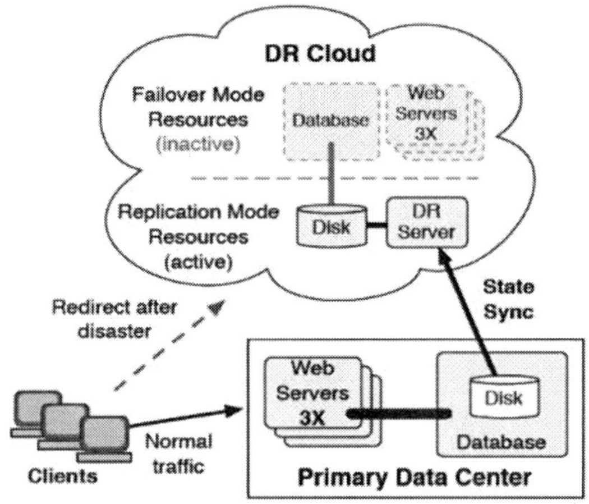

Figure 1. Conceptual Diagram of Disaster Recovery System on Cloud[1,2]

Figure 1 shows the basic concept of configuring a disaster recovery system in a cloud environment. The disaster recovery system activates resources to replicate critical data between the primary and disaster recovery centers normally. Server resources are allocated and maintained as a minimum resource. At disaster, The cloud system in the disaster recovery center allocate resources to the server using the resource provisioning function and operate the disaster recovery system to provide continuous business services to the customer

3. Building Disaster Recovery System for Cloud

3.1. Core / Context Analysis Framework for Disaster Recovery System

'Core/Context Analysis' is a analysis framework proposed by Jeffrey A. Moore[3,4]. Core means the core business of a corporation. Core business is an important part of enhancing a company's competitive advantage. Context is defined as all activities other than the core. And context task is not pursuing differentiation, but rather 'as efficient as possible in the standard way as possible' [5].

[Core / Context Analysis] is used as a framework for building a disaster recovery system in this research. The specified conditions in the Core are defined by 'The Guide to Using the Cloud Services in the Financial Services' printed by the Financial Security Institute[6]. In addition, the definition specified in mission-critical is selected through Business Impact Analysis (BIA).

The Business impact analysis is a series of analysis activities that define the tier matrix by analyzing the RTO / RPO for each business function and define the mission-critical (core or important task) according to the tier level connected to each business function. The results of business impact analysis can be defined a little differently depending on the consulting method, but the basic concept is applied equally[7,8].

Core task means that an IT system for processing, transmitting, receiving and delivering unique identification information or personal credit information. Context task means all other IT system.

And Mission-Critical task means processes shortfall creates serious and immediate risk. Non-Mission –Critical means all other processes.

Core-Mission-Critical task requires that a system that maintains direct relationships with customers in financial services and maintains competition with competitors. Context-Mission-Critical task requires that there is no direct relationship with financial services(A system that causes the business crisis when it does not work properly.). Core-Non-Mission-Critical task requires that a system that indirectly support financial services and require to maintain differentiation from competitors, but may be failed to deliver. Context-Non-Mission-Critical task requires that a System with low importance for corporate management.

Table 1. Core-Context Model for Disaster Recovery System for Cloud Service

Core/Context Analysis	Core	Context
Mission-critical	Legacy System	Public Cloud
Non-mission-critical	Private Cloud	Public Cloud

In table 1, the comparison between the mission-critical task and the non-mission-critical tasks is classified according to the importance of the task. For mission-critical operations, contracts must be made with cloud providers to enable continuous business services through redundancy or SLA recovery activities. Relatively non-mission-critical tasks are less impacted by the disability, so that contracts for disability recovery activities can be additionally provided through contracts that take economic considerations into account. Although the mission-critical and non-mission-critical tasks corresponding with the core conditions are classified according to the importance of work, they do not use the public cloud environment, so that they are applied to the maintenance of the legacy system environment and the private cloud environment. However, for mission-critical tasks, the transition to the cloud environment should be considered de-pending on the utilization of the business.

3.2. Core / context model based disaster recovery system

The method of constructing the disaster recovery system according to the system configuration type derived through the core / context analysis is differently configured. Generally, the scope of building a disaster recovery system will be different depending on the construction time and the operation time of disaster recovery system. However, since the mission-critical service has a great impact on the management of the financial company even, the disaster recovery system should be built in the advance.

- Definition of locations for providing disaster recovery services

'Mission-critical' services and 'non-mission-critical' services are in the same space from the view of Core system. 'Mission-critical' services and 'non-mission-critical' services that are associated with the 'Context' system and are also in the same space. However, the 'Core' system and the 'Context' system do not exist in the same space.

In the public cloud environment, the space of the 'mission-critical' service may be different from the space of the 'mission-critical'.

- Define when to provide disaster recovery services

Disaster of 'Mission-Critical' services can affect to whole financial business even in a temporary disruption situation so that disaster recovery systems should be con-figured and prepared in the advance. The 'Core' and 'Context' systems that provide 'non-mission-critical' services can be pre-configured and operated in accordance with the financial company's management policies and service priorities. And according to RTO(recovery time objective), the system may be newly configured and operated after a disaster occurs.

- Classification of services under the Public Cloud environment

'Context' systems with public cloud services have different 'mission-critical' and 'non-mission-critical' services. 'Mission-Critical' services should be provided for disability or disaster in the aspect of temporary interruption, while relatively 'non-mission-critical' services can be selected for low-level services in table 2.

Table 2. Disaster Recovery for Cloud Service

Service	SLA
'Mission-critical' service	Service recovery within the due time in case of a service failure Automatic Fail-Over in case of H/W failure Disaster recovery service (transition within RTO) Minimum data loss (data recovery within RPO), etc.
'Non-mission-critical' service	Service recovery within the due time in case of a service failure Automatic Fail-Over in case of H/W failure

3.3. Operate Disaster Recovery System with Core/Context Model

Disaster of 'Mission-critical' with 'Core' system and 'mission-critical' & 'Context' system can affect to whole financial business even in a disruption situation so that disaster recovery systems should be configured and prepared in the advance. How-ever, since the provision locations of the services are different from each other, the 'Core' and 'Context' systems do not cause the same disaster situation from the perspective of disaster.

'Mission-critical' services and 'non-mission-critical' services that are classified as 'Core' systems. And the services are directly affected when a disaster occurs. And the service continuity is maintained as shown in figure 2. It also connects with 'mission-critical' services using public cloud to provide continuous services.

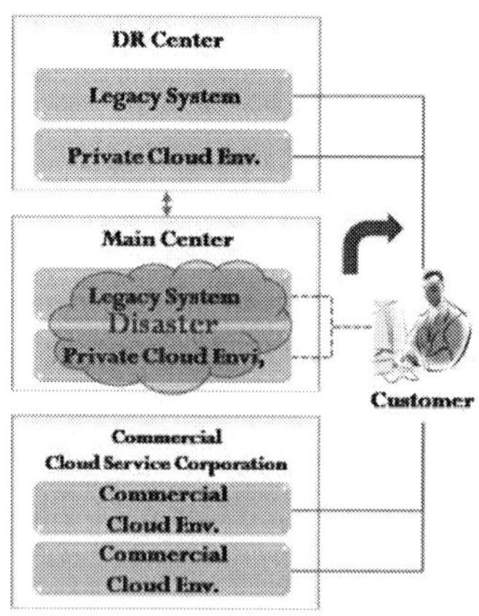

Figure 2. Configuration for Data Center Disaster

In the event of a disaster in a cloud service provider as shown in figure 3, business services resume within RTO and RPO through a disaster recovery service contracted with a cloud service provider. A disaster recovery center operated by a cloud service provider provides business continuity for business services with a disaster recovery service contract.

IOP Conf. Series: Journal of Physics: Conf. Series **1060** (2018) 012005 doi:10.1088/1742-6596/1060/1/012005

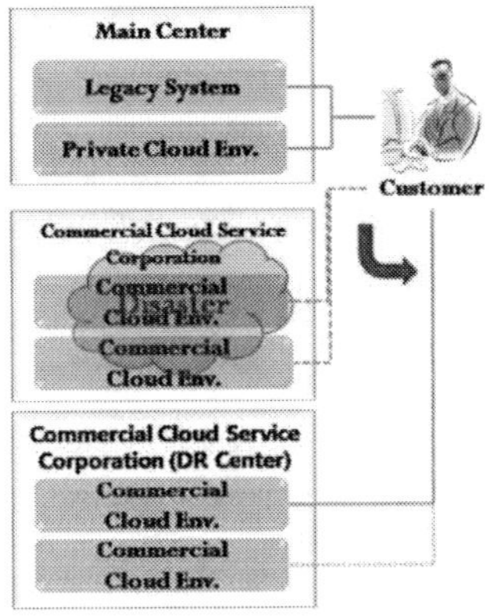

Figure 3. Configuration for Cloud Service Provider Disaster

If situations as shown in figure 2 and figure 3 occur, the disaster recovery system for each business service defined in the category is operated according to the operating procedure specified in tables 3 and table 4.

Table 3. Pre- and post-disaster Core/Context service Activities [Figure 2]

Category	Normal Operation	Disaster Recovery Operation
Core & Mission-critical	Disaster recovery system exists. The Disaster Recovery center is operated at a distance from the data center.	Start disaster recovery system within RTO. Resume service after data recovery according to RPO.
Context & Mission-critical	Disaster recovery system exists. Disaster recovery service contract through public cloud provider.	In case of own data center disaster, link with disaster recovery system Maintain existing services.
Core & Non-mission-critical	Disaster recovery system exists (possible). The Disaster Recovery center is operated at a distance from the data center.	If the disaster recovery system is operated, the disaster recovery system is operated within the RTO, Resume service after data recovery according to RPO.
Context & Non-mission-critical	Disaster recovery system does not exist.	In case of self-center disaster, link with disaster recovery system, Maintain existing services.

Table 4. Pre- and post-disaster Core/Context service Activities [Figure 3]

Category	Normal Operation	Disaster Recovery Operation
Core & Mission-critical	Disaster recovery system exists. The Disaster Recovery center is operated at a distance from the data center.	Integration of Disaster recovery system of public cloud center.
Context & Mission-critical	Disaster recovery system exists.	Operate disaster recovery system of public cloud center(RTO level assurance).

	Disaster recovery service contract through public cloud provider.	Resume target data recovery and according to RPO and business services.
Core & Non-mission-critical	Disaster recovery system exists (possible). The Disaster Recovery center is operated at a distance from the data center.	Integration of Disaster recovery system of public cloud center.
Context & Non-mission-critical	Disaster recovery system does not exist.	Rebuilding public cloud services. Resume business services.

4. Conclusion

Personally identifiable financial information dealt with in the financial industry requires the highest level of information security [9]. Therefore, adaptation of the cloud environment in the financial industry has many limitations. However, it is possible to derive information or financial services that can be handled in the cloud environment through business analysis of the financial industry.

In this paper, the criteria for classifying the business and financial information services of the financial industry are defined based on the [Core / Context Analysis] theory. In addition, those that can be deployed in the cloud environment and those that should be provided as legacy systems according to the newly defined [core / context classification criteria]. The disaster recovery system for financial services and financial business supported is designed and proposed in the cloud environment. In particular, an operational frame for the normal operation of financial information services for disaster recovery is proposed. Through this design, it will be possible to establish a financial system cloud system that can manage personal information and financial information efficiently and establish a disaster recovery system.

References

[1] Jongbeom Jeon, 'A Research for the New Disaster Recovery Method on Cloud Environment', 2014, Korea University

[2] Timothy Wood, Emmanuel Cecchet, K.K. Ramakrishnan, Prashant Shenoy, Jacobus van der Merwe, and Arun Venkataramani,'Disaster Recovery as a Cloud Service:Economic Benefits & Deployment Challeges', In proceedings of the Conference on Hot Topics in Cloud Computing (HotCloud),August 2010

[3] 'Living on the Fault Line, Revised Edition:Managing for shareholder Value in Any Economy', August 20, 2002, Harper Business, Geoffrey A. Moore

[4] 'Dealing with Darwin : How Great Companies Innovate at Every Phase of Their Evolution', 2005, Penguin, Geoffrey A. Moore

[5] 'CLOUD NO SHOUGEKI', 2009, TOYO KEIZAI INC., Makoto Shirota

[6] 'The guide for using financial service in the Cloud',2016.10.14,Financial Security Institute

[7] Omar H. Alhazmi , Yashwant K. Malaiya,'Evaluating Disaster Recovery Plan Using the Cloud',In proceeding of Reliability and Maintainability Symposium (RAMS), 2013

[8] Mr.A.Srinivas, Y.Seetha Ramayya, B.Venkatesh,' A Study on Cloud Computing Disaster Recovery', International Journal of Innovative Research in Computer and Communication Engineering, August 2013, Vol. 1,pp1383-1384

[9] 'Financial Supervisory Regulation',Amended on 2016.10.15

IOP Conf. Series: Journal of Physics: Conf. Series **1060** (2018) 012006 doi:10.1088/1742-6596/1060/1/012006

ISAR Target Recognition Based on Two-dimensional Locality Preserving Projection

Hong CAI[1], Le HAO[2],Yongzhi SU[1]

[1]NCO School,Space Engineering University, Beijing 102249, China
[2]Service Center For Foreign Ministry And Its Overseas Missions, Beijing 100010, China

Caihong72@qq.com

Abstract. It is well known that the relationship between different radar target data is often nonlinear, so the recognition rate will decrease if the traditional linear subspace methods are used to reduce the dimension of inverse synthetic aperture radar (ISAR) images. Recently, Locality Preserving Projections (LPP), a kind of manifold learning algorithm, is proposed for ISAR target recognition. But the ISAR images must be vectorized in LPP algorithm,which would result in the problem of singularity, and the local structural information from original images is easy to lose. In this paper, a novel ISAR target recognition algorithm is presented by using two-dimensional Locality Preserving Projections (2DLPP). The proposed 2DLPP algorithm is based directly on 2D ISAR images rather than 1D vectors as conventional LPP does. The simulated experimental results suggested that 2DLPP has better classification performance comparing to PCA,LDA and LPP.

1. Introduction

In recent years, automatic target recognition (ATR) based on inverse synthetic aperture radar (ISAR) images has received considerable attention from the researchers[1]–[4]. Different from the optical images which can describe the overall shape of the targets, ISAR images just represent the sparse distribution of scattering centers. Also, the dimension of the ISAR images is high, and the ISAR images of the same target will be changed when the target is in different distance, attitude and rotation rate. Therefore, it is important to adopt proper and robust techniques for dimension reduction and feature extraction in ISAR target recognition.

At present, some linear subspace methods, such as Principal Component Analysis (PCA)[5][6], Linear Discriminant Analysis (LDA)[7][8] and Independent Component Analysis (ICA)[9][10], are widely applied in radar target recognition domain. However, these methods only can find the linear structure of the observation data, but actually the ISAR data will be non-linear due to the complex movement of the target. For this reason, better results will be received if non-linear techniques are applied.

Nowadays, a number of research efforts have shown that many high-dimension observation data possibly reside on a non-linear sub-manifold which is controlled by a small number of implicit parameters. The idea of manifold learning provides us a new way to dimension reduction via analyzing the intrinsic dimension of high-dimension sampling data. Techniques that have been proposed for this purpose include Isometric Mapping (ISOMAP) [11], Locally Linear Embedding (LLE) [12] and Laplacian Eigenmap (LE)[13]. However, they yield maps that only defined on the training data points and it remains unclear how to evaluate the novel test data points. So these methods might not be used in ISAR image target recognition directly. In order to overcome the above shortcomings in manifold learning, He et al put forward a new algorithm called Locality Preserving Projections (LPP)[14] and applied it to face recognition successfully[15]. LPP is a linear subspace method derived from Laplacian Eigenmap, it not only has the advantages of linear subspace method, but also has the capability of manifold learning.

In 2010,Qiang He et al proposed LPP for ISAR target recognition[16], they pointed that the ISAR images do reside on a low-dimension sub-manifold embedded in the high-dimension ISAR image space, which is controlled by fewer parameters, such as attitude angle, cross-range scale and position, etc., and the better recognition performance is acquired with the low-dimension feature obtained by LPP, comparing to PCA and LDA.

However, there are two disadvantages in this application:(1)the 2D ISAR images must be transformed into 1D vector,this will result in the problem of singularity of matrices, when the number of samples is far smaller than the dimension of samples;(2) The image-to-vector transform will lose the local structural information embedded in the original ISAR images which may be more important for target recognition.

In order to avoid these problems, a novel ISAR target recognition algorithm is presented by using two-dimension Locality Preserving Projections (2DLPP) [17-19].The proposed 2DLPP algorithm is based directly on 2D ISAR image matrices rather than 1D vectors as conventional LPP does. The simulated experimental results suggest that the 2DLPP algorithm has better classification performance comparing to LPP algorithm.

2. Locality Preserving Projections (LPP)

LPP is the optimal linear approximations to the Laplacian Betrami operator on the manifold [14]. Therefore, although it is a linear method, it seems to recover the important aspects of the intrinsic non-linear manifold structure by preserving local structure.

Given a set of high-dimension data points $X=\{x_1, x_2,\ldots,x_m\}$ in high-dimension Euclidean space $\boldsymbol{R}^n$, find a transformation matrix $\boldsymbol{W}$ that maps these m points to a set of points $\boldsymbol{Y}=\{y_1,y_2,\ldots,y_m\}$ in $\boldsymbol{R}^l$ ($l<<n$), such that $y_i=\boldsymbol{W}^{\mathrm{T}}x_i$ is the one-dimension representation of original data vector x_i. The transformation matrix $\boldsymbol{W}$ will be obtained by minimizing the following objective function

$$J=\sum_{i,j}\left\|\boldsymbol{y}_i-\boldsymbol{y}_j\right\|^2\boldsymbol{S}_{ij} \tag{1}$$

In Equation (1), the local relationships among the input data vectors are described by similarity matrix $\boldsymbol{S}$, where the similarity relationship is defined as follows:

$$\boldsymbol{S}_{ij}=\begin{cases}\exp(-\|\boldsymbol{x}_i-\boldsymbol{x}_j\|^2/t), & e(\boldsymbol{x}_i,\boldsymbol{x}_j)=1\\ 0 & e(\boldsymbol{x}_i,\boldsymbol{x}_j)=0\end{cases} \tag{2}$$

In Equation (2),$e(x_i, x_j)$ is an indicator function designating whether x_i and x_j are neighbors and t is the heat kernel factor. The neighborhood of a given input vector x_i can be defined as the k-nearest vectors to x_i.

Following some simple algebraic steps, the above objective function reduced to follows:

$$\frac{1}{2}\sum_{i,j}\left\|\boldsymbol{y}_i-\boldsymbol{y}_j\right\|^2\boldsymbol{S}_{ij}=\mathrm{tr}(\boldsymbol{W}^{\mathrm{T}}\boldsymbol{XLX}^{\mathrm{T}}\boldsymbol{W}) \tag{3}$$

Where $\boldsymbol{L}=\boldsymbol{D}$-$\boldsymbol{S}$ is Laplacian matrix. The matrix $\boldsymbol{D}$ is a diagonal matrix of $m\times m$, its entries are $\boldsymbol{D}_{ii}=\sum_j\boldsymbol{S}_{ij}$. Matrix $\boldsymbol{D}$ provides a natural measure on the data points, the bigger the value $\boldsymbol{D}_{ii}$ is, the more "important" $\boldsymbol{y}_i$ is. Impose a constraint to the objective function as follows:

$$\boldsymbol{Y}^{\mathrm{T}}\boldsymbol{DY}=1\Rightarrow\boldsymbol{W}^{\mathrm{T}}\boldsymbol{XDX}^{\mathrm{T}}\boldsymbol{W}=1 \tag{4}$$

Finally, the transformation matrix $\boldsymbol{W}$ that minimizes the objective function is given by the maximum eigenvalue solution to the generalized eigenvalue problem:

$$\boldsymbol{XLX}^{\mathrm{T}}\boldsymbol{W}=\lambda\boldsymbol{XDX}^{\mathrm{T}}\boldsymbol{W} \tag{5}$$

3. Two-Dimension Locality Preserving Projections (2DLPP)

Consider a set of ISAR sample images $\{A_i\}_{i=1}^{k} \in \mathrm{R}^{m\times n}$,design a linear tranformation $y_i = A_i W$,which maps the original ($m\times n$)-dimension image space into an ($m\times d$)-dimension feature space. Here, $W \in \mathrm{R}^{n\times d}$ is the projection matrix.

The objective function of 2DLPP method is also equation (1):

$$\begin{aligned} J &= \sum_{i,j}\|y_i - y_j\|^2 S_{ij} = \sum_{i,j}\|A_i W - A_j W\|^2 S_{ij} \\ &= \sum_{i,j}\|(A_i - A_j)W\|^2 S_{ij} = W^{\mathrm{T}}(\sum_{i,j}(A_i^{\mathrm{T}}A_i - A_j^{\mathrm{T}}A_j)S_{ij})W \\ &= W^{\mathrm{T}}\sum_{i,j}(A_i^{\mathrm{T}}S_{ij}I_m A_i - A_j^{\mathrm{T}}S_{ij}I_m A_j)W \\ &= W^{\mathrm{T}}A^{\mathrm{T}}[(D-S)\otimes I_m]AW \\ &= W^{\mathrm{T}}A^{\mathrm{T}}[L\otimes I_m]AW \end{aligned} \tag{6}$$

Where A=[A_1, A_2,..., A_k],the definition of S,L and D is the same as LPP, $\otimes$ is definded as Kronecker product. Furthermore,in order to remove the arbitrary scaling factor in the embeding, a constraint is imposed as followings:

$$\begin{aligned} \sum_i D_{ii} X_i^{\mathrm{T}} X_i = 1 &\Rightarrow W^{\mathrm{T}}(\sum_i D_{ii} A_i^{\mathrm{T}} A_i)W \\ &= W^{\mathrm{T}}A^{\mathrm{T}}(D\otimes I_m)AW = 1 \end{aligned} \tag{7}$$

Then the minimization problem of the objective function becomes

$$\min_W W^{\mathrm{T}}A^{\mathrm{T}}(L\otimes I_m)AW \tag{8}$$

The transformation vector W is given by the minimum eigenvector solution to the generalized eigenvalue problem:

$$A^{\mathrm{T}}(L\otimes I_m)AW = \lambda A^{\mathrm{T}}(D\otimes I_m)AW \tag{9}$$

Through solving equation(9), the best projection Vector W can be obtained .

4. Proposed classification algorithm

In this section, we describe the details of our method.

4.1 Data pre-processing

In this study, we perform two steps to complete the pre-processing task. The first step is to extract the target from the background clutter or noise. Here, we assume that the signal-to-noise ratio (SNR) or signal-to-clutter is relatively high, so we can select a threshold to remove the background noise and clutter from the original image. Thus, those pixels with higher values than the threshold are considered as target, and the remaining pixels are set to zero. Nowadays, there are many methods to help us to determine the threshold, such as edge detection, histogram-based segmentation techniques and intensity-based method, and so on. But their performance is not always reliable [20].In this paper, we use the Constant False Alarm Rate (CFAR) algorithm to obtain an adaptive threshold.

The next step is to normalize the ISAR image intensity. As we all know that the radar echo intensity will be influenced by many factors, such as distance, receiver gain, and so on. For this reason, it is necessary for us to normalize the ISAR image intensity in order to remove the disadvantages arising from the varying intensity. In this paper, we use the whole energy to normalize the ISAR image. Suppose $I_o(x_i,y_j)$ is the intensity of a random pixel on the ISAR image, then the normalized intensity of this pixel can be expressed as:

IOP Conf. Series: Journal of Physics: Conf. Series **1060** (2018) 012006 doi:10.1088/1742-6596/1060/1/012006

$$I_{\text{n}}(x_i, y_j) = \frac{I_{\text{o}}(x_i, y_j)}{\sum_{i=1}^{M}\sum_{j=1}^{N} I_{\text{o}}(x_i, y_j)} \quad (10)$$

As a result, the influence of the varying intensity in recognition is removed.

4.2 Dimension reduction based on 2DLPP

Now the entire procedure of dimension reduction based on 2DLPP algorithm is summarized as follows:

1) *Create a graph G with n nodes*:The *i*th node denotes the ISAR image A_i. put an edge between the *i*th and *j*th node, if A_i is among the *k* nearest neighbors of A_j, or A_j is among the *k* nearest neighbors of A_i, and also A_i and A_j must have the same class label.
2) *Compute the weights*:If node *i* and node *j* are connected, put a similarity weight $\boldsymbol{S}_{ij}$ on it, otherwise zero. There are two variations for weighting the edges: to change the default, adjust the template as follows:

(1)*heat kernel*: If there is an edge between node *i* and node *j*, put $\boldsymbol{S}_{ij}$=exp $(-\|\boldsymbol{A}_i-\boldsymbol{A}_j\|^2/t)$, where *t* is a suitable constant, otherwise$\boldsymbol{S}_{ij}$=0.

(2)*simple-minded*: $\boldsymbol{S}_{ij}$=1,if node *i* and node *j* are linked by an edge, otherwise $\boldsymbol{S}_{ij}$=0.In our work, the heat kernel method is applied for its stability.

3) *Eigenmap*:Compute the eigenvectors and eigenvalues for the generalized eigevector problem: $\boldsymbol{A}^{\text{T}}(\boldsymbol{L}\otimes\boldsymbol{I}_{\text{m}})\boldsymbol{AW}=\lambda\ \boldsymbol{A}^{\text{T}}(\boldsymbol{D}\otimes\boldsymbol{I}_{\text{m}})\boldsymbol{AW}$.

Let the solutions of this equation are $\boldsymbol{w}_0,\boldsymbol{w}_1,\ldots,\boldsymbol{w}_{d-1}$,and their corresponding eigenvalues are λ_0, $\lambda_1,\ldots\lambda_{d-1}$.The final linear dimension reduction mapping is $\boldsymbol{A}_i\rightarrow\boldsymbol{y}_i=\boldsymbol{A}_i\boldsymbol{W}$, where $\boldsymbol{W}=(\boldsymbol{w}_0,\boldsymbol{w}_1,\ldots,\boldsymbol{w}_{d-1})$.

4.3 Classifier design

In this paper, the purpose is to evaluate the performance of the proposed dimension reduction algorithm, so the simple *k*-nearest neighbor classifier is used to complete the recognition. The *k*-nearest neighbor classifier is an important approach for non-parametric classification, its basic idea is: Suppose the training set has *c* classes ω_1, $\omega_2,\ldots$, ω_c, the number of each class is N_i, i=1,2,…,*c*. Calculate the Euclidean distance between the test sample $\boldsymbol{y}$ and the arbitrary train sample $\boldsymbol{y}_i^j$:

$$d_i^j = \left\|\boldsymbol{y}-\boldsymbol{y}_i^j\right\|, j=1,2,...,N_i \quad (11)$$

Where *i* represent the ω_i class, *j* represent the *j*th sample of ω_i class. Suppose the test sample $\boldsymbol{y}$ has *k* neighbors, the sample number of ω_1, $\omega_2,\ldots$, ω_c class is k_1, $k_2,\ldots$, k_c respectively. Define the distinguish function:

$$d_i(\mathbf{y}) = k_i, i=1,2,...,c \quad (12)$$

If the *k* nearest algorithm has no threshold, and an arbitrary class satisfies:

$$d_m(\mathbf{y}) = \max_{i=1,2,...,c} d_i(\mathbf{y}) \quad (13)$$

Then $\boldsymbol{y}$ is considered belonging to the *m*th class.In other words,for an unkown sample $\boldsymbol{y}$, its *k* nearest neighbors are selected, and $\boldsymbol{y}$ is labeled the class of most neighbors.

If the *k* nearest algorithm has threshold *r*,suppose

$$D = \min_i \|\mathbf{y}-\mathbf{y}_i\|, i=1,2,...,k \quad (14)$$

If $D \geq r$, $\boldsymbol{y}$ is considered not belonging to any class,otherwise the class label of $\boldsymbol{y}$ can be determined by *k* nearest method.

5. Experimental results

5.1 Data preparation

In order to demonstrate the proposed scheme, we made a classification experiment using four kinds of ISAR images: F15, F16,Jian10 and "Jaguar". These images are simulated based on high-frequency electromagnetic scattering theory[20] and stepped-frequency ISAR imaging principles [21-23].

The following radar parameters are selected: carrier frequency f_0=9*GHz*, stepped frequency step size Δf=3*MHz*, number of stepped frequency steps N=64, number of pulses M=256, pulse repetitive frequency f_p=15*KHz*.

Here,the ISAR images are generated from -45^0 to 45^0 yaw angle (0^{0}pitch and 0^{0}roll), -45^0 to 45^{0}pitching angle(0^{0}yaw and 0^{0}roll), -45^0 to 45^0 rolling angle (0^{0}pitch and 0^{0}yaw) respectively with a 3^{0}interval, 8 different rotate rates and 15 different distances are used. Thus, the number of these ISAR images for each target is 3600×3.These ISAR images are cut into N=35 in down-range direction and M=100 in cross-range direction. Figure 1 shows some of the ISAR images with different attitude angle, cross-range scale and position.

In this classification experiment, we select 600 samples randomly from F15, F16 and Jian10 respectively, yielding 600×3 train samples, and the test set A is include 900 samples which are selected from the remainder ISAR images of f15, f16 and jian10. The test set B has 300 samples from "Jaguar", where the situation of rejection recognition is considered. As a result, the training set and

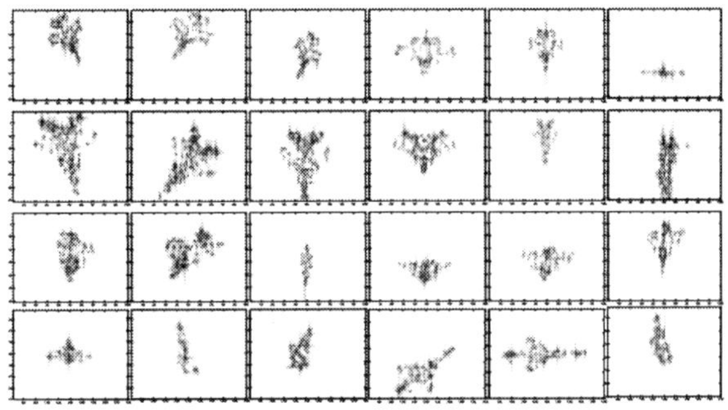

Figure 1. Some samples with different attitude angle, cross-range scale and position.

test set are entirely different. The training set is used to learn the transformation matrix $\boldsymbol{W}$, test samples are mapped into the feature subspace with the transformation matrix, and then the k-nearest neighbor classifier is used to complete the recognition.

5.2 Experimental result

Table 1.shows the recognition results. We can see that the higher recognition rate is acquired with the feature space obtained by 2DLPP algorithm comparing to PCA ,LDA and LPP. The recognition rate in original high-dimension ISAR image space is represented as "Baseline". In this simulation experimentation, the signal-to-noise ratio (SNR) of ISAR images is set to 10dB.

Table 1.Recognition performance with different method

Approach	Dims	Recognition Rate
Baseline	3500	69.1%
PCA	30	89.1%
PCA+LDA	2	85.2%
PCA+LPP	15	95.6%
2DLPP	35×6	98.6%

5.3 Algorithm complexity analysis

In this paper,the algorithm complex of LPP and 2DLPP is also be analysed.

1) Create adjacent graph G: the algorithm complexity is $O(N^2+\log N)+O((N(N-1)/2)(m\times n))$.
Where $m\times n$ is the pixel quantity of ISAR image, N is the number of ISAR images.
2) Compute the weights:
(1)*heat kernel*: The algorithm complexity is $O(K\times N\times m\times n)$,where k is the number of neighberhood.
(2)*Simple-minded*: In this case, the algorithm complexity is $O(k\times N)$.
3) Eigenmap:
(1)*LPP*: The algorithm complexity of $\boldsymbol{XLX}^T$ is $O(m\times n\times(k+1)\times N)+O((m\times n)^2\times N)$; The algorithm complexity of $\boldsymbol{XDX}^T$ is $O(m\times n\times N)+O((m\times n)^2\times N)$; The algorithm complexity of calculating the translation matrix $\boldsymbol{W}$ is $O((m\times n)^3)$;The total algorithm complexity is $O(2mn\times mnN+m^3n^3)$.
(2)*2DLPP*:
The algorithm complexity of $\boldsymbol{A}(\boldsymbol{L}\otimes \boldsymbol{I}_m)\boldsymbol{A}^T$ is $O(m\times(k+1)\times N)+O(m\times n\times N\times(k+1))+O(m\times N\times(n)^2)$;The algorithm complexity of $\boldsymbol{A}(\boldsymbol{D}\otimes \boldsymbol{I}_m)\boldsymbol{A}^T$ is $O(m\times N)+O(m\times n\times N)+O(m\times N\times(n)^2)$;The algorithm complexity of calculating the translation matrix $\boldsymbol{W}$ is $O((n)^3)$. The total algorithm complexity is $O(2N\times mnN+n^3)$.
Obviously,the algorithm complexity of LPP is much higher than 2DLPP.

6. Conclusion

In this paper, the idea of manifold learning is introduced into ISAR target recognition domain. Firstly, ISAR image subspace is obtained by 2D Locality Preserving Projections (2DLPP) algorithm, and then the k nearest neighbor classifier is applied to complete the recognition. The proposed 2DLPP algorithm is based directly on 2D ISAR images rather than 1D vectors as conventional LPP does. The simulated experimental results suggested that 2DLPP has better classification performance comparing to PCA,LDA and LPP.

References

[1] Hou Yingni, Yang Yuhao, Li Shiguo and Jiang Tao 2016 A study on ship classification based on ISAR images *J. Modern Radar* **38** 47-50
[2] Li Feng, Cao Jun, Ren Lixiang and Long Teng 2016 Micro-doppler extraction from ISAR image *J.ICASSP* 3101-05
[3] Bozkurt H and Erer I 2016 Multi scale bilateral filter enhanced ISAR target recognition *Signal Processing &Communication Application Conf.* 1993-96
[4] Anna Kurowska 2016 The preliminary survey of ship recognition algorithms using ISAR images *international Radar Symp.* 1-4
[5] Jolliffe I T 1986 *Principle component analysis*(New York: Springer-Verlag)
[6] Amit Kumar Mishra and Tshiamo Motaung 2015 Appication of linear and nonlinear PCA to SAR ATR *International Conf. Radioelektronika* 349-54
[7] Fukunaga K 1990 *Introduction to statistical pattern classification*(San Diego: Academic Press)
[8] Liu Hualin and Yang Wanlin 2007 Radar target recognition based on direct discriminant analysis using range profile *J. Chinese Journal of Radio Science* **22** 1021- 24
[9] Hyvarinen A, Oja E and Karhunen J *2001 Independent Component Analysis*(New York: Wiley)
[10] Bai Jinliang,Jiang Zhiye and Liu Jiaqi 2015 ISAR image target micro-doppler feature extraction and recognition method based on PCA and ICA *IET International Radar Conf.*
[11] Tenenbaum J B, Silva V D and Langford J C 2000 A global geometric framework for nonlinear dimensionality reduction *J. Science* **290** 2319-23
[12] Roweis S T and Saul LK 2000 Nonlinear dimensionality reduction by locally linear embedding *J. Science* **290** 2323-26
[13] Belkin M and Niyogi P 2003 Laplacian eigenmaps for dimensionality reduction and data representation *J. Neural Computation* **15** 1373-96
[14] He Xiaofei and Niyogi P 2003 Locality preserving projections *Proc.Conf. Advances in Neural Information Processing Systems*
[15] He Xiaofei, Yan Shuicheng and Hu Yuxiao 2005 Face recognition using Laplacianfaces *J.IEEE Transactions on Pattern Analysis and Machine Intelligence* **27** 328-40

[16] He Qiang,Cai Hong,Han Zhuang-zhi and Shang Chao-xuan 2010 ISAR target recognition based on non-linear manifold learning *J.ACTA ELECTRONICA SINICA* **38** 585-90
[17] Chen Sibao, Zhao Haifeng, Kong Min and Luo Bin 2007 2D-LPP:A two-dimensional extension of locality preserving Projections *J.Neurocomputing* **70** 912-21
[18] Hu Dewen,Feng Guiyu and Zhou Zonglan 2007 Two-dimensional locality preserving projection(2DLPP)with its application to palmprint recognition *J.Pattern Recognition* **40** 339-42
[19] Sun Lijuan,Yang Dan,Zhang Xiaohong and Sun Lili 2007 Face recognition based on SR-2DLPP *J.Application Research of Computers* **26** 2789-92
[20] Kyung-Tae Kim, Dong-Kyu Seo and Hyo-Tae Kim 2005 Efficient classification of ISAR images *J. IEEE Transactions on Antennas and Propagation* **53** 1611-21
[21] Luo Hong, Xu Xiaojian and Huang Peikang 1999 On the modeling and prediction of wideband radar signals *J. Acta Electronica Sinica* **27** 41-44
[22] Mun and Kok Leong 2002 *Stepped frequency imaging radar simulation*(Naval Postgraduate School: Monterey, California)
[23] Wang Yonglei,He Qiang and Caihong 2013 Simulation on ISAR 2-dimensional scattering point model based on 3DS *J.Fire Control &Command Control* **38** 18-20

IOP Conf. Series: Journal of Physics: Conf. Series **1060** (2018) 012007 doi:10.1088/1742-6596/1060/1/012007

Compatibility Studies of IMT System and Automotive Radar in the Frequency Range 71-78 GHz

Zhaojun Qian, Jiajia Chen* and Tan Wang

The State Radio Monitoring Center

*Email: chenjiajia@srrc.org.cn

Abstract. The millimeter waves have large available bandwidth and thus gigabit throughput can be easily achieved for future wireless communication systems. The frequency band has been identified as potentially suitable frequency band for IMT (International Mobile Telecommunications) systems. In this paper, consideration ECC and Huawei relevant paper, the coexistence interference of 71-78 GHz-band international mobile telecommunications (IMT) and automotive radar is analyzed based on actual geographical and ITU-R system characteristic. Specifically, the interferences of IMT system base stations (BS) and users (UE) to automotive radar are studied, considering high frequency band radio propagation clutter loss. This paper presents the Monte Carlo results in terms of the level of compatibility interference power vs. the distances between IMT and automotive radar in the straight trajectory by using Visualyse V7.9 simulation tool. The results show that in some case the IMT BS operating in the band 71-76 GHz may be well compatible with the automotive radar, the interference power value is mainly under protection criterion value. The provided results are useful for design of future 71/80 GHz IMT-2020 (5G) wireless communications systems and automotive radar system.

1. Introduction

The millimeter waves have large available bandwidth and thus gigabit throughput can be easily achieved for future wireless communication systems. The frequency band has been identified as potentially suitable frequency band for IMT (International Mobile Telecommunications) systems [1], in dense urban areas to provide large bandwidth and high volume business [2]. In China, the 26 GHz radio is approved by Chinese Ministry of Information Industry for local multipoint distribution system access network business, this band was carried out in China Mobile, China Telecom, China Unicom, China Netcom [3].WRC-19 agenda item 1.13 (IMT-2020) considers *identification of frequency bands for the future development of International Mobile Telecommunications (IMT),* between 24.25 GHz to 86 GHz. The bands of 71-76 GHz and 81-76 GHz are within the scope of the potential bands.

Considerate ECC and Huawei relevant paper [4][5], this contribution presents the initial compatibility studies of IMT system operating in the band 71-76 GHz with automotive radar in the adjacent band 76-77 GHz. In this included study, only single entry interference from an IMT base station (BS) and three user equipment (UE) to a receiver of automotive radar is addressed.

2 .Allocation information in the frequency range 66-81 GHz

The Radio Regulation [6] table of frequency allocations in the band 71-78 GHz is provided below for reference. Automotive radar is one kind of the radiolocation service and allocated in the band 76-81 GHz, see table 1.

Table 1. Allocation in the band 71-78 GHz allocation to services

Region 1	Region 2	Region 3
71-74	FIXED FIXED-SATELLITE (space-to-Earth) MOBILE MOBILE-SATELLITE (space-to-Earth)	
74-76	FIXED FIXED-SATELLITE (space-to-Earth) MOBILE BROADCASTING BROADCASTING-SATELLITE Space research (space-to-Earth) 5.561	
76-77.5	RADIO ASTRONOMY RADIOLOCATION Amateur Amateur-satellite Space research (space-to-Earth) 5.149	
77.5-78	AMATEUR AMATEUR-SATELLITE RADIOLOCATION 5.559B Radio astronomy Space research (space-to-Earth) 5.149	

3. Technical characteristics

3.1 Automotive radar parameters

According to Recommendation [7], the technical characteristics of automotive radar in the frequency band 76-77 GHz are summarized in Table 2.

Table 2. Technical characteristics of automotive radar in the band 76-77 GHz

Parameter	Radar A
Frequency	76-77 GHz
Bandwidth	1 GHz
Noise	-69 dB
Antenna height	0.7 m
Antenna deployment	Front of the vehicle
Antenna peak gain	25 dBi
Noise = RX sens (1 GHz) in dBm + RX-noise figure 15 dB (note 1)	-69 dBm (= -84 dBm + 15)
Antenna azimuth scan angel (degrees)	TX/RX:
Receiver noise figure (dB)	15

3.2 IMT parameters

According to the liaison statement from WP 5D to TG 5/1 [8][9], the typical characteristics of IMT system for sharing and compatibility analyses are listed in Table 3.

Table 3. Technical characteristics of imt bs in the band 71-76 GHz

Parameter	IMT (Base station)
Frequency	71-76 GHz
Channel bandwidth	200 MHz
Network loading factor	100 % for single-entry
Duplex method	TDD
BS TDD activity factor	100 % for single-entry
Antenna height	6 m
Antenna deployment	Below roof top
Downtilt	10°
Orientation	[-180°, 180°]
Antenna pattern	Refer to Rec. ITU-R M.2101
Element gain	5 dBi
Horizontal/vertical 3 dB beamwidth of single element	65° for both H/V
Horizontal/vertical front-to-back ratio	30 dB for both H/V
Antenna array configuration (Row × Column)	16×16 elements
Horizontal/Vertical radiating element spacing	0.5 of wavelength for both H/V
Array Ohmic loss	3 dB
Conducted power (before Ohmic loss) per antenna element	6 dBm/200 MHz
Spurious emissions	–13 dBm/MHz

3.3 unwanted emission marks of IMT BS in the band 76-77 GHz

For the band 71-76 GHz, the maximum BS transmit power , the following spectrum mask described in Doc. 5-1/36 [8] are then used, see Table 4.

Table 4. Spectrum masks for outdoor scenarios and $P_{TX} < 30.5$ dBm for BS in the frequency range 66-86 GHz

Frequency offset from "edge of transmission" Δf	Emission limit (BS)	Emission limit (UE)	Measurement bandwidth
$0 \leq \Delta f < 20$ MHz	–5 dBm/MHz	–5 dBm/MHz	1 MHz
20 MHz $\leq \Delta f <$ 400 MHz	–13.4 dBm/MHz	–13 dBm/MHz	1 MHz
$\Delta f >$ 400 MHz	Spurious domain limits*	Spurious domain limits	1 MHz

*Note: for BS and UE, the spurious emission level is set as being -13 dBm/MHz.

3.4 Propagation models

The analyses are based on the propagation models described in Recommendation ITU-R P.525-3 [10] is used to simulate the transmission loss from an IMT system include one BS and three UE to an automotive radar device.

IOP Conf. Series: Journal of Physics: Conf. Series **1060** (2018) 012007 doi:10.1088/1742-6596/1060/1/012007

3.5 Protection criterion for automotive radar

As recommended in Recommendation ITU-R M.2101[9], a protection criterion I/N of -6 dB is considered in the study.

4. Methodology

4.1Interference topology

Based on the information available in WP 5D and TG 5/1 documents, it can be assumed that IMT components operating in the 71-76 GHz will be deployed in an urban or a suburban environment.

Figure 2 illustrates a basic sketch of the interference scenario. It is assumed that there are an IMT BS operating in the band 75.8-76 GHz and an automotive radar in the adjacent band 76-77 GHz. The IMT BS is deployed at the side of the road, and serves the IMT user equipments (UEs) located in its serving sector. As recommended in Recommendation ITU-R M.2101, the range of angle for single serving sector is 120 degree. The automotive radar is installed on the front of a vehicle, which is traveling over a few meters away from the IMT BS in a straight trajectory.

Therefore, the out-of-band (OOB) and spurious emissions of IMT BS working in 75.8-76 GHz will affect the performance of automotive radar in 76-77 GHz.

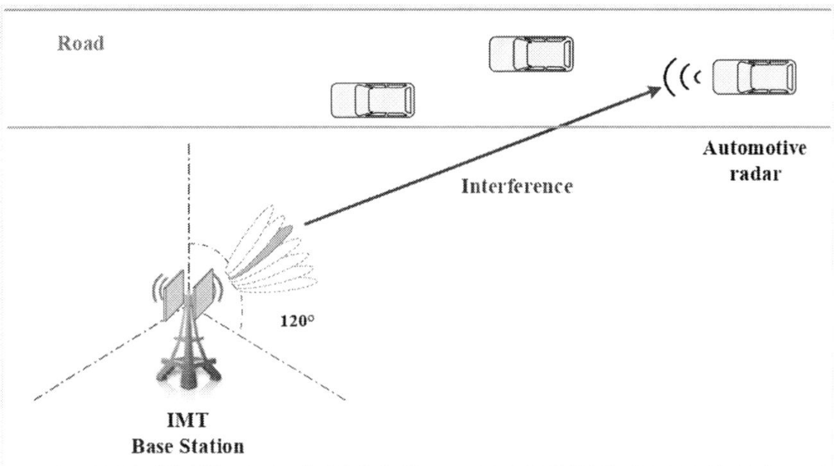

Figure 1. Illustration of compatibility interference between IMT BS and automotive radar

In the simulations, the scanning in azimuth of IMT BS is considered, to emulate the action that IMT BS is serving to the UE located in different geographical locations. Therefore, 1 000 snapshots are simulated for each victim positions.

4.2 Calculations

The interference power for IMT BS and UE at the receiver of automotive radar is given as:

$$I = Tx + G_{radar} + G_{IMT} - PL \quad (1)$$

where,

Tx:the emission power of IMT BS and UE, in dB;

G_{radar}:the antenna gain of automotive radar, in dBi;

G_{IMT}:the antenna gain of IMT BS and UE, in dBi;

PL:transmission losses for the path from IMT BS to automotive radar, in dB;

5. Results and analyses

This section presents the Monte Carlo results in terms of the level of compatibility interference power vs. the distances between IMT and automotive radar in the straight trajectory by using Visualyse V7.9 simulation tool.

IOP Conf. Series: Journal of Physics: Conf. Series **1060** (2018) 012007 doi:10.1088/1742-6596/1060/1/012007

In order to better assess the effect of the distances of IMT BS from the vehicle trajectory, i.e., R in Figure 1, this section shows the results for different distances of IMT BS from the vehicle trajectory. The horizontal scanning of automotive radar antenna is considered in the simulation. Two kinds of configurations on the distances of IMT BS from the vehicle trajectory are set, i.e., 4 m and 10 m, respectively. The corresponding results are given in Figure 2 and Figure 3.

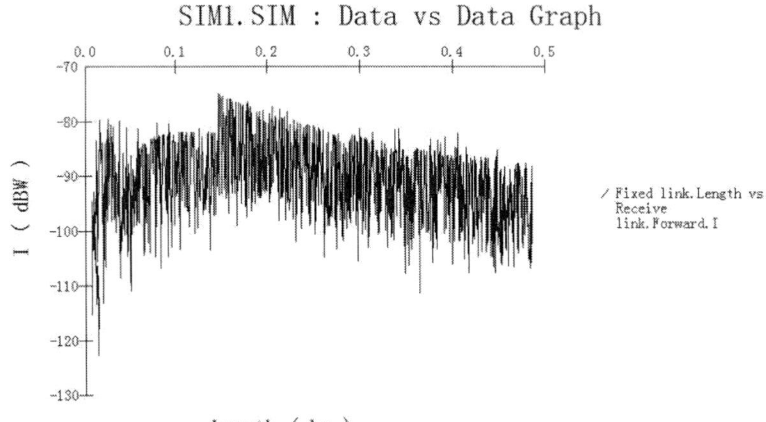

Figure 2. Compatibility interference of IMT BS at the automotive radar for different distances with 4m distances between IMT BS and the vehicle trajectory

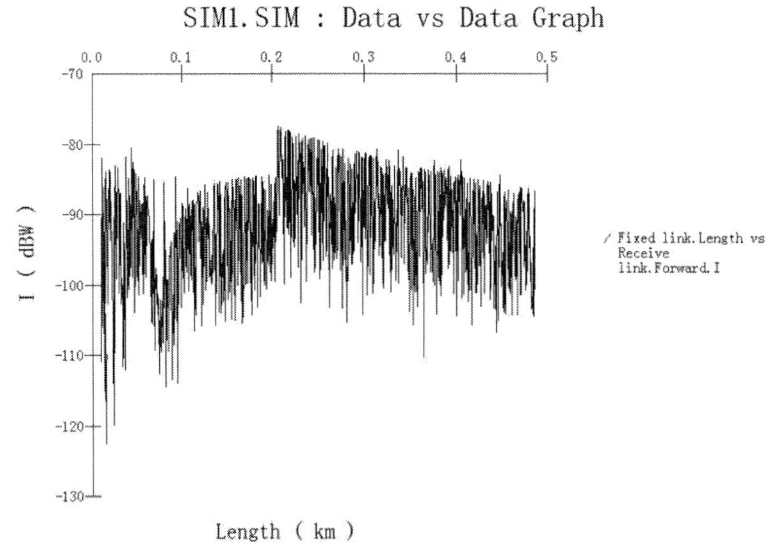

Figure 3. Compatibility interference of IMT BS at the automotive radar for different distances with 10m distances between IMT BS and the vehicle trajectory

It can be concluded that the interference will be reduced with the increase of the distances of IMT BS from the vehicle trajectory.

6. Summary

This initial study presents the potential of compatibility between IMT system and automotive radar operating in the band 76-77 GHz. The results show that in some case the IMT BS operating in the band 71-76 GHz may be well compatible with the automotive radar, the interference power value is mainly under protection criterion value. Obviously, if the antenna horizontal ±15° angle scanning of automotive radar, it seems that both impacts will be alleviated.

Acknowledgement
This work was supported by the National Science and Technology Major Project under Grants No. 2015ZX03002008.

Reference

[1] Bleicher A. The 5G phones future. IEEE Spectrum, 2013, 50(7): 15-16.
[2] Tan Wang, Gen Li, Jiaxin Ding, Qingyu Miao, Jingchun Li, and Ying Wang, "5G Spectrum: Is China Ready?", IEEE Communications Magazine, pp. 58-65, 2015.
[3] S. Y. Geng, Xing Li, Wei Hong, and X. W. Zhao, "Mm-wave 26 GHz channel characterization for future wireless communications," in 2016 International Conference on Control and Automation, 2016, pp. 475-480.
[4] "Initial compatibility studies of IMT-2020 and automotive radar in the frequency band 76-81 GHz for AI 1.13", revision 1 to ECC PT1(17)235rev1, Dec.2017.
[5] Contribution from ITU-R TG 5/1, "Compatibility studies of IMT system and automotive radar in the frequency range 76-81 GHz," 2018.
[6] Report ITU-R Radio Regulations, 2016.
[7] Recommendation ITU-R M.2057-1, "Systems characteristics of automotive radars operating in the frequency band 76-81 GHz for intelligent transport systems applications ," 2018.
[8] Liaison statement from ITU-R WP 5D to TG 5/1, "Spectrum needs and characteristics for the terrestrial component of IMT in the frequency range between 24.25 GHz and 86 GHz," 2017.
[9] Recommendation ITU-R M.2101-0, "Modelling and simulation of IMT networks and systems for use in sharing and compatibility studies," 2017.
[10] Recommendation ITU-R P.525-3, "Calculation of free-space attenuation," 2016.

IOP Conf. Series: Journal of Physics: Conf. Series **1060** (2018) 012008 doi:10.1088/1742-6596/1060/1/012008

Optimization of Image Enhancement Algorithm in Foggy Area

Zhang Hong-wei[1,2,a] and Zhang Jiu-gen[1,2]

[1]College of electrical engineering and control science, Nanjing University of Technology, Jiangsu Nanjing 211800, China

[2] Institute of Intelligent Building, Nanjing University of Technology, Jiangsu Nanjing 211800, China

[a]Corresponding Email:736159774@qq.com

Abstract. Nowadays, tourism scenic area gradually become people to relax, one of the best place to rest, the huangshan mountain, the Yellow River and other famous scenic spots, for example, however, at present, most of the scenic area in the design of the monitoring equipment does not consider the influence of the bad weather, so under the condition of rain day, under the influence of atmospheric scattering function monitoring equipment, unable to capture the image; At the same time, monitoring equipment working for a long time, more or less there will be a certain degree of degradation. Therefore, this article on the basis of defects based on the existing scenic spot to fog, mainly focus on a single scale fog image enhancement algorithm Retinex algorithm and multiscale Retinex algorithm concept, and gives the specific process of two kinds of algorithm of Matlab programming, through the analysis judgment, evaluation and prediction, and plenty of experimental data and theoretical support, through a large number of experiments, the results prove multiscale Retinex algorithm in image contrast to the scenic spots in the fog effect is best, in the process of use in the future to the scenic spot of fog, can get the desired effect.

1. Introduction

Nowadays, with the development of national economy and the improvement of people's living standard, the monitoring equipment of tourist scenic spots is getting more and more attention [1].

In the case of fog, haze, rain, snow and other severe weather conditions, the imaging effect of monitoring equipment is not significant due to the effect of atmospheric scattering effect [2]. Literature [3] proposed in many scenic areas, typical of Huang Shan scenic spot, the year of cloud mist, the weather is unpredictable, perhaps the first second cloud cover, but then the eyes of the sun; For example, in July and August, the Yangtze river basin is a rainy season, and most of the scenic spots are hazy with rain and fog.

However, the vast majority of the existing scenic spot monitoring equipment at the beginning of the design, did not consider the weather factors to the impact of its imaging, monitoring of the scenic spots workers cause serious interference. Therefore, according to the actual problem, realize the degradation of image processing, effectively improve the contrast, to improve the visual effect, to achieve real color of the scene, monitoring is of great significance to the security of the scenic spot. The experimental results show that the multi-scale Retinex algorithm is superior to the fog effect.

2. Image enhancement defog algorithm

2.1.Scale Retinex algorithm
The mathematical expressions of the single-scale Retinex algorithm are as follows:

$$ri(x,y)=logSi(x,y)-log[F(x,y)*Si(x,y)] \quad (1)$$

In equation (1), ri(x,y) is the brightness value of the coordinates at (x,y).(*) said convolution operation, F (x, y) surrounding the function, SSR method of convolution is forecast for light weight, it is used by a template center pixel with a weighted average of the other pixels in the template to replace the center pixel values, in order to eliminate the interference of light weight [4-5].

2.1.1. Simulation results of SSR at different scales. In figure 1, A-1 represents the original graph; The SSR simulation results of A-2 are presented; Simulation results of A-3; A-4 represents the simulation results.

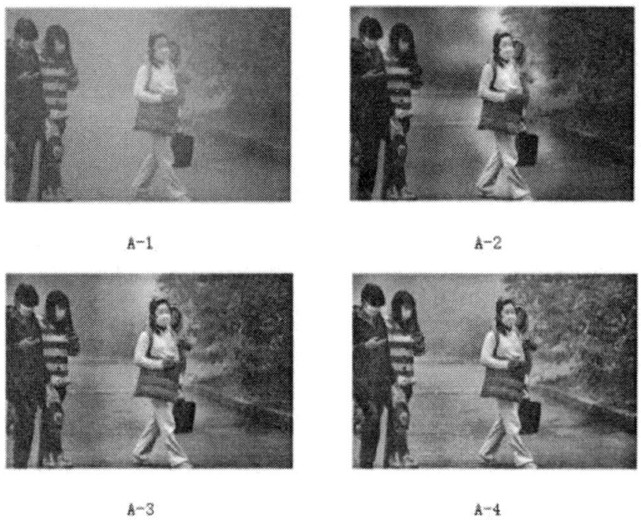

Figure 1. Simulation of SSR at different scales

Can be seen from the results, for SSR, the choice of different constant gaussian scale, will have different image enhancement effect, when the image can complete dynamic range compression, but the overall color distortion is serious; When the color distortion of the image decreases, the local details are more abundant. When, the overall color fidelity of the image is good, but the dynamic range compression completion is poor. In the SSR processing results of different scales, there is obvious "halo".

2.1.2. Single scale Retinex algorithm summary. By using gaussian function to deal with the input image in detail, the single-scale Retinex algorithm achieves good enhancement effect.

However, the choice of gaussian scale directly decides the image enhancement effect is good or bad, the dynamic range of inverse gaussian scale in image compression, but is proportional to the image of color constancy, SSR method is difficult to in the scope of the image compression and image color constancy both at the same time obtain good effect. To solve this problem, we introduce a multi-scale Retinex algorithm [6-8].

2.2. Multiscale Retinex algorithm
MSR is a weighted average of the results of different scales of SSR. It can also ensure the overall contrast and local details of the images have a good effect. Its expression is as follows:

$$R_{Mi}(x,y)=\sum_{k=1}^{K}W_k\{\log S_i(x,y)-\log[F_k(x,y)*S_i(x,y)]\} \quad (2)$$

In equation (2) , MSR is represented in the output of the ith color spectrum segment; The number of gaussian functions is denoted by K, and it also represents the scale of the surrounding function; The weight of each scale is represented by [9-10].

2.2.1. Simulation results of different scale MSR. In figure 2, A-1 represents the original figure, andA-2 represents the image after MSR simulation.

A-1 A-2

Figure 2. Simulation of different scale MSR

As can be seen from the processing results, MSR also has good local details as well as the overall contrast of the image, and there is no obvious "halo" generation [11-13].

3. Experimental results and conclusions

Whatever image processing method is adopted, the ultimate aim is to improve the quality of degraded images. Therefore, the evaluation of image quality is an indispensable link in the field of image processing, and also an important basis for evaluating the effectiveness of image processing methods.

This paper selects three basic parameters which are obvious in physical meaning and suitable for evaluation of image quality: brightness mean, variance and information moisture.

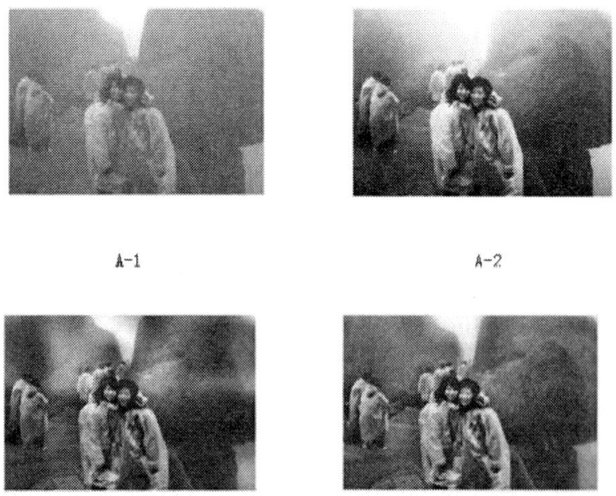

A-1 A-2

A-3 A-4

A-5 A-6

Figure 3. Image quality comparison of three basic parameters

In figure 3, A-1 represents the original graph; A-2 represents the image after the histogram equalization; The simulation results of the SSR are shown in A-3. The simulation results of SSR when A-4 is expressed; The simulation results of SSR when A-5 is expressed; A-6 represents the simulation result of MSR.

Table 1 elevator door open and closed systemObjective evaluation of simulation results

Enhancement method	A-1	SSR 25	SSR125	SSR 250	MSR
Mean brightness	188.5258	166.7752	170.6851	172.8772	139.4169
Nariance	3679.5	6888.6	6282.1	5982.5	10461
Information quantity	5.1930	5.6701	5.5961	5.5822	5.3574

It can be seen from table 1 that the images with SSR method are adopted, and the selection of different gaussian scales has different image enhancement effects. The scale of SSR can achieve dynamic range compression, but the overall color distortion is serious. The overall color sense consistency of the large scale SSR is good, but the dynamic range compression is poor, and in the processing results of the original figure 1, there is obvious "halo" generation.

SSR MSR algorithm because it contains a number of different scale, the advantages of the processed image is so use it to the improvement of the overall image contrast and local details shows two aspects that can achieve good results at the same time, although not completely eliminate the "halo", but effectively suppresses the production of "halo", in order to gain the better image processing effect. Therefore, this paper has a certain reference value for the research on the optimization of image calculation in foggy scenic spot.

4. Conclusion

This paper first introduces the existing basic principle and method of image enhancement to the fog, and in view of the single dimension Retinex algorithm and multiscale Retinex algorithm from the theoretical basis to the specific implementation expounded in detail. The experimental results show that compared with single scale Retinex algorithm of SSR method, the algorithm of multiscale

Retinex method of MSR in overall image contrast and local details shows in two aspects, can obtain good effect. According to the experimental results can not completely eliminate the "halo", the next step is to explore the formation of effective inhibition of the "halo", thus obtained a more ideal method of image processing effect.

References

[1] Yang Jing-rong. Research on the restoration algorithm of single foggy images [D]. Hefei: He Fei university of technology, 2013.

[2] Gonzalez. Digital image processing [M]. Beijing: electronic industry press,2007.

[3] Cheng Lei. Research on the rain and fog algorithm of video image [D]. Tianjin: tianjin university, 2011.

[4] Zhao Xiao-xia. Research on video image enhancement system based on RETINEX theory [D]. Beijing: China university of mining and technology, 2011.

[5] Lan Bell. Research and software implementation of single image defogging algorithm [D]. Chengdu: southwest jiaotong university, 2012.

[6] Zhang Jie. Research on image recognition algorithm of smoke video [D]. Tianjin university 2016.

[7] Tang Ning, Lu Yang. A fast single image defogging method based on physical model [J]. TV technology. 2015(09).

[8] Wang Yi-fan, Yin Chuan-li, Huang Yi-ming, Wang Hong-yu. Image defog based on bilateral filtering [J].Journal of Chinese image graphics. 2014(03).

[9] Zhang Ren-bin, Li Xing-xing, Xia Hong-li. Fast defogging algorithm based on the single image of the depth of field [J]. Computer engineering. 2014(02).

[10] Fan Feng-bing, Zhang Hong-ying, Wu Bin, Wu Yan-dong. A fast single image de-fog algorithm based on fog depth [J]. Journal of sichuan institute of technology (natural science). 2016(02).

[11] Zhao Chang-hui. Depth based on the structure of the PIM today to realize convolutional neural network approximation calculation [J]. 2017 (6) computer research and development.

[12] Xu Chao.The pedestrian detection method improved convolution neural network [J]. 2017 (6) computer application.

[13] Zhou Jun-yu. Convolution neural network applications in image classification and target detection in [J].2017 (13), computer engineering and application.

Camouflage Effectiveness Assessment Based on Fusion with Constant Color Background

ZHAO Dehui[1], WAGN Pengye[1] , ZHANG Hengbo[2] and DONG Zongge[3]

[1]Institute of Aeronautics and Astronautics Engineering, Air Force Engineering University, Xi'an 710038, China;
[2]National Defense Engineering Research Institute of Academy of Military Science of PLA, Beijing 100036, China;
[3]Department of Air Force Equipment, Beijing 100843, China

E-mail:w17729291493@163.com

Abstract. In the constant-color background, the structure feature is more influential in evaluation of the target camouflage effectiveness its surface is covered by the color that matches the background compared with the color feature. Against this, a method fusing the index of similarity and the index of difference is proposed. The index of similarity is designed based on texture, orientation and intensity features and described by saliency maps. The index of difference is designed based on target's own feature of straight-section edge and described by the weighted maximum entropy principle. These two features are fused to evaluate the camouflage effect. Experimental results show that our camouflage evaluation framework accords with the human visual perception mechanism.

1. Introduction

The technique of camouflage effect evaluation has a very important role in the design, implementation and optimization of the camouflage scheme[1].With the continuous development of computer technology, the camouflage effect evaluation method based on digital image features is gradually replacing the traditional subjective-based manual interpretation method due to its advantages of high processing efficiency, strong objectivity, and low occupancy rate of human resources[2,3].In recent years, the main contributions of relevant scholars in the field of objective camouflage effect evaluation are: Wang He designed approximate measurement indicators based on the tolerance near set theory to evaluate camouflage effect[4];Xue F extracted background-related characteristics and target's own characteristics to evaluate camouflage effect[5];Lin used feature's similarity based on image quality assessment techniques to judge the target camouflage effect[6];Lin also evaluated the effect of camouflage with the data obtained from eyeball motion[7]; Jia Qi introduced the human eye's visual attention mechanism to establish a corresponding evaluation model[8,9].The above methods all extract the most perceptible color information of the human visual system for the design of the effect evaluation indicator. In the case of the constant-color background(desert, jungle) and the target's surface is covered with a corresponding camouflage color, The effect of color brightness contrast of the target and the background will be greatly reduced in the evaluation of camouflage effects.

Therefore, the paper extracts characteristic which can reflect the surface texture (called context-related characteristic)to design the quantization index of similarity with visually significant characteristic; At the same time, taking into account the target being belong to Man-made geometry, which has its own outline edge (the target's own relevant feature) that is easily found by the human visual system. So we can extract the straight-line-segment edge features of the target and quantify the features as an indicator of the difference used the maximum entropy principle in effect evaluation.

Finally, a nonlinear fusion of the similarity index and the dissimilarity index is designed to evaluate the disguise effect of the target under the constant-color background. The results show that the comprehensive evaluation index is in good agreement with the professional interpretation results.

2. Differentiation indicators

2.1. Line-segment edge feature extraction

According to the human visual attention mechanism, the straight-line feature in the image is a typical rule that is easy to be visually noticed. It mostly appears in the surface texture and outline edge of the man-made objects in the image, and it is easier to be recognized by the human eye firstly in the natural scene mainly composed of irregular bodies[10,11]. Therefore, in the evaluation of camouflage effect, the number and length of long-straight-line segments existing in the camouflaged target edge are introduced as an important indicator for measuring the target's camouflage effect. The specific line feature extraction process is as follows:

STEP1: Input the image with the camouflaged target(a), using the edge detection algorithm to perform edge detection, and obtain the edge processing map(b).

STEP2: Setting a set of thresholds($(\mathrm{T}_1,...,\mathrm{T}_i,...), i \in N^*$) based on the sensitivity of the length of the straight-line segment in the human visual system with the Hough Transform ,straight line segments(b) under different thresholds (x ,unit: pixel) extraction;

STEP3: Statistics the number of edge segments (m_i) within thresholds $\mathrm{g}_i : (\mathrm{T}_i, \mathrm{T}_{i+1}]$ in the camouflaged areas and background areas .And note the maximum straight line length is n .

2.2. Feature quantization of line segments based on weighted maximum entropy

The principle of maximum entropy is a criterion in the learning of probability models. It is believed that the model with the largest entropy is the best model among the set of models that satisfy the constraints. This principle is widely used in the image segmentation. It is considered that the image features in the target region and the background region are similar, and when the entropy addition between the two regions reaches a maximum, the background and target segmentation are optimal[12,13].Inspired by this, the principle of maximum entropy is used in camouflage effect evaluation. The larger the total image feature entropy within the camouflaged region and the surrounding background region, the easier the camouflaged target is from the surrounding background, the worse the target camouflaging effect is, vice versa.

We get the number of edge segments(called m_i) in each threshold group from the previous section , $p(\mathrm{g}_i)$ indicates the threshold group (g_i) 's the probability of appearance. And it is defined as (O: target area, B: background area):

$$p_{O,B}(\mathrm{g}_i)=\frac{m_i}{\sum_i m_i} \tag{1}$$

The value has no effect on the target discovery probability, which is different from the characteristics of the general case processing. According to the study of Xu Shan et al, the relationship of the length of the edge segment and the probability of being discovered is a convex curve with a horizontal asymptote $P(n)=1$ and $P(1)\to 0$ .In order to meet the above requirements, we use functions ($P(x)=\log_n(x), x>1$) approximate expression of this relationship. Thus, the feature entropy is weighted and its weight coefficient is defined as:

$$w_i = P(\frac{T_i + T_{i+1}}{2}) / \sum P(\frac{T_i + T_{i+1}}{2}) \qquad (2)$$

Then, the feature entropies of the target area(O) and the background area(B) are respectively represented as:

$$\begin{cases} H_o(\mathrm{t}) = -\sum_i w_i p_o(\mathrm{g}_i) \ln p_o(\mathrm{g}_i) \\ H_B(\mathrm{t}) = -\sum_i w_i p_B(\mathrm{g}_i) \ln p_B(\mathrm{g}_i) \end{cases} \qquad (3)$$

The total entropy of the target and surrounding background is:

$$H(\mathrm{t}) = H_o(\mathrm{t}) + H_B(\mathrm{t}) \qquad (4)$$

The greater the total entropy value, the greater the distinction between the camouflaged target and the surrounding background in the image, and the worse the target camouflage effect to be evaluated. Vice versa.

3. Similarity Indicators

Relevant studies have shown that significance is an important measure of the degree of interest in human visuals, and the regions with higher significance are more likely to cause human visual attention. Therefore, the similarity test based on the saliency metric is reasonable to use in camouflage effect evaluation. The lower the region's significance is, the better the target's camouflage effect is. The specific flow of this model is shown in the following figure 1:

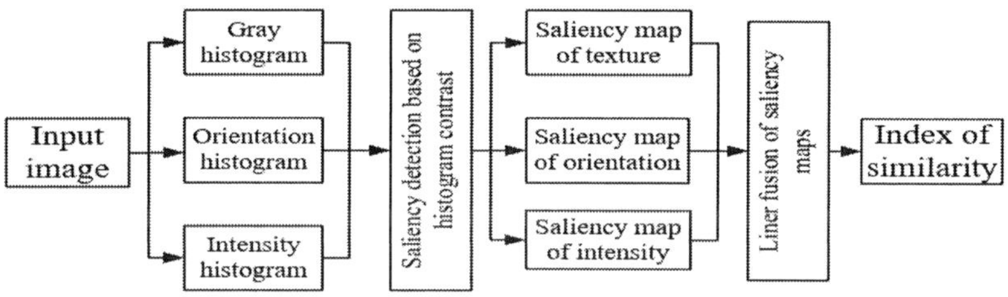

Figure. 1 The flow chart of saliency extraction algorithm

Feature extraction and quantification

When evaluating a target camouflage effect coated with a corresponding camouflage color under the constant-colour background, the gradient, texture, and the brightness are used to characterize the difference between the target and the background, due to the degraded effect to distinguish between the camouflaged target and the surrounding background.

Among them, the brightness and gradient information are expressed by the corresponding histogram. Since the texture information belongs to the area information, the expression can be quantified by a statistical method based on the image gray histogram.

Significant calculation

Significant detection based on histogram contrast is used to calculate the characteristic saliency values of texture, gradient and brightness. Due to space limitations, the specific calculations can be found in paper. From this, saliency based on texture, gradient, and brightness(S_1 , S_2 , S_3) can be obtained.

Significant fusion

The linear fusion of the saliency test results of different features, witch to create the saliency-based similarity indicator:

$$S = (S_1 + S_2 + S_3) / 3 \qquad (5)$$

4. Camouflage Evaluation Example

4.1. Comprehensive evaluation index establishment

In order to more fully evaluate the camouflage effect of the target, we fuse the similarity index and the difference index for the comprehensive evaluation index design. The former reflects the integration of the camouflaged target with the surrounding background, and the latter reflects that the camouflaged target distinguishes itself from the surrounding background.

Since the influence of the two on the target camouflage effect is opposite, the fusion between the two is not a linear additive relationship. We consider the characteristics of the two indicators and their relationship with the camouflage effect. A multiplicative strategy of index multiplication is used to express the "and" relationship between the two indicators by multiplying the indicators. The specific expression is defined as:

$$C = \frac{S}{S+c_1} \times \frac{0.1H(t)}{0.1H(t)+c_2} \tag{6}$$

Among them, c_1, c_2 are small constant values to avoid that the overall evaluation index is too large due to the small denominator. Because of the saliency, the size of the maximum entropy is inversely proportional to the disguise effect. Therefore, the greater the comprehensive evaluation index is, the worse the camouflage effect is.

The comprehensive evaluation process is shown in Figure 2.Determining the target area (target area to include the smallest circumscribed rectangle) disguised different ways in the image to be compared.

Considering that the background part far away from the target has relatively weak influence on the target detection result, eight adjacent rectangular areas of equal area around the target area are selected as the background area. The structural features of the target area and the surrounding background area are extracted and the length of the edge of the straight line segment is extracted. The saliency map is used to quantify the structural features and the improved maximum entropy principle quantifies the length of the line segment edge.

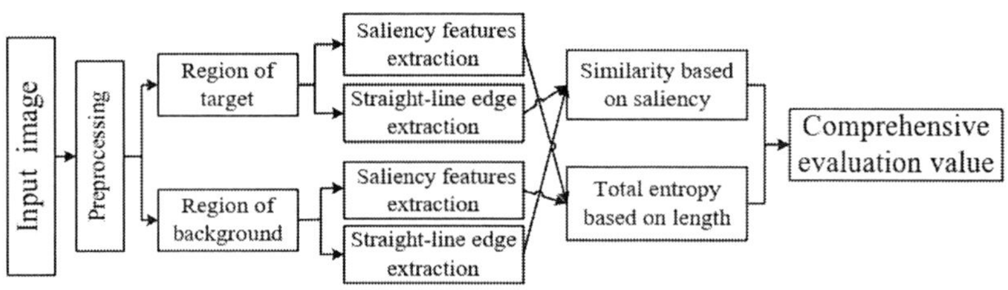

Figure. 2 Framework of comprehensive camouflage-effect evaluation

4.2. Example verification and result analysis

In order to test the effectiveness of the proposed comprehensive evaluation method for camouflage effect, this paper chooses to verify the method of covering different camouflage nets under a single green space background.

In a green plot, cover different systems of camouflage nets (numbers 1, 2, and 3) and use aerial photography to acquire experimental plots at the same height, as shown in the figure 3:

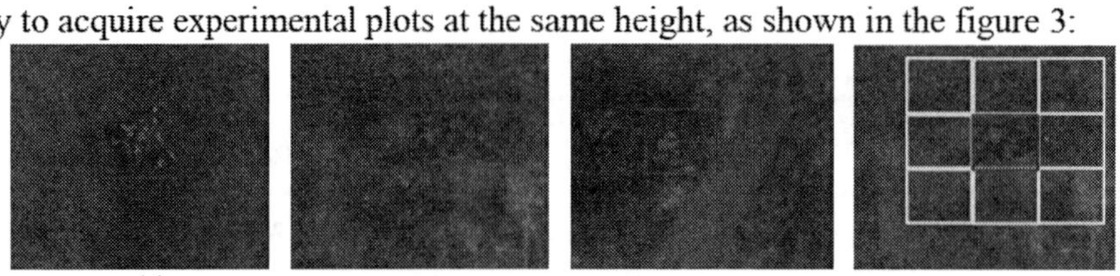

(a) (b) (c) (d)

Figure. 3 Camouflage net of (a) No. 1, (b) No. 2, and (c) No. 3; (d) Sketch map of "target+background"

Extract the characteristics of the edge of the target and its surrounding background, according to the actual situation, let the threshold ($T_1 = 1, T_i = T_{i+1} + 5$). Due to space limitations, some of the threshold values of the lower edge straight line segment are listed as follows in the figure 4:

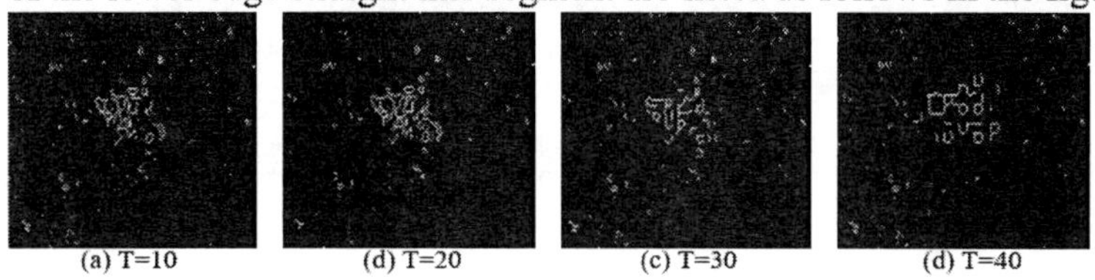

Figure. 4 Sketch map of straight-line edge of camouflage net NO.1 in different thresholds

The weighted maximum entropy theory is used to quantify the edges of the straight line segment of the target and the background region. The results are as as shown in the following table 1.

Table 1. The saliency based on different features of camouflage targets

Camouflage net	NO.1	NO.2	NO.3
H_o(t)	1.6356	1.6279	1.6288
H_B(t)	2.1541	1.8728	2.0004
H(t)	3.7897	3.5007	3.6292

According to the chart, the edge lengths, sparsity, and corresponding feature entropy values of the straight line segments of the target area and the surrounding background area in the images containing the camouflage network are different, which verifies that the straight line segment edges can be used for index design to reflect the difference between the target and the background.

Different camouflage networks correspond to different total entropy values, which indicates that the total entropy value can be used as a dissimilarity index and then a similarity index to evaluate comprehensively when evaluating the camouflage effect of the target.

The saliency map is obtained using the distinctive features of the disguised target, and the saliency of the target and the surrounding background is quantified. The specific results are as follows in the figure 5 and the table 2:

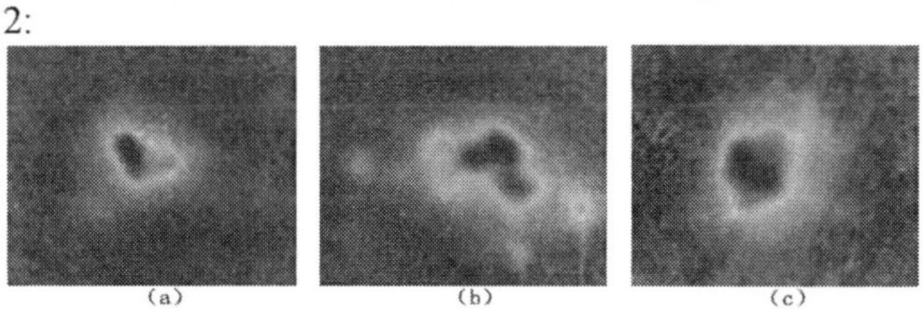

Figure. 5 Sketch map of saliency based on features fusion with camouflage net of (a) No. 1, (b) No. 2, and (c) No. 3

Table 2. The saliency based on different features of camouflage targets

Camouflage net	NO.1	NO.2	NO.3
Color	0.1607	0.1745	0.1562
Texture	0.4216	0.2291	0.2981
Intensity	0.3643	0.1962	0.3057
Orientation	0.3514	0.2561	0.2886
Features fusion	0.3791	0.2853	0.2975

As can be seen from the table 2, the three camouflage networks have little discrimination based on the saliency indicators of the color features, and cannot effectively evaluate the camouflage effect of different camouflage nets under the background of the green space; and the saliency detection results of the fused textures, directions, and brightness characteristics. The three kinds of camouflage nets in the value have a high degree of discrimination, and it is easier to describe the different targets more completely and reasonably than the saliency of a single feature.

Fusion indicators and similarity indicators were used to obtain comprehensive evaluation indicators, as shown in the following table 3:

Table 3. Index of comprehensive evaluation and comparison

Camouflage net	NO.1	NO.2	NO.3
Ours	0.9491	0.9392	0.9419
Visual result	Bad	Excellent	Good

From the table 3, we can see that the proposed method extracts features related to the image structure: texture, direction, edge line segments and brightness, and the results of reasonable quantification and fusion have good consistency with the interpretation results.

5. Conclusions

We present a new method for evaluating the camouflage effect of multiple targets in a single color background. The method highlights the role of structural features and considers both positive and negative aspects of similarity and difference. The test results prove the validity of the method, can replace the subjective artificial interpretation method to evaluate the target camouflage effect, and make up for the insufficiency of the existing methods to evaluate the failure under a single color background. This article still has the problem of noise interference in the process of extracting the edge line segment, and hopes to further solve it in the future research. The method has certain reference significance for the detection of camouflage effect based on infrared images and is reflected in the next research work.

References

[1] Yunhui Y, Zhan W, Dong Dewei, et al. Current Actuality and Development Tendency of Military Camouflage Technology[J]. China Mechanical Engineering, 2012,23(17): 2136-2141.(CHINA)

[2] ZHAO Dehui, WANG Pengye, DONG Zongge, LI Mingfeng.Optical Camouflage Effectiveness Assessment Fusing the Different Features[J]. Journal of air force engineering university(natural science edition).(CHINA)

[3] Guo Tong,Hua Wenshen,Liu Xun,Liu Xiaoguang,Cui Zihao.Comprehensive Evaluation of Optical Camouflage Effect Based on Hyperspectra[J].Laser & Optoelectronics Progress,2016,53(10):101002.(CHINA)

[4] Wang He, Bai Tingzhu. Evaluation of Optical Camouflage Based on Tolerance Nearness Sets Theory[J]. Acta Optica Sinica, 2012, 32(12): 1210001. (CHINA)

[5] Xue F, Wu F, Wang J, et al. Camouflage texture design based on its camouflage performance evaluation [J]. Neurocomputing, 2017.

[6] Lin CJ, Chang CC, Lee Y H. Evaluating camouflage design using eye movement data[J]. Applied ergonomics, 2014, 45(3): 714-723.

[7] Lin CJ, Chang C, Lee Y, et al. Evaluating camouflage design using eye movement data.[J]. Applied Ergonomics, 2014, 45(3): 714-723.

[8] Qi J, Xuliang L, Chao W, et al. Evaluation of engineering camouflage effectiveness based on human visual attention mechanisms[C]. International Conference on E-business and E-government, 2011: 1-4.

[9] Jia Qi, Lü Xuliang, the Chao Wu, et Al Camouflage E Evaluation of FF ectiveness the using Human Visual Attention mechanism [J] Journal of Applied Sciences 2011, 29 (. 3):. 294-298. (CHINA)
[10] Hou Yimin, Tang Yue, Sun Xiaoxue, Sui Wenxiu.Application of Gaussian-Rayleigh Mixture Model in Remote Sensing Image Segmentation[J].Acta Optica Sinica, 2015, 35(s1):s110004. (CHINA)
[11] Zhang Zhenjie,Hao Xiangyang,Liu Songlin,Cheng Chuanqi.Line Detection Based on Hough One-Dimensional Transform[J].Acta Optica Sinica,2016,36(4):0412005. (CHINA)
[12] Sun Xiaoyan, Dong Huailin, Wang Beizhan, Su chuangye. Region-based segmentation method of ships in remote sensing image[J]. Computer Engineering, 2008, 34(10):196-198. (CHINA)
[13] Chen L, Chen X, Wang S, et al. Foreign Fiber Image Segmentation Based on Maximum Entropy and Genetic Algorithm [J]. Journal of Computational Chemistry, 2015, 03(11): 1-7.

IOP Conf. Series: Journal of Physics: Conf. Series **1060** (2018) 012010 doi:10.1088/1742-6596/1060/1/012010

Topic Mining Based on the Heat of Micro-blog

Wang Siyao, Liu bo, Zhou Jiajun and Lin Fuqiang

Nation University of Defense Technology College of Computer Science
ChangSha, China

18608417754@163.com

Abstract. With the popularity of social networks, users interact with each other and comment on current events through online social network more and more frequently, how to extract the hot topic has become the focus of natural language processing research.In this paper, we propose a hot topic extraction method based on the popularity of micro-blog. First we use cross-entropy to define the heat of micro-blog according to the number of its comment and forwarding. Then combining the heat of micro-blog and word2vec model to assign weight for each word, and we apply bidirectional LSTM(Long Short Term Memory) to conduct document semantic coding and single-pass method for topic mining. Besides, We separately introduce three evaluation indicators to test the proposed method: UMI(Normalized Information), PMI(Point-wise Mutual Information)and Purity.We used crawlers to crawl over 10,000 micro-blogs in 15 hot topics in 2017 in Sina Weibo, the experiment results show that the proposed method performs better and has stronger robustness than the traditional topic detection method.

1. Introduction

Online Social Networks(OSN) have gradually become popular in daily life and become one of the most important media for users to interact with each other, comment on current events and keep track of topics, like Facebook, Twitter as well as Weibo. By 2016 there are more than 2.2 billion registered users in Facebook. In the second quarter of 2017, there are more than 2 billion active users in Facebook. How to effectively extract the topic of the user's published content has become the focus of the current natural language processing research.

Social media text data contain many new words, abbreviations and emoticons and the length of text is relatively short[1,2].This type of text data contains more noise,so the text matrix yield by topic model is extremely sparse and difficult to analysis and calculate.Commonly used probability topic model LDA and its variants[3].can display the semantic information of the texts through taking topic as the expression of the middle layer features.However, this model treats each word as an entity without considering the contextual relationships between words and the temporal dependencies.The machine-learning based approach is mainly used to process ordinary texts, using words as a basic attribute, represent texts as sets of words, and apply machine learning methods for matrix decomposition (LSI,NMF[4],topic clustering(like k-means, incremental clustering, hierarchical clustering, single-pass clustering, etc.)In addition, the development of deep learning provides a new research direction and technical method for the topic mining . Recently, word embedding and recurrent neural network (RNN) have become a research focus of deep learning in natural language processing.The word embedding method is to express the vocabulary as a dense real number vector on a low-dimensional space,in this way, the expression of lexical semantic features and the construction of language model can be realized.

However, the current study of hot topics mining in the OSN media mainly focus on the level of text, ignoring the interactivity of online social network:every text published by users hide a lot of social information.For example,the more the text was commented and forwarded, the more probability that text contain the hot topic.Therefore, based on the above ideas, we propose a heat-based topic mining method.In this paper,the main works are as followed:

Given that the word2vec model can not distinguish the importance of words in the text,we give each word a value based on the tweets popularity and reconstruct word embedding model.

Using LSTM to encode documents and characterize the dependencies between words, and we apply single-pass clustering method to mine the topic model.

2. Related Work

In this section, we review the prior study on the issue of hot topic mining. Social media text expression is not standardized, and there is a large number of short text messages, to solve such problems, Danushka and Liu[5,6] et.al proposed the use of search engines and hownet to expand the original text with the aim of weakening the low-frequency feature words on the clustering results. However, the methods are implemented by introducing a large number of external features,as a result the excessive time consumption is not suitable for large-scale expected research.

In terms of the expression of words, word embedding has become a very popular tool in recent years and is widely used in every aspects of natural language processing.Collobert & Weston et al. [7]proposed a multi-layer neural network model structure based on recurrent neural network for processing variable length word sequence input. In 2012, Eric H. Huang improved the method of C&W and proposed a word embedding training model based on cyclic neural network considering the local and global contexts[8]. In 2013, Google's Mikolov team proposed the word2vec word embedding model[9,10], expressing the vocabularies as a dense real number vector on a low-dimensional space, in order to achieve a lexical semantic feature expression. In 2014, Jeffrey and Socher fused the idea of local context information and global word co-occurrence matrix decomposition to further explore a global linear regression model Glove[11]. Other methods to consider the global context include adding text vectors such as PV-DM and PV-DBOW[12].models and introducing external knowledge base, but due to the particularity of words in natural language processing in terms of semantic expression and multi-word synonymy and so on, how to obtain high-quality representation of word features has always been an important topic in natural language processing and text mining. Sundermeyer et. al[13] explained how to build language models using LSTM neural networks. Kim[14] used CNN to complete the task of sentence classification with the pre-process of word vector.In 2015, Tang et.al[15] used the neural network model consisting of CNN and LSTM to conduct text topic mining.After experimental verification, neural networks composed of CNN and LSTM have achieved good results in topic mining.

With the deep application of deep neural network in the field of natural language processing, Encoder-Decoder framework[16] has been the mainstream method of text sequence modeling in recent years, It also has far-reaching effects on reading comprehension[17], text abstract[18], machine translation[19], automatic question answering[20]. Wang et al. Proposed an end-to-end deep learning framework for fusion question matching to model reading comprehension questions[21].The framework includes match-LSTM, a match expression model for questions and sentences, and a web-oriented Point Net for answer constraints that effectively enable reading comprehension on large datasets. The popularity of the Encoder-Decoder framework has led to many enhancements to codecs, and attention models[22] are the most powerful and most powerful enhancements available today.The attention model decodes the Decoder output to give a different weight to each input to the Encoder so that the weight of the input that is more important to the current output becomes larger.

3. Proposed Method

3.1. Definition of the heat of Micro-blog

Our thinking about the way of calculating blog heat comes from the description of cross-entropy in information theory:a small probability event contains a large amount of information, at the same time a large probability event contains a small amount of information.So the value of the information in event A is defined as formula(1) :

$$H(X) = E_p \log \frac{1}{p(x)} = -\sum_{x \in X} p(x) \log \ p(x) \quad (1)$$

Learning from the idea of calculating the value of information, assuming that the number of comments on a micro-blog m is r, the forwarding number is s, the definition of the p(x) is shown in formula(2)

$$P(m) = \frac{1}{r+s+1} \quad (2)$$

So we can calculate the cross-entropy as formula (a) as the heat of each blog. The heat of micro-blog is equal to the sum of the heat of the term in the micro-blog. Therefore, when there are N words in the micro-blog, the heat calculation formula of a single word is described as formula(3):

$$Heat(w) = \frac{Heat(m)}{N} \quad (3)$$

3.2. Word2Vec Model

Word embedding is a good way to express lexical features, and the vocabularies are expressed as dense real number vectors in low-dimensional space.Word2Vec is the most widely used word embedding technology, including two kinds of models:CBOW and Skip-Gram.The CBOW model uses the remaining words in the context to predict the probability of generating the target word, and Skip-Gram uses the target vocabulary to predict the probability of other terms in the context.Compared with the CBOW model, Skip_gram has higher semantic accuracy at the expense of higher computational complexity.This paper is based on the Skip-Gram model to improve the training of word vectors, so the prediction of contextual probability is defined as (4):

$$p(context(w) \mid w) = \prod_{\bar{w} \in context(w)} p(\bar{w} \mid w) \quad (4)$$

Word2vec applies a layered softmax function to improve computational efficiency,combined with the layered softmax function, equation (4) can be expanded to formula(5):

$$P(context(w) \mid w) = \prod_{\bar{w} \in context(w)} \prod_{j=2}^{l_{\bar{w}}} [\sigma(v_w^T x_{\bar{w},j-1})]^{y_{\bar{w},j}} [1 - \sigma(v_w^T x_{\bar{w},j-1})]^{1-y_{\bar{w},j}} \quad (5)$$

Where w denotes the target vocabulary, $\bar{w}$ indicates the context of the target vocabulary, l_w denotes the the path length of the context in the output layer hierarchy tree, v_w denotes the input word vector of target vocabulary. $x_{\bar{w},j}$ represents the output word vector at the corresponding level under a certain contextual vocabulary. $y_{\bar{w},j}$ is logistic output variable.

When training the model of word embedding, the robustness of the model is often enhanced by introducing some words that are not found in the corpus as negative samples,these negative samples can be pre-generated by negative sampling[23].The objective function of Skip-Gram model trained with negative sample is shown in formula(6):

$$L = \sum_{w \in C} \sum_{\bar{w} \in context(w)} \sum_{u \cup \{w\} Y N_{\bar{w}}} \{y_{wu} \log[\sigma(v_{\bar{w}}^T x_u)] + (1 - y_{wu}) \log[1 - \sigma(v_{\bar{w}}^T x_u)]\} \quad (6)$$

Where $N_{\bar{w}}$ is a negative sampling set under the vocabulary $\bar{w}$, y_{wu} is logistic output variable, $v_{\bar{w}}$ is the sum of word embedding of context, x_u denotes the parameter of the model.

3.3. Word2Vec based on the heat of Micro-blog

Given training corpus dictionary $vocab = \{t_i \mid i \in 1...N\}$ and document $d_i =< w_1, w_2,...w_j >$where N is the word vector dimension.We use Word2vec model to train the corpus to get the word vector,accumulate the word vectors in the text d_i to get the vector representation of the text d_i shown in formula(7)

$$R(d_i) = \sum_t word2vec(t) \quad where \quad t \in d_i \tag{7}$$

Next, we introduce the heat model to calculate the word weight in Word2vec model according to the heat of words,and the the weighted word vectors are accumulated to obtain a new vector representation of document d_i shown in formula (8)

$$weight\_R(d_i) = \sum_t word2vec(t) \times Heat_t(w) \tag{8}$$

3.4. Document Semantic coding based on LSTM

Automatic coding machine is a artificial neural network,using self-supervised learning to encode input samples ,in order to achieve the purpose of reducing the data sample dimension.It is mainly divided into two parts:Encoder and Decoder.The encoder mainly compresses the original data into the output code:

$$\phi \ : \mathrm{X} \rightarrow \mathrm{Z} \tag{9}$$

The decoder will restore the output code closely to the original data:

$$\varphi \ : Z \rightarrow X' \tag{10}$$

In order for the automatic coder to retain the primary information in the original sample, the optimization goal is set as formula(11):

$$\phi, \varphi = \arg \min_{\phi,\varphi} \| X - \varphi(\phi(X)) \|^2 \tag{11}$$

LSTM coding framework is divided into five layers,taking the word embedding expression of all the keywords in the document as input and the overall semantic embedding of the document as output. The details of these five levels are as follows:

Word embedding presentation layer:This layer is the input layer. Because some documents tend to have more lexicons and some unimportant words that have no direct effect on the expression of the document's theme,in this paper we only select the word embedding with larger heat value as input,and the word embedding for all words is obtained from word2vec model above.

Bidirectional LSTM hidden layer:Contains two LSTM hidden layers, forward and backward layer,at the same time, each word embedding is connected to both the forward and backward LSTM hidden layer unit,these two hidden layers are connected to the same output.The input word at the moment t is embedded as E_t ,the output to the forward LSTM hidden layer cell is h_{t-1}^f ,output to the backward LSTM hidden layer cell is h_{t-1}^b for the last moment.The output of the forward and backward layers at the current moment are shown in formula(12,13):

$$h_t^f = H(E_t, h_{t-1}^f, c_{t-1}, b_{t-1}) \tag{12}$$

$$h_t^b = H(E_t, h_{t+1}^b, c_{t-1}, b_{t-1}) \tag{13}$$

Where $H(\cdot)$ denotes the function of LSTM hidden layer, c_{t-1} indicates the status value of the Cell unit at the last moment, b_{t-1} denotes the offset at the last moment.

Bidirectional LSTM output layer:Each output cell is connected to both the forward and backward LSTM hidden layer cells at same moment.

Average pooling layer: We use average pooling to process the original eigenvalues and construct new features,as well as realizing the dimension reduction, enhancement and noise filtering of the original valid features.Through averaging all cell values over a certain range,local information can be taken into account,calculation is shown in formula(14), Where T is the length of the input word embedding sequence.

$$pool(g) = \sum_{t=1}^{T} \frac{g_t}{T} \tag{14}$$

Semantic encoding output layer:The result of the average pooling layer can be calculated by activating the function to get the final semantic coding vector of the entire document.The dimension of the semantic code vector is consistent with the dimension of the input word embedding in order to facilitate similar calculation.

3.5. Topic clustering

After applying LSTM to semantic encoding, each document exists in the form of a vector in the hidden subject space,and its dimension is much lower than that of the feature space in vector space model.Therefore, we use Single-Pass clustering algorithm to conduct document clustering. After that we get the number of clusters K (that is, the number of topics), and the results of the division of each document on the K topics in the document set.

4. Result Evaluation

Data set: We used crawlers to crawl over 10,000 micro-blogs in 15 hot topics in 2016 in Sina Weibo, and selected 200 micro-blogs from each topic, recording the number of each comments and forwarding on each micro-blog. As a result we take a total of 3000 micro-blogs as the experimental data set.

The data set need to be pre-processed before model training,including stop words, high and low frequency words, illegal characters.The word with the frequency of less than 5 in the corpus is considered as the low frequency word, and the word with the frequency more than 20% of the total number of words is considered as high frequency word.

4.1. Evaluation Measurement

For the results of the hot topic mining, we use the purity and the normalized information (UMI) as the evaluation measurement, and these two evaluation methods are applicable to the text data with label.

Purity:Purity is used to measure the proportion of correctly clustered documents in the total document. The greater the purity is, the better effect the topic clustering yields.The purity value is calculated as formula(15):

$$purity = \frac{1}{D}\sum_{i=1}^{k} \max_j | p_i \cap c_j | \tag{15}$$

Where D is the total number of micro-blogs, k is the number of topic clusters, p_i represents the set of words contained in the i-th cluster,c_j denotes the j-th micro-blog in the corpus.

Normalized mutual information (UMI):The NMI value is calculated as formula(16):

$$NMI = \sum_{t=1}^{K} \sum_{1<ijK} \frac{\log \frac{P(w_i) * P(w_j)}{P(w_i, w_j)}}{-\log P(w_i, w_j)} \tag{16}$$

Where K represents the number of topics, N is the first N words under the topic. $P(w_i, w_j)$ denotes the co-occurrence probability of word w_i, w_j. $P(w_.)$ is the probability of the word w under the topic k.

Point-wise Mutual Information,PMI is the most commonly used semantic coherence evaluation measurement.The results of PMI evaluation are often highly consistent with manual evaluation. The higher PMI scores, the stronger semantic coherence topic have.The PMI value is calculated as formula(17):

$$PMI(\varphi_k) = \frac{2}{V(V-1)} \sum_{1 \le i < j \le V} \log \frac{p(w_i, w_j)}{P(w_i) p(w_j)} \tag{17}$$

Where $p(w_i)$ is the probability of appearance of vocabulary w_i in the test document set. $p(w_i, w_j)$ represents the joint probability of vocabulary w_i and w_j in the test document set,V is the dimension of the vocabulary list. Therefore, in this paper, for each subject word distribution, we only select the first 10 words with the highest probability value to calculate PMI.

4.2. Result Analysis and comparison

We also implement the classic statistical-based TF*IDF, NTM and LDA algorithm to exact key words from micro-blog.We conducted several sets of comparative experiments.In the experiment, we set the number of topics as 6, 20, 40, 80 respectively for the experiment of purity, and we set the number of topics as 20, 40, 60, 80, 100 respectively for experiment of NMI.The dimension of word vector size is 300, the experimental results are shown in Figure.1 and Figure.2

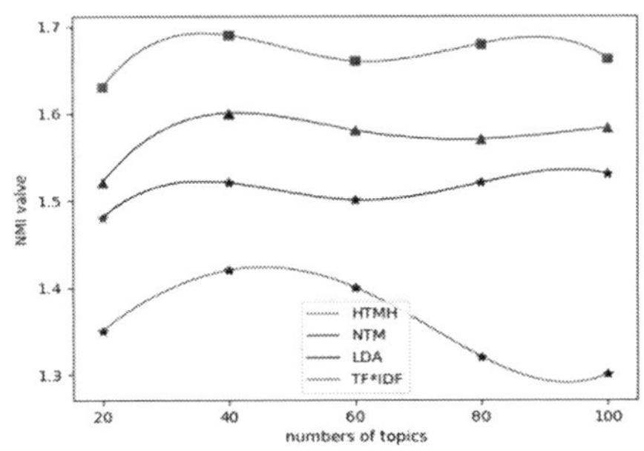

Figure.1 NMI value

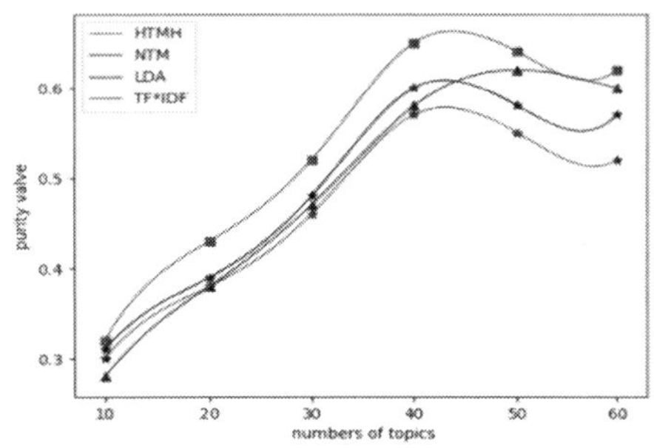

Figure2. Purity value

From the above two figures we can see,under the Sina Weibo corpus, the proposed HTMH(Hot Topic Ming Based on the Heat of Micro-blog) model has some improvement over other models in terms of purity and normalized mutual information,the purity reaches the maximum when the number of topics is 40,and the normalized mutual information (NMI) gets the highest accuracy when the number of topics is 20,This is also consistent with other models, HTMH has increased by 3% compared with TF * IDF, and increased by 5% compared with LDA.This is because word embedding technique is introduced as a semantic supplement.This makes the semantic relationship between words and subject strengthened.And in the follow-up training process, word embedding and topic model training promote each other, so we can identify the topic more accurately.

We can see from Figure.3, the HTMH model proposed in this paper has a generally higher PMI value.It shows that the extracted hot topic has a strong semantic coherence and as the number of subjects increases, the PMI value basically remains unchanged.The traditional LDA model has the lowest PMI because it does not consider the semantic reinforcement of documents and words,NTM is a neural network reconstruction of the LDA model, but it does not take into account the reinforcement of vocabulary and semantics.So it performs relatively poorly.

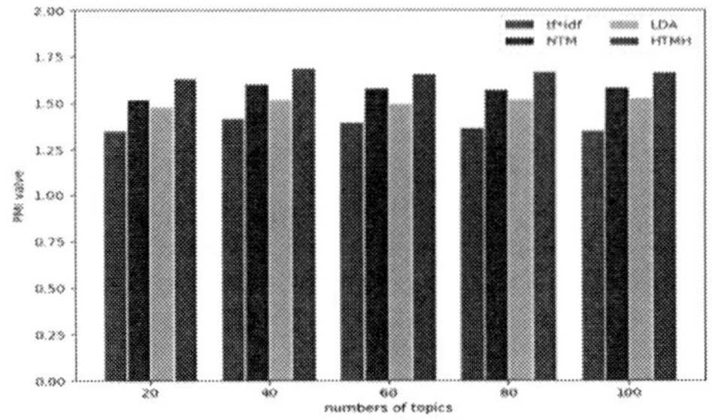

Figure3. PMI value

5. Conclusion

In this paper, we proposed a hot topic mining method based on micro-blog heat,based on the word2vec, we combined the idea of information theory and gave weight to each word. After using bidirectional lstm for semantic encoding, we applied single-pass clustering algorithm to conduct hot topic mining.

We then compared MTMH(Hot Topic Ming Based on the Heat of Micro-blog) model with TF * IDF, LDA and NTM algorithms,and introduced measurement of the topic mining quality,such as purity, NMI and PMI.Through the weibo corpus we proved that the HTMH model performs better on topic mining.

Future research focuses on the impact of sequence on the distribution of topics,in addition,collaborative training of feature expression of topic detection, emotion analysis and word embedding is also the trend of large-scale social media data analysis.

References

[1] Losada D E. The challenge of understanding the flow of sentiments in social media documents [C]//International Workshop on Search and Mining User-Generated Contents. ACM, 2011:1-2.

[2] Liu H. Mining social media: issues and challenges[C]// ACM Sigmm International Workshop on Social Media. ACM, 2011:1-2.

[3] Blei D M, Ng A Y, Jordan M I. Latent dirichlet allocation[J]. Journal of Machine Learning Research, 2003, 3:993-1022.

[4] Wang Q, Xu J, Li H, et al. Regularized latent semantic indexing[C]// International ACM SIGIR Conference on Research and Development in Information Retrieval. ACM, 2011:685-694.

[5] Bollegala D, Ishizuka M, Matsuo Y. Measuring semantic similarity between words using web search engines[J]. Computer Science, 2015:757-766.

[6] Liu Z, Yu W, Chen W, et al. Short Text Feature Selection for Micro-Blog Mining[C]// International Conference on Computational Intelligence and Software Engineering. IEEE, 2010:1-4.

[7] Collobert R, Weston J, Karlen M, et al. Natural Language Processing (Almost) from Scratch[J]. Journal of Machine Learning Research, 2011, 12(1):2493-2537.

[8] Huang E H, Socher R, Manning C D, et al. Improving word representations via global context and multiple word prototypes[C]// Meeting of the Association for Computational Linguistics: Long Papers. Association for Computational Linguistics, 2012:873-882.

[9] Mikolov T, Chen K, Corrado G, et al. Efficient Estimation of Word Representations in Vector Space[J]. Computer Science, 2013.

[10] Mikolov T, Sutskever I, Chen K, et al. Distributed Representations of Words and Phrases and their Compositionality[J]. Advances in Neural Information Processing Systems, 2013, 26:3111-3119.

[11] Pennington J, Socher R, Manning C. Glove: Global Vectors for Word Representation[C]// Conference on Empirical Methods in Natural Language Processing. 2014:1532-1543.

[12] Le Q V, Mikolov T. Distributed Representations of Sentences and Documents[J].2014, 4:II-1188

[13] Sundermeyer M, Schlüter R, Ney H. LSTM Neural Networks for Language Modeling[C]// Interspeech. 2012:601-608.

[14] Kim Y. Convolutional Neural Networks for Sentence Classification[J]. Eprint Arxiv, 2014

[15] Tang D, Qin B, Liu T. Document Modeling with Gated Recurrent Neural Network for Sentiment Classification[C]// Conference on Empirical Methods in Natural Language Processing. 2015:1422-1432

[16] Sutskever I, Vinyals O, Le Q V. Sequence to Sequence Learning with NeuralNetworks[J]. 2014, 4:3104-3112.

[17] Seo M, Kembhavi A, Farhadi A, et al. Bidirectional Attention Flow for Machine Comprehension[J]. 2016.

[18] Lopyrev K. Generating News Headlines with Recurrent Neural Networks[J]. Computer Science, 2015.

[19] Cho K, Merrienboer B V, Bahdanau D, et al. On the Properties of Neural Machine Translation: Encoder-Decoder Approaches[J]. Computer Science, 2014.

[20] Pengcheng Yin, Zhengdong Lu, Hang Li, et al. Neural Enquirer: Learning to Query Tables with Natural Language[J]. Computer Science, 2016.

[21] Wang S, Jiang J. Machine Comprehension Using Match-LSTM and Answer Pointer[J]. 2016.

[22] Bahdanau D, Cho K, Bengio Y. Neural Machine Translation by Jointly Learning to Align and Translate[J]. Computer Science, 2014.

[23] Mikolov T, Sutskever I, Chen K, et al. Distributed Representations of Words and Phrases and their Compositionality[J]. Advances in Neural Information Processing Systems, 2013, 26:3111-3119..

Healthy Building Management System Based on Business Intelligent

Yong Cao[1a], Chen Wang[1b], Jingshu Wei[1c], and Xiao Pan[2]

[1.] China Academy of Building Research, Beijing, 100013.
[2.] Sinosteel engineering design &research institute, Hebei, 050000

[a]cao.yong@cabr-cseb.com, [b]wang.chen@cabr-cseb.com, [c]jingshu.wei@cabr-cseb.com

Abstract. Huge account of building performance data are generated, collected and preserved in Building Automation Systems (BASs), which have been widely implemented with the development of smart building technology. Based on the key information which under covered in these data, building performance management plan can be created and improved. As a technology complex which concludes data warehouse, online analysis and data mining, business intelligence (BI) can extract, collect, and analyse large amounts of data to support decision-makers in making business decisions. With the assistance of BI based Decision Support System (DSS), desired information from massive building operation data can be found, which upgraded BAS as smart Building Management System. A business intelligence-based health building management system is introduced in this paper which can collect and store various performance data such as indoor environment quality (IEQ), water quality and energy consumption into one database. After processed with a predefined procedure, all preserved data can be processed and displayed with a user-friendly graphic interface. With the assistant of this BI based system, operating managers can make building management decision or revise management plan much more accurately as well as quickly according to real-time online monitoring data.

1. Introduction

With the development of economy and society, as well as the accelerated process of urbanization and industrialization, the demand of living quality and better environment is also increasing. On the other hand, in recent years, as a sign, smog diffuse in every winter, environmental pollution get much more public's focus. As a key factor of human health, clean air is highlighted more than ever

The large public buildings are always got air pollution problem because of crowded people and closed space. Research shows that if effective measures are taken to improve the indoor environmental quality, the efficiency of the internal staff can be increased by 15% to 20% [1]. However, there are still many problems in the indoor environment quality of large public office buildings: indoor environment assessment methods are not completed; indoor environmental quality intelligent monitoring technology is not perfect, lack of effective intelligent monitoring methods; lack of publicity system to show the AQI to residents, so that they cannot know whether the AQI has reached a healthy, efficient standard in their work place. Therefore, installing air quality monitoring systems in public buildings has become a necessary trend.

Building Automation Systems (BASs) are the decision support systems that support the energy management processes that consist of monitoring, analyzing, controlling, and optimizing energy usage.

Overall, BASs minimize energy consumption, and maximize productive conditions and energy efficiency [2]. In the process of the BASs, a large amount of monitoring data has been accumulated in various projects, such as indoor environment data, energy consumption data etc.

BAS comprise activities such as (1) consolidating energy related data from different sources, (2) using data access tools to analyze building performance, (3) visualizing energy related data, and (4) generating reports [3, 4]. The lack of a reference information model can be grasped in BASs with confusing user interfaces (that force users to throw away large amounts of data [5, 6]), and that are limited in terms of analysis capabilities (often forcing energy managers to use spreadsheets to analyze energy data) [7].

With the development of information technology, there is an urgent need for an integrated system to solve the contradiction between "excessive" data storage and lack of decision information. Therefore, the business intelligence (BI) has been introduced as an inevitable choice to improve the level of information.

2. Methodology

Data has become the second major factor in importance only to the business of talent. Business intelligence can integrates and analyses the external data from the company's operating market and the internal data from the company's resources, then extracts valuable information. Enterprise users can be assisted with BI to strengthen their management, making decisions promptly and scientific for future development, as well as discover new business opportunities. Corporate leaders and professional analysts are on longer the only users of business intelligence systems, all decision makers such as corporate managers, the functional persons within the company's internal departments, even customers, suppliers, and partners, can be treated as BI users.

Business intelligence technology can analyze various types of data such as structured, semi-structured, or unstructured data, also static historical data, and dynamic data streams. The business intelligence technology system is mainly composed of Data Warehouse (DW), On-line Analytical Processing (OLAP), Data Mining (DM), and Enterprise Information Portal (EIP). Partially composed, it integrates data discovery, management and decision support as a whole unit. It can integrate enterprise data resources on existing IT infrastructure and provide users with innovative applications and services.

Different from pure data analysis software. Business intelligence analysis software is more intelligent, standardized, and process-oriented towards business activities. Its procedure can be summarized as importing data sources, analyzing data indicators, and displaying data results. Finally, the BI analysis results are presented to end user through various forms such as ordinary reports and visual charts. From the perspective of different business users, the more easy-to-use business intelligence software is, the more popular it is. In order to improve the user experience, the user wants the software to be more convenient and the content of various types of data information is more intuitive. Therefore, the successful user experience needs to be realized through the visual interface of business intelligence analysis software.

3. Case Study

One of the most important features of healthy building operation management is the wide-distributed monitoring equipment, various systems, and high degree of coupling of monitoring data. Building management decision makers cannot inspect meters at real time. Due to lack of professional knowledge, it is difficult to understand the coupling relationship between indoor environment and equipment energy consumption. Therefore, implement business intelligence to guide the operation and management of healthy buildings would be a significant attempting.

Using business intelligence, building operation data can be analyzed, energy consumption can be optimized. The relationship between HVAC operation and indoor environment changing can be induct and proofed. According to the proofed relationship, optimized operating plan can be made in order to minimum energy consumption and improve indoor air quality. Therefore, the target of this system is to provide a path, which decision makers can either obtain information from the home page or use data analysis tools directly to query information from the data warehouse. Urgent information can be sent directly to the administrator's mailbox in a push mode, prompting administrators.

In this article, a healthy building management system based on business intelligence built in actual projects would be introduced as an example to illustrate the application of business intelligence technology in the entire system.

This building is a complex, contains office, commercial, and club. A healthy building management system has been implemented to collect real-time of indoor air quality data, as well as water quality of swimming pool, recovered gray water and portable water. With its assist, building managers can be advised to take appropriate measures to reduce or minimize indoor air pollution sources, in order to establish and promote the clean air plan to protect the health of residents.

The large area in public buildings, the more complex systems it is, especially HVAC and electrical systems. Ordinary methods is costly, inefficient and cannot manage every details in building. Only building automation (BA) system can meet the requirement of large-scale building management. By reserving data interfaces at the application layer. This system can be used as a building environment management function module for BA system. BA can collect the monitoring data in the device network, so that the BA system can be expanded into a building energy consumption and indoor environment quality parameter data management information system. Instead of logging multiple systems repeatedly, managers can monitor building operation in one system, to monitor real-time energy consumption and environmental quality parameters, data storage, historical reports and other works.

The general users can access this monitoring and management system through various ways such as personal terminals and mobile terminals, to find out real-time environmental qualities in the area. Property managers can conduct real-time monitoring of buildings through advanced authorizations, respond unexpected situations in accordance with routine plans, and realize the meticulous management of people-oriented building operations. On other hand, the workload of management and maintenance can be greatly reduced, work efficiency is improved, and management costs be dropped. Data analysts can use the senior administrator's authorization to log in to the System backstage for software maintenance and alarm management. Through a variety of alarming measures, every system can cooperates with each other and implements comprehensive management in the form of a network. Various emergencies can be handled quickly and decisively.

3.1. Graphical interface

Different from the traditional building management system, this system is for the general individuals who lack of professional knowledge. Therefore, it does not use simple data tables to pile up and display. Instead, these data is processed, analyzed, and then displayed to users in a most intuitive way.

For instance, the produce to show the indoor air quality in an office area is: all IAQ related data such as $PM_{2.5}$, PM_{10}, CO_2, and TVOC concentration shall be select according to an user-defined date range, then an AQI index be calculated in a pre-defined function, finally, this calculated number is mapped to a color shown in Figure 1.

IOP Conf. Series: Journal of Physics: Conf. Series **1060** (2018) 012011 doi:10.1088/1742-6596/1060/1/012011

Figure 1. The AQI of an office area

This procedure is automatic and can be real-time (based on-line data instead). Users can aware the most recently air quality only based on color and without any more thinking or mapping. When the color of a certain area gradually changes from yellow to red, it indicates that the indoor air quality has started to deteriorate, requiring the staff to turn on the equipment to ventilate and ventilate the air in time to avoid symptoms such as physical discomfort of the indoor staff.

3.2. Data analysis

In order to facilitate the needs of data statistics and analysis, the system includes a data analysis system. Users can query the change of building operating parameters, such as indoor air quality, water quality, energy consumption, etc., in any area at any time according to their needs. Among them, through the comparison of external weather data with indoor environmental parameters, the user can quickly understand whether the building operation is normal. The query interface is shown in Figure 2.

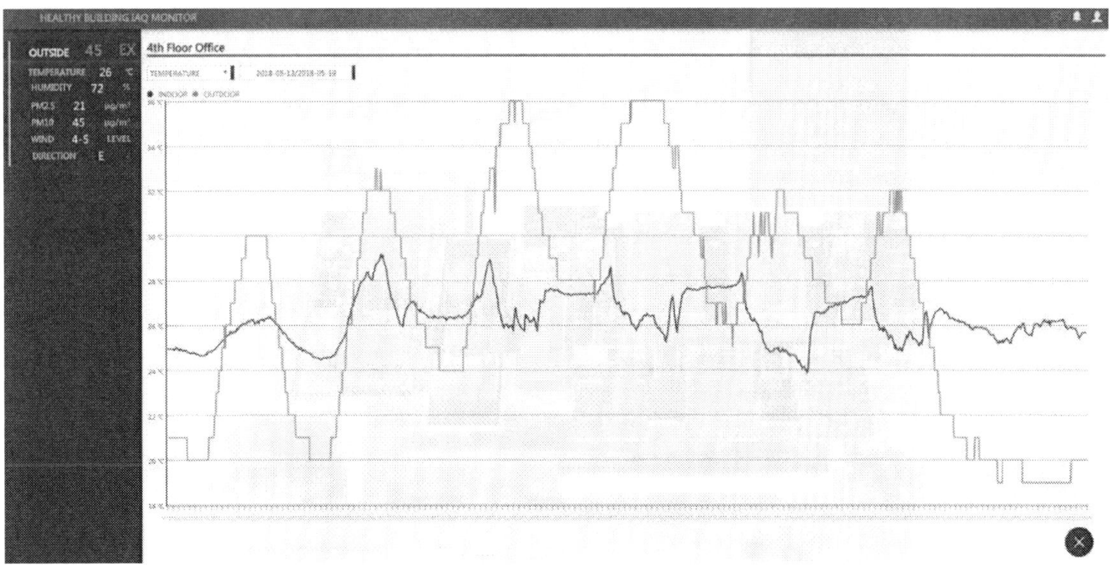

Figure 2. Data analysis

3.3. Equipment linkage

There are a lot of harmful gases in the exhaust gas from the car start. Since the air flow in the underground garage is limited by the limited entrance of the basement, it is difficult to achieve natural ventilation. Therefore, a comprehensive ventilation system must be set to promptly reduce the concentration of harmful substances. Dilute below the required specification.

Based on the data collected by the carbon monoxide concentration sensor installed in the garage and monitored by building automation system (BAS) and set up an automatic alerting value for the linkage and opening of the garage exhaust fan, the goal of saving labor costs and effectively ensuring air quality can be achieved.

3.4. Monthly report

Based on the collected building operational data, the system will automatically generate a public area environmental monitoring report monthly. In this report, not only the water quality of the month (turbidity, residual chlorine, pH, conductivity) but also the indoor environment quality data of the public areas, as well as the energy consumption in a Sankey chart are summarized to assist the managers in establishing a flexible combination of indoor comfort and building energy consumption. Equipment management system. The report interface is shown in Figure 3.

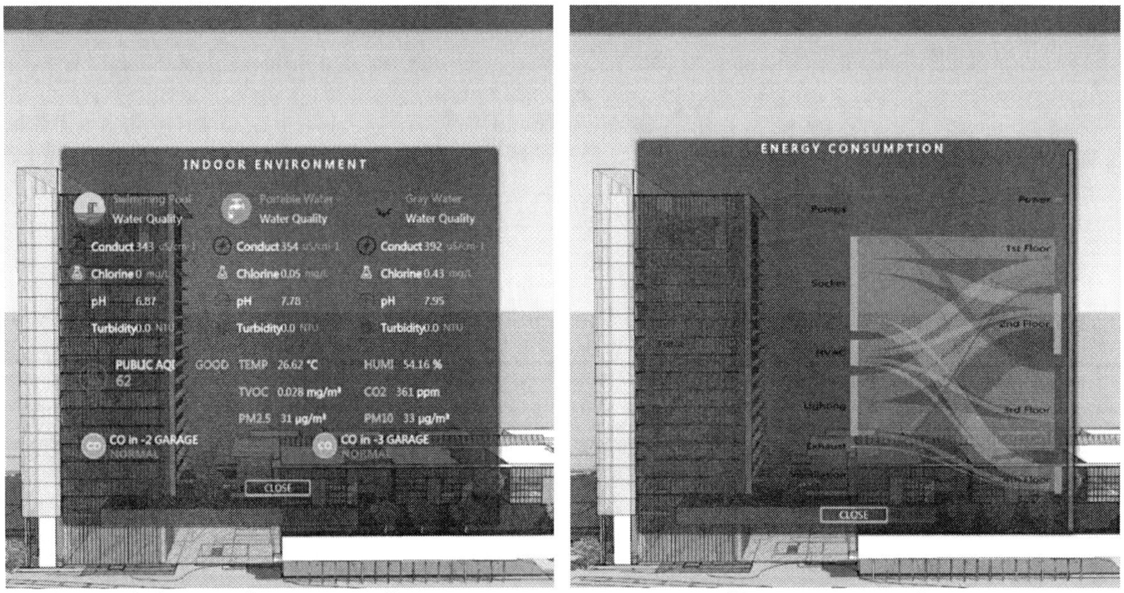

Indoor environment quality Energy consumption

Figure 3. Monthly report

3.5. Weather Information

In the case of Internet connection, the system can automatically obtain local weather data and provide real-time weather information services to remind residents to increase or decrease clothing according to weather changes and reduce the possibility of outbreaks of infectious diseases such as colds.

4. Conclusions

Building Automation Systems (BAS) have been widely implemented with the development of smart building technology. With its aid, building performance data can be generated, collected, managed and maintained automatically. Energy consumption analysis is performed by energy managers using BASs, which can be regarded as Decision Support Systems instantiated to the energy management domain [8].However, the key information are under covered by huge number of data. To solve this problem, business intelligence (BI) technology has been selected. BI technology can extract the desired information from massive building operation data in many ways, such as chats, colors, shapes, etc.

A real building management system for healthy building is introduced in this paper. This system can collect online indoor air quality index as well as water qualities. With persevered data interface, it can be invoked as a function module of BAS, transport data to BAS, to achieve equipment linkage. By implement BI techniques, all displayed data is processed, analyzed and user-friendly. Ordinary users who lack of professional knowledge can detect their surroundings much easier. After a month use, high evaluation are set for this system though there are further work to do.

Acknowledgments

This work is financially supported by the National Key R&D Program of China during the Thirteenth Five-year Plan: Key technologies for performance improving and renovation of existing public buildings (2016YFC0700707), as well as the Foundation for Youth of China Academy of Building Research (Data based 3D Visualization for Intelligent Energy Management System).

References

[1] International WELL Building Institue *WELL Building Standard-Certification Guidebook (Chinese)* v1 with Q3 2017 Addenda
[2] N. Motegi, M. A. Piette, S. Kinney, K. Herter, *Guide to Analysis Applications In Energy Information Systems,* Information Technology for Energy Managers, The Fairmont Press, Inc., 2004, pp. 145-155.
[3] P. Raghunathan, *Data Analysis and Decision Making: Using Spreadsheets and Pivot Tables To Get A Read On Energy Numbers*, Information Technology for Energy Managers, The Fairmont Press, Inc., 2004, pp. 157-169.
[4] A. Ahmed, J. Ploennigs, K. Menzel, B. Cahill, *Multidimensional building performance data management for continuous commissioning,* Advanced Engineering Informatics 24 (2010) 466-475.
[5] S. Tom, *Introduction to Web-based Information and Control Systems*, Information Technology for Energy Managers, The Fairmont Press, Inc., 2004, pp. 9-15.
[6] B. Gnerre, G. Cmar, *Dening the Next Generation Enterprise Energy Management System*, Web Based Energy Information and Control Systems: Case Studies and Applications, The Fairmont Press, Inc., 2005, pp. 403-434.
[7] B. L. Capehart, *Information Technology for Energy Managers*, The Fairmont Press, Inc., 2004.
[8] José Cavalheiro, *A Multidimensional Model for Building Energy Management,*

IOP Conf. Series: Journal of Physics: Conf. Series **1060** (2018) 012012 doi:10.1088/1742-6596/1060/1/012012

K-LRFMD: Method of Customer Value Segmentation in Shared Transportation Filed Based on Improved K-means Algorithm

Hong Li, Xiaosheng Yang, Yao Xia, Lujie Zheng, Guoqing Yang and Pan Lv

Zhejiang University, Hangzhou, 310027, China

lihong@zju.edu.cn

Abstract. The advent of information age has transformed the focus of enterprise marketing from product-centric to customer-centric, and customer relationship management becomes the core problem of enterprises. Accurate customer value classification results are an important basis for enterprises to optimize marketing resources allocation, and customer value classification is becoming one of the key issues that need to be solved urgently in customer relationship management. In the face of the fierce market competition of the vehicle-sharing industries, each shared transportation company has introduced more preferential marketing methods to attract more customers. In this paper, with the aid of the vehicle-sharing platform in a domestic university campus, we established a reasonable customer value evaluation model called K-LRFMD. K-LRFMD did some clustering analysis with the customers based on specific feature engineering and improved K-means algorithm. In this paper, we compare different customer value derived from K-LRFMD. The analysis can formulate the corresponding marketing strategy to provide personalized customer service for different customers.

1. Introduction

In recent years, a lot of capital is flocking to make China's shared transportation filed blustery. Ride-sharing enterprises such as ofo and mobike and car rental company DiDi accelerate competition of traffic market share. Its rapid ascent, universal attention, also explain that people's a great demand of the mass travel. How to use these massive customer data to establish effective communication among relevant business units (such as marketing and customer service) about the segmentation and focus limited resources on high-value customers to achieve the goal of maximizing corporate profits is the current very popular research topic.

Although some literatures [1, 2] introduce some customer segmentation applied to fields of banking, insurance, network account, etc. There is no literature involving shared transportation filed.

There are a lot of data accumulated in the shared transportation filed. However, these data with many attributes is huge, making it difficult to extract valuable information from them. In order to ensure the accuracy of customer segmentation, we must choose the correct and reasonable classification indicators and classification methods. Currently the most widely-value model identifies customers by three indicators (Recent Frequency Monetary), referred to as RFM model [3].Monetary represents the sum of the amount of product purchased over a period of time in the RFM model in [4]. Due to the factors such as driving distance, promotion activities and other factors affecting the business of vehicle-sharing operation, the Monetary is different for different customers with the same consumption amount. Therefore, this indicator does not apply to customer value analysis. [5] proposed AFH customer classification model based on the analysis of RFM model and introduced customer value matrix model. The model adds an indicator of the average consumption amount, but in essence it still doesn't take any factors such as discount coefficient and mileage into account. In terms of classification, clustering analysis is commonly used nowadays. Currently the most widely used clustering algorithm is the K-means algorithm [6, 7]. The effect of K-means algorithm proposed in [8, 9] is greatly influenced by factors such as number of clusters and initial cluster centres. Studies show that the influencing factors mentioned above are related to the specific case and subjective experience. Therefore, this paper replaces the original consumption amount(M) with the two indicators, such as the travel distance(M) accumulated by the customers in a certain period of time and the average value of the discount coefficient(D) enjoyed by the customers in a certain period of time. In addition, considering the membership mechanism of the vehicle-sharing platform, the length of the joining membership can affect the customer value to a certain degree, so the length of the membership (L) is added into the model as another indicator for differentiating the customers. In order to solve the two shortcomings associated with the initial value of K-means algorithm, this paper will set the number of clusters K as 5, and proposed an initial clustering centre selection strategy based on the specific case and competent experience. Finally, we evaluated the customer segmentation based on the actual application scenario.

In short, the model we developed called K-LRFMD has the following key features:

- Special feature engineering: Based on the traditional RFM model, we propose features that are more consistent with customer segmentation in the area of vehicle-sharing.
- Reasonable initial clustering centre selection strategy: We implement a maximum probability of selecting initial clustering centre strategy based on Euclidean distance.
- The vehicle-sharing platform is already in use of the model: Our model has been used in vehicle-sharing platform operated in campus of some universities.

The rest of this article is organized as follows. The second section is an overview of our modelling process. The third section details the method we propose. In the fourth part, we systematically evaluate the performance of the model based on the actual application scenario. The fifth section is discussion and limits of the model. Finally, the sixth section concludes.

2. Overview of building a K-LRFMD customer segmentation model

This section will do a brief introduction to the process of building a K-LRFMD customer segmentation model. In this paper, five indicators of customer relationship: length L, consumption time interval R, consumption frequency F, driving distance M and average discount coefficient D are taken as the customer value index of vehicle-sharing platform(see Table 1) and recorded as LRFMD model.

Table 1. Index meaning

model	L	R	F	M	D
K-LRFMD	The number of days from of Member Registration Time to the end of observation	The number of days from the the customers last driving to the observation window	The number of the time of customers drive in the observation window	Customer times accumulated mileage in the observation window	The average discount factor enjoyed by customers in the observation window

According to LRFMD, if the traditional binning method of attribute of RFM model is used, as shown in Figure.1, it is classified according to the average value of the property, which is bigger than the mean value is expressed as ↑, on the contrary is expressed as ↓, although the model can also identify the most valuable customers, but the number of the customer segmentation is too much, and increase the cost of targeted marketing. Therefore, this paper adopts clustering method to identify customer value. Based on the improved K-means algorithm, five indicators of LRFMD model are clustered to identify the most valuable customers.

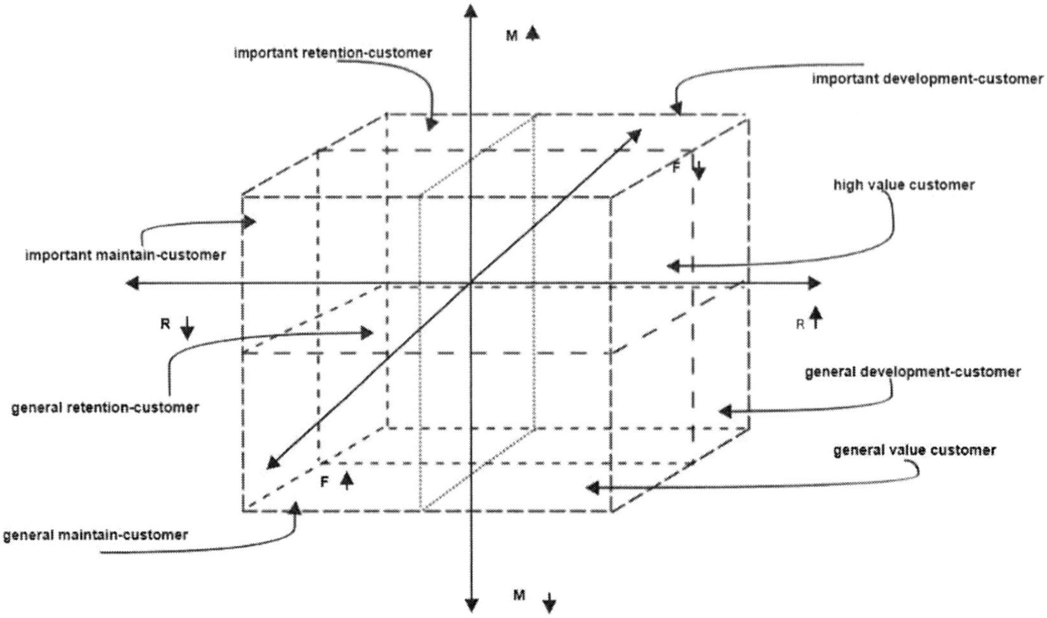

Figure 1. RFM Model Analysis

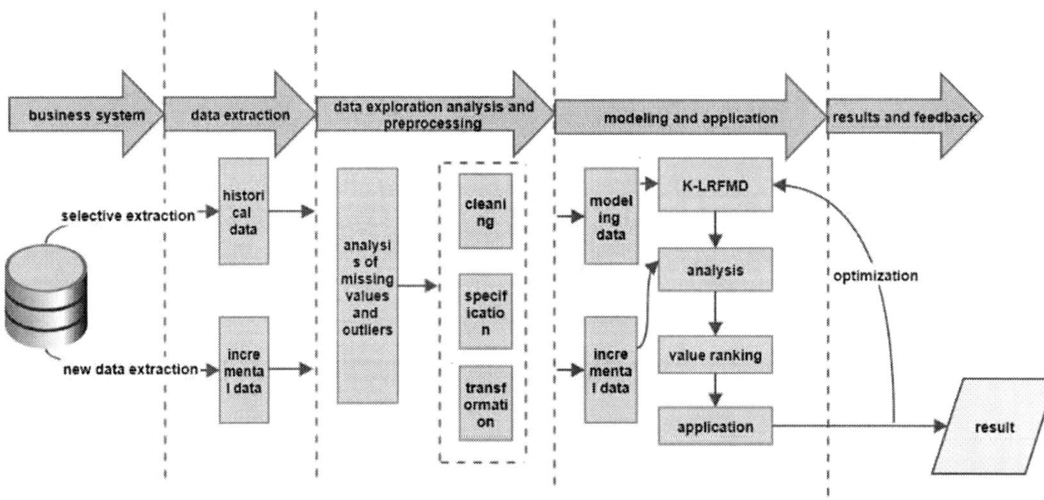

Figure 2. The Overall Process of Building K-LRFMD

The overall process of customer segmentation is shown in Figure.2 and the overall process of building K-LRFMD mainly includes the following steps.

- From the data sources operated by a vehicle-sharing platform in a university, do selective extraction and new data extraction to form historical data and incremental data, respectively.
- Perform Data exploration analysis and pre-processing of two datasets formed in step (1), including analysis of missing values and outliers, attribute specification, cleaning and transformation.

- Based on the modelling data of the completed data pre-processed in step (2), and the improved K-means algorithm, apply K-LRFMD model, and analyse the characteristics of each customer group to identify valuable customers.
- According to the results of the model, using different marketing tools to provide customized services.

3. K-LRFMD

In this section we will detail the specific processes and methods of building a model.

3.1. Data Extraction

For the ending time of November 30, 2017, the time window with a width of two years is selected as the analysis observation window, and the detailed data of all the customers are taken to form the historical data. For the subsequent added customer details, follow up the newest time point in the database as the end time, and extract the data in the same way as above to form incremental data.

From detailed information such as vehicle information and payment records in the database of vehicle-sharing platform in a university, the detailed data of all customers from November 30, 2015 to November 30, 2017 are extracted based on Last_Drive Time of the last driving date, for a total of 50715 records. Which contains the user ID, driving mileage, points, discounts and other 29 attributes.

3.2. Data Exploration and Analysis.

Data exploration and analysis is the analysis of the data missing value and outlier. Through the observation of the data, there is a record that the original data has a payment equal 0, a discount equal 0 and a total mileage more than zero. This phenomenon may generate because of free trial or redeeming points of the customer. The partial result of data exploration and analysis is in Table 2.

3.3. Data Preprocessing

This paper mainly uses the data cleaning, attribute specification and data transformation pre-processing method.

Table 2. Data Exploration Results Analysis Table

Attribute	Null value records	Maximum	Minimum
User id	0	47042	1939
Current miles	0	13710	0
...	...	...	...
Cost	11	78.1	-29.4
Car id	0	246	68

Data Cleaning Through data exploration and analysis, there exist missing value and abnormal value less than zero in the Cost and Money attributes. Due to the huge amount of raw data, such data occupy a relatively small proportion and have little effect on the result and therefore can be discarded. Specific treatment is as follows:

- Discard missing values
- Discard value less than 0 in the Cost and Money

Attribute specification There are too many attributes in the original data. According to the K-LRFMD model proposed in this paper, the six attributes associated with KLRFMD are selected, namely User id, Start time, Load time, Cost, Money, bonus.

Data transformation The data transformation is to transform the data into an "appropriate" format to accommodate the need for mining tasks and algorithms. In this paper, the data transformation method is used for attribute construction and data standardization. Since the original data does not directly give the five indexes of the K-LRFMD model, the five indexes need to be extracted through raw data. The specific calculation methods are as follows:

- L = Load time – Start time

The number of days from the Member Registration Time to the end of the observation window [unit: day]

- R = Last To End

The number of days from the time of customers last driving a shared car to the observation window while [unit: day]

- F = Drive Count

The number of times customers drive a shared car in the observation window [unit: day]

- M = SumCurrentMiles

Customer accumulated mileage in the observation window [unit: mile]

- D = Sum Bonus/Count

The average discount factor enjoyed by customers in the observation window [unit: None]

Table 3. K-LRFMD index range

Attribute	L	R	F	M	D
MAX	450	448	369	239968	111
MIN	1	1	1	0	0
AVG	257.39	168.54	14.57	9308.94	3.87

After five indicators are extracted, the data distribution of each indicator needs to be analysed. The range of the data is shown in Table 3.

From the data in the table, it can be found that the values of the five indexes vary greatly. In order to eliminate the impact of the magnitude of data, the data needs to be standardized. This paper uses Z-score approach and the result of some data processing is as shown in Table 4.

Table 4. Normalized data set

ZL	ZR	ZF	ZM	ZD
1.20882776	1.312803469	-0.185458018	-0.554793993	-0.526600821
0.956659445	1.497281062	-0.347784511	-0.554793993	-0.526600821
1.249094637	1.83109238	-0.510111005	-0.552708063	-0.322491976
0.872614763	1.463577819	-0.550692628	-0.554793993	-0.526600821
-1.581091175	-1.016106791	-0.550692628	-0.554793993	-0.526600821
1.033287101	1.625940291	-0.510111005	-0.554793993	-0.526600821
0.847914913	1.43860888	-0.510111005	-0.554793993	-0.526600821
1.230696342	-0.901523796	1.356643666	-0.091002535	2.085992401

3.4. K-means improved algorithm

K-means algorithm k-means clustering is a method of vector quantization, originally from signal processing, that is popular for cluster analysis in data mining. K-means clustering aims to partition n observations into k clusters in which each observation belongs to the cluster with the nearest mean, serving as a prototype of the cluster. This results in a partitioning of the data space into Voronoi cells [10, 11]. The problem is computationally difficult (NP-hard); however, there are efficient heuristic algorithms [12] that are commonly employed and converge quickly to a local optimum.

Improve K-means algorithm Two key features that make the K-means algorithm highly efficient are also seen as its biggest drawbacks:

- The number of clusters k is an input parameter. Choosing an inappropriate value of k may result in poor clustering results. This is why feature checking is needed to determine the number of clusters in a dataset.

- Convergence to a local optimal solution may lead to "counter-intuitive" erroneous results.

An important limitation of the k-means algorithm lies in its clustering model. The basic idea of this model is that we get spherical clusters separated from each other, in which the mean points tend to converge to the centre of the cluster. It is generally expected that the cluster sizes will be roughly equal, so assigning each observation to the nearest cluster centre (i.e. mean point) is the correct allocation.

Improve the initial clustering centre selection method The traditional k-means clustering algorithm obtains the last k centres through the initial central iteration. The relationship between the final clustering results and the initial clustering centres is still close, and different initial centres may get completely different results. [13] proposed the idea of data segmentation to choose the initial clustering centre. [14] proposed to determine the initial cluster centres based on distance estimation. In [15], K-means++ is proposed to determine the initial cluster centres based on the maximum probability. Although the K-means++ algorithm proposed in [16] can definitely initialize the clustering centre, it has one shortcoming in the scalability: its inherent ordering property: the choice of the next centre depends on Centre of choice. In response to this shortcoming, this paper changes the sampling strategy of each traversal based on K-means++, each traversal takes $O(k)$ samples, the sampling process is repeated about $O(logn)$ times, a total of $O(klogn)$ sample points obtained after repeated sampling, the set of constant factor approximation of the optimal solution, and then the $O(klogn)$ points are clustered into k points.

Table 5. Customer classification situation

ZL	ZR	ZF	ZM	ZD	num per
-1.26534934	-0.93586135	-0.24461487	-0.22321496	-0.68739483	1030 29.59%
0.56052897	0.81107607	-0.33850695	-0.35428804	-0.01618059	1415 40.65%
0.6649835	-0.90491723	4.04972263	4.07757598	-0.3143415	108 3.10%
0.3475817	0.87571874	-0.53503839	-0.46243334	2.44529302	376 10.80%
0.55734129	-0.75230912	0.89627725	0.84189442	-0.28001647	552 15.86%

Finally, these k points are sent to Lloyd iteration as the initial cluster centres. The actual experiments prove that $O(logn)$ sub-sampling is not needed; 5 times repeat sampling can get a better initial centre of clustering. The method is as follows:

- Randomly select $O(k)$ points from the set of input data points as the cluster centres and repeat the sampling five times to obtain a set of $O(5*k)$ sample points and then cluster them into k initial centres point;
- For each point x in the dataset, calculate the distance $D(x)$ between it and the nearest cluster centre (referring to the selected cluster centre), and select a new cluster centre based on the maximum probability criterion of Euclidean distance;
- Repeat step 2 until k clustering centres are found. In step 2, the distance between each data point and the nearest seed point (cluster centre) is calculated in turn, and a set D consist of $D(1),D(2),...,D(n)$. n represents the size of the data set.

In D, to avoid noise, you cannot directly select the element with the largest value. You should select the element with a larger value, and then use the corresponding

The number of clusters K value selection When using K-means algorithm, it is necessary to specify the number of clusters k. The algorithm is based on the field of customer value segmentation and can take k as 5 based on engineering experience.

Clustering results The K-means algorithm is used to cluster the standardized data including L, R, F, M and D, and the clustering results are shown in Table 5 and Figure.3.

4. Customer Value Analysis

According to the result of clustering, the characteristic analysis is shown in Figure.4.

Customer group 1 has the smallest R attribute; Customer group 2 has the largest R attribute; Customer group 3 has the largest L, F, M attributes and R attribute is smaller; Customer group 4 has the largest D, R attributes. According to the specific business analysis of vehicle-sharing platform, the characteristics of a group were evaluated by comparing the size of each index among the groups. For example, customer group 3 has the largest attribute L, F, M, and the lowest attribute R, so it can be said that L, R, F and M are inferior features in customer group 3. And so on, in order to sum up the advantages and disadvantages of each group features, the specific results shown in Table 6.

The chart analysed by the above characteristics shows that each customer group has significantly different performance characteristics. Based on this feature description, the present case defines five levels of customer categories: important maintain-customer, important development-customer, important retention-customer, general customer, Low value customers. Each customer category is as follows:

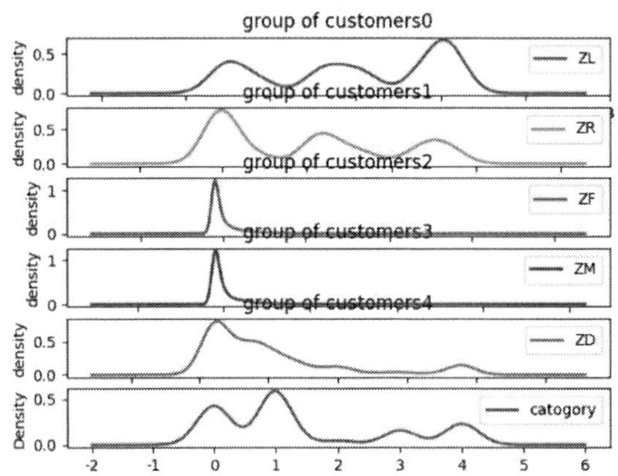

Figure 3. Cluster Diagram

Table 6. Customer Group Characteristics Description Table

Group category	Advantageous features	Weak features
Customer group 1	R L	
Customer group 2		R
Customer group 3	L R F M	
Customer group 4		R F M
Customer group 5		

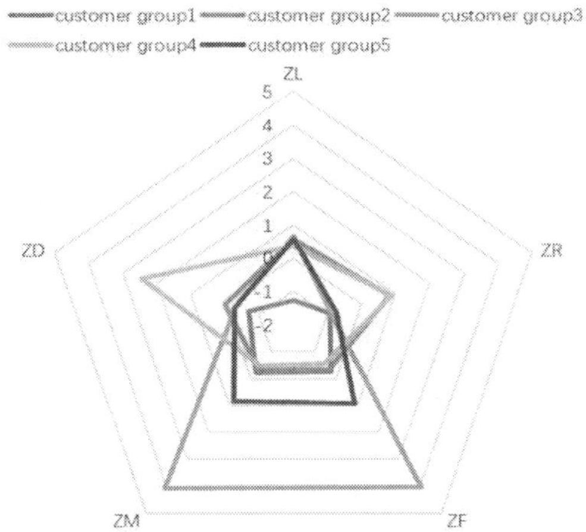

Figure 4. Analysis of Customer Characteristics

- Important maintain-customer: In general, such customers lower the average discount rate. Because the vehicle-sharing platform will hold some discount activities at certain time such as promotions. The lower average discount rate corresponds to more use times. The number of times that customer has used the driver recently (R) is low, the driving frequency (F) is high, and the driving mileage (M) is high, and the member time (L) is long. They are the ideal customer type and loyal customers of the vehicle-sharing platform, making the largest contribution to the operation of the vehicle-sharing platform, but the smallest proportion (3.10%).
- Important development-customer: Although these members have a short membership time (L), they have short R and driving range (M) and driving frequency (F) are relatively large, accounting for 29.59% of the total.
- Important retention-customer: These customers have a long membership time (L), a long R, but the total mileage (M) and driving times (F) are not low, accounting for (15.86%) which indicates the need for retain such customers. We should analyse why they have recently not use the service, and need to maintain more interaction with the customers.
- General customers and low-value customers: Such customers have not driven for a long time. Number (F) or mileage (M) less, accounting for (51.45%).

According to the characteristics of each customer type, customer value ranking of various customer groups, the results shown in Table 7.

Table 7. Customer Group Value Ranking

Customer group	Ranking	Ranking meaning
3	1	important maintain-customer
1	2	important development-customer
5	3	important maintain-customer
2	4	general customer
4	5	low value customers

5. Discussions and Limitations

The model proposed in this paper uses historical data modelling, with the change of time, the observation window of the analysis data is also changing. Therefore, it is recommended that the model should be run once a month considering the details of new customers and the actual situation of the business to judge clustering centre of new customers. Meanwhile, the characteristics of new customers are analysed. If the actual situation of incremental data and the results of the judgments vary widely, you need to focus on the business department to see the reasons for the changes and to confirm the stability of the model. If the stability of the model varies greatly, the model needs to be retrained to adjust.

6. Conclusion

This paper builds a reasonable customer value segmentation model K-LRFMD based on the shared transportation platform. Based on the improved K-means algorithm, Customers are clustered and divided into five types. We established customer segmentation table for customer value analysis, and put forward the corresponding marketing strategy. Empirical studies show that the proposed model and improved algorithm can effectively classify customers in shared transportation filed and differentiate between valueless and high value customers. The proposed K-LRFMD can also be extended to other areas such as aviation customer value segmentation, bank customer value segmentation and so on.

Acknowledgments

This work is partially supported by NSFC61602404, Zhejiang Provincial Natural Science Foundation LY16F020007, and the Fundamental Research Funds for the Central Universities 2017FZA5015.

References

[1] Li W, Wu X, Sun Y, et al. Credit Card Customer Segmentation and Target Marketing Based on Data Mining[C]// International Conference on Computational Intelligence and Security. IEEE, 2011:73-76.

[2] Jiang T, Tuzhilin A. Improving Personalization Solutions through Optimal Segmentation of Customer Bases[J]. IEEE Transactions on Knowledge and Data Engineering, 2009, 21(3):305-320.

[3] Wu J, Lin Z. Research on customer segmentation model by clustering[C]// International Conference on Electronic Commerce. ACM, 2005:316-318.

[4] Xu Xiangbin,Wang Jiaqiang,Tu Huan,et al. Customer classification of Ecommerce based on improved RFM model[J]. Journal of Computer Applications 2012,32(5):1439-1442.

[5] Wang K F. Applied Research on AFH Customer Classification based on Data Mining Technology[J]. Technoeconomics & Management Research, 2012.

[6] Berry M J, Linoff G. Data Mining Techniques: For Marketing, Sales, and Customer Support[M]. John Wiley & Sons, Inc. 1997.

[7] Berkhin P. A Survey of Clustering Data Mining Techniques[J]. Grouping Multidimensional Data, 2002, 43(1):25–71.

[8] Ngai E W T, Xiu L, Chau D C K. Application of data mining techniques in customer relationship management: A literature review and classification[J]. Expert Systems with Applications, 2009, 36(2):2592-2602.

[9] Zakrzewska D, Murlewski J. Clustering Algorithms for Bank Customer Segmentation[C]// International Conference on Intelligent Systems Design and Applications. IEEE Computer Society, 2005:197-202.

[10] MacKay, David. Chapter 20. An Example Inference Task: Clustering. Information Theory, Inference and Learning Algorithms. Cambridge University Press.2003: 284292. ISBN 0-521-64298-1. MR 2012999

[11] Aurenhammer, Franz (1991). "Voronoi Diagrams A Survey of a Fundamental Geometric Data Structure". ACM Computing Surveys. 23 (3): 345405.

[12] E.W. Forgy. Cluster analysis of multivariate data: efficiency versus interpretability of classifications. Biometrics. 1965, 21: 768769.

[13] Liu C, Zeng L, Zhang J, et al. An optimized K-means clustering algorithm for CMP systems based on data set partition[J]. Journal of Computational Information Systems, 2015, 11(13):4727-4738.

[14] Raed T. Aldahdooh, Wesam Ashour. DIMK-means Distance-based Initialization Method for K-means Clustering Algorithm[J]. International Journal of Intelligent Systems & Applications, 2013, 5(2074-904X):41-51.

[15] Bahmani B, Moseley B, Vattani A, et al. Scalable K-means++[J]. Proceedings of the Vldb Endowment, 2012, 5(7):622-633.

[16] Jdrzejowicz J, Jdrzejowicz P, Wierzbowska I. Apache Spark Implementation of the Distance-Based Kernel-Based Fuzzy C-Means Clustering Classifier[J]. 2016.

Study on the Generation of Fractal Image and Animation Based on MATLAB and Flash Platform

Xiaoming Xie

School of Electronics and Information Technology, Sun Yat-sen University, 132 Waihuan East Road, Guangzhou 510006, China

xiexm7@mail2.sysu.edu.cn

Abstract. In practice, the content of fractal graph generated by one kind of algorithm alone is relatively toneless. Combining different algorithms can not only break the limitation of single fractal object generation but also enrich the contents of scenes. In this paper, according to the needs of scene construction, more targeted algorithms are appropriately chosen. By combing two fractal algorithms of recursion and Iterator Function System (IFS for short), two static scenarios "Old Tree, Withered Grass and Snowy Night" and "Reed Land and Snowy Night" are built on MATLAB platform, which further enriches contents of the scene. Simultaneously, a trial has been made to experiment the generation of dynamic fractal graphs. Based on the fractal algorithm with parameters, a recursive windmill animation and a lunar eclipse animation based on recursion and IFS algorithm are generated on MATLAB platform, which enhances the creativity of the scene building of fractal animation. With respect to the selection of implementation platform, this paper compares the effects of fractal graph generation between MATLAB and Flash. The results show that, thanks to its powerful animation creation ability, Flash is undoubtedly more valuable than MATLAB in the implementation of relatively simple fractal animation effect in the light of speed and convenience. Nevertheless, in order to achieve richer and more realistic animation effects, the more powerful MATLAB platform is preferred. In general, it is necessary to choose the right platform according to actual requirements of creation so as to improve the development efficiency.

1. Introduction

As Mandelbrot [1] notes, fractal is the geometry of nature. Fractal is a new-type subject that has been developed in recent 30 years, and this word is translated from the English vocabulary "fractal", which means irregular, broken and numeral. In 1975 and 1982, a French mathematician of American nationality Benoit Mandelbrot, the creator of Fractal Theory, published two historical works: *Los objetos fractales: forma, azar y dimensión* and *the Fractal Geometry of Nature*. Since then, the fractal theory has taken shape.

The fractal theory is a strong theoretical tool for studying irregular objects in the natural world and engineering application. Overall, the fractal object has an uneven and irregular shape and such irregularity also coexists with regularity because of its self-similarity. With the development of this theory, the geometric characteristics of fractal object has no longer been limited to its self-similarity. Objects with self-affine transformation relationship between the whole and the part are also called as fractal. In general, the fractal object is featured by irregularity, self-similarity, self-affinity, self-affinity and smart structure [2].

The fractal image can simulate vividly all kinds of scenery in the nature. It furnishes a new perspective for human beings to know the nature and the world. With the rapid development of computer technology, the computing power of the computer is increasingly strong. The computer-aided fractal algorithm helps to generate complex textures and shapes, as well as producing fractal objects with physical forms and visual effects. The fractal geometry deals with the dot, line and facet of the figure from an overall perspective, so it is superior to traditional Euclidean geometry in terms of the description of nature.

At present, the fractal theory is widely applied to the fields of physics, chemistry, biology, medicine, geology and computer science [3]. In computer graphics, the fractal theory has broaden the channels and provided a new direction for computer graphics generation [4]. In aesthetics, the fractal geometry is applied to morphological creation of furniture [5], landscape design [6], architectural design [7] and textile industry [8]. In the age of digital image processing, the emergence of fractal theory has improved the compression ratio of image encoding, which is particularly important for image transmission and processing [9]. Moreover, the fractal theory also has some important applications in computer animation generative technology. It can help the computer to simulate vivid natural scenes and improve the efficiency of animation production. Therefore, it is widely applied in the field of film-making and game design.

This paper focuses on the application of fractal geometry in computer graphics. By combining recursive algorithm and IFS algorithm, corresponding fractal images are generated on MATLAB platform to build scenes and enrich the image contents. Besides the creation of static fractal images, this paper also explores the generating method of fractal animation and implements on MATLAB platform. Furthermore, in the selection of implementation platform, the effects of generating fractal images on platforms of MATLAB and Flash are compared.

2. Creation of snowy night image - combination of recursive and IFS algorithms

Different algorithms could simulate different natural objects in reality. The fractal image generated by one algorithm is very single in its content. If different algorithms could be used in combination in scene construction, a rather rich scene would be generated. Next, this study will apply different algorithms to generate corresponding fractal objects based on our needs and combine them in the same image to build natural scenes.

2.1. Creation of the scene "Old Tree, Withered Grass and Snowy Night"

This scene graph contains a row of withered grasses and a withered tree as its main scenery, in which the withered grasses are generated by recursive algorithm that is relatively simple among so many fractal algorithms, in consideration of the efficiency of program execution. Although IFS algorithm can be used to simulate high-fidelity natural scene, its implementation complexity is far higher than that of the recursive algorithm, and the IFS code is required to be obtained in the first place. If each withered grass is generated by calling IFS algorithm, time and memory resources will be consumed highly. In view of these problems of IFS algorithm, the part of withered grass is finished by recursive algorithm. As the withered grass is small in shape, it can just be generated at a low frequency of recursion. The final results of creation shows that the effects can be satisfied (Figure 1). As only one withered tree is required to be generated, IFS algorithm is used to achieve a vivid effect. The generation process of the scene graph and part of the codes (MATLAB) used are listed as follows. (1) Determine the hierarchy relationship of the scenery (also known as the layer relationship) to further sequence creation. In figure 1, the crescent is on the bottom layer, the fractal tree is on the second layer, the withered tree is on the third layer and the snow is on the top layer. (2) The image takes shape from the bottom layer. Firstly one moon is generated - a white circle is drawn. Then cover the white circle partly with another circle of the same background color to present a crescent effect. (3) Generation of a clump of grasses. The codes used for creating a clump of grasses are listed as follows.

```
function f=bulrush(ax,ay,bx,by,alfa)
c=0.05; PI=3.1415926; H=sqrt((bx-ax)*(bx-ax)+(by-ay)*(by-ay)); h=H/2; alpha=(alfa/180)*PI;
```

```
if bx<ax
   beta=atan((by-ay)/(bx-ax))+PI;
elseif bx==ax
   beta=PI/2;
else
   beta=atan((by-ay)/(bx-ax));
end
x=[ax,bx]; y=[ay,by]; plot(x,y,'Color',[rand() rand() rand()],'LineWidth',2);hold on;
if H>c
   gx=(bx+ax)/2; gy=(by+ay)/2;
   cx=gx+h*cos(beta+alpha); cy=gy+h*sin(beta+alpha);
   ex=bx+h*cos(beta+alpha); ey=by+h*sin(beta+alpha);
   x=[gx,cx]; y=[gy,cy]; plot(x,y,'Color',[rand() rand() rand()],'LineWidth',2); hold on;
   x=[bx,ex]; y=[by,ey]; plot(x,y,'Color',[rand() rand() rand()],'LineWidth',2); hold on;
   bulrush(gx,gy,cx,cy,alfa); bulrush(bx,by,ex,ey,alfa);
end
axis([-2 2 -2 2]); axis off;
for N=1:100
      ax=rand(); ay=0.035; bx=rand()*0.02-0.01+ax; by=rand()*0.07+0.07;
      bulrush(ax,ay,bx,by,15); hold on;
end
```

A total of 100 withered grasses is generated. “ax” and “ay” denote respectively as the horizontal and vertical coordinate of the withered grass position. The value of “ax” is produced by a random number, so the positions of the 100 withered grasses are all random. “bx” and “by” are related to the height of the withered grass. The height of each withered grass is also random and falls between 0.7 and 1.4.

(4) IFS algorithm is applied to generate fractal tree.

```
function f=tree_ifs_2(ax,ay)
figure('color',[1,1,1]);
x=[ax ay];
a1=[0.195 0.344;-0.488 0.443]; b1=[0.4431 0.2452];
a2=[0.462 -0.252;0.414 0.361]; b2=[0.2511 0.5692];
a3=[-0.058 0.453;-0.07 -0.111]; b3=[0.5976 0.0969];
a4=[-0.035 -0.469;0.07 -0.022]; b4=[0.4884 0.5069];
a5=[-0.637 0;0 0.501]; b5=[0.8562 0.2513]; r=rand(1,100000);
for N=1:100000
   if (r(1,N)>=0)&(r(1,N)<=0.25)
      x=x*a1+b1; plot(x(1),x(2),'Color',[0.80 0.75 0.44]); hold on;
   elseif (r(1,N)>0.25)&(r(1,N)<=0.5)
      x=x*a2+b2; plot(x(1),x(2),'Color',[0.80 0.75 0.44]); hold on;
   elseif (r(1,N)>0.5)&(r(1,N)<=0.75)
      x=x*a3+b3; plot(x(1),x(2),'Color',[0.80 0.75 0.44]); hold on;
   elseif (r(1,N)>0.75)&(r(1,N)<=0.95)
      x=x*a4+b4; plot(x(1),x(2),'Color',[0.80 0.75 0.44]); hold on;
   else
      x=x*a5+b5; plot(x(1),x(2),'Color',[0.80 0.75 0.44]); hold on;
   end
end
axis off
```

(5) Adorn the fractal image with snowflakes.

The snowflakes are produced by drawing "*" at the coordinate points which are generated by random numbers. The final effect is shown in figure 1.

Figure 1. The image of "Old Tree, Withered Grass and Snowy Night" generated by recursive and IFS algorithm.

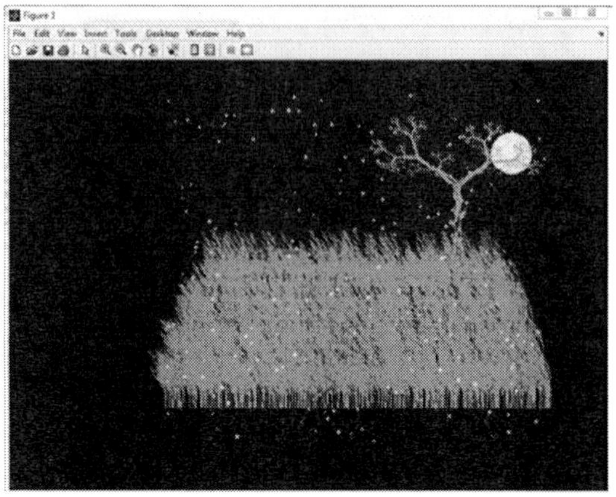

Figure 2. The image of "Reed Land and Snowy Night" generated by recursion and IFS algorithm.

2.2. Creation of the scene "Reed Land and Snowy Night"

An attempt of the 3D visual richness is conducted in the above situations and the reeds are further turned to a reed land. Such 3D visual presentation from close-by examples to those far off just requires, from the perspective of visual principle, a deceasing tendency of the reeds row by row, thereby producing effects of a piece of reed land. The codes for generating this image are just required to be modified from the code of reed part in figure 1 and the rest would be maintained. The final effect is shown in figure 2. The generating codes for reed land are listed as follows.

```
t=[500 400 300 250 200 120 120 120 120];
a_x=[1 0.96 0.92 0.88 0.84 0.76 0.76 0.76 0.76]; b_y_1=[0.06 0.03 0.03 0.02 0.02 -0.06 -0.06 -
0.06 -0.06]; b_y_2=[ 0.61 0.43 0.26 0.17 0.07 -0.06 -0.02 - 0.02 - 0.02];
n=0;
for N=1:9
    n=n+1;
    for N1=1:t(n)
        ax=((rand()*2)-1)*a_x(n); ay=-0.8; bx=rand()*0.02-0.01+ax;
        by=-rand()*b_y_1(n)- b_y_2(n);
        bulrush(ax,ay,bx,by,15); hold on;
    end
end
```

The vectors "t", "a_x", "b_y_1" and "b_y_2" are parameters used to control the number and height ranges of reeds in each row, in which "t" denotes the number of reeds in each row, "a_x" is used to control the range of reeds, "b_y_1" and "b_y_2" is used to control the height range of the reeds.

3. Generation of fractal animation based on MATLAB platform

3.1. Basic principle of fractal animation: fractal algorithm with parameters

Most fractal algorithms are required to contain parameters to control the appearance of the fractal objects. In other words, if one wants to change the appearance of the object generated by fractal algorithm, only the parameters of fractal algorithm are required to be modified. Taking IFS algorithm as an example, the appearance of the fractal object generated by IFS algorithm rests on IFS code. If the IFS codes are changed, so does the appearance of the fractal object. Such changes may be trivial or great, which is dependent on the role of the IFS codes play in the whole code block. Based on this principle, control of the continuous change of one key parameter in the algorithm can produce a fractal image with a series of continuous changes. If we consider these continuously changing fractal images as the animation frame and play them continuously, an effect of fractal animation could be produced. The following study is based on MATLAB platform. It generates several pieces of fractal animation respectively by recursive algorithm and IFS algorithm with parameters.

3.2. Basic ideas of generating animation on MATLAB platform

Generating continuously changing fractal image by continuously changing parameters also mean that a new piece of fractal image will be generated after a change of parameter each time. As the implementation of fractal algorithm may consume some time, the time interval between two adjacent images may be long. If the animation is generated by drawing firstly and displaying next, it will appear so discontinuous and affect the visual effect. Therefore, this study uses a method of generation frame by frame firstly and centralized playback next. After each piece of image is generated, the command "getframe" in MATLAB will be used to get each fractal image generated, namely, each frame of the animation. Then the frames will be saved via the command of "movie2avi" as an avi file. Finally, the effects of fractal animation will be seen continuously by playing this avi file. In addition, the play speed and loop playback may be set through relevant commands. Next this study will analyze in detail the design process and key codes of the fractal animation generated on MATLAB platform.

3.3. Windmill animation based on recursion

The animation effects of the windmill animation in this design are presented by controlling the continuous changes of the reed shape via the continuously changing parameters. In order to control the reed shape more intuitively and conveniently, the reed generated function is required to be modified. Specific details are shown as follows.

```
function f=bulrush_2(ax,ay,b,alfa_1,alfa_2)
   PI=3.1415926; alfa_1=alfa_1*(PI/180); bx=b*cos(alfa_1); by=b*sin(alfa_1);
   bulrush(ax,ay,bx,by,alfa_2);
end
```

Unlike the generated function of the above-mentioned reed bulrush (ax,ay,bx,by,alfa_1), bulrush_2 controls the shape of reed directly by passing its height (b) and angle (alfa_1) and both contribute to calculations of "bx" and "by" value, then the bulrush function will be called again to generate the reed. Such modification enables more intuitive and convenient control of the reed shape and simplifies the animation process of windmill animation. The generation process of the windmill animation is listed as follows.

(1) Generation of the windmill.

```
bulrush_2(0,0,1,0,-18); M(1)=getframe; r=0; dr=5;
for N=1:72
   clf; r=r+dr; bulrush_2(0,0,1,r,-18);
   if(r>=50)
      bulrush_2(0,0,1,45,-18);
   end
```

Similar to if(r>=50), the other 6 reeds will be generated respectively at 90°, 135°, 180°, 225°, 270° and 315°, and form a complete windmill when the precondition of "if" statement changes. So specific codes are not given here.The processes and effects of windmill generation are shown in figure 3.

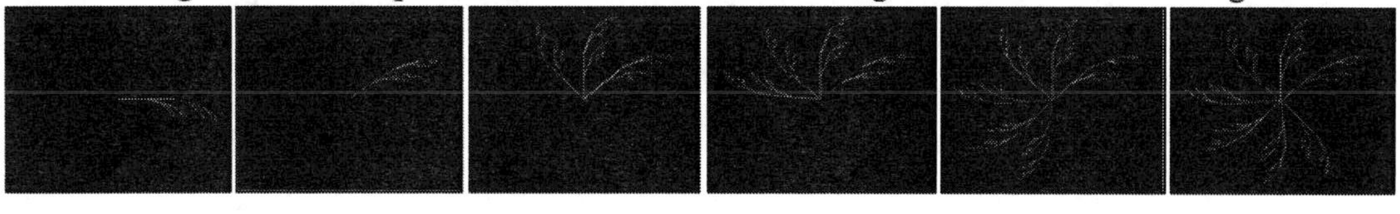

Figure 3. Windmill generation processes and effects.

(2) Rotation of the windmill.

Two parameters should be defined in this step. One controls the pendulum angle of small subbranches of the reed and the amplitude of swing ranges from -18 to 18. The other denotes the swing pace of the small subbranch and the subbranch swings 1° in each frame. The effects of windmill rotation are shown in figure 4.

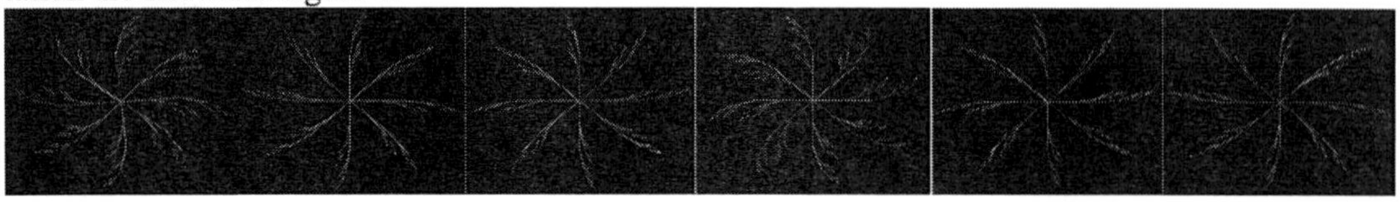

Figure 4. Windmill rotation.

(3) The windmill changes in size while rotating with smear.

Four parameters should be defined in this step. The first one controls the pendulum angle of small subbranches of the reeds and the amplitude of swing ranges from -18 to 18. The second one denotes the swing pace of the small subbranch and the subbranch swings 1° in each frame. The third one controls the length of the reeds, ranging from 0.028 to 1. The last one denotes the pace of length changing. The animation effects are shown in figure 5.

Figure 5. Size variation during rotation with smear.

(4) Generation of .avi files.

3.4. Lunar eclipse based on recursive and IFS algorithms

In the image of "Old Tree, Withered Grass and Snowy Night" (Figure 1), the crescent is formed by partial overlap of two circles and its shape changes with the movement of the circle-cover. If the process of continuous crescent changes is gotten in the form of frame, continuous playback helps to get the whole process of lunar eclipse. In addition, the continuous changes of parameters allow the reeds to render an effect of swinging with the wind. The generation process in shown as follows.

(1) Generate the 1st frame of animation.

In this step, we can define two vectors "ax" and "1", both are used to save the length information of each reed generated in the first frame. Only in this way can the reed of same length be generated at the same location in each of the following frames and the discontinuity between frames could be avoided.

(2) The process of lunar eclipse and swinging reeds.

There are four parameters defined in this step. The parameter "s" denotes the horizontal coordinate of the circle-cover, and "ds" denotes the pace of its change in each frame. The parameter "angel" is used to control the angle of reed, "da" is the pace of its change in each frame. In this animation, firstly the reeds swing to an angle of 105° and then swing back and forth between 103°and 105°. This effect is controlled by mark bit "flag". Once these parameters are defined, we can simply start to generate frame by frame through using a "for" loop.

(3) Generation of .avi files.

The final effects are shown in figure 6.

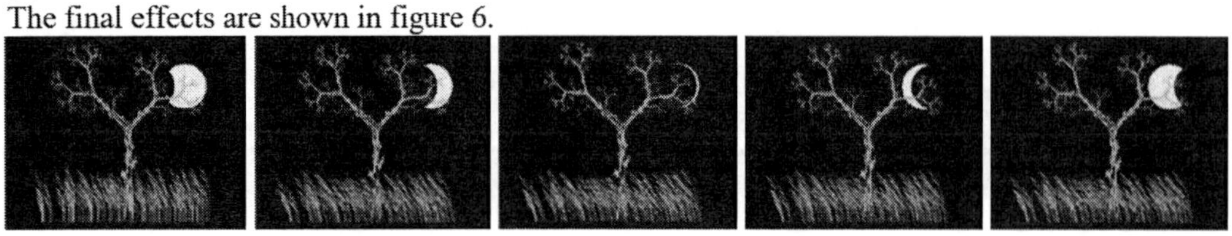

Figure 6. The effect of lunar eclipse animation.

4. The combination of Flash and fractal animation

As a powerful animation creation software, Flash is a rich animation creation tool that provides many advantages for animation creators. Although Flash is widely used in animation, the current research on the combination of Flash and fractal animation is still relatively rare. After Flash develops the function of supporting the copy of a movie clip, it has become very convenient to develop fractal animations on the Flash platform. The similarity between the part and the whole of fractals enables us to use the same small parts as the fractal overall shape to generate the final whole by the iteration of one layer after another. This design implements LS algorithm by writing script codes and simulates a plant's growth process from sparse to dense by creating components and adding simple shape to the components. The following is the script codes and analysis of the two LS algorithms with single-rule and multi-rule.

4.1. Implementation of single-rule LS algorithm

```
    function ls(w,ax,ay,t)
 {
     var t_0=0; _root.stack=[]; var stack_e=0; var x=ax; var y=ay; var w_temp=new String("");
     var alpha=-90; var alfa=25; var i=1;
     if(t>0)
     {
       for(i=0; i<w.length; i++)
       {
          ......//Generating the path under the rule: F -> F[-F]F[+F]F
       }
       this. ls(w_temp,x,y,t-1);
     }
     if(t==0)
     {
       for(i=0;i<w.length;i++)
       {
          switch(w.charAt(i))
          {
             case('F'): ......//step forward and draw a line
             case('['): ......//push the current direction and position into stack
             case(']'): ......//pop from stack
             case('-'): alpha=alpha-alfa; break;
             case('+'): alpha=alpha+alfa; break;
          }
       }
     }
 }
```

In the "function ls (w, ax, ay, t)", an array is used to implement the stack frame of the LS algorithm and creates objects of "String" for the strings so that character(s) is convenient for operation such as traversal, replace, connections etc. The animation effects are achieved mainly through the shape tweening of the component. By adding a component instance to the main scene with "attachMovie()", and adding the key frame to the new layer in the main scene, the function is added to the frame codes. The final effects are shown in figure 7.

Figure 7. A plant's growth process by applying Single-rule LS algorithm.

4.2. Implementation of multi-rule LS algorithm

"Multi-rule" in the multi-rule LS algorithm refers to the rule of multiple replacement. In the single-rule LS algorithm, replacement is required only when character 'F' appears in a string, while multi-rule requires to replace multi characters. Hence, the script codes only need to be partially modified. Firstly, add judgement statement in the string to be replaced according to the newly added rule. For example, in this multi-rule design, there are two replacement rules:

L→[LL[+L+L-[-R-R++R]]R]; R→[R-R++R]

Simply add the following statement to the first "for" loop in the single-rule function.

```
if(w.charAt(i)=='L'||w.charAt(i)=='R')
    {
      if(w.charAt(i)=='L')
      {
        w_temp=w_temp.concat("LL[+L+L-[-R-R++R]]R");
      }
      else
      {
        w_temp=w_temp.concat("[R-R++R]");
      }
    }
```

The others are similar to the previous single-rule. The final effect is shown in figure 8.

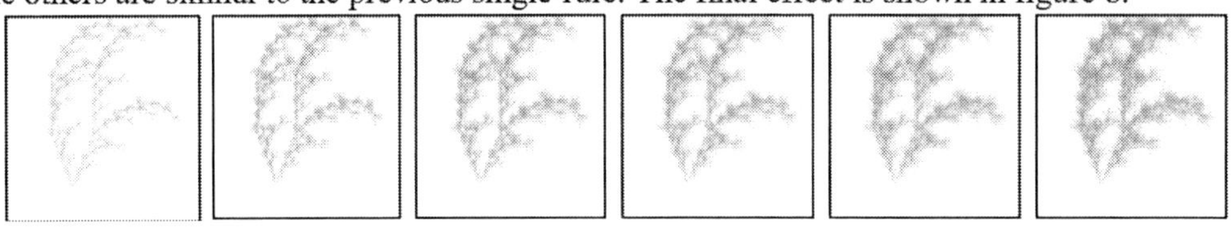

Figure 8. A plant's growth process by applying Multi-rule LS algorithm.

What can be concluded from the above figure 7 and figure 8 is that although it's easier to implement LS algorithm on Flash, deeper sense of recursion can't be realized because of the limitation of Flash computing power, thus the effect of plant simulation is not as expected. The main idea of LS algorithm is moving forward and drawing lines, which limits the animation effect of Flash and makes the effect monotonous and not obvious. From this perspective, the MATLAB platform's computing power is obviously much stronger than Flash because it is capable of deep recursion, and the simulation of plants on MATLAB is better than Flash in terms of fidelity.

5. Conclusion and discussion

In practice, the content of fractal graph generated by one kind of algorithm alone is relatively toneless. Combining different algorithms can not only break the limitation of single fractal object generation but also enrich the contents of scenes. In this paper, according to the needs of scene construction, more targeted algorithms are appropriately chosen. By combing two fractal algorithms of recursion and IFS, two static scenarios "Old Tree, Withered Grass and Snowy Night" and "Reed Land and Snowy Night" are built on MATLAB platform, which further enriches contents of the scene. Simultaneously, a trial has been made to experiment the generation of dynamic fractal graphs. Based on the fractal algorithm with parameters, a recursive windmill animation and a lunar eclipse animation based on recursion and IFS algorithm are generated on MATLAB platform, which enhances the creativity of the scene building of fractal animation.

With respect to the selection of implementation platform, this paper compares the effects of fractal graph generation between MATLAB and Flash. The results show that, the computing power of Flash is far less than that of MATLAB, so it might not be possible to realize the complex fractal algorithm on Flash. However, thanks to its powerful animation creation ability, Flash is undoubtedly more valuable than MATLAB in the implementation of relatively simple fractal animation effect in the light of speed and convenience. What's more, Flash has great flexibility, which can create vivid animation effect freely and also can make the realization of man-machine interactive operation more convenient. Nevertheless, in order to achieve richer and more realistic animation effects, the more powerful MATLAB platform is preferred. In general, MATLAB and Flash both have their advantages and disadvantages. Therefore, it is necessary to choose the right platform according to actual requirements of creation so as to improve the development efficiency.

Fractal geometry, with its unique charm, enormously enriches human's cognition of the world. It has extremely high theoretical value and has attracted many scholars' attentions in the academic circle. Nowadays, fractal theory has been applied in many fields such as signal processing, image coding, as well as new material research and development. In the future, fractal theory will achieve more fruitful results with further explorations being done. For example, in the field of architecture and home decoration, architect and building designer can use a complete fractal graph to decorate the whole surface of a wall or produce wallpaper and home ornamentation composed of fractal pattern. In the field of information security, fractal theory can contribute to information encryption technology as well. In the aspect of natural disaster warning, prediction accuracy of natural disasters such as earthquake and tsunami can be improved by using fractal theory. Besides, in the area of computer graphics, with the development of computer technology, the process of fractal image exploitation in computer will become more efficient and convenient, and the contents will be more colorful.

References

[1] Mandelbrot B (Translated by Wang J) 1991 Fractal - the geometry of nature *World Science* **11** 1–4

[2] Lou L 2010 *Communication Theory and Technology of Image* (Beijing: Tsinghua University Press) (in Chinese)

[3] Zhu H and Ji C 2011 *Fractal Theory and its Application* (Beijing: Science Press) (in Chinese)

[4] Chang T 2013 Application of Fractal Geometry in Computer Graphics *Computer CD Software and Applications* **17** 314–5 (in Chinese)

[5] Li S and Geng X 2018 Research on Furniture Shape Creation Based on Fractal Theory *Furniture and Interior Design* **1** 16–7 (in Chinese)

[6] Feng Y, Tian F and Tian C 2016 Fractal Geometry Aesthetics of Traditional Chinese Gardens - Rebellion against Western Euclidean Geometry Aesthetics *Huazhong Architecture* **1** 11–5 (in Chinese)

[7] Yan L and Yang C 2008 Application of Fractal Aesthetics in Architectural Design *Journal of Harbin Institute of Technology (Social Sciences Edition)* **10** 28–32 (in Chinese)

[8] Tang Y, Fang K and Shen L 2009 The Comparison between Fractal Patterns and Traditional Textile Printing Patterns *Journal of Textile Research* **30** 90–4 (in Chinese)

[9] Zhang A, He Y and Zhang J 2017 Image Compression Coding Algorithm Based on Wavelet and Fractal Theory *Computer Technology and Development* **27** 46–50 (in Chinese)

Is Group Means Imputation Any Better Than Mean Imputation: A Study Using C5.0 Classifier

Faizan U F Khan[1], Kashan U Z Khan[2] and S K Singh[1]

[1] Department of Computer Science and Engineering, IIT (BHU), Varanasi, 221005, India
[2] Department of Computer Engineering, Faculty of Engineering and Technology, Jamia Millia Islamia, New Delhi, 110025, India

fufahad.khan.cse14@iitbhu.ac.in

Abstract. Since most data-driven systems including classifiers require large amounts of complete data, the task of handling missing data has garnered much attention. If one of the variables under study in a dataset has some incomplete values, it is treated as a missing data problem. Various methods in the literature exist for dealing with missing data including complete case analysis, listwise deletion, single imputation and multiple imputations. Out of these, mean imputation remains a favourite for researchers due to its simplicity and ease of use, despite some glaring flaws. In this paper, we compare Mean imputation with a similar single imputation method – Group Means imputation – and present our results on nine real-world datasets with respect to classifier accuracy of the C5.0 classifier on the imputed dataset. We show that while Group Means imputation fares better on training data, the test set accuracies fall in favour of Mean Imputation, which deals with novel data in a much better fashion.

1. Introduction

In recent times, data-driven systems have seen a great boost in terms of application and have had a large impact in the field of computer technology. However, the data initially collected is rarely optimal and need to go through a data preprocessing phase before it can be utilized to mine knowledge. [1][2][3] Data Imputation is a preprocessing method that deals with the inconsistency of missing data. Missing data inconsistency is said to be present in a dataset when one or more variables under study have an absence of values [4].

Missing data can render decision-making systems including neural networks and Support Vector Machines useless as they require complete data to produce output. In such cases, it becomes essential to estimate the missing data which can be fed to the system to extract knowledge. Care should be taken while estimating the missing data as the goal while estimating data should be to make efficient and valid inferences about the sample data rather than to predict, estimate or retrieve missing data or to get the similar output that is with complete data. [5] Data Imputation has emerged as a valid candidate for the problem of missing data. The literature consists of various imputation methods including Mean/Mode Imputation, Group Mean Imputation, Predictive Mean Matching, Regression Imputation and Multiple Imputation using Chained Equations (MICE). [6][7] While many alternatives exist, Mean/Mode Imputation continues to be the most widely abused form of imputation, despite its various flaws. This is due to its simplicity and applicability to all cases, and the fact that it is time optimal.

In this paper, we test the efficiency of Mean Imputation versus a modified version of Mean Imputation – known as Group Mean Imputation – and see if it offers a better and simple alternative to Mean Imputation with respect to classifier accuracy. We use real-world datasets taken from the UCI Machine Learning Repository [8] and artificially introduce missing values to obtain imputed datasets. These datasets are then used to build a C5.0 classifier which is used to evaluate the quality of imputation of the different methods. [9]

2. Preliminaries

The objective of imputation is to replace the missing values by plausible values in order to exploit the information in the recorded variables in the incomplete cases for inference about population parameters. Complete data (Y) can be defined as a function of observed data (Y_{obs}), and missing data (Y_{mis}),

$$Y = (Y_{obs}, Y_{mis}) \tag{1}$$

2.1. *Missing data mechanisms*

Knowing the reason why the data are missing is beneficial in choosing the right technique to approximate missing data. [10] Generally, we have four missing data mechanisms [7][10], which dictate why the data is missing.

2.1.1. Missingness Completely At Random. This type of missingness does not bias the inferences. The probability of a value of a variable being missing is completely random and unrelated to any other values of other variables, whether missing or observed. There is no observable pattern in missing data and cases with complete data are indistinguishable from cases with incomplete data.

2.1.2. Missingness At Random. In this type of missingness, the cause of the missing data is related to other observed variables but unrelated to the missing values themselves. Also known as the *ignorable case.* We can predict the missing values based on observed values of other variables.

2.1.3. Missingness that depends on unobserved predictors. Missingness is no longer "at random" if it depends on information that has not been recorded and this information also predicts the missing values. This type of missingness must be explicitly modelled, or else bias can occur in the inferences

2.1.4. Missingness that depends on the missing value itself. The instance of the missing data depends on the unobserved data of the variable itself. Particularly hard to handle, it implies that the missing data mechanism is related to the missing values.

3. Mean Imputation and Group Mean Imputation

3.1. *Mean/Mode Imputation*

One of the most common methods of imputation, in this method, the mean of all values within the same attribute is calculated and then imputed in the missing data cells. If the attribute is nominal, mode substitution can be used instead. Mean substitution introduces bias in the parameter estimates as a result of loss in standard deviation, even though the mean of the data is generally preserved. [7] This is because we substitute all missing values with a single measure of central tendency.

$$A_i^j = \sum_{k \in (complete)} A_k^j / n_{|I(complete)|} \tag{2}$$

3.2. *Group-Means Imputation*

Group means substitution differs slightly from the normal mean/mode substitution. In this method, the missing values are replaced with the group (or class) mean of all known values of that attribute. [11] While the class wise imputation technique will impute multiple values, it will still underestimate the standard deviation and will lead to biased parameter estimates. As in normal Mean Imputation, an alternate measure of central tendency can be used if calculation of mean is not possible.

4. Experiment

The details of the experiment including the datasets used, the ratio and type of missingness and the classifier details are discussed further. The whole experiment was carried out in the R programming language.

4.1. Datasets

We applied the imputation methods on nine different datasets taken from the UCI Machine Learning Repository [8]. The details of the datasets are given in table 1.

Table 1. Datasets used in the experiments.

Dataset Name	Number of Categorical Attributes	Number of Ordinal/Continuous Attributes	Number of Cases
Breast Cancer Wisconsin	1	9	699
Car	7	-	1728
Credit	10	6	690
Cleveland	9	5	303
Ionosphere	1	34	351
Iris	1	4	150
Spambase	1	57	4601
Titanic	3	1	1313
Wine Quality	1	11	4898

For the *Wine Quality* dataset, we converted the continuous *quality* variable to an ordinal variable with three categories depicting the quality of the wine: *"good"* (>7), *"average"* (>4 & <7), *"bad"* (<4).

The whole dataset was divided in the training-test ratio of 70:30. Missing values were artificially introduced in each of the nine different datasets, with three different missing ratios of 30%, 40% and 50%. For each missing ratio, three different incomplete datasets were created, resulting in nine different variations of each dataset. The results were reported after aggregating the results of the imputed datasets to avoid sampling bias.

4.2. C5.0 classifier details

The C5.0 algorithm derives from the C4.5 algorithm, but is faster and less error prone. [9][12] It has some additions over the previous algorithm including generation of smaller decision trees, generation of less error prone rules, winnowing and the ability to incorporate boosting. In our experiments, we generated rule based models and set the number of boosting iterations to 10. Using adaptive boosting, we generate several classifiers rather than just one. When a new case is to be classified, each classifier votes for its predicted class and the votes are counted to determine the final class. We use the evaluation criteria similar to that proposed in [13] in our experiments. The results include the accuracy of the C5.0 classifier on both the training set and the test set. We used the R implementation of the C5.0 decision tree algorithm as provided in the R package *C50.* [14]

5. Results

Table 2. Classification accuracy of the C5.0 classifier with datasets having three different missing ratios. Bold values indicate a higher accuracy value than the other method.

		Training Set		Test Set	
Dataset	Missing Ratio	Mean	Group Means	Mean	Group Means

Breast Cancer Wisconsin	30	99.90	**100.0**	95.71	**96.03**
	40	99.33	**100.0**	**95.56**	95.24
	50	99.60	**100.0**	**96.51**	95.08
Car	30	82.47	**97.37**	72.32	**87.28**
	40	82.27	**97.33**	71.29	**87.86**
	50	82.07	**96.63**	71.29	**87.09**
Cleveland	30	85.43	**97.43**	**84.06**	76.92
	40	87.37	**98.53**	**83.73**	77.65
	50	87.37	**98.40**	**80.52**	77.65
Credit	30	95.43	**100.0**	79.48	**83.57**
	40	93.40	**100.0**	77.65	**82.13**
	50	92.80	**99.83**	77.65	**84.38**
Ionosphere	30	100.0	100.0	**94.66**	90.88
	40	99.73	**100.0**	**92.45**	91.82
	50	100.0	100.0	**95.28**	92.14
Iris	30	89.53	**98.70**	**97.78**	94.82
	40	89.50	**98.70**	**95.56**	91.85
	50	90.80	**98.40**	**94.07**	91.85
Spambase	30	91.43	**99.80**	**92.91**	91.43
	40	90.47	**99.77**	**93.17**	90.30
	50	91.07	**99.80**	**92.88**	90.49
Titanic	30	74.07	**96.30**	78.93	**79.19**
	40	73.37	**95.87**	76.99	**80.20**
	50	75.33	**95.90**	**79.19**	78.43
Wine Quality	30	81.97	**100.0**	**82.02**	75.65
	40	81.73	**99.93**	**82.18**	78.07
	50	84.83	**100.0**	**81.70**	77.80

The experiments showed varying results for both the imputation methods. We found Group Means Imputation to be almost exclusively better than normal Mean Imputation on all datasets when training set results were compared, as shown in table 2. The Group Means method posted training accuracies of above 96% for all datasets, even when the missing data ratio was 50%. However, this was not reflected in the test data accuracies for Group Means Imputation, which dropped significantly to about 77% for the *Wine Quality* and *Cleveland* datasets, where the training set accuracies were greater than 97%.

Mean Imputation on the other hand posted relatively low training set accuracies as compared to Group Means Imputation, but outperformed Group Means Imputation when test set accuracies were considered. It was better in six of the nine datasets for different missingness ratios. Group Means was better with respect to classification accuracy for the *Car*, *Credit* and *Titanic* datasets for all missingness ratios except for 50% missingness in the *Titanic* dataset. We observe that in all these cases, the number of categorical variables was larger than the number of continuous/ordinal variables, which may have been a factor for these observations. This might be due to the fact that we use mode substitution since mean is not defined for categorical variables.

For *Titanic*, *Spambase* and *Iris* datasets, Mean Imputation reported better accuracies with the test set than the training set, generalizing better on unseen data. This was not the case for Group Means which generalized very poorly for novel data.

6. Conclusion

Our study shows that while both Mean Imputation and Group Means Imputation use similar methodology to handle missing data, Mean Imputation is better suited for the task as it generalizes well for novel data. High classification accuracies for training set for Group Means Imputation is misleading and may result in worse results when dealing with novel data.

However, when the number of categorical variables was larger than the number of continuous/ordinal variables, Group Means Imputation outperformed normal Mean/Mode Imputation in both training and test set accuracies. This should be taken into account when selecting the better of the two methods, since Mode substitution due to a large number of categorical variables seems to reduce accuracy. The ratio of missingness had no observable correlation with the accuracies for both the methods. In addition, the size of the dataset was not a factor which may differentiate between the two methods.

References

[1] Larose DT 2014 *Discovering Knowledge in Data: An Introduction to Data Mining*. John Wiley & Sons
[2] Han J, Pei J and Kamber 2011 M. *Data Mining: Concepts and Techniques*. Elsevier
[3] Magnani M 2007 Techniques for dealing with missing data in knowledge discovery tasks. Obtido http://magnanim. web. cs. unibo. it/index. html.
[4] Bannon W 2015 Missing data within a quantitative research study: How to assess it, treat it, and why you should care. *Journal of the American Association of Nurse Practitioners* **27**(4) 230-2.
[5] Chhabra G, Vashisht V and Ranjan J 2017 A comparison of multiple imputation methods for data with missing values *Indian Journal of Science and Technol* **10**(19).
[6] Scheffer J. Dealing with missing data.
[7] Cheema JR 2014 A review of missing data handling methods in education research *Review of Educational Research* **84**(4) 487-508.
[8] Dua, D. and Karra Taniskidou, E. 2017 UCI Machine Learning Repository [http://archive.ics.uci.edu/ml]. Irvine, CA: University of California, School of Information and Computer Science.
[9] Pandya R and Pandya 2015 J. C5. 0 algorithm to improved decision tree with feature selection and reduced error pruning *International Journal of Computer Applications* **117**(16).
[10] Nelwamondo F, and Marwala T 2008 Key issues on computational intelligence techniques for missing data imputation-a review. *In Proc. of world multi conf. on systemics, cybernetics and informatics* **4** 35-40).
[11] Sim J, Lee JS and Kwon O 2015 Missing values and optimal selection of an imputation method and classification algorithm to improve the accuracy of ubiquitous computing applications. *Mathematical Problems in Engineering*.
[12] Quinlan JR 2014 C4. 5: programs for machine learning. Elsevier.
[13] Aljuaid T and Sasi S 2016 Intelligent imputation technique for missing values. *In Advances in Computing, Communications and Informatics (ICACCI)*, (pp. 2441-2445). IEEE
[14] Kuhn M, Weston S, Coulter N and Quinlan R. 2014 C50: C5. 0 decision trees and rule-based models. R package version 0.1. 0-21, URL http://CRAN. R-project. org/package C. 2014;50.

IOP Conf. Series: Journal of Physics: Conf. Series **1060** (2018) 012015 doi:10.1088/1742-6596/1060/1/012015

Research and Application on Personalized Recommendation of Potential Friends and Books Base on Collaborative Filtering

He Chengdong[1*] and Shu Chang[1]

[1] Lian Yi Technology Co., Ltd., Guangzhou 510000,Guangdong,China

*Corresponding author's e-mail address:heironhand@163.com

Abstract. The personalized recommendation of potential friends and book resources is important for improving the personalized service of digital campus. In this context, this paper proposes a recommendation method of potential friends and book classes based on collaborative filtering. The establishment of users profiles and the implicit rating prediction model by mining user demographic attributes and book borrowing features. It can improve social environment and utilization of library resources base on this method. The result shows that the model can effectively improve the quality of personalized service recommendation of potential friends and book resources.

1. Introduction

Construction of intelligence service and sharing platform break through the "Isolated Data Island". The platform integrates data of campus card, library borrowing, student achievement, attendance, entrance guard and network access log in a big data environment[1]-[3]. The key is mining and analyzing the rule of characteristic data of school students and managers in digital campus and smart campus.

One of the goals of the construction of digital campus platform and smart campus is to provide personalized services for students, and personalized recommendation of potential friends and book resources is an important part of personalized service[4][5]. Recommendation technology can be more intelligent to recommend the potential friends and learning resources to users with similar behaviour feature, so as to promote the development of the personalized service on campus.

2. Existing problems

At present, there exists few researches on recommending service for campus big data platform[6]. It does not form a deep mining for the campus users' basic attributes and real behavior data, and builds a model for interest prediction[7][8]. Therefore, recommendation technology applied to campus scene can alleviate information overloading, improve communication efficiency, and form personalized service. The method is to provide recommendation information based on users' basic attributes and interest preferences, and it also can effectively guide users to make the right decisions. The work we need to do, in this paper, is to establish a recommendation model for potential friends and book resources on campus, so as to improve the social interaction efficiency and the utilization rate of library resources.

Personalized recommendation technology usually adopts collaborative filtering algorithm that obtaining the user's eigenvector, calculating the similarity between users profiles, and recommend and predict[9][10]. Recommendation technology research mainly focuses on business service, but there are few researches on recommendation of digital campus resources. Therefore, it is necessary to research collaborative filtering technology applied to campus friends and Book Resources recommendation.

This paper proposes some solutions to the issues discussed above. Mining students' basic attributes and borrowing data, establishing user interest feature models, and using collaborative filtering algorithm to recommend friends and books. It is beneficial to improve the personalized service of the campus.

3. Existing problems

3.1. Student demographic attributes

The demographic attributes of the students are Student ID, Name, Sex, Birth date, ID, Native place, Department code, Professional code, class ID, role ID, Cell-phone number, E-mail, Account status, and School status in the management system of College Students' behavior analysis. This paper quantifies four factors: student ID(Grade), gender, ID card (Regions), department code (Department).

- Student ID(Grade):The first two numbers in the school number represent grade. Two methods can be used to quantify grade, for example, the sparse representation method 14 is expressed as (1000), 15 is (0100), 16 is (0010), and 17 is (0001). Another method of quantification is (1, 2, 3, 4), respectively.
- Sex: 'male' is quantified to be '1', 'female' is' 0 '.
- ID card(Regions): The first two numbers represent regions . The representation method is consistent with the Student ID(Grade) .
- Department code (Department): The representation method is similar to the above.

The table 1 indicates that the four feature quantization results are normalized and form user profiles matrix.

Table 1. User profiles matrix from demographic attributes

	F_1	F_2	F_3	F_4
U_1	H_{11}	H_{12}	H_{13}	H_{14}
U_2	H_{21}	H_{22}	H_{23}	H_{24}
...	...	...	...	...
U_i	H_{i1}	H_{i2}	H_{i3}	H_{i4}
...	...	...	...	...
Un	H_{n1}	H_{n2}	H_{n3}	H_{n4}

The feature vectors of user n and user m are defined as Hn=<Hnl, Hn2, Hn3, Hn4>, Hm=<Hml, Hm2, Hm3, Hm4>, respectively.

Pearson similarity method is used to calculate the similarity of user demographic characteristics.

$$sim(U_i,U_j)=\frac{\sum(H_{ik}-\overline{H_i})*(H_{jk}-\overline{H_j})}{\sum(H_{ik}-\overline{H_i})^2*(H_{jk}-\overline{H_j})^2} \quad (1)$$

H_{ik} , H_{jk} is the quantized values of user n and m corresponding to the feature k respectively. $\overline{H_i}$, $\overline{H_j}$ is average value of all the features by the user n and m.

3.2. User profiles

The library borrowing information includes Book ID, Borrower ID, Student ID, Staff ID, Barcode, Book name, Author, Press, Publication date, Library ID, Borrow date, and Return date in the management system of College Students' behaviour analysis.

Collaborative filtering usually mining users' potential interests, through analysing all rating data. But the books is special, they are clustered through the Chinese Library Classification that clusters books into n categories, and table 2 shows the form of book classification in Chinese Library Classification. Different classes of book can respond to interest of student. The cluster frequency borrowing is used to be converted to rating of items when lack of book rating and frequency borrowing of one book is one for a single user.

Table 2. Chinese Library Classification

Class	Name	Class	Name
A	Marx doctrine, Lenin doctrine, Mao Zedong thought and Deng Xiaoping theory	T-9	Industrial economy
B	Philosophy, Religion	TB	General industrial technology
C	social science general introduction	TD	Mineral engineering
D	Politics, Law	TE	Oil and gas industry
E	Military	TF	Metallurgical industry
F	Economics	TG	Metallization ,metal technology
G	Culture, Science, Education, Sports	TH	Machinery , instrument industry
H	Language, Writing	TJ	Weapons industry
I	Literature	TK	Power industry
J	Art	TL	Atomic Energy Technology
K	History, Geography	TM	Electrician technology
N	Natural science introduction	TN	Radio electronics, Telecommunications Technology
O	Mathematical science and chemistry	TP	Automation technology and computing technology
P	Astronomy, Earth Science	TQ	Chemical industry
Q	Bioscience	TS	Light industry, Handicraft industry
R	Medicine , Health	TU	Architecture Science
S	Agricultural Science	TV	Water conservancy engineering
T	Industrial Technology	U	Transportation
T-0	Industrial technology theory	V	Aeronautics, Astronautics
T-1	The status and development of industrial technology	X	Environmental Science, Safety Science
T-2	Institutions, Groups, Meetings	Z	Comprehensive books
T-6	Reference book		

In the context of implicit user rating in the frequency of user borrowing books, they are labeled as “not interested” when borrowing times are very few, On the other hand, it is marked as “very interested”. the higher the frequency is, the higher interest is, and tends to be stable. Therefore, in this paper, a nonlinear relation is used to indicate the frequency of books borrowing and their interest in book classes.

$$r(x_i) = \alpha(1-(1/\mathrm{e})^{\frac{x_i}{\mu}}) \tag{2}$$

x_i is borrowing frequency, it is marked as interested if above μ times. α is a parameter to determine the scope of the rating. In order to be consistent with the scoring mechanism, This paper sets up α=5, μ=5.

After the above transformation, the user rating matrix is indicated by table 3.

Table 3 User profiles matrix from implicit rating

	C_1	C_2	...	C_k	...	C_m
U_1	r_{11}	r_{12}	...	r_{1k}	...	r_{1m}
U_2	r_{21}	r_{22}	...	r_{2k}	...	r_{2m}
...	...	...	...		...	...
U_i	r_{i1}	r_{i2}	...	r_{ik}	...	r_{im}
...	...	...	...	...	...	
U_n	r_{n1}	r_{n2}	...	r_{nk}	...	r_{nm}

Pearson similarity method is used to calculate the similarity of user profiles.

$$sim(U_i,U_j)=\frac{\sum(r_{ik}-\overline{r_i})*(r_{jk}-\overline{r_j})}{\sum(r_{ik}-\overline{r_i})^2*(r_{jk}-\overline{r_j})^2} \tag{3}$$

Obtaining K adjacent users whom are recommended to the target user with higher similarity through comparison. Then prediction by the following formula model.

$$P_{i,C}=\frac{\sum_{j\in N} sim(\mathrm{i},\mathrm{j})*R_{j,C}}{\sum_{j\in N} sim(\mathrm{i},\mathrm{j})} \tag{4}$$

N represents an adjacent user set that has a rating on the item C, $R_{j,C}$ is rating by user j.

3.3. Personalized Recommendation Process

The process of personalized recommendation on friends and book classes is structured in four steps as follows. The figure 1 shows the flowchart of friends and book classes recommendation.

- Step 1: The related data is obtained through the server, and the required data is filtered and obtained.
- Step 2: The eigenvalue of labels are analyzed to get the user profiles between students and tags.
- Step 3: Obtaining K adjacent users whom are recommended to the target user with higher similarity between user profiles.
- Step 4: Top N items are recommended for predicting the missing rating of target user's feature.

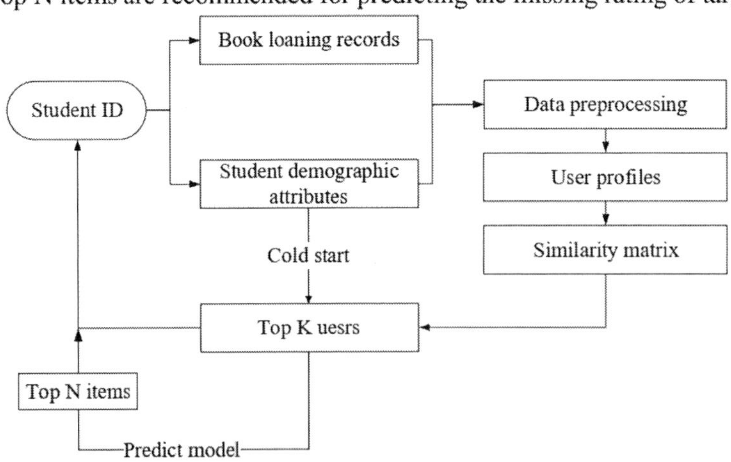

Figure 1. The flowchart of friends and book classes recommendation

4. Experimental results and analysis

The experimental data are 100 student's demographic information and book borrowing data. The data set that divided into training set and test set based on the ratio of 8:2 are filtered to get 1084 effective borrowing records. The mean absolute error (MAE), widely used in collaborative filtering recommendation, is used to measure the prediction accuracy.

$$MAE = \frac{1}{N}\sum_{i=1}^{N}\left|r_i - r_i^{'}\right| \tag{5}$$

N is the number of test samples, r_i and $r_i^{'}$ is real and prediction rating respectively.

The figure 2 is the prediction accuracy of the implicit rating of the book classes. The experimental results show that when the nearest users Top K is 20, the prediction accuracy of book classes implicit rating is stable at 1. Top N book classes are recommended for target user, and users with higher similarity as potential friends.

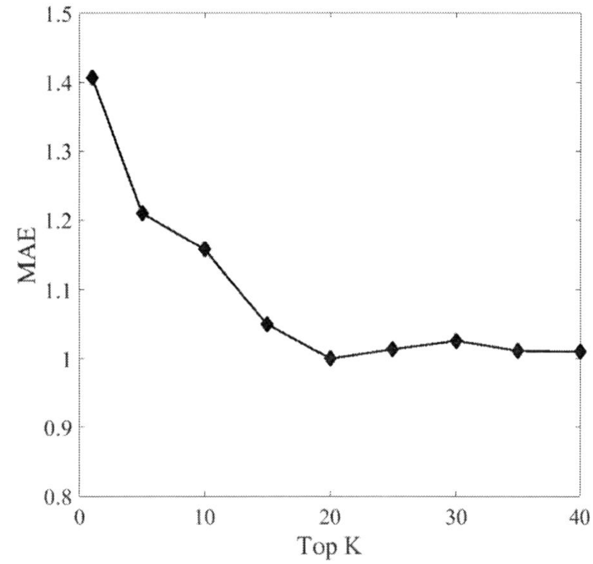

Figure 2. Prediction accuracy of book classes implicit rating

5. Conclusion

A personalized recommendation service based on the student demographic attributes and frequency borrowing of digital campus users is realized by data mining to obtain user profiles and establish potential friends and book classes recommendation models. In this paper, a recommendation collaborative filtering method using a potential friend and book classes is practical. It will improve the efficiency of social interaction and the utilization of library resources. Meanwhile, personalized recommendation service can also be applied to other scenes of digital campus, so as to promote the construction of smart campus.

References

[1] Zheng T, Sheng X L. Intelligent Campus Personalized Information Push Service Base on User Interest [J]. Journal of Zhejiang Shuren University. 2017, 17(1):15-19+27.

[2] Ding D, Li J, Wang H, et al. Student Behavior Clustering Method Based on Campus Big Data[C]// International Conference on Computational Intelligence and Security. IEEE Computer Society, 2017:500-503.

[3] Zhang N. A Campus Big-Data Platform Architecture for Data Mining and Business Intelligence in Education Institutes[C]// International Conference on Machinery, Materials, Environment, Biotechnology and Computer. 2016.

[4] Binas A, Charlin L, Levinshtein A, et al. Indigo Books and Friends Recommendation[J]. 2009.

[5] Gao B, Du S, Li X, et al. Research on the Application of Persona in Book Recommendation System[C]// Journal of Physics Conference Series. Journal of Physics Conference Series, 2017:012023.
[6] Hu H, Long D. A Study on Key Techniques of Wisdom Campus Information Recommendation Platform Based on Big Data[C]// International Conference on Intelligent Transportation, Big Data and Smart City. IEEE, 2016:56-58.
[7] Liang Z. Research on student behavior analysis and prediction method based on campus big data [D]. Xi'an University of technology, 2017.
[8] Lindsey R V, Khajah M, Mozer M C. Automatic discovery of cognitive skills to improve the prediction of student learning[C]. Advances in neural information processing systems, 2014: 1386-1394.
[9] Liu Q W. Research on Recommendation System Based on Collaborative Filtering [D]. University of Science & Technology China, 2013.
[10] Quan Z. Collaborative filtering recommendation based on user personality[C]// International Conference on Information Management, Innovation Management and Industrial Engineering. IEEE, 2014:1041-1046.

Data Processing and Analysis Method of Electronic Countermeasure Test Based on Rule Engine

Fen Liu[1], YongJie Zou[1], GuangYuan Ma[1] and GuangSun[1]

[1]Luoyang Electronic Equipment Test Center of China

715361425@qq.com

Abstract. The electronic countermeasure test data processing is the premise and basis of realizing equipment performance evaluation. At present, the uncertain problems of electronic counter- measure test data processing had been a major factor which would restrict the development of test evaluation means. This paper presented a graphical data processing and analysis method based on the procedure analysis of electronic countermeasure test data process. This method realized the custom design of electronic countermeasure data processing flow chart through flow analysis, procedure decomposition, graphical modeling, parameter configuration and flow resolving. The method based on rule engine used the rule modeling and resolving technology to realize the custom design of matching rules of electronic countermeasure test data; used the procedure data template to store the input, output and result data, which realized the traceability of results for data analysis and data processing based on the procedure.

1. Introduction

The technical approaches of evaluation for electronic countermeasure equipment performance test is the process for evaluating the overall performance of electronic countermeasure equipment, which used both test data and equipment technical indicators by signal parameter comparison, countermeasure behavior determination, data statistics, calculation of indicators, results evaluation and so on. The contents, standards and methods of evaluation for various types of equipment test are quite different due to the purpose, the object, the topic and so on. It is necessary to develop data processing programs in a targeted manner, which would impose a heavy burden on the evaluation for test. The uncertain problem in the data processing of equipment test had been a major factor which restricted the development of evaluation means construction of electronic countermeasure equipment test.

Data matching is the core function of data processing. The flexibility in matching rules setting is the key to determine the practicability of data processing. At present, the rule-based business system generally uses the technology of rule editing and rule resolving base on rule engine. Rule engine is a software component of explaining and executing rules, which completely separates business rules logic from the program code, expediently the edit and managing of business rules for users. Aiming at the uncertain problem in the data matching rules of data processing, This paper investigated the techniques of description, modeling and resolving on data matching rules for electronic countermeasure equipment test, referring to the technology of representation and executing of rule in rule engine of CLIPS[1].

Taking into account the flexibility of flow creation and the simplicity of user operation, this paper used graphical programming technology to graphically design the data processing to form a flow design window similar to the flow chart based on the analysis of test data processing flow of electronic countermeasure equipment, which would allow user add, view and modify the flow data processing. Aiming at the uncertain problems such as input data and matching parameters of electronic countermeasure equipment data processing, this paper realized the custom design of electronic countermeasure training data processing by configuring the input parameters of functional components of data processing flow.

Aiming at the problems that it is difficult to analyze and find cause in the evaluation of equipment performance by using data processing results, this paper realized the data analytical method of process-based electronic countermeasure equipment performance test evaluation based on the design of data processing and design of matching rules, through the way of designing the class of procedure data to store the logistic procedure data of data processing.

2. Rule Modeling and Resolving Technology

2.1. Rule description

The rule is a declarative description of transactions or activities. Operational standards, business processes, regulations, bylaw, and industry standards are all known as rules[2]. At present, there is no unified classification standard for rules in the academic circles. The classification of rules differs according to the research background or research object, but the classification methods are generally biased toward the business level. At the business level, the classification of rules refers to the classification of rules from a business perspective. The rule on business level is representatively summarized for five representative types, such as structural rules, integrity constraint rules, derivative rules, responsiveness rules and responsibility authority assignment rules. The description of a rule usually consists of three parts of type, condition and attribute. A rule may include multiple conditions or not. A condition indicates a dualistic operation or an unary operation, condition operators include "=", ">", "≥", "<", "≤", "AND", "OR", "NOT" and so on. The part of rule properties defined in this paper can be considered as a conclusion of conditional operations and the statistical action to be performed. The data matching rules mentioned in this paper are integrity constraint rules, which mainly refer to reconnaissance or interference signal matching rules of the electronic countermeasure equipment test and are used to determine the effect of reconnaissance or interference behavior. A matching rule may consist of a number of data items matching conditions.

2.2. Rule modeling

Rule modeling is a process that developers classifily map the rules at the business level to formats that system can identify, which currently mainly includes two kinds of directive encoding and declarative encoding. Directive encoding is a way of describing business rules with programming statements, which is more suitable for software systems with simple and stable rules and which can reduce the complexity of many implementations. However, the relational codes of software system need to be modified when rules change, which will lead to the software system not flexible. Declarative encoding is a way of describing the logical structure of rules during calculation rather than controlling its structure, which focuses on the task that rules will accomplish, rather than how to accomplish them. At present, typical forms of declarative rules describing are UML/OCL, SQL, CLIPS, XML and so on[3].

Aiming at achieving the statistics of matching results of data items and rules and learning from the modeling ideas of CLIPS's custom rules, The data matching rules described in this paper take the allowable match error values of electronic countermeasure equipment countermeasures activity signal parameters as the condition operands of the parameter matching rules, which is defined as data matching rules base.

2.3. Rule resolving technology

Referring to the rule resolving technology based on rule engine of CLIPS method, this paper presents a test data processing method of electronic countermeasure equipment based on rule engine and the technology approach is shown in Figure 1.

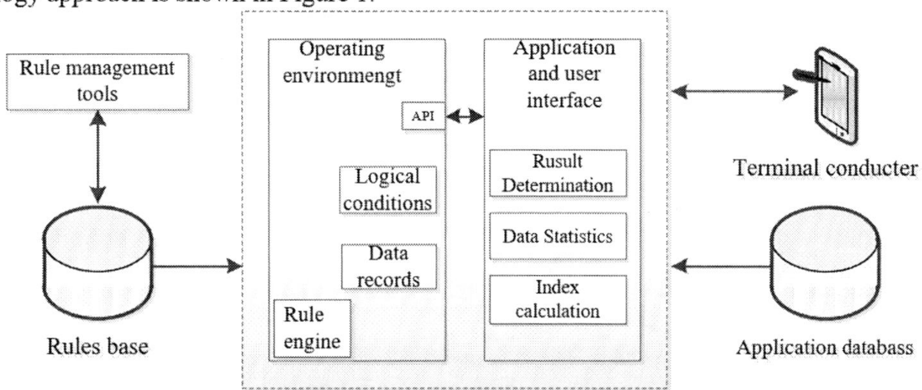

Figure 1. The technology approach of electronic countermeasure rule resolving

Based on defining data matching rules of all kinds of electronic countermeasure equipments, the method used rule management tools to design rules according to data processing requirements and rule engine to analyze and executing data matching rules designed by users.

3. The Graphical Modeling Method of Electronic Countermeasure Equipment Test Data Processing

3.1. Data processing analysis

As the difference between signal types of different professional electronic countermeasure equipment, there is a big difference between signal parameters and matching standards. Data items of radar countermeasure equipment data processing usually include carrier frequency, repetition frequency, pulse width, signal pattern, target interference level, target anti-jamming effect and so on; Data items of photoelectric countermeasure equipment data processing usually include wavelength, signal type, signal cycle, the target interference situation and so on; Data items of communication countermeasure equipment data processing usually include interference pattern, interference frequency, interference bandwidth, target interference situation and so on.

In order to facilitate the design of data processing rules and data analysis, the test data processing of electronic countermeasure equipment is usually divided into reconnaissance data processing and interference data processing on the basis of distinguishing radar countermeasure, communication countermeasure and photoelectric countermeasure. Reconnaissance data processing is the process that count the number of signal released by signal source, electronic countermeasure equipment reconnaissance signal number, electronic countermeasure equipment to correctly identify the signal and the signal parameters of matching results and so on according to the work records of signal sources and electronic countermeasure reconnaissance equipments. Interference data processing is the process that count the number of interference signal released by electronic countermeasure equipment, target equipment interference situation, the interference level, the interference time and so on according to the work records of electronic countermeasure and target equipments.

The basic steps of data processing of electronic countermeasure equipments mainly includes reading the signal data of target equipment one by one, reading the signal data of test equipment one by one, checking the signal parameters are correct one by one, count the matching error and matching result of each parameter, rating calculation equipment performance test evaluation index and so on[4], as shown in Figure 2.

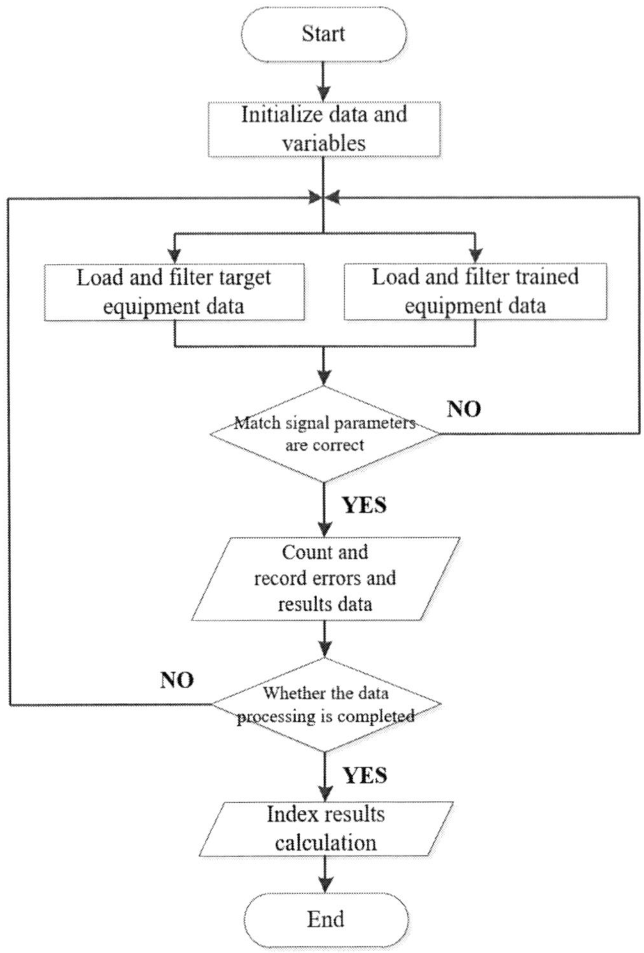

Figure 2. The basic steps of electronic countermeasure data processing

3.2. Process decomposition

Based on the analysis of data processing and basic steps of all kinds of electronic countermeasure equipments, this paper divides the logical structure of data processing into some functional modules such as red data, blue data, result data, data screening, screening conditions, data matching, matching rules, relationship connection and so on and users could configure parameters of each function module such as the input data, matching parameters, matching rules. Table 1 shows the various functional modules.

Table 1. Logical functional modules of electronic countermeasure data processing.

NO.	Name	Function
1	Red and Blue Data	Configure input and output the equipment data, select data items to be compared
2	Data Filter	Filter data based on the filter conditions
3	Filter Conditions	Configure the filter conditions
4	Data Matching,	Matching according to the data matching rules
5	Matching Rules	Configure the matching rules
6	Result Data	Count data matching results

3.3. Graphical modeling

Aiming at the uncertain problem of data processing, this paper used graphical modeling and visual technology to model the logic function modules of data processing as a process component based on the decomposition of logical structure[5], which taking fully account of flexibility of process creation and simplicity of operation, guided by the modeling theory based on business process, realizing the custom design of electronic countermeasure equipment test data processing. Users can realize the logical structure customization of data processing by selecting, dragging and adding the components, and solve the uncertain problems of data processing by configuring the input data and matching rules in the processing component. The logical structure of electronic countermeasure equipment test data processing was shown in Figure 3.

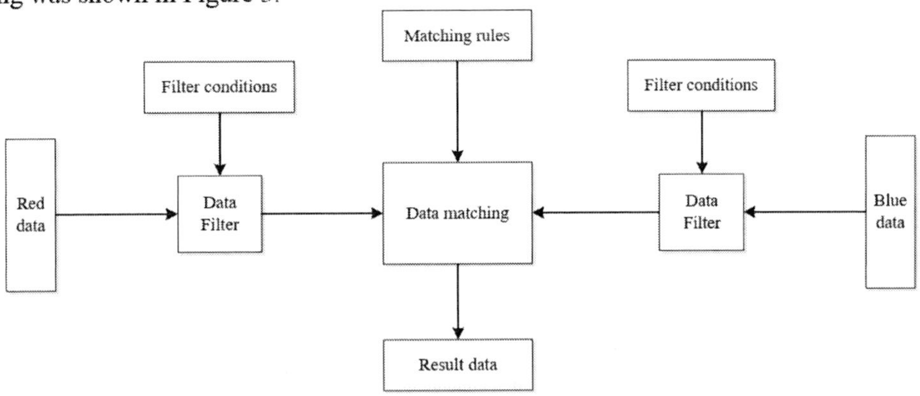

Figure 3. The logical structure of electronic countermeasure equipment test data processing

Users can bind the output data of data processing to the evaluation index according to the calculation requirement of electronic countermeasure equipment test evaluation index after designing of the logical structure customization and configuring of uncertain problems, and get evaluation results according to the calculation method of evaluation index at last.

4. The data analysis method of electronic countermeasure equipment performance test evaluation

At present, there were some problems in the performance test evaluation of electronic countermeasure equipment such as weak correlation between the result data and the original data, the untraceability of processing results, the data opaque of process procedure, the data analysis limitation by the evaluation index results and so on, which make it difficult to analysis and find problems in the evaluation of equipment performance test according to processing results[6]. In order to solve this problem, this paper presented a method of process-based data analysis based on the custom design of data processing and matching rules. Aiming at all the configured procedure logical nodes of the data processing custom design such as input data, data filter items, data matching items and so on, this method designed a formatted process data storage template class to store he input data, the output data and the statistical data of logical nodes of each matching rule so as to provide procedure data support for result data analysis during the procedure of rule engine resolving and executing matching rules. The procedure-based data analysis method of electronic countermeasure equipment test evaluation was shown in Figure 4.

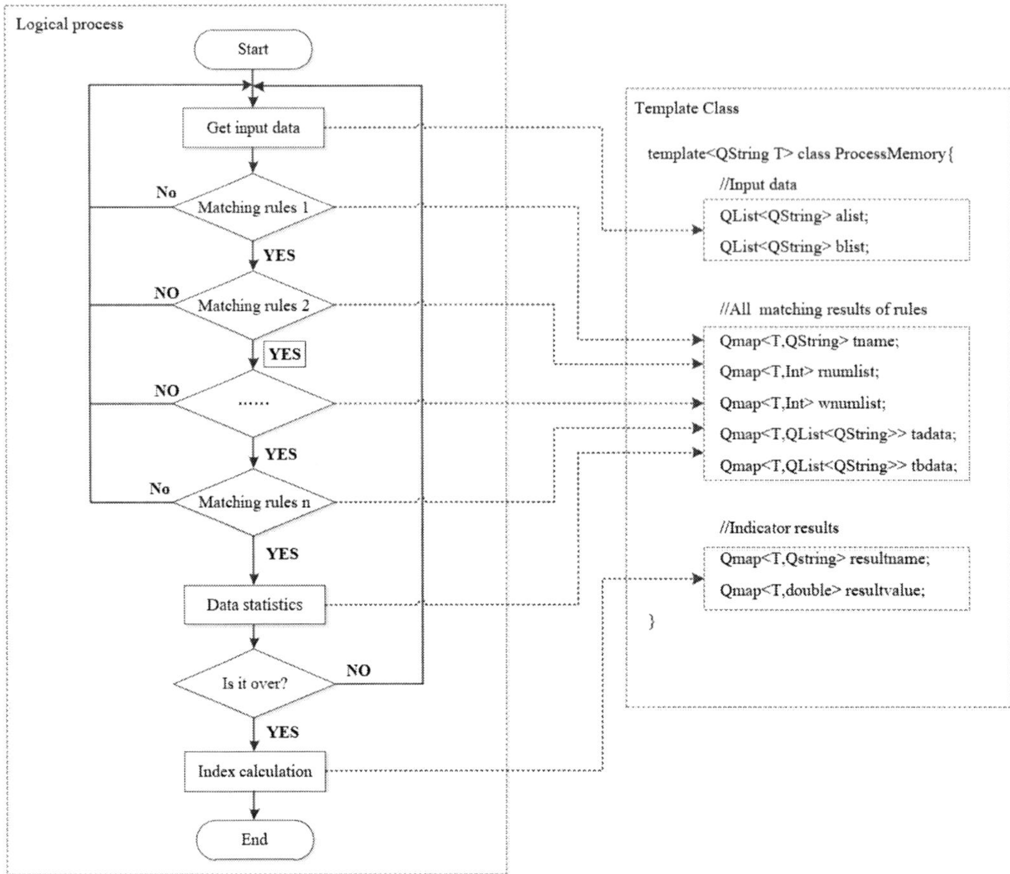

Figure 4. The process-based evaluation data analysis method of test evaluation

5. Conclusion

This paper researched on the modeling and resolving technology of rule engine-based rules, the processing decomposition and modeling methods based on graphical programming, the test evaluation requirement analysis of electronic countermeasure equipments and so on, and presented a rule engine-based data processing and analysis method of electronic countermeasure test, aiming at some uncertain problems in the test data process of different electronic countermeasure equipments such as input data matching parameters, matching standards ,etc. The method solved the uncertain problems of the data processing and the matching rules of electronic countermeasure equipment test, achieved the custom design of data processing, matching rules and data processing results. The procedure-based data analysis method enhanced the traceability of results.

References

[1] Weihui Wang, Guohua Geng, Mingquan Zhou. A Rule Repository Model for Rule-Driven Question-Answer-Based Web Applications. Proceedings-International Conference on Artificial Intelligence and Computational Intelligence, Sanya China, 2010: 17-22.

[2] Caro J L. Proposing a formal method for workflow modeling: Temporal Logic of Actions (TLA). International Journal of Computer Science: Theory and Application, 2014, 1(1):1-11.

[3] Nalepa G.J., Mach M.A. Business Rules design method of Business Process Management[A]. Proceedings of the International Multiconference on Computer Science and Information Technology[C], Polish Information Processing Society, IEEE Computer Society Press, 2009: 165-170.

[4] Goedertier S., Vanthienen J. Rule-based business process modeling and execution[A]. Proceedings of the IEEE EDOC Workshop on Vocabularies Ontologies and Rules for the Enterprise (VORTE 2005) [C], 2005:67-74.
[5] Royce, G.K. Integration of a Business Rules Engine to Manage Frequently Changing Workflow: A Case Study of Insurance Underwriting Workflow[A]. Proceedings of the 2007 Americas Conference on Information Systems[C], 2007: 495.
[6] Zhenchun Wei, Jianghong Han, Yang Lu. Hierarchical Modeling and Control of Discrete Event Control Systems based on Rule Description Method [C]// Proceedings of the 2006 IEEE International Conference on Mechatronics and Automation, June25-28, 2006, Luoyang, China(ICMA2006). USA:IEEE, 2006:2179-2183.

Threshold Decision Algorithm Based On The Maximum Interclass Variance For TDCS

Bingzheng Xu[1], Qinghua Ren[1,2] , Yang Liu[1] and Xiazhao Li[1]

[1.] Information and Navigation College in Air Force Engineering University, Xi'an 710077, China;
[2.]CETC key laboratory of aerospace information applications,Shijiazhuang 050081, China

E-mail: xubzheng@sina.com

Abstract. Transform domain communication system (TDCS) is a new type of wireless communication system, who can take the initiative to avoid interference. In this paper, Fractional Fourier Transforms (FRFT) is introduced into Transform Domain Communication Systems (FRFT-TDCS), which can effectively target linear frequency modulation (LFM). But the multicomponent linear frequency modulation (MLFM) is difficult to effectively eliminate, when using traditional threshold decision algorithm of TDCS. A novel threshold decision algorithm for FRFT-TDCS is proposed in this paper. Because the order of FRFT has no effect on noise, the threshold decision algorithm by using the idea of maximum interclass variance means minimum misclassification probability. Simulation results show that compared with the traditional hard threshold decision algorithm, the proposed algorithm can effectively eliminate MLFM interference and negative influence of the order of FRFT. Whether bipolar modulation or cyclic shift keying(CSK) modulation, the bit error rate(BER) of this threshold decision algorithm is always better than the traditional threshold decision algorithm; With the increase of signal to noise ratio(SNR), the BER performance of the two threshold decision algorithms increase gradually.

1. Introduction

In modern electronic warfare, anti-jamming communication technology is the key to battlefield perseverance. TDCS has been widely concerned in the communication field[1-4]. TDCS not only have the characteristics of Low Probability of Interception (LPI) and Low Probability of Interference (LPD), but also have unique ideas of actively avoiding interference by using spectrum holes. Traditional TDCS based on Fourier transform can effectively remove stationary interference but not to deal with non-stationary LFM interference effectively. In recent years, FRFT has been introduced into TDCS because of its better energy focusing property for LFM interference. And the related research on FRFT-TDCS has been carried out in the paper [5-8].

The anti-interference ability of TDCS depends on the ability of removing the interference. In addition, the setting of the threshold determines whether the interference can be completely eliminated. In the increasingly complex electromagnetic environment, the threshold decision algorithm has become one of the key technologies to improve the anti-interference ability of TDCS. In the aspect of amplitude spectrum generation, FRFT-TDCS removes the interference spectrum through the preset threshold, and the effect of interference rejection directly affects the anti-interference performance of

IOP Conf. Series: Journal of Physics: Conf. Series **1060** (2018) 012017 doi:10.1088/1742-6596/1060/1/012017

the system. At present, there are few researches on the threshold design of FRFT-TDCS. In paper [9], the interference spectrum is removed with 40% of the maximum spectral amplitude of the interference signal. The proposed method is simple in design and low in complexity. However, the effect of threshold decision algorithm is poor. In paper [10], a double-threshold amplitude spectrum generation method is proposed. Double thresholds are deduced from the probability of missed detection and the false alarm probability. The frequency spectrum between two thresholds is marked as 0.5, which improves the system throughput. However, The low-power access to the spectrum that may be interfered between the two thresholds which still leads to an increase of the system BER. In paper [11], deduced the relationship between the BER and the threshold value, the optimal threshold set by this method is obtained through one-dimensional search when the interference spectrum is determined. However, in most cases, we do not know whether a spectrum carrier is interference, which makes the applicability of the method has been more limited. The above methods are based on FFT-TDCS. However in FRFT-TDCS, MLFM interference in different orders of FRFT has different spectrum amplitude and position, which leads to it is difficult to effectively remove out the interference spectrum by the traditional TDCS threshold decision algorithm.

In this paper, a novel threshold decision algorithm is proposed, which aiming FRFT-TDCS to deal with MLFM poorly, by using the idea of maximum interclass variance meaning minimum misclassification probability. This algorithm can adaptively determine the threshold value, which not only can effectively eliminate the interference, but also can avoid the adverse influence by the order of FRFT.

The organization of this paper is as follows. The concept of LFM interference is introduced in Section 2. In Section 3, the principle of FRFT-TDCS is given. Threshold decision algorithm and its derivation are in 4 Section. The simulation result shows in 5 Section and paper is concluded in 6 Section.

2. LFM interference

LFM interference is a typical non-stationary interference signal whose instantaneous frequency varies linearly with time. MLFM interference can be expressed as:

$$x_i(t) = \sum_{i=1}^{L} A_i(j\pi k_i t^2 + j2\pi f_i t) \tag{1}$$

Where k_i , f_i , A_i respectively are frequency modulation rate, the initial frequency and amplitude of the MLFM interference

The method of paper [12] is used to estimate the parameters of the MLFM interference signal in the electromagnetic environment. The optimal order of the i component LFM interference signal can be calculated as:

$$p_i = \frac{2}{\pi} arc\cot(\text{-}k_i) \tag{2}$$

Figure.1 shows the time-frequency distribution of LFM interference. Figure.2 shows the LFM interference spectral distribution of the FRFT domain.

(a) Time-frequency distribution diagram of LFM interference

(b) Time frequency distribution of three component LFM interference

Figure.1. Time frequency distribution of LFM interference

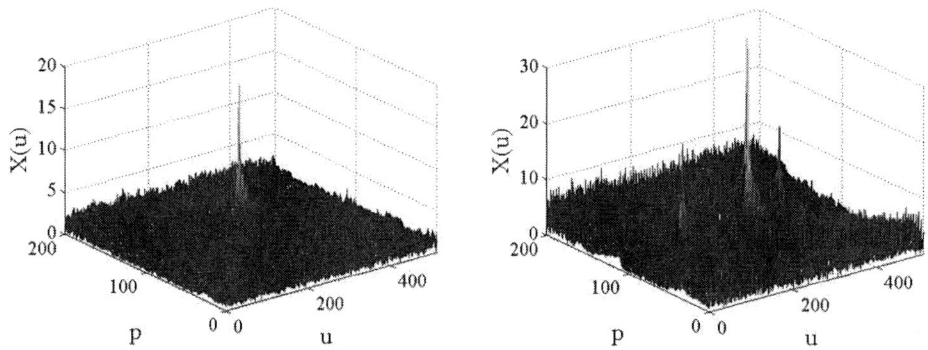

(a) Spectrum distribution in FRFT domain of LFM interference

(b) Spectrum distribution of three component LFM interference in FRFT domain

Figure.2. Spectrum distribution of LFM interference in FRFT domain

We can see that the optimal orders of different components of are not the same. Different components of the LFM interference only in its optimal order reach the maximum value. But other components are as far away from its optimal order, which resulted in decreased peak in FRFT domain, and tends to be stable.

3. Principle of FRFT-TDCS

FRFT-TDCS is a kind of multi-carrier communication system with strongly anti-jamming capability. It uses the "clean" spectrum to join the pseudo-random phase in the FRFT domain to generate the basis function with noise-like performance and multiple access performance. Working principle of FRFT-TDCS is shown in Figure.3.

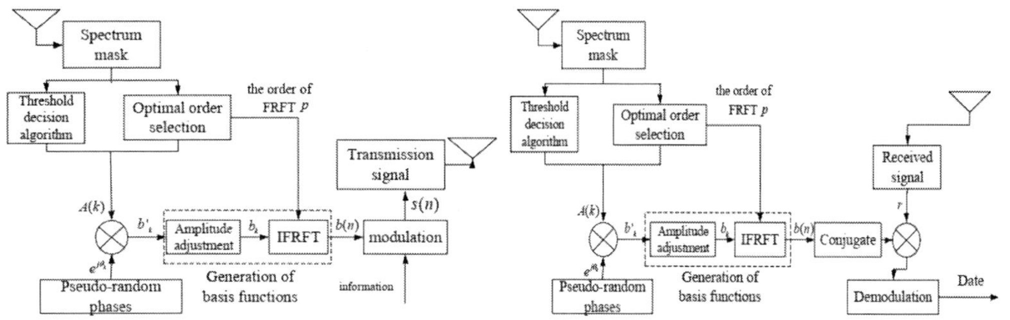

(a) Principle block diagram of transmitter **(b)** Principle block diagram of receiver

Figure.3. Principle block diagram of FRFT-TDCS

The transmitter take sample the electromagnetic environment in the FRFT domain to estimate the interference signal parameters; FRFT-TDCS removes the interference according to the threshold setting and selects the optimal order p to obtain the formed amplitude spectrum $A'(k)$; Then the amplitude spectrum $A'(k)$ multiplies the pseudo-random phase $e^{j\theta_k}$ to get b'_k . The basis function b_k which take amplitude adjustment should be obtained by inverse fractional Fourier transform (IFRFT) ; The time domain waveform of the basis function $b(n)$ can be obtained by the above steps; The information is modulated to produce a signal waveform. It is assumed that there are in the same electromagnetic environment for transmitter and receiver (e.g. Formation flying communication). Receiver will get the same basis function as transmitter, which can demodulate the received the signal waveform to get the data.

FRFT-TDCS can eliminate the interference spectrum in the FRFT domain, so that the spectrum of the generated base function is orthogonal to the interference signal spectrum. Threshold decision algorithm can judge the interference spectrum: the spectrum carrier which is less than the threshold value T is retained. In contrast, the spectrum carrier is higher than the threshold value T is removed. Then the amplitude spectrum $A(k)$ composed of 0 and 1 sequences is obtained.

$$A(k)=\begin{cases}1 & X(k)<T\\ 0 & X(k)>T\end{cases} \tag{3}$$

Therefore, the ability to eliminate the interference spectrum directly determines the anti-interference ability of FRFT-TDCS. If the threshold value is too high, the interference may not be eliminated, resulting in an increase in BER; if the threshold is too low, it may lead to eliminate the frequency band without interference, which can result in a reduction in the spectrum efficiency [13].

4. Adaptive threshold decision algorithm

First, it is proved in [14] that FRFT does not change the statistical properties of noise. Therefore, it can be considered that the amplitude of noise in the FRFT domain will not change with the change of transform order.

Then, let the phase points be N and FRFT transform order search points be M .$[0,1,2,\cdots,L-1]$ represent the amplitude of the FRFT domain. n_i stands for the number that amplitude is equal to i . $MN=n_0+n_1+n_2+\cdots+n_{L-1}$. The probability of that normalized amplitude is equal to i is $p_i=n_i/MN$. It can be drawn that:

$$\sum_{i=0}^{\text{L-1}} p_i=1, \quad p_i \gg 0 \tag{4}$$

Assuming there is a threshold value $T(k)=k,\quad 0<k<L-1$, the amplitude values are divided into two categories L_1 and L_2. The amplitude value L_1 consists of all search points $[0,k]$ and the amplitude value L_2 consists of all the search points $[k,L-1]$. If L_1 is a part of noise, but L_2 contains only the part of interference, k is considered as the optimal threshold k^*.

By using threshold k, the magnitude assigned to L_1 has probability of P_1:

$$P_1(k)=\sum_{i=0}^{k} p_i \tag{5}$$

Then we can calculate the value of the amplitude assigned to L_1 :

$$m_1=\sum_{i=0}^{k} iP(i/L_1) \tag{6}$$

According to Bayesian formula:

$$m_1=\sum_{i=0}^{k} iP(L_1/i)P(i)/P(L_1) \tag{7}$$

$$m_1=\frac{1}{P_1(k)}\sum_{i=0}^{k} iP_i \tag{8}$$

get the value of the amplitude assigned to L_2 :

$$\begin{aligned} m_2 &= \sum_{i=k+1}^{L-1} iP(i/L_2) \\ &= \frac{1}{P_2(k)}\sum_{i=k+1}^{L-1} iP_i \end{aligned} \tag{9}$$

The average amplitude value of the entire fractional domain three-dimensional distribution is given:

$$m_G=\sum_{i=0}^{L-1} iP_i \tag{10}$$

Based on the above formula, we can calculate the following conclusions:

$$P_1m_1+P_2m_2=m_G \tag{11}$$

$$P_1+P_2=1 \tag{12}$$

We define the interclass variance as follows:

$$\sigma_B^2=P_1(m_1-m_G)^2+P_2(m_2-m_G)^2 \tag{13}$$

Bringing equations (11) and (12) into (13), we can get the following equation:

$$\sigma_B^2=P_1P_2(m_1-m_2)^2 \tag{14}$$

It can be seen from the formula (14) that the bigger the difference between m_1 and m_2 is, the larger the value of σ_B^2 is. This phenomenon indicates that the interclass variance represents the measure of separability between classes [15] .Therefore, the optimal threshold is maximize the value k.

5. Simulation

Simulation parameter setting: the time value of LFM interference is $[-2,2]$; the sampling frequency f_s is 128Hz; the number of sampling points is 512;the dimension of FRFT domain coordinate is normalized to 4 (the same as the signal duration);the noise is 4dB. The interference signal is three-component LFM interference, whose frequency modulation rates are respectively 8, 12 and 20 Hz / s and the initial frequencies are 40, 15 and 5 Hz respectively. In addition, the initial phases of LFM interference are 0 rad. Corresponding to the order of 1.33, the interference power spectrum is showed in Figure.4.

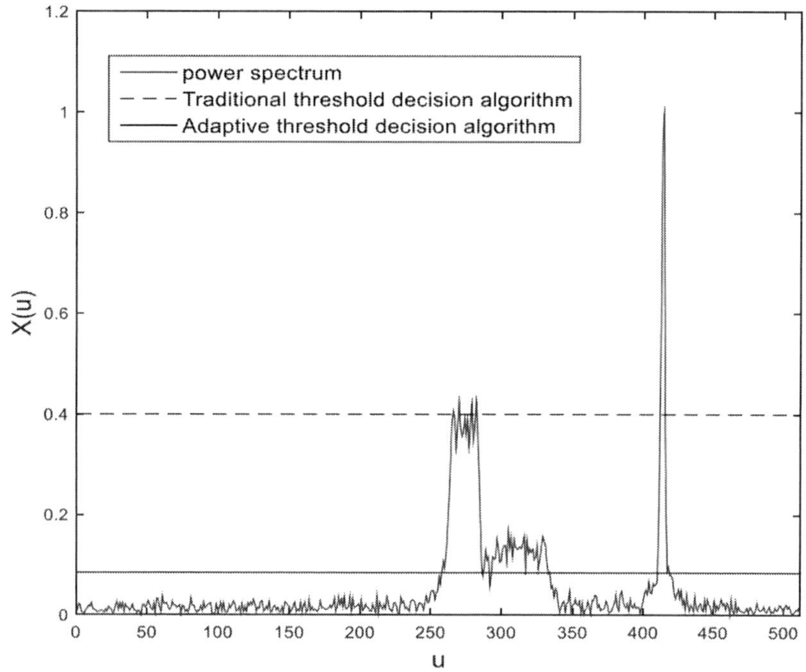

Figure.4. Power spectrum of interference with order 1.33

As shown in Figure.4, traditional threshold decision algorithm is set based on the 40% of maximum value of power spectrum[16]. For MLFM interference, the ability of the traditional threshold decision algorithm to remove interference is terrible. The traditional threshold algorithm leads to the decrease of FRFT-TDCS anti-jamming performance. However, the adaptive threshold decision algorithm in paper can ignore the influence of the transform order on the threshold setting, and can effectively eliminate the interference spectrum under different transformation orders.

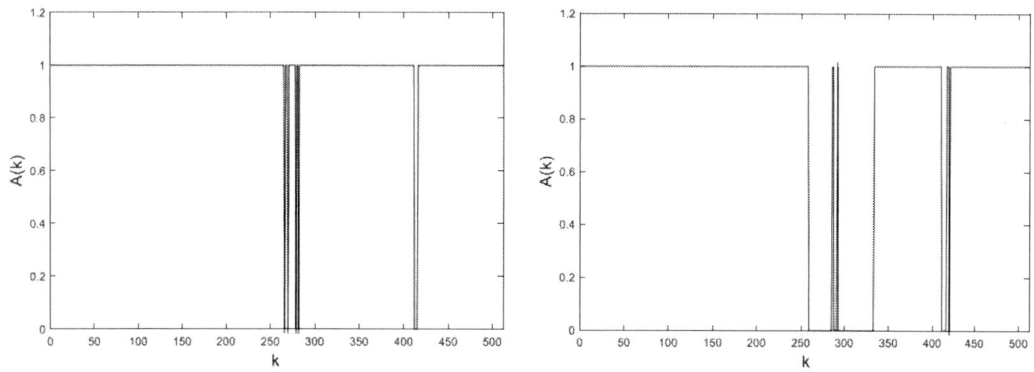

(a) Traditional threshold decision algorithm (b) Adaptive threshold decision algorithm

Figure.5. Amplitude spectrum of three component LFM interference under different threshold decision algorithms

The simulation results in Figure.5 show that the adaptive threshold decision algorithm is more accurate than the traditional threshold decision algorithm. The adaptive threshold decision algorithm proposed in this paper is suitable for the processing of MLFM interference.

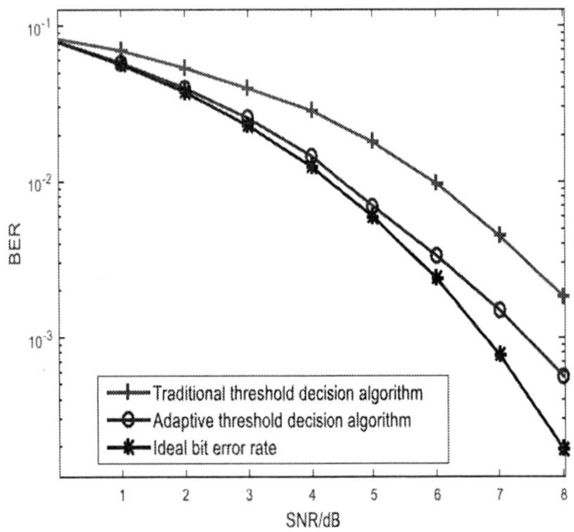

Figure.6. BER performance of different threshold decision algorithms under bipolar modulation

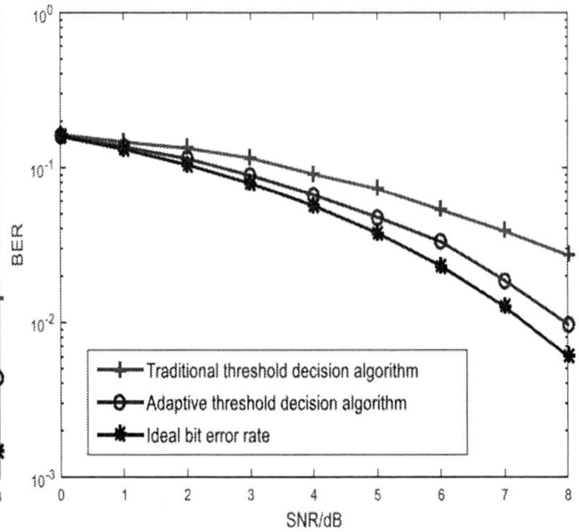

Figure.7. BER performance of different threshold decision algorithms under BCSK modulation

The BER performance of different threshold decision algorithms is shown in Figure. 6 and Figure. 7. Whether it is bipolar modulation or BCSK modulation, the improvement of BER is still very obvious. The BER of the adaptive threshold decision algorithm is closer to the ideal value (BER of the system without interference). With the increase of signal-to-noise (SNR), the difference of BER between the two threshold decision methods gradually increases.

6. Conclusion

Aiming that the MLFM interference is difficult to effectively eliminate for FRFT-TDCS. This paper proposes an adaptive threshold decision algorithm based on the maximum interclass variance. The algorithm adaptively determines threshold value in the three-dimensional FRFT domain. Simulation results show that compared with the traditional threshold decision algorithm , this algorithm can not only effectively eliminate the interference, but also avoid the influence of FRFT order on the threshold decision. The FRFT-TDCS anti-jamming performance has been greatly improved.

7. Appendices

This work is supported in part by University cooperation projects foundation of CETC key laboratory of aerospace information applications

References

[1] Chang C, Huan H, Xu J, et al. Multidimensional parallel combinatory transform domain communication system[J]. International Journal of Communication Systems, 2017.

[2] Wang Shu, Da Xinyu, Chu Zhenyong, et al. Magnitude weighting selection: a method for peak-to-average power ratio reduction in transform domain communication system [J]. IET Communications, 2015, 9(15): 1894-1901.

[3] Sharma M, Gupta R. Comparative analysis of various communication systems for intelligent sensing of spectrum[C]. Preceedings of the IEEE International Conference on Advances in Computing, Communications and Informatics (ICACCI), 2014: 902-908.
[4] Xie T, Da X, Chu Z, et al. Neural network based receiving technology in transform domain communication system[J]. Journal of Huazhong University of Science & Technology, 2014. (2): 96-100.
[5] Wang Chuandan. Zhang Zhongpei. Li Shaoqian. Interference mitigating based on Fractional Fourier Transform Domain Communication System [J]. Journal of Electronics (China), 2007, 24(2): 181-186.
[6] Shi Jun, Chi Yonggang, Sha Xuejun, et al. Fractional fourier domain communication system (FrFDCS): Analytic, modeling and simulation results [J]. Canadian Conference on Electrical & Computer Engineering, 2009:423-426.
[7] Yang Zeyi, Tao Ran, Wang Yue, et al. A Novel Multi-carrier Order Division Multi-access Communication System Based on TDCS with Fractional Fourier Transform Scheme [J]. Wireless Personal Communications, 2014, 79(2): 1301-1320.
[8] Yang Zeyi, Tao Ran, Wang Yue, et al. A Novel Multi-carrier Order Division Multi-access Communication System Based on TDCS with Fractional Fourier Transform Scheme[J]. Wireless Personal Communications, 2014, 79(2): 1301-1320.
[9] He Zhiqing. Design, modeling and Simulation of transform domain communication system [D]. Xi'an, Northwestern Polytechnical University, 2003.
[10] Xin L, Min J, Xiuhua L I, et al. Transform domain communication system based on double-threshold spectrum estimation[J]. Journal of Harbin Institute of Technology, 2013, 45(7):103-108.
[11] Xie T C, Da X Y, Chu Z Y, et al. Optimum threshold design of basis functions for transform domain communication system[J]. XI Tong Gong Cheng Yu Dian Zi Ji Shu/systems Engineering & Electronics, 2013, 35(10):2204-2207.
[12] QI Lin, TAO Ran, ZHOU Siyong, et al. Detection and parameter estimation of multicomponent LFM signal based on Fractional Fourier transform [J]. Science in china: Ser. F Information Sciences, 2004,47(2):184-198
[13] Wang Rouxi. Transform domain communication system research and simulation analysis of the key technology of [D]. Beijing, Beijing Institute of Technology, January 2016
[14] Luo peng. Research on non-stationary signal processing technology based on fractional Fourier transform [D] Tianjin, Tianjin University, 2011.
[15] Petrou M, Petrou C. 4. Image Enhancement[M]// Image Processing: The Fundamentals, Second Edition. John Wiley & Sons, Ltd, 2011:293-394.
[16] Sun Haixin. Research on transform domain communication system based on dynamic spectrum access and its key technologies [D]. [Ph.D. Thesis], Jilin University, 2012

2D-DOA Estimation for Multiple FH Signals Based on Unitary ESPRIT

Yin-song Yang[1], Ying Guo[1,2], Hong-guang Li[1], Kun-feng Zhang[1], Ping Sui[1] and Run-ze Dong[1]

[1] Information and Navigation College in Air Force Engineering University, Xi'an 710077, China;
[2] Science and Technology on Information Transmission and Dissemination in Communication Networks Laboratory, Shijiazhuang 050081, China.

E-mail: yangyinsong131@163.com

Abstract. In order to use the two-dimensional direction-of-arrival(2D-DOA) of the frequency hopping (FH) signals for network sorting, an FH signal 2D-DOA estimation algorithm based on Unitary ESPRIT is proposed in this paper. Firstly, the snapshot model of FH signal is established based on the structure of the planar antenna array. Then using the morphological filtering method to correct the time-frequency map of FH signals after the SPWVD transform to complete the effective hop extraction, so FH signals can be simplified as a narrow-band signal when studying one hop. Thirdly, the Unitary ESPRIT is exploited to estimate the 2D-DOA of FH signals, which transforms the received data from the complex field to the real domain by the unitary transformation to decrease the calculation complexity and reuses the data during constructing the real matrix to improve the accuracy of the estimation. Finally, we did three experiments to measure the performance of proposed algorithm from three aspects: signal-to-noise ratio, snapshot count and time-consuming. The simulation results show that the proposed algorithm has good performance.

1. Introduction

Frequency-hopping communication has been widely used in military communications because of its good anti-interference, low interception and strong networking capabilities, but it also poses a serious challenge for the communication countermeasure reconnaissance.[1] Network sorting is a key step in frequency hopping signal reconnaissance. It can sort out frequency hopping network platforms from the complex electromagnetic environment, provide support for implementing electronic interference with accurate interference.

Signal direction of arrival (DOA) plays an important role in frequency hopping signal network classification and has also been a hot area of research. [2][3] Belouchrani first proposed the concept of spatial-time-frequency, and then processing method based on spatial-time-frequency was used in non-linear frequency modulation signal estimation and blind source signal separation, the performance was better than the traditional method. In paper [4] and [5], Chen put forward a new idea for the FH signals DOA estimation by constructing the spatial-time-frequency distribution matrix of each hop and using the MUSIC algorithm, however, this algorithm needed to perform spectral peak search in the parameter space, which had high computational complexity, and it also did not consider the two-dimensional DOA situation, unable to locate the target of three-dimensional space. A spatial-

IOP Conf. Series: Journal of Physics: Conf. Series **1060** (2018) 012018 doi:10.1088/1742-6596/1060/1/012018

polarimetric-time-frequency distributions and ESPRIT algorithm [6][7] was proposed to estimate DOA and polarization of FH signals, it can not only avoid seeking the spectral peak, but also achieve the goal of three-dimensional space positioning, but the ESPRIT algorithm had complex eigenvalue roots and parameter matching problems, and the computational complexity was still high. Almidfa K et al. [8] compared DOA estimation performance of LS-ESPRIT, TLS-ESPRIT and unitary ESPRIT in W-CDMA communication, and the simulation results showed the unitary ESPRIT algorithm had the best estimation performance when SNR was greater than 0. Some scholars [9]-[12]have proposed the unitary ESPRIT algorithm and its improved algorithm into DOA estimation of MIMO radar, those algorithms transform the received data from complex operations into real operations, which overcomes the problems that ESPRIT algorithm needs parameter matching and has low estimation accuracy under the condition of low SNR and low snapshot data. The denoising method based on morphological filtering was proposed to obtain a clearer time-frequency map.[13]

Based on the above issues, the STFD&U-ESPRIT algorithm is proposed in this paper to estimate 2D-DOA of multiple FH signals. In section 2, a planar antenna array receiving model is established and an array snapshots model of the FH signal is deduced therefrom. Then the effective hops of FH signals from time and frequency domain is extracted and the spatial-time-frequency distribution matrix of this hop is established in section 3. In section 4, we present the detailed steps of the algorithm proposed in this paper. Finally, the simulation results and analysis are given in Section 5 and paper is concluded in 6 Section.

2. Snapshots model of FH signal

Assume that the hopping period of FH signal $s_n(t)$ is T_n, there are L hops within time of Δt in total. ω_{nl} and φ_{nl} represent the carrier frequency and initial phase of l-th hop respectively and the time of initial hop is Δt_{0n}. Then the $s_n(t)$ can be defined as:

$$s_n(t) = \upsilon_n(t)\sum_{l=0}^{L-1}\exp[j(\omega_{nl}t' + \varphi_{nl})]rect\left(\frac{t'}{T_n}\right) \tag{1}$$

Where $t' = t-(k-1)T_n - \Delta t_{0n}$, υ_n stands for the complex envelope of base band of signal $s_n(t)$, $rect$ is the unit rectangle pulse.

(a) The structure of array **(b)** The 2D-DOA of FH signals

Figure.1 The receiving model of FH signals

Array structure shown in Figure.1(a), the total number of array elements are $N \times M$, the bandwidth of the receiver $B = f_{\max} - f_{\min}$, the adjacent array spacing are d_1 and d_2, and the elements spacing $\max(d_1, d_2) < c/2f_{\max}$ (c denotes the speed of light), $f_{\max}$ and $f_{\min}$ denote the upper and lower limits of the receiver bandwidth, respectively. Suppose that K FH signals $s_k(t)(k = 1,2,...,K)$ impinge instantaneously onto the $N \times M$ array. The receiving model of FH signals as shown by Figure.1(b), let the pitch angle of the FH incident wave is θ_k ,the azimuth angle of the FH incident wave is φ_k , and the wavelength is λ. Where $\theta_k \in [0, \pi/2]$, $\varphi_k \in [0, 2\pi]$. Define the coordinate original point array element as the reference array element, then direction matrix of N elements on the X axis can be expressed as:

$$\boldsymbol{A}_x = [\boldsymbol{a}_1(\theta_1, \varphi_1), \boldsymbol{a}_2(\theta_2, \varphi_2), ..., \boldsymbol{a}_K(\theta_K, \varphi_K)] \tag{2}$$

Where $\boldsymbol{a}_k(\theta_k, \varphi_k) = [1, e^{j2\pi d_1 \cos\varphi_k \sin\theta_k/\lambda}, ..., e^{j2\pi d_1 (N\text{-}1)\cos\varphi_k \sin\theta_k/\lambda}]^T$. In a similar way, the direction matrix of M elements on the Y axis can be expressed as:

$$\boldsymbol{A}_y = [\boldsymbol{b}_1(\theta_1, \varphi_1), \boldsymbol{b}_2(\theta_2, \varphi_2), ..., \boldsymbol{b}_K(\theta_K, \varphi_K)] \tag{3}$$

Where $\boldsymbol{b}_k(\theta_k, \varphi_k) = [1, e^{j2\pi d_2 \sin\varphi_k \sin\theta_k/\lambda}, ..., e^{j2\pi d_2 (M-1)\sin\varphi_k \sin\theta_k/\lambda}]^T$. Then $N \times M$ array can be divided into M subarrays, so subarray 2 to M corresponds to the offset of subarray 1 along the Y axis. The difference between the direction matrices of adjacent subarrays is a fixed value $\Psi = diag(e^{j2\pi d_2 \sin\varphi_1 \sin\theta_1/\lambda}, e^{j2\pi d_2 \sin\varphi_2 \sin\theta_2/\lambda}, ..., e^{j2\pi d_2 \sin\varphi_K \sin\theta_K/\lambda})$ .Therefore, the flow pattern matrix of $N \times M$ array can be formulated as:

$$\boldsymbol{A} = \begin{bmatrix} \boldsymbol{A}_x \boldsymbol{D}_1(\boldsymbol{A}_y) \\ \boldsymbol{A}_x \boldsymbol{D}_2(\boldsymbol{A}_y) \\ \vdots \\ \boldsymbol{A}_x \boldsymbol{D}_M(\boldsymbol{A}_y) \end{bmatrix}_{NM \times K} \tag{4}$$

Where $\boldsymbol{D}_m(\bullet)$ represents a diagonal matrix constructed by m rows of the matrix. Assume the received data vector is $\boldsymbol{X}(t) \in R^{NM\times1}$, gaussian white noise data vector is $\boldsymbol{N}(t) \in R^{NM\times1}$, and FH source data vector is $\boldsymbol{S}(t) \in R^{K\times1}$, so snapshot vector model for array can be defined as:

$$\boldsymbol{X}(t) = \boldsymbol{AS}(t) + \boldsymbol{N}(t) \tag{5}$$

3. Extraction of effective hops and construction of space-time-frequency matrix

The frequency hopping signal is a broadband signal, and the carrier frequency of each hop jumps randomly, so flow pattern matrix of the plane array also changes with the carrier frequency hopping, however it can be simplified as a narrow-band signal when studying one hop. For the problem of reducing frequency hopping signal to narrowband signal processing, it has been studied by some scholars. Chen [4] extracts the effective hops of FH signal by the panoramic time-frequency map getting from polyphase filter group method. The effective hops of the FH signal are extracted by the combined time-frequency method of short-time Fourier transform (STFT) and Smoothed Pseudo Wagner Distribution (SPWVD).[5] Different improvements[6] [7] have been made to the combined time-frequency method by Zhang . He combines with time-frequency single-source to set the noise threshold, effectively remove the noise and a clear and robust time-frequency map is obtained, improving the performance of extracting the effective hops.

As mentioned in the above articles, we can see that getting a clear time-frequency map of the FH signal is the most crucial step, because it will directly affect the extraction accuracy of the effective hop, thereby affecting the performance of DOA estimation and network station sorting. In order to further improve the performance of 2D-DOA estimation, this paper uses the morphological filtering method to correct the time-frequency map of FH signals after the SPWVD transform to complete the effective hop extraction. The more details about process of morphological filtering see paper [13]. After getting the effective hops, we can establish the spatial-time-frequency distribution matrix of one effective hop of signal.

Cohen discrete time frequency distribution of signal $x_i(t)$ is expressed as:

$$\boldsymbol{D}_{x_i x_i}(t, f) = \sum_{\tau=-\infty}^{\infty} \sum_{l=-\infty}^{\infty} \varphi(l, \tau) x_i(t + l + \tau) x_i^*(t + l - \tau) e^{-j4\pi f \tau} \tag{6}$$

Where $\varphi(l,\tau)$ denotes kernel function. Therefore, the discrete cross-time-frequency distribution of signal $x_i(t)$ and $x_j(t)$ can be defined as:

$$D_{x_i x_j}(t,f)=\sum_{\tau=-\infty}^{\infty}\sum_{l=-\infty}^{\infty}\varphi(l,\tau)x_i(t+l+\tau)x_j^*(t+l-\tau)e^{-j4\pi f\tau} \tag{7}$$

Then the spatial-time-frequency distribution of signal $x(t)$ can be defined as:

$$D_{XX}(t,f)=\sum_{\tau=-\infty}^{\infty}\sum_{l=-\infty}^{\infty}\varphi(l,\tau)X(t+l+\tau)X^H(t+l-\tau)e^{-j4\pi f\tau} \tag{8}$$

Where $[D_{XX}(t,f)]_{ij}=D_{x_i x_j}(t,f)\,(i,j=1,2,\cdots,M)$ denotes the time frequency distribution between the output signals of each array. According to (5) and (8), the covariance matrix of time frequency domain can be written as:

$$E[D_{XX}(t,f)]=AD_{SS}(t,f)A^H+E[D_{NN}(t,f)] \tag{9}$$

4. 2D-DOA estimation of FH signal and network sorting

Suppose the number of frequency hopping sources are K, and the number of sources are less than the number of elements at any time-frequency point.

The unitary transformation matrix U_N is defined as follows:

$$U_{2n}=\frac{1}{\sqrt{2}}\begin{bmatrix} I_n & jI_n \\ \Pi_n & -j\Pi_n \end{bmatrix} \tag{10}$$

$$Q_{2n+1}=\frac{1}{\sqrt{2}}\begin{bmatrix} I_n & 0 & jI_n \\ 0^T & \sqrt{2} & 0^T \\ \Pi_n & 0 & -j\Pi_n \end{bmatrix} \tag{11}$$

Where Π_n is the antisymmetric unit matrix. Constructing unitary transformation matrix Q_N and Q_M based on the formulation (10) and (11). Then the real matrices Y converted from complex matrices X can be expressed as:

$$Y=(Q_M^H\otimes Q_N^H)\;X \tag{12}$$

Where $\otimes$ donates the Kronecker multiplication. Let $E=[\mathrm{Re}(Y),\mathrm{Im}(Y)]$, then we can obtain the $\tilde{R}_{EE}$, where $\tilde{R}_{EE}$ is the estimation value of array covariance matrix of the E. Then doing eigenvalue decomposition for $\tilde{R}_{EE}$, we can obtain the signal subspace E_S that is formed by the eigenvectors corresponding to the larger K eigenvalues, and noise subspace E_N whose dimensional is $NM-K$. Since each frequency hopping signal is irrelevant, there must be a full rank matrix T, s.t $E_S=DT$,where D is the Direction matrix of Y. We define the selection matrix can be expressed as:

$$J_1=[I_{(N-1)(N-1)},0_{(N-1)\times 1}] \tag{13}$$

$$J_2=[0_{(N-1)\times 1},I_{(N-1)(N-1)}] \tag{14}$$

According to the property of unitary matrix $U_N U_N^H=I$, we obtain the relationship between selection matrix and unitary matrix can be rewritten as $U_{N-1}^H J_2 U_N=(U_{N-1}^H J_1 U_N)^*$. Simultaneously we order that $\begin{cases} K_{\mu 1}=I_M\otimes\mathrm{Re}(U_{N-1}^H J_2 U_N) \\ K_{\mu 2}=I_M\otimes\mathrm{Im}(U_{N-1}^H J_2 U_N) \end{cases}$ and $\begin{cases} K_{v1}=\mathrm{Re}(U_{M-1}^H J_2 U_M)\otimes I_M \\ K_{v2}=\mathrm{Im}(U_{M-1}^H J_2 U_M)\otimes I_M \end{cases}$ .Considering the case of receiving K frequency hopping signals, then we have:

$$\begin{cases} \boldsymbol{K}_{\mu 1}\boldsymbol{D}\boldsymbol{\Omega}_{\mu} = \boldsymbol{K}_{\mu 2}\boldsymbol{D} \\ \boldsymbol{K}_{v1}\boldsymbol{D}\boldsymbol{\Omega}_{v} = \boldsymbol{K}_{v2}\boldsymbol{D} \end{cases} \tag{15}$$

Where $\boldsymbol{\Omega}_{\mu} = diag(\tan(\frac{\mu_1}{2}),...,\tan(\frac{\mu_K}{2}))$, $\boldsymbol{\Omega}_{v} = diag(\tan(\frac{v_1}{2}),...,\tan(\frac{v_K}{2}))$, and $\mu_k = \frac{2\pi}{\lambda}d_1\cos\varphi_k\sin\theta_k$, $v_k = \frac{2\pi}{\lambda}d_2\sin\varphi_k\sin\theta_k$, $k = 1,2,...,K$.

According to the relationship between $\boldsymbol{D}$ and $\boldsymbol{T}$, we can obtain:

$$\begin{cases} \boldsymbol{K}_{\mu 1}\boldsymbol{E}_s\boldsymbol{\psi}_{\mu} = \boldsymbol{K}_{\mu 2}\boldsymbol{E}_s \ \ \boldsymbol{\psi}_{\mu} = \boldsymbol{T}^{-1}\boldsymbol{\Omega}_{\mu}\boldsymbol{T} \\ \boldsymbol{K}_{v1}\boldsymbol{E}_s\boldsymbol{\psi}_{v} = \boldsymbol{K}_{v2}\boldsymbol{E}_s \ ,\ \boldsymbol{\psi}_{v} = \boldsymbol{T}^{-1}\boldsymbol{\Omega}_{v}\boldsymbol{T} \end{cases} \tag{16}$$

Merging the formulation (16), we obtain a new matrix can be written as:

$$\boldsymbol{P} = \boldsymbol{\Psi}_{\mu} + j\boldsymbol{\Psi}_{v} = \boldsymbol{T}^{-1}(\boldsymbol{\Omega}_{\mu} + \boldsymbol{\Omega}_{v})\boldsymbol{T} \tag{17}$$

Then doing eigenvalue decomposition of the $\boldsymbol{P}$, we get a diagonal matrix $\boldsymbol{\Sigma}$ composed of eigenvalues. In addition, $\boldsymbol{\mu}$ and $\boldsymbol{v}$ are defined as follows:

$$\begin{cases} \boldsymbol{\mu} = \frac{2}{\pi}\arctan(\mathrm{Re}(\boldsymbol{\Sigma})) \\ \boldsymbol{v} = \frac{2}{\pi}\arctan(\mathrm{Im}(\boldsymbol{\Sigma})) \end{cases} \tag{18}$$

Finally, the pitch angle estimation and azimuth angle estimation are obtained from the formulation (18), i.e.

$$\begin{cases} \tilde{\boldsymbol{\theta}} = \arcsin(\boldsymbol{\mu}^2 + \boldsymbol{v}^2)^{1/2} \\ \tilde{\boldsymbol{\phi}} = \arctan(\boldsymbol{\mu}^2 + \boldsymbol{v}^2)^{1/2} \end{cases} \tag{19}$$

From the formulation (19) derivation process, we can see that compared with the traditional ESPRIT algorithm, the unitary ESPRIT algorithm transforms the eigenvalue decomposition and the matrix operations of complex domain into the real domain, which greatly reduces the computational complexity. When constructing the real matrix $\boldsymbol{Y}$, the received data is repeatedly used that improves the estimation accuracy. In addition, 2D-DOA estimation can be automatically paired. After we get the 2D-DOA estimation, the network sorting of multi-FH signal can be achieved through the clustering analysis.

5. Simulation and analysis

Suppose that there are five frequency hopping signal FH1, FH2, FH3, FH4, FH5 in the space, the incident pitch angles are $\theta_1 = 15^\circ$, $\theta_2 = 30^\circ$, $\theta_3 = 45^\circ$, $\theta_4 = 60^\circ$, $\theta_5 = 75^\circ$ and the incident azimuth angles are φ_1=80°, φ_2=60°, φ_3=40°, φ_4=20°, φ_5=10° respectively, the hopping period is 10us, the normalize carrier frequency jumps between 0~0.7. 200 Monte Carlo experiments were performed, the overall root mean square error of DOA and overall estimated success rate were used as the performance criterion. The total root mean square error (RMSE) of DOA is defined as:

$$RMSE = \sqrt{\frac{1}{PK}\sum_{p=1}^{P}\sum_{k=1}^{K}[(\tilde{\theta}_{kp} - \theta_k)^2 + (\tilde{\varphi}_{kp} - \varphi_k)^2]} \tag{20}$$

Where K denotes the source number, P denotes the total Monte Carlo experiments, θ_k and φ_k denote the true values of pitch angle and azimuth angle of the k-th FH signal,

$\tilde{\theta}_{kp}$ and $\tilde{\varphi}_{kp}$ are p -th pitch angle and azimuth angle estimations of the k -th FH signal. The total estimated success rate η is defined as:

$$\eta = P_1 / P \tag{21}$$

Where P_1 denotes the number of successful experiments with DOA estimated deviations less than 1°, and P denotes the total number of experiments.

5.1 Experiment 1

In order to verify the impact of signal to noise ratio (SNR) on the performance of the algorithm, assume that the receiving array structure is 4×4 uniform array, $d_1 = d_2 = 0.5m$, and the number of snapshots are 600. Fig.2 shows the performance comparison of 2D-DOA estimation in proposed algorithm and ESPRIT algorithm when SNR increase from -5dB to 20dB.

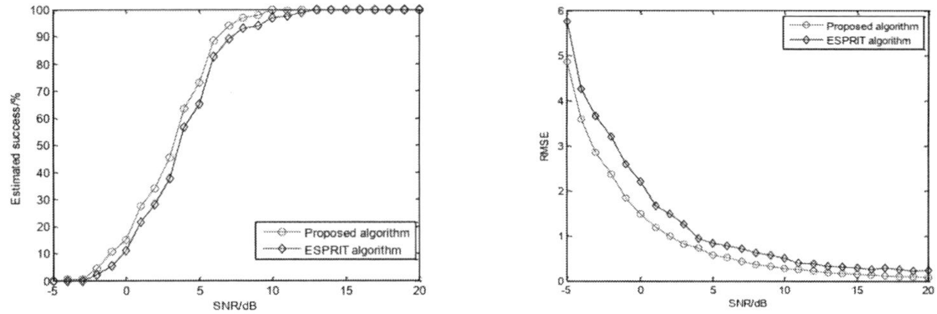

(a)The total success rate of Experiment 1 (b) The RMSE of Experiment 1

Figure.2 Performance comparison of 2D-DOA estimation

It can be seen from Figure.2 (a) that with the increase of SNR, the η of the proposed algorithm and the ESPRIT algorithm are gradually increased, the proposed algorithm is slightly larger; When the SNR reaches $9dB$, the η of the proposed algorithm is close to 100% while the ESPRIT algorithm needs to reach about $12dB$.

It can be seen from Figure.2(b) that with the increase of SNR, the RMSE of the proposed algorithm and the ESPRIT algorithm are gradually decreased; the RMSE of the proposed algorithm is obviously less than ESPRIT algorithm in general; When the SNR is greater than $11dB$, the RMSE of the proposed algorithm is close to the ESPRIT algorithm, and both tend to be stable.

5.2 Experiment 2

In order to verify the impact of the amount of snapshot data on the performance of the algorithm, assume that the number of snapshots per hop gradually increases from 400 to 2000, the structure of receiving arrays is 4 × 4, and the SNR are 0dB and 5dB, respectively, other experimental conditions are same with Experiment 1. Figure.3 shows the performance comparison of 2D-DOA estimation in proposed algorithm and ESPRIT algorithm.

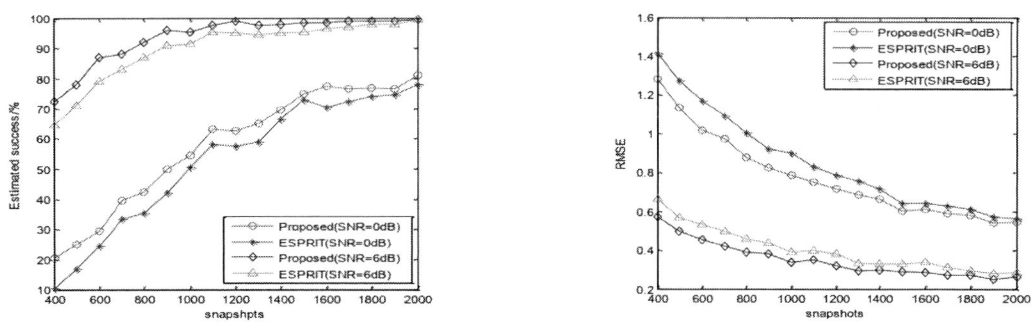

(a)The total success rate of Experiment 2 (b) The RMSE of Experiment 2

Figure.3 Performance comparison of 2D-DOA estimation

Figure.3(a) shows that under different SNR, the overall estimation success rates of both algorithms increase gradually with the increase of snapshots. However, under each snapshot, the success rate of proposed algorithm is better than ESPRIT algorithm. From the Figure.3(b), we can see that with the increase of snapshot data under different SNR, the total RMSE of the proposed algorithm and ESPRIT algorithm decrease gradually. In other words, the longer the frequency signal lasts, the more observation the sample gets and the better the estimated performance.

5.3 Experiment 3

In order to compare the algorithmic complexity of two algorithm, assume that the SNR is 5dB, the snapshot data of each hop is 3000, the structure of receiving arrays are 3×3 , 4×4 , 5×5 , 6×6 , 7×7 , 8×8 ,respectively, other experimental conditions are same with Experiment 1. The time required for the 2D-DOA estimation of the two algorithms is shown in Table 1.

Table.1 Comparison of two algorithms for 2D-DOA estimation time required(s)

Algorithm type	3×3	4×4	5×5	6×6	7×7	8×8
Proposed algorithm	0.572	0.951	1.515	2.210	3.802	5.407
ESPRIT algorithm	0.863	1.389	2.225	3.248	5.336	7.467

From Table.1, we can see that as the number of arrays increases, both the proposed algorithm and ESPRIT algorithm consume more time, but the ESPRIT algorithm required more. When the structure of array is the same, the proposed algorithm is less time consume than ESPRIT algorithm, and the time consume of proposed algorithm is about 0.65 times of ESPRIT algorithm.

6. Conclusion

The 2D-DOA information of the frequency hopping signals can be effectively used to complete multi-FH signal network sorting. The STFD&U-ESPRIT algorithm is deduced and explained in this paper to estimate the 2D-DOA information of multiple FH signals. Theoretical analysis and simulation results show that the proposed algorithm can reduce the computational complexity compared with the traditional ESPRIT algorithm, while the performance is still better. In addition, the proposed algorithm transforms data processing from complex to real domain. Therefore, it has higher applicability and more engineering practical value.

7. Appendices

This work was supported by the National Natural Science Foundation of China under Grant 61601500.

References

[1] Sha Z C, Liu Z M, Huang Z T, et al. Online Hop Timing Detection and Frequency Estimation of Multiple FH Signals[J]. ETRI Journal, 35,5(2013-10-01), 2013, 35(5):748–756.

[2] Belouchrani A, Amin M G. Time-frequency MUSIC[J]. IEEE Signal Processing Letters, 2002, 6(5):109-110.

[3] Belouchrani A, Amin M G. Blind source separation based on time-frequency signal representations[J]. IEEE Transactions on Signal Processing, 1998, 46(11):2888-2897.

[4] Chen L H, Zhang E Y. Directions of arrival estimation for multi frequency-hopping signals based on digital channelized receiver and spatial time-frequency analysis[J]. Journal on Communications, 2009, 30(10):68-74. (in Chinese)

[5] Chen L H. Directions of arrival estimation for multicomponent frequency-hopping signals based on spatial time-frequency analysis[J]. Systems Engineering & Electronics, 2011, 33(12):2587-2592. (in Chinese)

[6] Zhang D W, Guo Y, Qi Z S, et al. A joint estimation of 2D-DOA and polarization estimation for multiple frequency hopping signals[J]. Journal of Huazhong University of science & technology 2015, 09:84-88. (in Chinese)
[7] Zhang D W, Guo Y, Qi Z S, et al. A joint estimation algorithm of multiple parameters for frequency hopping signals using spatial polarimetric time frequency distributions[J]. Journal of Xi'an Jiao tong University, 2015, 49(8):17-23. (in Chinese)
[8] Almidfa K, Tsoulos G V, Nix A. Performance analysis of ESPRIT, TLS-ESPRIT and Unitary-ESPRIT algorithms for DOA estimation in a W-CDMA mobile system[C]// 3g Mobile Communication Technologies, 2000. First International Conference on. IET, 2000:200-203.
[9] Guimei Zheng, Baixiao Chen, Minglei Yang. Unitary ESPRIT algorithm for bistatic MIMO radar[J]. Electronics Letters, 2012, 48(3):179-181.
[10] Zheng G, Tang J, Yang X. ESPRIT and Unitary ESPRIT Algorithms for Coexistence of Circular and Noncircular Signals in Bistatic MIMO Radar[J]. IEEE Access, 2017, 4(99):7232-7240.
[11] Zhou Y, Fei Z, Yang S, et al. Joint Angle Estimation and Signal Reconstruction for Coherently Distributed Sources in Massive MIMO Systems Based on 2D Unitary ESPRIT[J]. IEEE Access, 2017, PP(99):1-1.
[12] Liu S Q, Wang B H, Li L J, et al. DOA estimation method for two-dimensional hybrid MIMO phased-array radar[J]. Journal of xi'an university of electronic science and technology (natural science edition), 2017, 44(3):157-163. (in Chinese)
[13] Gong X, Wang Z H, Wang C G. Parameter estimation of frequency hopping signal based on time-frequency analysis and morphological filtering. [J] Journal of Air Force Radar Academy, 2011, 25(4):243-246.(in Chinese)

Temporal and Spatial Analysis of Fake Base Stations

Yue Jia, Bo Lyu* and Yangyang Li

Innovation Center China Academy of Electronics and Information Technology

*blv@csdslab.net

Abstract. With the development of information technology, send spam messages by using the fake base station has become a new means of criminal and marketing advertisement. Fake base station usually dissembles itself as numbers which stand for banks, operators, state organs and forcible and sends fraud, pornography, gambling, advertising and other short messages to users. This action affects the normal communication and also has a serious violation of public security. Based on sample data sets of junk text messages opened by ChinaVis2017 Challenge, this paper develops a temporal and spatial trajectory analysis and visualization system for stopping fake base station. Moreover, this paper deeply excavates the temporal and spatial laws of the location and movement of fake base stations. Based on the analysis results of fake base station behaviour patterns, this paper provides several recommendations of tracking and arresting fake base station owners. This paper firstly introduces the realization scheme and key technology of spatial-temporal trajectory analysis and visualization system of fake base stations. Then, it analyses the temporal and spatial trajectory distribution, the content classification and the degree of harmful of fake base stations, and finally analyses the actual demand of fake base stations supervision and management with several practical solutions.

1. Introduction

Currently, fake base station is a means of telecom fraud high-tech equipment and it is mainly composed of hosts and notebook computers. It searches for a certain radius of the phone card information, and fraudulently imitates any other phone number sending fraud, advertising and other short messages to compromised users. When such device is running, the compromised users' mobile phone signal is forced to connect to the fake base station and they cannot connect to the public telecommunications network. Under this circumstance the normal mobile phone services cannot be performed [1].

As long as users are forced to connect to a fake base station, the fake base station can unlimitedly send any content of the message to the users. Those advertising short messages, fraudulent short messages and sensitive content messages could do great harm to security. Even if the spam message does not successfully lead to a crime, fake base station also affects the users' normal operator service as the user's mobile phone signal is forced to connect to the device of fake base station. Mobile phone users will temporarily off the public network in about 8 to 12 seconds and after that they may return to normal. However, some mobile phone users must restart their devices to re-enter the network. In addition, the fake base station will lead to mobile phone users frequently update the location, occupying the region's wireless network resources and the emergence of network congestion, affecting the normal communication of users. Harms as the fake base station may arise show as following:

a) Counterfeit bank customer service phone. For example, counterfeiting Industrial and Commercial Bank of China, fake base station will pretend as 95588 and sign the number at the end of the message. Short messages content might include prompting user password failure and a modify is needed by user to log in the site shown in the message, or indicating user to log on the site to exchange cash and other tricks.

b) Guide the user to log on to a phishing site. The fake base station can imitate the bank customer service number which is exactly the same number that the real bank customer service used. Many users will trust and click on the relevant link. At this time, users will be led to a fishing sit which is really similar to the bank's website.

c) Get users' confidential information. When logging on the phishing site, the fraud may guide users step by step to enter their information. Those information including bank card account number, mobile phone number and login password. If the fraudster wants to get more information about the victim, they will be on-line and offer more guidance to the users.

2. Theory and challenges

Fake base station occupies the public radio communication system network number, frequency resources, and so on, to disguise itself as the public radio communication network base stations neighbourhood. It uses public radio communication network to monitor signalling. In the normal system monitoring radio communication process, the phone searches for a base station signalling every five seconds, and interacts with a number of nearby base stations at the same time. According to the base station signal strength, the user selects the nearest and strongest base station. However, one of the easiest disguise methods is through the instantaneous signal power transmission, then the fake base station can take place of normal base station. After that, it can and automatically collect coverage within the public radio communication network mobile phone users mobile phone number and the current location information. Then, fake base station will earn trust from user and fraudulent use of other mobile phone numbers to these off-line mobile phone to send pre-set business customized advertising messages [2, 3].

Fake base station can be carried by a person or a vertical. So, it is hard to trace or arrest owners of fake base stations. Moreover, one fake base station can send as many as 50000 short messages per hour. Not only GSM users, but also 3G network and LTE network users may be compromised by fake base station. Moreover, there are two problems that make it impossible to use it as a sign to identify the fake base station trajectory. One problem is that a number of fake base stations may masquerade as the same number (for example, 95580 which is the service number for Postal Savings Bank of China, 10086 which is the service number of China Mobile, etc.), and another problem is that the same fake base station may use multiple camouflage numbers at the same time. Finding an efficiency searching system to tracing and locating fake base station becomes a crucial task for public security.

3. Data set analysis

The sample data set, with which this paper processed, has 63 days of junk text messages for Beijing, the capital city of China, from the data February 23th to April 26th of the year 2017. All messages are collected through user reporting mechanism. In total, there are 3358952 lines of data and each of them including message content, fake phone number, message receiving time, user reporting time and the longitude and latitude of the reporting location. The total number of different fake phone number is 11371.

In the part of the data preparation, in addition to match the data provided, we also use the following additional data sources: (a) Beijing city map data that provided by OpenStreetMap, to offset the existence of competition between open data and OSM data of manual correction; (b) in order to study the regularities of distribution of content of short messages, we use LBS API provided by Baidu to map the longitude and latitude coordinates in the data set. In the data cleaning section, the exception points of latitude and longitude beyond the geographical boundaries are eliminated, and data items beyond the scope of the study are eliminated [1].

In the cleaned data, the sender's telephone number disguised by the pseudo base station is the identification of a spatial-temporal trajectory, but there are two problems that make it impossible to

IOP Conf. Series: Journal of Physics: Conf. Series **1060** (2018) 012019 doi:10.1088/1742-6596/1060/1/012019

use it as a sign to identify the pseudo base station trajectory. The problem is that a number of pseudo base stations may masquerade as the same number (for example, 95580, 10086, etc.), and the problem two is that the same pseudo base station may use multiple camouflage numbers simultaneously. In order to pseudo base station trajectory visualization, each with the phone number labelled according to the spatial-temporal trajectory space of Euclidean distance for clustering, multi path differential behind the same period with the same numbers, so as to solve the problem. Then, by using the similarity of trajectories, the trajectories which are very close to the laws of time and space are aggregated into the same pseudo base station, so as to solve the second problem [4].

3.1. Content Categories

In order to analyse the spatial-temporal regularities of fake base station moving patterns and messages' occurring regulars, this paper analyses the word segmentation and the semantic of the short message and extract the key words of the content of the message. After that, the content of the message is classified and clustered according to the distance of the keyword. By applying above classification methods, the short messages will be divided into fake invoices, posing identity, pornographic, real estate advertising, contraband promotion, employment advertisements, illegal gambling, sales advertisements, education immigration and other types of ten categories.

3.2. Spatial Distribution Regularities

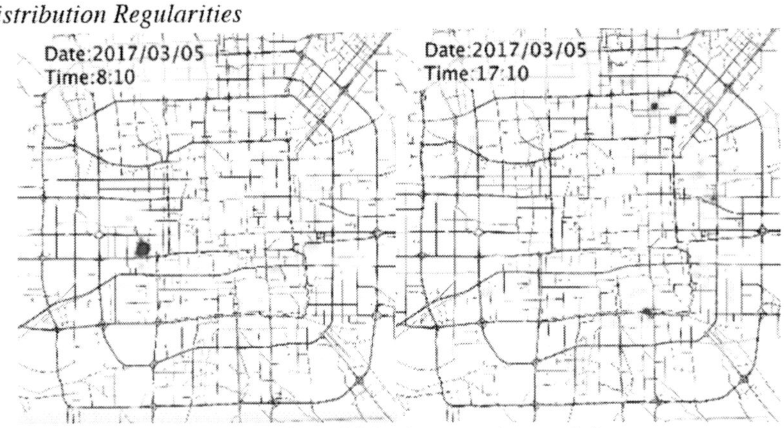

Figure 1: Samples of fake base station activity map.

From the Figure 1, we can see the temporal and spatial trends of the distribution of fake base stations. For example, on the day of March 5th, much more short messages were send by fake base stations, and more concentrated in a few hot spots (such as office buildings, shopping district, etc.). Vice the versa, one the day of April 5th, which is the tomb sweeping holidays, as most people rested at home or went out of town for tourism, the total number of messages has a significant reducing. Moreover, the distribution is more uniform. From the laws of the thermos gram, it can be concluded that the fake base station is very cautious and targeted when selecting the location and the sending time, especially in relation to the daily behaviour of the public. The thermal map is intuitive to see the distribution of fake base stations, but for the fake base station trajectory patterns, this paper provides more fine granular analysis means.

In raw data, the sender's telephone number disguised by the fake base station is the identification of a spatial-temporal trajectory. However, there are two issues that prevent us from using it as a marker to identify the fake base station trajectory. The first problem is that there are multiple fake base stations may disguise the same number (such as 95580,10086, etc.), the other problem is that the same fake base station may also use a number of camouflage numbers.

In order to visualize the trajectories of the fake base stations, each time-space trajectory marked with the telephone number is clustered according to the Euclidean distance in space, and the multiple trajectories behind the same number are identified in the same period to solve the problem. And then use the temporal and spatial similarity of the trajectory to gather the trajectories of the spatially close to the same fake base station to solve the problem.

Through the above spatiotemporal data analysis method, the action trajectory of the fake base station can be effectively extracted. On this basis, the team uses Processing, which is a flexible software sketchbook and a language for learning how to code within the context of the visual arts [5], to construct the space-time trajectory query and visualization system. The system supports the basic functions such as zoom out, track searching and tracking visualization. While supporting the map style selection, mouse latitude and longitude and other auxiliary functions. The user can enter the phone number in the interactive interface to retrieve the track and specify the visual colour of the track. In the visual trajectory, the traces of the thickness and saturation of the fake base station to send the number and frequency of text messages.

Figure 2:One fake base station forges different phone numbers

This paper founds that many trajectories have a high spatial and temporal similarity. As can be seen from the Figure 2, the same fake base station operator uses different cell phone numbers. To enlarge the map can be found (as shown on the right), the fake base station operator in parking layout of the north gate of the park and the South Gate of the fake base station, after a period of time by sending two mobile phone number, quickly leave the place.

Figure 3:One phone number forged by different fake base stations

In data analysis, the authors found that many fake base stations forged the same number to send text messages. In particular, the identity of the class like text messages like to use 10086, 95580, 95560 and other service class number. By means of spatial-temporal clustering can separate the trajectory of the same time distance, as shown on the left, the distinction between the two 95580 track of Chaoyang District and Haidian District.

Different fake base station hidden traces of the means are also different. The trajectory can be classified by calculating the distribution of the upper coordinate points of a trajectory. For example, some tracks are used to move around a place, or even stationary, and focus on sending messages at specific times (for example, rush hour), while the rest of the time hides the movement. While some fake base is to keep moving, usually along the fourth ring road along or send text messages as shown in Figure 3.

3.3. Temporal Distribution Regularities

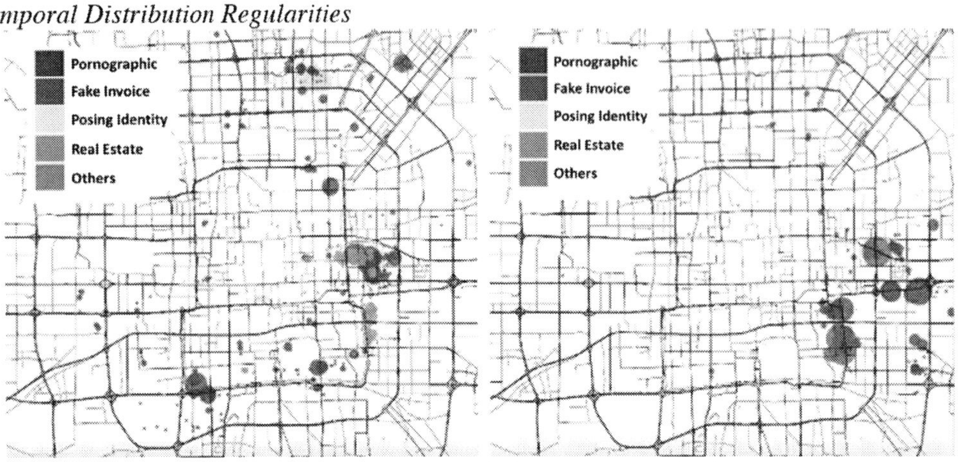

Figure 4:Samples of the temporal and spatial distribution of all kinds of short messages

In this paper, the macro situation of all kinds of text messages is displayed intuitively. As shown in Figure 4, the colour of scatter pictures represents the type of SMS, and the size of scatter points represents the number of messages. In February 23rd, for example, will be divided into four periods, each period from each of a representative point in time, as shown below, the morning pornographic messages as the main messages sent by the fake base, but mainly concentrated in the Sanhuan nearby bustling business district. The invoice information of the invoice is more evenly distributed, and exists from morning till night. February 23rd for the working day, the rush hour is also the fake base station to send short messages peak, real estate advertising and processing invoices, short messages concentrated in the vicinity of the financial district to send, sending time is short, but the number of sending large. In the evening, lurking a whole day porn Service short messages resurgence.

The X axis represents 24 hours a day, and the Y axis represents the number of days per month. Figure 5, Figure 6, Figure 7, Figure 8 are all under the same scale and use a tool named Echarts to generate [6].

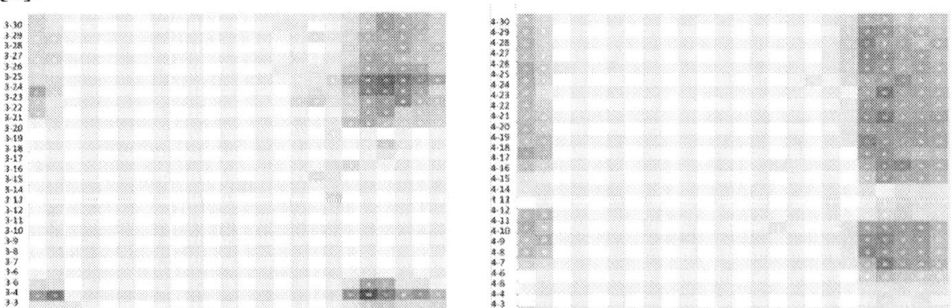

Figure 5:Temporal distribution of pornographic service message

From Figure 5, pornographic information is mainly active in the evening from 18 to 3 a.m.. April 3rd, 4 and 5 are tomb sweeping holidays. During this period, only a small amount of erotic information is active and distributed throughout the day. Another anomalous period was from March 6th to March 20th, when there was almost no information about sexual services. The analysis from the State Council Decree No. 666th "the State Council on amending some administrative regulations" published in March 1st and after the implementation of the national cultural market law enforcement departments to strengthen law enforcement, the Beijing section cracked several pornography cases, erotic services relevant information is greatly reduced. But after that time, porn messages have a clear trend of rebound.

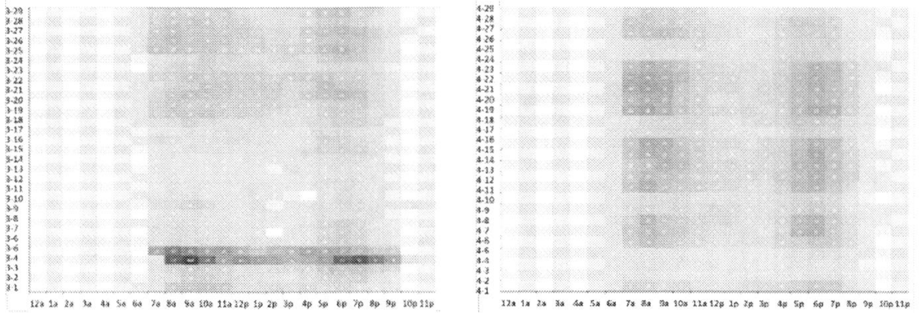

Figure 6:Temporal distribution of fake invoice message

Through the analysis chart of Figure 6, we can clearly find that the invoice information is mainly active from 6 to 21. There is a strong "cyclical", active working days, weekends more convergence. In addition, most dates in March were less active than in April. Abnormal 4-5 March rush invoice message, the paper believes that the overall amount of information in March 4-5 on the two day of the great, various types of message data grew, far more than other days of data for this abnormal surge in the amount of data, when analysing the behaviour is not too much to consider.

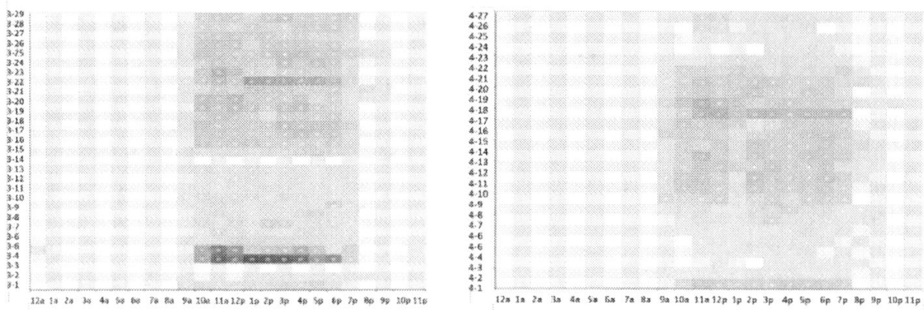

Figure 7:Temporal distribution of posing identity message

From Figure 7, pretending to be an identity class is mostly active during the day and more active. Interestingly, these messages did not appear in the two days of March 3rd and March 15th. March 15th is the consumer protection day, March 3rd is the fifth session of the twelfth CPPCC National Committee, the control is more stringent. On the tomb sweeping holidays, posing as identity class short messages active less.

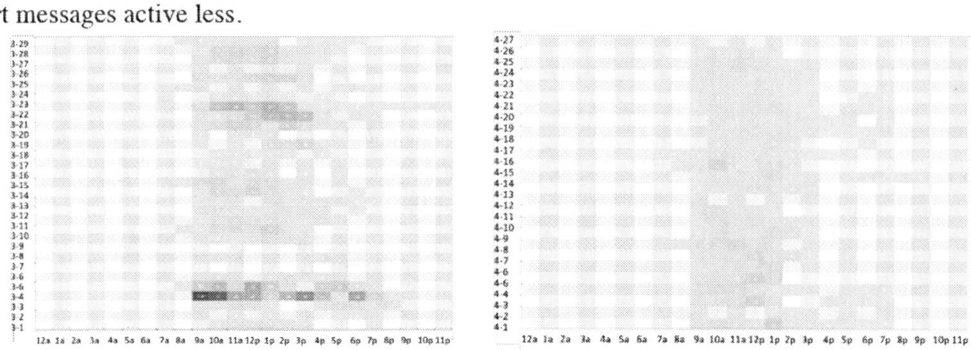

Figure 8:Temporal distribution of real estate advertisement

Through the analysis of Figure 8, it can clearly observe that the real estate advertising is mainly active in 9 to 13, during the tomb sweeping holidays, the active period is relatively small. Based on the above analysis can reveal various types of messages are distribution of their time, the whole tomb sweeping holidays. 3.15 is consumer protection day, holidays and important activities and policy release date, all kinds of messages that appear less, some types of information are not even all day long, have a great relationship crackdown should be with a police officer, but once the crackdown fell, all kinds of fake base short messages has rebounded, pornographic information is obvious.

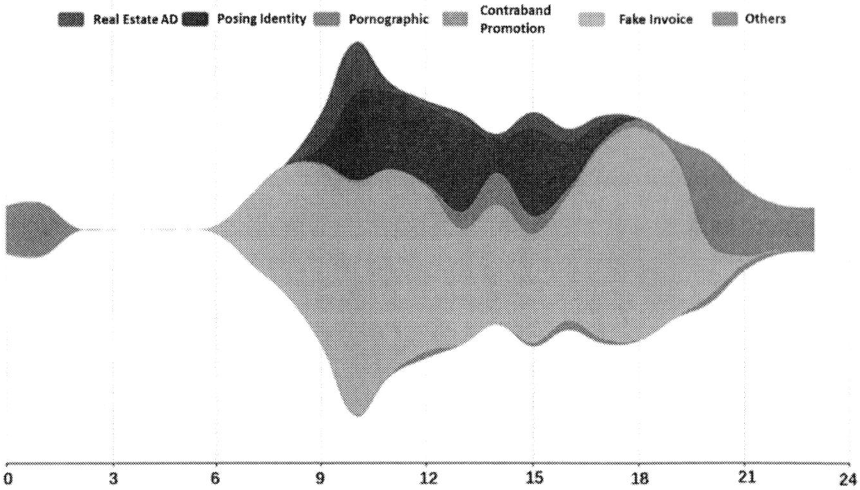

Figure 9:The proportion of different kinds of messages

In order to show the approximate time distribution and total amount of all SMS messages within a day, Figure 9 provides a view. The X axis represents 24 hours a day, and the Y axis represents the total number of each kind of message. From the Figure 9, the overall number of fake invoice message is the most, mainly distributed in the daytime; followed by a large number of pornographic messages are active in the night and noon; and after posing as a real estate advertising message is active in the work period; the remaining amount of other types of messages are scattered in all day long.

To more fully grasp the regularity and influence of the fake base short messages relation, the team used the Baidu LBS API will provide the original data is mapped to the latitude and longitude of the administrative district of Beijing City, according to the distribution of the fake base short messages to the administrative division of Beijing city as a unit of statistics, and the 10 class message division of harm in 5 degree, the degree of hazard level "as the worst impact degree and so on until five, according to the different factors of the content of messages, camouflage number, time distribution and geographical distribution, ten kinds of different text divided into 5 grade level of hazard system, the specific content is shown below.

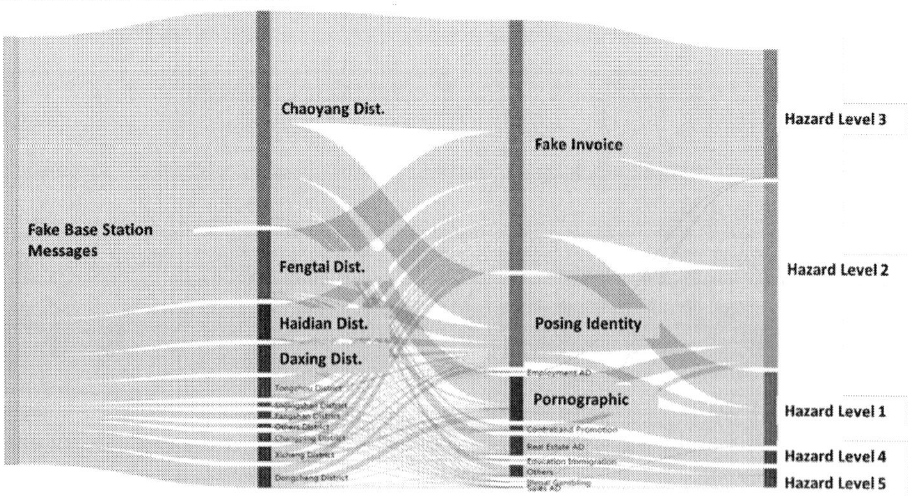

Figure 10: Distribution and hazard levels of pseudo base station messages

It can be obtained from Figure10 that fake base short messages are mainly distributed in the developed economy, office intensive six districts of the city, led by Chaoyang District, followed by Fengtai District, Haidian District, Dongcheng District, Xicheng District and Shijingshan District, but not with the population and land area is. From the Sankey diagram can be seen intuitively fake base message flows, such as short-term erotic services mainly from Chaoyang District, but the Dongcheng District erotic services short messages proportion also cannot be ignored, while the Haidian District erotic services short messages minimum proportion; Shijingshan District has no education immigration short messages.

4. Suggestions

In reality, fake base stations are usually in the form of vehicles, vehicles along the road, on the way to send text messages at fixed points. However, there is no license plate number in the existing fake base station system, which brings challenges to the fake base station traceability and the implementation of the arrest. The problem can be solved effectively by association analysis of fake base station data with traffic camera data. The existing traffic camera can recognize the license plate number accurately, and the moving track of a license plate number can be obtained by retrieving the license plate number in the system database. By using the method of correlation between time and space trajectories proposed in this paper, the license plate number of fake base station vehicles can be accurately locked. For example, as shown in Figure 3, in the 95580-B path, the vehicle from the starting from the East Fourth Ring Road, parking time in the north of the city of Victoria to normal running, turn from the North Sanhuan back through the road, location for the camera coverage area and easily from the vehicle monitoring a license plate number and running track.

According to the previous analysis, this paper can initially determine the highest probability of occurrence of different types of pseudo base stations in different periods, which can serve as a basis for priority attacks. For example, analysis found that after a period of hard strikes, pseudo base station activity will rebound substantially. According to previous data summary, in the size of holidays, important activities, important policies after the implementation of the rebound in the timing, sentinel arrest, you can raise the probability of capturing pseudo base station. Pornographic texts, for example, are expected to rebound after a period of hard strikes. According to the temporal and spatial patterns analysed from historical records, we can provide law enforcement personnel with a number of designated capture areas, and simultaneously with road monitoring linkage to improve the success rate of arresting.

5. Conclusion

Fake base station not only compromises the normal telecommunications order, endangering public safety, disrupt the market order, but also seriously damages the rights and interests of the masses of property, violations of personal privacy of citizens, which lead to serious social harm. According to the "People's Network" statistics, in each year there are nearly 100 billion frauds, gambling, marketing, winning and other text messages has been sent through the fake base station equipment, and the fake base station has become a major public hazard.

In the process of visual analysis, the team found that the time, space and content of the fake base station were highly reproducible. The fake base station movement route, the short message time and the text message content had a fixed behaviour pattern, which means that most of the fake The base station is controlled by a small number of operators. Using the means of trajectory correlation analysis and pattern analysis, the fake base station can be clustered into several operators, thus revealing the operating family behind the fake base station, providing the basis for the control of the fake base station.

6. Acknowledgement

This work is supported by National Key R&D Program of China (No. 2017YFC0803300), the Innovation Funds of CETC (Grant No.16105501) and the Joint Funds of CETC (Grant No. 20166141B08020101).

References

[1] Li X, Cao B, Criminal Regulation on a Case of Fake-base-station[J]. Journal of Guizhou Police Officer Vocational College, 2014.
[2] Zheng Y, Urban Computing: Tackling Urban Challenges Using Big Data[C]// Requirements Engineering Conference. IEEE, 2016:3-3.
[3] Javed M, Siddiqui A T, Transformation of Mobile Communication Network from 1G to 4G and 5G[J]. International Journal of Advanced Research in Computer Science, 2017, 8(3).
[4] Al-Dohuki S, Wu Y, Kamw F, et al. SemanticTraj: A New Approach to Interacting with Massive Taxi Trajectories[J]. IEEE Transactions on Visualization & Computer Graphics, 2016, 23(1):11-20.
[5] https://processing.org/, official website for processing.
[6] http://echarts.baidu.com/, official website for Echarts.

IOP Conf. Series: Journal of Physics: Conf. Series **1060** (2018) 012020 doi:10.1088/1742-6596/1060/1/012020

A Method of Content Discovery in CDN

Xinhua E[1] and Binjie Zhu[2]

[1]Faculty of Information Technology,Beijing University of Technology, Beijing,100124, China

[2]China Mobile Group Beijing Company Limited, Beijing, 100007,China

Elson1900@163.com

Abstract. The CDN management system distributes content objects to the edge of the Internet to achieve the user's near access. The content discovery is an important issue in content delivery network. The content t discovery algorithm is used to find the location of the content in the content distribution network. In this paper, we design a content resource discovery algorithm Global-hop, which optimizes the propagation process of topology change event messages. The events that cause the topology change are divided into two categories: one is the increase of nodes, and the other is the reduction of nodes. The reduction of nodes is divided into node withdrawal and node failure. Different from One-hop, the broadcast path of topology change event message in the Global-hop algorithm is tree-like broadcast. The broadcast topological information in the original algorithm divided nodes into several Slice, and each Slice was divided into several Unit. Each Slice has a Slice Leader, and each Unit also has a Unit Leader with other nodes being ordinary nodes. The message propagation delay of Global-hop algorithm is less than that of One-hop algorithm.

1. Introduction

Nowadays services such as Video-on-Demand and live TV programming available over Internet. Users want to have access to high quality video at any times. Content Delivery Network (CDN) is a solution that has been used to delivery traditional Web page content and Video content . Both content discovery and content routing belong to content request location routing, and the process of directing the user's request to the server is called content request routing. Content request routing in content distribution networks is divided into directional routing and location routing: the so-called directional routing directs content request into the server near the user, and when there is no requested content in the local server, the location of the requested content is found by the location routing system.

2. Related researches

In the early P2P network, the function nodes of the network are all peer to peer, and content distribution is stored in network nodes, but the centralized index is applied. When a node needs to search for contents of other nodes, it should send a look-up request to the index server, which is a centralized method of P2P content indexing [1-7]. Whereas the centralized index server is not applied in the distributed P2P content index. Content index distributes in network nodes which cooperate with each other to complete the content search process.

3. Content resource discovery

This paper designed a content resource discovery algorithm Global-hop, which adopted the structured

DHT to allocate and query the index. ID number N.ID can be achieved by the Hash value of IP information of each node in the network, N.ID=HASH(IP). The ID number of these nodes are arranged in sequence in the order of the ID space to form a node ring. The size of ID space is determined by the number of digits m of the Hash value, $T=2^m$. The T is the maximum number of ID represented by ID space, and m is the number of bits of the HASH value. Therefore, the node ring composed of all nodes ID in the network is a subset of the ID space, and the number of nodes in the node ring is equal to the number of host nodes. It is a node ring with 5 nodes. M is 3, so the maximum number of nodes that the ID space can represent is 8. The IP hash values of the 5 nodes are 0, 1, 2, 5, 7, and they form a node ring. The node followed closely by node n on the ring is called the descendant node of node n, and node n is called the precursor node of this node.

Content index tables are distributed in the nodes of the global resource discovery layer. The Hash function value of the resource name is used as the ID number of the resource, R.ID. The Hash value of the host in which the resource is located is used as the node ID number. The resource ID and node ID constitute the index entry <R.ID, N.ID>, and index entries of all resources make up the resource index table. According to the resource ID, the index item is placed in the corresponding node. The resource ID is used as the key of the index, and the index information is stored in the descendant node of the key. As shown in Figure 1, indexes with keyword 2 are stored in nodes with NID 2, indexes with keyword 4 are stored in nodes with NID 5, and indexes with keyword 6 are nodes with NID 7.

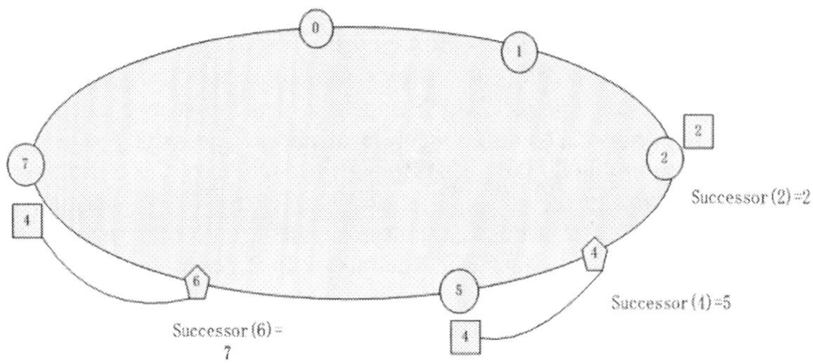

Figure 1. The storage of the index in the node ring

Index information tables are needed in index query. The information of all the nodes are stored in index information tables, so the node in which corresponding information is located can be routed. After the arrival of resource request, it searches the routing information table according to the ID of the resource index to get the ID of its storage node, and then requests the corresponding node according to the node information. As shown in Figure 2, the node 7 is searching for an index with the keyword 3. By looking up the routing table, the descendant node with keyword 3 is 5, and then the request is routed to a node with a NID of 5. Find the descendant node of the object, find the index to the descendant node, and find the node of the resource according to the index.

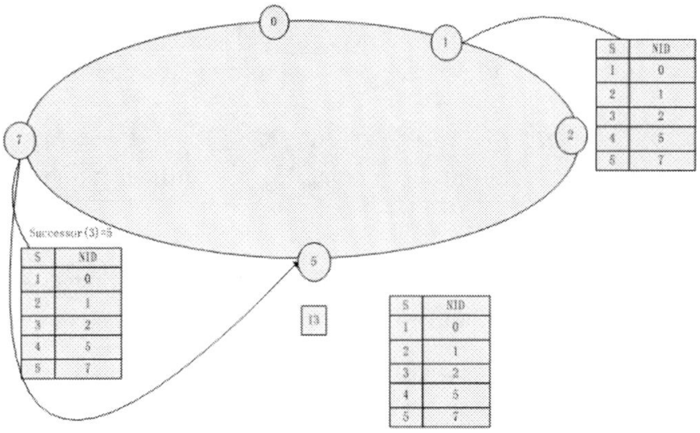

Figure 2. resource discovery

4. Topology information updating algorithm

The events that cause the topology change are divided into two categories: one is the increase of nodes, and the other is the reduction of nodes. The reduction of nodes is divided into node withdrawal and node failure. The node withdrawal is a node withdrawing from the network because of requirements like maintenance, etc, which is a prepared exit, while the node failure is a node failing, causing the node to be called a failure node, which happens at random.

When a node is added, the new node will get routing table information from an existing node in a network. The hash value of the new node IP is used as the NID number of the node. Insert node information into the routing information table based on the NID number. Broadcast the change information of topology with the descendant node of the new node functioned as the root node of the broadcast tree.

When the node withdraws, the withdraw node sends out the withdraw information to the descendant node, and its descendant node broadcast the change information of the topology functioned as the root node of the broadcast tree.

When the node fails, failure node will not take the initiative to send invalidation information, so nodes need to send Keep-Live information to determine that the neighbor node do not fail. Each node sends Keep-Live information to its predecessor node in a certain period. The predecessor node is regarded as a failure node when it does not reply, and a broadcast tree will be built with the node as the root node.

Divide all the nodes in a node ring into M groups with a L_g node in each group, Group Leader. The minimum NID node in each group is the L_g node of the group, and it consists of a broadcast sub-tree. When topology change information is found in a node, the information will be sent to the root nodes of M broadcast sub-tree, and will be sent to each node through broadcast sub-tree.

When a node N_c finds the topology change, node N_c will send the event information to all L_g nodes in the node ring, and L_g nodes will update the topology change onto nodes in each group. If L_g nodes send information to all the nodes directly, a large number of bandwidth will be consumed. Due to bandwidth constraints, each node is allowed to send information to M nodes. The discovery node N_c is only responsible to send information to M L_g nodes. $L_g$nodes will send information to M nodes in its group, and these M nodes will send information to another M nodes, and a broadcast sub-tree will be formed with L_g functioned as root node. A update system based on trees is formed hereby.

In each broadcast sub-tree, topology information is broadcast from root node in the group to each node. Firstly, root node send information to M descendant node. After receiving the time information, M descendant node will keep broadcasting.

Pseudo code of establishing message broadcasting path tree within a group is shown in Figure 3. The GL node in each group is the root node. N（i,j）is used to represent relevant position id in the tree. I represents the floor of the sub-node, and j represents the number of the leaf node in the floor.

When a node receives a message to build a tree, it should firstly get the tree coordinates of the node, and then calculate the coordinate N_n of M sub-node according to the coordination of the node, as shown in Formula.

$$N_n\ (i,j) = (i+1, (j-1)\cdot M+n),(1 \le n \le M) \quad (1)$$

When receiving coordinates of sub-node, calculate S_n the ordinal number of nodes by coordinate number, as shown in Formula 4-2.

$$S_n = \sum_{n=1}^{i-1} M^n + j + S_r \quad (2)$$

S_r is the ordinal number of root nodes. If the ordinal number is more than S_t, then the serial number is $S_c = S_n - S_t$, and S_t is the total serial number, otherwise S_c=S_n.

```
For each node get the tree createinformationdo
    {
Get_tree_id (information from father node);
        For n=1 to M do
            {
              Nn (i,j) = (i+1,(j−1)*M+n),(1≤n≤M);
              Sn = Σ_{n=1}^{i-1} M^n + j + Sr;
        IfSn > ST
                                  Sc = Sn − St;
              Else
                Sc=Sn;
                Mark the children nodeSc;
                Send information to children node;
            }
          End For
    }
End for
```

Figure 3. pseudo code for the process of building a message broadcast path tree

5. The performance analysis of the algorithm

Different from One-hop, the broadcast path of topology change event message in the Global-hop algorithm is tree-like broadcast. The broadcast topological information in the original algorithm divided nodes into several Slice, and each Slice was divided into several Unit. Each Slice has a Slice Leader, and each Unit also has a Unit Leader with other nodes being ordinary nodes. The message broadcast path is: SL- other SL- descendant or precursor nodes broadcasting in turn, which is a linear broadcast path essentially. The broadcast path in this method is: GL- sub-nodes in each layer, which is a tree-like broadcast way, and can broadcast faster. In the original method, messages should be sent firstly from the nodes of discovery event to SL nodes, and then from SL nodes to other nodes. The event messages pass from SL to UL nodes. 3 hops are needed in the previous several broadcast. The messages broadcast finally from UL node in a linear way. The node number of a Unit is U, so the hop of the time message broadcast all over a Unit is U/2, the total hops needed are 3+U/2, and progressive function set of hops is $\theta(n)$.

In the method designed in this paper, messages are sent from the message discovery node to M GL node directly, and do not need to be sent to other GL nodes through GL nodes in this group. Each node has same routing message, so it can make full use of the information and send it to other GL nodes.

When arriving GL nodes, event messages are sent to each nodes by message broadcast tree and with GL as root nodes. Therefore, the hops needed by Global-hop to broadcast topology change messages are H, as shown in Formula 4-1. The progressive function set of hops is $\theta \log_a(n)$.

$$H = \log_a(N - \frac{N-1}{a}) \quad (3)$$

Through calculation, the comparison of the two algorithms of message broadcast is shown in Figure 4. The algorithm 1 is Global-hop, and the algorithm 2 is One-Hop. Moreover, with the increase of the number of nodes in the network, the gap will be larger. In terms of broadcast time, the total broadcast time of the event messages of One-hop algorithm is divided into broadcast time and periodic waiting time, so apart from the broadcast time corresponding to each step, the total message delay also includes the periodic waiting time. While Global-hop does not need the periodic waiting time, so its total message broadcast delay only includes the broadcast time each hop corresponding to. Under the circumstances of the same network, the average broadcast time corresponding to each hop is the same, so the message broadcast delay of Global-hop is smaller than one-hop.

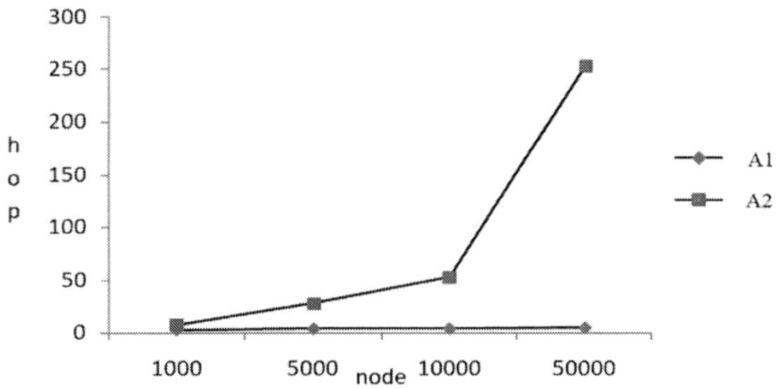

Figure 4. comparison of message broadcast hops of Global-hop and One-Hop

References

[1] SanaaSharafeddine, KarimJahed, Omar Farhat, et al. Failure recovery in wireless content distribution networks with device-to-device cooperation[J]. Computer Networks, 2017, 128(9): 108-122.

[2] Rami Halloush, Hang Liu, Lijun Dong, et al. Hop-by-hop Content Distribution with Network Coding in Multihop Wireless Networks[J]. Digital Communications and Networks, 2017, 3(1): 47-54.

[3] XiaoyingZheng, Ye Xia. Optimizing network objectives in collaborative content distribution [J]. Computer Networks, 2015, 91(14): 244-261.

[4] UttamMandal, PulakChowdhury, et al. Energy-efficient networking for content distribution over telecom network infrastructure[J]. Optical Switching and Networking, 2013, 10(4): 393-405.

[5] LazarosGkatzikis, VasilisSourlas, Carlo Fischione, et al. Low complexity content replication through clustering in Content-Delivery Networks[J]. Computer Networks, 2017,121(5): 137-151.

[6] Bernd Klasen. Efficient content distribution in social-aware hybrid networks [J]. Journal of Computational Science, Pages 209-218, 2013, 4(4): 209-218.

[7] Michele Mangili, Jocelyne Elias, Fabio Martignon, et al.Optimal planning of virtual content delivery networks under uncertain traffic demands [J]. Computer Networks, 2016, 106(4): 186-195.

Deep Learning in Encoding and Decoding of Polar Codes

Chao Yuan[1,a] ,Chang Wu[1,b,*] ,Dan Cheng[1,c] and Yang Yang[1,d]

[1]University of Electronic Science and Technology of china, Chengdu, China

[a]13076091292@163.com ,[b]changwu@uestc.edu.cn ,[c]danchenguestc@163.com ,[d]2654016764 @qq.com

Abstract. In this paper, we apply neural networks to the encoding and decoding of polar codes. Polar codes in 5G communication which meets the simple and ultra-low latency can also be improved. This paper aims to improve the polar codes to achieve simplicity and hardware friendliness. After trained by the neural network, we get a simple math formula to estimate the reliability of the encoding in the polar codes, which can achieve the same effect as GA (Gaussian Algorithm). In terms of decoding, by using NND (Neural Network Decoding) to decode polar codes, we achieve almost the same effect as the conventional SC (Successive Cancellation) algorithm. In order to achieve further improvements, we use the neural network to optimize BP (Belief Propagation) algorithm. Thanks to it, a small number of iterations can be enough to decode the polar codes, even with an iteration of one. The simulation shows that though the early stage of training is full of difficulties, the result can be used to achieve one-step and fool-type decoding.

1. Introduction

In 1943, W.Mcculloch and W.Pitts came up with an NN (neural network) idea to sequentially solve the problem by simulating the human brain architecture [1]. In 1990s, a variety ideas around the use of NN for decoding emerged. The concept of deep learning stems from the study of artificial neural networks. The advantage of deep learning is the use of unsupervised or semi-supervised feature learning and hierarchical feature extraction to efficiently replace manual feature capture. Deep learning is a new field in machine learning research, and it aims to establish a neural network that simulates analytic learning of human brain.

In 2008, Erdal Arikan first proposed the concept of channel polarization at the International Symposium on Information Theory (ISIT) conference [2]. Polar codes have a deterministic construction method and are the only known channel coding methods that can be rigorously demonstrated to "reach" the capacity of the channel.

However, the computational complexity over the code reliability estimation and decoding algorithms of polar codes is very huge, which requires strong mathematical foundation. At the same time, there exists a problem of big calculation which is linear with the block length. Thus, systems with different parameters can not be accepted, such as a system with different block length and code rate, and they are not even feasible for low latency, real-time encoder or decoder. As 5G control channel coding standard, Polar codes are required to gain low latency, but the traditional algorithm is not perfect to meet the requirement, so we have to improve the algorithm. In recent years, boosted by the great achievement people made on deep learning and parallel computing ability of computer, a series of problems the polar codes are facing can be solved by using a powerful deep learning network. This is the trend of development, and it is also the necessity of development.

This paper mainly aims to use deep learning networks to solve some of the problems of polar codes. The results shown below are based on my attempts, and more further work still needs more scholars to study together. In this paper, the polarized degree of polar codes, which is also the reliability of the polar codes, is calculated before decoding. The traditional way to calculate the reliability is DE (density evolution) and GA algorithms [3], these algorithms are complex and have large latency. By using the deep learning to train the polar codes , we can use curve fitting to conclude a simple formula to estimate the reliability, which can also be used to deal with the complex attached parts created by polarizing codes. Then, when we decode the polar codes, we can use deep learning to achieve fool-type, one-step decoding. If we have decoding training matrix in advance, once we receive the data, we only have to do simple multiplication and addition, but we can achieve the same effect with traditional SC decoding. In order to achieve better decoding results, we use the deep learning to modify the parameters in BP decoding algorithm, which really makes a difference.

The rest of this paper is organized as follows: The section II introduces the basic knowledge of deep learning and polar codes and the principles to be mastered. The section III shows the application of NN learning in decoding reliability estimation and simulation results. The section IV presents the application of NN when decoding the polar codes. The section V is the summary of this paper.

2. Introduction of polar codes and deep learning

For deep learning, the idea is to stack multiple layers, that is, the output of each layer is the input of next layer. In this way, the input information can be expressed hierarchically.

When decode the polar codes, we must first distinguish the reliability of N splitted channels, that is to distinguish which channel is reliable and which channel is unreliable [4]. There are three commonly used methods for measuring the reliability of each polarized channel: the Bhattacharyya Parameter method, the DE method, and the GA method.

Arikan gave a decoding algorithm of polar codes in his paper, that is, successive cancellation (SL) decoding algorithm [5]. For polar codes, when the code length tends to infinity, the channel polarization is more complete. SC decoding algorithm is a greedy algorithm, at each layer of the tree, we only search for the optimal path then we go to the next level, so the error can not be modified. SCL decoding algorithm is to improve the SC algorithm [6].

In addition, there are spherical decoding algorithms and BP decoding algorithms [7]. In the proposed BP decoder, The reliability of the transmission messages can be enhanced by increasing the checkout node. In particular, it is possible to correct the error messages transmitted from the frozen node. The coding of the polar codes can be divided into n stages, and at each stage, the size of processing sets (PS) is $N/2$. Each PS has two input and output nodes[8].

3. Deep learning and polarization endoding

1): Build a deep learning environment. Keras is a API of high-level neural network. It is written by pure Python language and based on Tensorflow, Theano and CNTK backend. As for the specific building environment, you can refer to the machine learning web: http://keras-cn.readthedocs.io/en/latest/.

2): Calculate Bhattachayya parameter of input data. Then the transition probability of the channel $w^N : x^N \rightarrow y^N$ is:

$$w^N(y_1^N \mid x_1^N) = \prod_i^N w(y \mid x) \tag{1}$$

Bhattacharyya parameter can be expressed as :

$$Z(w) \cong \sum_{y \in Y} \sqrt{w(y \mid x) w(y \mid x)} \tag{2}$$

IOP Conf. Series: Journal of Physics: Conf. Series **1060** (2018) 012021 doi:10.1088/1742-6596/1060/1/012021

Then, we need to calculate the Bhattacharyya parameter of several sets of input data. For example, the number of input bits N is: 16,32,64,128... 2^n ($n=\log_2 N$). The Bhattacharyya parameter corresponding to input data x_1^N is sorted as ascending order and recorded as z_1^N.

3): Convert the ordered subscript from 1 to N of input x_1^N to binary number b_i (i=[1,N]). Then build a model of deep learning and enter the training data: X_train=b_i ,Y_train=z_1^N.

The number of bits of input data is N. Generally speaking, we use the training networks with three layers(2N-4N-2N). You can also continue to correct the training network during the training process.

The NND flow chart as shown in Figure 1:

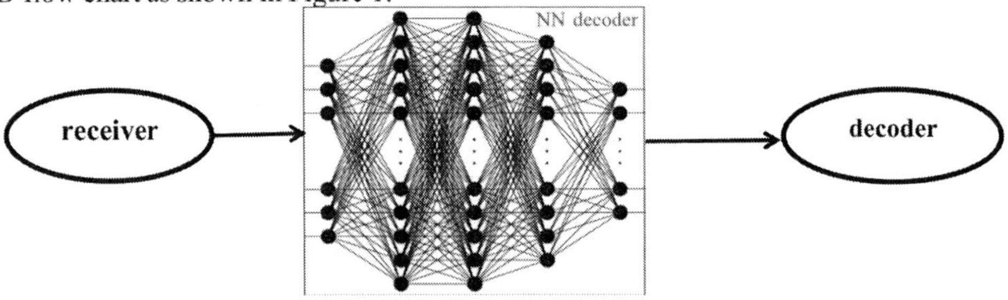

Figure 1. The NND flow chart about the NN of polar codes

By constantly adjusting the parameters corresponding to the deep learning model and training the data, and using the principle of β -expansions [9], a simple and efficient formula for estimating the reliability of polar code can be obtained:

$$k_i = \sum_{i=1}^{N} b_i \varphi^i + c \tag{3}$$

where k_1 is the estimated value of reliability, φ and c are constants and their values are not the same in different channel environments. In the Gaussian white noise channel environment, $\varphi = 1.1777$, $c = 1.1133$

4): How to use this formula.

For example, the code length of polar codes is $2^5 = 32$, and the binary expression whose subscript is 5 is 00101. A calculation of one of them is shown in formula (14). When all the estimated values are calculated, the value of the corresponding index position is sorted by descending order.

$$K_5 = \sum_{i=1}^{5} b_i(\sqrt{\phi})^i + c = 1*(\sqrt{\phi})^1 + 0*(\sqrt{\phi})^2 + 1*(\sqrt{\phi})^3 + 0*(\sqrt{\phi})^4 + 0*(\sqrt{\phi})^5 + \mathrm{c} \tag{4}$$

According to the simulation and BPE (Bhattacharyya Parameter Estimate), the comparisons are as follows Figure 2:

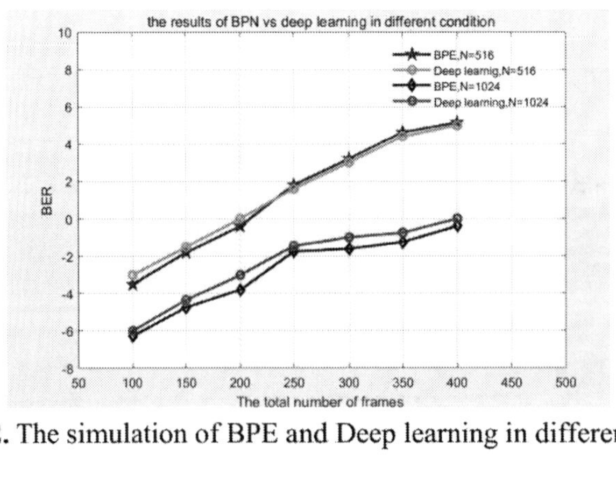

Figure 2. The simulation of BPE and Deep learning in different frames

The simulation data are shown as follows: There are two group data and their frames length are N=512 and N=1024 respectively. The total number of frames is M. The signal through the Gaussian white noise channel. The modulation method is BPSK. And the encoding efficiency of input is 0.5, in other words, the effective length of polar codes is 0.5N. We adopt successive cancellation list (SCL) to decode, where L=8. And we add the cyclic redundancy check (CRC) at the end, the length of CRC is 17. The signal-to-noise ratio (SNR) required for reliability of the BP method and deep learning method is compared under the circumstance of which the frame error rate is 0.001.

The simulation results show that: the DL (deep learning) method obtained by machine learning training and the BPE method need similar SNR under the same conditions, and the BPE is a little better than DL. But the calculation of DL is simple and easy to understand, and its calculation amount is much smaller than BPE. When the code length and bit rate are relatively large, the DL method can realize the instant implementation of encoder/decoder for low latency. But this method is not of universality. It needs to do machine learning in advance according to different channel environment and communication scenarios, and then correct the parameters involved in this method. Only when we get better parameters, this method can be comparable to the traditional algorithms.

4. Deep learning and polar code decoding

In this part, we want to use NN to decode noisy code-words. At the transmitter, k-bits information is encoded into code-words of length N. The encoded bits are modulated and transferred on the noisy channel. At the receiver, we can receive the code-words with noise. And the mission of decoder is to recover the corresponding information bits. Compared with iterative decoding, NN finds its estimated value through only one layer. It is a arduous and expensive task for deep learning to get good trained data. But using NN for channel encoding is special, because we deal with artificial signals. Therefore, we can generate as many training samples as possible. Moreover, the desired NN output, also denoted as label, is obtained for free because if noisy code-words are generated, the transmitted information bits are obviously known [10]. For simplicity, binary phase-shift keying (BPSK) modulation and additive white Gaussian noise channel (AWGN) are used. The other information of channels can be used directly, and the flexibility may be a particular advantage of NN-based decoding.

The examples of loss function are mean square error (MSE) and binary cross entropy (BCE):

$$L_{MSE} = \frac{1}{K}\sum_{i}(b_i - \hat{b}_i)^2 \quad (5)$$

$$L_{BCE} = -\frac{1}{K}\sum_{i}\left[b_i(\ln(\hat{b}_i)) + (1-b_i)\ln(1-\hat{b}_i)\right] \quad (6)$$

Where b_i is the input bit that is the encoded bit of polar codes, $b_i \in [0,1]$, $\hat{b}_i$ is the decoded bit estimated by NN, $\hat{b}_i \in [0,1]$.

This processing step can also be implemented as an additional layer without requiring any training parameters. Note that in this case, the noise variance must be known and provided as an additional input of NN. We use Keras3 as the advanced abstract front-end of keras3. It allows rapid deployment of NN from the very abstract perspective of the Python programming language, thus hiding much of the underlying complexity [11].

We train our NN decoder at so-called "epochs". At each epoch, the gradient of the loss function is calculated over the entire training set X by using Adam, a stochastic gradient descent optimization method [12]. Because the noise floor in our architecture will produce a new noise in each use, the NN decoder will never see the same input twice. For this reason, although the training set has a finite size of 2k code-words, we can train the data on a basic infinite training set by simply increasing the max training number of epoch (Mep). However, this makes it impossible to distinguish whether the the neural network has been improved through a large number of training samples or more optimization iterations.

The simulation result is shown in Figure 3. From the results, NN decoder is little better than the SC, but we need huge number of Mep which means we must spend a lot of time to train the NN. If the length of word-codes is long, we have to face the curse of dimension. So, this way of NN only fit to short word-codes. There is still a gap between the effect of direct decoding by traditional neural network and the decoding of polar codes. To achieve a better training effect, we study the combination of neural network and BP algorithm.

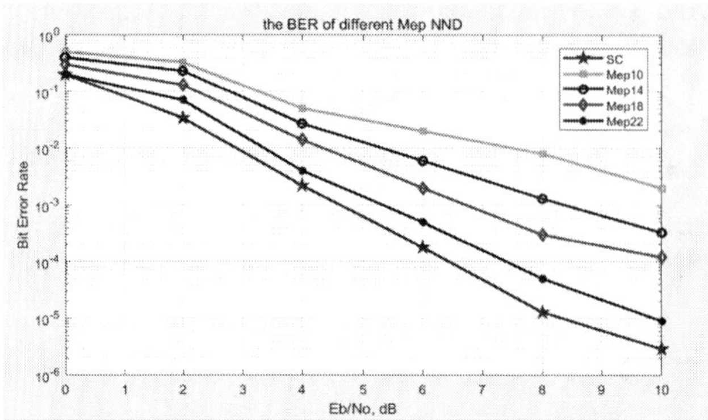

Figure 3. The simulation of deep learning on decode of polar codes with 64 bit-length codes with code rate r = 0.5 in AWGN channel

For the BP algorithm, in order to reduce the computation and make it easy to realize in hardware, we compare the LDPC decoding min-sum product algorithm (SPA). And we use the sum-min BP to train the neural network, because it is easy to realize.

The specific formulas are shown as follows (7-10) :

$$L^t_{i,j} = \alpha^t_i * sign(L^{t-1}_{i,j+1}) * sign(L^{t-1}_{i+n/2^j,j+1} + R^t_{i+n/2^j,j+1}) * \min(|L^{t-1}_{i,j+1}|, |L^{t-1}_{i+n/2^j,j+1} + R^t_{i+n/2^j,j}| + a^t_i \quad (7)$$

$$L^t_{i+n/2^j,j} = L^{t-1}_{i+n/2^j,j+1} + \beta^t_i * sign(L^{t-1}_{i,j+1}) * sign(R^t_{i,j}) * \min(|L^t_{i,j+1}|, |R^t_{i,j}| + b^t_i \quad (8)$$

$$R^t_{i,j+1} = \chi^t_i * sign(R^t_{i,j}) * sign(L^{t-1}_{i+n/2^j,j+1} + R^t_{i+n/2^j,j}) * \min(|R^t_{i,j}|, |L^{t-1}_{i+n/2^j,j+1} + R^t_{i+n/2^j,j}| + c^t_i \quad (9)$$

$$R^t_{i+n/2^j,j+1} = R^t_{i+n/2^j,j} + \delta^t_i * sign(L^{t-1}_{i,j+1}) * sign(R^t_{i,j}) * \min(|L^{t+1}_{i,j+1}|, |R^t_{i,j}| + d^t_i \quad (10)$$

where $\alpha, \beta, \chi, \delta, a, b, c, d$ are all the coefficients of each belief propagation.

Here, $L^t_{i,j}, R^t_{i,j}$ represent the messages at t-th iteration from the left-to-right and right-to-left based on log-likelihood ratio respectively. According to the above formulas, the neighbouring nodes can propagate their associated messages iteratively, and these messages can be updated by PS. After the BP decoder reaches the preset maximum iteration, we can determinate the decoding bits and output by the hard decision based on the overall LLR of the node in formula (11) :

$$\hat{u}_i = sign(LLR^{\max\_iter}_{i,1}) = sign(L^{\max\_iter}_{i,1} + R^{\max\_iter}_{i,1}) \quad (11)$$

The purpose of this paper is to use the neural network to correct the numerical value of each iteration of $\alpha, \beta, \chi, \delta, a, b, c, d$, thus, by making the BP algorithm in a specific channel environment achieve a better effect.

The specific calculation process is as follows:

1: Calculate the maximum log-likelihood ratio $LLR(r_i)$ of the signal received through channel.

2. Prepare some parameters. The frozen bits in the reliability result calculated by the neural network, are recorded as $frz = \{frz_1 \;\; frz_2 \;\; \cdots \;\; frz_{N-k}\}$. $\alpha, \beta, \chi, \delta$ are all initialized to 1, a, b, c, d are all initialized to 0. And we define the maximum number of iterations as T_{iter}.

3. Start BP iteration algorithm:

$$L^0_{i,j} = 0, R^0_{i,j} = 0 \quad (12)$$

$$\begin{cases} if:(j==1)\&(i\in frz)\rightarrow R_{i,j}^{t}=\infty \\ else: when(j==1+m)\rightarrow L_{i,m+1}^{t}=LLR(r_i) \end{cases} \quad (13)$$

$$\begin{cases} if(L_{i,m+1}^{t}+R_{i,m+1}^{t}\geq 0)\rightarrow \hat{x}_i=0, else\rightarrow \hat{x}_i=1 \\ if(L_{i,1}^{t}+R_{i,1}^{t}\geq 0)\rightarrow \hat{u}=0, else\rightarrow \hat{u}=1 \end{cases} \quad (14)$$

When the number of iterations is smaller than the maximum, we update the values of $L_{i,j}^{t}, R_{i,j}^{t}, \hat{u}, \hat{x}$ constantly.

It is time to stop iteration while the matrix meets the conditions: $\hat{u}G == \hat{x}$,where G is the generated matrix of polar code.

4. When it meets the conditions: $\hat{u}G == \hat{x}$, we record the values of $L_{i,j}^{t_s}, R_{i,j}^{t_s}, t_s$ is the iteration value at this time.

We use encoded u to derive the corresponding results: $u\rightarrow \overline{R}_{i,j}^{t_s}, \overline{L}_{i,j}^{t_s}$. $(\overline{L}_{i,j}^{t_s}, \overline{R}_{i,j}^{t_s}), (L_{i,j}^{t_s}, R_{i,j}^{t_s})$ can correct the value of $\alpha_s, \beta_s, \chi_s, \delta_s, a_s, b_s, c_s, d_s$ through the neural network training, where $t=t_s-1$. Keep training, when $t=0$, stop training.

Record the corrected value of $\alpha, \beta, \chi, \delta, a, b, c, d$ and decode. When $\hat{u}G==\hat{x}$, the decoding is finished. At this time, the whole decoding process has been finished.

The complete algorithm flow chart is as follows in Figure 4.

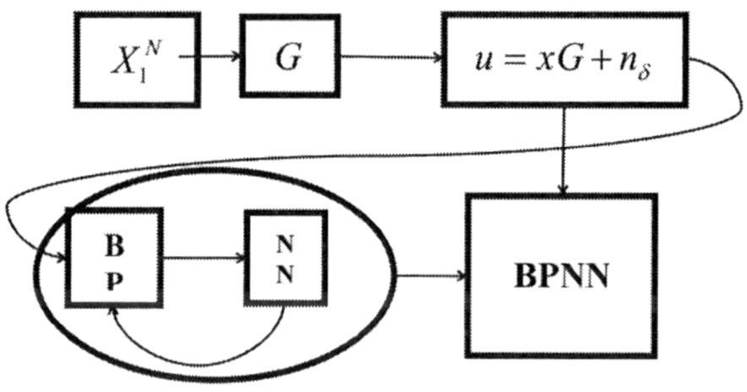

Figure 4. The complete algorithm flow chart of BPNN.

In this chart, G is generator matrix. δ_n is the channel condition. $F=\begin{bmatrix}1 & 0\\ 1 & 1\end{bmatrix}$, $\otimes$ represents the Kronecker product [13].

The result of simulations is depicted in Figure 5:

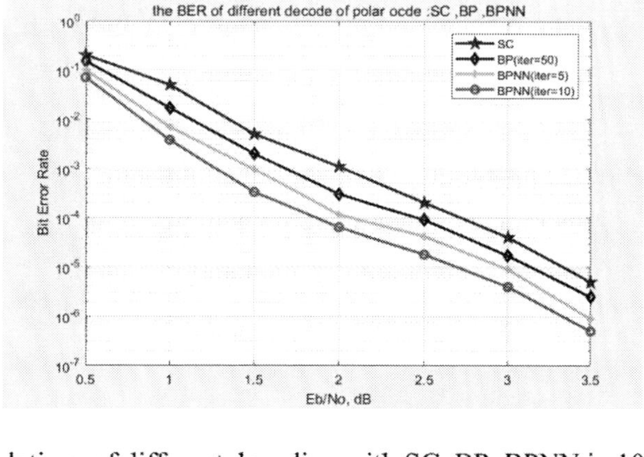

Figure 5. The simulations of different decoding with SC ,BP ,BPNN in 1024 bits of a frame (10000 frames) and r is 0.5, the modulation mode is BPSK, and the channel is AWGN.

By comparing the BP decoding with the BPNN (BP+NN) which is the BP through the neural network, we find that BPNN is better than BP especially at the larger SNR. The BPNN training work can be done well in advance, so the decoding of polar codes in BPNN decoding only needs a few times of iteration, even one iteration can realize decoding. It can reduce a lot of computation and lower time delay, and make the hardware implementation be easy at the meantime. BPNN also makes the encoding and decoding of polar codes be realized in one step, so that the polar codes are easy to implement in hardware and has ultra-small time-delay. So it is very suitable for 5G scene requirements.

5. Conclusion

Through the study of the neural network effects on polar codes' encoding and decoding, this paper has the following results. At the time of encoding, we obtain a simple and efficient estimation method with low latency by training the reliability of polar code. This method has no complicated calculation and is easy to understand and comparable to GE. When decoding, through the direct neural network decoding we can decode in one step. And this decoding has the same effect with SC decoding. When the requirement is not high or the channel environment is better, we can use the NN decoding. If the better bit error rate is needed, we can use BP+NN to achieve the BP decoding with one iteration by correcting the BP parameters by constant training. In conclusion, BPNN is a very suitable method. And the polar codes' encoding and decoding will be simple and practical after neural network training. It can implement by only one fool-style operation, and it can realize the high efficiency and low latency in hardware. This paper is based on special environment, and different environment needs neural network training. And with the length of code-words becomes larger and larger, the neural network becomes more and more difficult and it needs continuous study.

References

[1] W. S. McCulloch and W. Pitts, "A logical calculus of the ideas immanent in nervous activity,"The bulletin of mathematical biophysics, vol. 5, no. 4, pp. 115-133, 1943.

[2] E. Arikan, "Channel polarization: A method for constructing capacityachieving codes for symmetric binary-input memoryless channels," IEEE Trans. Inf. Theory, vol. 55, no. 7, pp. 3051-3073, Jul. 2009.

[3] I. Tal and A. Vardy, "How to construct polar codes," IEEE Trans. Inf.Theory, vol. 59, no. 10, pp. 6562-6582, 2013.

[4] N. Hussami, S. Korada, and R. Urbanke, "Performance of polar codes for channel and source coding," in Proc. IEEE ISIT, 2009, pp. 1488-1492.

[5] I. Tal and A. Vardy, "List decoding of polar codes," in Proc. IEEE Int.Symp. Inf. Theory, 2011, pp. 1-5.

[6] P. Trifonov, "Efficient Design and Decoding of Polar Codes," IEEE Transactions on Communications, vol. 60, no. 11, pp. 3221-3227,
November 2012

[7] Vikalo, H, and B. Hassibi. "On the sphere-decoding algorithm II. Generalizations, second-order statistics, and applications to communications." IEEE Transactions on Signal Processing 53.8(2005):2819-2834.

[8] B. Yuan and K. K. Parhi, "Architecture optimizations for BP polar decoders," in Proc. 38th IEEE ICASSP, 2013, pp. 2654-2658.

[9] Parry, W., "On the β-expansions of real numbers", Acta Mathematica Academiae Scientiarum Hungaricae, 11: 401-416, 1960.

[10] Gruber T, Cammerer S, Hoydis J, et al. On Deep Learning-Based Channel Decoding[J]. 2017.

[11] I. Goodfellow, Y. Bengio, and A. Courville, "Deep Learning," 2016, book in preparation for MIT Press. [Online]. Available: http://www.deeplearningbook.org

[12] D. P. Kingma and J. Ba, "Adam: A method for stochastic optimization," CoRR, 2014. [Online]. Available: http://arxiv.org/abs/1412.6980

[13] N. Hussami,S. B. Korada, and R. Urbanke, "Performance of polar codes for channel and source coding," in Proc. 2009 IEEE Int. Symp. Inform. Theory, (Seul, Korea), p. 1488-1492, 2009.

Efficient Method for Frozen Bits Encoding and Decoding of Polar Code

Yan-Haw Chen[1], Min-Yen Tsai[2], J.J. Wang[3], Zhe-Wei Kang[3] and Jian-Ting Pan[3]

[1]Dept. of Computer Science and Information Engineering, I-Shou University
[2]Dept. of Computer Science and Information Engineering, I-Shou University
[3]Dept. of Computer Science and Information Engineering, I-Shou University

The corresponding author's e-mail: wsx102218@gmail.com

Abstract. An efficient encoding and decoding of the Polar code utilizing algebraic method to find error patterns is presented in this study. The key idea behind the proposed decoding method is theoretically based on the existence of a one-to-one mapping between the frozen bits and correctable error patterns. This method would help reduce the decoding time for finding error patterns when decoding the (N=8, K=4) and (N=16, K=8) Polar Code. Furthermore, it would reduce the encoding by 91% with lookup table for G_3 and G_4 matrix and the encoding method can be used to the decoding of Polar Code with algebraic method. Ultimately, the proposed decoding algorithm for Polar codes can be made regular, simple, and suitable for software implementations.

1. Introduction

The encoding of Polar Code is a linear block coding method recently confirmed by Arıkan [1], which can not only correct the error but approach the Shannon-Hartley theorem. This caused the researchers' interest, the decoding of this code architecture is not complicated, the reliability of the data in the performance of error correction is excellent. Due to the complexity of encoding and decoding, it grows logarithmically with the increase of code length. If using Successive Cancelation (SC) algorithm [2-4], the decoding complexity increases little. However, the SC decoding algorithm used in the long code decoding process, its error correction performance is worse than Turbo code and LDPC code. Recent research literature based on the lack of SC decoding algorithm for method improving [5-9]. For example, performance of algorithms such as List SC (SCL), Stack SC (SSC), Cyclic Redundancy Checker-assisted SCL (CRC-SCL). These methods are also implemented as hardware circuits [10], and these decoding technical documents only focus on the soft decision mode to discuss the error correction performance. Since the standard form of Polar Code is Non-systematic code [1], and in [11] it is mentioned that systematic code has a lower bit error rate (BER) than non-systematic codes. Therefore, the coding method in this paper, use the method of look-up table and algebra scheme to correct errors. The error correction algorithm of Polar Code will use the characteristics presented by the frozen bit. Since the frozen bit must be zero after decoding without error, a simple error correction rule can be established by using this feature and solving the data errors that may occur in real communications.

2. Experiment Method

There will be divided into three parts to illustrate: generation matrix, encoding, decoding and error correction.

2.1. Generation Matrix

The definition of a generation matrix is as follows:

$$G = \begin{bmatrix} 1 & 1 \\ 0 & 1 \end{bmatrix} \tag{1}$$

The definition of expansion matrix of generation matrix is as follows, using Kronecker product to produce matrix of encoding and decoding:

$$G^{\otimes k} = \begin{bmatrix} 1 * G^{\otimes k-1} & 1 * G^{\otimes k-1} \\ 0 * G^{\otimes k-1} & 1 * G^{\otimes k-1} \end{bmatrix}, \mathrm{k} \geq 2, \mathrm{k} \in Z^{+} \tag{2}$$

In this paper, it will use $G^{\otimes 3}$ and $G^{\otimes 4}$:

$$G^{\otimes 3} = \begin{bmatrix} 1 * G^{\otimes 2} & 1 * G^{\otimes 2} \\ 0 * G^{\otimes 2} & 1 * G^{\otimes 2} \end{bmatrix} = \begin{bmatrix} 1 & 1 & 1 & 1 & 1 & 1 & 1 & 1 \\ 0 & 1 & 0 & 1 & 0 & 1 & 0 & 1 \\ 0 & 0 & 1 & 1 & 0 & 0 & 1 & 1 \\ 0 & 0 & 0 & 1 & 0 & 0 & 0 & 1 \\ 0 & 0 & 0 & 0 & 1 & 1 & 1 & 1 \\ 0 & 0 & 0 & 0 & 0 & 1 & 0 & 1 \\ 0 & 0 & 0 & 0 & 0 & 0 & 1 & 1 \\ 0 & 0 & 0 & 0 & 0 & 0 & 0 & 1 \end{bmatrix} \tag{3}$$

$$G^{\otimes 4} = \begin{bmatrix} 1 * G^{\otimes 3} & 1 * G^{\otimes 3} \\ 0 * G^{\otimes 3} & 1 * G^{\otimes 3} \end{bmatrix} \tag{4}$$

Since the generation matrix of the construction of Polar Code is related to the generation of Reed-Muller (RM) code, the generator matrix has the characteristics of the upper triangular, which also means that the encoding matrix is inverse matrix.

2.2. Encoding

The vector of 8 bits original data is as follows:

$$D = (d_0, d_1, d_2, d_3, d_4, d_5, d_6, d_7) \tag{5}$$

The vector of 8 bits frozen bits and its elements must be zero is as follows:

$$F = (f_0, f_1, f_2, f_3, f_4, f_5, f_6, f_7) = (0,0,0,0,0,0,0,0) \tag{6}$$

The vector of 16 bits data with D and F is as follows:

$$\begin{aligned} M &= (f_0, f_1, f_2, d_0, f_3, f_4, d_1, f_5, f_6, d_2, d_3, d_4, f_7, d_5, d_6, d_7) \\ &= (m_0, m_1, m_2, m_3, m_4, m_5, m_6, m_7, m_8, m_9, m_{10}, m_{11}, m_{12}, m_{13}, m_{14}, m_{15}) \end{aligned} \tag{7}$$

2.2.1. Original Method. The original encoding and decoding method is to use matrix operations to get the result. The operation is as follows:

$$\begin{aligned} M' &= G^{\otimes 4} M^T \\ &= (m'_0, m'_1, m'_2, m'_3, m'_4, m'_5, m'_6, m'_7, m'_8, m'_9, m'_{10}, m'_{11}, m'_{12}, m'_{13}, m'_{14}, m'_{15})^T \end{aligned} \tag{8}$$

This will take 256 times AND operations and 240 times XOR operations. The lower triangular part of $G^{\otimes 4}$ does not need to be calculated, so only 136 AND operations and 120 XOR operations are needed actually.

2.2.2. Look-up Table Method. Before discuss about the look-up table method of encoding and decoding of Polar Code, it need to establish the table T by the matrix multiplication of $G^{\otimes 3}$ and $X_j = (x_0, x_1, x_2, x_3, x_4, x_5, x_6, x_7)$. X_j is a binary combination of 0 to 255. X' is the result of $G^{\otimes 3}X$.

$$\begin{aligned} X' &= G^{\otimes 3} X \\ T[j] &= \bigoplus_{i=0}^{7} x_i, 0 \leqq j \leqq 255 \end{aligned} \tag{9}$$

Splitting M into two parts M_0, M_1:

$$M_0 = (m_0, m_1, m_2, m_3, m_4, m_5, m_6, m_7) \tag{10}$$

$$M_1 = (m_8\,, m_9\,, m_{10}\,, m_{11}\,, m_{12}\,, m_{13}\,, m_{14}\,, m_{15})$$

and forming two new 8*8 matrix $P_0\,, P_1$ taken from the parts of $G^{\otimes 4}$, in fact, it is the result of Kronecker product of $1 * G^{\otimes 3}$ ($G^{\otimes 3}$Matrix) and $0 * G^{\otimes 3}$ (Zero Matrix):

$$P_0 = \begin{bmatrix} 1 & 1 & 1 & 1 & 1 & 1 & 1 & 1 \\ 0 & 1 & 0 & 1 & 0 & 1 & 0 & 1 \\ 0 & 0 & 1 & 1 & 0 & 0 & 1 & 1 \\ 0 & 0 & 0 & 1 & 0 & 0 & 0 & 1 \\ 0 & 0 & 0 & 0 & 1 & 1 & 1 & 1 \\ 0 & 0 & 0 & 0 & 0 & 1 & 0 & 1 \\ 0 & 0 & 0 & 0 & 0 & 0 & 1 & 1 \\ 0 & 0 & 0 & 0 & 0 & 0 & 0 & 1 \end{bmatrix}, P_1 = \begin{bmatrix} 0 & 0 & 0 & 0 & 0 & 0 & 0 & 0 \\ 0 & 0 & 0 & 0 & 0 & 0 & 0 & 0 \\ 0 & 0 & 0 & 0 & 0 & 0 & 0 & 0 \\ 0 & 0 & 0 & 0 & 0 & 0 & 0 & 0 \\ 0 & 0 & 0 & 0 & 0 & 0 & 0 & 0 \\ 0 & 0 & 0 & 0 & 0 & 0 & 0 & 0 \\ 0 & 0 & 0 & 0 & 0 & 0 & 0 & 0 \\ 0 & 0 & 0 & 0 & 0 & 0 & 0 & 0 \end{bmatrix} \tag{11}$$

The original the encoding and the decoding formula of Polar code are as follows:

$$M' = (m'_0\,, m'_1\,, m'_2\,, m'_3\,, m'_4\,, m'_5\,, m'_6\,, m'_7\,, m'_8\,, m'_9\,, m'_{10}\,, m'_{11}\,, m'_{12}\,, m'_{13}\,, m'_{14}\,, m'_{15})^T \tag{12}$$

The above M' matrix can be split into four sub-matrixes as follows:

$$\begin{aligned} M'_A &= P_0 {M_0}^T = (m_{A0}\,, m_{A1}\,, m_{A2}\,, m_{A3}\,, m_{A4}\,, m_{A5}\,, m_{A6}\,, m_{A7})^T \\ M'_B &= P_0 {M_1}^T = (m_{B0}\,, m_{B1}\,, m_{B2}\,, m_{B3}\,, m_{B4}\,, m_{B5}\,, m_{B6}\,, m_{B7})^T \\ M'_C &= P_1 {M_0}^T = (m_{C0}\,, m_{C1}\,, m_{C2}\,, m_{C3}\,, m_{C4}\,, m_{C5}\,, m_{C6}\,, m_{C7})^T \\ M'_D &= P_0 {M_1}^T = (m_{D0}\,, m_{D1}\,, m_{D2}\,, m_{D3}\,, m_{D4}\,, m_{D5}\,, m_{D6}\,, m_{D7})^T \end{aligned} \tag{13}$$

Each row of P_{0d}, P_{1d} can be presented array $P_{0d}[], P_{1d}[]$ with $(255\,, 85\,, 51\,, 17\,, 15\,, 5\,, 3\,, 1)$ and $(0\,, 0\,, 0\,, 0\,, 0\,, 0\,, 0\,, 0)$, respectively.

$M_0\,, M_1$ is presented value m_{0-7}, m_{8-15} in decimal. And then using the table and look-up table mentioned earlier, The result of m_{C0-7} must be zero, and the result of m_{D0-7} is same as m_{B0-7}. So neither of these two parts need to be calculated:

$$\begin{array}{lll} m_{A0} = T[P_{0d}[0]\,\&\,M_0] & m_{B0} = T[P_{0d}[0]\,\&\,M_1] = m_{D0} & m_{C0} = T[P_{1d}[0]\,\&\,M_0] \\ m_{A1} = T[P_{0d}[1]\,\&\,M_0] & m_{B1} = T[P_{0d}[1]\,\&\,M_1] = m_{D0} & m_{C1} = T[P_{1d}[1]\,\&\,M_0] \\ m_{A2} = T[P_{0d}[2]\,\&\,M_0] & m_{B2} = T[P_{0d}[2]\,\&\,M_1] = m_{D0} & m_{C2} = T[P_{1d}[2]\,\&\,M_0] \\ m_{A3} = T[P_{0d}[3]\,\&\,M_0] & m_{B3} = T[P_{0d}[3]\,\&\,M_1] = m_{D0} & m_{C3} = T[P_{1d}[3]\,\&\,M_0] \\ m_{A4} = T[P_{0d}[4]\,\&\,M_0] & m_{B4} = T[P_{0d}[4]\,\&\,M_1] = m_{D0} & m_{C4} = T[P_{1d}[4]\,\&\,M_0] \\ m_{A5} = T[P_{0d}[5]\,\&\,M_0] & m_{B5} = T[P_{0d}[5]\,\&\,M_1] = m_{D0} & m_{C5} = T[P_{1d}[5]\,\&\,M_0] \\ m_{A6} = T[P_{0d}[6]\,\&\,M_0] & m_{B6} = T[P_{0d}[6]\,\&\,M_1] = m_{D0} & m_{C6} = T[P_{1d}[6]\,\&\,M_0] \\ m_{A7} = T[P_{0d}[7]\,\&\,M_0] & m_{B7} = T[P_{0d}[7]\,\&\,M_1] = m_{D0} & m_{C7} = T[P_{1d}[7]\,\&\,M_0] \end{array} \tag{14}$$

The subsequent operation is completed as follows:

$$\begin{array}{llll} m'_0 = m_{A0} \oplus m_{B0} & m'_1 = m_{A1} \oplus m_{B1} & m'_2 = m_{A2} \oplus m_{B2} & m'_3 = m_{A3} \oplus m_{B3} \\ m'_4 = m_{A4} \oplus m_{B4} & m'_5 = m_{A5} \oplus m_{B5} & m'_6 = m_{A6} \oplus m_{B6} & m'_7 = m_{A7} \oplus m_{B7} \\ m'_8 = m_{B0} & m'_9 = m_{B1} & m'_{10} = m_{B2} & m'_{11} = m_{B3} \\ m'_{12} = m_{B4} & m'_{13} = m_{B5} & m'_{14} = m_{B6} & m'_{15} = m_{B7} \end{array} \tag{15}$$

The look-up table method will take 16 times AND operations and 8 times XOR operations. In addition to 16 times the look-up table.

2.3. Decoding and Error Correction

Decoding and encoding use the same matrix for multiplication, so the same can be through the look-up table solution. Similarly, each bit must also be calculated as a decimal number in order to look up the table. Assumed that the encoded M' codeword is transmitted with a single bit error or not with error, and then the decoded codeword will be presented by $M'_e = M' + e_i$, i represents the location into which a single-bit error is injected.

$$\begin{gathered} M' \xrightarrow{\text{inject a single bit error}} M'_e \\ M'_0 = (m'_0\,, m'_1\,, m'_2\,, m'_3\,, m'_4\,, m'_5\,, m'_6\,, m'_7) \\ M'_1 = (m'_8\,, m'_9\,, m'_{10}\,, m'_{11}\,, m'_{12}\,, m'_{13}\,, m'_{14}\,, m'_{15}) \end{gathered} \tag{16}$$

Through program verification of all codewords generated when a single bit error is injected, one-to-one property of frozen bit and single bit error can be found. Before correcting error, it need to decode M_e.

$$M_{eA} = P_0 M_0'^T = (m_{e0}, m_{e1}, m_{e2}, m_{e3}, m_{e4}, m_{e5}, m_{e6}, m_{e7})^T$$
$$M_{eB} = P_0 M_1'^T = (m_{e8}, m_{e9}, m_{e10}, m_{e11}, m_{e12}, m_{e13}, m_{e14}, m_{e15})^T \quad (17)$$

Where P_0 is same as $1 * G^{\otimes 3}$ ($G^{\otimes 3}$Matrix).

Next, M'_{eA} is obtained the same procedure, the table and rules of a single bit error correction is as follows:

(1) If $m_{e0} = 1$ Then codeword has a single bit error and it can find the error location through the following Table 1..
Else If All frozen bits $= 0$ Then codeword has no error.
Else $m_{e0} = 0$, but other frozen bits $\neq 0$ Then codeword has least two bits error.

(2) Table 1. Frozen bit for a single bit error table. Through a specific frozen bit position, this table can tell which bit was wrong in encoding.

Table 1. Frozen bit for a single bit error table.

Frozen location Error location	m_{e0}	m_{e1}	m_{e2}	m_{e4}	m_{e8}	**Frozen location Error location**	m_{e0}	m_{e1}	m_{e2}	m_{e4}	m_{e8}
m'_0	1	1	1	1	1	m'_8	1	1	1	1	0
m'_1	1	0	1	1	1	m'_9	1	0	1	1	0
m'_2	1	1	0	1	1	m'_{10}	1	1	0	1	0
m'_3	1	0	0	1	1	m'_{11}	1	0	0	1	0
m'_4	1	1	1	0	1	m'_{12}	1	1	1	0	0
m'_5	1	0	1	0	1	m'_{13}	1	0	1	0	0
m'_6	1	1	0	0	1	m'_{14}	1	1	0	0	0
m'_7	1	0	0	0	1	m'_{15}	1	0	0	0	0

However, it is not worthwhile to handle only one bit error. Therefore, this study aims at the adjustment of data bit and frozen bit, and proposes a modified look-up table method.

2.4. Modified Look-up Table Method

Modifying data vector D ,frozen bit vector F and 16 bits data vector M:

$$D = (d_0, d_1, d_2, d_3, d_4, d_5)$$
$$F = (f_0, f_1, f_2, f_3, f_4, f_5, f_6, f_7, f_8, f_9) = (0,0,0,0,0,0,0,0,0,0)$$
$$M = (f_0, f_1, f_2, d_0, f_3, f_4, d_1, f_5, f_6, f_7, f_8, d_4, f_9, d_5, d_6, d_7) \quad (18)$$
$$= (m_0, m_1, m_2, m_3, m_4, m_5, m_6, m_7, m_8, m_9, m_{10}, m_{11}, m_{12}, m_{13}, m_{14}, m_{15})$$

The frozen bit of 2 bits error frozen bit is calculated as an integer number F_Z, it will be used as an index to establish the table T_f, and correspond the error correction value.

$$T_f[Fz] = (error\ correction\ value) \quad (19)$$

When the Fz value is 1, 2, 3, 9, 18, 27, 65, 73, 97, 130, 146, 195, 219, there are a problem of the same frozen bits at different 2-bit error position. Therefore, if the value of Fz has the value mentioned above, it will need to check 2 times.

3. Result and Conclusion

3.1. Original method v.s. Look-up table

Table 2. Origin method v.s. Look-up table.

	Origin method	Look-up table
AND	272 times	53 times
XOR	240 times	16 times
OR	-	32 times
Shift	-	37 times
Look-up	-	32 times

The results from the previous mathematical operation are shown in Table 2. The look-up table method saves a lot of AND and XOR operations, which is still cost-effective to the original matrix operation coding, although it needs to be built before use. This will be conducive to the current Internet of Things (IoT) trends and needs.

3.2. Error Correction and Detection

It is possible to correct a 1-bit error and detect at least two bit errors by frozen bit. About error correction rules can refer to *2.3.* Section. If the decoding needs to correct two bits error, it will to decrease k/n rate.

References

[1] E. Arıkan, "Channel combining and splitting for cutoff rate improvement," *IEEE Trans. Inform. Theory*, vol. IT-52, pp. 628–639, Feb. 2006.
[2] E. Arıkan, "Channel polarization: A method for constructing capacity achieving codes for symmetric binary-input memoryless channels," *IEEE Trans. Inform. Theory*, vol. 55, pp. 3051–3073, 2009.
[3] E. Arıkan, "Source polarization," in Proc. IEEE Inform. Theory, Austin, TX, USA, June 2010.
[4] E. Arıkan and E. Telatar, "On the rate of channel polarization," in Proc. IEEE Symp. Inform. Theory, Seoul, South Korea, pp. 1493–1495, 2009.
[5] N. Hussami, S. B. Korada, and R. Urbanke, "Performance of polar codes for channel and source coding," in Proc. IEEE Int. Symp. Inform. Theory, pp. 1488–1492, June 2009.
[6] S. B. Korada, "Polar codes for channel and source coding," Ph.D. thesis, EPFL, Lausanne, 2009.
[7] E. Abbe and E. Telatar, "MAC polar codes and matroids," in Proc. Inform. Theory and App. Workshop, pp. 1-8, Jan. 2010.
[8] E. Hof, I. Sason, and S. Shamai, "Polar coding for reliable communications over parallel channels," in Proc. IEEE Inform. Theory Workshop, Aug. 2010.
[9] H. Mahdavifar and A. Vardy, "Achieving the secrecy capacity of wiretap channels using polar codes," *IEEE Trans. Inform. Theory*, vol. 57, no. 10, pp.6428-6443, October 2011..
[10] C. Leroux, I. Tal, A. Vardy, and W. J. Gross, "Hardware architectures for successive cancellation decoding of polar codes," ICASSP 2011, May 2011.
[11] Erdal Arıkan, "Systematic Polar Coding," *IEEE Commun. Lett.*, vol. 15, no. 8, pp.860-862, August 2011.

BDAP: A Big Data Analysis Platform Based On Spark

Zhixiang WANG [1,a], Yao Bu[1], Demeng BAI [2], Bin WU[1,b] and Jiafeng QIN[2]

[1] Key Laboratory of Intelligent Telecommunications Software and Multimedia, Beijing University of Posts and Telecommunications Beijing, Beijing, 100876;
[2] State Grid ShanDong Electric Power Research Institute, Jinan, 250000

Email: [a]Wangzhixiang513@gmail.com, [b]wubin@bupt.edu.cn

Abstract. Big data analysis platform consists of a serious of application and technologies for gathering, cleaning, storing, and analysing, which contribute to manage large-scale information, get analysis result and make decision precisely. However, traditional data analysis platforms meet new challenges as both continued growth in both scale and complexity in real data, existing data analysis platforms rarely have comprehensive functions for users. In response to these challenges, the BDAP (Big Data Analysis Platform) is proposed. BDAP supports large-scale data storage, parallel ETL process, data mining, statistical analysis, text mining and power grid data processing which are based on Spark. This demo introduces two application scenarios, transformer power prediction and video classification are presented. Experimental results show that BDAP is efficient and effective for data mining.

1. Introduction

Data analysis platform consists of a serious of application programs and technologies for gathering, cleaning, storing, analysing and providing access to data, which contribute to manage large-scale information, get analysis result and make decision precisely. Big data analysis platform research has attracted more and more attention, the application of Hadoop and other big data processing technology are introduced in paper [1]. From the perspective of open source tools, both MapReduce (based on mahout project) and Spark (based on the MLlib) are focused on the realization of the algorithm implemented parallel. However, they are all single algorithm package, not integrated as one.

Current big data analysis platforms are each have different emphasis. Datix [2] implements an extensible network analysis platform, which proposes intelligent pre-partition storage scheme to lessen the time for pre-processing such as filtering and querying. Tsinghua cloud set up Kloud [3] platform to achieve the table auxiliary index design, which improve the table process efficiency. Big-Provision [4] provides a platform for comparing the accuracy and performance of the same algorithm in different frameworks. Relatively enterprise projects have PDMiner [5] and BC-PDM [6]. These integrations of some basic data mining pre-processing and algorithms, but these are not comprehensive enough, and the correlation between components is too strong to modify easily.

In response to these challenges, the BDAP is proposed which based on Spark. BDAP provides not only multiple data pre-processing and data mining algorithms, but also individual business-specific components. Platform framework also has been significantly improved, the combination of OSGI and Tomcat can build low coupling and scalable workflow engine easily.

Technical background of BDAP includes Hadoop [7] and Spark [8]. HDFS (Hadoop Distributed File System) is a distributed file system based on Hadoop framework. Hadoop includes a variety of tools including HBase [9], Pig, Hive, and Sqoop, which cover distributed databases, distributed data warehouses, and data transfer. Apache Hadoop YARN [10] is a new MapReduce [11] framework refactored by Hadoop, which can be called as new MapReduce. Alluxio is a high performance, fault-tolerant, memory-based open source distributed storage system which is compatible with features of Hadoop MapReduce and Apache Spark. Workflow engine [12] can determine the information transmission routes by the role of the division of labour and different conditions, content levels and other core solutions. Workflow engine includes the process of node management, flow management, process management and other important features. JBPM, Activiti and Shark are mature workflow engines. Storm is a more mature real-time processing framework now, which always coordinates with Kafka.

The rest of this paper is organized as follows: The architecture and the features of BDAP are described in Section 2. Section 3 present the experimental results. Section 4 introduces the application scenarios of this platform. Finally, we conclude and present directions of future work in Section 5.

2. Platform Overview

The overview architecture of BDAP is presented in Figure 1. The platform consist of four main layers and feature of each layer is described as followed:

- Cloud Platform Layer provides cloud computing environment and data storage.
- Algorithm Layer provides the core data processing components of this platform, including ETL, data statistical analysis and some commonly used data mining algorithms.
- Service Layer provides operational functions such as operating flow engine, scheduling engine, progress monitoring, system management, data visualization analysis and so on.
- Interface Layer is based on the web application of the user terminal; it provides good experience for the user to set the data processing flow or the other service. As shown in Figure 2.

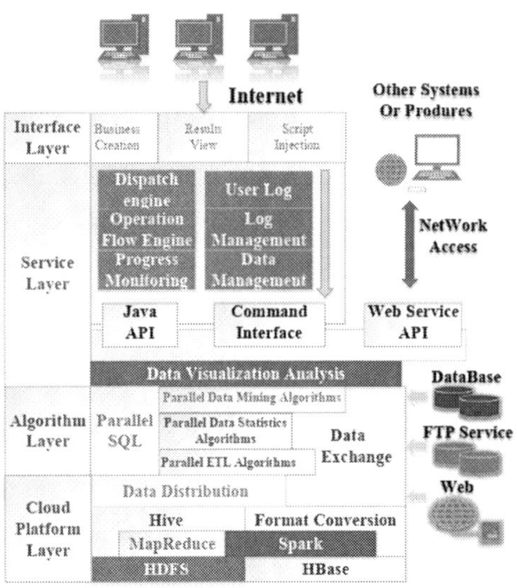

Figure 1. BDAP architecture

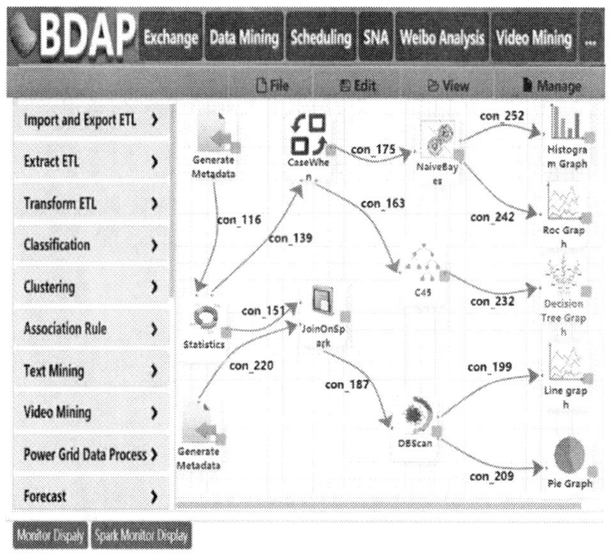

Figure 1. BDAP architecture **Figure 2.** User Interface of BDAP

And BDAP has also these following features:

2.1. Fully Functions

BDAP can handle various data including structured, semi-structured and unstructured text data. Compared with MLlib, as shown in Table 1, it's clear that BDAP basically covers functions of MLlib.

IOP Conf. Series: Journal of Physics: Conf. Series **1060** (2018) 012023 doi:10.1088/1742-6596/1060/1/012023

Table 1. Comparison of algorithms for MLib and BDAP

	MLlib	BDAP
Classification algorithm	Text Bayesian/ decision tree/ neural network/ SVM/ linear regression/logistic regression	Text Bayesian/naive Bayes/SVM/ linear regression/logistic regression/C45/CART/ CHAIID/neural network/automatic classifier
Clustering algorithm	Kmeans/LDA	Kmeans/LDA/DBScan/Brich/Clara/Automatic clustering
Association rules	FpGrowth	Apriori/FpGrowth/Time series association rules/ Association rules based on information entropy

2.2. Low Coupling

In BDAP, functions are highly modular. Import, pre-processing operation, various algorithms and results-presentation are all integrated into independent components.

Workflow and scheduling flow is one of the features of BDAP. Workflow is manually set up by users, it is connected by some processing components and directional arrows, which are dragged by the user to the "canvas". As shown in Figure 3, the operation of the workflow is: Import Metadata → Data Type Checking → Field Type Conversion → where → SVM → Roc.

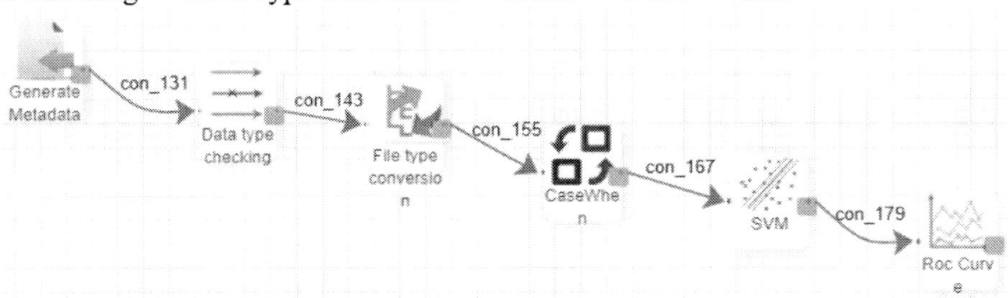

Figure 3. Workflow presentation

Workflow engine is the core module of the system batch processing, mainly because of the OSGI. OSGI will be integrated with Tomcat to call the service, BDAP use OSGI to achieve service management. OSGI-based service is a scalable and flexible plug-in architecture, every algorithm component is independent and has good scalability and low coupling.

2.3. Visualization

Visualization of the platform can improve the user experience, users can participate in changing parameters, processes, monitoring and results presentation.

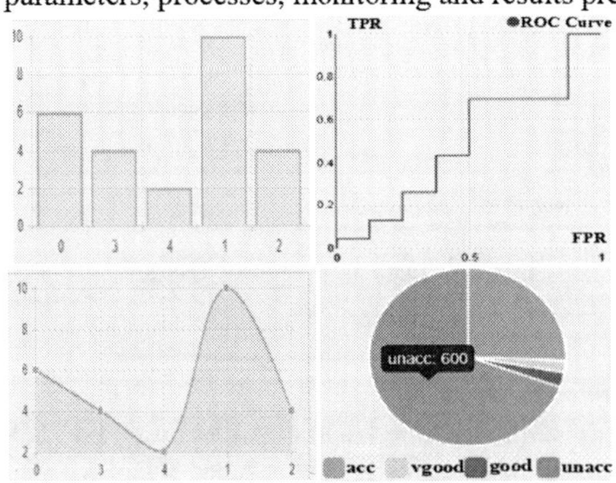

Figure 4. Results

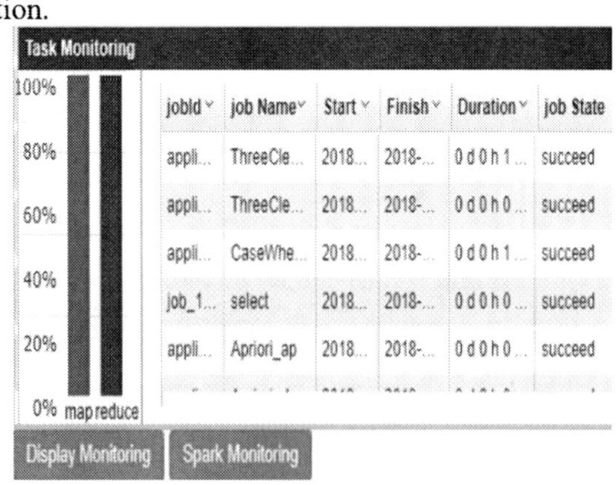

Figure 5. Monitoring

IOP Conf. Series: Journal of Physics: Conf. Series **1060** (2018) 012023 doi:10.1088/1742-6596/1060/1/012023

Results Presentation: The output components enable the visualization of the results presentation. As shown in Figure 4, BDAP currently supported output results are text output, pie chart, histogram, line graph, Roc curve, table and decision tree.

Monitoring: A workflow starts running with a monitoring window, including job name, start time, end time, duration time and job status. The performance of algorithms is intuitively shown as Figure 5.

3. Performance

Experimental environment introduction: The Spark cluster consists of 31 slave nodes and 1 master node. The server is dell R720. CPU is Intel(R) Xeon(R) CPU E5-2620 v4 @ 2.10GHz. Each machine has 2 processors;64G running memory. The operating system is 64-bit centos6.5. JDK version is jdk1.7. Parallel computing framework is hadoop 2.6.0 and Spark 1.5.1.

3.1. ETL and Hive

In order to test the large-scale data, this paper uses the randomly generated 25-dimensional data sets sized as 100M, 1G, 2G, 5G respectively.

The SQL statements required in OperGenColumn component are as followed:

- alter table t1 add columns (c string); update t1 set column26 = column1 + column2;

The SQL statements required in the Attribute exchange component are as followed:

- insert into newt1 select column2, column1, column3, column4... column25, column26, from t1;

It is obvious that even a simple function of the components requires a number of SQL statements. Hive maps structured data into a table and it is SQL statement that completes the steps. So more complex functions need more SQL statements, it takes time to convert SQL statements into MapReduce, which explains why ETL is superior to Hive in Figure 6.

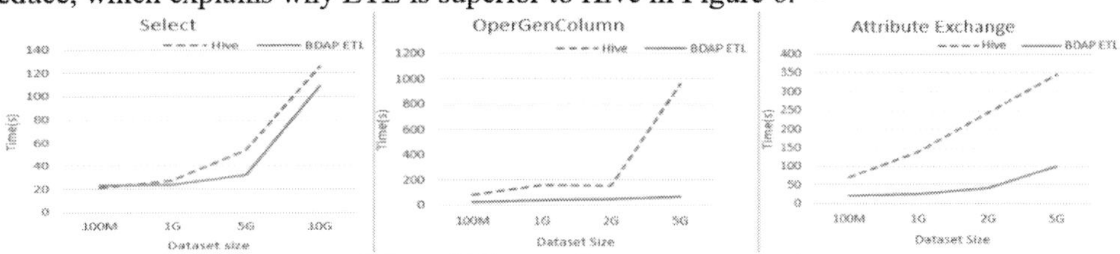

Figure 6. ETL and Hive performance comparison.

3.2. Algorithm Performance

MLlib is a commonly used machine learning algorithm library. MLlib and BDAP have been compared in aspect of function comprehension. Now, we compare the performance of the algorithm and take the BP algorithm in BDAP as an example. This algorithm adjusts the learning rate dynamically, increases the momentum factor, uses the small batch gradient descent method, trades the cross entropy cost function. The BP algorithm is compared with the multi-level perception classifier (MLPC) in MLlib.

In order to test large-scale data, this paper use data replication to generate the data based on data set covtype. The data scale is half million, 1 million, 2 million, 5 million and 10 million respectively. Result is shown in Figure 7.

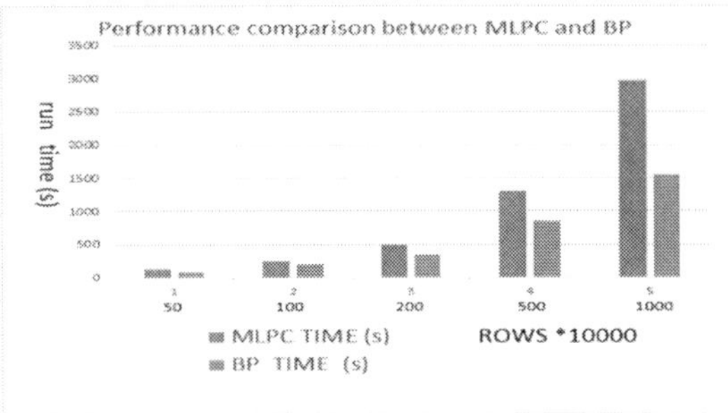

Figure 7. Running time comparison between BP and MLPC

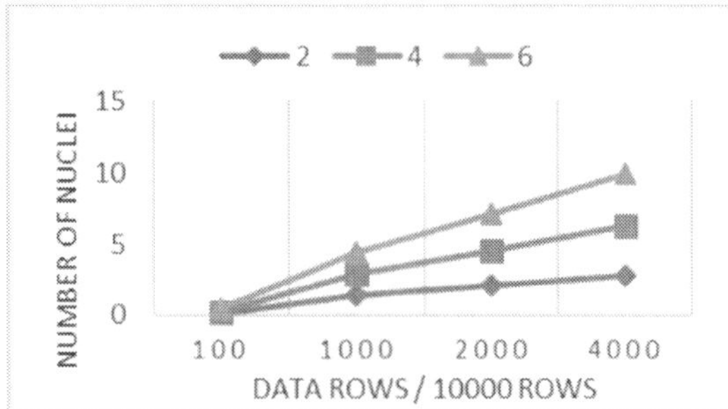

Figure 8. Speedup ratio of Spark-BIRCH in different nucleus and different data line

Table 2. Compared With The Experimental Results.

	a9a		Ijcnn1		skin_nonskin		Webspam		covtype	
	Time	Acc	Time	Acc	Time	Acc	Time	Acc	Time	Acc
MBSSVM	46	0.841	30	0.904	98	0.773	97	0.602	71	0.695
PMDP-SVM	35	0.844	13	0.983	37	0.999	95	0.991	102	0.960
LIBSVM	130	0.842	115	0.986	2109	0.995	29472	0.992	83631	0.961

In addition, BDAP provides a PMDP-SVM [13] (Partial-Model Dividing Parallel Support Vector Machine) model, which combines the characteristics of data distribution. PMDP-SVM combined with the Spark parallel framework, the clustering algorithm is introduced to support parallel training, and proposed a local model partitioning algorithm based on sample distribution. In this part, we evaluate the SVM on the platform by experiments, mainly focus on the performance of the training time and the accuracy of prediction. The experimental results in Table 2 show that the classification performance of PMDP-SVM is close to global optimal solution of LIBSVM, the prediction accuracy is higher than the Spark framework SVMWithMiniBatchSGD provided by MLlib tool.

In general, the speedup ratio of large data sets is proportional to the number of CPU cores. With the increase in the number of CPU cores, the speedup ration also increases. In the case of a certain number of CPU cores, more data points lead to higher speedup ratio. With the extension of data, both of the value of speedup ratio and trend of increase is more obvious. It is shown in Figure 8.

4. Application Scenarios

4.1. Power Grid Data

The EMS data is an important type of various power grid data, relating to the efficiency of power grid running. The changes in each value of EMS data never appear in one kind of value, and is often associated with changing trend of other kinds of value. Mining potential internal relationship between the EMS data has great significance in improving the data processing.

Based on BDAP, we use the improved algorithms and design a workflow to analyse the association rule of the EMS data, including EMS data pre-processing, calculation EMS data threshold, association rules analysis and etc. Figure 9 shows the workflow and we get the number of rules form the dataset through the workflow is far more than the algorithm based on clustering as shown in Figure 10.

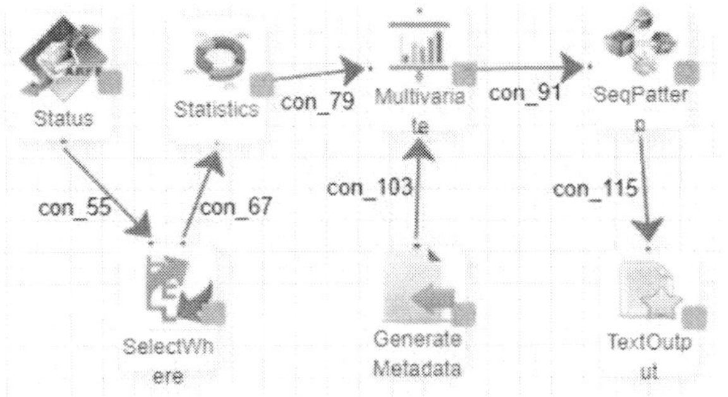

Figure 9. Power grid data analysis example

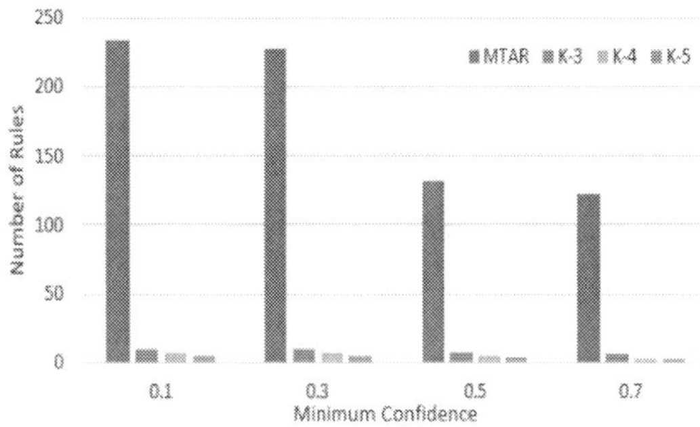

Figure 10. Comparison of the results

4.2. Video Classification

In this part, we use the SVM classification to classify and predict. We use data set KTH to train and test, KTH includes 6 types of human actions: walking, jogging, running, boxing, waving and clapping. 25 subjects respectively in 4 scenarios, 6 types of video recorded, S1 is a small range of outdoor sports scenes, S2 is a large range of outdoor with certain scale transformation scenes, S3 outdoor scenes wearing different clothes, S4 is the indoor scenes.

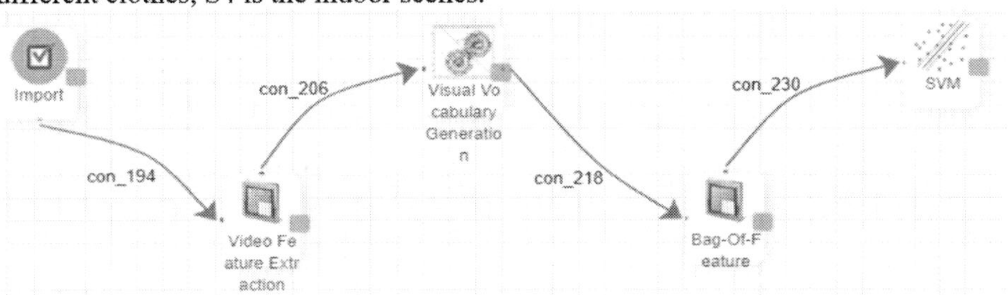

Figure 11. Workflow of video classification

Figure 11 shows the process of building information. We convert extracted original visual features into Bag of Features vector, and then use the SVM component in BDAP to train and predict.

5. Conclusions And Future Work

In this paper, a big data analysis platform (BDAP) based on Spark and Hadoop is proposed. In this platform, functions are comprehensive. The coupling degree between the components is low. Some operations such as adjusting parameters, establishing processes, monitoring, and results presentation are all visible. BDAP set up concepts of workflow and scheduling flow, workflow can save and delete. In performance, the performance of the algorithm of BDAP is better than that of Hive and MLlib. Furthermore, we analyse 2 application Scenarios, one is on power grid data, the second is on video classification.

In future work, we plan to optimize the authority management in order to let the different users can operate different data. Data exchange progress should be real-time monitored. We will set up more analytical applications around the Storm, such as real-time fault prediction analysis, real-time video mining, real-time text processing, etc.

6. Acknowledgment

This work is supported in part by the National High-tech R&D Program (863 Program) of China (2015AA050204) and State Grid Science and Technology Project (520626170011).

References

[1] Jie Song, Zongzhe Sun and Keming Mao et al. 2017. *Journal of software*. MapReduce big data processing platform and the research progress in algorithm. 28(3). 514-543

[2] Sarlis D, Papailiou N, Konstantinou I, et al. 2015. Datix: a system for scalable network analytics. *Acm Sigcomm Computer Communication Review*. 45(5). 21-28

[3] An Zhuo. 2012. *Research and implementation of large data analysis platform based on P2P scalable architecture* (Beijing: Tsinghua University)

[4] Li H, Lu K, Meng S. 2015. Bigprovision: a provisioning framework for big data analytics. *IEEE Network*. 29(5). 50-56.

[5] Qing He, Fuzhen Zhuang and Li Zeng et al. 2014. PDMiner: parallel distributed data mining tool platform based on Cloud Computing. *Chinese science*. 44(7). 871-885

[6] Yu L, Zheng J, Shen W C, et al. 2012. BC-PDM: data mining, social network analysis and text mining system based on cloud computing. *ACM SIGKDD International Conference on Knowledge Discovery and Data Mining*. 1496-1499.

[7] Mckenna A, Hanna M, Banks E, et al. 2010. The Genome Analysis Toolkit: a MapReduce framework for analyzing next-generation DNA sequencing data. *Genome Research*. 20(9). 1297-303.

[8] Chatzimilioudis G, Costa C, Zeinalipouryazti D, et al. 2016. Distributed In-Memory Processing of All k Nearest Neighbor Queries. *IEEE Transactions on Knowledge & Data Engineering*. 28(4) 925-938.

[9] George, L. 2011. *HBase: The Definitive Guide*. (Sebastopol: O'Reilly Media, Inc).

[10] Vavilapalli V K, Murthy A C, Douglas C, et al. 2013. Apache Hadoop YARN: yet another resource negotiator. *Symposium on Cloud Computing, ACM*. 5. 1-16.

[11] Dean J, Ghemawat S. 2004. MapReduce: Simplified Data Processing on Large Clusters. *Conference on Symposium on Opearting Systems Design & Implementation*. 10. 107-113.

[12] Jain A, Ong S P, Chen W, et al. 2015. FireWorks: a dynamic workflow system designed for high-throughput applications. *Concurrency and Computation: Practice and Experience*. 27(17). 5037-5059

[13] Chang Liu, Bin Wu and Yi Yang. et.al. 2016. Multiple Submodels Parallel Support Vector Machine on Spark. *IEEE International Conference on Big Data*. 945-950.

IOP Conf. Series: Journal of Physics: Conf. Series **1060** (2018) 012024 doi:10.1088/1742-6596/1060/1/012024

Data Analysis of Soccer Athletes' Physical Fitness Test Based on Multi-View Clustering

Bin Jiang[1], Hua Sun[2], Wanjian Bai[1], Hongmei Li[1], Yong Wang[1], Hongwei Xiong[3] and Ning Wang[2]

[1]State Grid Shangdong Electric Power Company, Shandong, China
[2]Shandong Luneng Sports & Culture Company, State Grid Shandong Company, Shandong, China
[3]Shandong Luneng Software Technology Co. Ltd., Shandong, China

{ls_sunht}@163.com

Abstract. Data mining is a process that uses machine learning to extract valuable knowledge from a large number of data. The sportsman has a long period of exhaustion of human body motor function. The changes of physical function signals in each period are not obvious, there is a lack of dynamic association between the body function data, which makes the connection between the data not close. Traditional data mining methods are based on association rules mining technology [1], which excavates the most relevant data attributes in the data. Once the athletes' body function data appear more obvious faults, the correlation is weakened, which will cause the deep excavation of the disease is inaccurate. In this paper, we take athletes' different body function test data as different views of body performance, adopts a multi view clustering technology to analyse the association between test items, and further optimizes the physical fitness test index. The comprehensive evaluation of athletes based on physical data is transformed into a clustering problem, which can effectively solve the athletes' physical status evaluation.

1. Introduction

For competitive sports, the ultimate goals of sports training are to enhance excellent sports performance, and athlete's physical ability is the most basic factor to improve athletic ability. Physical fitness testing is a basic way for coaches to understand the physical status of athletes [1][2]. Coaches should regularly perform physical fitness tests for athletes, according to different test standards, calculate the performance of each athlete's physical fitness test, and then evaluate their physical condition based on their own experience, and make corresponding training plan and guide training accordingly [3]. However, with the accumulation of test data [1], it is becoming more and more difficult to manage and analyse these data by manual processing. By using the traditional data processing and database management functions of computers, we can solve the management problems of physical test data, help coaches to manage athletes, translate performance scores, and manage historical data, so as to improve the efficiency of data processing [4]. However, the scientific evaluation and prediction of sports physical status cannot be solved, and the knowledge hidden behind the data resources cannot be found.

Analysis of data is local or the surface characteristics of the traditional data analysis and processing method, which cannot be predicted information about the overall features of the data hidden in the data and then describe the trend of its development, and the physical condition of the evaluation and prediction of concern is precisely the important information hidden behind the data. Data mining technology can extract valuable, unknown, hidden and potentially useful knowledge from a large number of original data [5]. Therefore, it is very suitable for the analysis of athletes' physical test data.

The existing main methods usually use classification and association rules [6][7] in data mining to analyse athletes' physical fitness test data, on the one hand, try to find out the relationship between physical fitness test items, and to optimize test indexes. On the other hand, the classification prediction model is set up, which is used to evaluate and predict the physical status of athletes. However, with the increasing number of athletes' physical fitness test data, the dimension of data is also increasing. At the same time, the classification method belongs to supervised learning, which requires a lot of artificial annotation data, so the traditional data processing method is difficult to meet the needs of practical applications. Therefore, this paper adopts a multi view clustering method to mine the importance degree of each physical fitness test index, and cluster and evaluate athletes' physical state.

2. Existing Problems

Taking football as an example, the physical test of athletes generally includes 15 indexes, such as aerobic, anaerobic, jumping, and lower limb joints. By analysing the physical test data, the coach could find out the gap between each athlete and formulate the training program. However, with the accumulation of physical test data, the traditional processing and analysis methods of data cannot meet the actual needs, the following two points are particularly important.

2.1. Optimization of Physical Energy Test Index

At present, the 15 indicators of physical fitness in football are the coaches based on many years' practical experience, mainly involving jumping, Li, physiological and biochemical, lower extremity joints and so on. Due to the limitations of equipment, time, climate and other factors, it is very difficult to carry out the test of all 15 projects regularly. Moreover, practice shows that there is a correlation between some physical fitness items in the 15 test items, that is, the improvement of physical fitness test results can lead to the improvement of one or more physical fitness test results. For the above reasons, the coaches put forward the further optimization of physical fitness test indicators. So, what standards are optimized and optimized? First of all, we should make clear the relationship between the various physical testing projects, and provide a scientific basis for further optimization of physical energy indicators.

2.2. Assessment of Physical Status

The purpose of physical training is to improve athletes' physical quality through training, so that athletes can achieve the best physical fitness during competition and play the best competitive performance. The traditional method is highly relying on the experience of the coaches, to adjust and strengthen the best physical fitness of the athletes during the preparation of the competition. This requires the coaches to have rich experience in training and fully understand the physical state of the athletes. Obviously, this increases the pressure of the coach's decision making and the possibility of making a mistake. Now, the physical fitness test indicators have a quantitative standard, can we analyse the physical state of the athletes by these data and evaluate their physical status. And a long time test has accumulated rich test data. These data are important historical data for the physical status changes of each athlete. Through these historical data, we can analyse athlete's physical state development and change rule, predict athlete's physical state in a certain period, and provide scientific basis for coaches to make training plan.

3. The Proposed Method

During the training process, the coach can evaluate the physical status of each athlete by careful analysis of the athletes' physical test data. Taking football as an example, the coach usually divides the physical status of athletes into three states, which are poor, general and good according to the test data.

This process of evaluation is a process that coaches judge according to experience, and it is difficult to describe exactly. The clustering analysis method in data mining through data mining model potential from the characteristics of the known training data, through each data clustering to different clusters have an effective description or model, and then use this model to the unknown new data division. Therefore, the coach's comprehensive evaluation of the athletes can be regarded as a potential clustering model. In the actual life, it is necessary to evaluate the real physical fitness of the athletes by testing many different physical indexes. In order to fully test the athlete's physical state, need to consider a number of fitness test index, in this paper, a different view of the data we will a fitness test index as the physical condition of reaction of athletes, by using a method of multi view clustering [8] of athlete's physical state clustering and evaluation. The algorithm is proposed by Cai [9], which can make use of multiple physical test indicators and learn the importance of different physical test indicators through weights.

3.1. Multi-view item clustering

Let $X^{(v)} \in R^{M\times d^{(v)}}$ denote the v-th test index data matrix of physical fitness data, $G^{(v)} \in R^{M\times K}$ would be the clustering indicator matrix of the v-th test index data, $F^{(v)} \in R^{d^{(v)}\times K}$ could be the centroid matrix for the v-th test index data, K could be the number of clusters.

As many different test index data shows that the physical state of the same athlete, thus the clustering indicator matrix $G^{(v)}$ of different test index data should be unique. This multi-view clustering method belongs to hard clustering methods, so the clustering indicator matrix $G \in R^{M\times K}$ should satisfy the 1-of-K coding mode.

The RMKMC algorithm used the structured sparsity-inducing norm, $L_{2,1}$-norm, which could be robust to data outliers and achieve stable results with different initializations. Using those techniques, the loss function of RMKMC method can be represented as:

$$\min_{F^{(v)},G,\alpha^{(v)}} \sum_{v=1}^{V} (\alpha^{(v)})^{\gamma} \left\| X^{(v)} - G(F^{(v)})^T \right\|_{2,1} \tag{1}$$

$$s.t. G_{ik} \in \{0,1\}, \sum_{k=1}^{K} G_{ik} = 1, \sum_{v=1}^{V} \alpha^{v} = 1 \tag{2}$$

where $\alpha^{(v)}$ is the weight factor for the v-th test index data and γ is the parameter which controls the distribution of weights for different test indexes data, V is the number of test indexes data matrix.

In order to effectively solve this object function, Cai [9] proposed an efficient optimization algorithm, which rewrite the formula (1) as:

$$J = \min_{F^{(v)},D^{(v)},G,\alpha^{(v)}} \sum_{v=1}^{V} (\alpha^{(v)})^{\gamma} H^{(v)} \tag{3}$$

Where

$$H^{(v)} = \mathrm{Tr}\{(X^{(v)} - G(F^{(v)})^T)^T \mathrm{D}^{(v)} (X^{(v)} - G(F^{(v)})^T)\} \tag{4}$$

Through this algorithm, we could alternatively update $F^{(v)}$, G, $D^{(v)}$, as well as $\alpha^{(v)}$, and repeat the process iteratively until the proposed objective function is convergence. For more detail of optimizations about RMKMC algorithm can be seen in [9]. Using the unique cluster indicator matrix G, the physical test data of each athletes could be clustered into K groups.

4. Experiments

In this section, we will validate our motivation, which applies the RMKMC method to the real soccer athletes' physical fitness test data. Our data is collected by the Shandong Luneng football club in the physical test of the athletes. Because of the privacy of physical data, most of the data are encrypted or processed by Hash.

4.1. Experimental Setup

In this subsection, we compare the results of our approach with some baseline methods, includes simple multi-view K-means clustering(NKMC), affinity and Propagation (AP). We also neglect the weight of each type of physical fitness test data, and we degenerate the RMCMC method into a simple version, called simple MKMC (SMKMC), which means each test index data having same importance. We compare RMKMC approach with this simple version method. Before doing the multi-view clustering algorithm, each physical fitness test index data should be normalized, we making all the values of each physical fitness test index data in the range [-1; 1]. When we implement SMKMC method, we simply concatenate all of the normalized data matrix as input for the classic K-means method.

In addition, RMKMC has a parameter γ to control the weight distribution between all physical fitness test index of data. We search this parameter as described in [9]. We use the K-fold cross validation to select the best number of clustering, and we set the number of clustering K=5. Since the performance of all the clustering algorithms are highly depending on the initializations, we repeat 50 times using random initialization for all the clustering methods and show the average results.

4.2. Clustering Results Comparisons

In this subsection, we compared different clustering results on soccer athlete data set. In this experiments, we use the NMI, ACC and Purity as our metrics to measure the clustering result of different methods, those metrics are commonly used in most of clustering methods. As shown in Table 1, the RMKMC method obtain the best clustering results.

Table 1. Comparisons of different methods with RMKMC approach

Method	NMI	ACC	Purity
K-means	0.6612	0.5621	0.6432
AP	0.6536	0.5955	0.6422
Co-Reg	0.7151	0.6632	0.7258
SMKMC	0.7633	0.6939	0.7812
RMKMC	0.7901	0.7798	0.8602

Compared with the SMKMC method, which just concatenate all of the normalized data matrix as input for the classic K-means method, the RMKMC method have the excellent clustering performance. This results shows the RMKMC method can make use of multiple physical test indicators and learn the importance of different physical test indicators through weight.

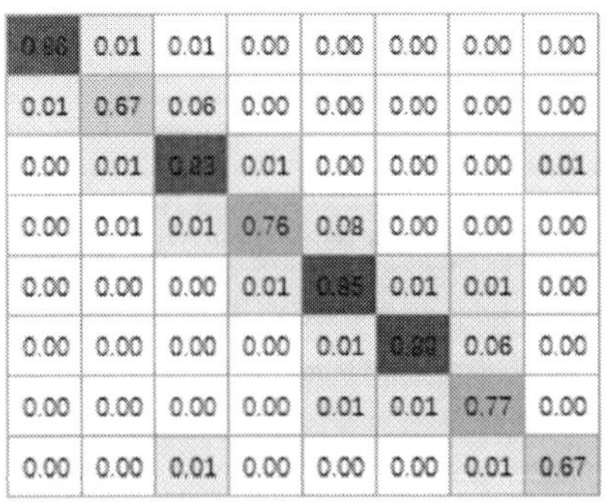

Figure 1. The calculated average clustering accuracy confusion matrix for Soccer Athletes' data sets.

We plot the confusion matrices of RMKMC in terms of clustering accuracy in Figure 1. From both tables and figures, we can see that our proposed methods consistently beat the base line method on the data set.

Through the analysis of the clustering results of the athletes' physical fitness test data, we communicate with the coaches and the medical staff. The results show that the mean VO2max for 56.2ml/kg/min, the maximum heart rate 194.3/min, 4 min after exercise blood lactic acid 12.9mmol/L, anaerobic threshold heart rate 173/min, immediately after exercise heart rate 189.7/min, 3 min after exercise heart rate 130.1/min, anaerobic threshold the speed of 4.01m/s. The development of aerobic capacity should refer to the anaerobic threshold heart rate or anaerobic threshold, and the best training results can be achieved by training at different intensities.

5. Conclusions

The physical fitness test is a basic way to understand the physical condition of athletes for the coach. The traditional data mining method is based on association rules mining technology, which excavates the most relevant data attributes in the data. However, once the body appeared obvious fault, the relevance between data is weakened, will result in inaccurate association mining. In this paper, we regard different physical fitness test data as different views of athletes' body performance, and use a highly robust multi view clustering method to cluster and evaluate athletes' physical state. Finally, the experimental comparisons show the robustness and effectiveness of the algorithm.

References

[1] Pelegrini A, Silva D A S, Petroski E L, et al. Health-related physical fitness in brasilian schoolchildren: data from the Brazil sport program.[J]. *Revista Brasileira De Medicina Do Esporte*, 2011,17(2) :92-96.

[2] Kamangar F. Heterogeneous image feature integration via multi-modal spectral clustering[C]// *Computer Vision and Pattern Recognition. IEEE*, 2011:1977-1984.

[3] Yu H, Zheng D, Zhao B Y, et al. Understanding user behavior in large-scale video-on-demand systems[J]. *ACM SIGOPS Operating Systems Review*, 2006, 40(4):333-344.

[4] Russo S G, Neumann P, Reinhardt S, et al. Impact of physical fitness and biometric data on the quality of external chest compression: a randomised, crossover trial.[J]. *Bmc Emergency Medicine*, 2011,11(1):1-9.

[5] Piatetsky-Shapiro G, Piatetsky-Shapiro G, Smyth P. The KDD process for extracting useful knowledge from volumes of data[M]. *ACM*, 1996.

[6] Rani B, Kaur P, Bansal A. Analysis of student physical fitness data using data mining algorithm[J]. *International Journal of Applied Engineering Research*, 2012, 7(11):1659-1662.

[7] Cui G X, Liang L I, Wang K K, et al. Research and improvement on Apriori algorithm of association rule mining[J]. *Journal of Computer Applications*, 2010, 30(11):2952-2955.

[8] Yan D, Huang L, Jordan M I. Fast approximate spectral clustering[C]// *ACM SIGKDD International Conference on Knowledge Discovery and Data Mining. ACM*, 2009:907-916.

[9] Cai X, Nie F, Huang H. Multi-view K-means clustering on big data[C]// *International Joint Conference on Artificial Intelligence. AAAI Press,* 2013:2598-2604.

IOP Conf. Series: Journal of Physics: Conf. Series **1060** (2018) 012025 doi:10.1088/1742-6596/1060/1/012025

The Study of Customer Split-Flow Based on Data Mining

CHEN Jun-De[1] and ZHANG De-Fu[2]

[1]Huawei Technologies Co.Ltd, Shenzhen, Guangdong, China
[2]School of information science and engineering, Xiamen University, China

chen2wo@126.com

Abstract. In this paper, we established the business preference model to identify who often applied the low-value businesses in service hall, and the channel adapter model was built, which was used by data mining, and we made this model determine the suitability customers for them to apply the low-value business by electronic channels. Three categories were divided that fit the specific recommendation, required patience to guide, and not suitable recommended. Finally, according to the actual business data, we validated the model and obtained the satisfactory results for customer split-flow.

1. Introduction

With the growth of the business for people who handle their affairs in physical business hall and hotline channel, that is necessary to divert low-value business users to other channel such as electronic business hall, induce them to handle the business in online business hall directly and not crowd the high value business in physical hall.

Data mining is a new technology to meet the needs of extracting information from massive databases in information society. It can help decision-makers extract the implicit, unknown, non trivial and potentially valuable information or knowledge [1], Classification problem is the critical problem, and classification algorithm is the key technology for data mining, it has a wide application in many fields, such as pattern recognition, data analysis, image processing and so on. Such as Sujinlong, et al. [2] pretreated a video image, and used the method of fuzzy neural network to implement target identification. Based on the customer data of game, Tzu-Liang (Bill) Tseng, et al. [3] classified the customer. Michael Y. Hu et al. [4] extracted the features by the method of neural network etc. Classification problem is becoming a very active research topic in the field of data mining, and this split-flow for business hall could be regarded as the classification problem as also.

For the split-flow, the customers should be classified and suitable customer was required to be distinguished firstly. By applied the classification algorithm of data mining, we could distinguish and perform the split-flow for the users who often deal with the existing physical hall and hotline channels for the low value business. From three aspects of customer, business, channel, the three-dimensional adaptation model was built, which could reduce the operating pressure in physical channel, and increase the service efficiency in electronic channels, improve the user satisfaction.

2. Frequent Classification Algorithm

2.1. FRI Algorithm

Fuzzy Rules Induction (FRI) is a self-organizing data mining and non-parametric method based on Group Method of Data Handling (GMDH) [5]. FRI algorithm could automatically extract the fuzzy rules from data, and describe the complex systems in the form of IF-THEN by close natural language (such as good, general, poor). FRI Uses the black-box approach to analyze and dispose the relation of input and output variables in system, and automatically extract the fuzzy rules from the data by GMDH technology, this algorithm could give a forecast for the system, and is a kind of effective way to describe the behavior of the system.

2.2. Artificial Neural Network (ANN)

2.2.1. ANN Algorithm Principle. Artificial neural network (ANN) is a new intelligent handling method in the process of imitating human brain to solve problem, through a lot of simple handling units called neurons constitute, it constitutes a nonlinear dynamic system, and the basic neuron model is shown as Figure 1.

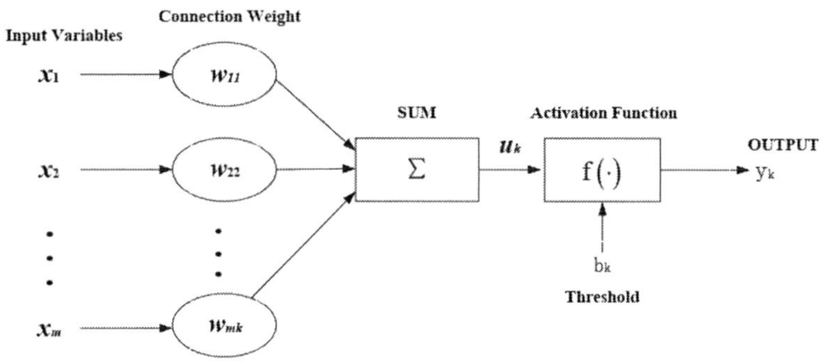

Figure 1. Artificial Neuron

2.2.2. ANN Classification. The fuzzy rules of if-then form are the most basic knowledge of complex decision-support techniques. The classical theoretical methods for obtaining fuzzy rules include the rough set theory proposed by Pawlak and the fuzzy model proposed by Takagi and Sugeno[6], ChenHui [7] proposed that the rough sets and fuzzy sets has certain limitations, such as fuzzy set has unthoroughness, and rough sets has generality, etc. HanMin, et al. [8] proposed a classification method for multi-attribute fuzzy rules based on cloud-neural network, based on these analyses the rule classification method of neural network was studied in the paper. For discrete properties, the front parts of the rule classification are usually represented as Boolean functions. For example, the sample has two discrete properties (x_1, x_2), and the value of x_1 maybe be $x_{11}, x_{12}, x_{13}, x_{14}$; The possible value of x_2 maybe be x_{21}, x_{22}. there are 2*4 species, and the format of the extracted classification rules is shown below.

$$\begin{array}{l} IF(x_1 = x_{12}) \vee (x_1 = x_{11} \wedge x_2 = x_{22}) \vee ((x_1 = x_{13} \wedge x_1 = x_{14}) \wedge (x_2 = x_{21})) \\ \quad THEN\ \ t = 1 \quad ELSE\ \ t = 0 \end{array} \tag{1}$$

2.3. Decision Tree

The common algorithms of decision tree are ID3 and C4.5, the C4.5 algorithm is based on the ID3 algorithm and proposed by Quinlan in 1993. As a typical classification algorithm, decision tree[9] has been used in the financial, medical and other industries widely.Similar to the above algorithm, the data pretreatment should be performed before the compute of decision tree, the continuous attribute was needed to be discretized, and the training set should be formed for the decision tree, and the basic theory of C4.5 algorithm is clear and simpler.

IOP Conf. Series: Journal of Physics: Conf. Series **1060** (2018) 012025 doi:10.1088/1742-6596/1060/1/012025

3. Algorithms Selection

3.1. FRI Algorithm

So, through analyze the algorithm principle and characteristics for artificial neural network, decision tree, FRI, we have a basic understanding for the advantages and disadvantages of each algorithms. It should be pointed out that, in view of the different problems or scene, there will be different method to solve it, which requires a scientific approach to analyze and study. When faced the general classification problem, we would estimate which kind of algorithms are done relatively well or suitably in advance.

Therefore, the use of test and comparison would be a feasible method to select the suitable algorithm, and this could be implemented well in the UCI data sets. Took the balance-scale dataset for example, we selected 500 samples orderly to constitute the training set, and 125 samples to constitute the test data. Used the above algorithms to calculate respectively, we acquired the average error rate was 9.4% in training set, and 6.4% in test set. Then, we chose other part respectively for testing set, and the rest of the four parts as a learning set, this dataset was run repeatedly five times. The average error rate in the test set is 13.44%. The extraction rule for L class is

$$\begin{aligned} &IF \quad NOT-A_1 \ \& \ C_1 \ \ OR \ \ NOT-B_1 \,\&\, D_1 \\ &THEN \ \ L \end{aligned} \tag{2}$$

The signification of above expression (2) is that if the weight both the left and right side is not 1, or the distance between left and right is not 1, then it belongs to L class. Similarly, the Vote dataset has 435 samples, we constituted the 348 samples as the learning set, 87 samples as the test set, we acquired that the average error rate in the test set is 3.45%.

The other data sets were processed similarly, by contrasted the test results for the above algorithms of FRI, decision tree and ANN, we could get the different accuracy of these algorithms and select the relatively good algorithm to apply. In this case, we selected the decision tree to distinguish the customer and implement the business split-flow.

4. Model Building

4.1. Channel adaptation model

4.1.1. Model Preparation. Both the domestic and foreign scholars have some study in the channel preference of customer. Such as Siwon Cho, et al.[10] studied the preference of physical or virtual channel for the customers. Chiang, et al.[11] used the artificial neural network to predict the customers' selection behavior for different channels. Kevin , et al.[12] first used AHP to estimate the customers' channel preference, etc.

The Channel optimization analysis should firstly analyze the urgent split-flow business to confirm the target customer group, and on this basis, the channel adaptation model would be built to determine whether the customer was worth to recommend or not, through the relevant measures were taken to appropriate customer, the goal of improving customer satisfaction was able to be achieved. Through the establishment of the customer - business - channel adaptation model, it helps the manager take pointed split-flow measures to target customer groups and guide the customers to use remote electronic channels.

As is shown in the below table 1, according to the business value, the customer business was divided into three levels of high, middle and low, and at the same time, the relevant customer channel was divided into three categories of Suitable recommendation, Need to be patient guidance and not Suitable recommendation as also. The detail was as follows.

Table 1. Adaptable split-flow strategy

Customer-business	Customer-channel	measures and strategies
High	Suitable recommendation	Priority diversion guidance

High	Need to be patient guidance	Suggest manual or SMS recommendation
High	Not suitable for recommendation	Temporarily not diverting
Median	Suitable recommendation	Priority diversion guidance
Median	Need to be patient guidance	Suggest manual or SMS recommendation
Median	Not suitable for recommendation	Temporarily not diverting
Low	Suitable recommendation	Priority diversion guidance
Low	Need to be patient guidance	Suggest manual or SMS recommendation
Low	Not suitable for recommendation	Temporarily not diverting

4.1.2. Data Preprocessing. Through the comprehensive analysis for the data of customer channel, the key fields of the channel adaptation model are determined, including users age, type, Average electricity fee for nearly three cycle, Number of arrears, Transfer failure times within a year, Cash payment times within a year, contract content, whether the electronic bill was opened, the IVR times within a year, total 9 fields were extracted and these fields were represented by $X_i\left(i=1,2,K\ ,9\right)$ respectively. The extracted fields were calculated twice, and the data was preprocessed at the same time, including the missing value processing, abnormal value processing, data regulation and dimension reduction, etc.

*4.1.3. Model Process.*By analyzed the proportion of channel business, the service staff could obtain which was the most urgent business type for the channel split-flow. The business preference model was used to identify who often applied the low-value businesses in service hall, and the channel adapter model was used to channel split-flow, such as the low-value business was guided to electronic channels and so on.

As the chart shows below, the channel adapter model and business preference model were built separately. Among the target customers, the channel adaption model was calculated. The customer channel was divided into three categories, and each category was recommended to the specific channel. The customer - business preference model was calculated as also, and the prediction probability was calculated for the business preference. Then, we used the model in the actual business data and verified it. The last was the result analysis. Detail was listed as Figure 2.

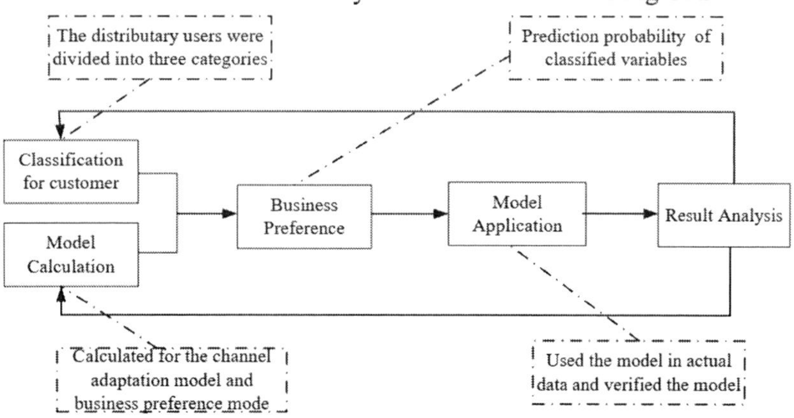

Figure 2. The Chart of Modeling Process.

In this way, the model could distinguish the customer who transacted low-value business in physical business hall and hotline channel. Here, the algorithm of decision tree was used in the model, and the typical algorithm of decision tree is ID3 and C4.5. The C4.5 algorithm was improved from ID3 and inherited the advantages of ID3 algorithm. Compared with other classification algorithms, such as statistical methods and neural networks, the C4.5 algorithm had the following advantages: the generated classification rules are easy to understand and the accuracy is higher.

The C4.5 decision tree model was used to distinguish whether the customers were suitable to transact the business by electronic channels or not, and the calculation process was shown in Figure 3. Then, it will be guided and recommended to the electronic channels for the suitable user categories, in this way, the purpose of channel diversion was achieved.

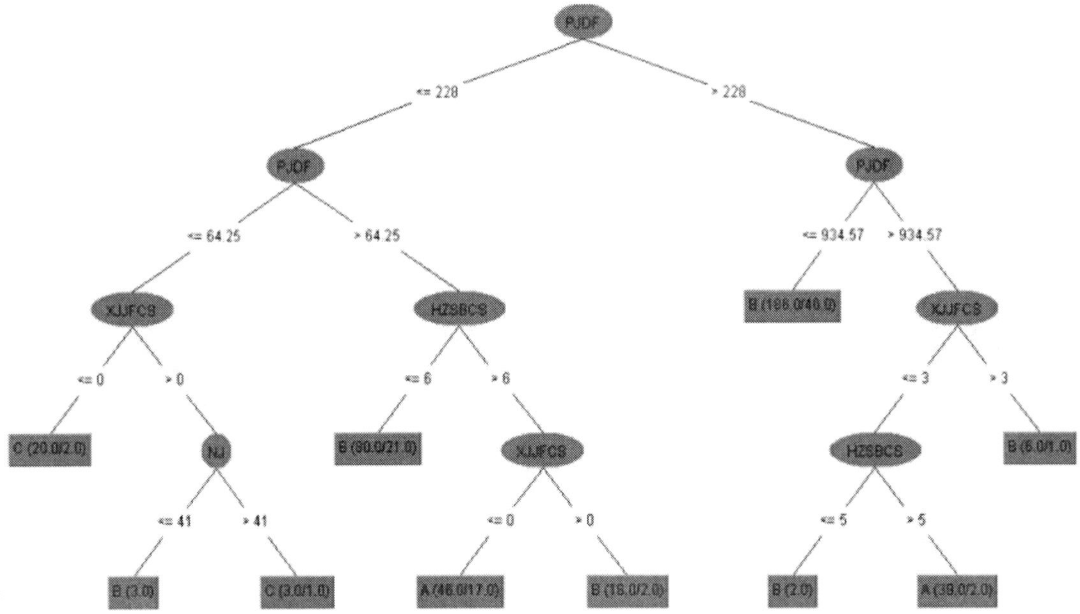

Figure 3. C4.5 decision tree

L is the final leaf node, the recommended proportion of each leaf node is the probability fit to recommend the electronic channel, below table 2 is the detailed result

Table 2. The User Classification Result

USERS	Attr	Suitable recommendation	Need patient guidance	Not suitable recommendation
USER 1	......	87%	13%	0%
USER 2	......	70%	25%	5%
USER 3	......	20%	78%	2%
USER 4	......	0%	17%	83%
......	......	......	......	......

The classification categories include the recommended for electronic channels, need patience guidance of electronic channels, and not suitable to recommend for electronic channels, at the same time, the probability of each category was calculated as above.

As to the different channel, different customer has the different preference for the channels, and table 3 is the result of preference for different customer channel. Such as the large users, this group was favor of the experience platform, and the most of them would like to transact the business by experience platform, the probability is 47%.

Table 3. Calculation Result of Customer Preference

Matrix of Customer-Channel	BOSS Self	coop	SMS	Online sales	Experience platform	Hotline
Large users	51%	51%	41%	35%	47%	84%
……	……	……	……	……	……	……
Family users	29.5%	29%	12.5%	8.5%	31.5%	56%
Indiv-com users	25.9%	30%	10.8%	9.5%	34%	70%

5. Conclusion

The model of customer split-flow is established to divert customers' pressure on reducing physical channel operations, improve overall service efficiency and enhance customer satisfaction. It consists of two parts, one is the channel adaptation model, and the other is the business preference model. First of all, the customers who often deal with the low value business are identified, and according to the business preference model, the customers are divided into three categories of high, medium and low as also. Then, the channel adaptation model is established to determine whether the customers are suitable to electronic channels or not, this divided categories are suitable for recommendation, need for patient guidance and not recommended. Finally, the model is validated on the actual business data, and the satisfactory split-flow results are obtained.

6. Acknowledgment

The authors wish to thank the project chance provided by Guangzhou Power Supply Bureau, and thank Mr. Zhang Liang-jun, the chairman of Guangzhou TipDM Intelligent Technology Co., Ltd., for valuable discussion and contribution to the successful delivery of the project.

References

[1] Zhang Liangjun, Chen Junde, Liu Mingjun, Chen Rong. Data mining: practical case study [M]. Beijing: China Machine Press, 2013: 10-1
[2] Su kanadaki, HERBERT H C IU, FERNANDO T, Embedded Design in Neural Network and Optical Flow Based High-speed Target Recognition & Tracing [J]. China Journal of Highway and Transport 2015, 28 (11): 113 – 121
[3] Tzu-Liang(Bill)Tseng, Chun-CheHuang, Rough set-based approach to feature selection in customer relationship management [J]. Omega 35 (2007) 365–383.
[4] Michael Y. Hu , Murali Shanker , G. Peter Zhang, Ming S. Hung Modeling consumer situational choice of long distance communication with neural networks[J]. Decision Support Systems 44 (2008) 899–908
[5] He Changzheng. Self-organizing data mining and economic forecasting [M]. Beijing: Science Press, 2005.
[6] Leng G, McGinnity T M, Prasad G. An approach for on-line extraction of fuzzy rules using a self-organising fuzzy neural network[J]. Fuzzy Sets and Systems,2005, 150(2): 211-243.
[7] ChenHui, Li De-Yi, Shen Cheng-Zhi. A Cloud Model Applied to Controlling inverted pendulum [J]. Journal of Computer Research& Development, 1999, 36 (10): 1180-1187.
[8] HanMin,LiZheng.Multi-attribute fuzzy rules classification based on cloud-neural network[J].Control and Decision,2009, 24(6): 933-936.
[9] Lin Zhang, et al. Efficient privacy-preserving classification construction model with differential privacy technology[j]. Journal of Systems Engineering and Electronics,2017(01)

[10] Siwon Cho, Jane E. Workman Effects of need for touch, centrality of visual product aesthetics and gender on channel preference for apparel shopping[J].Journal of Global Fashion Marketing,2015,Vol.6(2),pp.120-135.
[11] Chiang W Y K, Zhang D, Zhou L. Predicting and Explaining Patronage Behavior toward Web and Traditional Stores Using Neural Networks: A Comparative Analysis with Logistic Regression[J].Decision Support System,2006,42(2):18.
[12] Kevin Chiang, Zhen Li. An analytic hierarchy process approach to assessing consumers' distribution channel preference [J]. International Journal of Retail & Distribution Management, 2010,Vol.38(2)2010.

IOP Conf. Series: Journal of Physics: Conf. Series **1060** (2018) 012026 doi:10.1088/1742-6596/1060/1/012026

An Improved MUSIC Algorithm for DOA Estimation of Non-Coherent Signals with Planar Array

Liu Yaning[1, a], Fu Juntao[1, b], Ran Xinghao[1, c] and Ming Le[1, d]

[1]Air and Missile Defence College, Air Force Engineering University

[a] 13231126178@163.com; [b]762050486@qq.com; [c]1275465401@qq.com; [d]2390043575@qq.com

Abstract-To reduce the computational complexity of two-dimensional (2-D) direction of arrival (DOA) estimation of non-coherent signals, this paper proposes a semi-real-valued MUSIC algorithm with planar array. Firstly the real part of the received signal covariance matrix is derived and performed an eigen-decomposition to obtain the noise subspace. Then use the reduced-dimensional MUSIC and the least-square method to evaluate the 2-D DOA estimates. The proposed algorithm can avoid the high computational cost and have very close performance to conventional 2-D MUSIC algorithm. Simulation results verify the availability of the algorithm.

1. Introduction

Recently, two-dimensional (2-D) direction of arrival (DOA) has become a research hotspot of array signal processing.It can not only be applied to antenna, but also be a key parameter estimation technique of location of signal source.

The commonly used signal array models of 2-D DOA estimation include L-shape array, planar array and parallel linear array [1-3]. The conventional 2-D DOA estimation algorithms like MUSIC[4], ESPRIT[5], Capon[6] are complex.Therefore, more researches are focused on how to reduce the complexity of these algorithms. An ESPRIT algorithm with one single snapshot is derived in [7] to achieve the 2-D DOA estimation. [8,9] use one-dimensional MUSIC and Capon algorithm to reduce calculation amount based on the characteristic of Kronecker.

According to [10], the planar array can do better in resisting disturbance and more accurate estimation can be obtained.Based on the planar array, this paper proposes a semi-real-valued MUSIC algorithm which can significantly reduce calculation amount. The proposed algorithm can avoid the high computational cost and have very close performance to conventional 2-D MUSIC algorithm. What's more, the automatic pairing of angle can be realized.

2. Signal model

As shown in Figure.1, we consider a planar array (PA) with N elements on each row and M elements on each column. The inter-element spacing is d. Suppose that K narrowband source signals impinge on the PA from directions $(\theta_1,\phi_1),(\theta_2,\phi_2),...,(\theta_k,\phi_k)$,respectively. θ_k and ϕ_k represent azimuth and elevation of the kth signal source $\theta_k \in [-90°,90°]$,$\phi_k \in [-90°,90°]$. α_k and β_k represent the angle between the incident direction and X axis and Y axis.

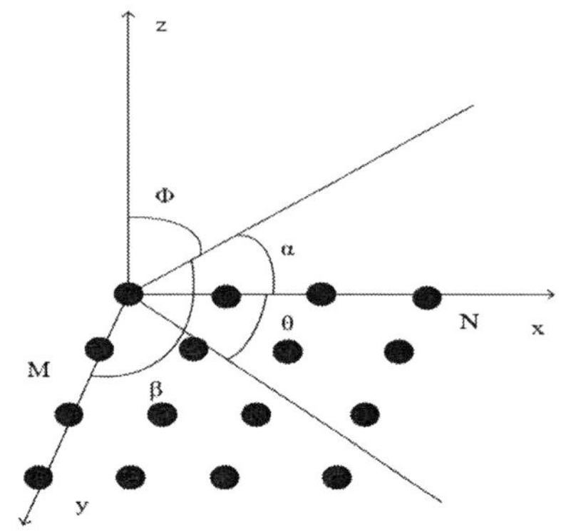

Figure.1 signal array model

The received noise is white Gaussian noise which is uncorrelated with source signals. The output of the element is given by

$$y_{k,nm}(t) = \sum_{k=1}^{K} \delta_{k,nm} s_k(t - \tau_{k,nm}) + n_{nm}(t) \tag{1}$$

where $y_{k,nm}(t)$ is the output of the nth row and mth column element. $\delta_{k,nm}$ is the gain of the element and $\tau_{k,nm}$ is the delay relative to the coordinate origin. $s_k(t) = \exp(2\pi f_k t)$ represent the kth signal with the frequency f_k. $n_{nm}(t)$ is white Gaussian noise with a mean value of 0 and a variance of σ^2.

According to the signal model, we can obtain the delay as:

$$\tau_{k,nm} = (n\cos\theta_k \sin\phi_k + m\sin\theta_k \sin\phi_k)\cdot d/\lambda \tag{2}$$

As shown in fig1, we can have the following relation:

$$\begin{aligned} \cos\alpha_k &= \cos\theta_k \sin\phi_k \\ \cos\beta_k &= \sin\theta_k \sin\phi_k \end{aligned} \tag{3}$$

Put (3) into (2),

$$\tau_{k,nm} = (n\cos\alpha_k + m\cos\beta_k)\cdot d/\lambda \tag{4}$$

We can know from (3) that:

$$\begin{aligned} \theta_k &= \arctan(\cos\beta_k / \cos\alpha_k) \\ \phi_k &= \arcsin(\cos^2\alpha_k + \cos^2\beta_k)^{\frac{1}{2}} \end{aligned} \tag{5}$$

Define the steering vector of the kth signal is a_k. The output of the array can be expressed by:

$$\boldsymbol{Y}(t) = \sum_{k=1}^{K}\sum_{m=0}^{M}\sum_{n=0}^{N} (\boldsymbol{a}_n(\alpha_k) \otimes \boldsymbol{a}_m(\beta_k))\boldsymbol{s}_k(t) + \boldsymbol{N}(t) \tag{6}$$

where

$$\begin{aligned} \boldsymbol{a}_n(\alpha_k) &= (1, \exp(j2\pi d/\lambda)\cos\alpha_k, \ldots \exp(j2\pi(N-1)d/\lambda)\cos\alpha_k)^T \\ \boldsymbol{a}_m(\beta_k) &= (1, \exp(j2\pi d/\lambda)\cos\beta_k, \ldots \exp(j2\pi(M-1)d/\lambda)\cos\beta_k)^T \end{aligned} \tag{7}$$

3. Conventional 2-D MUSIC algorithm

2-D MUSIC is widely used to estimate azimuth and elevation of the signals. The covariance matrix of received signals can be described as:

$$\boldsymbol{R} = \frac{1}{L}\sum_{l=1}^{L} \boldsymbol{Y}(t_l) * \boldsymbol{Y}(t_l)^H \tag{8}$$

where L is the number of snapshots

Do the eigenvalue decomposition and (8) can be written by:

$$R = E_S U_S E_S^H + E_N U_N E_N^H \tag{9}$$

where U_S is a diagonal matrix whose diagonal elements contain the largest K eigenvalues and U_N stands for a diagonal matrix whose diagonal elements contain the smallest $MN\text{-}K$ eigenvalues. E_S is the matrix composed of the eigenvectors corresponding to the largest K eigenvalues of R , while E_N represents the matrix including the rest eigenvectors.

According to [11], the 2-D MUSIC spectrum function can be constructed as:

$$P_{2D\text{-}MUSIC} = \frac{1}{a(\alpha,\beta)^H E_N E_N^H a(\alpha,\beta)} \tag{10}$$

$$\alpha \in (-90°, 90°], \beta \in (-90°, 90°]$$

Using the characteristic of Kronecker, (10) can be described as:

$$P_{2D\text{-}MUSIC} = \frac{1}{[a_n(\alpha)\otimes a_m(\beta)]^H E_N E_N^H [a_n(\alpha)\otimes a_m(\beta)]} \tag{11}$$

This algorithm can effectively estimate 2-D DOA, but searching the peak in the whole space costs lots of calculation amounts. Therefore, we proposed an improved algorithm to solve the problem.

4. Improved 2-D MUSIC algorithm

4.1 2-D DOA estimation via semi-real-valued MUSIC(srv MUSIC)

Define the denominator in [11] is Φ :

$$\Phi = [a_n(\alpha)\otimes a_m(\beta)]^H E_n E_n^H [a_n(\alpha)\otimes a_m(\beta)] \tag{12}$$

From (7),(10),(11), we can obtain that $\mathbf{a}(\alpha,\beta) = \mathbf{a}_n(\alpha)\otimes \mathbf{a}_m(\beta)$, $a_n(\alpha) = a_n(-\alpha)$, $a_m(\beta) = a_m(-\beta)$ .So:

$$\begin{aligned}
\mathbf{a}(\alpha,\beta) &= \mathbf{a}_n(\alpha)\otimes \mathbf{a}_m(\beta) = \mathbf{a}_n(-\alpha)\otimes \mathbf{a}_m(-\beta) = \mathbf{a}(-\alpha,-\beta) \\
\mathbf{a}(\alpha,\beta) &= \mathbf{a}_n(\alpha)\otimes \mathbf{a}_m(\beta) = \mathbf{a}_n(\alpha)\otimes \mathbf{a}_m(-\beta) = \mathbf{a}(\alpha,-\beta) \\
\mathbf{a}(\alpha,\beta) &= \mathbf{a}_n(\alpha)\otimes \mathbf{a}_m(\beta) = \mathbf{a}_n(-\alpha)\otimes \mathbf{a}_m(\beta) = \mathbf{a}(-\alpha,\beta)
\end{aligned} \tag{13}$$

It can be known from (13) that the search domain of 1-D DOA can be halved, $\alpha \in [0, 90°]$, $\beta \in [0, 90°]$. Therefore,the domain of 2-D DOA can be reduced to $\frac{1}{4}$. According to [12], the noise subspace can be obtained by the eigenvalue decomposition of the real covariance matrix $\mathrm{Re}(R)$.

According to the characteristic of Kronecker, $a_n(\alpha)\otimes a_m(\beta)$ can be transformed into $[a_n(\alpha)\otimes I_M]\, a_m(\beta)$.(12) can also be expressed as:

$$\Phi = a_m(\beta)^H [a_n(\alpha)\otimes I_M]^H E_n E_n^H [a_n(\alpha)\otimes I_M] a_m(\beta) \tag{14}$$

Define $P(\alpha) = [a_n(\alpha)\otimes I_M]^H E_n E_n^H [a_n(\alpha)\otimes I_M]$. It's easy to calculate $a_m(\beta)^H a_m(\beta) = M$. So (14) can be transformed to:

$$\Phi = M \bullet \frac{a_m(\beta)^H P(\alpha) a_m(\beta)}{a_m(\beta)^H a_m(\beta)} \tag{15}$$

According to Rayleigh-Ritz theory, the minimum of (15) is $\lambda_{\min}(\alpha)$ which is the minimum eigenvalue of $P(\alpha)$ .It can be described by:

$$\min \frac{a_m(\beta)^H P(\alpha) a_m(\beta)}{a_m(\beta)^H a_m(\beta)} = \lambda_{\min}(\alpha) \tag{16}$$

So the estimation of α can be obtained by:

$$\bar{\alpha} = \arg\max \frac{1}{\lambda_{\min}(\alpha)} \quad \alpha \in [0, 90°] \tag{17}$$

The estimation of k received signals' α are recorded separately as $(\bar{\alpha}_1, \bar{\alpha}_2, ..., \bar{\alpha}_k)$.Then estimate β.

Define $\varphi_k = angle(\boldsymbol{a}_y(\beta_k))$, Combining with (7), we can know that:

$$\varphi_k = [0, (2\pi d/\lambda)\cos\beta_k, ...(2\pi(M-1)d/\lambda)\cos\beta_k]^T \tag{18}$$

According to the least-square method , define

$$\boldsymbol{C} = \begin{bmatrix} 1 & 0 \\ 1 & 1 \\ \mathrm{M} & \mathrm{M} \\ 1 & M-1 \end{bmatrix} \tag{19}$$

Then solve the minimum of $\|Cd_k - \varphi_k\|^2$, and the result is $\boldsymbol{d}_k^T = (\boldsymbol{C}^T\boldsymbol{C})^{-1}\boldsymbol{C}^T\varphi_k$. β_k can be sovled by following formula:

$$\beta_k = \arcsin(\mathbf{d}_k(2)/(2\pi d/\lambda)) \tag{20}$$

4.2 steps of algorithm

Step 1: Solve the covariance matrix $\boldsymbol{R}$ of received signals and derive the real part $\mathrm{Re}(\boldsymbol{R})$. Do the eigenvalue decomposition on it to obtain the real number domain noise subspace.

Step 2: Construct $\boldsymbol{P}(\alpha)$ based on the characteristic of Kronecker and solve the minimum eigenvalue $\lambda_{\min}(\alpha)$. Put it in (17) to solve the estimation of α.

Step 3: Put the estimation of α in (16) and obtain the eigenvectors. Then following (18)-(20), use least-square method to get the estimation of β.

Step 4: Obtain the 2-D DOA estimation according to (5).

4.3 Analysis on computation complexity

Analyses the computation complexity according to the steps above and compare it with conventional 2D-MUSIC. Assume the number of signals is K, the snapshots are L, the search points are n. Total calculation is $O\{(MN)^2L + (\mathrm{MN})^3/4 + n(MN-K)(M^2N+M^2)+10M\}$. And the calculation of 2D-MUSIC is $O\{(MN)^2L + (\mathrm{MN})^3 + 2n(MN-K)(M^2N+M^2)+N^2\}$.

To see more intuitively, specific values are given to draw the Figure.2. Assume $N = M$, $K = 3$, $L = 200$, $n = 3\times10^3$.It can be seen from Figure.2 that the algorithm in this paper can significantly reduce the computation complexity compared with 2-D MUSIC.

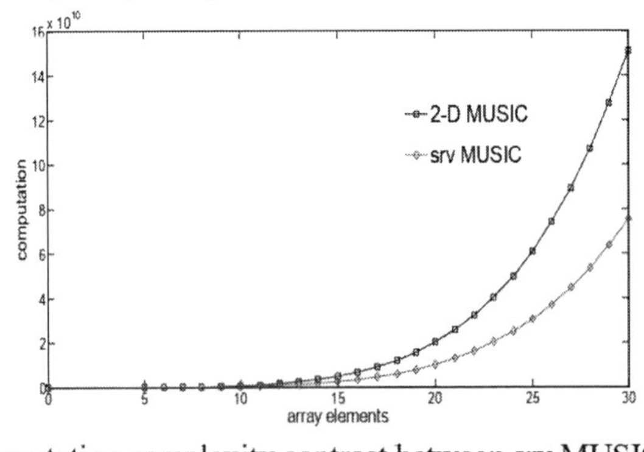

Figure.2 Computation complexity contrast between srv MUSIC and 2-D MUSIC

IOP Conf. Series: Journal of Physics: Conf. Series **1060** (2018) 012026 doi:10.1088/1742-6596/1060/1/012026

5. Simulation results and analysis

5.1 Algorithm evaluation

To evaluate the algorithm, simulation is carried out just as following. We present 500 Monte Carlo simulations to assess the angle estimation of the algorithm. Define root mean squared error(RMSE). We adopt the array with $N = M = 8$, and d is $\lambda / 2$. Let SNR is 20dB and $L = 100$. Assume that there are two non-coherent source signals located at angles $(\theta_1, \phi_1) = (10°, 15°), (\theta_2, \phi_2) = (25°, 35°)$ .Figure 3 presents angle estimation results of the algorithm for all targets. It can be seen that 2-D DOAs can be clearly observed. So it is indicated that the algorithm has high accuracy in 2-D DOA estimation.

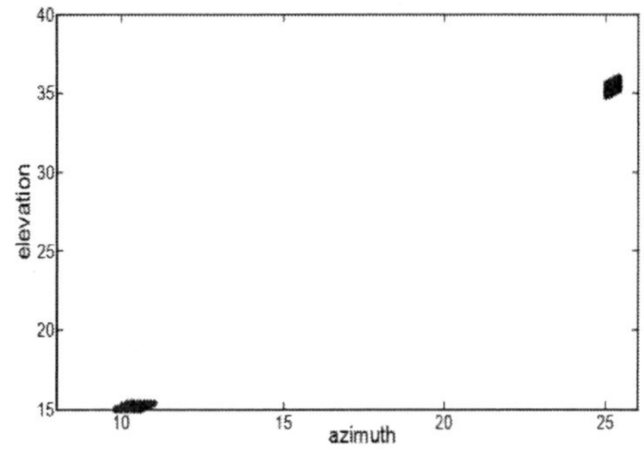

Figure.3 DOA estimation with the algorithm for two targets

5.2 RMSE versus SNR

The number of Monte Carlo tests is 500. The number of snapshots is set as 100. There are both 8 array elements on x axis and y axis. The range of SNR is from -4dB to 20dB with interval 4dB. Compare the proposed algorithm with 2-D MUSIC, and the results are shown in Figure.4.

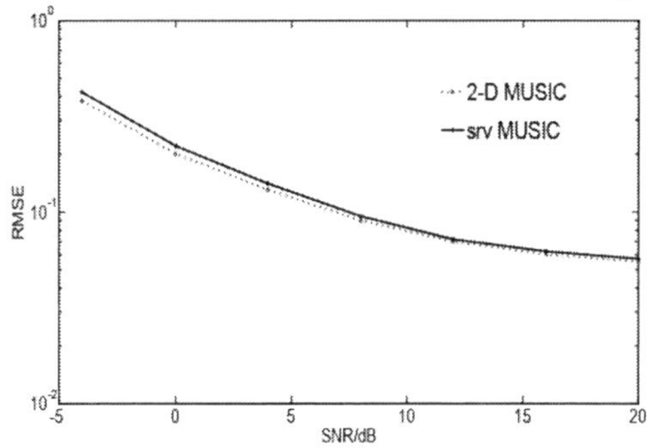

(a)The change curve of RMSE of θ with SNR

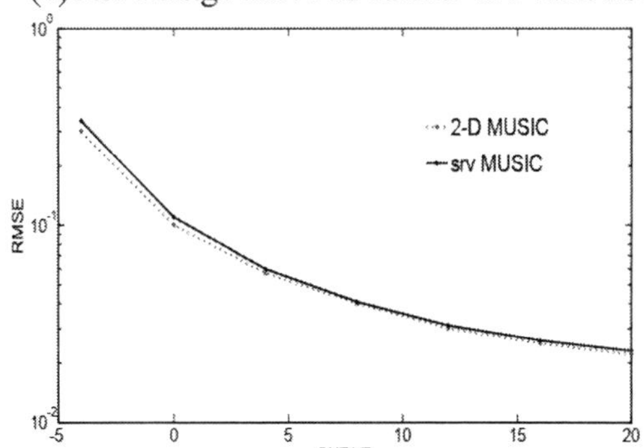

(b)The change curve of RMSE of ϕ with SNR

Figure.4 RMSE versus SNR

It can be seen that the algorithm in this paper is a little worse than 2-D MUSIC when the SNR is low. But the difference will be narrowed with the SNR being higher, and the estimation accuracy will be mainly the same. What's more, the RMSE curves become lower and lower with the SNR increasing. It indicates that the estimation performance becomes better, however, the trend of becoming better slows down.

6. Conclusion

Based on the conventional 2-D MUSIC, this paper proposes an improved MUSIC algorithm to reduce the computation complexity. Compared with 2-D MUSIC, the search domain can be halved. Therefore, the calculation amounts can be reduced significantly. Through simulation verify, the proposed algorithm can have very close performance to 2-D MUSIC. All in all, this algorithm is an efficient 2-D DOA estimation algorithm.

References

[1] N. T, H. M. K. L-shape 2-dimension arrival angle estimation with propagator method[J]. IEEE Trans. on Antennas and Propagation. 2005, 53(1):1622-1630.
[2] N. Y, T. K. S. 2-D unitary matrix pencil method for efficient direction of arrival estimation[J]. Digital Signal Processing. 2006, 16(6):767-781.
[3] He Z Q, Liu Q H, Jin LN. Low complexity method for DOA estimation using array covariance matrix sparse representation[J]. Electronics Letters. 2013, 49(3):228-230.
[4] Schmidt R O. Multiple emitter location and signal parameter estimation[J]. IEEE Trans.on Antennas and Propagation. 1986, 34(3):276-280.
[5] Capon J. High-resolution frequency wavenumber spectrum analysis[J]. proceedings of the IEEE. 1987, 57(8):1408-1418.
[6] Cai J J,Li P,Zhao G Q.Two-dimensional DOA with reduced-dimension MUSIC[J].Journal of Xidian University. 2013, 40(3):81-86.
[7] Zhang X, Xu D. Angle estimation in MIMO radar using reduced-dimension Capon[J]. Electronics Letters. 2010, 46(12):860-861.
[8] Zhang X, Xu L. DOD and DOA estimation in MIMO radar with reduced-dimension MUSIC [J]. IEEE Communications Letters. 2010, 14(12):1161-1163.
[9] Heidenreich P, Zoubir AM, Rubsamen M. Joint 2D DOA estimation and phase calibration for uniform rectangular arrays[J]. IEEE Transactions On Signal Processing. 2012, 60(9): 4683-4693.
[10] Liu X D,Zhou X L,Xiao J G.Single Snapshot DOA Estimation Algorithm Based on Spatial Smoothing[J].Journal of Detection & Control. 2015,37(06):66-70.
[11] Yan F, Jin M, Liu S. Real-valued MUSIC for efficient direction estimation with arbitrary arrays[J]. IEEE Trans. on Signal Processing. 2014, 62(6):1548-1560.
[12] Golub GH, Loan CF. Matrixs computations[M]: The John Hopkins University Press. 1996.

IOP Conf. Series: Journal of Physics: Conf. Series **1060** (2018) 012027 doi:10.1088/1742-6596/1060/1/012027

BTM Topic Modeling Approach to Named Entity Linking

Menglu You [1], Hong Yang [1], Zhengkui Lin[1], Bopeng Liu[1] and Chao Li[1]

[1]School of Computer Science and Technology, Dalian Maritime University

dalianjx@163.com

Abstract. Named entity link is the process of linking a given named entity in a document to an unambiguous entity in a knowledge base, including the merging of synonymous entities, disambiguation of ambiguous entities, etc. Firstly, the paper construct the named entity knowledge base by using the offline version of Wikipedia. We improve the MPM co word measure which only considers the co-occurrence frequency without considering the frequency of individual words so that it can calculate word co-occurrence coefficient. Secondly, we propose a method named entity BTM link based on topic model. Finally, the named entity in the given text is linked to an ambiguous named entity in the knowledge base.

1. Introduction

Named Entity Linking (NEL) refers to the process of linking a named entity that has been identified in the text to a specific real entity in the knowledge base[1]. The core problem of NEL is the semantic disambiguation problem of the named entity. The main reason is that there is a problem of entity ambiguity in the text written in natural language, that is, the diversity of entity semantics and expression forms, so that a named entity may correspond to Multiple real-world entities[2]. In order to solve the problem that the link quality of the named entity link technology is not high, this paper proposes a named entity link method based on BTM theme model.

2. Related work

The naming entity is a text entity with explicit semantic information in the document[3], including naming aliases, pronouns, nouns, etc., which are included in the text. The named entity link (NEL) links the named entity in the document to a Known, unambiguous exogenous knowledge base in the process. The named entity link can be divided into two main steps: candidate entity generation and candidate entity sorting, that is, by selecting the scope of the candidate entity by the candidate entity, and selecting the most appropriate context entity as the result of the final link within this range.

Sun Zhen et al. [4] summarized the research techniques for named entity recognition in their papers, and listed the main methods for named entity recognition, including rule-based methods, statistics-based methods, and hybrid methods. Xiaohui Yan et al. [5] proposed a BTM model, which can play a good role in dimension reduction, and can also use the topic features to express the characteristics of the document. It helps the semantic modeling of short texts. This article comprehensively discusses the methods of constructing the knowledge base proposed in the literature [1,2], uses the features of Wikipedia, improves the specific implementation technology, and focuses on the knowledge base construction and word co-occurrence measurement based on Wikipedia. The research of entity semantics modeling proposes an improved method of named entity semantic modeling based on BTM and a named entity linking method based on BTM topic model. The effectiveness of the method is verified by experiments.

3. Knowledge Base Construction Based on Wikipedia and Word Coverage

3.1. Word covariance measure

Word is now widely used in the field of text analysis and information retrieval. Its core idea is that the frequency of co-occurrence of words and words reflects the semantic association between words to some extent. The methods used to collect the common words are common matrix method, Point Mutual Information (PMI), conditional probability method and T / Z fraction method.

The process of word co-occurrence is divided into the following four steps: document preprocessing, processing, projection, calculation.

(1) Document preprocessing

The words of the word library are divided into sentences and the number of each sentence is recorded, and the sentence is divided by words. Each occurrence of the word marks the number of the sentence and its number at the sentence position, and then deletes the word to mark part of speech.

(2) Processing

After the text is preprocessed, the word frequency of each word can be counted and the corresponding (tf * 2) matrix is generated. The first column in the matrix is the number of the sentence where the word is, and the second column is the position number of the sentence where the word is.

(3) Projection

After the text has been processed, the word can be projected onto the two-dimensional coordinates according to the word matrix. The # S axis represents the number of the sentence, and the P-axis indicates the position number of the word in the sentence.

(4) Calculation

The Pasteur coefficient (BC) is an approximate calculation of the overlap of the two statistical samples, as shown in Equation (1).

$$BC(a, b) = \sum_{k=1}^{N} \left(\sum a_k b_k\right)^{1/2} \quad (1)$$

In Equation (1), N is the number of partitions, Σak and Σbk are a, b samples appear in the k partition. The calculation is based on a fixed length of the calculation, not suited to a variety of sentences, based on the Pasteur coefficient, Ji Qi [6] proposed formula (2), by calculating the average distance between the two words, and thus dynamically adapt to other types of sentence.

$$MPM(a, b) = \sum_{k=1}^{N} \left[\frac{\sum a_k \sum b_k}{avgD(a_k, b_k)\,(1+ \mid \sum a_k - \sum b_k \mid)} \right]^{1/2} \quad (2)$$

N is the number of sentences, Σak and Σbk words a and the number of words b appear in the sentence k, avgD (ak, bk) is the average distance between the word a and the word b in the sentence k. When the frequency of the occurrence of a word is large, and the other occurs very little, it will lead to the measurement of the word is not ideal. Olga Kolesnikova et al. [7] used conditional probabilities to fix the deviations and solve the above problems. Based on Equation (2), we introduce the conditional probability difference of word a and word b to fix the deviation. The formula is shown in Equation (3) as follows:

$$LMPM(a, b) = \sum_{k=1}^{N} \left[\frac{\sum a_k \sum b_k}{lavgD(a_k, b_k)\,(1+ \mid P(a_k|b_k) - P(b_k|a_k) \mid)} \right]^{1/2} \quad (3)$$

In Equation (3), N is the number of sentences, Σak and Σbk are the numbers of words a and b in the sentence k, and ΣakΣbk represents the number of occurrences of words a and b simultaneously. P (ak | bk) is the probability that the word a appears in the case where the word b appears in the sentence k, and lavgD (ak, bk) is the average distance between the word a and the word b in the sentence k.

3.2. Experimental and experimental results analysis

Because of the correlation of semantics, this paper compares the semantic relevance with other word covariance methods to judge whether the covariance of extracted words is good or bad. In order to evaluate the experimental results, the RMSE coefficients are obtained by combining the semantic correlation with the combination words, and the formula is shown in Equation (4) as follows:

$$RMSE = \left[\frac{1}{n}\sum_{i=1}^{n}(learn_i - test_i)^2\right]^{1/2} \tag{4}$$

Where n is the logarithm of the data set, $learn_i$ is the semantic relevance of the i-th word pair, and $test_i$ is the data set. Where the result of the relational distribution is in the range of [0,1] and the result of the distance distribution is in [0, +), in order to make the distance distribution within the range of [0,1],so use the Equation (5) as follows:

$$r = 1 - \frac{d}{\max(D)} \tag{5}$$

Where d is the distance value, r is the uniform correlation, max (D) is the largest distance in the result set, and the larger the r value is, the less the result is.

The data set used in this experiment is the CSAW and Tweets datasets, with 307 articles (107 from CSAW and 200 from Tweets). The experimental results are shown in Table 1 and Table 2 below.

Table 1. Comparison results on csaw.

	PMI	CP	TS	BC	MPM	LMPM
Jaccard	0.869	0.231	0.51	0.287	0.249	0.243
Dice	0.138	0.553	0.569	0.364	0.124	0.125
Cosine	0.504	0.811	0.790	0.542	0.434	0.442
E-Distance	0.385	0.781	0.434	0.409	0.288	0.286
KL-	0.217	0.186	0.623	0.525	0.504	0.501

Table 2. Comparison results on Tweets.

	PMI	CP	TS	BC	MPM	LMPM
Jaccard	0.956	0.434	0.907	0.398	0.324	0.321
Dice	0.945	0.817	0.851	0.462	0.273	0.275
Cosine	0.451	0.660	0.748	0.533	0.509	0.506
E-Distance	0.429	0.582	0.717	0.786	0.265	0.267
KL-	0.212	0.345	0.334	0.160	0.145	0.14

As shown in Table 1 and Table 2 above, Jaccard, Dice, E-Distance, E-Distance calculates semantic relevance LMPM on CSAW documents, and Jaccard, Cosine and KL-Divergence are best studied in Tweets documents. LMPM is best The The general trend shows that LMPM is better than other computational methods in word covariance.

4. Naming Entity Semantic Modeling Based on BTM Theme Model

4.1. Semantic Modeling Based on BTM Theme Model

The whole model of BTM is a three-tiered model of word pairs, themes and words. The word pair - subject obeys the Dirichlet distribution, the subject - the word follows the Multinomial distribution, which is layered like LDA. The BTM model is generated as follows:

(1)For each topic z:

(2)Describe the word distribution $\varphi z \sim Dir(\beta)$ under this particular z,

(3)For a whole corpus to describe a global theme distribution $\theta \sim Dir(\alpha)$

(4)For each word pair B (wi, wj) in the corpus, we do the following:

(a) Extract a subject $z \sim Mult(\theta)$ from θ

(b) From the subject z to be drawn from the two words wi, $wj \sim Mult(\varphi z)$

Through the BTM model generation process, the union probability of word pair b (wi, wj) is shown in Equation (6) as follows:

$$P(b) = \sum_z P(z)P(w_i \mid z)P(w_j \mid z) = \sum_z \theta_z \phi_{i|z} \phi_{j|z} \tag{6}$$

The probability formula for the entire corpus is shown in Equation (7) as follows:

$$P(B) = \prod_{(i,j)} \sum_z \theta_z \phi_{i|z} \phi_{j|z} \tag{7}$$

Figure 1 shows the graph representation of BTM.

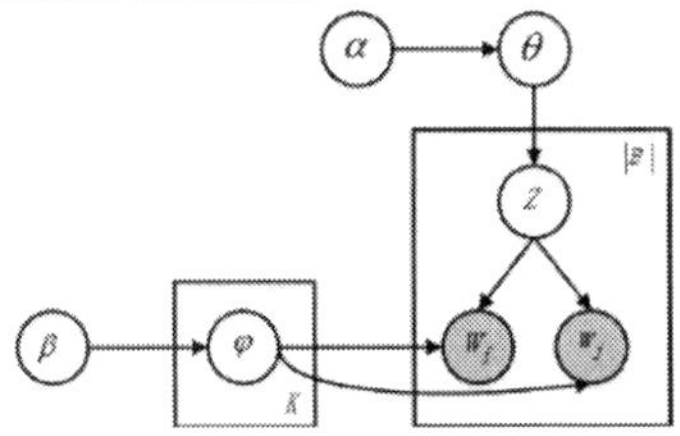

Figure 1. Graphical Representation of BTM.

Which is the subject distribution of BTM corpus, φ is the probability distribution of words under a subject, | B | is the number of word pairs in corpus, wi, wj is two different words in a word pair. However, BTM cannot directly obtain the topic distribution of the document. In order to infer the topic distribution of the document, it is considered that the topic probability of the document is equal to the expected value of the topic probability of the word pair generated from this document.

$$P(z \mid d) = \sum_b P(z \mid b)P(b \mid d) \tag{8}$$

P (z | b) in Equation (8) is estimated from the parameters of BTM, and P (z | b) can be obtained by Bayesian formul as shown in the following Equation (9).

$$P(z \mid b) = \frac{P(z)P(w_i \mid z)P(w_j \mid z)}{\sum_z P(z)P(w_i \mid z)P(w_j \mid z)} \tag{9}$$

The empirical probability distribution of the word pairs in the document is taken as the value of P (b | d), nd (b) is the value of P (z) = θz, P (wi | z) = φi | z). Is the number of occurrences of the word b in document d, as shown in the following Equation (10).

$$P(b \mid d) = \frac{n_d(b)}{\sum_b n_d(b)} \tag{10}$$

In BTM there are two a priori coefficients of α and β, α is the corpus - the a priori coefficient of the subject, β is the a priori of the subject-word, and LMPM can be used as β. B is calculated as Equation (11).

$$\beta_{LMPM}^{w_i,w_j} = \beta_{SYM} + 0.1 * LMPM(w_i, w_j) \tag{11}$$

When βLMPM wi and wj are applied to BTM, a new BTM model of βBTM is obtained, as shown in the following Equation (12).

$$\varphi = \frac{(n_{k,-b}^{w_i} + \beta_{LMPM}^{w_i,w_j})}{\sum_{t=1}^{V}(n_{k,-b}^{w_t} + \beta_k^{w_t})} \times \frac{(n_{k,-b}^{w_j} + \beta_{LMPM}^{w_i,w_j})}{\sum_{t=1}^{V}(n_{k,-b}^{w_t} + \beta_k^{w_t})} \tag{12}$$

Based on semantic modeling process:

(1) Preprocess the document, remove the stop word, stem extraction.

(2) The corpus word pairs are generated for the processed documents and the LMPM coefficients of the word pairs are calculated.

(3) Randomly arrange a topic zi, i [1, | B |] for each word pair, this procedure is the initialization of MARKOV CHAIN.

(4) Calculate the next state based on the previous state and state transition probability.

(a) The subject zb is generated according to P (z | z-b, B, α, βLMPMWi, Wj).

(b) Update nz, nWi | z and nWj | z.

(5) (4) iterates N times until the Markov Chain is stable and writes down the current value of the subject of each word pair.

(6) Calculate φ using Equation 4.7 and use Equation 4.3 to calculate the subject distribution ψ for each document.

Next, the algorithm is used to describe the improved BTM semantic modeling process.

algorithm 1: Semantic Modeling Based on BTM Model
enter: Document, number of topics, super parametersα, β_{sym}, Number of iterations N
Output: θ, φ and ψ algorithm: Document preprocessing, access to the word on the corpus B LMPM Coefficient of Calculated Words to Corpus Initialize, randomly for a word on a subject for i = 1 to N for b∈B a). According to P (z | z-b, B, α, βLMPMWi, Wj) for the word pair b distribution theme z_b b). Update nz, nWi | z and nWj | z. Calculate θ, φ and ψ according to Equations 4.8 and 4.11 and 4.3

4.2. Analysis of experimental and experimental results

This experiment uses Gibbs-LDA, BTM and methods for comparison. Using the Tweets data set in UDI-TwitterCrawl-Aug2012, there are 140,000 Tweets files, each with 450 songs, each randomly selected 200 files. More than 90,000 Tweets data were used for Gibbs-LDA, BTM and betaB experiment. In the experiment, the number of subjects was K = 5, 10, 15, 20, 25, 30, the number of iterations was 1000 and the iterations were started after 800 iterations, once every 40 times, α was 0.05, But for βBTM it is necessary to recalculate the βLMPM w_i, w_j between the word pairs according to β and using Equation (3.5).

In this paper, we use the probability distribution of Gibbs-LDA, BTM and βBTM in different subject cases as the input feature matrix of similarity calculation. The subject is used as the text feature, the similarity is calculated by JS distance and the similarity is used as KNN algorithm The measurement standard (KNN is the algorithm in WEKA), using Precision and F1 to measure the experimental results of the three models.

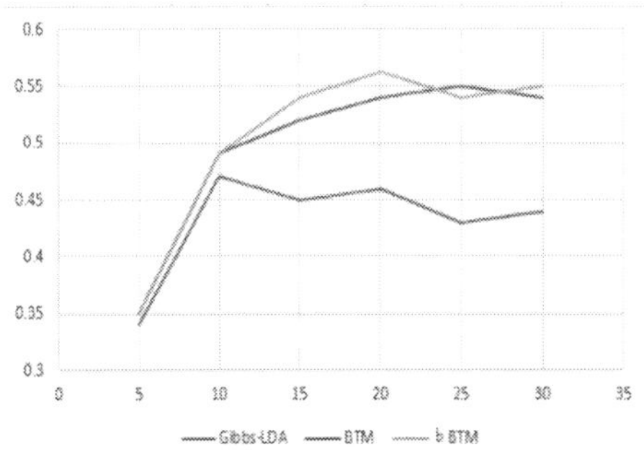

Figure 2. Gibbs-LDA, BTM and βBTM different topic Precision.

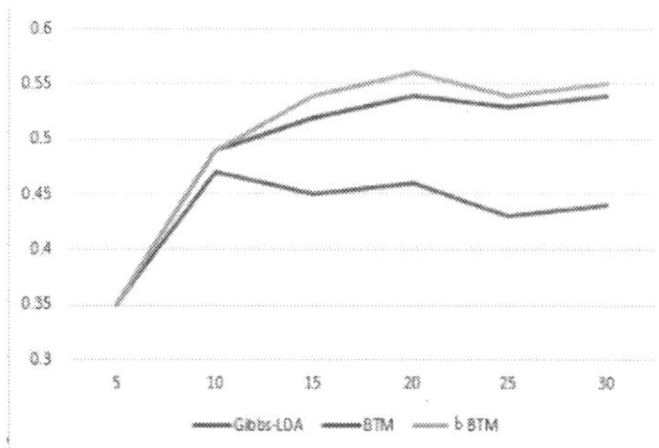

Figure 3. Gibbs-LDA, BTM and βBTM different topic F1.

Figure 2 is the accuracy of Gibbs-LDA, BTM and βBTM under different topics, Gibbs-LDA has the maximum accuracy when the number of subjects is 10, and the accuracy of BTM is 25 when the number of subjects is 25, βBTM achieves maximum accuracy when the number of subjects is 20. The overall trend of βBTM is better than Gibbs-LDA and BTM.

In Figure 3, under the same subject, Gibbs-LDA, BTM and βBTM F1 value, Gibbs-LDA in the subject number of 10 when the F1 value to the maximum, BTM in the subject number of 20, F1 value to the maximum, βBTM in the subject number of 20, F1 value to the maximum. The overall trend of βBTM is better than Gibbs-LDA and BTM.

5. Naming Entity Linking Method Based on BTM Theme Model

5.1. Naming Entity Linking Process Based on BTM Theme Model

After constructing the model by βBTM theme modeling method, the named entity link system can get the distribution of words and named entities in the subject space respectively.

The named entity link algorithm is as follows:

algorithm 2: Name the entity link method
(1). According to the synonyms in the knowledge base, all the named entities in e are Entity merging, using standard naming entities to represent, get a description of the standard, = {e1, e2, e3, …, en};
(2). For each named entity entity ei ∈ e 'in e', according to the ambiguous vocabulary in the The candidate entities are pruned to obtain the candidate entity entity set Cei = {ei, 1, ei, 2, ei, | Cei | = 0, that failed to find the corresponding entity in the knowledge base, that the named , Return to the label NIL; when | Cei |> 0, perform steps 3) to 5);
(3). Using the formula (3.3) to calculate the degree factor of each pair of pairs in wd, we can The subject distribution of named entities Te;
(4). The subject probability of the document is equal to the expected value of the subject βBTM in the θ and φ to infer the subject of the document distribution Td;
(5). For each named entity ei ∈ e 'in e', the subject distribution Td of its document is ei, j the theme distribution Tei, j cosine similarity, through the training set to learn the When the large value is less than λ, returns NIL, which is greater than λ, returns the closest

5.2. Analysis of experimental and experimental results

In this paper, the experimental environment is Ubuntu 16.04 32bit system, Inter (R) Core (TM) i5-3470 CPU @ 3.20GH, RAM 4G, the parameters α, βsym initial value of 0.05,0.01, using Gipps sampling method, iterative The number of 1000 and the iteration 800 after the start to save, save every 40 times. In this paper, we use the named entity linking method of Gibbs-LDA, BTM and βBTM in different subject cases.

First, the training set of data training to get the optimal λ, Table 3 for the use of Tweets training set and based on the βBTM model named entity link, in different cases, the accuracy rate, recall rate, F1 worthy of the situation.

Table 3. Precision, recall and F1 under different number λ

λ	Accuracy	Recall rate	F1
0.1	65.72%	56.82%	60.95%
0.3	67.91%	59.26%	63.29%
0.6	68.65%	62.71%	65.55%
0.8	67.51%	59.16%	63.06%
0.9	66.03%	57.12%	61.25%

From the experimental results, with the growth of λ, F1 value increases, when λ grows to 0.6, F1 reaches the maximum, λ> 0.6, F1 value begins to decrease, indicating that in the Tweets data set, λ is 0.6 , The name of the entity link the best results.

Figure 4 shows the accuracy of the Tweets data set under different themes. Figure 5 shows the recall rate for the Tweets dataset under different themes. Figure 6 shows the value of the Tweets dataset under different themes.

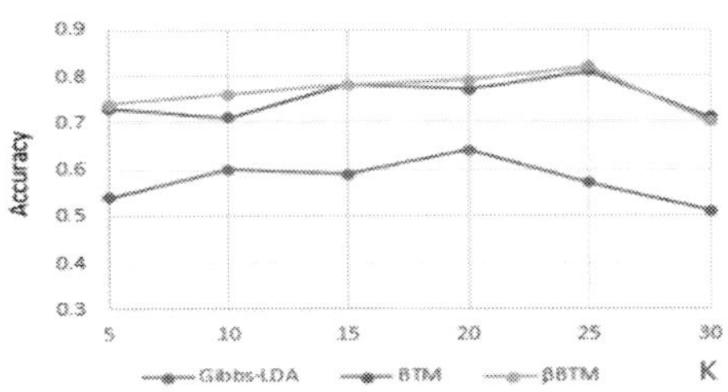

Figure 4. Precision-value under different number of Latent Topics

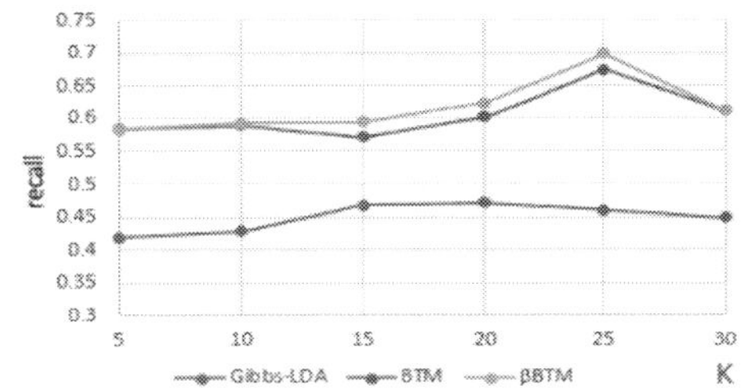

Figure 5. Recall-value under different number of Latent Topics

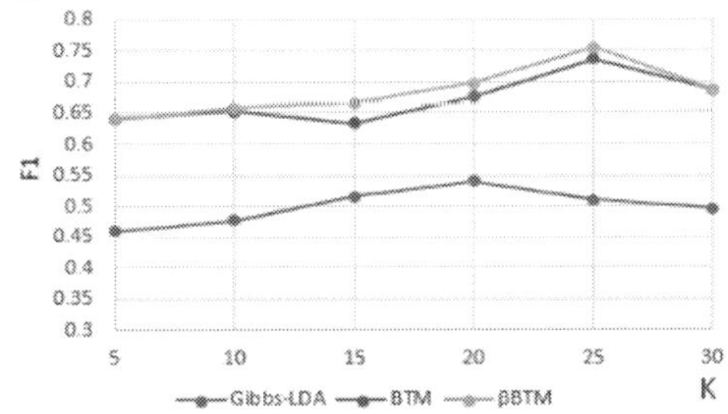

Figure 6. F1-value under different number of Latent Topics

From the experimental results, βBTM achieves the best accuracy, recall and F1 values when the number of subjects is 25; BTM achieves the best accuracy, recall and F1 values when the number of topics is 25. Gibbs-LDA achieves the best accuracy, recall and F1 values when the number of subjects is 20. From the overall trend, the βBTM link is better than BTM and Gibbs-LDA.

6. Summary

In this paper, a named entity linking method in short text is deeply researched and analyzed. A named entity linking method based on βBTM theme model is proposed. After naming entities in the document, semantic modeling can be obtained. The distribution of the subject space is calculated by calculating the cosine similarity of each candidate entity of the named entity in the named entity of the named entity, and selecting the closest named entity as the final link. However, there are still deficiencies. In the construction of the knowledge base, this article only uses some of the knowledge in Wikipedia, and no further research and discussion has been conducted on other external knowledge bases such as YAGO and Freebase.

References

[1] Baoxing Huai,Tengfei Bao,Huanshu zhu,et al.A named entity linking method based on the probability theme model[J].Software journal.2014,25(9):2076-2087.
[2] Gang Zhao,Ji Wu,Dingding Wang,et al. Entity disambiguation to Wikipedia using collective ranking[J]. Information Processing and Management.2016. pp:1247–1257.
[3] Hachey B, Radford W, Nothman J,et al. Evaluating entity linking with Wikipedia[J]. Artificial Intelligence,2013,194:130–150.
[4] Sun Zhen,Wang Huilin. Review of research progress on named entity recognition[J].New Technology of Library and Information Technology.2010.
[5] Xiaohui Yan, Jiafeng Guo, Yanyan Lan,et al. A Biterm Topic Model for Short Texts. IW3C2. ACM,2013.
[6] Ji Qi, Yukio Ohsawa. Matrix Plane Model: A Novel Measure of Word Co-occurrence and Application on Semantic Relatedness. ICDMW.2015.pp:1246-1253.
[7] Olga Kolesnikova. Survey of Word Co-occurrence Measures for Collocation Detection. Computaciony Sistemas. 2016, pp. 327–344.

A Method of Fine - grained Emotion Analysis Based on Semantic Analysis Network

Bopeng Liu [1], Hong Yang [1], Zhengkui Lin[1], Menglu You[1] and Zhaohui Wang[1]

[1]School of Computer Science and Technology, Dalian Maritime University

dalianjx@163.com

Abstract. Sentiment Analysis also known as the Tendency Analysis, it is the emotional color of the subjective text analysis, processing, induction process. The traditional text emotion analysis research mainly faces the chapter and the sentence level, realizes the corresponding emotional polarity judgment. Based on the characteristics of Chinese language commentary language expression, this paper uses a semantic analysis based on the combination of attribute words and emotional words to identify the extraction method, and the extraction method and the method of attribute words combined with a method based on semantic analysis Network Review Fine - grained Emotion Analysis.

1. Introduction

With the increase of the number of comments on e-commerce sites and commodity attributes, the traditional commentary emotional analysis has been unable to meet the needs of shops or consumers. Buyers want to get products that match their own preferences from the product reviews. The merchant wants to get the commentary's comments on the product from the comments, that is, the buyer's comments show the degree of preference for the goods and improve their competitiveness.

2. related work and problems

A class of goods have a lot of attribute, fine-grained emotional polarity classification focuses on finding these attributes from the comments and judging the emotional tendencies of consumers for each attribute. In practice, fine-grained emotional polarity classification is significant for consumers to improve consumer behavior, product performance and service. The main research work focuses on extracting keywords and emotional words.

The product attributes are divided into explicit and implicit attributes through Mingqing Hu, Bing Liu's [1] method. Explicit sphrases with explicit attributes as attribute words. The implicit attribute does not appear in the comment as a noun or noun phrase that describes the performance of the purchased item, which is obtained by way of a semantic understanding of the statement before or after it. Kobayash [2] build product attributes for cars and games, each of which is represented by a triplet <Subject, Attribute, Value> and then manually tagged it.

In general, the research of emotion analysis is mainly analysis of the emotional polarity of coarse-grained users, and the research on fine-grained emotional polarity classification for product or service attributes is relatively few. Therefore, this paper based on the literature [3] to further study the network assessment of fine-grained emotional analysis.

3. Fine - grained emotion analysis corpus pretreatment

Fine - grained emotional analysis corpus preprocessing mainly includes Chinese word segmentation and part of speech annotation. This paper adopts the Language Technology Platform (LTP) of Harbin Institute of Technology to carry out data preprocessing research. In terms of part-of-speech, LTP uses 863 parts. Examples of processing results are shown in Table 1 below. The English meaning is "Often to eat, It's super good, cost-effective, Complete variety, taste delicious. Everyone is having a good time".

Table.1 Part of the word, part of the annotation, to stop the word processing results examples.

Comments before corpus processing	Comments After the corpus preprocessing
Often to eat, It's super good, cost-effective, Complete variety, taste delicious. Everyone is having a good time	eat/v | super/b | good/a | cost/d | effective/a | complete/n | taste/n | delicious/a | everyone/d | eat/v | have/d | good time/a

4. A Method of Joint Extraction of Attribute Words and Emotional Words Based on Semantic Analysis

Emotional words in the modified attribute words expressed in a more flexible way, it's speech and syntax are very rich. In view of these characteristics of network commentary, this paper combines the semantic and grammatical extraction rules put forward by Jiang Tengjiao and Tang Ming-feng [4]-[6] to extract a new attribute word in combination with word frequency and word distribution ratio extraction method.

4.1 Candidate attribute words - emotional words on the extraction method

In this section, the attribute word - emotional word pair is defined as a set of <attribute words, emotional words>. The main goal of the semantic role annotation is to identify and mark the semantic roles associated with the predicate to find the attribute words that the predicate is modified. However, the role of the general emotional word in the comment sentence is not only as a predicate, but also may be other elements of the element, at this time, you must expand the semantic role label. The meaning of each label is shown in Table 2.

Table.2 Marked Names and Meanings in Semantic Role Marking.

Annotated	Explanation	Annotated	Explanation
A0	agent	**A1**	Accept the matter
A2	Indirect role object	**TMP**	time
DIR	direction	**DIS**	Chapter mark
EXT	range	**LOC**	position
MOD	General modification	**NEG**	negative
PRP	proposal	**ADV**	Adverbs

The semantic analysis of the commentary sentences in the LTP system can be briefly described by the following examples.

example: "The taste of Bashu hot pot is getting better and better now, and the long-term development prospect is good." This sentence is marked by LTP, as shown in Figure 1 below.The English meaning is "The flavor of the Batu hotpot is now getting better and the long-term development prospects are promising"

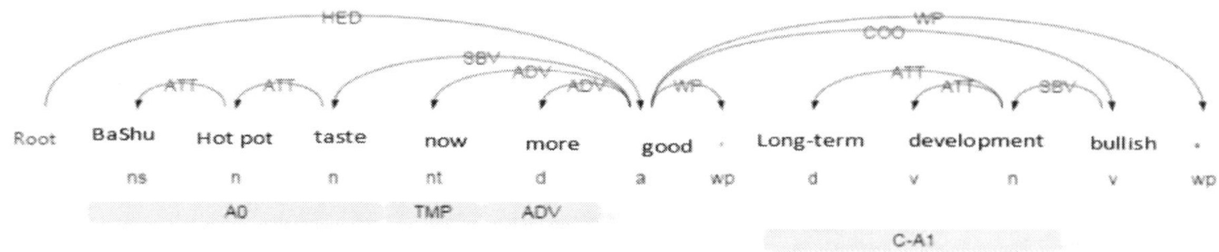

Figure.1 Semantic analysis and annotation of example 1.

In Figure 1, the word is separated by words and words, and the result of the annotation is expressed below the word. The interdependence between the words is shown in the figure's arc; the semantic role annotation results are shown below the part of speech.

Jiang Tengjiao in the literature proposed predicate attribute words meet the pre-rules 1, pre-rules 2 and pre-rules 3, respectively, the corresponding conditions of use and attribute word priority:

(1) non-psychological verb as a predicate, the priority of A0 → A1 → A2.

(2) psychological verbs as predicates, the priority of A1 → A2 → A0.

(3) the three in the same sentence, the attribute word for a number of A0, A1 or A2 combination.

In the commentary, the emotional words can be presented as many corners, including adverbs, predicates, adverbs of the object, and attributive elements. Thus, the rule of extract attribute are shown in Table 3.

Table.3 Extraction of Attribute Words - Eight Rule Tables of Emotional Word

rule	Predicate state	Attribute word - emotional word pair
rule 1	When the emotional word is a predicate in a semantic role	<The predicate word, the emotional word>
rule 2	When the emotional word is a predicate object, adverbial, or complement	<The word of the predicate + predicate
rule 3	An adverbial adverb or adverb	<Predicate A0 + predicate + object, emotional word>
rule 4	The above rules have not yet found the emotional word of the word	The latest term of emotional words, emotional words

In order to facilitate the use of LTP platform, the system proposed the definition of 12 kinds of dependencies, used to analyze the sentence structure and the word collocation relationship, as shown in Table 4.

Table.4 LTP dependency relationship name and meaning table

Annotated	Structural relationship	Annotated	Structural relationship
ATT	Fixed relations	**DE**	“A”structure
QUN	Quantitative relationship	**DI**	“the”structure
COO	Constellation	**DEI**	“get”structure
APP	Cohabitation	**BA**	“hold”structure
LAD	Pre-attached relationship	**BEI**	“cover”structur e
HED	core	**SBV**	s-v relation

IOP Conf. Series: Journal of Physics: Conf. Series **1060** (2018) 012028 doi:10.1088/1742-6596/1060/1/012028

4.2 Attributes and emotional words of the characteristics of the calculation method

With the candidate word pair, we can calculate the alignment probability of the attribute word to the emotional word according to the formula (1).

$$P(\omega_o \mid \omega_t) = \frac{count(\omega_o, \omega_t)}{count(\omega_t)} \tag{1}$$

Similarly, according to formula (2) to calculate the emotional word between the words to the probability of alignment.

$$P(\omega_t \mid \omega_o) = \frac{count(\omega_o, \omega_t)}{count(\omega_o)} \tag{2}$$

The number of times expressed and multiplied in the formulas (1) and (2) above. Through the probability of alignment and the noun / noun phrases and adjectives / verbs within the existence of an emotional relationship between the specific out, the formula as shown in (3)

$$Association(w_o, w_t) = (\mu * p(w_o \mid w_t) + (1-\mu) * p(w_t \mid w_o))^{-1} \tag{3}$$

The formula (3) is a harmonic factor that adjusts the two alignment probabilities that are set in this article.

In the commentary, the attribute word is the product itself; in addition, the most likely to be the attribute word in the commentary corpus is the evenly distributed candidate word with high frequency. To this end, we have selected five commentary corpus (sports, politics, humanities, science and technology, news commentary) in the field that is not the same as the topic "hot pot" in this paper.

The corpus of the comment in front of the text is written as C. In C, each comment can be treated as a separate class, and C contains n useful comments, that is, by calculating the information entropy of the comment, we can get the complexity of the whole system, but also can show the attributes The distribution of words in the document. The formula is as shown in (4):

$$IE(t_i) = -\sum_{j=1}^{n} p(d_j, t_i) \log p(d_j, t_i) \tag{4}$$

Equation (4) shows the probability that the candidate word appears in the comment, as shown in equation (5):

$$p(d_j, t_i) = \frac{tf_{ij}}{\sum_{j=1}^{n} f_{ij}} \tag{5}$$

Which means that the word frequency of the candidate attribute word in the jth comment, if only in a comment, then, then, so that, in order to simplify the calculation, the probability of inaccuracy is 0, so in equation (5) Add the constant factor and sum in the denominator with the previous denominator. The new probability formula is (6):

$$p(d_j, t_i) = \frac{tf_{ij}}{\sum_{j=1}^{n} f_{ij} + 0.0001} \tag{6}$$

"Person", "things" and other frequent nouns do not belong to the members of the word. That is, the frequency of occurrence of the attribute word is not necessarily high. On the contrary, the frequency of low vocabulary may also be a property word. In order to dig out more of the attribute words in the comments as much as possible, this paper is calculated by the corpus of different fields, combined with the document frequency. The specific formula is (7):

$$Ds(t_i)_j = \begin{cases} \dfrac{\log(1 + df_{in})}{\log(1 + df_{out\_j})} & if\ df_{out\_j} \neq 0 \\ \alpha \times \log(1 + df_{in}) & otherwise \end{cases} \tag{7}$$

In equation (7), the frequency of the document in the commentary corpus is the frequency of the document in the corpus of the jth (5 in total) domain. (4) and the formula (7) to find the characteristics of the word word Indicator formula (8) as shown in the formula:

$$I(t_i) = \overline{Ds(t_i)} \times IE(t_i) \quad (8)$$

In formula (8) $\overline{Ds(t_i)} = \sum_{j=1}^{m} Ds(t_i)j \Big/ m$

For the recognition of emotional words, this paper combines the word frequency of words and the distribution of words in the corpus to calculate the characteristic Indicator of candidate emotion words. The formula is (9):

$$I(o_i) = \log(1 + df_i) \times D_i \quad (9)$$

$$D_i = \frac{\overline{tf_i}}{\sqrt{\sum_{j=1}^{n} (tf_{ij} - \overline{tf_i})^2 \Big/ (n-1)}} \quad (10)$$

In the formula (9), we show the frequency of the candidate words in the corpus. The formula (10) indicates the distribution of the candidate words in the corpus. Is the frequency of the candidate word in the j'th comment, indicating the average word frequency in the commentary corpus.

4.3 The Method of Exact Extraction of Attribute Words and Emotional Words

We can extract the emotional words and attribute words accurately and combine with the random walk algorithm of undirected bimodal and restart to construct a probabilistic model. The probability of extraction is higher than that of the former Set the words of the threshold as emotional words or attribute words. The algorithmic algorithm is described as follows:

Suppose that s is the starting node of random walk, q is the current node, node q either returns to the starting node with probability c (ie, restart probability), or moves along probability (1-c) along the edge of the graph Adjacent nodes, A is a standardized adjacency matrix, to calculate the relationship between the nodes in the graph and the node s (11):

$$\vec{r} = c\tilde{A}\vec{r} + (1-c)\vec{e} \quad (11)$$

Equation (11) is the starting vector, when the initial node of the walk is s, otherwise.

The bipartite graph G (V, E, A) is shown in Figure 2. V denotes the set of points in the bimodal, representing the emotional word, representing the attribute word. E represents the edge set in graph G, and if there is an emotional relationship between the attribute word in the graph and the emotional word, it is represented by an edge.

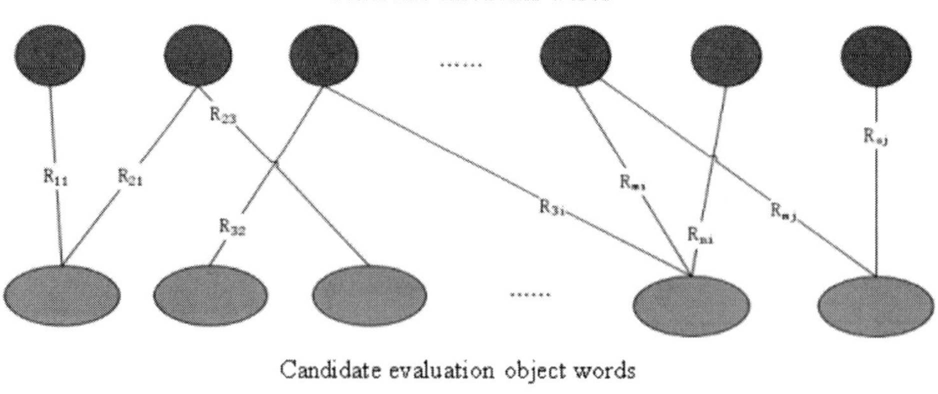

Figure. 2 An example of a dichotomous graph of the relationship between the modeled emotional word and the attribute word.

A represents the emotional relationship between the two, calculated by the formula (3). The formula (12) represents the process of calculating the affective word and the attribute word, and iterates the formula (12) until the energy value state is stable.

$$E(t) = \lambda \times R \times E(o) + (1-\lambda) \times I_t$$
$$E(o) = \lambda \times R \times E(t) + (1-\lambda) \times I_o \quad (12)$$

In the formula (12), the energy values of the attribute word and the emotional word are expressed by the sum. R denotes the relationship weight between the relation matrix, the i-th candidate attribute word and the j-th candidate emotion word. The characteristic vector representing the candidate attribute word is calculated by the formula (8). The characteristic vector representing the candidate emotion word is calculated by the formula (9). Is a harmonic parameter.

4.4 Analysis of experimental and experimental results

In this paper, the sample data set is marked by manual annotation to verify the accuracy of the extracted results, and compared with other extraction methods.

The parameters in equation (12) determine the relationship between the characteristics of words and the emotional relationship between words. The exact rate, recall rate, and F-measure value are shown in Figure 3.

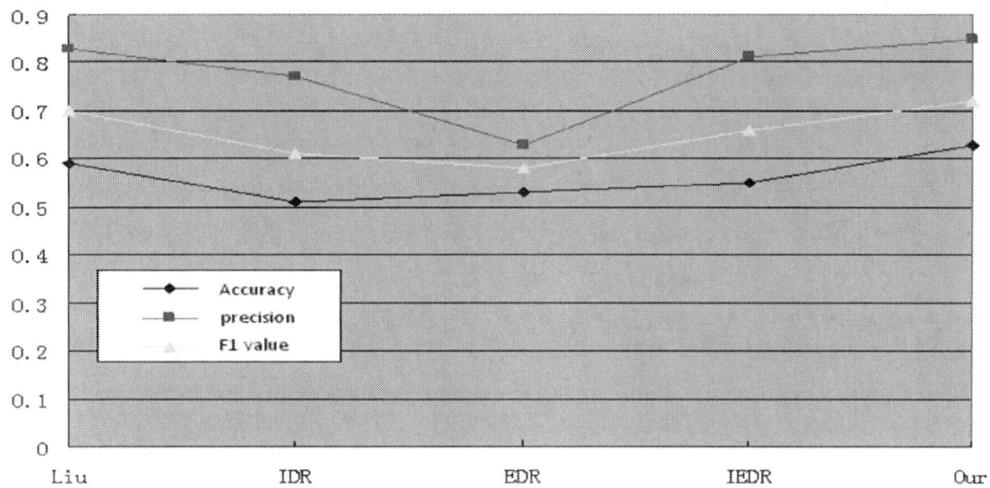

Figure. 3 The graph different λ in the three values on the dataset

Through the curve of Figure 3 can be seen when the effect of about 0.4 when the best.After the parameter values are determined, we compare our experimental results with Hu, IDR, EDR, IEDR methods (Our method). The experimental results are shown in Table 5.

Table.5 Five methods in the hot pot restaurants on the results of the experimental comparison

meathod	Accuracy	Recall rate	F1 value
Liu	0.60	0.85	0.71
IDR	0.51	0.77	0.61
EDR	0.53	0.63	0.58
IEDR	0.55	0.81	0.66
Our	0.63	0.85	0.72

From the comparison of the results of Table 5, with the hot pot shop comments as experimental corpus, respectively, the five groups of experiments, we can see in this method the highest accuracy. We found that the Hu method does not consider the distribution of words, domain correlation, candidate emotional words and other information, but only consider the importance of candidate attribute words, resulting in low accuracy. Therefore, the three important factors influencing the extraction of attribute words include the distribution of words, the information of field related information and the word information of emotional words.

5. Fine - grained Emotional Tendency Calculation Method for Attribute Words

In this paper, we find that if we have both the attribute words and the emotional words related to it in the sentence in the commentary corpus, then there is a specific relationship between the two. The commentary in the commentary is an attribute word or a category to which it belongs.

5.1 Emotional Review Unit Construction

After the pretreatment of the test corpus, the attribute words affect the word pairs, the degree adverbs and the negative words, and construct the corresponding emotional evaluation unit according to the form given below:

Emotional comment unit = <comment unit ID, store name, attribute word, emotional word, degree word, negative word>

(1)The extraction of degree adverbs

An adverb is an adverb that limits or modifies an adjective or adverb. In this paper, 219 degree words are selected in the knowledge network. Divided into six levels, the weight added to the degree adverbs. The degree of adverb and its weight setting examples and extraction algorithms are simplified. Table 6 and Table 7.

Table.6 Degree adverb and its weight setting example

Degree adverb extremely weight setting instance
exceed(1.5), very(1.25), extremely(2), most(2), owe(0.5)......, slightly(0.8)

Table.7 Degree Adverb Extraction Algorithm

```
Algorithm illustrate:Look up the degree adverb in a given sentence
Algorithm input:CString
Algorithm output:Ctring
Algorithm description:
For each Paragraph in Document:
 For each Line Paragraph
  For each Group in Line:
   For each Word in Group:
    if word in senDict:
     senWord = (Position in sentence,Emotional tendencies,Emotional intensity)
     LastSenWordPosition = 0 ##The position of an emotional word in a sentence
     for i in range(senWord[0],LastSenWordPosition,-1):
      if Group[i] in degreeDict:
         DegreeWord = ((Position in sentence,Modification strength))
     LastSenWordPosition = senWord[0]
```

IOP Conf. Series: Journal of Physics: Conf. Series **1060** (2018) 012028 doi:10.1088/1742-6596/1060/1/012028

5.2 Emotional Aggregation and Emotional Tendency Classification

In the process of analysis and research, this paper divides the comment sentence into seven categories: price, service, side dish, soup, seasoning, environment and location, and returns the extracted attribute words to the corresponding classification. The tendency of calculating the corpus in all the emotional tendencies of aggregated corpus is the analysis of coarse-grained emotion. The emotional tendencies of the unit or sentence are emotional analysis. Among them, the emotional unit of the emotional value of the formula is:

Emotional unit emotional value = negative word -1 * degree word weight * emotional word weight

Because the negative words and the degree of adverbs in the sentence in the local differences for the emotional unit expressed by the degree of emotional will have different effects, when the negative word is located before the adverb adverb, the negative feelings of the negative word 0.5, when the negative word in the degree adverb after its Emotional weight take -1.

5.3 Analysis of experimental and experimental results

According to the fine-grained emotion analysis process, combined with the semantic analysis of the combination of attribute words and emotional words extraction method, the construction of emotional units, while the emotional value of the calculation, the data visualization of the results shown in Figure 4 and Figure 5 shows.

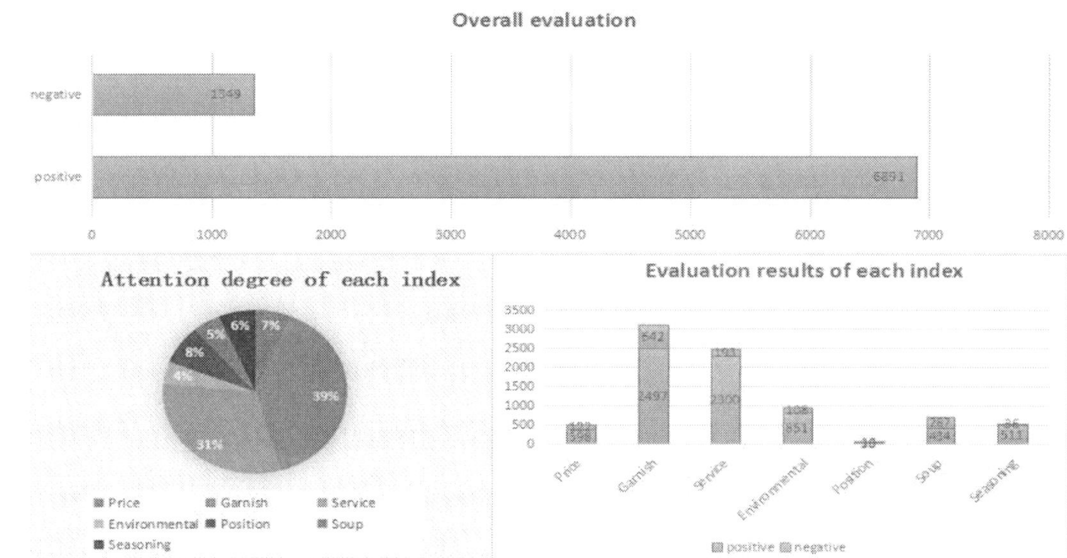

Figure.4 Results of emotional analysis 1

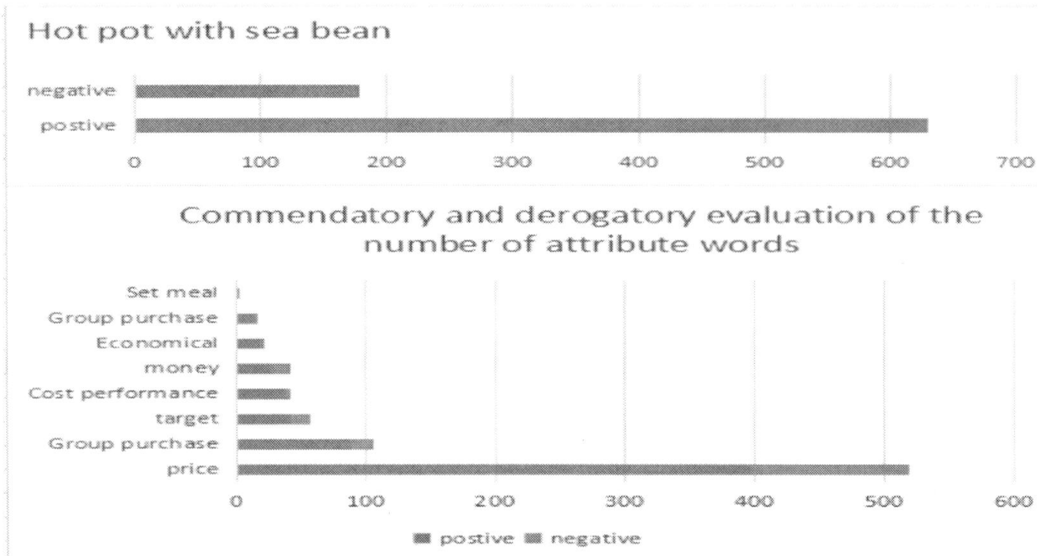

Figure.5 Results of emotional analysis 2

6. Summary and Prospect

In this paper, based on the semantic analysis of attribute words and emotional words jointly extract candidate attribute words and emotional word pairs, and calculate the word frequency and document ratio characteristics of attribute words and emotional words and phrases to further accurate calculation, and then the extraction method and orientation The method of attribute word combination puts forward a semantic analysis based on fine-grained sentiment analysis method for web commentary. Finally, it verifies the effectiveness and superiority of this method over traditional methods. However, due to the diversity of natural language expressions, sometimes subjective views of critics do not appear clearly in sentences, but are concealed in the comments with hidden language. Further mining of implicit attribute words and emotions helps to more fully understand the author's subjective evaluation of goods or topics.

References

[1] Hu M, Liu B. Mining opinion features in customer reviews[C]. AAAI. 2004: 755-760.
[2] Kobayashi N, Inui K, Matsumoto Y. Collecting evaluative expressions for opinion extraction[M]. Lecture Notes in Artificial Intelligence, 2004(3248):596-605.
[3] ShiHanXiao.Research on Fine-grained Sentiment Analysis[D]:(Doctoral Dissertation). Suzhou: Soochow University, 2013.
[4] JIANU Tend Jiao WAN Chang-Xuan LIU De-Xi LIU Xi-Ping LIAO Uuo-Qiong
[5] Extracting Target-0pinion Pairs Based on Semantic Analysis[J]. Journal of computer science, 2017, 03: 617-633.
[6] TangMingShuang. Sentiment Analysis of Online Review Based on Machine Translation Model[D].University of Electronic Science and technology, 2016.

The R&D Efficiency Evaluation of the University Teachers Based on A Two-Stage DEA Model

Xuehao Sun[1], Zhengkui Lin[1], Lei Zhang[1], Yu Lian[1] and Lei Ding[1]

[1]School of Computer Science and Technology, Dalian Maritime University

dalianjx@163.com

Abstract. This paper measures university teacher's R&D efficiency based on a two-stage DEA model. Firstly, this paper analyzes traditional DEA model and proposes a two-stage DEA model to overcome the disadvantage of traditional DEA, and then builds a new input-output index system to evaluate teacher's R&D efficiency. The new model is applied to analyze the R&D efficiency of 75 teachers of a university in 2014. The result show that professors, female teachers and engineering teachers have larger efficiency.

1. Introduction

At present, many scholars, at home or abroad, have made some progress in related researches of the performance of university R&D input-output. Kim[1]applied DEA model to evaluate 38 university R&D, and utilized all kinds of input-output variables to ensure technology and scale efficiency. And then, in a concrete case, he discussed how to use a way of organizational change to improve the performance of inefficiency university R&D. On the basis of bibliometrics methods, Abramo et al.[2] applied DEA method to evaluate university R&D efficiency of cross-discipline in Italy. In this method ,the input values are researcher's levels and research funds, the output values are quantity, quality and contribution levels of scientific publications. During 2000-2009 year, Feng, Chen and Tian[3] applied DEA method to evaluate the integral development of university R&D efficiency in our country. Then, they used Malmquist exponent to analyze dynamic change of university R&D efficiency. The result suggested university input-output as a whole is efficient in our country. And there is a trend that R&D efficiency in eastern region is the highest and that in central region is higher than that in western region. There are two times from 1999 to 2002 and from 2003 to 2006, respectively. Tian and Miao[4] applied DEA method to evaluate R&D efficiency in universities of out country, aimed to make references for R&D administrators making resources allocation policies and how to improve R&D efficiency of universities.

By the existing literature, we found DEA model is one of the widest application methods. But, at present, multi-stage DEA model is not yet applied to evaluate the university R&D efficiency. Based on above studies, this paper constructs two-stage DEA model and two-stage index system of university R&D efficiency, and utilizes empirical data of a university to analyze university R&D efficiency. Hence, we get the corresponding results.

2. A two-stage associated DEA model

Based on the traditional model, the two-stage associated DEA model is improved, the improved content is as follows:

Suppose there are n DMUs, each of DMUs may fall into two successive sub-processes, i.e., sub-process 1 and sub-process 2. As you can see in Figure1, x_{ij} and y_{rj} represent the i-th input of DMU j and the r-th output, respectively. Intermediate product z_{pj} represent the p-th output of sub-process 1 and the p-th input of sub-process 2, respectively. Where $i = 1,2,\text{L} ,m$, $r = 1,2,\text{L} ,s$ and $p = 1,2,\text{L} ,q$.

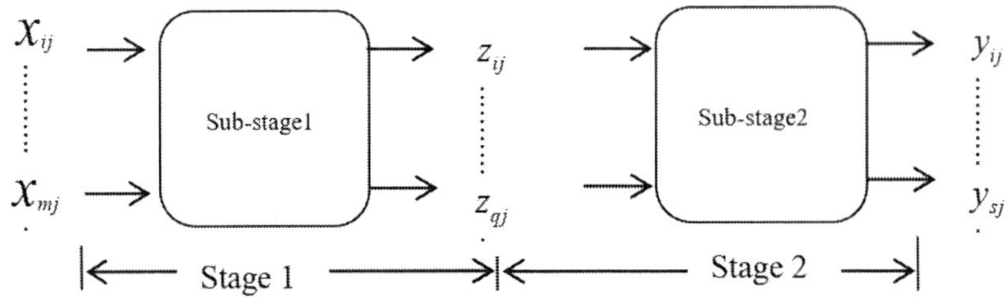

Figure 1. Productive Process of DMU

In a method of the efficiency of measured two-stage productive process, the interconnected relation between sub-processes is not considered, namely, all kinds of sub-processes regard as separate productive processes[5]. Suppose θ_k represents the overall efficiency of DMU. θ_k^1 represents the efficiency of sub-process 1 and θ_k^2 represents the efficiency of sub-process 2. Applied CCR-DEA model of constant returns to scale, the overall efficiency θ_k of DMU k can be calculated in this paper as follows:

$$\theta_k = max \sum_{r=1}^{s} u_r y_{rk} / \sum_{i=1}^{m} v_i x_{ik}$$
$$\text{s.t.} \begin{cases} \dfrac{\sum_{r=1}^{s} u_r y_{rj}}{\sum_{i=1}^{m} v_i x_{ij}} \leq 1, j = 1, \cdots, n \\ u_r, v_i \geq \varepsilon, r = 1, \cdots, s; i = 1, \cdots, m \end{cases} \tag{1}$$

Where ε represents non-Archimedean infinitesimal. $\theta_k = 1$ represents that DMU k is relative efficiency, while $\theta_k < 1$ represents that DMU k is inefficiency. On the same theory, we can construct two CCR-DEA model as follows:

$$\theta_k^1 = max \sum_{p=1}^{q} \pi_p z_{pk} / \sum_{i=1}^{m} v_i x_{ik}$$
$$\text{s.t.} \begin{cases} \sum_{p=1}^{q} \pi_p z_{pj} / \sum_{i=1}^{m} v_i x_{ij} \leq 1, j = 1, \cdots, n \\ \pi_p, v_i \geq \varepsilon, r = 1, \cdots, q; i = 1, \cdots, m \end{cases} \tag{2}$$

$$\theta_k^2 = max \sum_{r=1}^{s} u_r y_{rk} / \sum_{p=1}^{q} \pi_p z_{pk}$$
$$\text{s.t.} \begin{cases} \sum_{r=1}^{s} u_r y_{rj} / \sum_{p=1}^{q} \pi_p z_{pj} \leq 1, j = 1, \cdots, n \\ u_r, \pi_p \geq \varepsilon, r = 1, \cdots, s; p = 1, \cdots, q \end{cases} \tag{3}$$

Where θ_k^1 represents the efficiency of sub-process 1, and θ_k^2 represents the efficiency of sub-process 2 in measured DMU k.

In a measured method of the efficiency of the formulas 5.2 and 5.3, the interconnected relation between sub-processes is not considered, and this method do not accord to facts. The outputs of sub-process 1 is the inputs of sub-process 2, and the production of sub-process 1 has important influence for the efficiency of sub-process 2. By using separate DEA model, we can easy to calculate the results that may be hard to understand.

The intermediate product has kind of a dual role in productive process. A weighed mean $\sum_{p=1}^{q} \pi_p z_{pj}$ of intermediate product z_{pj} should be consistent in the formula (2) and the formula (3).

IOP Conf. Series: Journal of Physics: Conf. Series **1060** (2018) 012029 doi:10.1088/1742-6596/1060/1/012029

If u_r^*, v_i^* and π_p^* are optimal multipliers of DMU k, then

$$\begin{cases} \theta_k = \sum_{r=1}^{s} u_r^* y_{rk} / \sum_{i=1}^{m} v_i^* x_{ik} \leq 1 \\ \theta_k^1 = \sum_{p=1}^{q} \pi_p^* z_{pk} / \sum_{i=1}^{m} v_i^* x_{ik} \leq 1 \ A_i' \\ \theta_k^2 = \sum_{r=1}^{s} u_r^* y_{rk} / \sum_{p=1}^{q} \pi_p^* z_{pk} \leq 1 \end{cases} \tag{4}$$

From the computational process of the formula (4), we can find that as follows:

$$\theta_k = \frac{\sum_{r=1}^{s} u_r^* y_{rk}}{\sum_{i=1}^{m} v_i^* x_{ik}} = \frac{\sum_{p=1}^{q} \pi_p^* z_{pk}}{\sum_{i=1}^{m} v_i^* x_{ik}} \times \frac{\sum_{r=1}^{s} u_r^* y_{rk}}{\sum_{p=1}^{q} \pi_p^* z_{pk}} = \theta_k^1 \times \theta_k^2 \tag{5}$$

For the formula (5) hold, a key is that we can get optimal multipliers u_r^*, v_i^* and π_p^*.

$$\theta_k = max \sum_{r=1}^{s} u_r y_{rk} / \sum_{i=1}^{m} v_i x_{ik}$$

$$\text{s.t.} \begin{cases} \frac{\sum_{r=1}^{s} u_r y_{rj}}{\sum_{i=1}^{m} v_i x_{ij}} \leq 1, j = 1, \cdots, n \\ \sum_{p=1}^{q} \pi_p z_{pj} / \sum_{i=1}^{m} v_i x_{ij} \leq 1, j = 1, \cdots, n \\ \sum_{r=1}^{s} u_r y_{rj} / \sum_{p=1}^{q} \pi_p z_{pj} \leq 1, j = 1, \cdots, n \\ u_r, v_i, \pi_p \geq \varepsilon, r = 1, \cdots, s; i = 1, \cdots, m; p = 1, \cdots, q \end{cases} \tag{6}$$

The formula (6) is a fraction programming model. By using ChamesCooper transform, we can translate the formula (6) into linear programming as follows:

$$\theta_k = max \sum_{r=1}^{s} u_r y_{rk}$$

$$\text{s.t.} \begin{cases} \sum_{i=1}^{m} v_i x_{ik} = 1 \\ \sum_{r=1}^{s} u_r y_{rj} - \sum_{i=1}^{m} v_i x_{ij} \leq 0, j = 1, \cdots, n \\ \sum_{p=1}^{q} \pi_p z_{pj} - \sum_{i=1}^{m} v_i x_{ij} \leq 1, j = 1, \cdots, n \\ \sum_{r=1}^{s} u_r y_{rj} - \sum_{p=1}^{q} \pi_p z_{pj} \leq 1, j = 1, \cdots, n \\ u_r, v_i, \pi_p \geq \varepsilon \\ r = 1, \cdots, s; i = 1, \cdots, m; p = 1, \cdots, q \end{cases} \tag{7}$$

The formula (6) and the formula (7) integrate the constraint condition of the formulas (1)-(3). The associated relation of between sub-process 1 and sub-process 2 is considered. By dual transform, we have get a dual model of the formula (7) in this paper as follows:

$$min\ \theta_k$$

$$\text{s.t.} \begin{cases} \sum_{j=1}^{n} \alpha_j x_{ij} + \sum_{j=1}^{n} \beta_j x_{ij} \leq \theta_k x_{ik}, i = 1, \cdots, m \\ \sum_{j=1}^{n} \alpha_j y_{rj} + \sum_{j=1}^{n} \gamma_j y_{rj} \leq y_{rk}, r = 1, \cdots, s \\ \sum_{j=1}^{n} \beta_j z_{pj} - \sum_{j=1}^{n} \gamma_j z_{pj} \geq 0, p = 1, \cdots, q \\ \alpha_j, \beta_j, \gamma_j \geq 0, j = 1, \cdots, n \end{cases} \tag{8}$$

On the basis of the formula (7) or the formula (8), we can calculate the overall efficiency of DMU k. In the same theory, we also can construct the associated DEA model that measures θ_k^1 and θ_k^2. Where θ_k^1 represents the efficiency of sub-process 1 and θ_k^2 represents the efficiency of sub-process 2. For under the same standard to compare θ_k^1 with θ_k^2, we also construct the associated DEA model that measures the efficiency of sub-processes. Under keeping the constant overall efficiency of evaluated DMU, we get the efficiency of sub-processes in evaluation DMU k. A formula (9) represents the associated DEA model that measures the efficiency of sub-process 2.

$$\theta_k^2 = max \sum_{r=1}^{s} u_r y_{rk}$$

$$s.t.\begin{cases} \sum_{p=1}^{q} \pi_p z_{pk} = 1 \\ \sum_{r=1}^{s} u_r y_{rj} - \theta_k \sum_{i=1}^{m} v_i x_{ij} = 0 \\ \sum_{r=1}^{s} u_r y_{rj} - \sum_{i=1}^{m} v_i x_{ij} \le 0, j = 1, \cdots, n \\ \sum_{p=1}^{q} \pi_p z_{pj} - \sum_{i=1}^{m} v_i x_{ij} \le 1, j = 1, \cdots, n \\ \sum_{r=1}^{s} u_r y_{rj} - \sum_{p=1}^{q} \pi_p z_{pj} \le 1, j = 1, \cdots, n \\ u_r, v_i, \pi_p \ge \varepsilon \\ r = 1, \cdots, s; i = 1, \cdots, m; p = 1, \cdots, q \end{cases} \tag{9}$$

Applied above model, we can calculate the efficiency of sub-process 2. And then, by the formula $\theta_k^1 = \theta_k / \theta_k^2$, we can get the efficiency of sub-process 1.

3. The construction of DEA model input-output index system

An input-output analysis of DMU is a foundation to construct evaluation index system. Based on multianalysis of the input-output relation of between universities and teachers, an evaluation index system of the teacher's R&D input-output is constructed. This method considers problem domain and requirements of evaluation index system.

3.1 The structure of teacher's R&D process

Teacher's R&D activities include R&D project, government sponsored research, academic paper and so on. By the wages and crossing research project expenditure, teacher's R&D is supported. And teachers have win over the more government sponsored research expenditure. The final achievements of teacher's R&D work are academic paper. This also is an important aspect that consider the efficiency of teacher's R&D input-output. Therefore, as show in figure 2, teacher's R&D work may fall into two sub-processes, i.e., R&D project and corresponding academic paper.

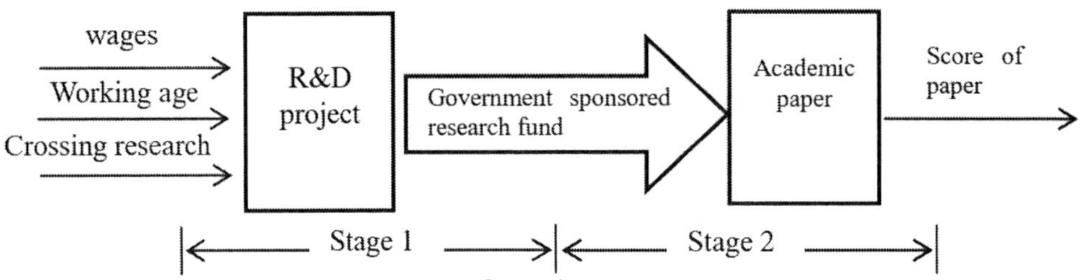

Figure 2. Process of Teacher Research Work

3.2 The construction and illustration of model input-output index system

Considering teacher's R&D characteristic as well as the availability of data, we select several indexes in this paper as follows:

(1) The input variables of production system and that of stage 1.

① The salary x_1 includes wages and end of year performance pay.

② The working age x_2 is teacher's work seniority. The longer working age is, the longer teacher work on their professional skills. In general, teachers have better R&D resource.

③ The crossing research project fund x_3 is teacher's project fund from firm.

(2) The output variables of production system and that of stage 2.

The score of paper y_1 is the total score of papers in a whole year. y_1 represents teacher's achievements in papers. In this paper, papers can fall into four grades. Every grade has different score.The score of the highest grade paper is 120, and the lowest is 30.

(3) Intermediate variables of production system

Government sponsored research fund z_1 is that teachers get fund from national and provincial departments. Government sponsored research is both the output of sub-process 1, and the input of sub-process 2. In this paper, project grades can fall into four grades. On the basis of different scores of four grades, we can calculate amount among different projects.

4. Empirical analysis

4.1 Data acquisition

The empirical data of this paper was drawn from the same university teacher's management system. Where the data includes 75 teacher's wages, crossing research project fund, government sponsored research fund, score of paper, personal information and so on in 2014. Statistical description of average input-output indexes shown in table 1.

Table 1. Statistics Teacher Research Sample Input-output in 2014

Variables	average	Standard deviation	minimum	maximum
x_1	201196	59088	101164	419544
x_2	16.5	8.7	2	37
x_3	896610	2033059	6250	14185700
z_1	661324	1412749	1000	10500000
y_1	183	165	4	691

4.2 The result analysis

In this paper, we apply two-stage associated DEA model to calculate the overall efficiency of teacher sample data and the efficiency θ_k, θ_k^1 and θ_k^2 of sub-processes. We get the results shown in tale 2.Computational process mainly applies the formula (5.7) and the formula (5.9) to measure every teacher's R&D and the efficiency of paper, then we apply $\theta_k^1 = \theta_k / \theta_k^2$ to calculate the efficiency of R&D projects.

Table 2. Score of Input-output Efficiency

DMU	θ_k^1	θ_k^2	θ_k	DMU	θ_k^1	θ_k^2	θ_k
1	0.17500	0.00600	0.00105	39	0.12100	0.00300	0.00036
2	1.00000	0.00100	0.00100	40	0.07800	0.00900	0.00070
3	0.20600	0.00100	0.00021	41	0.02800	0.02500	0.00070
4	0.27800	0.00100	0.00028	42	0.41700	0.00100	0.00042

From the result of table 2, we can find that individual teacher has difference in both R&D projects and academic papers, It may relate to university teacher's feature attributes and concentration. Faced the increasingly R&D competition, what measures should apply the administrative department of university R&D take to analyze efficiency. For a manifestation of a sample of teachers in different production stages, the efficiency average of two-stage associated DEA is a boundary line by which 75 teachers are divided into four classes(as show in figure 3). On the basis of this foundation, we can analyze different type of teachers to take proper measures.

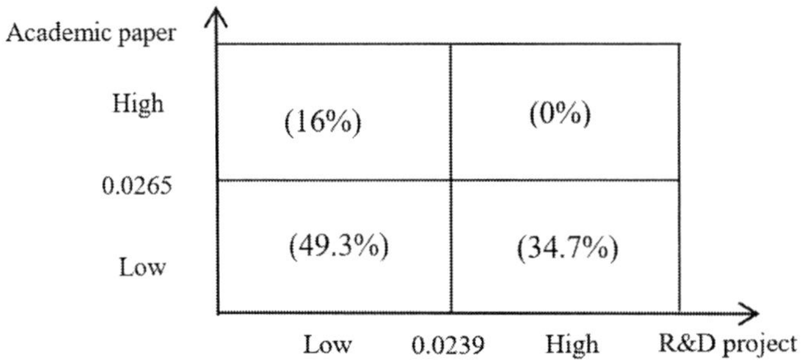

Figure 3. Research and Academic Efficiency Matrix

In addition to this, this paper uses teacher's feature attributes to cross-analysis from another point of view. There are many feature attributes such as professional title, department category, sex and so on.

As shown in Table 3, we get the analysis results.

Table 3 Cross-analysis of Research and Academic Efficiency

	attributes	θ_k^1Average	θ_k^2Average	θ_kAverage
Total	-	0.239	0.0265	0.00067
Professional title	Professor	0.267	0.029	0.00068
	Associate professor	0.18	0.021	0.00065
	Lecturer	0.035	0.013	0.000455
sex	Male	0.22	0.0132	0.000627
	Female	0.307	0.0756	0.00081
The large classification of subjects	Culture	0.148	0.335	0.000598
	Administration	0.16	0.0173	0.000613
	Marine engineering	0.267	0.0056	0.000516
	Engineering course	0.278	0.0179	0.000822
	science	0.212	0.004	0.000848

By above analysis results and the result in Table 3, we can get the tentative conclusions that are about the relation of between input-output efficiency of university teacher's R&D and teacher's feature attributes.

From aspect of professional title, professors are higher than other teacher's profession title in two-stage and input-output efficiency of overall R&D. On the one hand, it might be because professors have a certain social status. Professors can obtain sufficient resources, and have abundant academic knowledge and higher R&D level. On the other hand, professors generally are doctoral supervisor or master tutor, and team leader. Many excellent young teachers, doctors and masters share achievements with professors in R&D. So professors have higher efficiency of R&D.

From aspect of sex, female teachers are higher than male teachers in two-stage and input-output efficiency of overall R&D. This is a interesting phenomenon that it might be because female teachers are more carefully than male teachers in projects and papers. It also might be because female teachers have heavier work pressure than male teachers. On the basis of above reason, female teachers have put her heart and soul into R&D and project application.

From category of college, the teachers in engineering course and marine engineering are higher than the average efficiency of all the teachers in the first stage R&D input-output efficiency; The teachers in culture are much higher than the average efficiency of all teachers in the second stage R&D input-output efficiency; The teachers in engineering course and science are higher than the average of all the teacher in overall R&D input-output efficiency. Compared with social science, natural science has a relatively large proportion in natural and enterprise's projects. Natural science can directly generate the productive forces, so it gets serious attention by many people.

5. Conclusions

From subjective and objective points of view, this paper applies two-stage associated DEA model to construct the relation between university R&D resource inputs and university teacher's R&D outputs, it is convenient to evaluate university teacher's R&D. This paper applies teacher's feature attributes to cross-analyze teacher's R&D input-output efficiency an teacher's feature attributes. We can obtain some advise that assist university R&D administration to a certain degree.

Reference

[1] Kim S. A Comparison Study on University Research Efficiency Using DEA Analysis: focused on A University Case[J]. 2013, 15(1):249-258.
[2] Abramo G, D'Angelo C A, Pugini F. The measurement of Italian universities' research productivity by a non parametric-bibliometric methodology[J]. Scientometrics, 2008, 76(2):225-244.
[3] Feng Guangjun, Chen Wei, Tian Jinfang. Analysis of Scientific Research Efficiency in Universities in China Based on DEA-Malmquist Method——Experiential Research from 30 Provincial Panel Data[J]. Modern Finance and Economics-Journal of Tianjin University of Finance and Economics, 2012(9):61-73.
[4] Tian Dongping, Miao Yufeng. Evaluation of Scientific Research Efficiency in Universities in China Based on DEA[J]. Journal of Design Art, 2005, 24(4):6-8.
[5] Wu Desheng. Research on several theories and methods of data envelopment analysis[D]. University of Science and Technology of China, 2006.

Analysis of Simulator Sickness and Performance in Virtual Training

Jue Qu[1], Wei Wang[2], Shuzhi Yuan[2], Fangzheng Li[2] and Rui Cai[2]

[1] School of Aeronautics, Northwestern Polytechnical University, Xi'an 710072
[2] China Air and Missile Defense College, Air Force Engineering University Xi'an, Shaanxi, 710051, China

qujue402@sina.com

Abstract. Simulator sickness is inevitable during virtual training. At the process of missile lifting task, the problem "how long and how to train" is presented in virtual training of troop, crossed and uncrossed parallax stereoscopic display are set up as different virtual training environment. To analysis the change rules of sickness and performance along with time and different parallax types, subjects are asked to operate the box loading vehicle for 1 hour, and the score of simulator sickness questionnaire is recorded every 10 minutes. The result shows performance is better in crossed environment than which in uncrossed. There is no significant difference in operational performance over time; all subjects feel slight simulator sickness after 10 min training and obvious feeling after 30min. Suggestions are proposed that virtual training should be in crossed parallax environment and keep in 30 min.

1. Introduction

With the development of virtual reality technology, virtual training has attracted widely attention in military, traffic, medical treatment, entertainment and education owing to its safety and economy. Virtual training plays an important role due to the strict security requirements and exorbitant fees of actual military equipments. Virtual training has many advantages, and also many problems, of which the most common is simulator sickness. Exposed in 3D virtual environment for a long time will lead people to dizziness, vertigo, nausea, vomiting and other symptoms of motion sickness, called simulator sickness.

In most cases, people have a better performance of location estimation, object recognition, spatial memory etc. in the 3D virtual environment than in 2D[1], but it will cause more visual fatigue[2]. Continuous operating in the virtual environment, operational performance will decline with the onset of simulator sickness[3-4], the human body will be damaged. Article [5] shows that when keep driving operation on the car simulator for 10 minutes, the participants will feel sick and the operating performance will decrease significantly. If exposed to virtual 3D environment for more than 1 hour, human eye will gradually adapt to the contradictions between regulation effect and convergence effect, resulting in esotropia[6], although it helps to adapt to the virtual environment, it is harmful in the real environment.

For 3D virtual display, many researches focus on visual fatigue and comfort. The study on virtual training also focuses on the performance and despise fatigue, which can not guide the practice of "how long and how to practice" in virtual training. In this paper, we study the relationship between operational performance and simulator illness in different parallaxed (cross and non-cross) virtual 3D environment, and find the best virtual training duration and training method under the condition of simulator illness disturbance.

2. Causative factor analysis of simulator sickness

Domestic and overseas scholars have done a lot of researches from the aspects of visual imaging principle, EEG, virtual object movement about the causative factor analysis of simulator sickness[5,7,8], one of the most important factors is contradictions between binocular regulation and convergence[6,7,9,10,11]. More than 80% of external information is received by eyes, and about 90% of the brain energy is used to deal with external information and form awareness[12], article [13] argued that the natural of stereopsis fatigue is the brain fatigue caused by brain cognitive overload. When human sensory conflict exceeds the body central physiological tolerance limit, it will lead to a series of vegetative nervous disorder, resulting in a variety of short-term symptoms[7], that is, simulator sickness.

In addition, simulator sickness is also associated with a variety of factors: The ability of people to perceive the depth information of objects is different, there are about 12% of people who are weaker in the depth perception[14]. Individuals with better stereo vision ability will have better performance in the virtual environment, but more likely to emerge symptoms of simulator sickness [10], and similarly, the elder are less likely to emerge symptoms of simulator sickness than young people because of eye movement system recession [15]; article [16]shows that the faint shaking of virtual images caused by head shaking will lead to adverse reactions in human body. The sensitivity of different gender to simulator sickness is also different, and women are more likely to feel discomfort in virtual environments [10, 15, 17]; human reading speed and object recognition ability are better in a bright virtual environment than low-brightness environment[18]. Parallax variation frequency can affect visual fatigue more than parallax size, vertical movement of virtual image is more likely to cause fatigue than horizontal movement . Distance is also an important factor, it is comfortable to watch 3D TV when viewing distance is 3 times as large as the screen height[6]. Besides these reasons, simulator sickness is also affected by a number of equipment factors. At present, 3D display technology is varied, the most widely used is the optical perspective helmet display (HMD), computer screen display and projection display. Among them, helmet display technology is the earliest and has been widely used in various virtual reality (VR) fields, such as virtual maintenance and fault diagnosis. But researches show that helmet display is the most likely way to cause visual fatigue[19]. Computer screen display, such as 3D TV, is the most closest to the civilian, but 3D immersion is not as good as the other two ways and cannot give a good solution to the problem of visual fatigue, this technology has not been applied widely; projection stereoscopic display technology is most likely to achieve, such as 3D movie, but is susceptible to the polarized glasses. In addition, the images mutual crosstalk between left and right eye, screen resolution, image delay, flicker, etc. can also lead to visual fatigue.

3. Stereo vision and the virtual stereo vision

We can see the three-dimensional views, the most important foundation is physical structure of the eyes. When observing the same object there will be a slight angle difference due to the distance between two eyes, the retina will receive the same object with two different images, and then the brain will integrate two different images into a stereoscopic vision, form three-dimensional visual perception.

Depth perception of objects in stereoscopic vision is mainly determined by the parallax. As shown in Figure.1, assuming M is the fixation point, the M point is clearly imaged by the cooperative action of binocular regulation and convergence, although point A and point B (farther away from the fovea) is not clear, it can still be imaged. In order to see objects clearly, eyes will adjust the refractive power, it's known as regulation effect, refers to the phenomenon of adduction or extorsion of both eyes when looking near or distant targets[9]. View angles form the basis of depth perception, and the angle difference between them is called parallax, when $\alpha-\beta>0$, it is called cross parallax; $\gamma-\beta<0$, it is called uncross parallax.

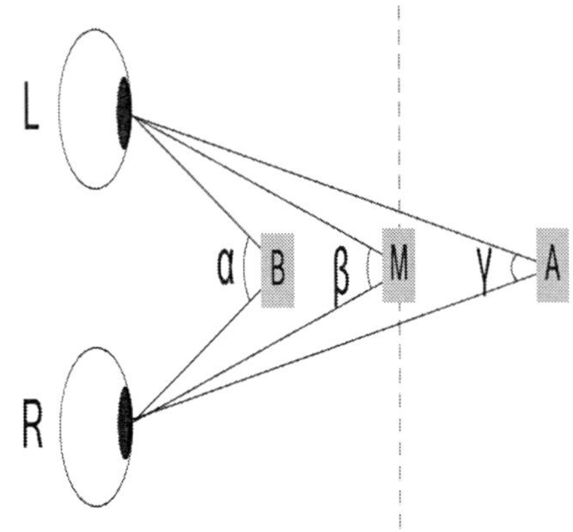

Figure 1. Crossed and uncrossed parallax

We do not notice unclear objects beyond the eyes focus because our eyes move continuously at a speed of 3 times per second, putting focus on the surrounding environment objects selectively. In this case, the regulation and convergence effect of eyes is consistent, namely the adjustment will increase along with the increase of convergence, decrease with the decrease of convergence. But in the virtual environment, adjustment and convergence will appear inconsistent in some cases. The principle of virtual 3D vision is the two cameras show different images to left and right eyes respectively in order to meet the three-dimensional visual perception. Finally, the image will be displayed on the screen, in order to see the image, the eyes will focus on the screen, but the virtual 3D object is presented in front of the screen or behind the screen, convergence located on the virtual 3D Object. In Figure. 1, we assume that point M is the screen position, A and B is the virtual 3D object, and the accommodation acts at point M, the converge acts at point A or B. Therefore, there is a phenomenon of inconsistent between regulation and convergence, leading to visual fatigue.

Studies have shown that cross parallax causes more visual fatigue than uncross parallax[20]. However, some studies have also shown that the effects of different parallax types have no significant differences[21]. There is few studies focus on the effect of different parallax types on operational performance. Article [22] argued that cross-parallax can stimulate people's stress reaction and raise the cortical arousal level to pay more attention to virtual scene, but whether this can improve operational performance need further research.

The movement of virtual objects includes two-dimensional motion and depth motion, depth motion refers to movement that virtual object close to or away from the observer, depth motion is more likely to lead to visual fatigue than two-dimensional motion, and the faster the speed of depth movement, the easier the eyes feel fatigue[9,14], fast depth motion means the rapid change of the parallax, and the rapid change of parallax and frequent change in the direction of depth movement are easy to cause visual fatigue[6]. Research [23]shows that the comfortable scope for parallax is 35 '- 40', and when the variation exceed 1 degrees, it will be unbearable. In this paper, we take virtual training system of box loading as background and focus on the depth of movement.

4. Virtual training experiment

4.1. Subjects

Subjects are eight male soldiers, corrected visual acuity were 1.0 or more, at the age of 22 to 23, scatter plot stereoscopic examination shows that subjects have passed the 60 "parallax test, belonging to the normal parallax crowd. Before the experiment, all participants have involved in different missile lifting training with actual equipment, but haven't operate in the simulator before, all subjects can be regarded as novice.

4.2. Experiment design

Optoma LC1 stereoscopic projector matched with stroboscopic polarized glasses is used in the experiment. The distance between subjects and the screen is 2.5m, and the sight distance between subjects and the virtual equipment is set as 10m. The projection area is shown in Figure.2, taking the eye spacing of 0.06m as the benchmark, the maximum parallax of 3D virtual image is 5.5 degrees, the minimum is 0, maximum speed of swing arm is 9 degrees per second, and the maximum parallax change rate is about 0.04 degrees per second by calculation.

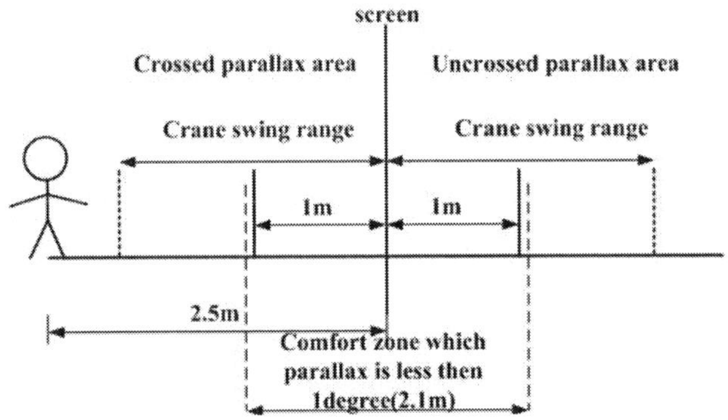

Figure 2. Sketch map of virtual images

Subjects were asked to hoist the missile launch box from one vehicle to another vehicle. During the hoisting process, the main task of the participants was to control the crane to place the hook in the specified position, and the auxiliary staff would perform the grasping operation. The experimental performance index is the depth error between the position of the parking hook with the standard position, as shown in Figure.3.

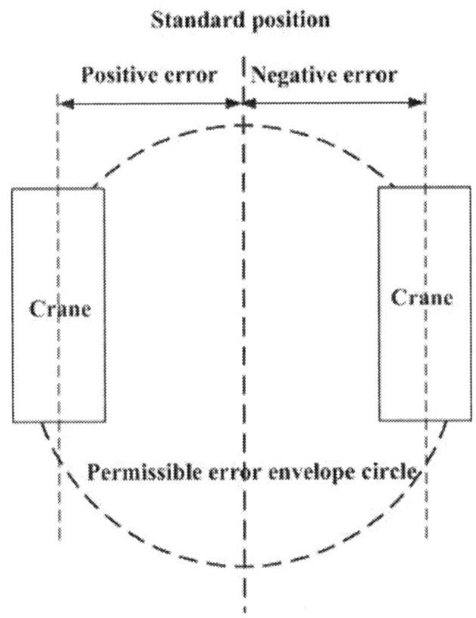

Figure 3. Schematic map of depth deviation counting

When training for ten minutes, pause for one minute to conduct Simulator Sickness Questionnaire (SSQ). The questionnaire was mainly used to check the adverse symptoms of the subjects, includes nausea, Ophthalmic discomfort, lost direction, and further refined into 16 kinds of index, each index has 4 different degrees (no, mild, moderate, severe), corresponding to the weight coefficient from 0 to 3 respectively, the indexs and weights shown in Table 1, The final score is calculated as a weighted sum of two weights.

Table.1 Details of simulator sickness questionnaire

number	index	weight		
		Nausea(N)	Ophthalmic discomfort(O)	Lost direction(D)
1	General discomfort	1	1	0
2	Fatigue	0	1	0
3	Headache	0	1	0
4	Visual fatigue	0	1	0
5	Difficulty focusing	0	1	1
6	Salivation increase	1	0	0
7	Sweating	1	0	0
8	Nausea	1	0	1
9	Difficulty concentrating	1	1	0
10	Fullness of the head	0	0	1
11	Blurred vision	0	1	1
12	Dizziness eyes open	0	0	1
13	Dizziness eyes closes	0	0	1
14	Vertigo	0	0	1
15	Stomach awareness	1	0	0
16	Burping	1	0	0
	Total	A	B	C

Score
N=A×9.54
O=B×7.58
D=C×13.92
Simulator sickness degree=(A+B+C)×3.74

Each experiment was conducted for 1 hour, with an SSQ test and an error recorded every 10 minutes. 4 people were asked to take part in cross-parallaxed (animation out the screen) experiment, and the other 4 were asked to take part in uncross-parallaxed (animation in the screen) experiment. During the experiment, keep the curtain closed, to prevent the background light from affecting the experiment.

5. experiment results and analysis

5.1. Experiment results

The average operational error of 8 subjects is shown in Figure 4. The degree of simulator sickness is shown in Figure.5 and Figure.6.

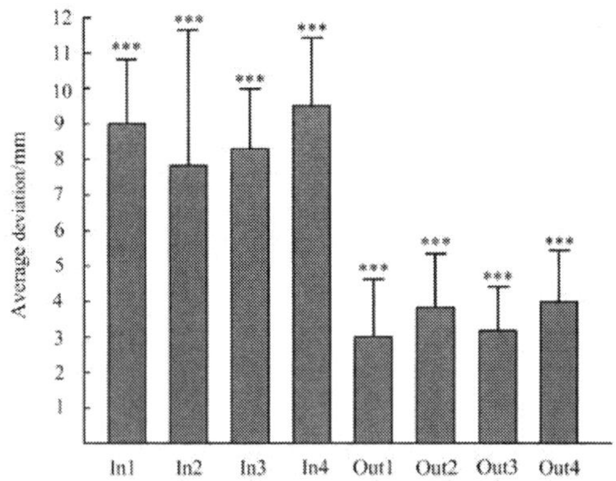

Figure. 4 Average deviation(error bars represent standard deviation)

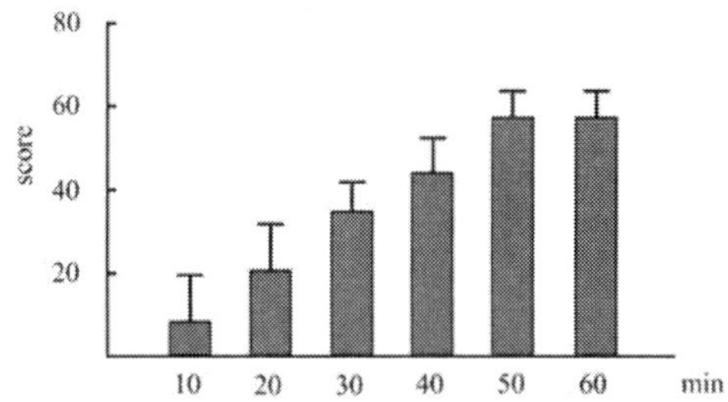

Figure.5 Average score of uncrossed experiment(error bars represent standard deviation)

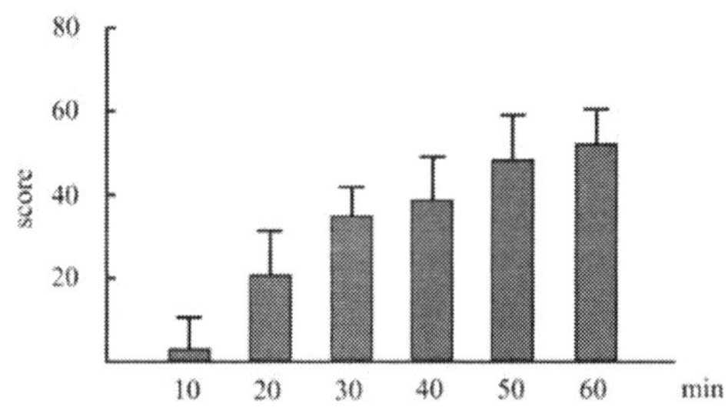

Figure.6 Average score of crossed experiment(error bars represent standard deviation)

As shown in Figure.4, Operational performance of cross-parallax and uncross-parallax is significantly different. Analysis of variance shows that there is no significant difference between four subjects in two groups（F_{in} (3, 20)=0.547, P=0.655; F_{out} (3, 20)=0.728, P=0.547）, the operational performance in two groups has no significant difference over time（F_{in}(5, 18)=1.49, P=0.241; F_{out}(5, 18)=0.926, P=0.487）. There are significant differences in the operational performance of the two groups（F(1, 46)=86.0, P<<0.01）.

The longer the training time, the higher the SSQ scores of the eight subjects, that is, the simulator sickness became more and more serious. The analysis of variance shows that the time effect of SSQ scores is significant（F_{in}(5, 18)=24.973, P<<0.01）; F_{out}(5, 18)=18.212, P<<0.01）. ANOVA of SSQ scores between animation in and out at different timing are shown in Table.2. We can see that there are no significant differences in two groups at all timings.

Table.2 ANOVA of SSQ scores between animation in and out the screen at different timing

	F（1, 6）, $F_{0.05}$（1, 6）=5.987	P
10Min	0.711	0.431
20Min	0	1
30Min	<0.001	1.0
40Min	0.760	0.417
50Min	2.288	0.181
60Min	1.130	0.329

After training for 10min, there were two in cross-parallaxed experiment, one in uncross-parallaxed experiment appeared the symptoms of lost direction such as head bilge and vertigo. After training for 20min, all subjects appeared the symptoms of lost direction, two subjects appeared ophthalmic discomfort symptom in each group. After training for 30 min, only one subject didn't appear ophthalmic discomfort. 50 min later, there is one subject appeared nausea symptom (sweating) besides lost direction and ophthalmic discomfort.

5.2. Experiment analysis

Although the virtual training of cross-parallax and non-cross-parallax increases visual fatigue with training time, the effect of the two types of parallax on operator fatigue is not very different, which is consistent with the conclusion of article [20], and inconsistent with article [21]. Experiment simulate the real load training process, the emphasis is focused on the analysis of the effects of different types of parallax on the fatigue of virtual training, so the maximum rotational angular velocity of the boom is set at a slower value (9 ° / s) for real safety reasons. So for the operator, the two main factors (parallax amplitude and parallax change rate) that affect the simulator sickness are invariant in the two groups and the main operating area is set in the proposed comfort range. The results show that the effect of parallax type on the fatigue of virtual training is not significant.

There is significant difference in the operation performance of two groups. The article [22] suggests that the human eye is much more sensitive to 3D effects in cross parallax than uncross parallax, which can be interpreted as cross-parallax makes people feel more immersed, that is, people in a virtual environment mentally while the body is in real environment[24]. Some researches show that the enhancement of immersion can improve the emotional arousal level and it will further improve the operational performance [24], but it will increase more fatigue[15].

6. Conclusions

In this paper, the following conclusions are obtained by setting up the virtual training of cross parallax and uncross parallax:

(1)In the process of virtual training, the training time has great influence on fatigue of the trainee. Although the operating performance will not be reduced significantly after 30 minutes, fatigue will increase rapidly due to simulator sickness. Mild degree simulator sickness takes 1 to 2 hours to disappear gradually, and there are extreme cases need 48 hours to eliminating [25]. So it is suggested that the virtual training time is less than 30min.

(2)Cross-parallax makes people feel more immersed. In this paper, we focus on the effect of depth motion of virtual objects on operation performance and fatigue. In order to achieve better training effect, stereo vision should be set as cross parallax in virtual training with a large number of depth motion.

At present, the influence of the parallax amplitude and parallax change rate on the fatigue and the operation performance in the 3D virtual environment is a hot research topic. Because of the large number of simulation diseases, the various hardware devices, the experimental variables are difficult to control, and the experimental research background is different, resulting in a lot of uniform conclusions. In this paper, there are some variables do not be effective controlled, such as the environmental brightness is controlled by closing the curtain in natural light conditions, and experiment takes the loading process of a certain kind of box as background, which is highly targeted, universality needs to be further studied.

References

[1] JOHN P McIntire, PAUL R Havig, ERIC E Geiselman. Stereoscopic 3D displays and human performance: A comprehensive review[J]. Displays, 2014, 35, 18-26.
[2] M Emoto, Y Nojiri, F Okano. Changes in fusional vergence limit and its hysteresis after viewing stereoscopic TV[J]. Displays, 2004, 25, 67–76.
[3] STANNEY K, Kingdon K, Graeber D, et al. Human performance in immersive virtual environments: effects of duration, user control, and scene complexity[J]. Human Perfor -mance , 2002, 15(4), 339–366.
[4] S Park, M J Won, S Mun, et al. Does visual fatigue from3D displays affect autonomic regulation and heart rhythm[J]. International Journal of Psychophysiology, 2014, 92, 42-48.
[5] BYUNG ChanMin, Soon CheolChung, Yoon KiMin, et al. Psychophysiological evaluation of simulator sickness evoked by a graphic simulator[J]. Applied Ergonomics, 2004, 35, 549-556.
[6] TAKEHIKO Bando, Atsuhiko Iijima, SumioYano. Visual fatigue caused by stereoscopic images and the search for the requirement to prevent them: A review[J]. Displays, 2013, 33, 76-83.
[7] SHI Guowei, GUO Xiaochao, XIONG Duanqin, et al. The Cause of Binocular Vision Related Simulator Sckness in Pose-exposure Virtual Environments[J]. Journal of Computer-Aid Design & Computer Graphics, 2013, 25(4): 474-479.
[8] JING Li, MARCUS Barkowsky, PATRICK Le Callet. Visual discomfort of stereoscopic 3D videos: Influence of 3D motion[J]. Displays, 2014, 35, 49-57.
[9] LI Shuyin, WAN Mingxi, LI Xinxiao, et al. Visual perception in virtual environment[J]. Journal of Image and Graphics; A, 2000, 5(11): 557-561.
[10] BRIAN Allen, TAYLOR Hanley, BAS Rokers, et al. Visual 3D motion acuity predicts discomfort in 3D stereoscopic environments[J]. Entert -ainment Computing, 2016, 13, 1-9.
[11] PANG Yu. Study on visual comfort enhancement of stereoscopic video based on perceiving distance of the viewer[D]. Jilin: Jilin university, 2015.
[12] RAICHLE M E, SNYDER A Z. A default mode of brain function: a brief history of an evolving idea[J]. NeuroImage, 2007, 37(4): 83–90.
[13] SANGIN Park, MYYOUNG Ju Won, EUI Chul Lee, et al. Evaluation of 3D cognitive fatigue using heart–brain synchronization[J]. International Journal of Psychophysiology, 2015, 97, 120-130.
[14] KELLY S Hale, KAY M Stanney. Effects of low stereo acuity on performance, presence and sickness within a virtual environment[J]. Applied Ergonomics, 2006, 37, 329-339.
[15] MARIANNA O, DANIELA W, THOMAS M, et al. Viewing experience of 3DTV: An exploration of the feeling of sickness and presence in a shopping mall[J]. Entertainment Computing, 2013, 4, 71-81.
[16] JEFF Johnson. Designing with the mind in mind: simple guide to understanding user interface design rules[M]. Beijing: POST & TELECOM PRESS, 2011.
[17] WU Miaoli. The Gender Sensibility Test and Analysis in a Driving Simulator Platform[J]. Research and Exploration In Laboratory, 2013, 32(9), 189-192.
[18] ARSALAN Heydarian, JOAO P Carneiro, David Gerber, et al. Immersive virtual environments versus physical built environments: A benchmarking study for building design and user-built environment explorations[J]. Automation in Construction, 2015, 54, 116-126.
[19] SHARPLES S, COBB S, MOODY A, et al. Virtual reality induced symptoms, effects(VRISE): Comparison of head mounted display(HMD), desk top and projection display systems[J]. Displays, 2008, 29, 58–69.
[20] LEE S, Jujung Y, HOSIK S, et al. Visual discomfort induced by fast salient object motion in

stereoscopic video[C]. SPIE Stereoscopic Displays and Virtual Reality Systems XXII, 2011, 7863: 786305.
[21] LI J, Barkowsky M, WANG J, et al. Study on visual discomfort induced by stimulus movement at fixed depth on stereoscopic displays using shutter glasses[C]. 17th International Conference on Digital Signal Processing(DSP),IEEE, 2011:1-8.
[22] JIN Shuai, XING Lidong, QIAN Zhiyu. The Association between Depth Information of 3D Movie and Viewer Fatigue[J]. Chinese Journal of Biomedical Engi -neering, 2014, 33(2): 15 —21.
[23] TAM W, SPERANZA F, YANO S, et al. Stereoscopic 3D-TV: visual comfort[J]. Broadcasting IEEE Transactions on, 2011, 57(2), 335–346.
[24] WITMER B, SINGER M. Measuring Presence in Virtual Environments: A Presence Questionnai -re[J]. Presence: Teleoperators and Virtual Envi -ronments, 1998, 3, 225-240.
[25] WU Miaoli. Sickness Sensibility Research of Individual Participants in Driving Simulator Platform[J]. Experiment Science and Technology, 2013, 11(5): 9-12.

IOP Conf. Series: Journal of Physics: Conf. Series **1060** (2018) 012031 doi:10.1088/1742-6596/1060/1/012031

Comparison Study of Visual Search on 6 Different Types of Icons

Yongqiang Liang[1], Wei Wang[1], Jue Qu[1] and Jie Yang[1]

[1]Air and Missile Defense College, Air Force Engineering University Xi'an, Shaanxi, 710051, China

1102268250@qq.com

Abstract. In order to study the similarities and differences of the visual search features on different types of icons in the human-computer interface, using digital icons, English icons, Chinese icons, abstract icons, graphical icons and image icons as experimental materials. Combined with eye tracking devices, a visual search experiment was designed adopting E-Prime programming, and a comparative study was conducted from 3 aspects of reaction time, correct rate, and eye movement characteristics. The results of single factor analysis of variance indicated that: (1) The difference among 6 different types of icons in reaction time of visual search was statistically significant, and the reaction time was ranked as digital icons > English icons > Chinese icons > abstract icons > graphical icons > image icons. (2) In addition to the abstract icons, the other 5 types of icons in the visual search showed no obvious error. (3) As for the eye movement characteristics of 6 different types of icons, there were significant differences in fixation duration, fixation times, saccade amplitude and saccade speed, however, pupil diameter and blink frequency were not significantly different. The research results have reference value for the icon design and usability evaluation in human-computer interface.

1. Introduction

Visual display is the main form of the information transmission and expression of the graphical user interface in the human computer interaction. Icons, as an important components of graphical user interface, play an important role in the software system interface, web page navigation interface, radar search interface, and aircraft display interface. At present, icon design mainly uses small-pictures that are intuitive and easy for the user to understand to represent specific meanings or functions, and are used for information identification and pointing interaction[1]. Reasonable icon design has a significant impact on the accuracy and efficiency of human-computer interaction, in view of the comparison of the visual search characteristics of different types of icons, it is of great significance to provide ergonomic basis for the design and evaluation of human-computer interface.

At present, the icons in the graphical user interface are mainly in the form of characters, graphics, images, etc. Among them, the characters are suitable for expressing abstract concepts, the images are more efficient in conveying entities and their context information, and the graphics are between the two forms of visual expression[2].

Dual Coding Theory holds that knowledge is stored in two forms which are language and imagery, and there are two independent cognitive subsystems, which are used to deal with these 2 forms of information[3]. Cognitive psychology points out that the process of human information processing is from input to processing and then to output, people recognize characters mainly by comparing their semantic information, and for graphics and images, they are used to compare their visual differences and implicit semantic information[4].

Researchers have studied the visual search features of different types of icons. Isherwood[5] proved that the search efficiency and perception performance of graphic icons were better than the textual icons, and the subjective feelings of the participants in graphic icons were more beautiful. Ghayas et al. [6] studied the recognition rate of mobile phone icons for users of different ages and found that the recognition rate depends on the degree of familiarity of the participants with the icon, and pointed out that the icon design should follow the principle of consistency. Schröder[7] examined the influence of concreteness and complexity on the semantic transparency of icons, and found that complexity had a significant effect on response time, while concreteness had a greater impact on correctness. Passini et al.[8] found that the response time and accuracy were positively correlated with the icon visual degree(concreteness) in the match test of icon semantics and appearance. Johanna et al.[9] considered that the semantic distance is a critical factor in icon design for in-car infotainment systems, the research results showed that the primed product comparisons method could be efficiently utilised for finding an optimised set of icons for time-critical applications out of a larger set of icons. Federica et al.[10] investigated the extent to which a sample of Italian users comprehended safety pictorials used on agricultural machinery, the results indicated that many of the pictorials adopted for agricultural machinery investigated in this stud yielded low comprehension scores.

The above research results provided an important theoretical basis for the design and evaluation of the icon in the human-computer interface, but the research object had not yet included all types of icons. Based on the above research results, this study used the icon types commonly found in 6 types of graphical user interfaces, including the Arabic numeral icons (abbreviated herein as digital icons), English icons, Chinese icons, abstract icons, graphical icons, and image icons. And visual search experiments were designed with eye tracking technique.

2. Experimental design

2.1. Experimental environment and equipments

The experiment was carried out in a specialized ergonomics laboratory, and the laboratory space was spacious, the temperature and humidity were suitable, the light intensity could be adjusted, no odor, no noise and other external interference factors. The experimental equipment included a laptop computer, a RED5 eye tracker, the highest sampling frequency is 500Hz, and the host display resolution was 1280 × 1024 pixel, and the screen brightness was adjusted to 300cd/m2, the distance between the eyes of the participants and the screen was about 50cm, the center of the screen and the participants' eyes were at an equal level. The experiment used the standardized psychological experiment system E-Prime software to compile and automatically record the preservation of experimental data, the eye movement data was processed by BeGaze software.

2.2. Participants

Students at the Air Force Engineering University in China (N = 12) participated in this experiment, including 8 males and 4 females, aged 18 to 22 years old (SD = 3.08). All participants were in good health, and had normal or corrected-to-normal vision, no color blindness, color weakness or other characteristics. All participants have reached CET-4, and have more than 3 years of experience using computers.

2.3. Experiment materials

The experimental materials consist of 6 different types of icons, which are made up of Arabia digital coding, English words, Chinese words, abstract graphics, image and physical images, each of the icons has 64 (8×8) total 384 experimental elements.

Table 1 shows a group of experimental materials of different types of icons. In the experimental material: the Arabia digital icon is made up of two digits. The English icon is made up of high-frequency words with the number of letters 4 to 5, and the Chinese icon is composed of high-frequency words including two Chinese characters (word frequency is higher than 200 times / million times). The abstract icons and graphical icons mainly come from various types of symbols and common display interface graphics, and the image icons are mainly physical photos with the same semantics as Chinese and English icons. The experimental icon size is 30×30 pixels. In order to avoid the color interference, all the experimental icon materials are processed into grayscale images using Photoshop except the image icon retain the source color.

Table 1. A set of experimental materials

digital icons	English icons	Chinese icons	abstract icons	graphical icons	image icons
04	tank	坦克	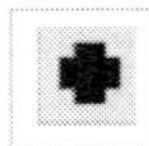		

2.4. *Experimental Procedures*

The experiment required the participant to have a good mental state. Before the formal experiment, the participants needed to read the introduction guide first, and then through 3 practice operations to ensure familiarity with the experimental procedure.

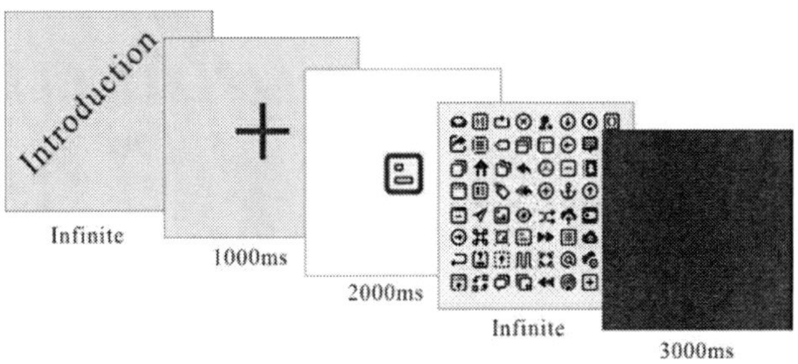

Figure 1. Experimental process

Figure 2. Abstract icon experimental stimulus

As shown in Figure 1, when the experiment started, the participants needed to read the experimental introduction until understood the content of the introduction, then pressed the keyboard' s any key, next, a cross graphic was displayed in the center of the computer screen, after 1000 ms, the screen automatically entered the next interface. At this time, the target icon was displayed in the center of the computer screen for a duration of 2000 ms, and then the computer screen automatically presented experimental stimuli (as shown in Figure 2 is the experimental stimuli of the abstract icon). It can be seen that the experimental stimuli are divided into four areas: the upper left, the lower left, the upper right, and the lower right. Participants needed to find the target icon in the 8×8 icon array as soon as possible, and then pressed the keyboard number key according to the area where the icon was located to end a search task, the corresponding rules for the experimental interface area and the numeric keys were: upper left=4, lower left=1, upper right=5, and lower right=2. After a search task was completed, the computer screen was displayed with a black screen, after 3000 ms, the next operation experiment was automatically entered, and the above process was repeated until all 10 search tasks were completed. All 18 subjects were tested separately until the end of the experiment.

3. Experiments results

3.1. Reaction time and correct rate

Each participant in the experiment completed visual search tasks for 6 different types of icons. After processing a few extreme data, the average response time and average correct rate of each type of icon were collected and calculated. As shown in Figure 3, it can be seen that the order of reaction time is as follows: image icon > graphical icon > abstract icon > Chinese icon > English icon > digital icon, in addition to the low correct rate of the abstract icons, the other types of icons are more than 95% correct rate in visual search tasks.

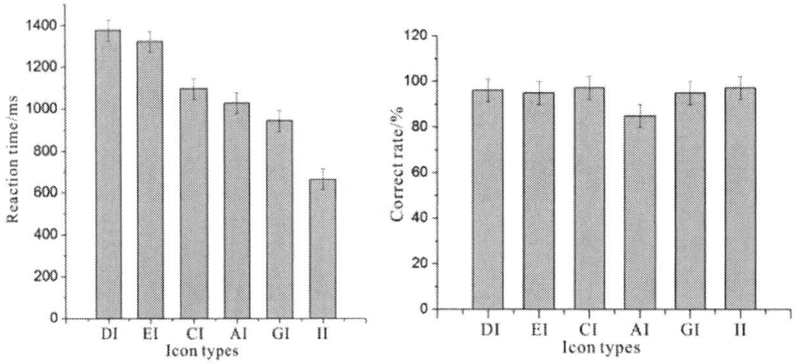

Figure 3. Reaction time and correct rate of different types of icons

Note: In Figure 3, DI = digital icon, EI = English icons, CI = Chinese icons, AI = abstract icons, GI = graphical icons, II = image icons, the short form in the following text stands for the same meaning.

The results of single factor analysis of variance (ANOVA) found that there were significant differences in the reaction time among 6 different types of icons ($F(5,85)=5.50$, $p<0.001$), further paired t test was analysed on the reaction time of the 6 different types of icons. The results are shown in Table 2.It can be seen that the digital icon vs English icon $t(17)=3.972$, $p>0.05$, the digital icon vs Chinese icon t(17)=2.126, p>0.05, and the digital icon vs abstract icon $t(17)=3.576$, $p>0.05$ have no significant difference in reaction time, however, digital icon vs image icon $t(17)=3.287$, $p<0.05$ and digital icon vs image icon $t(17)=2.785$, $p<0.001$ have a significant difference in reaction time. Similarly, it can be seen that the paired t-test difference between any other two different types of icons.

In addition to the low correct rate of the abstract icon (AI), the correct rate of the other 5 types of icons in the visual search is greater than 95%, and mainly due to mistakes in the experiment operations. Therefore, in this study, we did not perform the paired t-test on the correct rate, but believed that there was a significant difference between the correct rate of the abstract icon and the other types of icons, and the latter did not show obvious errors in the visual search experiment.

Table 2. Paired t-test of reaction time

	EI	CI	AI	GI	II
DI	3.792	2.126	3.576	3.287*	2.785**
EI	---	4.128	2.049*	5.419*	3.683**
CI	---	---	4.598	2.369*	2.979**
AI	---	---	---	5.211	3.210**
GI	---	---	---	---	1.875*

In the table: * indicated that the paired t-test results were significant at the 0.05 level; ** indicated that the results were significant at the 0.01 level.

3.2. Eye Movement characteristics analysis
Collecting eye movement characteristic data of participants during the visual search of each type of icon. Selecting the fixation duration, total fixation counts of a single search task, saccade amplitude, pupil diameter, blink frequency and saccade velocity' s average of the participants in the visual search experiment as evaluation indexes of eye movement characteristics. The calculation results are shown in Figure 4 and Table 3.

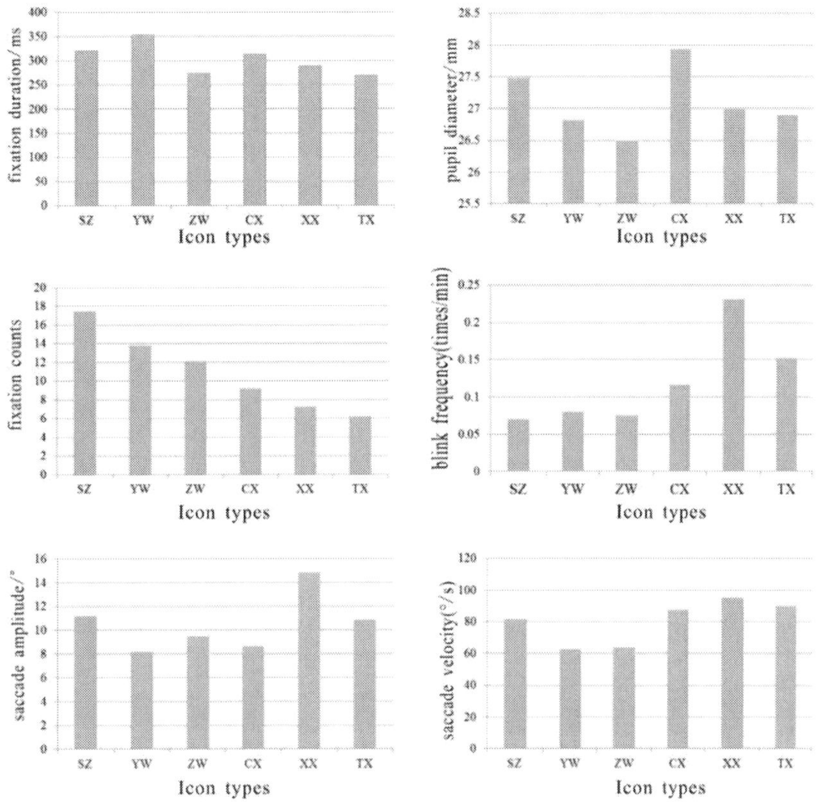

Figure 4. Eye movement characteristics of 6 types of icons

It can be seen that for the 6 types of icons, the average fixation duration for searching for English icons is the longest, while searching for the Chinese icons is the shortest. Similarly, it can be seen that the average pupil diameter when participants searching for abstract icons is significantly larger than searching for the other 5 types of icons, and the average blink frequency and saccade amplitude of searching graphical icons are significantly larger than the other 5 types of icons. The average saccade velocity is smaller when searching for the English icon and the Chinese icon, and the saccade velocity are higher when searching for the abstract icon, the graphical icon and the image icon. In addition, the participants shows a decreasing trend in the average fixation counts of gazes when searching for the 6 types of icons (from digital icons to image icons).

Table 3. The results of eye movement characteristic

	FD	*FC*	*SA*	*PD*	*BF*	*SV*
DI	320	17.5	11.18	27.48	0.070	81.52
EI	354	13.8	8.18	26.82	0.080	62.83
CI	274	12.1	9.48	26.50	0.075	63.74
AI	313	9.2	8.63	27.94	0.116	87.25
GI	288	7.3	14.86	26.99	0.231	95.32
II	269	6.2	10.82	26.90	0.152	90.21
F	1.88	26.66	8.05	2.29	34.82	3.79
P	0.134	0.000	0.001	0.771	0.000	0.000

Note: In Table 3, FD = fixation duration, FC = fixation counts, SA = saccade amplitude, PD = pupil diameter, and BF = blink frequency, SV = saccade velocity.

As shown in Table 3, single factor variance analysis shows that there are significant difference in fixation duration ($F(5,85)=1.88$, $p<0.05$), fixation counts ($F(5,85)=6.06$, $p<0.001$), saccade amplitude ($F(5,85)=8.05$, $p<0.001$) and saccade velocity ($F(5,85)=3.79$, $p<0.001$) among 6 types of icons in visual search experiment, while the differences in pupil diameter (F $(5,85)=2.29$, $p>0.05$) and blink frequency ($F(5,85)=4.82$, $p>0.05$) are not significant.

3.3. *Analysis of visual search strategy*

The eye movement scanning path diagrams of 6 different types of icons were analyzed, and the eye movement characteristics such as the fixation points distribution and the fixation counts of searching for each type of icon can be directly observed through the scan path diagram (as shown in Figure 5). It can be seen that when searching for the digital icons, the English icons, the Chinese icons, and the abstract icons, participants have significantly more fixation counts than abstract icons and image icons. While searching for digital icons and English icons, the participant's fixation duration (the size of the dots in the figure) is significantly greater than the other 4 types of icons. Similarly, it can be seen that the saccade amplitude (spacing of fixations) of the participants when searching for the graphical icons and image icons is significantly greater than the searching for the other 4 types of icons, and the fixations distribution of searching for these 2 types icons is more sparse.

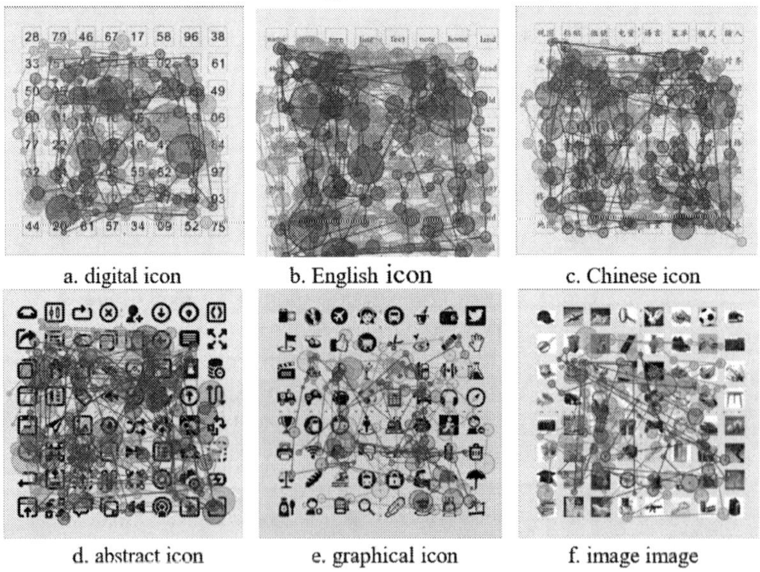

a. digital icon b. English icon c. Chinese icon

d. abstract icon e. graphical icon f. image image

Figure 5. Eye movement path

IOP Conf. Series: Journal of Physics: Conf. Series **1060** (2018) 012031 doi:10.1088/1742-6596/1060/1/012031

4. Discussion

4.1. Analysis of visual search performance of different types of icons

This paper studied the difference in reaction time and correct rate of different types of icons in the visual search experiment. The experimental results not only confirmed the conclusions of some scholars but also had new discoveries. The analysis of specific conclusions and causes were as follows:

The difference among 6 types of experimental icons in this study was considered to be the difference in visual degree (or concreteness), This paper holds that for Chinese people, their visual degree (or concreteness) sequence is: image icon > Image Icon > Abstract icon > Chinese icons > English icons > Digital icons, the sequence is exactly the opposite of reaction time in their visual search experiments. The reaction time in the visual search experiment mainly depends on the difficulty of the feature extraction of the search target by the participants, it is considered that the more concrete of icons in visual search tasks, the easier the feature extraction is. The semiotic theory held that characters was more advanced than images in the symbol system , the more advanced symbols required more complex cognitive processing in the brain, the longer the reaction time was [11]. We believed that the higher the visual degree of icon types, the closer the cognitive mechanism needed for visual search was to human instincts, and therefore the less reaction time. The paired t-test showed that there was a greater possibility of significant differences in reaction time between the two types of icon materials which had a larger distance in visual degree (eg, digital icons and image icons). This also indicated that the reaction time in the visual search had a great relationship with the visual degree of the target icon.

The visual search task of this study required relatively low visual cognition of participants. Therefore, this study initially assumed that all 6 types of icons will not show obvious errors in the visual search task (the correct rate is greater than 95%). However, the experimental results showed that the correct rate of abstract icons was less than 85%, while the correct rates of the other 5 types icons were greater than 95%. And the participants reported that the visual search for abstract icons might be forgotten and confused, further analyzing the experimental data, it was found that 90% of the participants in searching for abstract icons were wrong if the reaction time was greater than 10s (the average reaction time of the abstract icons was 3.82s). This also indicated that after more than 10 seconds, the participants' memory of the abstract icon would become blurred, according to the feature matching theory, participants might confuse some icons with similar appearance characteristics and cause misoperation. The other 5 icons can all be associated with a semantic concept, so only if the corresponding functional memory area of the brain has been activated, this kind of memory is often long-term, so the visual search task can hardly go wrong.

4.2. Analysis of visual search strategies for different types of icons

Visual search strategies include system search, random search, and semi-systematic search. This study analysed the visual search strategies of participants of 6 types icons experimental materials through eye movement path and other eye movement characteristics data. The experimental results showed that the participants tended to adopt semi-systematic search strategy to search for coding symbolic class experimental icon materials (digital icons, English icons, and Chinese icons), as for graphic symbolic class experimental icon materials (abstract icons, image icons, and image icons) tended to choose random search strategy. In the interview with the participants, it was also learned that when searching for digital icons or English icons, the participants would use a search strategy that matches the characteristics of the head-end feature (such as the first digit of the digital icons, the first letter of English icons), this semi-systematic search strategy could reduce the number of repeated searches, while searching for graphic symbolic class icons, the search efficiency could be effectively improved by adopting a random strategy.

4.3. Analysis of visual perception performance for different types of icons

Analysing the visual perception performance of different types of icons based on reaction time, correct rate, and eye movement characteristics. The experimental results of the reaction time confirmed the difference in the perceived performance of the 6 different types of icons, in section 4.2 it was analysed that might be related to the visual degree of icon materials. The correct rate of experimental results indicated that abstract icons are more likely to be misoperated in visual search than other types of icons. Therefore, it could be considered that the visual perception performance of abstract icons was poor, and it should be avoided to use abstract icon as much as possible in the actual graphical user interface design.

The eye movement characteristics data indicated that the average fixation duration of the digital icons, the English icons and the abstract icons was longer, which could reflect the visual processing of these 3 types of icons in the visual search was more difficult. As it was showed in Figure 4, it could be seen that the average fixation counts had a downward trend from the digital icon to the image icon. It could be considered that the average fixation counts was negatively related to the icon's visual degree, which also indicated that the higher the visual degree, the higher the visual search efficiency of experimental icons materials. The shift of the fovea of the visual system causes saccade. We held that the saccade amplitude can reflect the range of the visual processing sampled area. The experimental data indicated that the English icon and the abstract icon had the smallest saccade amplitude, while the graphical icon had the largest saccade amplitude, which also proved that the visual perception performance of the graphical icon was better. The pupil diameter can reflect the psychological workload of the participants during the experiment. The experimental results indicated that the pupil diameters of the participants when searching for the digital icon and the abstract icon experimental material were larger. This also indicated that these 2 experimental materials might cause higher psychological workload in visual search tasks. Some scholars have proved that when the visual cognitive workload was high, the brain needed to pay more attention to the fixation and saccade activities, and the blinking frequency would be reduced accordingly. The experimental data indicated that when searching for graphical icon, the participant' s blink frequency was the highest, which was consistent with the above conclusion.

To sum up, we believed that the graphical icons had the best visual perception performance among the 6 icon types, and had the highest visual search efficiency in human-computer interaction, while the abstract icon had the worst visual perception performance in visual search task. Digital icons were suitable for expressing content that needed to be quantified and had unique advantages in expressing quantity concepts. While the English icons and the Chinese icons had advantages for expressing abstract concepts, its visual perception performance was related to the language used by the operator, so the design of icons in the human-computer interface should also take into account the type of users.

5. Conclusion

In this paper we comprehensively studies and compares the search performance and eye movement characteristics of 6 common types icons in the graphical user interface in visual search experiment. The research results provide ergonomic evidence for the selection and evaluation of icon types in human-computer interaction graphical user interface, which is of great significance for improving the interaction efficiency of the graphical user interface. Next stage, we will combine with EEG and other physiological measurement devices to further study the visual cognition mechanism of different types of icons.

References

[1] Moshagen, M, Thielsch, M T. Facets of visual aesthetics[J]. International Journal of Human-Computer Studies, 2010, 68(10): 689-709.

[2] Mestres-Missé A, Münte Thomas F, Rodriguez- Fornells A. Mapping concrete and abstract

meanings to new words using verbal contexts[J]. Second Language Research, 2014, 30(2): 191-233.

[3] Wang Shuzeng. Frame of pattern recognition model based on cognitive psychology[J]. Geomatics and Information Science of Wuhan University, 2002, 27(05): 543-547.

[4] Barber H A, Otten L J, Kousta S T, Vigliocco G. Concreteness in word processing: ERP and behavioral effects in a lexical decision task[J]. Brain & Language, 2013, 125(1): 47-53.

[5] Isherwood S. Graphics and semantics: The relationship between what is seen and what is meant in icon design[C]// Proceedings of Engineering Psychology and Cognitive Ergonomics. San Diego: EPCE, 2009: 19-24.

[6] Ghayas S, Sulaiman S, Khan M, Jaafar J. The effects of icon characteristics on users' perception[C]// Proceedings of Advances in Visual Informatics. Selangor: IVIC, 2013: 652-663

[7] Schröder S, Ziefle M. Effects of icon concreteness and complexity on semantic transparency: younger vs. older users[C]// Proceedings of Computers Helping People with Special Needs. Linz: ICCHP, 2008: 90-97.

[8] Passini S, Strazzari F, Borghi A. Icon-function relationship in toolbar icons[J]. Displays, 2008, 29(5): 521-525.

[9] Silvennoinen J M, Kujala T, Jpp J. Semantic distance as a critical factor in icon design for in-car infotainment systems.[J]. Applied Ergonomics, 2017, 65:369-381.

[10] Caffaro F, Mirisola A, Cavallo E. Safety signs on agricultural machinery: Pictorials do not always successfully convey their messages to target users[J]. Applied Ergonomics, 2017, 58:156-166.

[11] Amor T A, Luković M, Herrmann H J, et al. Influence of scene structure and content on visual search strategies[J]. Journal of the Royal Society Interface, 2017, 14(132): 20170406.

A Comprehensive Integration Method Based on Unbalanced Data Classification Problems

Yu Lian [1, *], Di Zhang[1], Zhengkui Lin[1], XueHao Sun[1] and WenLi Yu[1]

[1]School of Computer Science and Technology, Dalian Maritime University

[*]525812069@qq.com

Abstract. This paper first introduces the characteristics of non-equilibrium data and the main problems brought by its classification. Based on this, it proposes a comprehensive integration method based on non-equilibrium data classification problems, and the method improvement background and strategy. The construction of base classifiers, and the selective integration methods of base classifiers are described in detail. Finally, the method proposed in this chapter is verified by experiments. The experimental results show that the proposed method is effective.

1. Introduction

In recent years, with the explosion of knowledge brought about by the development of informatization, the problem of classification of non-equilibrium data sets has attracted more and more attention from researchers. The so-called non-equilibrium data set refers to the fact that there are far more samples of certain classes in the data set than other classes[1], in which a class with many samples is called a majority class, and a class with a few samples is called a minority class. The classification of unbalanced data sets is very common in practical applications, such as fraudulent identification of fraudulent card transactions, oil well eruption of satellite image detection, failure prediction of telecommunication equipment, medical diagnosis[2], bankruptcy prediction, radar image surveillance, financial loan management, and fraud. Detection, customer loss prediction, malicious arrears user identification, network intrusion detection[3], transaction fraud identification, biological data identification and text classification and so on. The problem of sending class has one thing in common, that is, the minority information is the focus of our attention. For example, in the monitoring of credit card transaction fraud, the number of fraud users is a minority and it is our target; in the diagnosis process of cancer patients, Cancer patients are the minority and are the objects of our concern. The problem of classification of non-equilibrium data sets exists widely in various fields. Therefore, the research on the classification of non-equilibrium data has a wide range of application prospects and practical significance.

2. related work and problems

At present, there are three classification strategies to study non-equilibrium data: First, from the perspective of the data level by changing the balance of the class distribution of the data; Second, from the perspective of improving the traditional algorithm; Third, through the data level and algorithm level Combine this with an improved perspective.

The most common method to change the balance of data distribution from the data level is re-sampling. The SMOTE (Synthetic Minority Over-sampling Technique) algorithm proposed by

Chawla et al[4]. determines the K-nearest neighbors for each minority class sample, and then randomly fetches the points on the connection line between the sample and its neighbors to generate a duplicate minority sample. The advanced under-sampling method mainly includes Kubat and Matwin [5] unilateral sampling method and Wilson's Editing method.

The method of processing unbalanced data at the algorithm level mainly includes ensemble learning methods. Fan et al[6]. proposed that the AdaCost algorithm is one of the representative methods of cost-sensitive learning. Joshi et al[7]. showed that the AdaCost algorithm has high accuracy and recall rate for a few class samples. The most commonly used methods in ensemble learning are the Boosting algorithm and the Bagging algorithm. Considering that SMOTE algorithm is a mainstream sampling learning algorithm, and it is effective in the preprocessing of non-equilibrium data, some scholars have combined it with ensemble learning in the classification of non-equilibrium data, such as SMOTE. SMOTEBoost algorithm combined with Boosting. In 2002, Zhou et al[8]. demonstrated theoretically the validity of selective integration for the first time, and proposed a selective ensemble learning method based on real-valued coding genetic algorithm (GASEN, Genetic Algorithm based Select Selective Ensemble). GASEN method can achieve better classification accuracy with fewer neural networks.

3. Improvement of Unbalanced Data Classification Integration Method

3.1. The characteristics of unbalanced data sets and their main problems

Non-equilibrium data refers to the situation in which certain types of sample sizes are much larger than those of other types of data in some data sets. In practical problems, such classification problems are widespread. For example, in the analysis of customer churn, churn customers are often categorized as rare categories because there are usually only a few changes to service providers, and in the case of cancer diagnosis, samples of afflicted diseases also account for a very small percentage of total data.

The main characteristics of non-equilibrium data sets are that the number of samples in different categories differs greatly and sometimes they are heavily skewed; second, the distribution of different categories is non-equilibrium, resulting in the inability of the characteristics of rare samples to be detected. It is also the main cause of the low classification accuracy of a few samples. In addition, non-equilibrium data can easily cause other problems in the classification process, mainly including:

(1) Data fragmentation problem

(2) Summarizing deviations

(3) Noise problem

(4) Evaluation index problem

3.2 Improvement of Bagging Integrated Classification Algorithm Based on Unbalanced Data Sets

3.2.1. Algorithm Improvement Background and Strategy. Bagging algorithm is based on the simplest and most intuitive integrated approach to training set. Its own characteristics determine that it is suitable for designing multiple base classifiers in parallel. At the same time, multiple training sets with independent sample distributions are generated, and then learning is performed on different training sets to generate different base classifiers.

The traditional Bagging algorithm in some classifiers such as C4.5 decision tree, while classifying non-equilibrium data sets, can achieve high classification accuracy, but the classification accuracy of a few of them is not high, and In the classification of non-equilibrium data, people tend to pay more attention to the classification results of a few classes. Therefore, it is necessary to improve the traditional Bagging algorithm in order to improve its classification accuracy for a few classes of data on an unbalanced data set.

Combined with the implementation principle of the traditional Bagging algorithm, refer to the main strategies currently adopted to deal with the classification problem of unbalanced data sets. Improvement strategy:

(1) In order to ensure that the distribution of the data samples in the training data set is consistent with the overall distribution, the classification accuracy of the base classifier for a few classes of data on an unbalanced data set is improved, According to the literature[1], Firstly, the data set is changed based on the SMOTE algorithm idea and the undersampling method, and the unbalanced data is changed into relatively equalized data. Then the traditional Bagging algorithm is substituted into the base classifier to weaken the minority and majority classes.

(2) In order to change the traditional Bagging algorithm using simple voting or averaging method to integrate the classification results of the base classifier, due to the failure to fully consider the problem of the classification accuracy caused by the contribution of the base classifier in the classifier integrated system, we proposed a A Classifier Selective Integration Strategy Based on Hybrid Genetic Algorithms .

3.2.2. Classifier Construction was Improved Based on Bagging Algorithm. Definition: $X = \{x_1, x_2, ..., x_n\}$ is the set of total sample instances, $Y = \{y_1, y_2, ..., y_n\}$, Y is the attribute collection, and Type is the sample category array. X_{min} is a collection of a few class samples in X, X_{max} is a collection of most class samples in X.

(1) the few classes was identified

Traversing X, N records the number of samples included in each category, and stores the sample data contained in each category during the traversal; computes the mean value of Nvg=AVG(N), and the variance $square = Square(N)$; For the second classification problem, concave for each category $type[i]$ if the number of samples $N_i < avg$ and $|N_i - avg| > square$ then this class is a minority class, recorded as a positive class; otherwise it is a majority class, a negative class. If it is a multi-classification problem, the few classes that are most concerned are recorded as positive classes, and the rest are negative categories.

(2) Using a mixed sampling method to change the data set, changing the unbalanced data to relatively balanced data

① text oversample was fixed: for a few class samples $X_{min}[i]$; within each sample, $L = (avg - X_{min})/X_{min}$ samples need to be reconstructed; set the parameter K value in the K-adjacent algorithm, that is, select the nearest neighbor, using the K nearest neighbor algorithm Find each sample X, the nearest K of the same sample, $j_1, j_2, ..., j_k$ calculate the attribute difference value attr of each X_i and the K samples, set U to be a random number between $(0,1)$, and the new minority data was reconstruct $Newins = X_i + u * attr$.

② Undersampling was Defined: Reduce $T = (X_{max} - avg)/X_{max}$ samples for each sample in most classes. In order to ensure that the data characteristics of the original majority samples can still be preserved after the undersampling, the method proposed in the reference literature sets Avg_{max} as the mean value point of the majority of the samples; the distance between each sample of the majority class and Avg_{max} is calculated, and The distance is arranged from the largest to the smallest; Set the counter $time = 0$, traverse the sorted data sample set, each time a data sample, Time plus 1, if and only if Time is an integer multiple of 1/T, deleting this Class data.

(3) Substituting the new sample set into the traditional Bagging algorithm trains several base classifiers.

3.3. Base classifier selective integration based on hybrid algorithm

According to the literatue[9], a hybrid algorithm based on genetic algorithm and particle swarm optimization for multi-subsidiary stratification is proposed. Combined with 1.2.2, the base classifier selective integration algorithm based on hybrid genetic algorithm (Hybrid Genetic Algorithm Based) is proposed. Selective Ensemble, HGASEN) is described as follows:

(1) The training set is generated through the bootstrap method and the training subset Di is obtained. After training with Di, the base classifier Li is obtained. Using a hybrid genetic algorithm to select the base classifier L :

① First initialize N subgroups randomly, $GA_i, i = 1,2,...,N$, Each subgroup independently runs its own genetic algorithm;

② For the best individuals in each population formed in the selection, the DPSO was used to evolve elite groups.When initializing the velocity of the elite group particle, the direction of the particle's velocity in each dimension is limited to be positive or negative at the same time, that is, the velocity interval is used at the same time. $[0, \mathrm{vjmax}]$ or $[-\mathrm{vjmax}, 0]$.

③ After a certain amount of algebra, it is judged whether the stop criterion is satisfied.The difference between this algorithm and other genetic and particle swarm algorithms is that the adoption of genetic algorithm as the main body ensures the superior global search ability of the algorithm; secondly, the use of multiple population structures can better maintain the diversity of the population and can Accelerated evolution is completed; and using dissipative particle swarm optimization algorithm to perform mutation and local search on elite individuals, the convergence speed is fast and the precision is high; finally, a hierarchical structure is used to separate the individuals responsible for global search at the bottom layer from the individuals responsible for local search at the top layer. Not only can speed up the search speed, but also can avoid premature convergence and fall into a local optimal solution.

4. Experiment and result analysis

4.1 Initial data set class balance characteristics

In this paper, we use the lower three groups of UCI imbalance datasets for comparison experiments. In the experiment, a few classes are defined as positive classes, and most classes are defined as negative classes.

The data set car_evaluation, the data has 6 features per sample, a total of 4 categories, the test selected 'Vgood' as a positive class, and the rest as a unified class; The data set balance-scale, the data has 4 characteristics per sample, a total of 3 categories, the test selected "B" as a positive class, the rest as a unified class;The data set breast_cancer, the data has 9 features per sample, and there are 2 categories. In the experiment, 'recurrence-events' is selected as the positive category, and the rest are unified as the negative category.

Table 1. Class Balance Characteristics of Data Sets before Processing

data set	Characteristic dimension	Positive sample number	Negative sample size	Positive and negative sample ratio
Car_evaluation	6	65	1663	0.05
Balance_scale	4	49	576	0.17
Breast_cancer	9	85	201	0.42

Table 2. Data set Class Balance Characteristics After Processing

data set	Characteristic dimension	Positive sample number	Negative sample size	Positive and negative sample ratio
Car_evaluation	6	896	831	1.08
Balance_scale	4	294	288	1.02
Breast_cancer	9	170	201	0.85

As shown in Table 1,Table 2, the data processed according to the method in 3.2.2 of this chapter has basically achieved class balance.

4.2 Classifier classification performance comparison

The experimental uses the C4.5 decision tree was used as a Bagging base classifier. The accuracy rate and recall rate were selected as evaluation indicators. The classification performances of the traditional Bagging classifier integration algorithm, GASEN algorithm and the HGASEN algorithm proposed in this chapter were experimentally compared.

In the experiment, the initial weights of each base classifier are the same. The hybrid particle swarm optimization algorithm uses real-number coding. The integrated AUC value is used as the fitness function value of the genetic algorithm and the particle swarm algorithm. The crossover probability pc is 0.6 and the mutation probability pm is 0.2. There are 30 subgroups and 10 individuals in each subgroup. The optimal particles in each subgroup constitute the upper elite subgroup with 30 particles. Iterated for 50 rounds of iterations. Each round of genetic algorithm and particle swarm algorithm runs for 20 generations. The number of individuals in each substitution substratum group is set to 2.The experimental results are shown in Table 3 and Table 4 below.

Table 3. Comparison of Classification Performance of Classifier

Data set	Classification algorithm	Precision rate	Recall rate
Balance_scale	Bagging	0.695	0.731
	GASEN	0.738	0.750
	HGASEN	0.751	0.762
Car_evaluation	Bagging	0.933	0.931
	GASEN	0.942	0.946
	HGASEN	0.948	0.951
Breast_cancer	Bagging	0.763	0.766
	GASEN	0.789	0.782
	HGASEN	0.795	0.789

Table 4 .Comparison of Classification Performance of Minority Samples

Data set	Classification algorithm	Precision rate	Recall rate
Balance_scale	Bagging	0.200	0.020
	GASEN	0.400	0.167
	HGASEN	0.475	0.364
Car_evaluation	Bagging	0.860	0.880
	GASEN	0.878	0.889
	HGASEN	0.879	0.901
Breast_cancer	Bagging	0.750	0.318
	GASEN	0.783	0.354
	HGASEN	0.779	0.458

4.3 The result analysis

As can be seen from Table 3 and Table 4, the improved algorithm proposed in this chapter is superior to the GASEN and Bagging classifiers in classifier recall and precision, and it is lower than the minority class recall rate. The GASEN and Bagging classifier accuracy rate on the breast_cancer data set is lower than the GASEN difference. This chapter focuses on the classification accuracy of a few classes. The recall rate is the focus of attention, and the improved algorithm has better classification effect in a few classes.

5. Conclusions

The improved algorithm presented in this paper is superior to the GASEN and Bagging classifiers in the classifier recall and precision, and the GASEN and Bagging classifiers breast_cancer data set is better than GASEN in the minority class recall. The difference is that this chapter focuses on the classification accuracy of a few categories. The recall rate is the focus of attention, so the improved algorithm in this paper has a better classification effect in a few categories.

6. Summary and Prospect

This paper proposes that the integration algorithm for non-equilibrium data sets is based on the hybrid genetic algorithm, although to a certain extent by raising the difference between the classifiers participating in the integration, the final classification effect is raised. However, it still has some limitations. Further studies are mainly considered from the following three perspectives:

(1) Base classifier selection.
(2) Choose a better base classifier fusion method.
(3) Integrate based on heterogeneous algorithms.

References

[1] Haque M N,Noman N,Berretta R,et al. Heterogeneous Ensemble Combination Search Using Genetic Algorithm for Class Imbalanced Data Classification.[J].Plos One. 2016, 11(1):e0146116

[2] Zung M, Zou B, Wei F, et al. Effective prediction of three common diseases by combining SMOTE with Tomek links technique for imbalanced medicaldata[C]//Online Prediction and Computing Science (ICOACS) IEEE International Conference of . 2016:225-228

[3] Rodda S,Erothi U S R.Class imbalance problem in the Network Intrusion Detection Systems[C]//Electrical Electronics and Optimization Techniques (ICEEOT) International Conference on .IEEE, 2016:2685-2688.

[4] Chawlal NV,Lazarevic A,Hall LO.SMOTEBoost:improving prediction of the minority class in boosting[C]//The 7th Eu-ropean Conf on Principles and Practice of Knowledge Discovery in Databases . 2003:107-119

[5] Kubat M,Matwin S. Addressing the Curse of Imbalanced Data sets:One-Sided Sampling.[C]//Proc of the 14th International Conference on Machine Learning. 997:179186

[6] Fan W, Stolfo S J, Zhang J,et al.Adacost:Misclassification Cost-Sensitive Boosting[C]// Proceedings of the Sixteenth International Conference on Machine Learning .1999:97105

[7] Joshi M,Kumar V,Agarwal R.Evaluating Boosting Algorithms to Classify RareClasses:Comparison and Improvements[A]. First IEEE International Conferenceon Data Mining[C]. San Jose CA, 2001

[8] Zhou Z H, Wu J X, Tang W.Ensembling several networks: Many could be better than all. Artificial Intelligence. 2002 , 137(1-2):239:263

[9] Jin M, Lu H. A multi-subgroup hierarchical hybrid algorithm based on genetic algorithm and particle swarm optimization [J]. Control Theory and Application.2013,30(10):1231-1238

IOP Conf. Series: Journal of Physics: Conf. Series **1060** (2018) 012033 doi:10.1088/1742-6596/1060/1/012033

The Extraction of Emotional and Attribute Words in Perspective Sentences

Naijun Cao[1], Zhengkui Lin [1], Hong Yang [1], Huancheng Zhang[1] and ZenghuiQiu[1]

[1]School of Computer Science and Technology, Dalian Maritime University

m18742521258_1@163.com

Abstract: The article focuses on the research on the classification criteria of emotional words and attribute words. Firstly, the semantic similarity and its representative method based on distance, the Euclidean distance calculation method and the cosine distance calculation method are introduced. Based on this, a comparison experiment of semantic similarity between emotional words and attribute words based on Euclidean distance and cosine distance was carried out. According to the experimental results, a method for determining the classification criteria of emotional words and attribute words based on the distribution rules of words and cosine distances was proposed.

1. Introduction

In recent years, online shopping reviews have become one of the most important sources of information for online consumer evaluation [1]. The consumer sentiment of the online shopping review information on the purchased goods has an important influence on the purchase decision of subsequent potential consumers, and the online shopping commentary in the online shopping commentary extracts the emotional words and attribute words in the online shopping review. The key to the premise and foundation of the sentiment analysis is to select the appropriate classification feature model , in other words, the classification criteria, to support the extraction of emotional words and attribute words. This dissertation focuses on the semantic similarity calculation method of emotional words and attribute words and the related research on the classification methods of emotional words and attribute words based on the distribution law of words and cosine distances.

2. Distance-based semantic similarity calculation method

2.1. Word distance and semantic similarity

Dekang Lin thinks that the similarity of any two words depends on their Commonality and Differences, and then from the perspective of information theory, it gives the definition formula of semantic similarity:

$$\mathrm{Sim(A, B)} = \frac{\log \mathrm{p(common(A,B))}}{\log\mathrm{p(description(A,B))}} \quad (1)$$

Among them, the numerator represents the amount of information required to describe the commonality of A and B; the denominator represents the amount of information required to fully describe A and B.

At present, there are two main methods for calculating the semantic similarity of words: one is to use the statistical information method to solve the word context information, and the other is to calculate the word similarity based on distance. Simple correspondence, and then through the semantic dictionary, the concept of related words organized in a tree-like structure, by computing the distance between nodes in the tree to solve the semantic similarity between words.

$$\mathrm{Sim}(w_1, w_2) = \frac{k}{\mathrm{Dis}(w_1, w_2)} \quad (2)$$

Among them, Dis(W1, W2) is the distance of the tree represented by W1, W2, and k is the proportional coefficient.

In distance-based word similarity calculation methods, common distance calculation methods mainly include Euclidean distance, cosine distance, Manhattan distance, and the like. This paper focuses on the performance comparison of semantic similarity between Euclidean distance and cosine distance calculations.

2.2. Euclidean distance

Euclidean distance, or euclidean metric, is a widely used definition of distance, which refers to the true distance between two points in m-dimensional space, or the natural length of a vector (that is, the point to the origin Distance) [2], calculated as follows:

$$\mathrm{distance}(X, Y) = \sqrt[2]{\sum_{i=1}^{n}(X_i - Y_i)^2} \quad (3)$$

Among them, X and Y are n-dimensional vectors, X_i is the i-th component of X, and Y_i is the i-th component of Y.

2.3. Cosine distance

Cosine distance is also called cosine similarity. The cosine distance is calculated as follows[3][4]

$$\mathrm{sim}(X, Y) = \frac{\sum_{i=1}^{n}(X_i * Y_i)}{\sqrt{\sum_{i=1}^{n}(X_i)^2} * \sqrt{\sum_{i=1}^{n}(Y_i)^2}} \quad (4)$$

Among them, X and Y are n-dimensional vectors, Xi is the i-th component of X, and Yi is the i-th component of Y.

The cosine distance represents the angular difference between the two vectors and is applied to the word vector. Since the word vector has semantic meaning, the cosine distance between two word vectors can be used to represent the difference between the semantics expressed by the two word vectors.

3. Experiment Based on Distance Emotional Words and Attribute Words

3.1. Experimental Design

Both Euclidean distance and cosine distance can be used to calculate the semantic similarity between words.[5] In order to obtain better classification of affective words and attribute words and classification criteria of emotional words and attribute words, this paper develops Euclidean distances and cosine distances. The similarity degree calculation of the emotional words and attribute words is compared with the experiment.

The experiment is based on the previous viewpoints. Each word and its corresponding word vector in the viewpoint are used as the basic corpora;[6] then the basic corpora are paired with the word pairs, and after removing the duplicates, formula 1.1 and formula are used respectively. 1.2 Calculate the Euclidean distance and cosine distance between pairs of words to obtain experimental corpus; Finally, select several representative pairs of words from experimental corpus and determine distance-based classification criteria of emotional words and attribute words.[7][8][9]The properties of the manually labeled word pairs and their corresponding Euclidean distance and cosine distance calculation results are shown in Table 1 and Table 2 respectively.

Table 1.examples of Euclidean distance

Word pair	Word pair nature	Euclidean distance
System - Bluetooth	Attribute word - attribute word	3.545893
Function - Signal	Attribute word - attribute word	4.362586
Cache - Satisfaction	Attribute word – non attribute word	3.250933
Good-big	Emotional Words - Emotional Words	3.385302
…….	…….	……
Good- appearance	Emotional Words – non Emotional Words	3.351089

Table 2.examples of Cosine distance

Word pair	Word pair nature	Cosine distance
Signal - Touch Screen	Attribute word - attribute word	0.3550653294881015
Screen - Battery	Attribute word - attribute word	-0.0213554490291685
Touch screen - old man	Attribute word – non attribute word	0.5220542183771003
Beautiful - beautiful	Emotional Words - Emotional Words	0.7212404951757746
……	……	……
Decision - Touch Screen	Emotional Words – non Emotional Words	0.6637506361054051

3.2. Experimental analysis of distribution rules of emotional words and attribute words based on Euclidean distance

(1)Experimental process and experimental results

The maximum Euclidean distance of all word pairs in the experiment corpora obtained according to the experimental scheme is 16.096646, which is a study of attribute word-attribute word pair, attribute word-non-attribute word pair, emotional word-emotion word pair and emotional word-non-emotion word pair. The distribution of the law, first of all, the Euclidean distance in experimental corpora is divided into [0.0, 1.6), [1.6, 3.2), [3.2, 4.8), [4.8, 6.4), [6.4, 8.0], [8.0, 9.6], [9.6,11.2), [11.2,12.8), [12.8,14.4], [14.4,16.096646]10 intervals, and then statistically count the number of attribute words-attribute word pairs and attribute words in 10 intervals- The number of non-attribute word pairs, the number of emotional word-emotion word pairs, and the number of emotional word-non-emotion word pairs. For convenience of statistics, the symbol "a" represents the attribute word-attribute word pair, and the symbol "b "On behalf of attribute words - non-attribute word pairs, symbol "c" represents emotional word-emotion word pairs, symbol "d" represents emotional word-non-emotional word pairs, and finally, based on statistical results, the number of pairs of words plotted Summarize its distribution pattern. The algorithm described above is described in Table 3 below.

Table 3. Euclidean distance segmentation is used to calculate the attribute statistics

Process: For segmented statistics for word pairs that have been calculated for Euclidean distance, the formal representation of this algorithm takes one of the intervals as an example.
Symbol Description: Dic_0 _0_1_6_a: Euclidean distance in the range [0.0, 1.6) and the word pair property is the statistical result of the attribute word-attribute word

Dic_0 _0_1_6_b: The Euclidean distance is in the range [0.0, 1.6) and the word pair property is the statistical result of attribute word-non-attribute word

Dic_0 _0_1_6_c: Euclidean distance in the range [0.0, 1.6) and the word pair property is the statistical result of the attribute word-attribute word

Dic_0 _0_1_6_d: Euclidean distance in the range [0.0, 1.6) and the word pair property is the statistical result of the attribute word-non-attribute word

a : The word pair property is attribute word-attribute word; b : The word pair property is attribute word-non-attribute word

c : The word pair nature is emotional word-emotion word; d : The word pair nature is emotional word-non-emotion word

Vector_dis : experimental corpora after artificially marking word pair properties and calculating Euclidean distance

Results : statistical results of the nature of various word pairs in 10 intervals

Process:

```
Input →vector_dis
FORline INvector_dis
     Get the word pair for each line and write it as word-word
     Get the word pair attribute in each row, and write it as words-mark
     Get the corresponding Euclidean distance for word pairs in each row, written as word-distance
IF words-mark == a and word-distance∈[0.0,1.6)THEN
   Word-word → dic_0 _0_1_6_a
ELSE IF words-mark == b and word-distance∈[0.0,1.6)THEN
   Word-word → dic_0 _0_1_6_b
ELSE IF words-mark == c and word-distance∈[0.0,1.6)THEN
   Word-word→ dic_0 _0_1_6_c
ELSE IF words-mark == c and word-distance∈[0.0,1.6)THEN
   Word-word→dic_0 _0_1_6_d
END IF
END FOR
Output→results
```

According to the above algorithm, the number of word pairs of different word pair properties in 10 Euclidean distance intervals is counted. The results are shown in Table 4 below:

Table 4. Euclidean distance distribution data

Euclidean distance range	Attribute Word - Attribute Word Pair Number	Attribute Word - Non-attribute Word Pair Number	Emotional Words - Emotional Word Pairs	Emotional Words - Non-Emotional Word Pairs
[0.0 , 1.6)	2080	88	1174	840
[1.6 , 3.2)	1230	943	902	160
[3.2 , 4.8)	860	490	1730	172
[4.8 , 6.4)	22	34	27	1530
[6.4 , 8.0)	990	1860	780	2315
[8.0 , 9.6)	103	2127	166	1025
[9.6 , 11.2)	157	1000	157	34
[11.2 , 12.8)	1200	161	132	193
[12.8 , 14.4)	109	136	1900	500
[14.4 , 16.0]	400	300	400	300

(2)Analysis of results

On the one hand, since the range of Euclidean distances is [0,+∞), the maximum Euclidean distance of experimental corpora obtained at present is only 16.096646, which will lead to incompleteness of the distribution rules. With the continuous expansion, although the maximum value of the Euclidean distance may change, the overall trend of the distribution may also change, but the analysis of the distribution of the nature of word pairs will not have much impact.

On the other hand, according to statistical data, for the attribute word-attribute word pair and emotional word-emotion word pair, although there is a gap in the overall distribution, the difference is not significant, that is, different pairs of words are different. Since the number of Euclidean distances is equivalent, it is not possible to use a range of Euclidean distances as a classification criterion (threshold) for distinguishing attribute words, non-attribute words, emotional words, and non-emotion words.

3.3. Experimental Analysis of the Distribution Rules of Emotional Words and Attribute Words Based on Cosine Distance

(1) Experimental process and experimental results

The experimental process is similar to 3.2. Firstly, the cosine distances in the experiment corpora are divided into [-1.0, -0.8), [-0.8, -0.6), [-0.6, -0.4), [-0.4, -0.2). , [-0.2, 0.0), [0.0, 0.2), [0.2, 0.4), [0.4, 0.6), [0.6, 0.8), [0.8, 1.0] 10 intervals; then statistics on 10 intervals Word-attribute word pair number, attribute word-non-attribute word pair number, emotional word-emotion word pair number and emotional word-non-emotion word pair number, for convenient statistics, also note the symbol "a "Proxy attribute word-attribute word pair, symbol "b" for attribute word-non-attribute word pair, symbol "c" for emotional word-emotion word pair, and symbol "d" for emotional word-non-emotion word pair Finally, according to the statistical results, draw the distribution of word pairs and summarize their distribution rules. According to the above algorithm, the number of word pairs of different word pair properties in 10 cosine distance intervals is counted.

According to the algorithm, the number of word pairs of different word pair properties in 10 cosine distance intervals is counted.

(2) Analysis of experimental results

On one hand, according to the cosine characteristics, since the range of the cosine distance is [-1.0, 1.0], different from the Euclidean distance, the distribution rule reflected by the experimental results will not only occur because of the change in the number of experimental corpora. The fundamental changes, on the contrary, with the continuous expansion of the scale of experimental corpus, the above-mentioned distribution characteristics will become stronger and stronger;

On the other hand, according to the statistical results in Table 4, the following rules can be found: The distribution law of cosine distances for word pairs of different natures is used as a classification basis for emotional words and attribute words.

4. Classification criteria of emotional words and attribute words based on word-to-cosine distance distribution

Based on the different comparison of emotional words and analysis of the words, the distance can be used to calculate the words to be classified. Knowing the semantic similarity of emotional words and attribute words, and comparing and analyzing the distribution characteristics of cosine distances of different nature words, the classification criteria of emotional words and attribute words are determined .

As this: the process of determining the classification criteria for emotional words and attribute words can be described as follows:

1. Establish basic attribute word lexicons, emotional word lexicons, and other word lexicons (that is words that are neither attributed nor sentimental);

2. For words of unknown nature, semantic similarity is calculated with the words in the dictionary of attribute words. If the value of semantic similarity is within the threshold of the attribute word-attribute word pair ([-1.0 , -0.4)), When the attribute word dictionary is 80% of the number of words, it is determined that the keyword belongs to the attribute word, otherwise the semantic similarity is calculated with the selected emotional word, if the value of the semantic similarity is in the threshold Of the emotional word-emotion word pair ([0.6 , 1.0]), it is determined that the word is an emotional word, otherwise it is other words;

3. After the selection of emotional words, attribute words, and other words, the basic emotional word dictionary, basic attribute word dictionary, and basic other word dictionary are expanded to continuously improve the accuracy of the classification criteria. In terms of the extraction of attribute words and sentiment words, this paper proposes a semantic similarity calculation based on experimental corpora, and empirically researches the classification rules of semantic similarity based on Euclidean distance, cosine distance, emotional words, and attribute words. The emotional word-attribute word extraction method is based on corpus rules and matches emotional words and attribute words. For the emotional words that do not match the attribute words, the design algorithm extracts the implicit attributes of the words and compares them with the manually labeled results. The rate reached about 75%. However, because the subject involves a wide range of knowledge, coupled with my limited time and energy, and the following aspects can be further studied and carried out: cross-domain word contribution rate. For the value of the contribution of words to sentence semantics, the training corpus of this article is completed by manual annotation. However, the subjectivity of this part of people is relatively large. At present, there are no certain norms and standards for this aspect, resulting in possible final experimental results. Cannot be applied to all fields of language, for the word to sentence semantics contribution rate across the domain to achieve is the next major research content

5. Summary and Prospect

In terms of the extraction of attribute words and sentiment words, this paper proposes a semantic similarity calculation based on experimental corpora, and empirically researches the classification rules of semantic similarity based on Euclidean distance, cosine distance, emotional words, and attribute words. The emotional word-attribute word extraction method is based on corpus rules and matches emotional words and attribute words. For the emotional words that do not match the attribute words, the design algorithm extracts the implicit attributes of the words and compares them with the manually labeled results. The rate reached about 75%.

However, because the subject involves a wide range of knowledge, coupled with my limited time and energy, and the following aspects can be further studied and carried out: cross-domain word contribution rate. For the value of the contribution of words to sentence semantics, the training corpus of this article is completed by manual annotation. However, the subjectivity of this part of people is relatively large. At present, there are no certain norms and standards for this aspect, resulting in possible final experimental results. Cannot be applied to all fields of language, for the word to sentence semantics contribution rate across the domain to achieve is the next major research content.

References

[1]https://baike.baidu.com/item/%E6%AC%A7%E5%87%A0%E9%87%8C%E5%BE%97%E5%BA%A6%E9%87%8F/1274107?fr=aladdin&fromid=2809635&fromtitle=%E6%AC%A7%E5%BC%8F%E8%B7%9D%E7%A6%BB.

[2]https://baike.baidu.com/item/%E4%BD%99%E5%BC%A6%E7%9B%B8%E4%BC%BC%E5%BA%A6/17509249?fr=aladdin.

[3] Xiaobo Tang , Guangchao Liang. Summary of Research on Fine-grained Sentiment Analysis [J]. Joining of CI Books.2017,61(5):132-140.

[4] Turney P D, Littman M L.Measuring praise and criticism: Inference of semantic orientation from association[J].ACM Transactions on Information Systems(TOIS) . 2003,21(4):315 — 346.

[5] Zhichun W, Juanzi Li, Zhigang W, et al. Cross-lingual knowledge linking across wiki knowledge bases [C].In Proceedings of the 21st internationalconference on World Wide Web, 2012: 459-469.

[6] Ku L W, Wu T H, Lee L Y, et al. Construction of an evaluation corpus for opinion extraction[C].Ntcir.2005:513-520.

[7] Pang B, Lee L, Vaithyanathan S.Thumbs up: Sentiment classification using machine learning techniques[C]. In: Proceedings of Association for Computational Linguistics Conference on Empirical Methods in Natural Language Processing. ACL'02, 2002:79 — 86.

[8] Sajib S, Ng V. Mine the Easy, Classify the Hard: A Semi-Supervised Approach to AutomaticSentiment Classification[C].In:Meeting of the Association for Computational Linguistics.2009:701-709.

[9] Huancheng Zhang. Online shopping review informationResearch on Fine-grained Sentiment Analysis Method [D].Dalian Maritime University, 2016.

The Layout Optimization Design of Command Control Cabin Display and Control Panel

Shuzhi Yuan[1, *], Wei Wang[1], Jue Qu[1] and Chao Wang[1]

[1]Air and Missile Defense College, Air Force Engineering University Xi'an, Shaanxi, 710051, China

[*]1664267629@qq.com

Abstract. According to the type and quantity of the control panel of the weapon and equipment control panel, the operation difficulty is big, the repetitive movement, the blocking posture, the accumulated task load, Easily induced Musculoskeletal disorders, the impact of task performance, this paper comprehensive ergonomics criteria and the characteristics of human upper limb movement, and through the human in the loop simulation equipment. In the operation, the operation time data of a certain type of equipment operation interface is collected and analyzed, and the positioning movement time of each area of the equipment interface is obtained. Distribution map, to provide data basis for the model, the establishment of the control panel layout model based on linear programming and Fitts' law, and the use of LINGO software 0-1 programming model to solve the programming. The results show that the operating loss distance and operation time are optimized by 28% and 16% respectively, and the optimization effect is remarkable.

1. Introduction

In man-machine system operation, the human-machine interface is the concept space of information and force exchange between the user and the machine. The layout of the manipulating interface plays a vital role in the interaction efficiency. In the early days, the research on the layout of the manipulative interface was mainly based on the qualitative optimization of the interface between the ergonomic guidelines. In 1993, Sanders and McCormick proposed four basic principles for user interface layout: importance, frequency of use, functional similarity, and order of use. It forms the basis for research and is widely used in the design of user interfaces for various products and systems [1]. In 1995, ESJUNG detailed the ergonomics criteria affecting layout, refined into ergonomic criteria affecting relative position decisions, and ergonomic guidelines affecting absolute position decisions, and proposed CSP (Constraint Satisfaction Problem) techniques for solving Multi-constraint satisfaction problem [2]. In 2003, Grady T. Holman optimized the control panel design using the linear programming method from the perspective of ergonomics for the first time, and demonstrated that the ergonomic design based on ergonomics significantly reduced skeletal muscle disorder and its severity. In 2008, Maurício Alves C designed a nuclear power plant virtual control panel, which has significantly improved the efficiency of operation and training due to its flexible layout design and its low cost.[3] In 2013, Isaac José Antonio Luquettidos Santos reconfigured the alarm panel of the nuclear control room with a human factors questionnaire and an operator action analysis method based on a cognitive function model to improve the safety level of the control room of a nuclear power plant [4]. In 2015, Cem Alppay used ergonomics guidelines and user perspectives and preferences to quantitatively evaluate interface layout principles through interviews with professional pilots, and creatively presented a paper prototype method that allows pilots to optimize display devices on the instrument panel. The layout forms the methodology for optimizing aircraft instrument layout [5].

In recent years, the domestic operation control interface layout design mainly focuses on the rail vehicle driver console, aircraft cockpit, industrial control room and other fields. In 2014, Chen Dejun and Fang Weining proposed an optimization design model for console layout based on factors of human factor and geometric position matching. The particle swarm algorithm was used to solve the model and optimized the comfort operation requirements of the rail vehicle driver console [6]. In 2010, Wang Rui and Zhuang Damin used anthropometric data and Lagrange equations to establish an upper limb operating dynamics model, which was successfully applied to the ergonomics analysis of cabin manipulator layout [7]. In 2011, Deng Li proposed an optimization model for workspace layout based on human posture driving, and optimized the layout of manipulators in the right-hand area of the driller's console of the oil rig [8]. In 2013, Niu Song proposed the use of genetic algorithms to optimize the layout of the civil cockpit, solved the quantification of layout principles, and optimized the layout of a certain aircraft cockpit. The results showed that the genetic algorithm was suitable for layout optimization [9].

At present, there are relatively few researches on human upper body reachability probability and positioning movement characteristics of the man-machine interface. There are few mathematic models to apply the human upper limb reachability probability and positioning movement characteristics in the man-machine interface layout optimization model. Although Grady T. Holman's study considers the loss of the operating distance, it is based on the fact that foreign body data does not meet domestic equipment requirements, and then does not consider the optimization of the operating time. The ergonomic guidelines for ergonomics are only qualitatively applied.

Based on the improved Grady T. Holman's model, this paper proposes a prediction model for the layout optimization design of command and control console console panel based on linear programming (LP) and Fitts' law. The LINGO software is used to program the objective function of the linear programming. Through the iterative optimization of the decision variables, the best combination of the decision variables with the shortest loss distance and the shortest operating time is finally sought, and the overall layout scheme is optimized.

2. Control panel layout optimization design model

2.1. Description of the layout problem of the equipment control panel

After a long-term investigation, it is found that in the stressful task, the layout of the equipment interface could not well meet the movement characteristics of the human upper limbs. It is easy to cause repetitive operation, muscle fatigue damage, which leads to low operation efficiency and high operation error rate, and reduces the efficiency and reliability of human-machine interaction. How to improve the ergonomic design level of the human-computer interaction interface has become an urgent issue that needs to be resolved. Based on human upper limb movement characteristics and ergonomic guidelines, this paper improves the effectiveness of operating distances and operating time, and meets the basic ergonomic requirements such as frequency of use, importance, order of use, and spatial compatibility, and fundamentally addresses poor human-machine matching. Human-computer interaction problems.

2.2. Technical methods and modeling basis

2.2.1 Linear programming. Linear Programming (LP) is an important branch of mathematical planning and has been widely used in military operations, national defense construction, and engineering technology. The problem of linear programming is to find the maximum or minimum value of linear objective function under linear constraint conditions. Decision variables, constraints, and objective functions are the three elements of linear programming. The standard form is [10]:

$$\min c^T x$$
$$st.\quad Ax = b, x \geq 0, \tag{1}$$
$$c \in R^n, b \in R^m \quad A \in R^{m \times n}$$

2.2.2 Fitts' law. Fitts'law is widely used in the field of human-computer interaction and provides a scientific basis for the design of human-computer interaction interface. It is an important quantitative prediction model [11]. It is used to study how long it takes for a pointing device to locate a target object in a positioning movement. This article selects the 1989 MacKenzie revised formula. The MacKenzie model is the most commonly used, and proved to be the best in accordance with the experimental results, mainly because of:

The model is better fitted to the results observed by the experiment, and the information theory implied by the Fitts' law is accurately simulated. The expression is as follows:

$$MT = a + b\log_2\left(\frac{A}{w} + 1\right) = a + b \bullet ID \tag{2}$$

Fitts'law is a continuous, fast, and purposeful movement model. It shows that the time required to use a pointing device to reach a goal is related to the following two factors:

1) The distance between the current position and the target position of the device (A). The longer the distance, the longer it will take.

2) Target size (W), the greater the goal, the shorter the time spent.

2.3. Mathematical model based on linear programming and Fitts' law

The above various layout optimization models have achieved good results in terms of ergonomics criteria, minimum distance, etc., but they still cannot meet the high standards of equipment requirements. Operation time is the main factor restricting the efficiency of the equipment. In the layout problem, it is necessary to consider the distance between the optimization operation and the optimization of the operation time. It can be seen from the mathematical expression of Fitts' law that the operating time is not only affected by the operating distance, but also related to the orientation of the operation and the target size. In addition, the general algorithm cannot effectively solve the multi-objective linear programming model. This article uses LINGO programming to solve such problems efficiently.

According to a commonly used type of equipment control panel size design panel layout as shown in Figure 1. The panel consists of 15 small panels, represented by panel j (j=1,2,...,15) , each panel size is 147x92 (unit: mm), panel numbers are arranged from left to right. The ellipse represents the operator, the circle represents the initial position of the operator's hand, and the dashed line length represents the horizontal distance of the hand to each small panel and is recorded below the small panel, as shown in Figure 1.

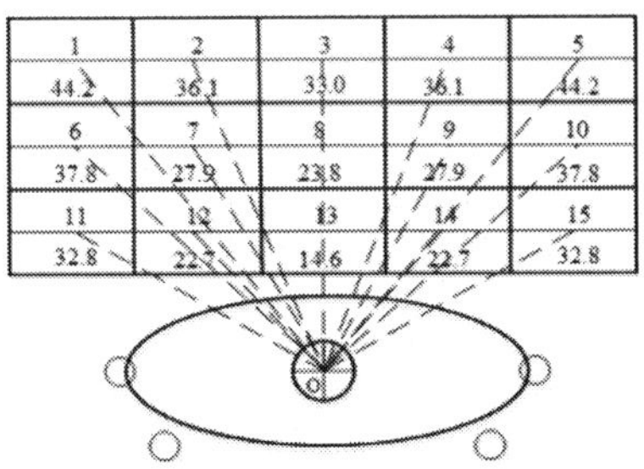

Figure.1 Each panel area from the distance operator

Based on the 50th percentile Chinese male body measurement size, the normal and maximum operating area of the following level working face is constructed. The dashed arc indicates where the tip of the thumb can reach when the arm is straight, such as the arc 3; the middle line represents the normal operating area, the reachable area of the tip of the thumb when the arm is placed normally and the upper arm is in a natural droop, such as the arc line 2; the solid inner line represents the reach of the center of the elbow in the horizontal state of the upper arm, such as the arc 1. According to the RULA criterion, when the elbow rotates around the shoulder joint normally and the wrist moves normally, the area within the arc 1 will be 100% efficiently used, and the probability of reaching the upper sagittal plane closer to the human body will decrease correspondingly. Applying the same criteria, the effective use probability of arc 3 is about 70%, and the effective use probability outside this area is less than 30%. Figure 2 shows the probability distribution of effective use of small panels by 50-percentile Chinese men.

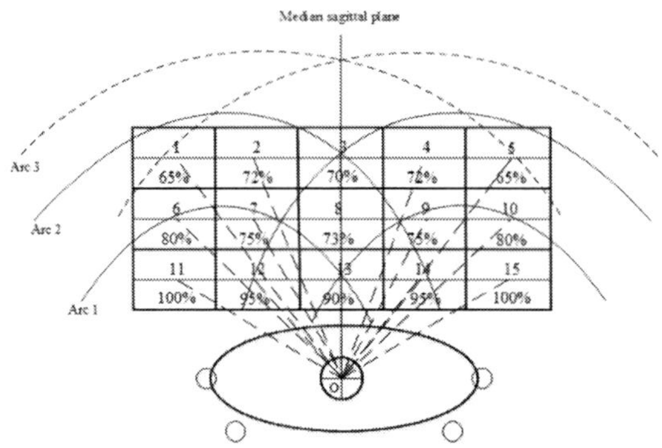

Figure.2 The effectiveness of upper limb movement area up to a probability distribution

2.3.1 Objective function. From the above analysis, it can be seen that the effectiveness of the positioning and operating movements of the upper limbs varies among different regions. In order to optimize the effective operation distance, the loss of upper limb movement distance caused by invalid operation is minimized, and the operation time is optimized. Apply the human upper limb movement law and Fitts' law and the ergonomic layout guidelines of the console to establish the following layout optimization model, the objective function is:

$$\begin{cases} \min F(x) = (f_1, f_2) \\ f_1 = \sum_{i=1}^{m}\sum_{j=1}^{n} C_{ij} X_{ij} \\ f_2 = \sum_{i=1}^{m}\sum_{j=1}^{n} E_{ij} X_{ij} \end{cases} \tag{3}$$

In the formula, the optimization goal $\min F(x)$ is to minimize the loss distance f_1 and operating time during operation f_2; C_{ij} is the loss distance cost variable, which represents the distance loss where function module i is placed at position j, $C_{ij} = f_i^{'}\Delta\beta_j = f_i^{'}(1-P_j)S_j$, $\Delta\beta_j$ is the operating loss distance at panel j, $f_i^{'}$ is the frequency of use of functional module i, P_j is the reachability probability of panel j, S_j is the distance from the starting point O to panel j; E_{ij} is the operating time cost variable, which represents the required positioning movement time when the functional module i is placed at position j, $E_{ij} = f_i^{'}MT_j = f_i^{'}\left(a_j + b_j \log_2\left(\frac{A_j}{W_j}+1\right)\right)$, MT_j is the operating time for positioning movement starting point to panel j, $f_i^{'}$ is the frequency of use of functional module i, A_j and W_j are the distance from the starting point of operation to panel j and the width of panel j respectively; X_{ij} is a decision variable for judging whether the function module i is placed in the position of the panel j, which is 1, if not 0; $f_i^{'}$ is the frequency of use of module i.

2.3.2 0-1 planning model constraints. (1) Control panel model basic constraints. The manipulation panel model consists of two basic constraints, and each panel j can only place one function module i, that is: $\sum_{i=1}^{n} X_{ij} = 1, i = 1,2,\cdots,n$; each function module i can only occupy one panel j, that is: $\sum_{i=1}^{m} X_{ij} = 1, i = 1,2,\cdots,m$. The constraints are as follows:

$$st.\begin{cases} \sum_{i=1}^{n} X_{ij} = 1, & j = 1,2,\cdots,n \\ \sum_{j=1}^{m} X_{ij} = 1, & i = 1,2,\cdots,m \\ X_{ij} = \{0,1\} & \end{cases} \tag{4}$$

(2) Constraint conditions based on Ergonomics

To facilitate the application of man-machine efficiency criteria, define variables S_{ij}, I_{ij}, *PV*, *CR*. S_{ij} represents the interaction between functional modules i and j, as Sij={3,5}, the representation module j can be placed on the right and right side of the i, as shown in Figure 3. I_{ij} represents a module that can be placed between i and j. *PV* stands for a module that can be placed in the main visual area, *CR* represents a module that can be placed in a comfortably accessible area.

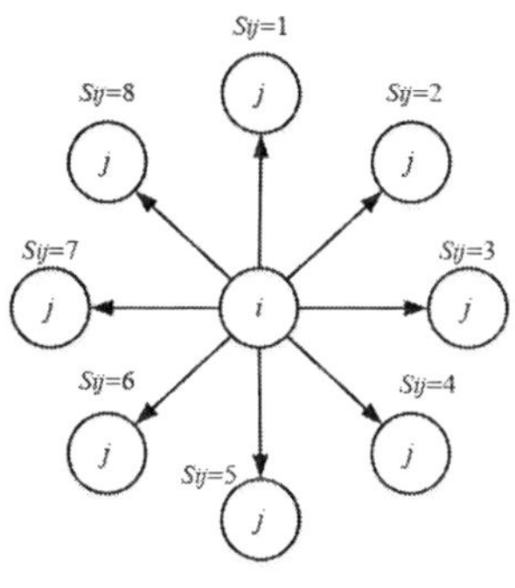

Figure.3 Between the orientation of the function-module constraints schematic

1) Ergonomic criteria affecting relative position decision

Spatial compatibility: According to the relationship between controllers i and j, determine the relative position between them.

Functional grouping: Controllers and displays with similar functions should be arranged together.

Sequence-of-use: The completion of a task requires multiple controllers to be used simultaneously, and the controllers should be arranged in order of operation (from top to bottom, from left to right).

2) Ergonomic criteria affecting absolute position decision

Frequency-of-use: High-frequency controllers should be placed in the most accessible range of limbs.

Importance: More important controllers and displays should be arranged in the best field of vision and the most accessible area of the body.

IOP Conf. Series: Journal of Physics: Conf. Series **1060** (2018) 012034 doi:10.1088/1742-6596/1060/1/012034

3. Study on the motion characteristics of the upper limb

In this paper, the virtual simulation environment is used to collect experimental data by using the body motion controller Leap Motion and the virtual simulation software Unity3d [12]. By collecting and analyzing the operation time data of a certain type of equipment control panel by people in the loop simulation equipment operation simulation experiment, the positioning movement time allocation maps of each area of the equipment interface are obtained, and the data basis is provided for the model establishment.

3.1. Experimental design of control panel data measurement

The purpose of the experimental design is to measure the operating time of the upper extremities when performing the positioning exercise, and to provide data basis for the objective function f_2. The experiment is carried out on the 736mmx276mm equipment control panel. It is divided into 15 small panels of the same size. Each small panel size is 147mmx92mm, and the center location is respectively arranged with 1 trigger button S1 ,S2, ... S15, the sequence number order is the same as that in Figure 1, as shown in Figure 4:

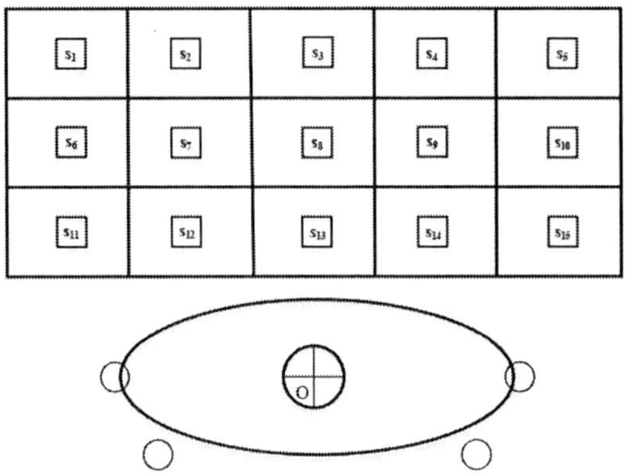

Figure.4 Experimental data test function modules

3.2. Experimental steps

The virtual simulation scene is designed by using the virtual simulation software Unity3D. The equipment console in the virtual environment is consistent with the actual equipment size, and the Leap Motion is used to read the hand data and complete the operation task. The experimental scene is shown in Figure 5.

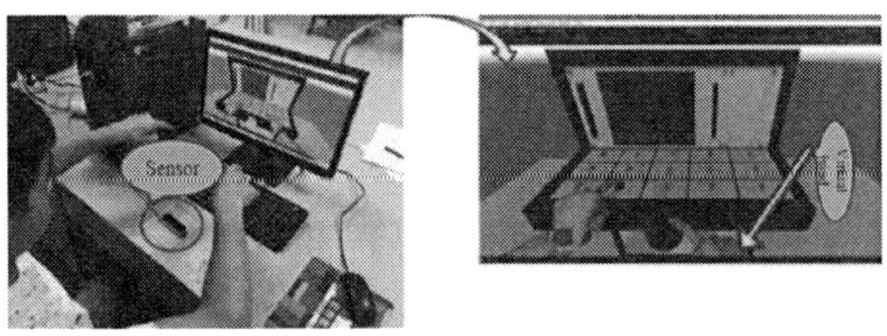

(a) human in the loop (b) Virtual experiment environment

Figure.5 Index finger positioning exercise test renderings

The white sphere in the virtual environment is the starting point of the finger movement, that is, the position of the O point in Figure 4. When the index finger triggers the sphere, the program starts timing. The subject rapidly leaves and clicks the button on the console. The button triggers red, the program ends timing, and a positioning motion is completed. The time data is recorded in the text file. Before the experiment, 10 students with normal eyesight and computer skills were invited to take part in the experiment, and the ability of virtual hand control in virtual environment was repeatedly practiced, so as to achieve a quick and accurate proficiency of one click task. After each click of the task is completed, the index finger returns to the white ball position to start the next button. To avoid interference between the left and right hands, the right hand hits the right three column, and the left hand clicks the left two column. Click order follows from left to right, near and far. The experimental process satisfies the experimental requirements of Fiits' law.

4. Case study and analysis

4.1. case study

Taking the man-machine control console of the command and control cabin of certain weapon equipment as the experimental research object, as shown in Figure 6. The functions of the control devices are grouped. Correspondences of the functional modules are shown in Figure 6. The size of the control panel is 736mm x 276mm, which is divided into 15 areas, and the size of each small panel is 92mm x 147mm.

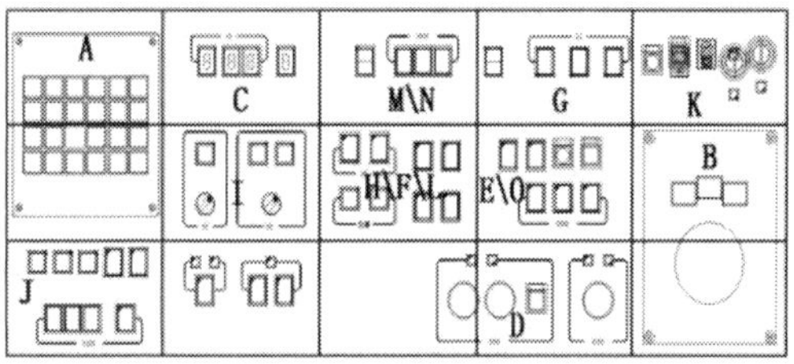

Figure.6 Certain allegations cabin console original layout

The author invites ten operational trumpeters of a certain type of weaponry equipment to fill out in detail the frequency of use and the degree of importance of each module of the control panel of the command and control panel. Finally, the experts are asked to check the experimental data and obtain reliable experimental data. It lays a solid foundation for the control panel layout optimization design model. Through the analysis of the model, LINGO can be used to solve the problem. Layout of the control panel is performed by integrating the global optimal solution of the objective functions f_1, f_2 and the ergonomic constraint relationship, the final layout optimization result of Figure 7 can be obtained.

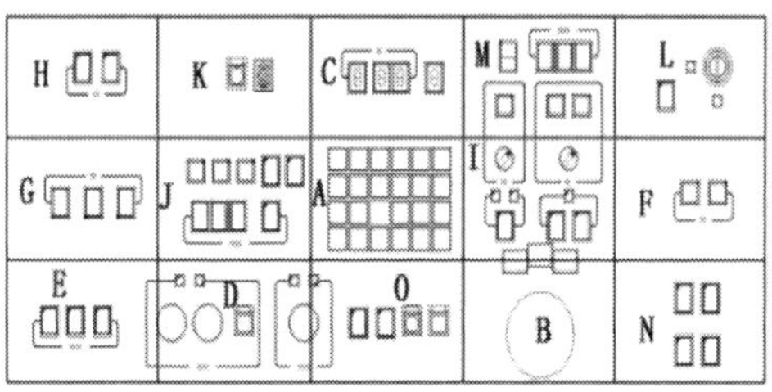

Figure.7 Optimize the layout of the final results

4.2. Layout analysis

The blue curve in Figure8 represents the distribution of operation loss distance and operation time in the panel area of the panel layout before the optimization in Figure 6. The red curve represents the distribution of operating distance loss and operating time MT of each panel area in Figure 7 after optimization of layout results. The new layout reduced the total operating loss distance from 487.60cm to 348.64cm, and the optimization ratio reached 28%, as shown in Figure 8(a). The operating time is reduced from 38976ms to 32804ms, and the optimization ratio reaches 16%, as shown in Figure8 (b). From this we can see that the 0-1 linear programming model of this paper has played a significant role in optimizing the operation loss distance and operation time of the control panel layout.

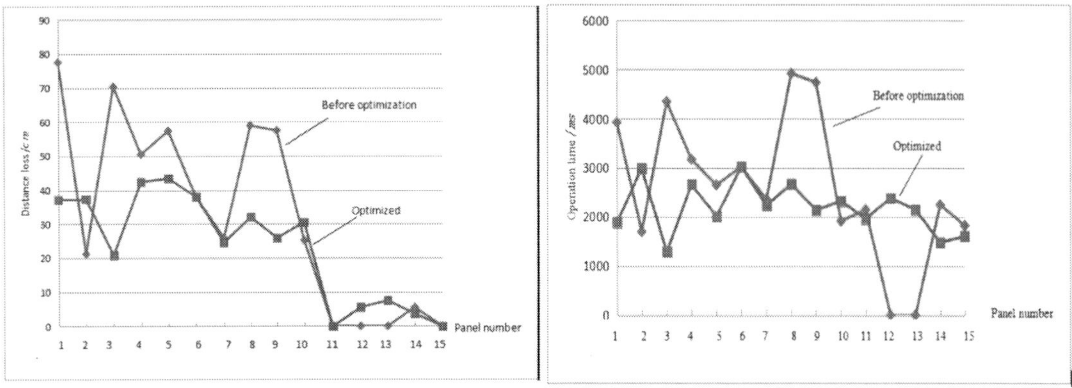

(a)Trying to distance optimization (28%) (b)Time optimization view (16%)

Figure.8 Layout Optimization after images

5. Conclusion

(1) In this paper, the upper body reachability probability distribution map of the upper limbs is obtained by performing upper body reachability statistics for each area of a certain type of equipment operation panel.

(2) The upper limb positioning movement characteristics research experiment is designed. Through the simulation of the upper limb positioning movement in the loop simulation equipment operation, the operation time data of a certain type of equipment manipulation interface is collected and analyzed, and the positioning movement time distribution maps of various areas of the equipment interface are obtained.

(3) A 0-1 linear programming model of equipment operation panel is established and solved. The layout results play a very good optimization effect in minimizing distance loss and operation time. The final layout results meet the frequency of use, importance, sequence of use in the ergonomics criteria.

The research method can provide theoretical and methodological basis for the future design of the new ground-to-air missile weaponry equipment manipulator interface and other industrial control room control panel multi-objective design, optimize operation efficiency in multiple directions and improve the fluency and accuracy of the operation.

References

[1] Sander, McCormick. Human Factors in Engineering and Design[M]. 7th Ed. New York: Mcgraw Hill press,1993: 456-484.

[2] E.S. JUNG, S. PARK and S.Y. CHANG. A CSP technique-based interactive control panel layout[J]. ERGONOMICS, 1995, 9(38):1884-1893.

[3] Maurício Alves C. Aghina, Antônio Carlos A. Mól, Carlo Alexandre F. Jorge, and Cláudi M.N.A. et al. Virtual Control Desks for Nuclear Power Plant Simulation Improving Operator Training[C]. //Published by the IEEE Computer Society. Brazil, 2008:6-9.

[4] Isaac José Antonio Luquettidos Santos, Marcos Santana Farias, Fernando Toledo Ferraz, et al. Human factors applied to alarm panel modernization of nuclear control room[J]. Journal of Loss Prevention in the Process Industries, 2013, 26: 1308-1320.

[5] Cem Alppay, Nigan Bayazit. A ergonomics based design research method for the arrangement of helicopter flight instrument panels [J]. Applied Ergonomics, 2015, 51:85-101.

[6] CHEN De-jun, FANG Wei-ning, QIN Yong-zhen, et al. Optimizing model and algorithm for human-machine interface layout of metro train drivers desk. JOURNAL OF THE CHINA RAILWAY SOCIETY, 2014, 36(11):40-47.

[7] WANG Rui, ZHUANG Da-min. Layout Optimization of Cockpit Based on Human Comfort, ACTA ARMAMENTARII, 2008, 29 (9): 1149-1152.

[8] DENG Li. Working space research & application based on the human body Posture drive Product design method [D]. Southwest Petroleum University, 2011.
[9] NIU Song. The Layout Optimization Study For Civil Aircraft Cockpit Ergonomics Design[D]. Nanjing University of Aeronautics and Astronautics, 2013.
[10] LI Jing-hua. Research on infeasible interior-point algorithms in linear programming[D]. Xi'an Technological University, 2014.
[11] WEI Xian-min. Fitts' law research in human computer interaction[J], Manufacturing Automation, 2012, 34 (3): 82-83.
[12] HU Hong, CHAO Jian-gang, YANG Jin, et al. Key Point Model for Hand Pose Estimation Based on Leap Motion[J], Journal of Computer-Aided Design & Computer Graphics, 2015,27 (7): 1211-1216.

Sharing Study of IMT-2020(5G) and SRS in 25.5-27GHz

Jiajia Chen[1, *], Zhaojun Qian[1], Tan Wang[1] and Xi Li[2]

[1]State Radio Monitoring Center, Beijing, China
[2]Beijing Credit Top Company Limited, Beijing, China

chenjiajia@srrc.org.cn

Abstract. Based on the WRC-19 1.13 issue research framework and the compatibility analysis requirements for the fifth generation of mobile communication candidate bands above 6GHz in China, this paper studied the coexistence between IMT-2020(5G) and Space Research(space to ground) system(SRS) in the same running bands 25.5-27GHz. The interference of IMT-2020(5G) system to the downlink of the SRS is analysed by the aggregated interference assessment method. The research shows that, IMT-2020(5G) would not interfere SRS harmfully by the certain isolation distance. The results can provide a technical basis for the actual deployment of IMT-2020(5G) and protection of SRS system in the frequency bands 25.5-27GHz in the future.

1. Introduction

At present, with the rapid development of the international mobile system, the spectrum demand of IMT-2020(5G) will become more. In order to effectively improve the data transmission rate and system capacity, meet the need of 5G communication, the IMT system needs to excavate the frequency spectrum resources of high frequency bands above 6GHz. 25.5-27GHz band is relatively low and continuous large bandwidth, regarded as IMT system's candidate frequency by some countries and regions in the world. This frequency band is also one of the most important candidate frequency bands of IMT system in China. However, the band has been widely used in earth exploration satellite system (space to ground), Inter-satellite system, Space research system (space to ground). It is an important band for developing and planning large capacity broadband satellite. China also has more than one satellite systems running and planning to be established in this band. So the planning and deployment of IMT system in this band would have an impact on satellite communication. The compatibility analysis between them is urgently needed.

Accounting the domestic research need of the fifth generation mobile communication candidate bands allocation, the framework of WRC-19 1.13 issues and ITU-R M.2101, investigating the typical parameters of satellites operating and planning in 25.5-27GHz in China, the research between IMT system and Non-GSO SRS system was carried out. Based on the typical deployment model and typical parameters of IMT, this paper analysed the impact from IMT system to SRS Non-GSO system and preliminarily concluded that the two systems can coexist and share the same bands by certain isolation distance.

2. The allocation situation in 25.5-27GHz

The 25.5-27GHz band allocation given by the 2016 edition of ITU <Radio Regulation>[1]is shown in Table 1. In the 2014 edition of radio frequency allocation in China, the frequency allocation is the same as that of ITU RR in 25.5-27GHz. The SRS is allocated in the frequency band 25.25-27.5 GHz as primary service.

Table 1. Frequency allocation in the 25.25-27.5 GHz frequency range

International Table		
Region 1	**Region 2**	**Region 3**
25.5-27 EARTH EXPLORATION-SATELLITE (space-to-Earth) FIXED INTER-SATELLITE MOBILE SPACE RESEARCH (space-to-Earth) Standard frequency and time signal-satellite (Earth-to-space)		

Part of the band 25.5-27 GHz is allocated to SRS (space-to-Earth). IMT system would interfere SRS Non-GSO earth station if the IMT system is deployed around the SRS earth station when the two systems use the same 25.5-27GHz frequency band. The interference model is shown in figure 1. The protection distance is needed and this paper analysed how long the protection distance is needed and whether the two systems can share the frequency band of 25.5-27GHz.

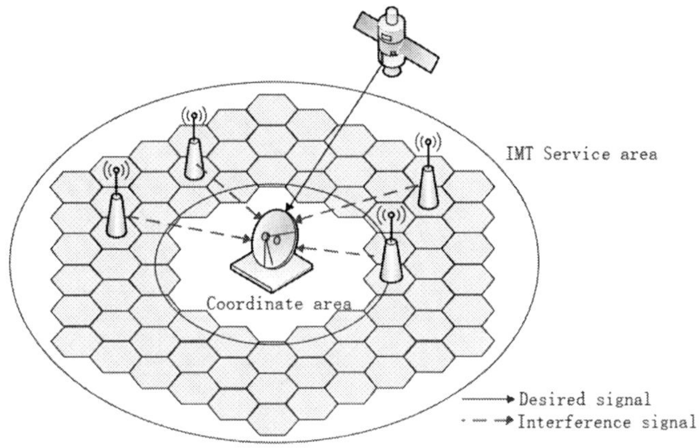

Figure 1. Scenario for SRS earth station interference from IMT-2020

3. Technical characteristics

3.1. Technical and operational characteristics of IMT systems operating in the 25.5-27.0 GHz frequency range

The technical and operational characteristics of IMT-2020 are given in the following subsections.

3.1.1. IMT-2020 base station. All the parameters and values are calculated from Recommendation ITU-R M.2101-0[2] with input parameters from the liaison statement from WP 5D to TG 5/1. The composite antenna gain pattern for the base station generated from both the documents is shown in figure 2.

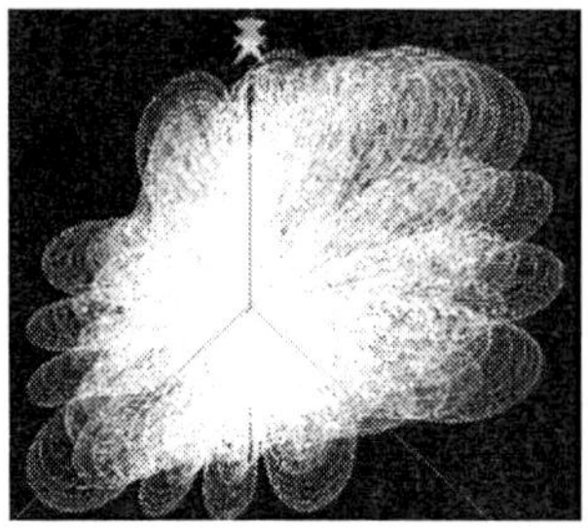

Figure 2. The 3D polar plot on the composite antenna gain pattern for base station according to Recommendation ITU-R M.2101-0

Table 2 shows the parameters of IMT-2020 used in this study. These parameters, for bands between 24.25 GHz and 33.4 GHz, are taken from ITU-R P.2101.

Table 2. IMT-2020 base station/UE parameters

Parameters	Base Station	User Equipment
Antenna height(m)	6	1.5
Antenna array configuration (Row × Column)	8x8	4x4
Sectorization	Single sector	
Downtilt and pointing	-10°to UE	To BS
Antenna pattern	ITU-R M.2101	
Element gain（dBi）	5	
Horizontal/vertical 3 dB bandwidth of single element	65° for both H/V	
Horizontal/vertical front-to-back ratio (dB)	30° for both H/V	
Antenna polarization	Linear ±45°	
Horizontal/Vertical radiating element spacing	0.5 of wavelength for both H/V	
Array Ohmic loss (dB)	3	
Conducted power per antenna element (dBm/200 MHz)	10	
BS maximum coverage angle in the horizontal plane	120°	
Deployment area	50 km^2	
Network loading factor	50%	
TDD activity factor	80%	20%
Density	30BS/km^2（urban）,10BS/km^2（suburb）	
Transportation Model	ITU-R P.452, ITU-R P.2108-0	
Hotspot coverage ratio(Ra), Built-up area ratio(Rb)	Ra=7%(urban), 3%(suburb), Rb=100%	

3.1.2. IMT-2020 user equipment. All the parameters and values are calculated from Recommendation ITU-R M.2101-0. The composite antenna gain pattern for the user equipment is shown in Figure 3.

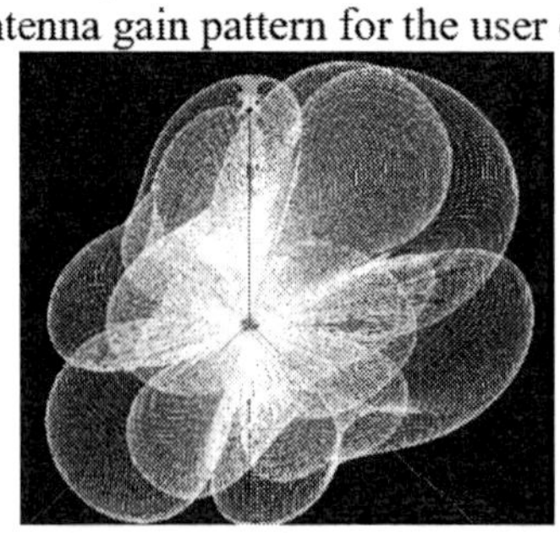

Figure 3. The 3D polar plot on the composite antenna gain pattern for user equipment according to Recommendation ITU-R M.2101-0

IOP Conf. Series: Journal of Physics: Conf. Series **1060** (2018) 012035 doi:10.1088/1742-6596/1060/1/012035

3.2. *Technical and operational characteristics of SRS*

The characteristics for the SRS systems for non-GSO satellites in the band 25.5-27 GHz can be found in Table 3.

Table 3. Non-GSO SRS system in the band 25.5-27 GHz

Simulation Parameters	**Value**
Carrier frequency (MHz)	26 250
Necessary bandwidth (MHz)	300
earth station antenna radius (m)	35
earth station antenna gain toward satellite (dBi)	78
earth station antenna radiation diagram	ITU-R S.465-6[3]
earth station receiver noise temperature (K)	70
Minimum elevation angle（degree）	5/10
earth station antenna polarisation	RHCP/LHCP

The composite antenna gain pattern for the SRS earth station is shown in Figure 4.

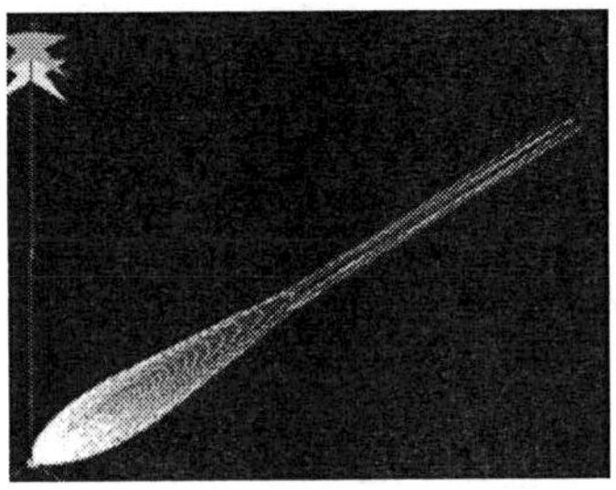

Figure 4. The 3D polar plot on the antenna gain pattern for SRS earth station according to Recommendation ITU-R M.465-6

Recommendation ITU-R SA. 609[4] contains the protection criteria for radio communication links for manned and unmanned near-Earth research satellites. In the frequency range 20-30 GHz, harmful interference can occur if the total time during which the power density of noise-like interference or the total power of CW-type interference in any single band or in all sets of bands 1 MHz wide, is greater than -156 dB(W/MHz) at the input terminals of the receivers for a period exceeding 0.001% of the time for manned missions, and 0.1% of the time for all other near-Earth space research missions.

3.3.Propagation models for sharing and compatibility studies in the 25.5-27.0 GHz frequency range

For the study of single-entry interference, the propagation model used is Recommendation ITU-R P.452-16[5] with a 50 metre resolution terrain. No clutter is considered.

For the study of aggregate interference, the propagation model in Recommendation ITU-R P.452-16 with a 50 metre resolution terrain is used for the interference path. The interference path also includes clutter loss calculated in accordance with the statistical model for terrestrial paths method described in Recommendation ITU-R P.2108[6].

4.Technical analysis

This study analyses the interference from IMT stations (base stations and user equipment) to Non-GSO SRS systems.

4.1Interference topology from IMT base station / UE to SRS

According to the parameters provided by WP 5D, the IMT BSs can be deployed in four scenarios: outdoor suburban open space hotspot, outdoor suburban hotspot, outdoor urban hotspot and indoor.

Compared with the other three scenarios, when IMT BSs are deployed indoor, the building entry loss will introduce significant attenuation to the IMT signal. So the aggregate interference power will be smaller than that in the other three scenarios.

As for the outdoor suburban open space hotspot scenario, the density of BSs is 0 or 1 BS/km2. However, the densities of BSs in outdoor suburban hotspot and outdoor urban hotspot scenario are respectively 10 and 30 BS/km2. When the BSs are deployed in different scenarios they follow the same distribution, fewer BSs are most likely to lead to smaller aggregate interference power.

Comparing the outdoor suburban hotspot and the outdoor urban hotspot scenarios, the latter has a larger BS density and Ra, so the aggregate interference will be larger.

As was stated above, the interference from BSs deployed in the outdoor urban hotspot scenario to the EESS systems is most likely to be the worst.

Recommendation ITU-R M.2101 provides the methodology how to model the IMT network and use Monte-Carlo simulation to evaluate the aggregated interference to other systems.

In the case of the aggregated interference, there are some options to deployment of IMT base stations and UEs.

The topology of equally-spaced IMT macro base stations located around the SRS earth station is assumed. And the IMT micro base stations which may interfere with the SRS earth station are located in the macro base station area. Taking the outdoor urban IMT scenario as an example, the number of IMT macro base stations is assessed according to the separation distance and the base station inter-site distance as following:

The radius of the ith ring is:

$$D(i)= Dprotection + (i-1)* Dintersite \quad (1)$$

The number N(i) of IMT macro base stations located on the ith ring is assessed according to the corresponding distance D(i) and the base station intersite distance range as following:

$$N(i) = pi / (arc \sin (Dintersite/ (2*D(i)))) \quad (2)$$

It should be noted that IMT-2020 networks will only be deployed in a hotspot area and not as seamless coverage. Some areas will not deploy IMT systems and the real deployed IMT base station numbers will be less than N(i), see Figure 5.

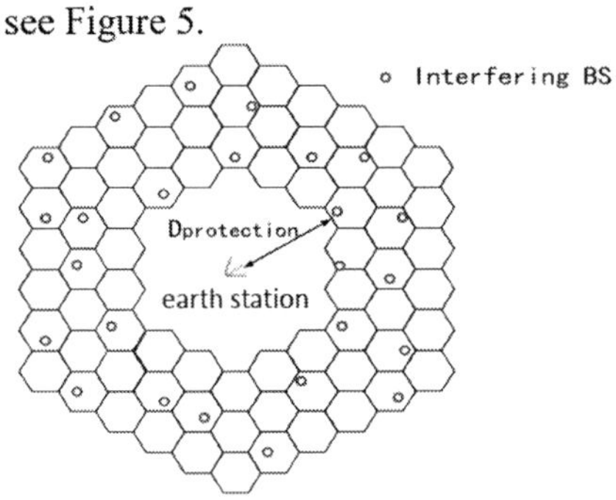

Figure 5. Aggregated IMT base stations scenario

4.2. Calculation of aggregate interference

A Monte-Carlo simulation was developed in order to assess the aggregate interference from multiple base stations and UEs deployed at a given distance from the EESS earth station. In a single snapshot of the Monte-Carlo simulation, the aggregate interference from IMT base stations or user equipment can be calculated as follow.

$$I_{BS/UE} = 10\log\left(\sum_{n=1}^{n=N_{BS/UE}} 10^{[PD_{tx}-OL_{tx}+G_{tx}(\theta_{tx})-L_n+G_{rx}(\theta_{rx})]/10}\right) \quad (3)$$

where:

IBS/UE: Aggregate interference from all IMT base stations or UEs;

NBS/UE: Total number of base stations or UEs;

n: index of the base station or UE;

PDtx: Transmit station signal power density;

OLtx: Transmit station array ohmic loss;

Gtx(θtx): Transmit station antenna gain in direction of receive station;

Ln: propagation loss between the base station and UE of index n and the SRS station (based on Recommendation ITU-R P.452 and ITU-R P.2101);

Grx(θrx): Receive station antenna gain in direction of transmit station taking into account the off-axis angle.

The aggregate interference from IMT base stations and UEs can be calculated as follow.

$$I_{aggr} = 10 * \lg(TDDFactor_{BS} * 10^{\frac{I_{BS}}{10}} + TDDFactor_{UE} * 10^{\frac{I_{UE}}{10}}) \tag{4}$$

where:

Iaggr: Aggregate interference power from IMT base stations and UEs;

TDDFactorBS: TDD activity factor of base station;

TDDFactorUE: TDD activity factor of user equipment;

The time percentage value used in Recommendation ITU-R P.452 is set as a fixed value, which is equal to the time percentage in interference criteria. In the simulation, an aggregate interference value is calculated in each snapshot and the linear average value is generated from all snapshot. The additional margin could be calculated by subtraction linear average of aggregate interference value from interference criteria.

5.Simulation results of interference from IMT system to SRS

As was analysed in section 3 and 4, take a typical big city in China as reference, set the simulation area to about 50 km2 and it was assumed that the IMT base stations are deployed in the suburban outdoor hotspot scenario. The SRS earth station is located in the centre of the simulation area and ensures that the distance between the SRS earth station and the nearest IMT micro BS is the 2 000 m. The IMT network loading factor is 20%. According the simulation assumptions above, the simulation results are as follow.

5.1.Interference caused by IMT BSs

The simulation results of the interference from IMT BSs to SRS earth station are shown in Table 4.

Table 4. Interference caused by IMT BSs to SRS earth station

earth station elevation (degrees)	Dprotection (m)	Aggregate interference (dB(W/MHz))	Interfering signal power at the input terminals of the receivers to be exceeded for a period not exceeding p% of the time (dB(W/MHz))	Interference margin (dB)
5	2 000	-166.7	-156 p=0.1	11.7
10	2 000	-170.1		14.1
5	2 000	-161.0	-156 p=0.001	15.0
10	2 000	-164.2		18.2

If the separation distance is set to 2 000 m, the simulation results meet the protection criterion of an SRS earth station in the 25.5-27 GHz frequency range.

5.2.Interference caused by IMT UEs

The simulation results of the interference from IMT UEs to SRS earth station are shown in Table 5.

Table 5. Interference caused by IMT UEs to SRS earth station

earth station elevation (degrees)	Dprotection (m)	Aggregate interference (dB(W/MHz))	Interfering signal power at the input terminals of the receivers to be exceeded for a period not exceeding p% of the time	Interference margin (dB)

			(dB(W/MHz))	
5	2 000	-176.5	-156 p=0.1	20.5
10	2 000	-178.4		22.4
5	2 000	-174.9	-156 p=0.001	18.9
10	2 000	-176.7		20.7

If the separation distance is set to 2 000 m, the simulation results meet the protection criterion of an SRS earth station in the 25.5-27 GHz frequency range.

5.3.Interference caused by IMT stations (BSs and UEs)

The simulation results of the interference from IMT stations to SRS earth station are shown in Table 6.

Table 6. Interference by IMT stations to SRS earth station

earth station elevation (degrees)	Dprotection (m)	Aggregate interference (dB(W/MHz))	Interfering signal power at the input terminals of the receivers to be exceeded for a period not exceeding p% of the time (dB(W/MHz))	Interference margin (dB)
5	2 000	-167.1	-156 p=0.1	11.1
10	2 000	-169.7		13.7
5	2 000	-165.1	-156 p=0.001	9.1
10	2 000	-167.5		11.5

If the separation distance is set to 2 000 m, the simulation results meet the protection criterion of an SRS earth station in the 25.5-27 GHz frequency range.

6.Conclusion

According to the interference protection criteria and the characteristic parameters of reference, the IMT system and SRS earth station can coexist when the protection distance is 2000m when using the same frequency band 25.5-27GHz, and the interference margin was above 10dB.

The next work should be to clarify the interference protection distance between the IMT system and the SRS system earth station under the actual geographical and terrain conditions, and to delimit the isolation area.

Acknowledgment

This work was supported by the National Science and Technology Major Project under Grants No. 2015ZX03002008.

Reference

[1] ITU-R. Radio regulation of the International Telecommunication Union(2016 ed.)

[2] ITU-R. "Modelling and simulation of IMT networks and systems for use in sharing and compatibility studies", Recommendation ITU-R M.2101-0,2017

[3] ITU-R. "Reference radiation pattern for earth station antennas in the fixed-satellite service for use in coordination and interference assessment in the frequency range from 2 to 31GHz", Recommendation ITU-R S.465-6,2010.

[4] ITU-R. "Protection criteria for radiocommunication links for manned and unmanned near-Earth research satellites ", Recommendation ITU-R SA.609-2, 2006.

[5] ITU-R. "Prediction procedure for the evaluation of interference between stations on the surface of the earth at frequencies above about 0.1 GHz", Recommendation ITU-R P.452-16,2015.

[6] ITU-R. "Prediction of Clutter Loss", Recommendation ITU-R P.2108,2017.

IOP Conf. Series: Journal of Physics: Conf. Series **1060** (2018) 012036 doi:10.1088/1742-6596/1060/1/012036

Supervised Ensemble Machine Learning Aided Performance Evaluation of Sentiment Classification

Sheikh Shah Mohammad Motiur Rahman[1, a], Md. Habibur Rahman[1, 2, b], Kaushik Sarker[1, c], Md. Samadur Rahman[3, d], Nazmul Ahsan[3, e] and M. Mesbahuddin Sarker[3, f]

[1]Department of Software Engineering, Daffodil International University, Dhaka, Bangladesh
[2]Department of Information and Communication Technology, Mawlana Bhashani Science and Technology University, Santosh, Tangail, Bangladesh
[3]Institute of Information Technology, Jahangirnagar University, Dhaka, Bangladesh

[a]motiur.swe@diu.edu.bd; [b]mdhabibur.r.bd@ieee.org; [c]kaushik.swe@daffodilvarsity.edu.bd; [d]engrsamadcse@gmail.com; [e]ahsan171118@gmail.com, [f]sarker@juniv.edu

Abstract. Text vectorization, features extraction and machine learning algorithms play a vital role to the field of sentiment classification. Accuracy of sentiment classification varies depending on various machine learning approaches, vectorization models and features extraction methods. This paper represents multiple ways of evaluations with the necessary steps needed to achieve highest accuracy for classifying the sentiment of reviews. We apply two n-gram vectorization models - Unigram and Bigram individually. Later on, we also apply features extraction method TF-IDF with Unigram and Bigram respectively. Five ensemble machine learning algorithms namely Random Forest (RF), Extra Tree (ET), Bagging Classifier (BC), Ada Boost (ADA) and Gradient Boost (GB) are used here. The key findings in this study is to determine which combination of vectorization models (Bigram, Unigram) along with feature extraction method (TF-IDF) and ensemble classifier gives the better performance of sentiment classification.

1. Introduction

Sentiment Analysis is the process to identify the user centric visualization based on user feedbacks. The main concept of sentiment analysis is to analyze the written text by the user and to extract their feedbacks [1]. Sentiment Analysis is used in different areas of healthcare, politics and movie industries [2]. In the industry of healthcare, it is used to find out the quality of patient care at any hospital or health care centre [2]. In the area of politics, the analysis of public's sentiment towards candidate of political parties [3], to predict the results of election [4]. In the industry of movie, it is used to find out the reaction of people who have watched the movie from their opinions and posted reviews [5]. To perform the analysis, researchers are using various machine learning algorithms which carry out an optimal way to find out the result from people's sentiment. Two types of machine learning algorithm are used in sentiment analysis, one is supervised and another is unsupervised. In supervised learning, the dataset is labeled during preprocessing or during analysis then trained the machine to obtain a meaningful output that helps to make the decision [6]-[7]. In unsupervised learning, it's getting harder to obtain reasonable result because of unlabeled data. That means unsupervised learning not provide any labeled data to machine. Various clustering methods are used to solve this issue [6]. Again the robustness of supervised learning can be enhance by applying the averaging, boosting and stacking techniques. These are named as ensemble machine learning techniques [8]-[9]. Thus, ensemble machine learning techniques are also known as supervised learning techniques. In this study ensemble machine learning techniques are used to perform classification of sentiment based on labeled data. Mainly three levels are used to investigate the sentiment classification such as aspect level, document level and sentence level [10]. The whole presented expressions and the references aspect within a document is focused on aspect level classification. In document level classification, each and every single document contain either a positive or negative reviews. Whether a sentence is positive or negative can be found out in sentence level classification [10].

This study focused on the aspect of document level classification. The structure of this paper is as follows. Section 2 is provided with the Literature Review based on the sentiment analysis research activities held recently. Section 3 is described the research methodology. Section 4 evaluates the performance to show that results meet up with design and objectives. Section 5 concludes the paper.

2. Literature Review

Zhang et al. [11] have considered word2vec and SVMperf method. They have found 87.10% and 90.30% classification accuracy for word2vec and SVMperf respectively. They use Chinese comment on clothing products dataset and the feature selection methods lexicon-based and part-of-speech have used. Martin et al. [12]. have used voting and stacking concept with machine learning algorithms. Implementing the document classification and sentiment classification they use Naïve Bayes, Support Vector Machine, Bayesian Logistic Regression (BLR) and C4.5. Using the MuchoCine(MC) and Spanish Corpus dataset they found the accuracy 88.5%, 88.31%, 88.27% and 87.66% respectively for the above algorithms. Xia et al. [13] have found 82.7%, 86.1% and 84.85% classification accuracy for Naïve Bayes, Support Vector Machine and Maximum Entropy respectively. The mostly use Cornell movie review dataset, document and multi-domain sentiment dataset. For selecting features set, they have considered part-of-speech(POS) and the word-relation based feature sets. Matsumoto et al. [14] have worked with document level sentiment analysis and for sentiment classification they use support vector machine and apply on Internet Movie Database (IMDb) and Polarity Dataset. They have found the classification accuracy 83.7% for Unigram, 80.4% for Bigram and 84.6% for Unigram + Bigram using only support vector machine learning algorithm. Tripathy et al. [6] have considered document label classification with CountVectorizer for vectorization along with TF-IDF to transform text to numeric values. The have used Naive Bayes (NB), Support Vector Machine (SVM), Random Forest and Linear Discriminant Analysis (LDA) machine learning algorithm for sentiment classification. They found 89.5%, 94.0%, 95.0% and 92.0% accuracy form Naive Bayes (NB),Support Vector Machine (SVM), Random Forest, and Linear Discriminant Analysis (LDA) respectively. They use aclIMDb [15] and polarity movie [16] review dataset. However highest accuracy into 100% is not found for any classification algorithm yet. This study try to found the highest accuracy for classifying sentiment of reviews. Random Forest (RF), Extra Tree (ET), Bagging Classifier (BC), Ada Boost (ADA) and Gradient Boost (GB) machine learning algorithms are used to test the accuracy level.

3. Methodology

The process of sentiment classification are two types: binary sentiment classification and multiclass sentiment classification [4]. Each document doc_i in DOC, where DOC = {doc_1 , doc_2 , doc_3 ,..., doc_n} is classified in binary classification. It is labeled as S, where, S is a predefined set of review category like S = {1,0} = {Positive, Negative}. Each document doc_i is classified in multi class sentiment analysis as a level in S* where, S* = {strong positive, positive, neutral, negative, strong negative} [6]. Binary sentiment classification is considered in this study. The overall system is described in figure 1.

3.1 Dataset Used
The dataset used for our experiment is "Large Movie Review Dataset" collected by Andrew Maas [17] [18]. This dataset consists of a collection of 50,000 movie reviews from IMDb [15]. The movies are covered by the category comedy, funny, action, science fiction, romantic, horror, fantasy and adventure. At most 30 reviews are considered as per movie. The movie reviews are classified into equal number of positive and negative reviews. Neutral reviews are ignored. Based on the user rating out of 10, less than or equal 4 is classified as a Negative review and greater than or equal 7 is classified as positive reviews.

3.2 Vector Creation and Feature extraction
Generate a vectorized documents by n-Grams of tokens (unigram and bigram) methods from each textual reviews. A term n-Gram is defined as a series of consecutive tokens of length n. Feature extraction is an important method to determine the robust key features from large number reviews [19] [20]. Feature extraction methods TF-IDF is used in this research which has implemented on both of unigram and bigram methods [21].

3.3 Evaluation with Ensemble Machine Learning
After that, the performance of classifying the sentiment of reviews from vectorized documents which are generated using the Unigram, Bigram, Unigram + TF-IDF and Bigram + TF-IDF models on dataset is being evaluated. Five ensemble machine learning approaches (Bagging Classifier, Random Forest, Extra Tree, Ada Boosting, Gradient Tree Boosting) are applied to classify the sentiment of reviews and to calculate the accuracy of classification [22].

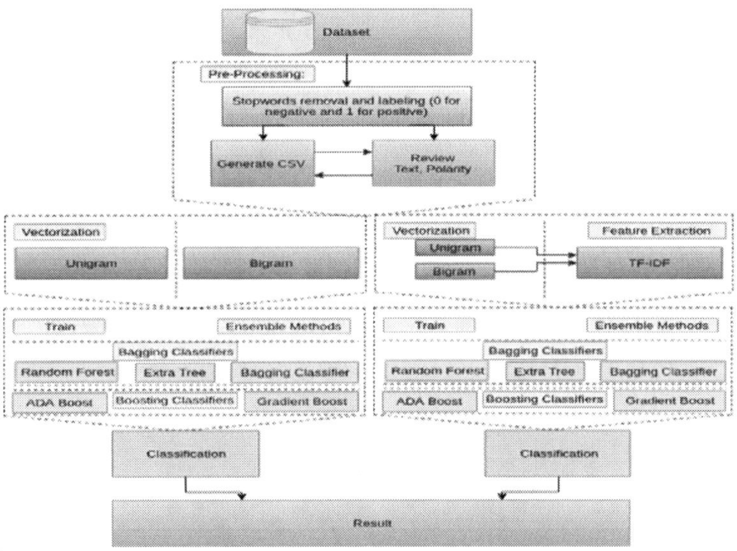

Figure 1. Overview of System Architecture

4. Performance Evaluation and Result Analysis
This section analyzes the accuracy level and the overall performance of the proposed model. Bigram and Unigram vectorization models and a well established feature extraction methods (TF-IDF) have been implemented and evaluated in this study. Here ensemble machine learning algorithms are considered [23]. Several feature extraction methods are applied for evaluating the performance [24].

4.1 Evaluation Parameter
To evaluate the performance of the classification of sentiment of reviews, there are some metrics are used which are as outlined in Table 1.

Table 1. Evaluation Parameter for Measuring the Accuracy [5] [6].

Evaluation Parameter	Description	Formula/Visualization
Precision	It measures the exact value of the result provided by the classifier.	Positive Precision = TP/(TP + FP) Negative Precision = TN/(TN+FN)
Recall	It measures the thoroughness of the result provided by classifier.	Positive Recall = TP/(TP + FN) Negative Recall = TN/(TN+FP)
F1-Score	It defines the harmonic mean of exactness and completeness.	F1-Score = (2*Precision* Recall) / (Precision + Recall)
Accuracy	It defines the ratio of correctly classified data to total number of data.	Accuracy = (TP + TN) / (TP + TN + FP + FN)

TP : True Positive represents the reviews are positive and also classifier classified as positive; FP: False Positive represent the reviews are positive but classifier classified as negative; TN: True Negative represents the reviews are negative and also classifier classified as negative; FN: False Negative represent the reviews are negative but classifier classified as negative; [5]

Figure 2 depicts the results obtained from the Unigram and Bigram vectorization model with five ensemble machine learning techniques. The blue and maroon color represents the accuracy of Unigram and Bigram vectorization model respectively. We obtain 99.404%, 100% , 98.664%, 80.426% and 81.722% accuracy by using Random Forest (RF), Extra Tree (ET), Bagging Classifier (BC), ADA Boost (ADA) and Gradient Boost (GB) respectively for Unigram vectorization model.On the other hand, the accuracy for Bigram vectorization model obtained from Random Forest(RF), Extra Tree(ET), Bagging Classifier(BC), ADA Boost (ADA) and Gradient Boost(GB) classifiers are 99.404%, 100%, 98.754%, 80.474% and 81.76% respectively represented in figure 2.

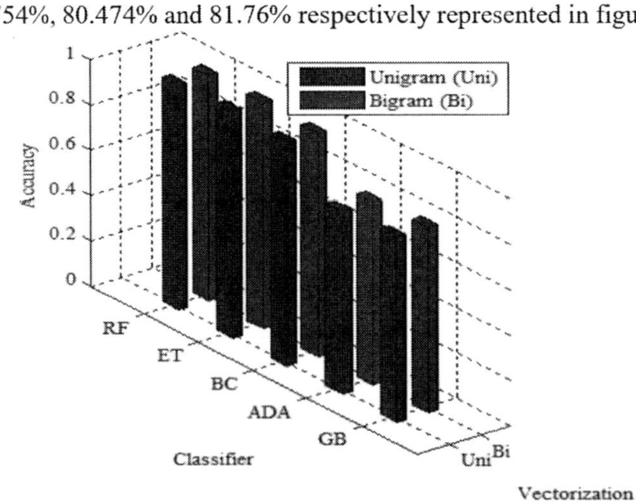

Figure 2. Accuracy for Unigram and Bigram Vectorization Model using RF, ET, BC, ADA and GB Classifiers

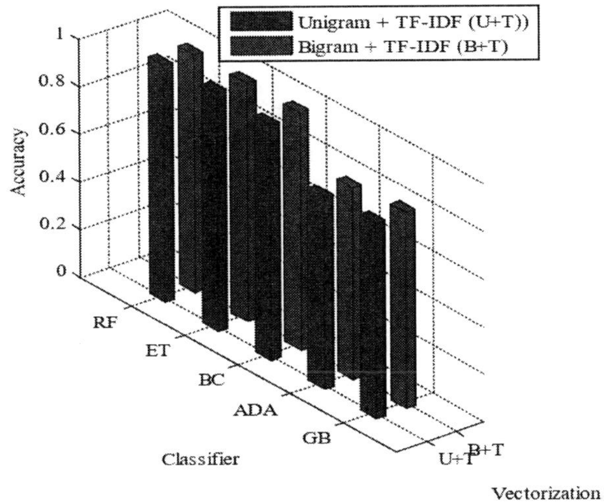

Figure 3. Accuracy for Unigram+TF-IDF and Bigram+TF-IDF Vectorization Model using RF, ET, BC, ADA and GB Classifiers

The results obtained from the Unigram + TF-IDF and Bigram + TF-IDF vectorization model with five ensemble machine learning techniques represented in figure 3. The accuracy of Unigram + TF-IDF and Bigram + TF-IDF vectorization model represented respectively by blue and maroon color. The accuracy of classification obtained from Random Forest(RF), Extra Tree(ET), Bagging Classifier(BC), ADA Boost (ADA) and Gradient Boost(GB) classifiers are 99.38%, 100%, 98.892%, 80.814% and 82.024% respectively using Unigram + TF-IDF technique.On the other hand, we observe that Random Forest(RF), Extra Tree(ET), Bagging Classifier(BC), ADA Boost (ADA) and Gradient Boost(GB) algorithms provide the accuracy of classification 99.47%, 100%, 98.914%, 80.516% and 82.258% respectively using Bigram + TF-IDF technique. All of the experimental result with the results of evaluation parameters are comparatively shown at table 2.

Table 2. Comparison among the combination of Experimental Results.

Classifier + N-Gram	Precision	Recall	F1-Score	Accuracy
RF+Uni	0.99	0.99	0.99	99.4%
ET+Uni	**1.00**	**1.00**	**1.00**	**100%**
BC+Uni	0.99	0.99	0.99	98.7%
ADA+Uni	0.81	0.80	0.80	80.4%
GB+Uni	0.82	0.82	0.82	81.7%

Classifier +N-Gram	Precision	Recall	F1-Score	Accuracy
RF + Bi	0.99	0.99	0.99	99.4%
ET + Bi	**1.00**	**1.00**	**1.00**	**100%**
BC + Bi	0.99	0.99	0.99	98.8%
ADA+Bi	0.81	0.80	0.80	80.4%
GB + Bi	0.82	0.82	0.82	81.8%

Classifier +N-Gram	Precision	Recall	F1-Score	Accuracy
RF+Uni+TF-IDF	0.99	0.99	0.99	99.38%
ET+Uni+TF-IDF	**1.00**	**1.00**	**1.00**	**100%**
BC+Uni+ TF-IDF	0.99	0.99	0.99	98.9%
ADA+Uni+ TF-IDF	0.81	0.81	0.81	80.8%
GB+Uni+ TF-IDF	0.82	0.82	0.82	82.0%

Classifier+N-Gram	Precision	Recall	F1-Score	Accuracy
RF+Bi+TF-IDF	0.99	0.99	0.99	99.5%
ET+Bi+TF-IDF	**1.00**	**1.00**	**1.00**	**100%**
BC + Bi+ TF-IDF	0.99	0.99	0.99	98.9%
ADA+Bi+TF-IDF	0.81	0.81	0.80	80.5%
GB+Bi+ TF-IDF	0.83	0.82	0.82	82.3%

5. Conclusion

The combination of text vectorization models without and with feature extraction method TF-IDF are broadly experimented with five ensemble machine learning algorithms. Table 2 represent evaluation parameters and accuracy obtained and Figure 2 and 3 represent the accuracy obtained from Unigram, Bigram, Unigram + TF-IDF and Bigram + TF-IDF respectively after classification using five Ensemble Machine Learning classifiers. We found the maximum classifier can provide highest accuracy from the combination of Bigram + TF-IDF. And Extra Tree provide 100% accuracy for all combinations, ADA Boost for Unigram +TF-IDF and rest three classifiers Random Forest, Bagging Classifier and Gradient Boost algorithm for Bigram + TF-IDF combination.

References

[1]. Arif, M. H., Li, J., Iqbal, M., & Liu, K. Sentiment analysis and spam detection in short informal text using learning classifier systems. Soft Computing, 1-11, (2017).
[2]. Elnagar, A., Khalifa, Y. S., & Einea, A. Hotel Arabic-Reviews Dataset Construction for Sentiment Analysis Applications. In Intelligent Natural Language Processing: Trends and Applications (pp. 35-52). Springer, Cham, (2018).
[3]. Wang, H., Can, D., Kazemzadeh, A., Bar, F., & Narayanan, S. A system for real-time twitter sentiment analysis of 2012 us presidential election cycle. In Proceedings of the Association for Computational Linguistics System Demonstrations. pp. 115-120, (2012).
[4]. Liu, B. Sentiment analysis and opinion mining. Synthesis lectures on human language technologies, vol. 5(1), pp. 1-167. (2012).
[5]. Sorostinean, M., Sana, K., Mohamed, M. O. H. A. M. E. D., & Targhi, A. Sentiment Analysis on Movie Reviews, (2017)
[6]. Tripathy, A., & Rath, S. K. Classification of sentiment of reviews using supervised machine learning techniques. International Journal of Rough Sets and Data Analysis (IJRSDA), 4(1), 56-74, (2017).
[7]. Gautam, G., & Yadav, D. Sentiment analysis of twitter data using machine learning approaches and semantic analysis. In seventh international conference on Contemporary computing (IC3), pp. 437-442. IEEE, (2014).
[8]. Pedregosa, F., Varoquaux, G., Gramfort, A., Michel, V., Thirion, B., Grisel, O. & Vanderplas, J. Scikit-learn: Machine learning in Python. Journal of machine learning research, 12(Oct), 2825-2830., (2011).
[9]. Ensemble methods, http://scikit-learn.org/stable/modules/ensemble.html.
[10]. Feldman, R. Techniques and applications for sentiment analysis. Communications of the ACM, vol. 56(4), pp. 82-89, (2013).
[11]. Zhang, D., Xu, H., Su, Z., & Xu, Y. Chinese comments sentiment classification based on

word2vec and SVMperf. Expert Systems with Applications, vol. 42(4), pp. 1857-1863, (2015).

[12]. MartíN-Valdivia, M. T., MartíNez-CáMara, E., Perea-Ortega, J. M., & UreñA-LóPez, L. A. Sentiment polarity detection in Spanish reviews combining supervised and unsupervised approaches. Expert Systems with Applications, vol. 40(10), pp. 3934-3942, (2013).

[13]. Xia, R., Zong, C., & Li, S. Ensemble of feature sets and classification algorithms for sentiment classification. Information Sciences, vol. 181(6), pp. 1138-1152, (2011).

[14]. Matsumoto, S., Takamura, H., & Okumura, M. Sentiment classification using word sub-sequences and dependency sub-trees. In Pacific-Asia Conference on Knowledge Discovery and Data Mining, pp. 301-311, Springer, Berlin, Heidelberg, (2005).

[15]. Large Movie Review Dataset, http://ai.stanford.edu/~amaas/data/sentiment/.

[16]. Movie Review Data, http://www.cs.cornell.edu/people/pabo/movie-review-data/.

[17]. Tang, H., Tan, S., & Cheng, X. A survey on sentiment detection of reviews. Expert Systems with Applications, vol. 36(7), pp. 10760-10773, (2009).

[18]. Types of language models, https://nlp.stanford.edu/IR-book/html/htmledition/types-of-language-models-1.html.

[19]. Croft, W. B., Metzler, D., & Strohman, T. Search engines: Information retrieval in practice vol. 283. Reading: Addison-Wesley, (2010).

[20]. Manning, C. D., Raghavan, P., & Schütze, H. (2008). Introduction to information retrieval (Vol. 1, No. 1, p. 496). Cambridge: Cambridge university press.

[21]. Tf-idf, http://www.tfidf.com/

[22]. Maas, A. L., Daly, R. E., Pham, P. T., Huang, D., Ng, A. Y., & Potts, C. Learning word vectors for sentiment analysis. In Proceedings of the 49th annual meeting of the association for computational linguistics: Human language technologies-volume 1 (pp. 142-150). Association for Computational Linguistics, (2011).

[23]. Language Modeling, https://web.stanford.edu/class/cs124/lec/languagemodeling.pdf.

[24]. Ahmad, F. K. Comparative Analysis of Feature Extraction Techniques for Event Detection from News Channels' Facebook Page. Journal of Telecommunication, Electronic and Computer Engineering (JTEC), vol. 9(1-2), pp. 13-17, (2017).

Airport Detection on Remote Sensing Images Using Fater Region-based Convolutional Neural Network

M M Zhu[1,*], Y L Xu[1], S P Ma[1] and H Q Ma[1]

[1]Air Force Engineering University, Xi'an, China

E-mail: *ming_paper@163.com

Abstract. An airport detection method on remote sensing images based on transfer learning and hard example mining is proposed. We use Faster region-based convolutional neural network framework with end-to-end advantage instead of the sliding windows and artificial features in the existing traditional methods. Transfer learning is used to solve this problem that the number of airport remote sensing data itself is limited. Hard example mining is used to make full use of hard examples. As a result, it can improve the ability to discriminate objects and make training more efficient. The experimental results show that the proposed method can detect different types of airports accurately under complex backgrounds. Compared with other airport detection methods, the proposed method has higher detection rate 92.6%, lower false-alarm rate 15% and shorter average processing time 0.2s.

1. Introduction

The extensive acquisition and use of high-resolution remote sensing images have greatly promoted the development of airport detection in remote sensing images. Due to the complex backgrounds and different shapes of airport, the real-time and accuracy of airport detection is faced with significant challenges. Traditional methods are mainly divided into two kinds: edge line [1] and regional segmentation [2]. The former has the characteristics of high speed and low complexity, but it is easy to be disturbed. The performance of the latter is improved but there is a problem that the sliding window overlaps redundancy. Recently, the deep learning has provided new ideas for it: many scholars use the convolutional neural network (CNN) to identify airports and runway areas [3]. However, the above methods only use CNN's powerful identification ability. The acquisition of airport proposal boxes still depends on the manual methods (edge line and regional segmentation), the limitations of traditional methods still exist.

The design of the region-based convolutional neural networks (R-CNN) [4] framework has made a huge breakthrough in the object detection and opened the upsurge of object detection based on deep learning. However, due to the large amount of hard disk space required by training and feature repeated extraction, its speed of training and testing is very slow. Therefore, the successor of R-CNN, Fast R-CNN, is proposed [5], which greatly improve the accuracy and speed. Due to Selective Search algorithm (SS) [4], the detection speed is still very low. Faster R-CNN [6] adopts regional Proposal Network (RPN) to extract the proposal boxes, while RPN and the subsequent detection network, Fast R-CNN, share the convolution layer, which greatly improves the speed of object detection. The training of models based on deep learning often requires lots of learning samples, while the number of remote sensing images about airport is very limited. It's difficult to train network models directly.

However, the theory of transfer learning [7] provides an effective means to solve the airport detection problem with limited data.

Therefore, this paper abandons the previous method of sliding window plus features extracted manually and takes the Faster R-CNN as the basic framework. We reuse the model trained from other natural image databases after corresponding modification and improvement. At the same time, hard example mining is used to make training more efficient.

2. Methodology

2.1. Overview of Faster R-CNN

As shown in Figure 1. Faster R-CNN network takes as input an entire image and as output predicted probability value and detection boxes. The network is mainly divided into two parts: region proposal networks (RPN) and a detection network with region of interest (RoI) pooling layer, fully connected layers and loss layers. The shared convolutional layers are a common part of two networks. Since the object detection is unified into a deep network framework, the whole framework has end-to-end characteristics.

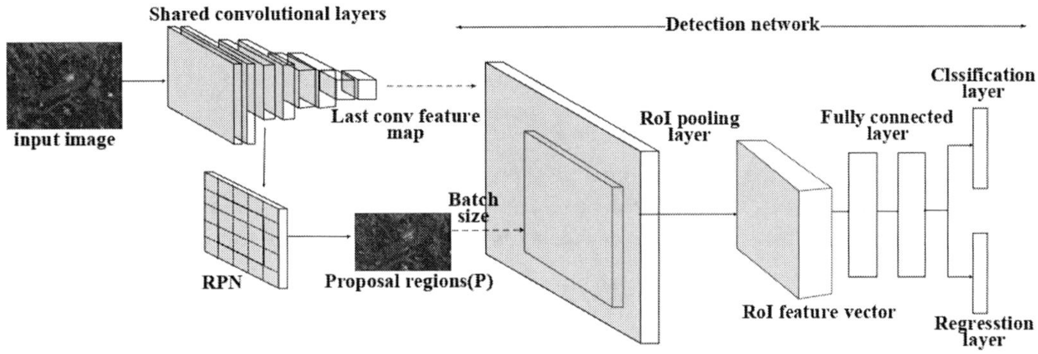

Figure 1. The architecture of Faster R-CNN.

2.1.1. Region Proposal Network. Region Proposal Network (RPN) uses a 3×3 sliding window to convolve the feature map. Each sliding window is mapped to a lower dimensional feature. A box-regression layer and a box-classification layer take the feature as input, as shown in Figure 2.

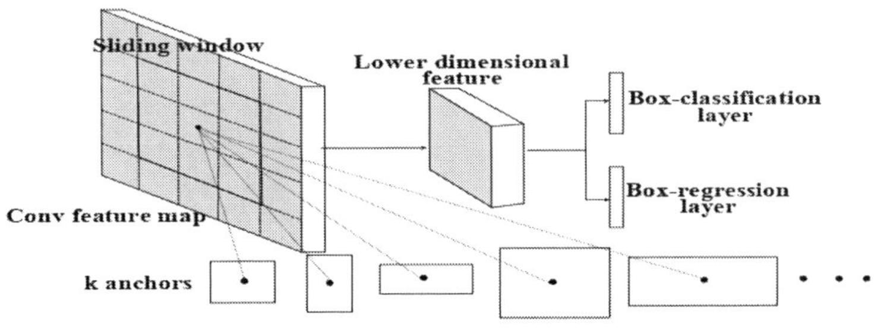

Figure 2. The structure of RPN.

In order to meet the multi-scale objects, the proposal boxes of multiple scale and aspect ratio are considered. These proposal boxes are parameterized by anchors. 3 scales and 3 aspect ratios are used at each position.

To train RPNs, a binary class label is assigned to each anchor. A positive label is assigned to two kinds of anchors: the anchor whose Intersection-over-Union (IoU) overlap with a ground-truth box is the highest or the anchor whose IoU overlap with any ground-truth box is higher than 0.7. A negative label is assigned to an anchor whose IoU ratio is lower than 0.3 for each ground-truth box. The multi-task loss is used in the training, the definition of loss function is described in literature [6].

A bounding-box regression layer is used to predict a new bounding box for the detection. As shown in Figure 3, the green box indicates the ground-truth bounding box, the yellow box indicates the proposal bounding box and the red box indicates the predicted box after the regression. Specific details can be found in literature [4].

Figure 3. Bounding-box regression.

2.1.2. Non-maximum Suppression. Some proposal boxes highly overlap with each other. To reduce redundancy, non-maxium suppression (NMS) is adopted based on their scores. The specific implementation process of NMS is:

(1) The classification score of the same target is ranked from high to low.

(2) Select the proposal box with the highest score and calculate the IoU value with other proposal boxes. If the IoU exceeds the threshold, then reject the proposal box with lower score.

(3) Select the highest score from the remaining proposal boxes and repeat step (2).

(4) Repeat step (3) until all proposal boxes are traversed. The above process is essentially an iterative-ergodic-elimination process. We can set the threshold of IoU to significantly reduce the number of proposal boxes.

2.1.3. RoI Pooling Layer. To reduce the calculation redundancy, the region of interest (RoI) pooling layer matches the spatial location of the proposal region and the corresponding feature map from the same input image, and it does not repeatedly input to the same network for the calculation. At the same time, due to the different sizes of the feature maps corresponding to proposal regions but the subsequent full-connection layer requires a fixed-size input, the RoI pooling layer uses max pooling to convert the feature maps with different sizes into a fixed-size feature map.

In order to share computation between proposal regions, we sample stochastic gradient descent (SGD) mini-batches hierarchically. For each mini-batch B, N images are sampled and then B/N proposal regions are sampled from each image.

2.2. Transfer learning

In order to extract image features efficiently, the manual design features with limited capability and poor robustness are discarded, and the convolutional neural network is used to extract features. Although the number of airport remote sensing data itself is limited, there are the common low-level and intermediate-level visual features between remote sensing images and natural images. So transfer learning can solve this problem through large database pre-training and small database fine-tuning. We use the pre-training network VGG16 for ImageNet classification to initialize the weights of the CNNs and then use the airport database for fine-tuning training. The experiment finally proved that it is very effective for transfer learning.

In order to match the proposed model with airport detection, according to the characteristics of the size and shape of the airport and the experimental verification, we set 3 scales and 3 aspect ratios in our network, as shown in Table 1:

Table 1. The settings of anchor.

	Anchor scales	Aspect ratios
settings	$(128^2, 256^2, 384^2)$	(1:1, 2:3, 3:2)

2.3. Hard example mining

Datasets always contain most of the easy examples and a few hard examples. Simple examples have little significance for training, while complex examples with high diversity and loss have a greater impact on classification and detection. Making full use of complex examples can improve the ability to discriminate objects. In fact, the imbalance between easy examples and hard examples is not a new challenge. The idea of dataset bootstrapping [8], typically called hard example mining, has been applied to the training of most detection tasks. Even the recent object detection model A-Fast-RCNN [9] that generates examples with occlusions and deformations using generative adversarial nets can be regarded as an example mining method. For airports with a small amount of data, hard example mining becomes more important. Therefore, referring to the idea of hard example mining, the overall structure of proposed method is shown in Figure 4:

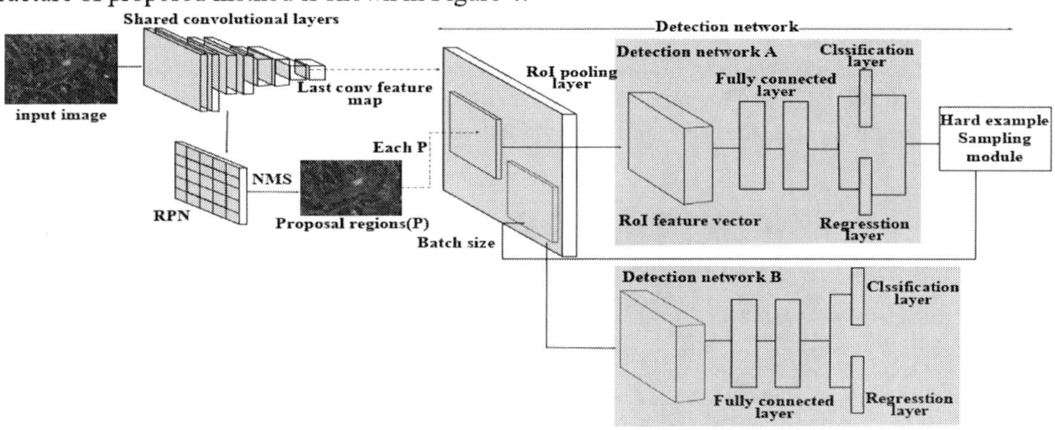

Figure 4. The architecture of proposed method.

The original detection network is duplicated into two networks, which are represented simply by the network A and B respectively and share the network parameters. The network A only has forward pass, while the network B is a standard detection network that has both forward and backward passes. Firstly, the network A takes as input the feature maps corresponding to proposal regions and performs a forward and computes loss values of all input proposal regions. Then the hard example sampling module selects B/N hard examples for which the current network performs worst by sorting the input proposal regions by loss. Finally, the above hard examples are used as input to the network B for normal model training. The input of network A is all the proposal regions P from N image instead of mini-batch B, while the input of network B is mini-batch B. We use $N=2$ (which results in $P \approx 4000$) and $B=128$.

2.4. Training

The entire networks are implement using the Caffe framework. The shared convolution layers are initialized by a pre-training model, the weights of other layers are initialized randomly with a zero-mean Gaussian distribution with standard deviation 0.01. we use a basic learning rate of 0.001, a momentum of 0.9 and a weight decay of 0.0005. We use alternating optimization strategy to train that allows for sharing convolution layers between the RPN and detection network, which is as follows:

1. The RPN is initialized with a pre-training model VGG16, the weight parameters are fine-tuned end-to-end.

2. The detection network is also initialized by the pre-training model VGG16 and trained by using the proposal regions of step-1 RPN.

3. We use the step-2 detection network to initialize RPN, fix the shared convolutional layers and only fine-tune the layers of RPN.

4. The shared convolution layers are fixed, other layers of the detection network are fine-tuned again.

3. Experiment

The proposed method is trained and tested in real remote sensing images that cut from Google Earth from all over the world. The sizes of all the images are 1000×600. The data set includes 600 images, in which 400 images contain airports and the remaining images do not contain airport but mainly background scenes such as railways, highways, bridges, and buildings. 240 airport images and 120 non-airport images are randomly selected to form a training set, and the remaining images are composed of a test set. The images are horizontally flipped with probability 0.5 to enhance the data. The computing platform consisted of a CPU of Intel i7-7700 and a GPU of NVIDIA GTX1080.

3.1. Detection results

Some detection results are shown in Figure 5, which includes various airport background types such as roads, rivers, and mountains. The second row of Figure 5 shows some airport images with unique shapes. There results prove that our model is robust to complex backgrounds and various airport structures.

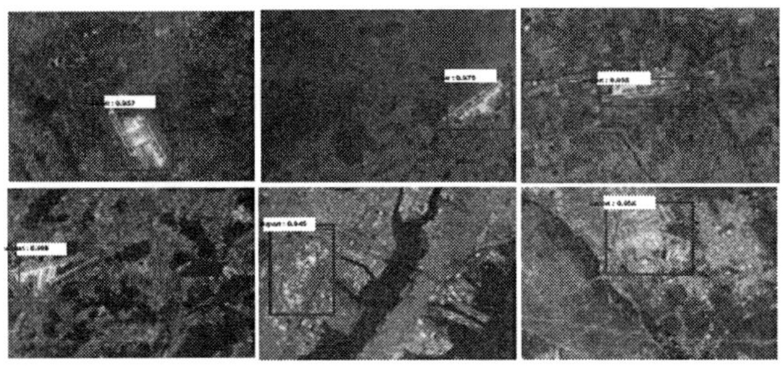

Figure 5. Detection results.

3.2. Comparison with other methods

To demonstrate the efficiency of our method, the existing airport detection methods are selected for comparative experiments. The detection rate (DR), false-alarm rate (FAR) and average processing time of each method on the test set are recorded and compared. All the results in the experiment were the average of multiple random experiments, as shown in Table 2. The literature [10] is an edge-based airport detection method, the literature [11] is an airport detection method based on regional segmentation, and the literature [3] is an airport detection method based on convolutional neural network.

Table 2. Performance comparison among different methods.

Method	DR /%	FAR /%	Average processing time /s
literature [10]	65	22.5	2.4
literature [11]	71.8	27.5	>100
literature [3]	81.1	35	21.3
Ours	92.6	15	0.2

According to the Table 2, the proposed method has the highest detection rate, the lowest false-alarm rate and the shortest average processing time. The reason is that the previous methods adopt the way of sliding window plus manual design feature, the calculation efficiency of sliding window is very low and the robustness of manual design features is not strong. The proposed method integrates feature extraction, acquisition of proposal boxes and object classification and regression into an end-to-end deep network framework, makes full use of CNN's powerful feature expression capabilities and allows for sharing convolutional layers between the RPN and detection network. Therefore, the proposed method has the best performance.

4. Conclusion

This paper proposes an airport detection method based on transfer learning and hard example mining. In this method, RPN and the non-maximum suppression are used to extract proposal boxes. The RoI pooling layer avoids extracting features repeatedly. Transfer learning is used to solve this problem that airport data is limited. Hard example mining is used to make full use of hard examples and make training more efficient. Both RPN and detection network share the convolution layer, which greatly improves the efficiency of airport detection. The results show that the proposed method is better than other methods. It has strong theoretical and practical significance for the real-time and accuracy of airport detection.

References

[1] Budak U, Halici U and Sengur A 2016. Efficient Airport Detection Using Line Segment Detector and Fisher Vector Representation. *J. IEEE Geoscience and Remote Sensing Letters.* **13**(8) 1079-83
[2] Zhu D, Wang B and Zhang L 2015. Airport target detection in remote sensing images: A new method based on two-way saliency. *J. IEEE Geoscience and Remote Sensing Letters.* **12**(5) 1096-1100
[3] Zhang P, Niu X and Dou Y 2017. Airport detection on optical satellite images using deep convolutional neural networks. *J. IEEE Geoscience and Remote Sensing Letters.* **14**(8) 1183-87
[4] Girshick R B, Donahue J and Darrell T 2014. Rich Feature Hierarchies for Accurate Object Detection and Semantic Segmentation. *J. CVPR.* 580-587
[5] Girshick R B 2015. Fast R-CNN. *J. CVPR.* 1440-48
[6] Ren S Q, He K M and Girshick R 2017. Faster R-CNN: towards real-time object detection with region proposal networks. *J. IEEE Transactions on Pattern Analysis and Machine Intelligence.* **39**(6) 1137-49
[7] Pan S J, and Yang Q 2010. A Survey on Transfer Learning. *J. IEEE Transactions on Knowledge and Data Engineering.* **22**(10) 1345-59
[8] Felzenszwalb P F, Girshick R B and Mcallester D A 2010. Object Detection with Discriminatively Trained Part-Based Models. *J. IEEE Transactions on Pattern Analysis and Machine Intelligence.* **32**(9) 1627-45
[9] Wang X, Shrivastava A and Gupta A 2017. A-Fast-RCNN: Hard Positive Generation via Adversary for Object Detection. *J. CVPR.*
[10] Qu Y, Li C and Zheng N 2005. Airport Detection Base on Support Vector Machine from A Single Image. *J. Fifth International Conference on Information, Communications and Signal Processing.* 546-549
[11] Tao C, Tan Y and Cai H 2011. Airport Detection from Large IKONOS Images Using Clustered SIFT Keypoints and Region Information. *J. IEEE Geoscience and Remote Sensing Letters.* **8**(1) 128-132

Study of Wisdom Campus Overall Architecture Model in Vocational Colleges Based on Service Oriented Architecture

Wang Xuejun

Shandong Labor Vocational and Technical College,Ji'nan 250014,China

17142660@qq.com

Abstract. Wisdom campus is a high-level stage of digital campus construction and needs to be improved in addressing the issues of information silos and legacy systems. Based on the SERVICE ORIENTED ARCHITECTURE-based vocational schools wisdom campus overall architecture model contains reusable services and well-defined, standards-compliant interface, through services to achieve different platforms, different languages on the exchange between the system can be targeted at vocational colleges to provide business processes , To provide fast and flexible changes. The model of "platform + application + service" proposed by the model and service-oriented thinking can better solve the problem of information silos and legacy systems and provide an effective method for upgrading the Informatization level of vocational colleges.

1. Introduction

With the further development of the Internet and the gradual popularization and application of the Internet technology of a new generation, the international education informationization ushers in new opportunities for development. The traditional campus will gradually move toward the smart campus through the digital campus. After three stages of development, such as networking, digitization and informationization, the education industry in China began to gradually enter the stage of intelligent education. The smart campus is one of the landmark products of the combination of digital campus and wisdom education. It is a hot spot in information construction and reform in education industry in recent years. How to accurately grasp the connotation of "smart campus" and give full consideration to the emerging new technologies and make a far-sighted planning and construction of "smart campus" is a new topic for educators. At present, on the whole, our country has made a major historical breakthrough in the construction of education informationization, but at the same time it faces a series of outstanding problems. First, the information infrastructure is still at a low level, can't effectively meet the development needs of various types of education at all levels; Second, a large number of digital campus system, the system independent of each other, there are isolated islands of information and applications; third is the construction of university information Lack of complete unity, covering a comprehensive, efficient and stable, scalable application framework. [1-3]

Wisdom campus generally refers to the new information technology that integrates school teaching, research, management and other types of campus resources and application system efficiently through the application of new technologies such as the Internet, Internet of Things and social software so as to realize intelligent management and service Campus construction mode. Therefore, the existing basic platform should be combined with new technologies such as cloud computing, big data and Internet of Things to build a complete, unified, comprehensive, efficient, stable and scalable application framework to solve the problems existing in the informatization of vocational colleges. To "smart campus" become a hot issue of information technology in vocational colleges. [2,4].

The construction of the smart campus in vocational schools is still in its infancy, fulfilling the requirements of a basic digital campus, but far from being intelligent, especially in the field of big data. Mainly reflected in the various departments, the data between the colleges did not achieve sharing, fragmented, resulting in "information island" phenomenon, leading to different data when the leadership decision-making, affecting the effectiveness of decision-making and pertinence. [3] Wisdom Campus "Information Island" and other problems and the solution to the remaining problems [5,6].

2. Enterprise-class SERVICE ORIENTED ARCHITECTURE

Service-Oriented Architecture is an enterprise-class service-oriented architecture, enables the distributed deployment, composition, and use of loosely coupled coarse-grained application components over the network as required. SERVICE ORIENTED ARCHITECTURE, with its loosely coupled nature, allows colleges to add new services or update existing services in a modular fashion to address new business developments that can be served through different channels, with existing or existing Application as a service to protect the existing IT infrastructure investment. SERVICE ORIENTED ARCHITECTURE can respond quickly to business changes, take advantage of investments in existing applications and application infrastructures to address new business needs and present a framework that supports organic business [7]. SERVICE ORIENTED ARCHITECTURE framework for enterprise-class applications provided by the distributed, high reliability, advanced solutions. The Web-based software industry's public standards, independent of the operating system and server-independent "cross-platform," make it "write once, run anywhere" and is the best fit for running on the Internet with powerful "business logic" Application [8].

3. Vocational college wisdom campus model

The vocational school campus structure model from the wisdom of vocational colleges and universities hardware infrastructure, public support platform, intelligent software applications, integrated information services, scientific decision support, security protection, operation and maintenance services management, uniform standards from the bottom up Design, high-performance cloud computing as a data storage, mining, analysis of the core technology, the use of big data as a basis for organizational management decision support, make an overall plan and top-level design, construction and Education, management, service information as one of the key core database of school education information cloud service platform to form an information system suitable for the planning and development goals of the college and form an informationized learning environment and supporting service system for all to enjoy quality educational resources, Promote the teaching of online courses, e-resource sharing of book documents, virtual simulation experiment center and other supporting information construction, and promote the reform of mixed education teaching mode. We will promote the construction of information resources such as digital book resources and digital teaching resources, carry out extensive training in enhancing teachers 'informational instructional capabilities, and continuously improve teachers' information literacy. In the campus network infrastructure, data center infrastructure, information security and other aspects continue to develop, built into the office automation, educational administration, research management, student management, personnel management, enrollment and employment, logistics services, "Palm University" and other information subsystems One-stop online service hall to achieve the universal use of online resources in the teaching and learning process, to provide a solid information protection for the improvement of education and teaching quality, and to create a high-level intelligent campus model framework featuring information collection and resource sharing .

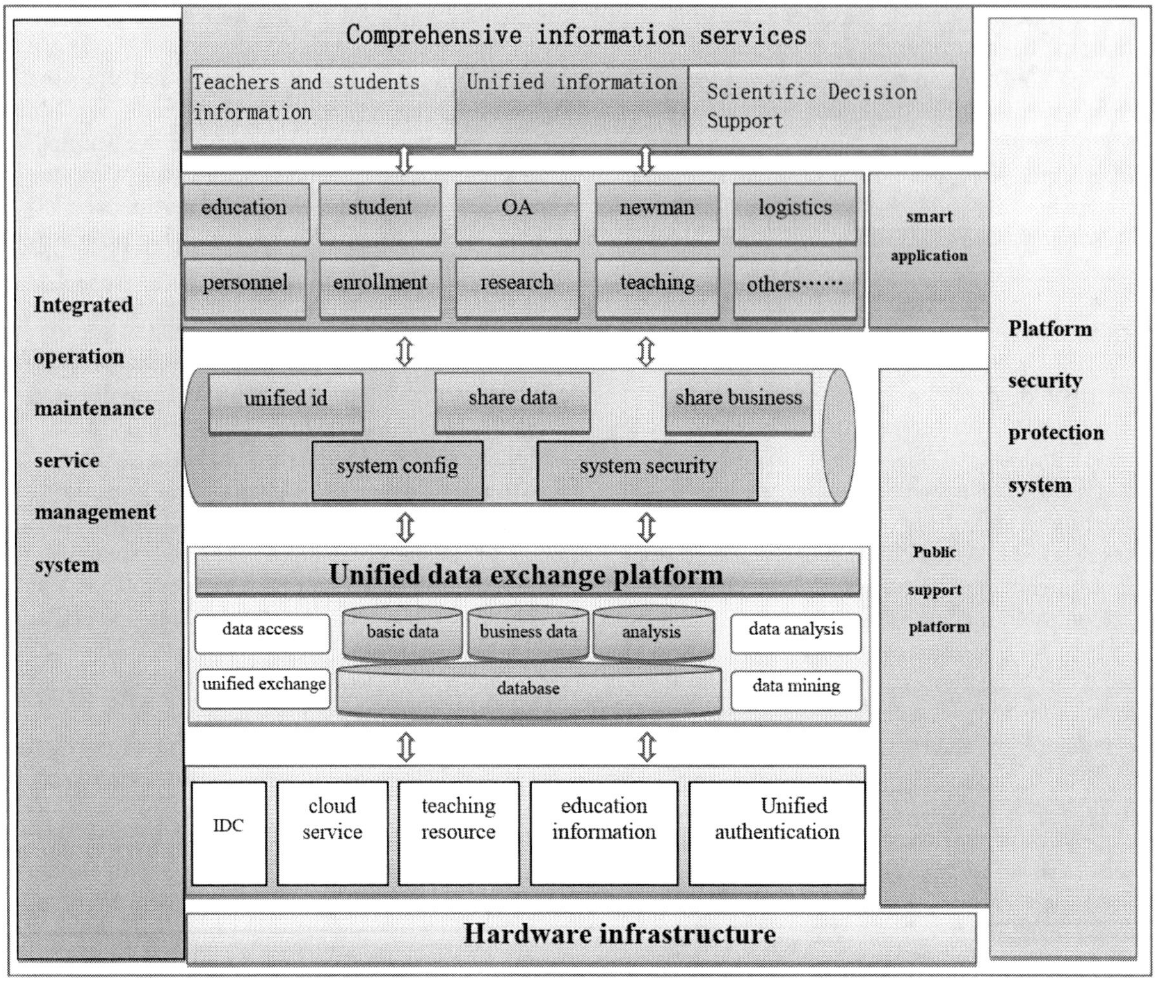

Figure. 1"smart campus" integrated service model

The "smart campus" integrated service model is shown in Figure.1 mainly includes seven levels: hardware support environment, software support environment, information resource center, application system public support platform, application system layer, user authentication and user layer, information service layer, and all applications are strictly Follow the relevant technical standards and norms.

The platform adopts the N-level architecture system, and builds the hardware and network layers, information resource center, public support platform, application layer, user authentication and user layer from the bottom up, all layers are based on the services provided by the lower layers. Platform through heterogeneous application integration technology, integration of existing information systems in various departments, through the interface integration and single sign-on technology for the end user to provide a unified interface.

(1)Infrastructure layer: Wisdom campus hardware and software support system, including network resources, hardware servers, storage, support software.

(2) Wisdom campus application system based platform: including educational information hub platform, unified identity authentication platform, public data center platform.

(3) Application system layer: It is for all kinds of information management system and various types of information service system for teachers and students in school departments. It can be divided into management center, resource center and service center.

4)Information service layer: for all levels of leadership, administrators, teachers, students provide a variety of personal business operations services, queries, reports, statistical analysis, decision-making.

5)Information Security System: In the platform construction process, give full consideration to the various levels of security measures and security technology, through hardware and software technology and security management tools to ensure the system in a safe and stable environment to run. Through the computer room management, internal and external network isolation, data encryption, access control and other security mechanisms to achieve legal access to data and information.

6)Information Standards / Management / Security System: In the platform construction process, full reference to various national technical specifications and industry standards. The standard specification system is an important guarantee for the normal operation of the system and contains two meanings: data standardization and management standardization. Data standardization refers to the establishment of a standardization system for spatial data and related business data. Management standardization refers to the formulation of work standards and assessment standards for all relevant responsible parties so as to improve the daily work system.

4. SERVICE ORIENTED ARCHITECTURE-based vocational schools wisdom campus architecture model

SERVICE ORIENTED ARCHITECTURE framework for enterprise-class applications provided by the distributed, high reliability, advanced solutions. The Web-based software industry's public standards, independent of the operating system and server-independent "cross-platform," make it "write once, run anywhere" and is the best fit for running on the Internet with powerful "business logic" application. Based on the SERVICE ORIENTED ARCHITECTURE architecture system of intelligent campus and its application system should follow the following SERVICE ORIENTED ARCHITECTURE is shown in Figure.2 technologies, including service component architecture related technologies and service data objects related technologies and so on.

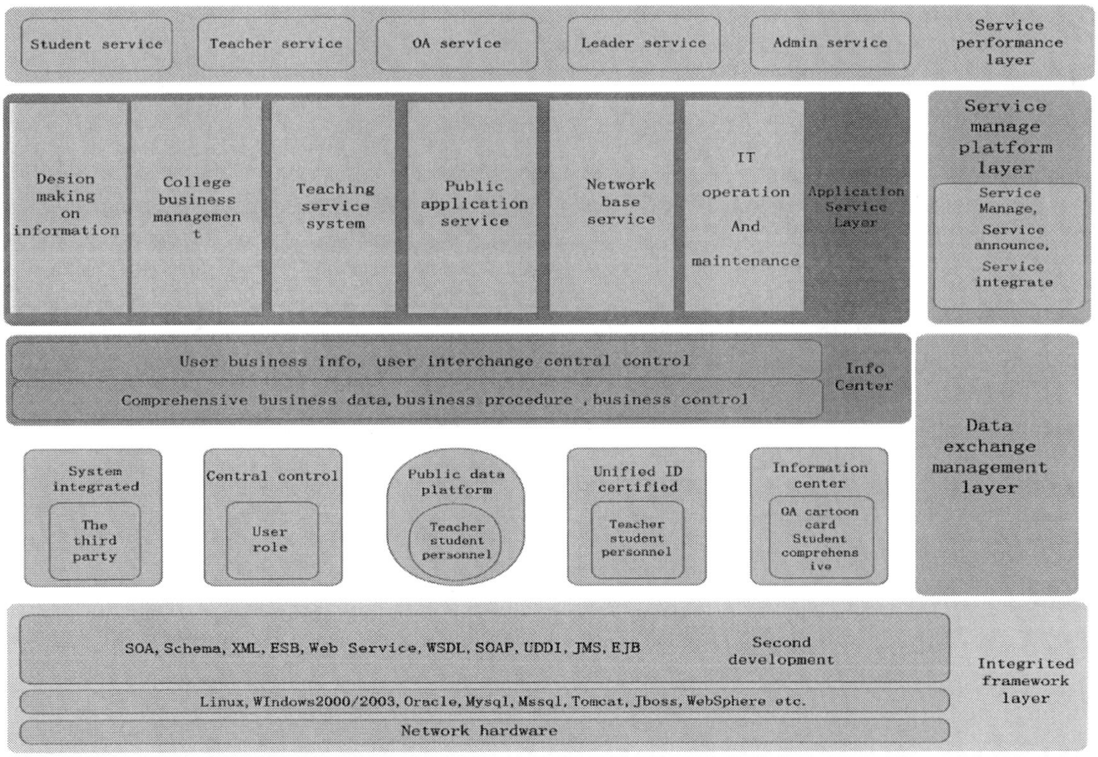

Figure. 2 SERVICE ORIENTED ARCHITECTURE

(1) object-oriented component technology

Object-oriented component technology is a completely hardware and operating system-independent development environment that focuses on developing reusable components that make up the business "business object" of an application that enables the smooth setup of distributed applications.

The application platform module is relatively independent and the interface is clear. The internal business process upgrade and transformation are independent of other modules. All modules are designed based on the pluggable interface of components and provide data support for the secondary development and analysis of the college in the future.

(2) application server cluster

According to the college's situation, put forward the use of load balancing and Session High Availability program:User front-end with hardware load balancing; backend, deploying multiple Application Servers and enabling Application Server clustering.

(3) performance indicators

The main factors affecting system performance include: network environment (campus network and Internet access), hardware configuration (server and client), system software (operating system, database management system and WEB service system), system architecture design, data structure design , Core algorithm design, deployment design and so on. Fully consider the scientific design of the overall system architecture, the rationality of the data structure design, the advanced design of the core algorithm, the flexibility of deployment design to ensure that the overall system performance is superior and fully meet the design requirements.

5. Application experiment

Based on the above research model, based on the vocational college wisdom campus application platform for SERVICE ORIENTED ARCHITECTURE vocational school campus intelligent campus model for application experiments.

5.1. application experiment

(1) example for one-stop comprehensive service hall is shown in Figure.3

To achieve the core business system interoperability, convergence and integration, built a large-capacity high-performance storage as a support, access convenient, stable and reliable high-speed, to meet the teaching, teaching, research, office workers, enrollment, employment, party, real estate , Operation and maintenance and other application subsystems digital campus unified service platform. For the integrated office, teaching, library management, personnel management, asset management, scientific research management, real estate management, card and other systems do a good job of data interface to achieve the sharing of campus information resources for the "Palm University", integrated school intelligence analysis and other campus applications Provide basic data services for the majority of teachers and students to provide school calendar, class tables, leading schedule, payroll, duty watch, phone book, book search, multimedia management, rules and regulations and other convenient one-stop comprehensive services.

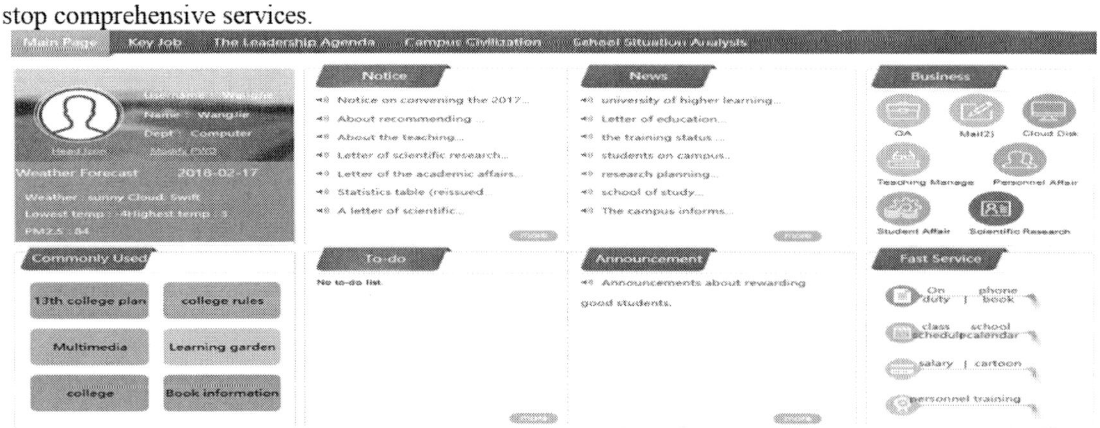

Figure.3 one-stop comprehensive service hall

5.2. Experimental Conclusion
The vocational school wisdom campus data center extracts, merges and screens the business data in the operation of the business system, forms a complete and independent thematic database for cross-departmental business and personal comprehensive business, and provides data encapsulation and application service interface. To solve the problem of "information silos" caused by the isolation of isolated business systems, unified data standards, application of standard interface standards, reduction of data exchange costs, inventory of school data assets, statistical inquiry and data mining, exploration of big data analysis The application in school provides a scientific basis for school development decision. It can be seen from the application effect that it can promote the development of informationization of vocational colleges with high precision and fast speed, so it is better suited to solve the "information silos" and the legacy problems in vocational colleges.

6. Conclusion
Service-Oriented Wisdom The overall architecture of the campus system integrates resources, data, information and application processes into service-based smart campus construction through service-oriented integration through information service so that they are related to each other and data are shared. It also helps to reorganize the organization and business processes and effectively coordinate personnel and resources so as to improve the overall efficiency in teaching, research, management, office and learning. In order to achieve the IT and business integration, to meet individual needs, and its rapid response ability to respond flexibly to changes in demand, support for education reform and innovation [9].

The application of SERVICE ORIENTED ARCHITECTURE-based intelligent campus overall structure model in vocational colleges makes it possible to greatly improve the sharing of information resources inside and outside the school. The unified management service platform for high-performance education information cloud and a sound data system can make educational administration, teaching and research , Office, academic, enrollment, employment, party affairs, real estate, operation and maintenance and other application subsystems digital campus education and management information level significantly increased. It can be believed that the overall campus intelligent campus model based on SERVICE ORIENTED ARCHITECTURE forms the rationality of the smart campus with SERVICE ORIENTED ARCHITECTURE architecture and application architecture, which provides a unique reference model for the construction of a smart campus in a vocational college. It will certainly be implemented in a wider area Get more widely used.

References:
[1] 863 Smart City (first phase) project group. Technical white paper of Smart City(2012)[EB/OL].www.docin.com/p-417808631.htm,2016-06-18.
[2] http://cdn.nmc.org/media/2015-nmc-horizon-report-HE-CN.pdf,2016-07-10.
[3] Jim Rudd et al. Education for a Smarter Planet: The Future of learning [EB/OL] .http://www.redbooks.ibm.com/redpapers/pdfs/redp4564.pdf,2012-09-09.
[4] http://www.csipiot.com/xingyechanpin/shiyanshi/68.html,2015-03-29.
[5] The Eclipse foundation[EB/OL]. （2011） .http://www.eclipse.org/modeling/emf/.
[6] Fausto Ibarra. The Enterprise Service Bus: Building Enterprise SERVICE ORIENTED ARCHITECTURE. BEA systems,2004
[7] Annrai O'Toole. Web Service-Oriented Architecture: The Best Solution to Business Integrate. Copyright(c) Cape Clear Software Inc,2003
[8] Boris Lublinsky.SOA Agents:Grid computing meets SOA.Dec11,2008. https://www.infoq.com/articles/lublinsky-soa-grid
[9] Martin Keen. Patterns: Serial and Parallel Processes for Process Choreography and Workflow. http://www.ibm.com/redbooks/,2004.04

Neighbor Trigger for Hovering Information

X Y Shuai[1]

[1] College of Computer Science and Technology, Taizhou University, Taizhou, China

E-mail: xyshuai@163.com

Abstract. Hovering information is generated by specific area in a mobile ad hoc network(MANET) to be attached to the anchor area for some time. MANET is a self-organizing and dynamic topology network without infrastructure. Some nodes maybe not receive the hovering information due to possible partitioning of the network in some areas with sparse nodes, on the other hand broadcast packets lead to broadcast storm problem and high latency in high traffic density network. To improve the reachability and deduce the number of broadcast of hovering information, the adaptive probabilistic flooding scheme is proposed by Andreas Xeros. But the calculation of probability p is too complex. So, this paper proposes a new adaptive probability flooding scheme based on random graph and neighbour trigger (algorithm of neighbour trigger, ANT) for floating the information in the hovering area. When a new node enters the range of hovered node and become a neighbour of the node that has received the hovering information, the informed node checks its neighbour list to decide whether to trigger the broadcast program with probability p. I analyse the performance of the proposed algorithm. The results show that the ANT outperforms other candidate hovering information algorithms in terms of the number of broadcast messages.

1. Introduction

Messages such as emergency messages, commercial advertisements must continue to be attached to the anchor area for some time to transmit the messages to new network nodes that enter the area. To float message within the anchor area, hovering information is proposed.

1.1 Mobile ad hoc

MANET [1]is a temporary dynamic network include many wireless mobile nodes, and it does not use any fixed network infrastructure. Node is free to move randomly and can exchange message directly with another mobile node within the range of radio transmission. When a destination that is located outside source node's radio radius, packets are stored and forwarded through a sequence of intermediate nodes. The all nodes in a MANET are required to participate in the relay of packets on behalf of other nodes.

VANET is vehicular MANET that has appeared over the past few years. VANET can provide safety and convenient services through wireless communications between vehicles that brings people with new functions and applications such as emergency services. However, VANET is confronted with many great challenges due to fast changing network topology. It is an important problem to study an efficient and reliable solution for packets dissemination in a VANET [2].

1.2 Hovering information

Hovering information is an information dissemination scheme in MANET. It was first proposed by A. Villalba and D. Konstantas [3], and a more elaborate and definition are provided by [4]. Alfredo A. Villalba Castro defined formally hovering information later in [5].

Some application, such as emergency messages in a MANET, as well as traffic accident and traffic congestion in the VANET, can generate a piece of floating information. The area around the source node is called hovering area. It is very important for each node residing in the hovering to be informed of the emergency to take effective and valid ways. Every node in hovering area stores and rebroadcasts the hovering information to suspend the information in the special zone. Hovering information transmits from one mobile node to another through ad hoc connectivity. The information which is floating on the anchor area is available to nodes that are entering the area. The hovering information is applicable to MANET, especially to VANETs.

2. Related works

A few researches have been done on disseminating information in MANET. This section reviews the epidemic routing and adaptive probability flooding protocol of hovering information in MANET (VANET).

2.1 Epidemic routing

In a MANET or VANET, due to a source and a destination are not always direct connect there must be a routing (such as DSR and DSDV, et al.) to deliver this messages through intermediate nodes. But these routings are limited to fully connected networks. In fact, there are disconnected portions of MANET (VANET) due to the mobility of the nodes and limitations in radio range. The presence of such scenarios where the connection path from the source to the destination is not always available.

According to these requirements and limitation, epidemic routing based on epidemic algorithm [6] was proposed by Vahdat and Becker [7] to solve the problems caused by such scenarios. The goals are to 1) provide message delivery with high probability, 2) minimize message latency, 3) minimize the total resource consumption. Epidemic routing relies on nodes called carriers which receive the message then store it in their buffers. When carrier comes into the range of another node through node mobility, two nodes exchange message. The message spread into additional island of nodes, so achieve high delivery ratios. Epidemic routing (host S sends a message to host D) is shown in figure1.

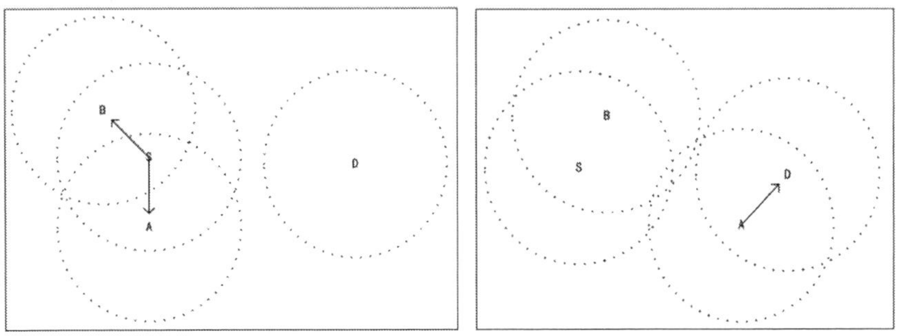

Figure 1. Epidemic routing.

2.2 Adaptive probability flooding(APF)

Andreas Xeros [8] proposed an algorithm using adaptive probabilistic flooding to float information in VANET. APF used epidemic route inside the anchor area and probabilistic flooding outside the anchor area. Vehicles outside the anchor area received the hovering information then rebroadcast the packet with probability p. The p is given by equation (1).

$$p = e^{-\frac{d^2}{2\sigma^2}} \quad (1)$$

The σ is a design parameter representing the standard deviation, *d* represents the distance from vehicles to the hovering area. The σ must be updated in time based on the connectivity of the network in the hovering area.

Because of possible partitioning of the network in some areas with sparse nodes, some nodes maybe not receive the hovering information. The epidemic routing approach can alleviate the problem. Those informed vehicles can store and forward the message serving as information bridges. With the mobility of informed vehicles, the hovering information may be relayed to the partitioned uninformed areas. The scheme increases the achieved reachability. However, epidemic routing within the anchor area will generate a lot of broadcast packets lead to broadcast storm problem and high latency in high traffic density network. In high traffic density network, the VANET is almost fully connected. So, no each vehicle floods the received critical messages are required inside the hovering area and no each vehicle replicates the message are required outside. Probabilistic flooding within the hovering area is applied in paper [9].

Adaptive probability flooding protocol reduces large number of redundant messages of hovering information based on epidemic routing. Redundant broadcasts increase the communication channel burden and contention and lead to large delay of packet delivery.

3. Algorithm of neighbor trigger (ANT) for hovering information

Adaptive probability flooding protocol reduces large number of redundant messages of hovering information based on epidemic routing and increases the achieved reachability of message. However, the probability p in the protocol is related to σ. The parameter σ is computed online and complexly. In order to reduce the complexity of p in APF, I propose a new adaptive probability flooding based on random graph and neighbor trigger. The probability p is defined by equation (2).

$$\mathrm{p} = \frac{1}{n} \tag{2}$$

The n is the number of nodes that received the hovering information in a connected portion. Random geometric graph (RGG) is a stochastic graph with metrics. RGG has been used in model of large networks, such as MANET. Each node randomly and independently placed in RGG and can be connect with each other only within the distance r (radio radius) [10]. The threshold for connectivity is $p = \log n/n$ and the threshold for a giant component is $p = 1/n$ in the ER random graph G (n, p) [11].

3.1 Algorithm of neighbor trigger

Each node has a neighbor list to store its current neighbors. The nodes in the communication range can exchange neighbor information with each other through the neighbor discovery procedure. The algorithm is given as follow.

```
Procedure Discover neighbor
    Send hello message to find neighbor;
    Exchange (id,flag); /*if received the hovering information then flag=1 else flag=0*/
    Update owe neighbor node list;
    If has a new neighbor
    Then
      Return newneighbor=1;
    Else
      Return newneighbor=0;
   Endif
Endprocedure
```

When a new node enters the range of hovering node, the node checks its list to decide whether to trigger the broadcast program. The process is shown in figure 2.

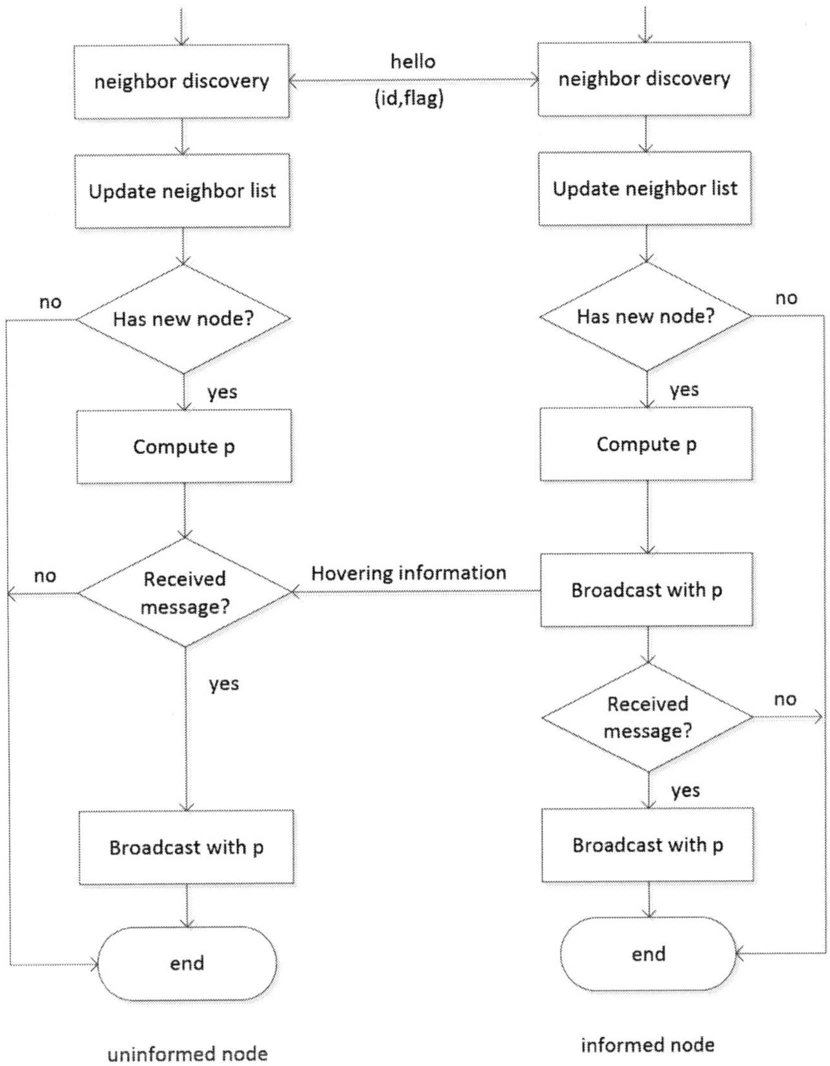

Figure 2. Process of ANT

The procedure ANT is shown as follow.

Procedure ANT
  Discover neighbor ();
    Compute (p); /*compute probability p */
  If *newneighbor* then
    Replication (p); /*relay the received hovering information with p*/
  Endif
  If received a replication then
    Rebroadcast (p); /*relay the received hovering information with p*/
  Endif
Endprocedure

Hovering node traverses owe neighbor list to count the number of informed node to compute the probability p. The procedure is shown as follow.

Procedure Compute(p)

n=1;
While (neighbor list is not null)
If *flag* then
n=n+1;
Endif
Endwhile
p=1/n;
Endprocedure

3.2 Performance analysis

I analysis the number of message of the ANT in the scenario descripted in paper [8]. The radium of hovering area is 500m, the radio radius of node is set to 150m-180m. I vary the number of nodes N= {5, 10, 15, 20}. Comparison of the ANT with APF in terms of the number of message is shown in figure 3.

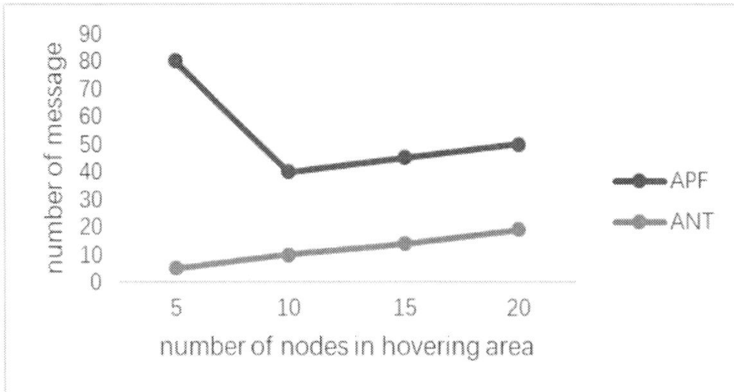

Figure 3. Comparison of the number of message.

Figure 3 shows the number of message broadcast by ANT is less than that of APF.

4. Summary

In this paper, I proposed a scheme for hovering information in an anchor area. The proposed scheme is based on epidemic routing and RGG. When an uninformed node enters the range of hovering node, the replication procedure is triggered with probability p. Any node will rebroadcast the hovering information with probability p when it receives replication. The result of analysis shows the ANT outperforms APF in terms of the number of message.

In the future, we will use more realistic simulations of scenario and network. Use RGG to establish more accurate model for analysis and optimization.

Acknowledgement

This research was supported by the Taizhou University Foundation for the Talents (QD2016035(702065)).

References

[1] Subir Kumar Sarkar 2012 Ad Hoc Mobile Wireless Networks Principles, Protocols and Applications (Boca Raton :CRC Press)

[2] D Tian, Jianshan Zhou and Yunpeng Wang 2016 An Adaptive Vehicular Epidemic Routing Method Based on Attractor Selection Model Ad Hoc Networks 36 p 465

[3] Alfredo Villalba and D. Konstantas 2006 Towards hovering information Lecture Notes in Computer Science 4272 p 161
[4] Giovanna Di, Alfredo Villalba and D. Konstantas 2007 Dependable Requirements for Hovering Information the 37th Annual IEEE/IFIP International Conference on Dependable Systems and Networks
[5] Alfredo Villalba, Giovanna Di and D. Konstantas 2008 Hovering information- Self-Organising information that Finds its Own Storage International IEEE Conference on Sensor Networks, Ubiquitous and Trustworthy Computing (TaiWan) 112 p 193
[6] A. Demers, D. Greene and C. Hauser 1987 Epidemic Algorithms for Replicated Database Maintenance Proceedings of Annual Acm Symposium on Principles of Distributed Computing Vancouver Acm 22 1
[7] Amin Vahdat and David Becker 2000 Epidemic routing for partially connected ad hoc networks http://cseweb.ucsd.edu/~vahdat/papers/epidemic.pdf
[8] Andreas Xeros, Marios Lestas and Maria Andreou 2010 Adaptive probabilistic flooding for information hovering in VANETs the second IEEE Conference on Vehicular Networking (USA) p 239
[9] Andreas Xeros, Themis Constantinides and Marios Lestas 2013 Adaptive Information Hovering in VANETs: Robustness and Enhancements International Conference on Connected Vehicles and Expo (USA) p 622
[10] J. Diaz, D. Mitsche, X. Perez 2008 On the Connectivity of Dynamic Random Geometric Graphs the 19th annual ACM Symposium on Discrete Algorithms (USA) 8 p 601
[11] Milan Bradonjiić and Will Perkins 2014 On Sharp Thresholds in Random Geometric Graphs the 18th International Workshop on Randomization and Computation (Germany) p 500

Affective Learning Reaction Model Based on Smartphone and Sensors

Kwang Sik Chung [2,*] and Sungho Chin [1]

[1] Dept. Of Computer Science, Korea National Open University, Jongno-gu, Dongsung-Dong, Seoul, Republic of Korea
[2] Cloud Center, LG Electronics, Yeongdeungpo-gu, Seoul, Korea

*E-mail : kchung0825@knou.ac.kr

Abstract. As we daily use smartphones as like as televisions and cars, smartphones are usually used to access and communicate with learning contents and lecturers. And through smartphone, various types of multimedia contents can interact with users, since there are many types of sensors equipped with smartphone. Information and sensor technology are very precious that users can be detected their movement, emotion and learning status. Especially, most of learners (above 90% of Korea National Open University students) access to learning contents with their smartphones in the mobile learning environment. But, there is few methods, technology and service model for learners' learning status tracking, only focusing degree and learning concentration level. In this paper, we propose learning reaction model and criteria for learning state. For simple approach, we propose sensor-based metrics with smartphone and the learning emotional reaction model between learners and learning contents is based on learner's learning actions and learner's sage state for a smartphone. Proposed Affective Learning Reaction Model(ALRM) consists of learner's learning emotional state, learners' personal environment state and learner's learning activity patterns. Each state has both mandatory and optional information. Mandatory information is important metrics to decide learners' learning emotion state. And Optional information is complimentary information. Those metrics have their own value and sum of them decides learner's learning state. With the ARLM, personalized and modified learning contents can be delivered to a learner.

1. Introduction

On e-learning environment, learning contents are delivered to a learner and the learner should do everything by him/herself. After delivering learning contents to a learner, e-learning system or service does not track the learner's learning activity and does not feed learning activity results and personalized learning compliments back to the learner. These situation makes learners loss study interest or learning concentration and students drop out of school or university. And, in these kinds of e-learning environment, learning progress and study achievement or assessments totally depend on learners or learners' self-controls. In order to make up for these weakness, we need agents or tools to track learner's learning activity and learner's learning emotional state. If we have learner's learning emotional state, then we can modify learning contents and make personal learning environment according to personal learning state[1,2].

But, nowadays most learning devices have various kinds of sensors and seamless network communication capability. With these devices, sensors and network communication environment, e-learning system can track a learner's learning activity and feed personalized learning contents back to the learner.

And in digital textbooks, sensors can initiate learners' learning interactions with learning contents and easily track learners' learning interactions. These kinds of e-learning service are studied and proposed in previous works[2]. But, without learning emotional model, consistency of personalized learning contents and personalized learning environment can not be guaranteed. If a learner studies one learning subject with an e-learning system, he/she can be provided learning contents related the subject and learning environment change makes another types or delivery methods of the learning contents and the modified learning contents are provided with to the learner. In this situation, without learning emotional model, learners' learning environment could be misunderstood and misestimated, and learning contents modification rules could not be choose consistently. This means that we need consistency learning emotional model that supports learning contents modification rules.

We propose Affective Learning Reaction Model as learning emotional model that considers learner emotion, learner's learning activity, and smartphone attitude. And we also study on Learning Reaction Analysis Engine to analyze learner's activity, learner's smartphone attitude, bio signals and Learner's attitude, and infer learners' emotion state. The remainder of paper is organized as follows. In chapter 2, we re-view previous affective computing and emotion analysis research for e-learning learning. The proposed learners' Affective Learning Reaction Model and learner activity-related information (learner's activity, learner's smartphone attitude, bio signals and Learner's attitude, and infer learners' emotion state) are described in chapter 3. We introduce the architecture and function modules of Learning Reaction Analysis Engine and Affective Learning Reaction Model are defined and described in chapter 4. Finally, we conclude in chapter 5.

2. Related Works

In [3], various emotion states metrics are proposed and dealt with by intelligent tutoring system. But only bio-signal based information is focused. Behavior of learner is useful to analyze learner's emotion state. In this paper, we involve those kinds of bio information. In [4], problems of technology usage for the learner's learning environment is discussed and affective emotion based approach is suggested for programming courses at university students. In [5], affective computing platform for e-learning is proposed and analyze learners' emotional information. But, emotional information is limited within bio-signal information. In [6], the relationship between learners' perceptions of affective learning, instructors' attractiveness and instructor evaluations is revealed and shows positive correlation with task attractiveness, social attractiveness, affective learning and instructor evaluations.

These kinds of information metrics are useful to broaden emotional model for learners, teachers and learning contents. In [7], emotion is defined as a critical role in the learning activity and learner's emotion analysis researches and studies are reviewed, since learning emotion impacts on learners' motivation and learning achievement. But, there is no way or methods to measure and collect the related data (sensed learner's activity or environment). In [8], Path model of emotions and learning analytics behavior and cognition is proposed. In Path model of emotions and learning analytics behavior and cognition, the detailed analysis model is proposed only for learner's behaviors. Except for learner's behavior, the other data or situation data could be used for analysis of learner's learning emotion.

Most of the previous works have focused on bio-signal information or learner's actions or activity, and simple emotion models are used to analyze and to track for learner's emotion state with only one or two sensors within smartphone. This paper tries to collect various learning emotional information with various sensors and use broad emotion model for a learner. Especially, various sensed data from smartphone would be collected and analyzed. In order to propose the analysis methods, learner's emotional model and emotional metrics criteria are proposed.

3. Affective Learning Reaction Model and Learner Activity Analysis System

Proposed Affective Learning Reaction Model has various information about a learner. Some information is directly related with a learner or learning activity. And some information is not directly related with a learner nor learning activity. We define information directly related with a learner as Direct Affective Learning Reaction In-formation (DALRI) and information indirectly related with a learner as Indirect Affective Learning Reaction Information (IALRI). DALRI can be directly analyzed and used for a learner state, learning interest level and learning concentration level. But, an indirect learner state will be inferred from IALRI analysis. Combination of a learner state and an indirect learner decides affective learning reaction state of the learner. According to the affective learning reaction state, appropriate learning con-tents delivery strategy for a learner can be provided to Automated Tutoring Engine [2].

3.1. Affective Learning Reaction Model

Learning Reaction Analysis Engine needs three types of learner's learning information that can be classified into personal environment state information, learning activity pattern information and learning emotional state information. Learning Reaction Analysis Engine decides learners' learning emotional state with personal environment state information, learning activity pattern information and learning emotional state information.

Personal environment state information, learning activity pattern information and learning emotional state information are stored in format of XML at learners' e-portfolio DB and periodically changed. Personal environment state information, learning activity pattern information and learning emotional state are like as bellows;

- Learner's activity
  - learning activity time(LAT): activity level of learning duration time(learning activities time per one session of learning contents)
  - learning interaction(LI) : initiative level of interaction with learning contents(learning interaction numbers time per one session of learning contents)
- Learner's smartphone attitude
  - smartphone declined angle(SDA) : decline angle detected by gyroscope sensor in smartphone
  - learner's face shape of smartphone camera(FSC) : learner's concentration level on learning contents
  - display touch frequency(DTF) : number of smart phone touch according to learning contents(basic touch number predefined by contents designers)
  - learning sector stay time(SST) : time to concentrate on learning contents(basic duration time predefined by contents designers)
  - shaking frequency(SF) : time to make irrelevant learning(basic touch number predefined by contents designers)
  - smartphone movement velocity(SMV) : learner's movement velocity
  - smartphone location(SL) : learner's location (in-building, in-car, in-class room, etc.)
  - illumination changes ratio(ICR) : illuminations of learners' environment and illumination change state
  - environment noise with microphone sensor(ENM) : learner's environment state for learning
- Bio signals
  - heart bit(HB) : emotional data
  - temperature(TP) : emotional data
  - hands movement and shaking frequency(HMS) : emotional data
- Learner's attitude
  - another APP launching types and execution time(AET) : number and execution time of other Apps launched on smartphone
  - activity tracking information(ATI) : learner's smartphone usage type

Learner's learning activity patterns means learner's concentration about smartphone contents.

Without other information, smartphone contents may be game application, video, or digital textbook for learning. Learner's learning activity patterns consists of Learner's activity(learning activity time, learning interaction, Learner's smartphone attitude, display Touch frequency, sector stay time, shaking frequency, smartphone location, illumination changes ratio change state, environment noise with microphone sensor), Learner's attitude(activity tracking information). Learner's personal environment state information means learner's environment that makes learning emotion accuracy of the learner. Learner's personal environment state information consists of Learner's smartphone attitude(smartphone declined angle, smartphone movement velocity, smartphone location, illumination changes ratio change state, environment noise with microphone sensor). Lastly, learner's learning emotional state information consists of Learner's activity(learning activity time, learning interaction), Learner's smartphone attitude(smartphone declined angle, display Touch frequency, sector stay time, learner's face shape of smartphone camera), Bio signals(Heart bit, Temperature, Hands movement and shaking), Learner's attitude(another APP launching types and execution time). Learner's learning emotional state information decides learner's learning emotion that resent learning difficulty, learning interests, and learning achievement. In Table 1, each information has their own value and, according to the value sum, learner emotion state can be decided.

Table 1. Learner's Information of LRAE Model.

Learner's Info.	Learner's Information	Value
Learner's activity	learning activity time	appropriate time(specified by learning content designer) / below Appropriate time
	learning interaction	appropriate frequency(specified by learning content designer) / below frequency
Learner's smart phone attitude	smartphone declined angle	between 30° ~ 75° / not-between 30° ~ 75° degree
	learner's face shape of smartphone camera	60 % ~ 100 % of face / not-[60%~100%] of face
	smartphone movement velocity	not-move or above-10km per hour / 1~10km per hour
	smartphone location	on the street / in a vehicle / in a building
	illumination changes ratio change state	frequently / occasionally / not-change-dark / not-change-light
	environment noise with microphone sensor	below 40 dB / between 40 dB ~60 dB / above 60 dB
	display touch frequency	appropriate frequency(specified by learning content designer) / below appropriate frequency / above appropriate frequency
	sector stay time	appropriate frequency(specified by learning content lecturer) / below appropriate frequency / above appropriate frequency
	shaking frequency	frequently (decided by learning content designer) / occasionally / not-change-dark / not-change-light
Bio signals	Heart bit	70~90 bits per min / above normal heart bit
	Temperature	35~37°C / above normal temperature
	Hands movement and shaking	appropriate movement / frequent movement
Learner's attitude	another APP launching types and execution time	loner than digital textbook launching time / shorter than digital textbook launching time
	activity tracking information	appropriate digital textbook foreground running time(designed by learning contents designer) / bellow digital textbook foreground running time

3.2. Affective Learning Reaction Information Processing Module

In order to process affective learning reaction information(ALRI) and decide learner's emotion state, three stages of process are defined. The first process stage analyzes learner's learning emotional state information. From the analysis of learner's learning emotional state information, Learning Reaction Analysis System knows whether a learner is studying or not. At the second stage, Learners' personal environment state information are analyzed and Learning Reaction Analysis System knows whether a learner concentrate on the subject of digital textbook or not. At the last stage, learner's learning activity patterns information(module) are analyzed and Learning Reaction Analysis System decides whether a learner's environment is appropriate for a learning activity or not.

At the first stage, mandatory information of learning emotional state information is smartphone declined angle, learner's face shape of smartphone camera, and another APP launching types and execution time. And optional information is learning activity time, learning interaction, display touch frequency and sector stay time. The mandatory information must satisfy the defined level. If mandatory information of learning emotional state information dose not satisfy the defined level, the learner would not learn or concentrate on the other subject or the other activity. If the defined level of mandatory information of learning emotional state information satisfies all three conditions, then the learner would be learning the subject of digital textbook application. The three conditions are as follows;

- Condition 1 of learning emotional state: smartphone declined angle is between 30° ~ 75°,
- Condition 2 of learning emotional state : learner's face shape of smartphone camera is between 60 % ~ 100 % of face,
- Condition 3 of learning emotional state: another APP launching types and execution time is shorter than digital textbook launching time,

At the second stage, mandatory information of learners' personal environment state information is smartphone movement velocity, illumination changes change ratio state, environment noise with microphone sensor. The mandatory information must satisfy the defined level. If mandatory information of learners' personal environment state information dose not satisfy the defined level and mandatory information of learning emotional state information would satisfy the defined level, the learner would not concentrate on subject. If mandatory information of learning emotional state information would satisfy the defined level and mandatory information of learners' personal environment state information would satisfy all three conditions, then the learner would be learning the subject of digital textbook application. The three conditions are as follows;

- Condition 1 of learners' personal environment state : smartphone velocity is between 0 km ~ 2 km,
- Condition 2 of learners' personal environment state : illumination changes change ratio is between 0 times/min. ~ 10 times/min.,
- Condition 3 of learners' personal environment state : another APP launching types and execution time is below 40 dB

At the third stage, mandatory information of learner's learning activity patterns information is learning activity time, learning interaction, display touch frequency, shaking frequency and sector stay time. And optional information is smartphone location, illumination changes ratio change state, environment noise with microphone sensor, and activity tracking information. The mandatory information of learner's learning activity patterns must satisfy the defined level. If mandatory information of learner's learning activity patterns dose not satisfy the defined level and mandatory information of learning emotional state information would satisfy the defined level, the learner would not concentrate on subject. If mandatory information of learning emotional state information would satisfy the defined level and mandatory information of learner's learning activity patterns information would satisfy all fours conditions, then the concentration of the learner would be high and the learner would be learning the subject of digital textbook application. The four conditions are as follows;

- Condition 1 of learner's learning activity patterns : learning activity time is appropriate time(specified by learning content designer),
- Condition 2 of learner's learning activity patterns : learning interaction appropriate frequency(specified by learning content designer),
- Condition 3 of learner's learning activity patterns : display touch frequency is appropriate frequency(specified by learning content designer),
- Condition 4 of learner's learning activity patterns : shaking frequency and sector stay time is appropriate frequency(specified by learning content lecturer)

4. Conclusion

We propose Affective Learning Reaction Model(ALRM). ALRM consists of learner's learning emotional state, learners' personal environment state and learner's learning activity patterns. Each state is inferred by Learning Reaction Analysis Engine. Learner's learning emotional state is defined as learner's emotional state for learning con-tents and environment. Learner's learning emotional state information consists of learning activity time, learning interaction, smartphone declined angle, display touch frequency, sector stay time, learner's face shape of smartphone camera, Heart bit, Temperature and Hands movement and shaking. Learners' personal environment state is defined as learners' environment information for learning. Learners' personal environment state information consists of smartphone declined angle, smartphone movement velocity, smartphone location, illumination changes ratio change state and environment noise with microphone sensor. Learner's learning activity patterns is defined as learners' learning activity. Learner's learning activity patterns information consists of learning activity time, learning interaction, display Touch frequency, sector stay time, shaking frequency, smartphone location, illumination changes ratio change state, environment noise with microphone sensor and activity tracking information.

All of the information are produced by smartphone or mobile devices equipped with various sensors. Various sensors collect learner's learning activities information and learner's environment information. With those information, learner's learning emotional state is decided and sent to ATE.

ATE decides new learning scenario and reconstructs strategy according to ALRM. In order to decide learner's learning emotional state, ATE applies conditions of three analysis stage. First stage for learning emotional state information, mandatory information is defined as smartphone declined angle, learner's face shape of smartphone camera, and another APP launching types and execution time. Second analysis stage for learners' personal environment state information, mandatory information is defined as smartphone movement velocity, illumination changes ratio change state, environment noise with microphone sensor. Third analysis stage for learner's learning activity patterns information, mandatory information is defined as learning activity time, learning interaction, display touch frequency, shaking frequency and sector stay time. And threshold value is defined for each mandatory information and threshold value satisfaction condition means learning positive state of a learner. These kinds of information from various aspects of a learner increases accuracy of estimation for learner's learning emotional state. We have a plan to define and design XML rule for ALRM.

Acknowledge

This work was supported by 2017 Korea National Open University Research Fund.

References

[1] Chung, K., Kim, Y., Lee, C., Jung, S 2014 KNOU Smart Learning: Beyond the Future KNOU Learning Environment *28th proceedings of AAOU*, AAOU 2014, 2014.

[2] Kwang Sik Chung, Learning Reaction Analysis Engine for Interactive Digital Textbook Platform *FutureTech 2017: Advanced Multimedia and Ubiquitous Engineering* pp 459-465, 2017.

[3] Mehdi Malekzadeh, Mumtaz Begum Mustafa and Adel Lahsasna, A Review of Emotion Regulation in Intelligent Tutoring Systems *Educational Technology & Society*, 18 (4), 435–445,

2015.

[4] Alvin Prasad, Mohammed Farik 2015 Integration Of Innovative Technologies And Affective Teaching & Learning *In Programming Courses International Journal of Scientific & Technology Research* Vol. 4, Issue 10, Oct 2015.

[5] Khaled El-Abbasy, Anastassia Angelopoulou1, and Tony Towell 2015 Affective Computing to Enhance E-Learning in Segregated Societies", *2015 Imperial College Computing Student Workshop (ICCSW 2015)*. Editors: Claudia Schulz and Daniel Liew; pp. 13–20, 2015.

[6] Irem E. AYDIN 2012 Relationship between Affective Learning, Instructor, Attractiveness and Instructor Evaluation in Videoconference based Distance Education Courses *The Turkish Online Journal of Educational*, volume 11 Issue 4 247-252, 2012.

[7] Bart Rienties and Bethany Alden Rivers 2014 Measuring and Understanding Learner Emotions: Evidence and Prospects", 2014.

[8] D.T. Tempelaar, B. Rienties, B. Giesbers, "In search for the most informative data for feedback generation: Learning Analytics in a data-rich context." Computers in Human Behavior, 2014.

Role of Trust, Customer Satisfaction, and Perceived Effectivenes of E-Commerce Institutional Mechanism to Repurchase Intention: Systematic Literature Review

Rifa Rizka Anisah[a] **and Agus Achmad Suhendra**[b]

Department of Industrial Engineering, Telkom University

[a]rifarizka@gmail.com, [b]agus@telkomuniversity.ac.id

Abstract. Development of all e-commerce company in the world increase significantly each year, every company innovate their strategy especially in the transactional mechanism which sensitivity level to the customer is very high. Every e-commerce company definitely needs a repurchase intention from their customer, then they needed the right strategy to achieve it. To make the good perception of the customer in case to keep the customer's trust and their satisfaction high, e-commerce company establish transaction mechanism which involves the third party to decrease transactional risk activity. The paper uses systematic literature review (SLR) as a methodology in reviewing literature is searched through published articles systematically such as; Emerald Insight, ISSN, and Elsevier. The paper will be classified of each research on primary studies by decision variables, variables, and methods. The paper is written in only English-languages with specific keywords to find in the online database. From the literature review results, it is known that there is influence between the role of trust, customer satisfaction, and perceived effectiveness of e-commerce institutional mechanism (PEEIM) to repurchase intention in blanja.com Indonesia with using structural equation modelling (SEM) method.

1. Introduction

E-commerce industries always update technology to take the public's attention so the market will trust earnestly and adopt new lifestyle of e-commerce adoption. The developing of internet user in Indonesia increase significantly since 2016 reach 132,7 million people from all of 256,2 million people in Indonesia, in 2014 with only 88,1 million internet user [1]. That shows us enhancement of internet user in Indonesia and it becomes golden opportunity for e-commerce industries to develop their product and change the habit of market in shopping. Previous research by wearesocial.com show us e-commerce statistic development in 2016, amount of internet user who had e-commerce transaction reach 24,74 million people in Indonesia then they have spent US$ 5,6 billion (Rp 74,6 trillion) to shop in e-commerce site [2].

2. Background of Information

The Developing of e-commerce industries change shopping habit in the market. Google Indonesia and Indonesia E-Commerce Association (idEA) work together and had a research about the reason of market habit in e-commerce shopping. 72% for e-commerce customer because of faster, 66% more efficient because it doesn't need to directly visit the store, 61% because it is easier to differentiate between products in e-commerce features, 58% because it is easier to access the goods [3]. We can conclude from these 5 reasons that ease and comfort become the basic reason for shopping in e-commerce. There are top 5 e-commerces site in Indonesia, these are Lazada, Bukalapak, Tokopedia, Blibli, and Blanja (tekno.liputan6.com) and three of them (Bukalapak, Tokopedia, and Blanja) are customer to customer (C2C) model of e-commerce with escrow payment system while Lazada and Blibli are business to customer (B2C) model e-commerce [4]. Google trend empirical evidence that Blanja has more traffic rate than Tokopedia with 27:25 scale while trustedcompany.com empirical evidence about customer satisfaction with review indicator that Tokopedia has better rating than Blanja with 3.1:2.6 scale. It becomes a big question when blanja has a good traffic rating with low rating of customer review. The results of customer review in Blanja certainly in accordance with their experience. We can hypotize with the low review of customer, it affects customer trust in Blanja. In addition, the role of payment mechanism strategy by company is very important which raises the perception from each customer in choosing e-commerce company. Thus, to make repurchase intention from customer in the same e-commerce company.

3. Related Work

Many types of researches in industrial engineering especially industrial management, including mechanism strategy to develop the vision of a company. Furthermore, there is a little comprehensive research about perceived effectiveness of e-commerce institutional mechanism (PEEIM) which moderate the role of interactivity and trust to repurchase intention. This part describes to some papers that there are in field of these roles.

This study discusses the importance of a trust, interactivity, peeim and CMC in e-commerce process. In this study it helps other practitioners to strategize and develop business processes to retain customers in e-commerce. This research was measured and analyzed by empirical modeling and survey data with PLS-SEM equation.

The results of this study is a positive relationship between the customer satisfaction and trust in the seller causing a repurchase intention. The perceived effects of PEEIM values show that the effects are moderate and significant. For further research, researchers can expand their research into other e-commerce.[5]

4. Research Method

Research paradigm is a concept or a structure diagram about perspective explanation from researcher in theory used [6]. Paradigm research is used by researcher to confirm research approach, methods and techniques to collect and analyze data that can explain the problems in study. In this study, the paradigm used is positivism paradigm because analyzing of statistical data obtained through the distribution of questionnaires on the respondents [7].Repurchase intention is influenced by independent variable that are trust in seller and customer satisfaction, then influenced by moderating variable of perceived effectiveness of E-Commerce Institutional Mechanisms (PEEIM).

Table 1 below, shows that the research is a quantitative research. Quantitative research is a research based on positivism that is used to examine in a particular population or sample [8]. Sampling technique is generally done randomly, data collection using research instrument, data analysis is quantitative statistic with the aim to test the hypothesis that has been established. This research is a deductive research, because it aims to test theory in certain context, that is theory about dependent variable, repurchase intention influenced by independent variable that is trust in seller and customer satisfaction, then influenced by moderating variable of perceived effectiveness of E-Commerce Institutional Mechanisms (PEEIM).This study included into correlation research. Correlation research is a research conducted by researchers to determine the level of relationship between two or more variables, without making any changes, additions or manipulations to existing data [9].The author does not intervene data but only collect data from the specified sample. Unit analysis is individual, because the data analysis is done individually by the author. The study design based on the time of execution is divided into two, cross-sectional and longitudinal. This study used cross-sectional research, where data collection was done in one period, then the data was processed, analyzed, and drawn its conclusion [10].

Table 1. Research Characteristics

No.	Characteristics of the Study	Type
1	Based on Paradigm	Positivism
2	Research Approach	Deductive
3	Methodology	Quantitative
4	By Destination	Causal
5	Type of Inquiry	Correlation
6	Based on Research Engagement	No Interfering Data
7	Based on the Analysis Unit	Organization
8	Based on Timeline	Cross Section

Research variables are basically anything that shaped what is set by the author to be studied so that obtained information about it and draw conclusions [8]. In this study, variables used include:

1) Independent Variable (Independent Variable). This variable directly affects the dependent variable, from positive or negative influences. The independent variable in this research is about trust in seller and customer satisfaction.
2) Dependent Variable. The dependent variable is the variable that is the main concern of the author, where this variable is directly used to answer the research problem. Variables are the result, because of the independent variables [10] . The dependent variable in this research is about repurchase intention.
3) Moderating Variables (Moderating Variable). Modating variable is a type of variable that has an influence on the nature or direction of relationships between variables. The nature or direction of the relationship between the independent variables with the positive or negative dependent variables in this case depends on the moderating variable. Therefore, the moderating variable is also called the contingency variable. The moderating variable in this study is Perceived Effectiveness of E-Commerce Institutional Mechanisms (PEEIM).

The operational variable is determined by the researcher to be studied to get information and can be drawn a conclusion. Based on the independent variable, dependent variable and moderating variable that will be discussed in this research, researcher made indicators for each variable in accordance with the problems of e-commerce.

In the Perceived Effectiveness of E-Commerce Institutional Mechanisms (PEEIM) moderating variable, it has dimension of trust in protection when conducting a transaction with e-commerce where the indicator is the existence of the mechanism of protection when making a transaction, The existence of trust against the protection of third parties when transact, The confidence of personal data is not utilized by other parties and the belief of a party that has a duty to protect in case of a transaction error.Operational Variable that exist in this research in seen in table 2.

Table 2. Variable Operationalization

Variable	Dimension	Indicator	No	Refference
PEEIM	Trust in protection	The existence of a protection mechanism when making a transaction	1	Bao et el (2016)
		There is trust in the protection of third parties when making transactions	2	Bao et el (2016)
		The confidence of personal data is not utilized by others	3	Bao et el (2016)
		Confidence in the presence of a party that has a duty to protect in case of a transaction error	4	Bao et el (2016)
Trust in Seller	Competence	Consumer trust in quality and service	5	Bao et el (2016)
		Reliable trust	6	Bao et el (2016)
		The belief that the seller meets the needs and demands of the customer	7	Bao et el (2016)
	Integrity	Honesty from seller	8	Bao et el (2016)
		High integrity of the seller	9	Bao et el (2016)
Customer Satisfaction	Re-purchase	Make a buyback	10	Wijoseno, Ariyanti (2017)
		Expectations exceeded	11	Wijoseno, Ariyanti (2017)
	Word-of-Mouth	Recommendations to others	12	Bao et el (2016)
Repurchase Intention	Made a re-purchase	Short-term period	13	Bao et el (2016)
		Long-term period	14	Bao et el (2016)
		Consumers are willing to make online transactions	15	Wijoseno, Ariyanti (2017)

This paper aims to collect and invertigate variables that has been considered in trust, customer satisfaction, perceived effectiveness of e-commerce institutional mechanism (PEEIM), and repurchase intention. This paper also investigate research trends to generate research agenda and identify the most used methods in related topics. Accuracy is needed while design the research question so there will be related match between paper and topic to be researched by give more attention to important component in each criteria of PICOC [11]. Table.3 shows the (PICOC) structure of the research questions.

Table 3. Summary of PICOC

Population	Random Indonesia people, Trust, Customer Satisfaction, Perceived Effectiveness of E-Commerce Institutional Mechanism, Repurchase Intention
Intervention	Proposing effectiveness strategy of repurchase intention
Comparison	n/a
Outcomes	Results of repurchase intention effectivity strategy
Context	Studies in market that analyze trust, customer satisfaction, and perceived effectiveness of e-commerce institutional mechanism

With referring to PICOC criteria above then further can be identified specific research questions. The following research questions (RQs) are raised that shows in Table 4. From the research questions in Table 4, at this early stage we can search for journals based on characteristic, because each study case of repurchase intention has it own characteristics based on its case study (Q2). We also filter journals based on specific keywords to get the related journals (Q1). After that, we can indentify variables that most used considered variables with related topics (Q3) and also the methods that the most used to solve this problem (Q4). The next step is state the variables that must be considered in future research (Q5). From this explanation, we conclude that research question (Q3 until Q5) are the main research questions. While research question (Q1 and Q2) used to identify summary of the referenced journal.

Table 4. Research Question on Literature Review

ID	Research Question	Motivation
RQ1	Which journal is the most significant in reviewing repurchase intention in e-commerce	Identify the most significant journal which reviewing repurchase intention in e-commerce
RQ2	What kind of research topics which are often discussed repurchase intention in e-commerce	Identify the research trends in repurchase intention in e-commerce
RQ3	What kind of variables considered are used most often for repurchase intention in e-commerce	Identify the most used factors for repurchase intention in e-commerce
RQ4	What kind of methods are used most often for repurchase intention in e-commerce	Identify the most used methods for repurchase intention in e-commerce
RQ5	What kind of variables which must be considered for the next research in e-repurchase intention in e-commerce	Identify variables considered for future research in repurchase intention in e-commerce

5. Conclusion

This paper presented a systematic literature review on role of trust, customer satisfaction, and perceived effectiveness of e-commerce institutional mechanism to repurchase intention. In this study conducted systematic literature review to get research opportunities and research gap which will be examined in subsequent research. The research model is empirically tested using survey data analyzed with partial least squares structural equation modelling (PLS-SEM). The paper hipotyzed that there is influence between the role of trust, customer satisfaction, and perceived effectiveness of e-commerce institutional mechanism (PEEIM) to repurchase intention in blanja.com Indonesia.

References

[1] Indonesia Internet Service Provider Association 2017 *Infografis Penetrasi & Perilaku Pengguna Internet Indonesia* Retrieved from https://www.apjii.or.id/ Jakarta p 6
[2] Pratama A H 2017 *Perkembangan Pengguna Internet di Indonesia Tahun 2016 Terbesar di Dunia* Retrieved from https://id.techinasia.com/pertumbuhan-pengguna-internet-di-indonesia-tahun-2016 Jakarta
[3] Erlangga M 2014 *Nilai Pasar E-commerce Indonesia Diprediksi Capai $25 Miliar di Tahun 2016* Retrieved from https://dailysocial.id/post/idea-e-commerce-indonesia-2016 Jakarta
[4] Jeko I R 2015 *5 Situs E-commerce Terbaik yang ada di Indonesia* Retrieved from https://www.liputan6.com/tekno/read/2322305/5-situs-e-commerce-terbaik-di-indonesia
[5] Bao H, Li B, Shen J and Hou F 2016 *Repurchse Intention in The Chinese E-Marketplace* Vol 116
[6] Noor 2011 *Metodologi Penelitian Skripsi Tesi Disertasi dan Karya Ilmiah* (Jakarta : Kencana) p 37
[7] Abdillah and Jogiyanto 2015 *Partial Least Square (PLS) Alternatif Structural Equation Modeling (SEM) dalam Penelitian Bisnis* (Yogyakarta : Andi) p 129
[8] Sugiyono 2013 *Metoda Penelitian Pendidikan Pendekatan Kuantitatif, Kualitatif dan r&d* (Bandung : Alfabeta) p 142
[9] Arikunto 2007 *Manajemen Penelitian* (Jakarta : Rineka Cipta) p 89
[10] Indrawati 2015 *Metode Penelitian Manajemen dan Bisnis Konvergensi Teknologi Komunikasi dan Informasi* (Bandung : Aditama) p 37
[11] Jesson J, Mateson L and Lacey F M 2011 *Doing Your Literature Review Traditional and Sistematic Techniques* Vol 2 London pp 56-80

Optimization Method of Relay Network Deployment Using Multi-UAV for Emergency Communication

Azaliya D. Ibrah, Liu Chuang*, Lv Na and Zhu Meng-yuan

School of informational and navigation, Air Force Engineering university, Xi'an 710077.China

*lc515626@formail.com

Abstract. The unique advantages of UAV with good maneuverability, ease of deployment and control can effectively meet the needs of emergency communications network construction. In order to solve the problem of effective communication security in large-area rescue areas, the idea of using multiple UAVs to construct an emergency relay network was proposed. Under the given mission area, the deployment of this network was studied. First of all, in order to fit the practical application, consider the channel fading to model the UAV coverage area, and propose a UAV relay network model based on connectivity probability. Then, according to the needs of emergency communication scenarios, from the perspective of simplifying the algorithm time complexity and guaranteeing the accuracy of the solution, an algorithm based on simulated annealing for deployment of UAV relay network is proposed. The simulation results show that the proposed algorithm has strong effectiveness and stability and can be applied to the deployment of UAV relay network in the context of emergency communications.

1. Introduction

In recent years, with the development of technology, uavs have greatly improved in terms of flight altitude, endurance time and load capacity. UAVs have the unique advantages of good maneuverability, low cost, ease of deployment and control, flexible communication networking, and easy updating of communication devices(e.g.[1],[2]). As an aerial platform, UAVs can be used to build an emergency communication relay network. Following the link, the user's communication coverage in the task area is achieved. Unlike previous missions that use a small number of uavs to complete single-hop or multi-hop relay missions, under large-scale disaster scenarios such as earthquakes and tsunamis, multiple uavs are needed to achieve effective communication protection for the entire rescue area. Emergency communications relay networks, and the current research in this area is less[3]. In commercial applications, at the Mobile World Congress in 2017, China Mobile and Huawei have cooperated to launch a tandem unmanned aircraft emergency communication high-altitude base station. This kind of uav base station can be quickly lifted into space within 5 minutes to solve the coverage radius of dozens. The 4G signal coverage problem within kilometers [4].From a practical point of view, the use of multiple uavs to build a large-scale emergency relay network is still in the exploratory phase. There are still many theoretical issues that need to be resolved in the form of network, network protocol design, and network planning.

In this paper, according to the actual application scenario, the idea of using relay UAV to construct emergency communication network is proposed, and the deployment of relay network is studied. This paper is organized as follows. In paper Section II we review related work .In Section III we formulate the problem,. A random deployment algorithm based on SA is designed in Section IV. In Section V we evaluate the algorithms via simulation. We summarize the results in Section VI.

2. Scenes and Modeling

Under normal circumstances, the coverage area of the relay UAV is determined by the altitude of the flight, the height of the coverage area and the maximum propagation distance of the relay signal, and the maximum propagation distance of the relay signal is related to the channel condition of the relay link. A single drone coverage model is shown in Figure 1.Without considering the propagation loss, the paper gives the relational expression of the geometric model of the high altitude UAV relay platform coverage area in [5]. The coverage radius r is

$$r = \sqrt{\frac{R(d^2 - h^2)}{R+h}} \tag{1}$$

The title is set 17 point Times Bold, flush left, unjustified. The first letter of the title should be capitalized with the rest in lower case. It should not be indented. Leave 28 mm of space above the title and 10 mm after the title. Among them, R is the radius of the earth, the height of the UAV is h, and the distance traveled by the d relay signal.

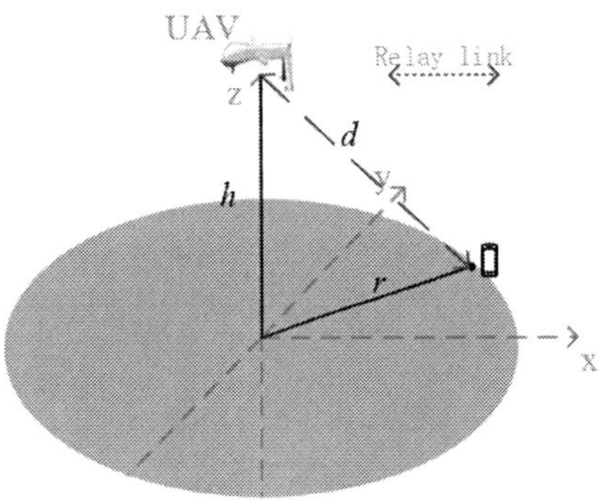

Figure 1. a relay UAV coverage model

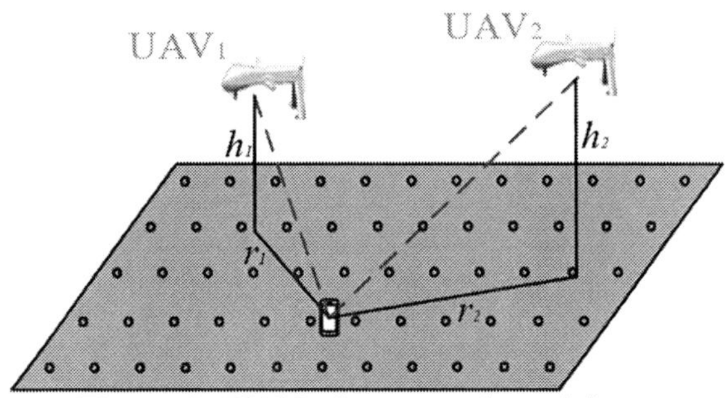

Figure 2. UAV relay network model

It can be seen from formula (1) that the coverage radius is related to the propagation distance of the relay signal, and the wireless signal propagation process is usually affected by three factors: path loss, multipath fading and shadow fading, resulting in maximum propagation of the relay link. The distance is usually uncertain. The uav relay link signal attenuation model can be expressed as:

$$L_f = (4\pi d/\lambda)^2 L_s L_m \tag{2}$$

In the formula: , respectively, the path loss based on the shadow and multipath random fading. According to the literature [6], the composite fading distribution of the fade propagation of the UAV relay link is:

$$f_{L_f}(l) = \int_0^{\infty} f_{L_m}(l|x) f_{L_m}(x|L_0)dx = \frac{2}{\Gamma(m)\Gamma(m_s)}\left[\frac{mm_s}{(4\pi d/\lambda)^2}\right]^{\frac{m+m_s}{2}} l^{\frac{m+m_s-2}{2}} \cdot K_{m_s-m}\left[2\sqrt{\frac{mm_s}{(4\pi d/\lambda)^2}l}\right], l>0 \tag{3}$$

Affected by propagation loss, the connectivity between the user and the relay node is indefinite and presents a probability distribution. Define the connectivity of ground users and relay nodes:

$$P_{con} = \int_0^{A/\gamma_0} f_{L_f}(l)dl = F_{L_f}(d)\Big|_{l=A/\gamma_0} \tag{4}$$

In the formula: P_t transmit power; G_t, G_r transmit antenna gains; ΔL extraordinary weather such as atmospheric refraction, rain fade, polarization, feeder loss and other factors; $F_{L_f}(d,l)$ to accumulate the probability density function for compound fading, substituting it into formula (4) yields:

$$d_{\max} = F_{L_f}{}^{-1} \bullet P_{con} \tag{5}$$

Substituting equation (5) into equation (1) yields the relationship between the connectivity probability and the coverage radius as:

$$P_{con_1} = \frac{1}{F_{L_f}^{-1}} \sqrt{\frac{r_1^2(R+h)}{R} + h_1^2} \tag{6}$$

The research in this paper is to build a scenario for emergency communications networks under large-scale disaster areas such as earthquakes and tsunamis. To study the convenience of the problem, it is assumed that all users are at the same altitude above ground, and the task area covered by the relay network is $M \in R^2$. In order to effectively evaluate the coverage performance of the relay network, the discretization method is used to divide the mission area into discrete points with equal spacing $M' \subset M$. The total number of discrete points is $|M'|$, The distance between adjacent discrete points is determined by the accuracy of the problem to be solved. Taking into account the different user density in different locations of the mission area, resulting in different network coverage requirements, set the weight of each discrete point p ($0 \le p \le 1$),It represents the size of the user density near the discrete point. Figure 2 shows the UAV relay network schematic.

Assume that the relay UAV node set is $U = \{u_i | i = 1,2,...,n\}$,n is the number of relay UAV, One relay UAV node in three-dimensional space $u_i(x_i, y_i, z_i)$. Then according to formula (6), any discrete point in the task area $m(x, y)$ .The probability of connectivity with this UAV is:

$$P_m = \frac{1}{F_{L_f}^{-1}} \sqrt{\frac{d(u_i,m)^2(R+h)}{R} + z_1^2} \tag{7}$$

$d(u_i,m) = \sqrt{(x-x_i)^2 + (y-y_i)^2}$ indicates the coverage radius of the relay UAV to the discrete point m, z_1 is the height of the UAV. Assume that there are n relay nodes in the network, and the connectivity rate between m and these UAV is $P(m_1)$, $P(m_2)$,…, $P(m_n)$,the probability of defining the location of point m to connect to the relay network is:

$$P_{con}(m) = 1 - \prod_{i=1}^{n}(1 - P(m_i)) \tag{8}$$

Definition 1. Effective coverage point: when $P_{con}(m) \ge P_{con}(th)$ call this point a valid coverage point, denoted as $N(x, y)$

$$R_{area}(U) = \frac{\sum p(x,y) \bullet N_{(x,y)}(U)}{|M'|} \tag{9}$$

The relay node coverage model shown in Equation (9) is often referred to as a probability coverage model. Compared with the traditional geometric model with constant coverage radius, the probability coverage model has more realistic characteristics. However, in the relay network deployment process, the probability coverage model is more complex than the geometric coverage model, and it is often difficult to adopt the geometric analysis method. Solve the optimal deployment location of relay nodes.

3. Random Deployment Algorithm Design

In the solving of the UAV relay network coverage optimization problem, the vector formed $U = \{u_i | i = 1,2,...,n\}$ by the position of the relay node set is used as the input parameter of the SA algorithm, and the area of the network effective coverage area is used as the optimization targetIn a three-. dimensional plane, randomly deploy n relay nodes, the search space dimension of the algorithm $q = 3n$, the population position vector can be expressed as $X_i = (x_{i1}^1, y_{i1}^2, z_{i1}^3, x_{i2}^1, ..., z_{in}^3)$. Each element is in turn represented by the x-axis, y-axis, and z-axis coordinates of the relay node.

Although there is no doubt about the ability of SA[7], it is guaranteed by a strict annealing plan. Specifically, it is a sufficiently high initial temperature, a slow annealing speed, a large number of iterations, and a sufficient number of disturbances at the same temperature. Therefore, the efficiency of SA has always been a major obstacle to the practical application of the algorithm. In order to solve the problem of deployment of UAV relay network, aiming at the defects of SA itself, this paper improves on the basis of SA and proposes an algorithm for deployment of UAV relay network based on simulated annealing(RND-ISA).The following key steps:

1) Population individual initialization. The population of individuals in the algorithm is a vector X consisting of q elements. The SA algorithm usually adopts a completely random initial population. Although this method can maintain good population diversity, it is easy to generate individuals with poor fitness, which is not conducive to algorithm convergence. Based on the major small discrete points right, the initial population distribution of weight values of several large area, so close to the initial population the right to re-value the larger node can effectively save the original running time of SA algorithm to improve the accuracy.

2) Y generated neighborhood solutions. In this paper, the elements of the current solution X are randomly changed during each iteration to obtain the neighborhood solution Y, and each element is selected from the corresponding sub-domain. The neighborhood solution Y is not only related to the current solution X, but also related to the current temperature T and the number of iterations m. The large variation of Y in the initial iteration is beneficial to maintain the diversity of solutions. In the later iterations, the small change of Y is beneficial to the convergence of the algorithm.

3) Acceptance probability *P*. The acceptance probability of the PSA algorithm is related to the size of the weight. By controlling the weight, the acceptance probability of a certain target can be increased or decreased. PSA is a probability *P* to accept the new solution.

$$P = \min(1, \prod_{i=1}^{N} \exp\{\frac{(R_{area}(X) - R_{area}(Y))}{T_i}\}) \tag{10}$$

4. Simulation Analysis

This section analyzes the optimization of the algorithm under conditions of moderate shadow fading. The weights of the discrete points are generated by using random parameters. The actual application can be set according to the population density. The distribution of relay nodes is shown in Figure 3. In the area of dotted lines in the figure, disaster relief areas are included. In order to compare the performance of the proposed algorithm, the basic SA algorithm, ant colony algorithm (ACO)[8] and artificial fish school algorithm (IFO) [9]were used in the same simulation experiment environment.

Table 1. PERFORMANCE COMPARISON OF 50 INDEPENDENT EXPERIMENTS

Algorithm	Coverage	variance
SA	91.13%	0.0557
CAO	93.33%	0.0710
IFO	92.61%	0.0371
RND-ISA	96.23%	0.0186

In order to verify the effectiveness and stability of the proposed algorithm, the optimization of network coverage is compared under the same task area and the number of relay nodes. The experimental results are shown in Table 1.It can be seen from Table 1 that the coverage optimization performance of the proposed algorithm is obviously better than the other three algorithms, and the variance of 50 results is smaller and the stability is higher.

Figure 4 shows the experimental fluctuations of 100 experiments. The exhaustive algorithm [10] (brute force, BF) is used to traverse the search for the combination of all the solutions during the experiment to find the optimal solution, as a comparison to verify the loss of the solution accuracy of the IPSA algorithm and computational advantage time, a straight line as shown in Figure 4. It can be seen that the accuracy of the improved SA algorithm is obviously higher than that of the basic SA algorithm and is closer to the optimal solution.

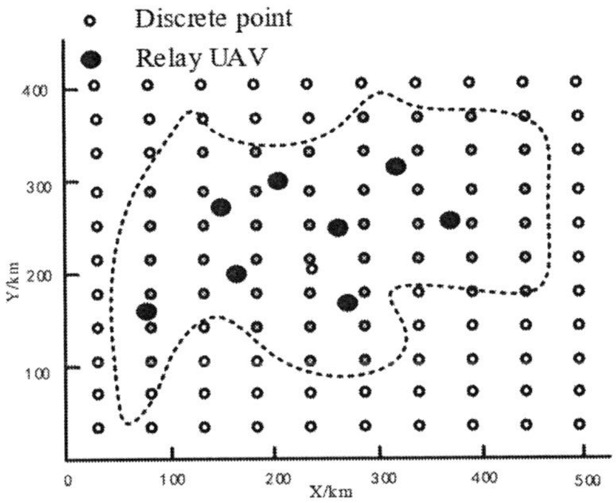

Figure 3. Schematic diagram of UAV deployment results

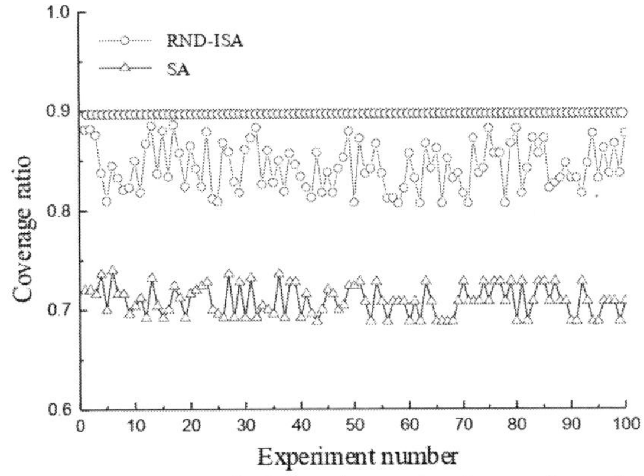

Figure 4. 100 independent experimental results

In order to verify the weight of different task regions and discrete points, the effectiveness of the proposed algorithm is proved. Using equal interval method to randomly generate multiple task areas within the area, using discrete points to represent the size of different task areas. The number of identical discrete points is averaged over 50 times to verify that the coverage of the task area changes and calculate time changes.

Figure 5 shows the relationship between the number of discrete points and the network coverage rate and the number of relay nodes. It can be seen from the figure that the greater the number of relay nodes, the higher the network coverage, but it is worth noting that with the relay With the increase in the number of nodes, the change in network coverage is not significant. This shows that selecting a suitable number of relay uavs for a given mission area is of great significance for optimizing network coverage performance, blindly increasing the number of uavs does not effectively increase network coverage.

Figure 6 shows the comparison of the calculation time at discrete points. It can be seen that as the task area increases, the calculation time of the four algorithms is increasing. The calculation time of the RND-ISA algorithm is much shorter than that of the CAO and IFO algorithms. When the number of discrete points is 120 or more, the average calculation of the RND-ISA is performed. The time is about three times that of the algorithm in this paper. By comparison, the proposed algorithm is more applicable to the deployment of relay networks in emergency communication scenarios.

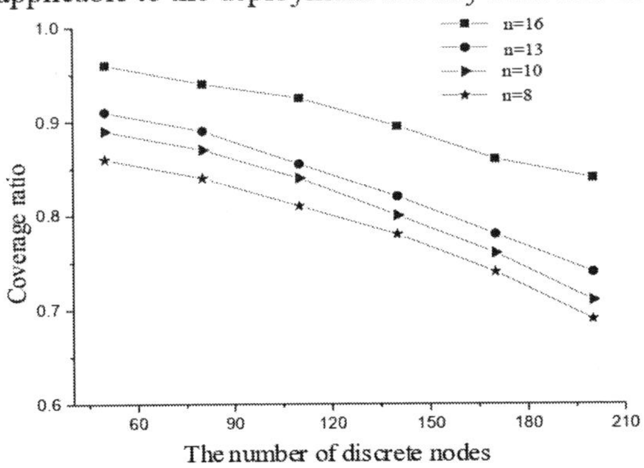

Figure 5. Schematic diagram of coverage changes

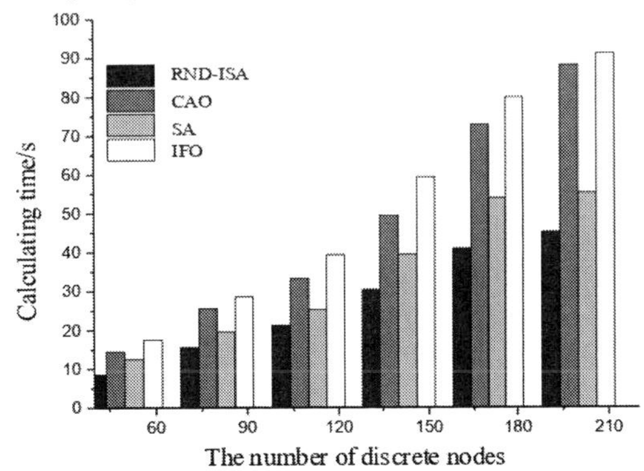

Figure 6. Schematic diagram of calculation time

5. Summary

In this paper, based on the background of relay network construction in disaster rescue scenarios, this paper proposes an emergency communications-oriented UAV relay network and studies the deployment of the network. From the perspective of practical application, a high altitude relay uav coverage model based on connectivity probability is proposed. From the scene requirements, an SA-based UAV relay network deployment algorithm is designed. The simulation results show that the algorithm has strong effectiveness and stability, and is more suitable for deployment of UAV relay networks in the context of emergency communications.

Acknowledgments

This work was supported by the National Natural Science Foundation of China under Grant No.61472443 and Research on the Application of Swarm Intelligent Induced Emergence in Aeronautic Swarm.

References

[1] Bupe P, Haddad R, Rios-Gutierrez F. Relief and emergency communication network based on an autonomous decentralized UAV clustering network[C]// Southeastcon. IEEE, 2015:1-8.
[2] Gao T, Lang F, Guo N. An Emergency Communication System Based on UAV-assisted Self-Organizing Network[C]// International Conference on Innovative Mobile and Internet Services in Ubiquitous Computing. IEEE, 2016:90-95.
[3] Kim S, Oh H, Suk J, et al. Coordinated trajectory planning for efficient communication relay using multiple UAVs[J]. Control Engineering Practice, 2014, 29(19):42-49.
[4] Jia Qunfeng. On the Problems and Countermeasures of Emergency Communication Support in China[J]. Network Security Technology and Application, 2018(1).
[5] Sun Shuai. Statistical model of UAV relay platform coverage area[J]. Engineering Technology: Abstract Edition, 2016(10): 00100-00100.
[6] Blasi L, Barbato S, D'Amato E. A mixed probabilistic-geometric strategy for UAV optimum flight path identification based on bit-coded basic manoeuvres[J]. Aerospace Science & Technology, 2017, 71.
[7] Senouci M R, Mellouk A, Aitsaadi N, et al. Fusion-based surveillance WSN deployment using Dempster–Shafer theory[J]. Journal of Network & Computer Applications, 2016, 64(C):154-166.
[8] Lee J W, Lee J J. Ant-Colony-Based Scheduling Algorithm for Energy-Efficient Coverage of WSN[J]. IEEE Sensors Journal, 2012, 12(10):3036-3046.
[9] Yu S, Wang R, Xu H, et al. WSN nodes deployment based on artificial fish school algorithm for Traffic Monitoring System[C]// Iet International Conference on Smart and Sustainable City. IET, 2011:1-5.
[10] An K, Sun K, Huang W, et al. Joint optimization of node selection and power allocation in wireless sensor networks[J]. 2014:2211 2215.

IOP Conf. Series: Journal of Physics: Conf. Series **1060** (2018) 012043 doi:10.1088/1742-6596/1060/1/012043

Research on Human-Computer Interaction Design of Self-service Public Facilities for Aging

Wen-Ting ZHENG and Qing-Jiang LIU

Initial Design Lab, No.155-2, Innovation Road, Shenyang City, China

zhr_zhengwt@163.com

Abstract. With the continuous improvement of the level of modernization, the public facilities that provide self-service have become more and more frequently appearing in people's life. On the basis of traditional human-computer interaction technology, the rise of multichannel human-computer interaction technology provides a pluralistic interaction mode between human and machine, and provides more choices for human-computer interaction design of self-service public facilities than before. However, the technical advancement should not be the primary term which is taken in consideration for the self-service public facilities. The purpose is to serve the public. The main features of the self-service public facilities are studied in this paper. This paper puts forward the design principle of "maximum compatibility", expounds the influence of aging social development trend on the popularization of self-service public facilities, and studies the design method of human-computer interaction to adapt to the current trend of aging.

1. Introduction

Led by science and technology, human-computer interaction achieves sustainable development, which brings endless challenges for designers. First of all, interdisciplinary characteristics of human-computer interaction design are becoming increasingly prominent, and multidisciplinary knowledge of different fields brings multiple barriers to designers; secondly, the factors of Human-Machine-Environment system influence each other and interact with each other, that the process is very complex, it is difficult to analyse the effects of multiple variables completely; thirdly, the speed of technology innovation is very fast, there is a contradiction between the advance and the applicability of human-computer interaction.

The purpose of self-service public facilities' establishment is to serve people, to make people satisfaction in a fluid way of interaction, and to achieve their goals. It's the ultimate goal of human-machine interaction design for self-service public facilities. The eyes of designers can be blindfolded easily by high-tech. Designers who call for advanced technology, have always put the cool experience into their design. They have ignored the original intention of human-computer interaction. Of course, it is very nice to experience so cool in Science and Technology Museum or exhibition hall or of some other entertainment facilities, there is no ground for blame. These aspects need such experience to stimulate the curiosity of the users and to impress the users. Our design can be beautiful and the method of interaction can be very new, but sometimes the experience can not be good.

The study of human-computer interaction is a very important process in design. The study of human-computer interaction can be divided into three categories: image research, experimental research and evaluation research.[1] The research on human-machine interaction of self-service public facilities is quite difficult, because of the complex user base. There are different gender, age and cultural background. Even if a large number of user studies have been done, the result may still be unsatisfactory.

Therefore, the design principle of "maximum compatibility" should be followed for the self-service public facilities. The so-called "maximum compatibility" is designed to meet the needs of as many people as possible. In the design of self-service public facilities, we expect more people to achieve the purpose of service by using, rather than adopting the "average" design method to our design.

2. Aging

2.1. *The trend of population aging*

In the human-computer interaction design of self-service public facilities, the problem of social aging can not be ignored. The elderly are one of the disadvantaged groups in the self-service public facilities. The self-service public facilities for the elderly can best reflect the two words of "public service". Allowing more people to use is one of the criteria for judging whether a design is successful or not.

The age distribution of the three times' census recently in China is shown in *Figure 1*. In 2010, the Sixth Population Census showed that 0-14 years old population accounted for 16.60%, 15-59 years old accounted for 70.14%, 60 years old and above accounted for 13.26%, of which 65 years and over accounted for 8.87%. Compared with the Fifth Population Census in 2000, the proportion of the 0-14 years old has fallen by 6.29%. The proportion of the 15-59 years old increased by 3.36%. The proportion of aged 60 and above increased by 2.93%, and the proportion of aged 65 and over increased by 1.91%. Compared with the Fourth Population Census in 1990, the proportion of the 0-14 years old dropped by 11.9%, and the proportion of aged 65 and over increased by 3.3%. [2]-[3]

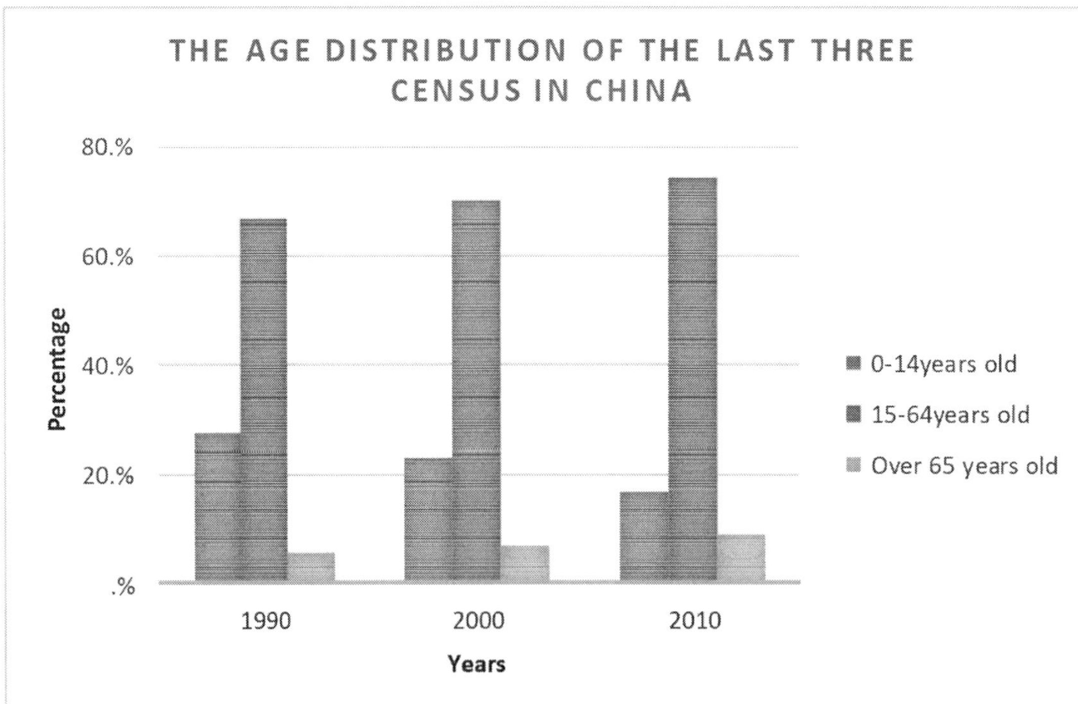

Figure 1. Figure with the age distribution of the last three census in China

2.2. *The relationship between aging and self-service public facilities*

The changes in the age structure of the population in China indicate that the fertility rate in China remains low, and the aging process is gradually accelerating. By 2010, the average life-span of our country had reached 75 years old. With the rapid economic and social development, the people's living standards and medical conditions are increasing, and the life-span of the population will be further increased. The average life of China's population is expected to reach 85 years old until 2050. [4]

Aging is a worldwide problem. Although all countries are introducing policies to deal with the negative effects of aging, maybe the average age or the proportion of elderly may be lowered, but the number of elderly population will not be decreased in the future.

People born in 60s and 70s, lack of knowledge and potential to accept new things. In addition, the social situation of The Empty Nest Man and Left-behind children is becoming more and more prominent, and many old people are faced with their own participation in social activities without the help of others. Therefore, in the design of self-service public facilities, we should pay attention to the influence of aging.

In fact, more than a dozen years ago, the study of computer technology for the elderly was started more than a decade ago. In the United States, AARP's Older Wiser Wired plans to provide education for the elderly and to provide guidance for the designers. There are also many plans and research projects in the European Union that provide support for computer technology for the elderly. [5]Our country is also putting forward relevant service policies for the market of the elderly. There is a report about the research of the human factors affecting the needs of the elderly, which issued by the National Research Council on aging,in which it defines the following definitions: elderly is the gradual change of physiological and psychological functions in a different way. As the age increases, the average ability to distinguish between the visual and auditory senses decreases, while the average intensity and speed of the response are decreased. Losing at least one kind of memory function, decreasing the sense of flexibility, slowing down the speed of "stimulus encoding", and getting difficulty in obtaining complex thinking skills. As age increases, visual function, including static object discrimination, dark adaptation, adjustment ability, contrast sensitivity and indirect vision, has decreased from the average level (Czaja, 1987)[5]

Through the survey, it is found that in the current trend of aging, users are faced with the following problems in the use of self-service public facilities.

One of the problems is the fast technological innovation. In today's rapid development of science and technology, many facilities had not been accepted completely and used by most users, then they have been updated. Take a simple example. The operating interface of the self-service device is changed from the initial button to the resistive touch screen, and then to the capacitive touch screen. Products of the previous generation have not yet been fully popularized, and ones of the new generation have been in blossom everywhere. Above all, there is no difference in appearance between the resistive touch screen and the capacitive one, but the way of working is quite different from each other. The resistive touch screen works with a sense of pressure, and the capacitive one works by using the human current. But many people do not understand the difference, especially for older users, with limited technical information. As a result, people often see that someone click the screen of the self-service equipment hard all the time, but there is not any response on the interface, making it mistaken for the machine to be broken. The fact is that using a fingernail to knock on a capacitive screen will not work. Of course, many people touch the resistance screen for half a day, until the atmosphere's strength is used to wake up the screen. Though there are some tips on the interface, most of them can't play a part in short time, because users can't really understand the real meaning of the tips, especially for the elderly users,who is difficult to adapt to the high speed technology revolution.

Another problem is diversification of the self-service device. In many places, such as stations, hospitals and cinemas, there is beginning to build self-service public facilities. Because there is still no industry standard, and different designs are designed to show their design features, design presents a variety. For the users, a variety of self-service public facilities are not good for making a Classical conditioning in use. Every time you accept a public service, you have to focus on what you want to do next. The more diversified the function of the self-service public facility is, the more complex the interaction process will be. Using a self-service device, because it is not familiar with the process, especially for the elderly users, it takes more time than young adults in order to achieve the same goal.

The last not the least problem is the disorder for interaction. The construction sites of self-service public facilities are mostly crowded places, and voice interaction is often not effective. Harsh or dark has a great influence on visual interaction. Especially for middle-aged and elderly users, the visual and hearing levels of most people will decline with aging. It is the reason that many people do not want to use self-service equipment. In the design of the human-computer interaction interface of the self-service public facilities, it is not only for the elderly, but also simple and easy to use. It is the criterion of interactive interface design.

So how do you make the "maximum compatibility" design? Since old people can't accept new things, is it possible to avoid high-tech and new-tech deliberately in the design? This recognition has gone into a misunderstanding. Although the old people have poor ability to accept new things, in fact, the old people can't accept new technology, the farther away from the new technology. The more disconnected to the times, the worse it gets. Not only bring the social service burden, but also create a sense of frustration to the heart of the old man.

3. The method

The purpose of the new interactive design for the elderly is to extract the understanding and acceptance of the old people from the new technology.

3.1. Overcoming the psychological cognition of "I do not understand"

It looks so complicated that it makes people stand there even like a clown, that is why the most people is afraid to try to use it. In the development of the society, many social problems resulted in that the elderly has become the "Empty Nest Man", the children has become the "Left-behind children",in the Empty Nester City, there are very few opportunity for the elderly to communicate with people, so they gradually lost the confidence to communicate with others, think of themselves as the social burden from cognition, and know nothing. So before the design and construction of the self-service public facilities, it is necessary to research on the potential motivation of the old people, how to inspire the elderly and how to attract them to use actively,that will help the elderly gradually overcome the psychological cognition, “I don't understand”, to establish their confidence to contact with fresh things, and to improve the participation of the elderly social activities.

Studies have shown that increasing the compensatory function of the brain by stimulating of new things can prevent brain dysfunction and reduce the incidence of cognitive impairment and Alzheimer's disease in the elderly. Therefore, in the design of human-computer interaction of self-service public facilities, we should try to avoid overly deep description, which is simple and easy to understand, and is the difference between service facilities and artworks.

3.2. Building the effective operation guidance

Through investigation, an interesting discovery are often found. Standing at the head of the long queue, a person has been working on the self-service equipment for a long time, and always fails to achieve the goal, which leads to the impatience of the people behind the queue, even the sound of grievance. The operation is repeated but unsuccessfully for some reason, most of which is the failure of interactive design. The guidance of the operation does not give an effective guidance. Lucky people to complete the operation with the help of others, some people give up. The geriatric feeling of frustration is stronger than the young. How dose it succeed in building effective operation guidance? Take the self-service registration machine of the hospital as an example. There will be a department option on the front page of the self-service registration machine, and a department will be selected.

After entering the department information, there are doctors' messages. It sounds like very easy to make a choice. But why do you still have a lot of duplication?

3.2.1. To be described from the user's point of view. Doctors are familiar with human structures and medical names, but direct guidance is more effective for patients with very different cultural backgrounds. Different departments of the hospital's setting is not exactly the same, it is hard to determine which department should the patient choice, and not as good as ask them straightly "what's the matter with you?". It may be easier to describe the patient's choice to be treated by a patient's perspective.

3.2.2. Priority of important information. As for the description of the doctor's information, in a limited area, it is a good choice to present the important information what people want to know most.

For example, doctors' professional direction, professional title, queuing number and cost are more concerned for patients, and resume information is not that important, such as published academic papers, honors and so on.

3.2.3. Clear the language. The "next" and "continue" are common operation guides, but such guidance is the most meaningless. For example, when a doctor has been chosen, "the next step" button has appeared.It is better to show directly"Are you sure to determine the medical treatment and pay for registration fee 10 yuan? You can scan the two-dimensional code to pay for it." Some guides like "sign in" and "sign up" are always wonders for the user. So clear the language and avoid these confused words.

3.2.4. Let the user do as little as possible. We should make full of use of the computing power and the big data system of the machine, and try to use the default items as much as possible. Users make a choice by their own, not only increases the psychological pressure of users, but also reduces the efficiency of interaction.In a word, the options provided to the user can not be ambiguous, incomprehensible, and intersecting. Such effective guidance is a successful design, and it is also the original purpose of the user experience and human-computer interaction design.

4. Summary

Human services are being replaced by self-service machines in an increasing fields, which liberates more and more labor force. It is a tendency to be unable to hold back the popularization of science and technology that machines work on the behalf of human beings, and become intelligent. There is no problem in pursuing the development of science and technology, but it is worth thinking about the "degree" of the application of science and technology in our life. What is the role of self-service machines playing in the human society of the future? How do machines participate in human social life? Is the purpose to make machines serve humans or replace humans with technology? There are many problems like this. As long as science and technology are developing continuously, human beings will not be able to avoid these problems.

Human-computer interaction usually lacks the temperature of human communication. Indifference and apathy are not what we expect. Followed the design principle of "maximum compatibility", it is in order to meet the needs of people for the premise, not designed for technology, because ultimately we hope that human machine dose not to say "Hey,do not I look great? ", but "Excuse me, what can I do for you?"

Public facilities are also different from individual goods. Social factors are considered as one of the most important studies of self-service public facilities. As the Times change, social factors will not remain the same. There is no absolute criterion for human-computer interaction, only the same purpose.

References

[1] Wei Liu, Xiu-Gan Yuan 2008 *Design and evaluation of human-computer interaction* (Beijing: Posts and Telecommunications Press) p65
[2] National Bureau of Statistics 2011 *Main data communiqu for the sixth national census in 2010*

(http://www.stats.gov.cn/tjsj/tjgb/rkpcgb/qgrkpcgb/201104/t20110428_30327.html) p1
[3] National Bureau of Statistics 2001 *Main data communiqu for the sixth national census in 2000* (http://www.stats.gov.cn/tjsj/tjgb/rkpcgb/qgrkpcgb/200203/t20020331_30314.html) p1
[4] Wei Liu, Xiu-Gan Yuan 2008 *Design and evaluation of human-computer interaction* (Beijing: People's Medical Publishing House) p15
[5] More referencesShneiderman B and Plaisant C 2004 *Designing the User Interface: Strategies for Effective Human-Computer Interaction* (US: Pearson Education) p152

Node Placement in WSN for Rail Track Monitoring System

Anqi Lv[b], Cuiran Li[a], Jianli Xie and Wenqian Guo

School of Electronics & Information Engineering Lanzhou Jiaotong University Lanzhou, China

e-mail:[a] licr@mail.lzjtu.cn, [b]2810282717@qq.com

Abstract. Railway environmental monitoring is essential to ensure the safe operation of trains. WSN has great potential in environmental detection, space exploration, smart home, military and other fields. In this paper, The WSN monitoring network is built by deploying relay/sensor nodes along two sides of the rail, to realize the railway environmental monitoring. The problem of "energy hole" caused by uneven energy consumption of nodes in linear WSN is discussed, and a non-uniform distance correlation deployment strategy is designed. In the strategy, the rectangular monitoring region is divided into multiple triangular region. And, the relationship among the number of sensor nodes, the area of unit triangle, the distance between the triangle vertices (relay nodes) and the energy consumption of relay nodes is analyzed. The simulation results have showed that the proposed algorithm has the smaller energy consumption gap between relay nodes, compared with uniform and non-uniform uncorrelated deployment algorithms. It ensures the stability of the whole network energy consumption, and has better performance in prolonging the network lifetime.

1. Introduction

Railway technology has been developing rapidly, and the speed of high-speed train has reached 350km/h, so the environmental monitoring system on both sides of rail is important to ensure the operation of trains. WSN has the characteristics of simple deployment, low maintenance cost, flexible networking, and relay et al. The WSN-based rail monitoring system can be used to collect environmental data efficiently and reliably, so as to make correct protection measures for train operation. The nodes in WSN are usually battery powered and cannot be replaced. The data collected by sensor nodes will been transmitted to the sink node through multi-hop and forwarding, the network lifetime depends on the condition of nodes' energy consumption . Particularly, the WSN along the rail is a linear structure, which leads to uneven energy consumption of nodes, and shortened the network lifetime. The node which is closest to the sink node transmits largest data and its energy will be used up first. The phenomenon is called "energy hole".

There has been a lot of research on solving the "energy hole" and prolong the network lifecycle. A new cluster mechanism is suggested in [1], to balance the network energy consumption. In [2], the mobile relay nodes are introduced to share the load of nodes near the base station, thereby prolonging the network lifetime. Two node deployment schemes are proposed in [3], in which the effect of node density and distance between nodes on the network lifecycle is analyzed. In [4], the expected cost of energy consumption is defined. And, the expected cost is reduced by adjusting the communication radius of sensors. In addition, an adaptive radius adjustment algorithm is put forward in [5], and the network energy consumption is decreased by adjusting the transmission power. An efficient and equidistance deployment method is suggested in [6], for linear node placement. In [7], the density and location of relay node are derived, by balancing the power consumption between relay node and sensor node. In [8], a non-uniform node placement algorithm is proposed, which prolongs the network lifetime by adjusting the number of nodes proportionally according to area of the monitoring region, and the network lifetime is prolonged.

The existing node deployment algorithm does not take into account the influence of changing distance between adjacent relay nodes on the network energy consumption, which restricts the further performance improvement. In the paper, a non-uniform distance correlation deployment strategy is proposed, in which the relationship of the unit monitoring region, the number of sensor nodes, the distance between the triangle vertices (relay nodes) and energy consumption of relay node is analyzed. The proposed strategy optimizes the transmission distance of relay node, and avoids the phenomenon of "energy hole" caused by uneven energy consumption. In addition, it prolongs the network lifetime.

2. Node Placement Model

The node placement along the railway track is divided into two types: uniform deployment and non-uniform deployment. In the uniform deployment, the distance between adjacent relay nodes on the same side is equal, and the number of sensor nodes in unit monitoring region is the same, shown in Figure 1.

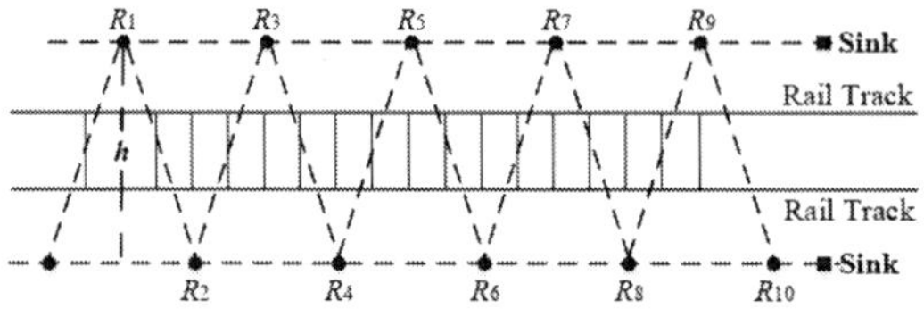

Figure 1. Uniform deployment.

In our non-uniform distance correlation deployment strategy, the distance between relay nodes is not equal, and the distribution of the relay nodes along two sides of the track follows the "from far to near" placement, depicted in Figure 2.

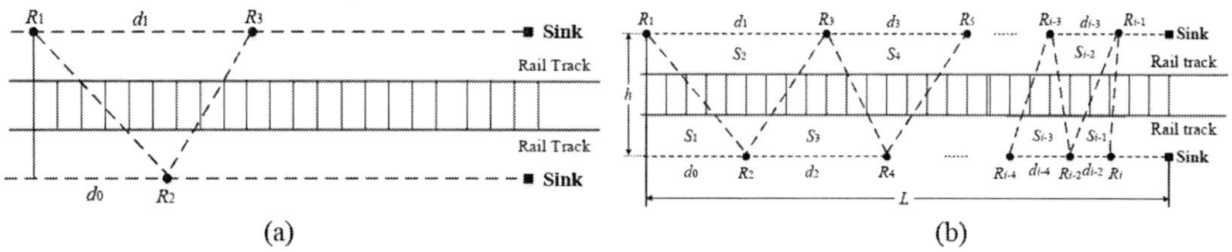

Figure 2. Non-uniform distance correlation deployment.

In Figure 2, the monitoring region along the track is a rectangular with length of L and width of h. Let R_1, R_2,...be the relay nodes, S_1, S_2,...is the area of unit triangle, N_1, N_2,...represents the number of sensor nodes in unit triangle, and d_0, d_1,...is the distance between relay nodes. In the node placement model, the data collected by sensor nodes is transmitted by relay nodes along one side of the rail track, i. e., $R_1 \rightarrow R_3 \rightarrow$...Sink, and $R_2 \rightarrow R_4 \rightarrow$...Sink.

3. Non-uniform Distance Correlation Deployment Algorithm

The phenomenon of "energy hole" is due to the excessive energy consumption gap between relay nodes in WSN. When the date acquisition rate of sensor node is the same for each unit monitoring region, the energy consumption gap among relay nodes will be reduced and the network lifetime is prolonged. In order to ensure the same information acquisition rate, the number of sensor nodes in unit monitoring region N_i and the area of unit monitoring region S_i need to meet the recursive equation, (1)

$$\begin{cases} N_2 / N_1 = S_2 / S_1 = d_1 / d_0 \\ N_3 / N_1 = S_3 / S_1 = d_2 / d_0 \\ \vdots \\ N_i / N_1 = S_i / S_1 = d_{i-1} / d_0 \end{cases} \tag{1}$$

Assume that the data acquisition rate of sensor node in unit time is k, then the energy model of relay node can be expressed by [9]

$$E_T = \begin{cases} kE_{elec} + k\varepsilon_{fs} \cdot d^{\alpha} & d < d_0 \\ kE_{elec} + k\varepsilon_{amp} \cdot d^{\alpha} & d \geq d_0 \end{cases} \tag{2}$$

And

$$E_R = kE_{elec} \tag{3}$$

where E_T is used to represent the energy consumption for transmitting data, E_R is used to represent the energy consumption for receiving data, of relay node, the value of α is 2 in free space model, or 4 in multipath fading model, d is the distance between transmission and receiving, E_{elec} is the energy consumed by transmitting circuit, ε_{fs} is the power amplification factor in free space model, ε_{amp} is the power amplification factor in multipath fading model, and d_0 is the distance threshold, expressed by

$$d_0 = \sqrt{\varepsilon_{fs} / \varepsilon_{amp}} \tag{4}$$

E_1 is the energy consumption of relay node R_1 per unit time, which is written as

$$E_1 = E_{r1} + E_{t1} = 2N_1kE_{elec} + N_1k\varepsilon_{fs} \cdot d_1^2 \tag{5}$$

Where E_{r1} is the energy used during receiving the data from region S_1 and E_{t1} is the energy used during transmitting the data to R_3. And, the energy consumption of relay node R_2 per unit time is given by

$$E_2 = 2N_2kE_{elec} + N_2k\varepsilon_{fs} \cdot d_2^2 \tag{6}$$

The energy consumption of relay node R_3 per unit time consists of the energy for receiving the data from region S_3, the energy for receiving the data from R_1, and the energy for transmitting the data to R_5, which is written as

$$E_3 = 2(N_3+N_1)kE_{elec} + (N_3+N_1)k\varepsilon_{fs} \cdot d_3^2 \tag{7}$$

Then, for j=1,2,…, the energy consumption of relay node per unit time is obtained as

$$\begin{cases} E_{2j\text{-}1} = 2(N_{2j-1} + N_{2j-3} + +\cdot N_1)kE_{elec} + (N_{2j-1} + N_{2j-3} + +\cdot N_1)k\varepsilon_{fs} \cdot d_{2j-1}^2 \\ E_{2j} = 2(N_{2j} + N_{2j-2} + +\cdot N_2)kE_{elec} + (N_{2j} + N_{2j-2} + +\cdot N_2)k\varepsilon_{fs} \cdot d_{2j}^2 \end{cases} \tag{8}$$

Introduce variable $M_{2j\text{-}1}$ and M_{2j}, (j=1,2,…,), which is defined as

$$\begin{cases} M_{2j-1} = \dfrac{N_1}{S_1}\sum_{i=1}^{j} S_{2j-1} = \dfrac{N_1}{d_0}\sum_{i=1}^{j} d_{2j-2} \\ M_{2j} = \dfrac{N_1}{S_1}\sum_{i=1}^{j} S_{2j} = \dfrac{N_1}{d_0}\sum_{i=1}^{j} d_{2j-1} \end{cases} \tag{9}$$

Substituting equation, (9) in equation, (8), then energy consumption E_i is

$$E_i = 2M_ikE_{elec} + M_ik\varepsilon_{fs} \cdot d_i^2, \quad i\text{=}1,2,\cdots \tag{10}$$

When the relay nodes have equal energy consumption, the whole network is in equilibrium state. If d_0 and d_1 are given (The values of d_0 and d_1 are determined by simulation experiment, see Section 4), d_2 can be calculated by

$$d_2 = \left(d_0d_1 + 2\frac{(d_0 - d_1)}{d_1} \cdot E_{elec} / \varepsilon_{fs} \right)^{1/2} \tag{11}$$

Furthermore, for j=1,2,…, the distance between relay nodes is given by

$$\begin{cases} d_{2j+1} = \left(\dfrac{N_1}{M_{2j+1}} d_1^2 - 2\left(1 - \dfrac{N_1}{M_{2j+1}}\right) E_{elec} / \varepsilon_{fs} \right)^{1/2} \\ d_{2j+2} = \left(\dfrac{N_1 d_1}{M_{2j+1} \cdot d_0} d_2^2 - 2\left(1 - \dfrac{N_1 d_1}{M_{2j+1} \cdot d_0}\right) E_{elec} / \varepsilon_{fs} \right)^{1/2} \end{cases} \quad (12)$$

Let L be the length of monitoring rail track, the sum of distances between relay nodes need to satisfy the constraint equation, (13)

$$\min\left(\sum_j d_{2j}, \sum_j d_{2j+1} \right) \geq L, \quad j=0,1,2,\cdots \quad (13)$$

In the non-uniform distance correlation deployment model, when the distance between relay nodes satisfies equation, (11) ~ (13), the energy balanced WSN can be obtained.

It is assumed that E_0 is the initial energy of the relay node, t_i is the lifecycle of relay node i, and T is the network lifetime, which is usually regarded as the lifetime of the earliest dead node due to battery power consumption. And, t_i and T is expressed respectively by

$$t_i = E_0 / E_i \quad i = 1,2,\cdots \quad (14)$$

$$T = \min\{t_i\} \quad i = 1,2,\cdots \quad (15)$$

4. Simulation Results

In simulation experiments, the monitoring region along the track is a rectangular with length of L and width of h. The relay nodes are placed in non-uniform distance correlation deployment strategy. The values of the simulation parameters are provided in Table 1[9].

Table 1. Parameters and values

Parameters	Values
L(m)	1000
h(m)	2.5
ε_{amp}(pJ/bit/m^4)	0.0013
ε_{fs}(pJ/bit/m^2)	10
E_{elec}(nJ/bit)	50
E_0(J)	2
k	5

Firstly, we need to determine the value of d_0 and d_1. Figure. 3 illustrates the relationship between d_0 and d_1.

Figure 4 shows the number of relay nodes and sensor nodes required for different values of d_1. It can be seen that when d_1=275m, the number of nodes required is the least. We know the number of sensor nodes in unit triangle N_i obtained by equation, (1), might not be an integer. In the actual deployment, top integral function is needed for N_i.

Figure 5 gives a comparison of theoretical energy consumption and the actual one of relay nodes with various values of d_1.

Figure 6 reveals the relationship between different values of d_1 and network lifetime. Obviously smaller values of d_1 generate the longer network lifetime. However, as in Figure 4, the number of nodes required for smaller values of d_1, is not the least. Therefore, the number of nodes should be taken into account, in addition to the network lifetime, in determination of values of d_0 and d_1.

The comparisons of energy consumption and relay node lifecycle for three node deployment algorithms are shown, respectively, in Figures 7 and 8. In the figures, the number of relay node is 16, with d_0=171m, d_1=275m. It can be seen that compared with uniform and non-uniform uncorrelated deployment [10, 11], the proposed non-uniform distance correlation algorithm has the least energy consumption gap among relay nodes, and the longer node lifecycle.

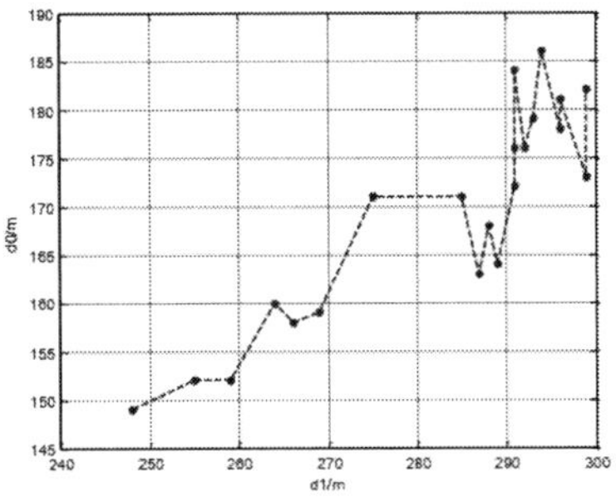

Figure 3. Relationship between d_0 and d_1 required

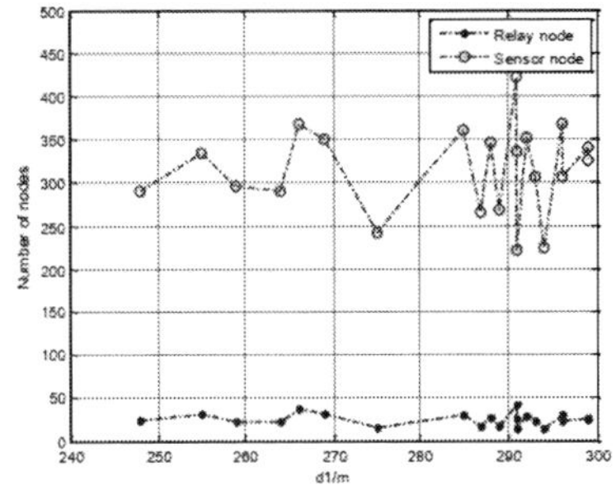

Figure 4. Number of relay nodes and sensor nodes

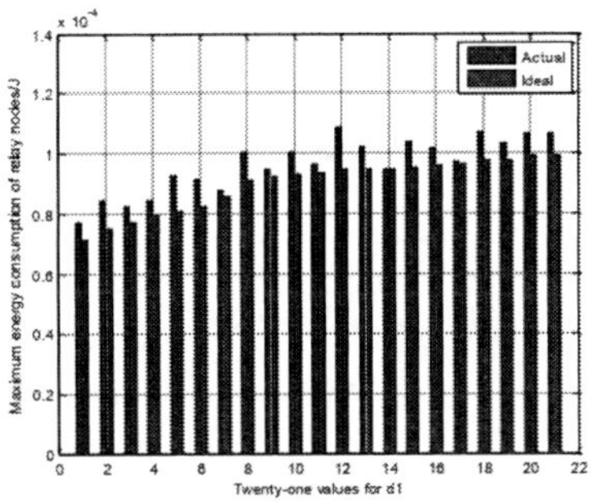

Figure 5. Comparison of theoretical and actual energy consumption of relay nodes for various values of d_1

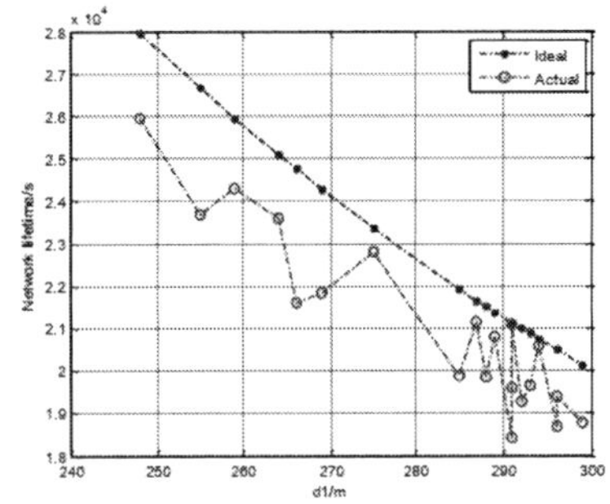

Figure 6. Relationship between different values of d_1 and network lifetime

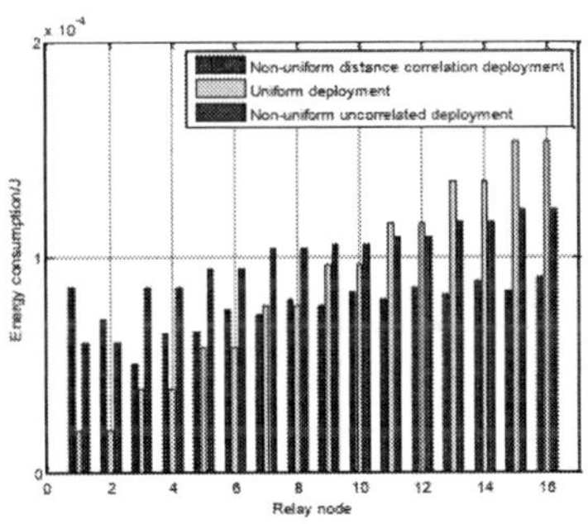

Figure 7. Comparisons of energy consumption for different deployment strategies

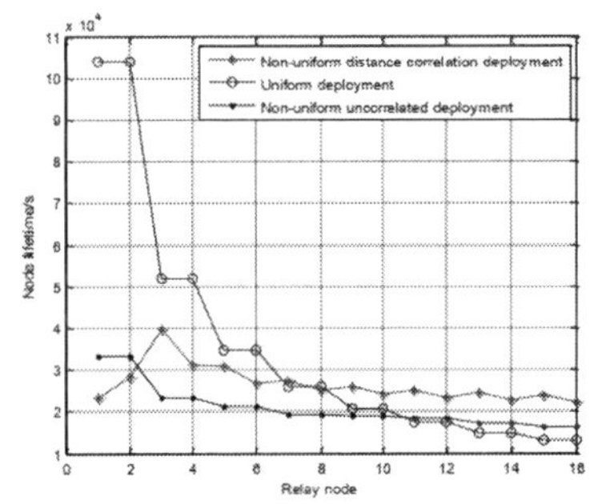

Figure 8. Comparisons of relay node lifecycle for different deployment strategies

5. Conclusion

The WSN along the rail track is a linear structure, which leads to uneven energy consumption of relay nodes, and shortens the network lifetime. The non-uniform distance correlation deployment strategy is proposed in this paper, to overcome the "energy hole". Compared with uniform and non-uniform uncorrelated deployment algorithms, the proposed deployment strategy has the least energy consumption gap among relay nodes, and the longer node lifecycle.

For future research, we need to investigate the effect of different path loss factors on energy consumption of relay node in energy model, so that the proposed strategy can be applied to the actual rail environment monitoring system.

Acknowledgment
This work is partially supported by the National Natural Science Foundation of China (61661025, 61661026), and Foundation of A hundred Youth Talents Training Program of Lanzhou Jiaotong University.

References

[1] Venniraselvi G and Manoharan R 2014 A survey of energy efficient unequal clustering algorithms for wireless sensor networks *International Journal of Computer Applications* **79** pp 1-4
[2] Wang W, Srinivasan V and Chua K C 2005 Using mobile relays to prolong the lifetime of wireless sensor networks International Conference on Mobile Computing and NETWORKING Cologne Germany ACM pp 270-283
[3] Hong L and Xu S 2010 Energy-Efficient Node Placement in Linear Wireless Sensor Networks International Conference on Measuring Technology and Mechatronics Automation Changsha China IEEE 2 pp 104-107
[4] Dong P and Gao S 2008 Adjustment of Transmission Radius in Linear Wireless Sensor Networks International Conference on Wireless Communications NETWORKING and Mobile Computing Dalian China IEEE pp 1-4
[5] Elleuchi M, Boujeleben M and Jmal W, et al 2017 Energy aware linear sensor placement scheme for water pipeline monitoring Computer Systems and Applications Agadir Morocco IEEE pp 1-8
[6] Wang S G, Yang L, Liu M C and Automation S O 2016 Adaptive radius adjustment algorithm of wireless sensor networks coverage Transducer & Microsystem Technologies 35 pp 137-139
[7] Lu K, Liu G and Mao R, et al 2011 Relay node placement based on balancing power consumption in wireless sensor networks Iet Wireless Sensor Systems 1 pp 1-6
[8] Guo A M and Zhong J 2012 Realization of subprime network energy consumption equilibrium in linear wireless sensor network Computer Engineering 38 pp 93-92
[9] Hu Y, Niu Y G and Zou Y Y 2017. A zone-based unequal multi-hop clustering algorithm in wsns. control & Decision 32 pp 1695-1700
[10] Lv X, Li J and Shi, T,et al 2016 Topology analysis based on linear wireless sensor networks in monitoring of high-speed railways Chinese Control and Decision Conference YinChuan China IEEE pp 1797-1802
[11] Duan J, Shi T and Lv X, et al 2016 Optimal node deployment scheme for WSN-based railway environment monitoring system Control and Decision Conference YinChuan China IEEE pp 6529-6534

Survey & Research on Mobile Terminals in Classroom Teaching

Li Li

School of Humanities and Social Science, Guangzhou Civil Aviation College, Guangzhou, Guangdong, China, 510403

E-mail: lili@caac.net

Abstract: Under Internet Plus environment, mobile terminal, taking mobile phone as a representative, is gradually participating in classroom teaching. By conducting a survey on 1,598 vocational college students, current classroom teaching is revealed and problems are exposed. Students were tested from 5 modules (mobile phone reliance; selection of the curriculum; subjective understanding; reasons and purpose; suggestions and solutions) with 10 specific questions. Some interesting and aspiring factors have been discovered, which illustrated there was a gap between the cognition of using mobile terminals in classroom teaching between students and teachers. Strategic analysis and solutions are proposed based on the survey. School management needs to be enhanced and the curriculum design should be optimized. There is a high demand in innovative classroom management from students and co-building a new class agreement in mobile terminal environment is highly recommended. There is also room for classroom teaching structure rebuilding; blended teaching module such as MOOC, SPOC and etc. provides more effective solutions. Meanwhile the mobile terminals should be redefined by teachers and teachers should guide the students to use mobile terminals positively and constructively in classroom to assist and promote classroom teaching. With in-depth investigation and big data analysis, the paper tries to figure out how mobile terminals currently role-played in classroom teaching and brings out practical suggestions.

1. Introduction

Representing by mobile phone, mobile terminal is penetrating into our daily life with its smartness, effectiveness and convenience. According to the Annual Report on Society Mentality of China 2015 co-released by Institute of social China Academy of Social Science and Social Sciences Academic Press, average time spent on smart phones of college students in China is around 5 hours 17 minutes, which occupies 22% of a day. [1] In education area, mobile terminals entering classroom teaching becomes a global phenomenon. Competing with mobile phones to seize students' interest develops into a worldwide problem and brings new challenges to educators. With China's entering into the era of Internet Plus, colleges and universities are building and upgrading smart campus from digital campus based on internet and information technology innovation. Mobile phone, as a representative of mobile terminal, gradually integrates with classroom teaching and becomes an indispensable part of college classroom. Chen Huaizhi once pointed out: the abuse of mobile phone took huge crash and challenges to classroom teaching organization and management in universities. Some college teachers even considered mobile terminals as "monsters in class". [2]

IOP Conf. Series: Journal of Physics: Conf. Series **1060** (2018) 012045 doi:10.1088/1742-6596/1060/1/012045

2. Survey

To better understand the status quo and how mobile phones affect classroom teaching, a survey was conducted on students enrolling in 2016 in a vocational college. Altogether 1,598 questionnaires were distributed and received 1,315 back. There were 1,296 effective questionnaires, which was 82% of the total. There were 5 modules in this questionnaire as shown below in Table 1, with detailed questions included in each module.

Table 1 Modules of the Survey

Module	Detailed questions
1. Mobile phone Reliance	The frequency of using mobile phone in class
	Specific time section of using mobile phone in class
	Time spent on mobile phones in class
	Students' mentation without mobile phones
2. Selection of the curriculum	In what subject class students use mobile phone more
3. Subjective understanding	How students estimate the behavior of using mobile phone in class
	How students evaluate the influence of using mobile phone in class
4. Reason and purpose	The reason why students use mobile phone in class
	What do students use mobile phone to do in class
5. Suggestion and solutions	How students suggest to solve the problem

3. Findings

3.1. Mobile Phone Reliance

3.1.1 The frequency of using mobile phone in class

Table 2 Frequency of using mobile phone

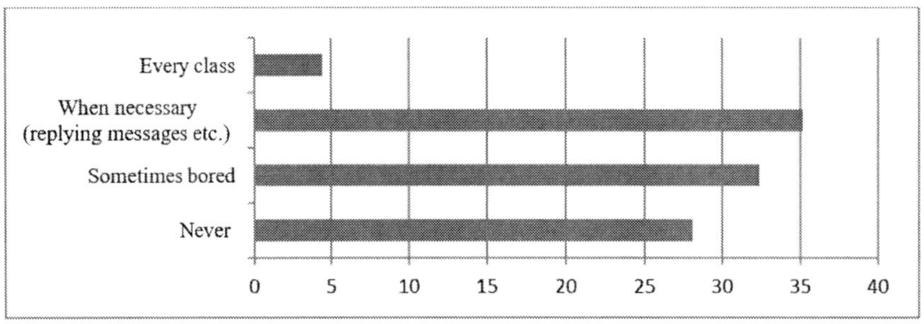

As shown in the Table 2, 28.06% of the students never use mobile phones in class which tells that the majority would use mobile phones in class. The ratio of using mobile phones every class is rather small, only 4.41%. For the rest part, 35.16% will use their mobile phones when necessary, for example checking or replying messages. The other 32.37% will use mobile phones for fun when they feel bored. This two figures weight almost alike.

3.1.2 Specific time section of using mobile phone in class

IOP Conf. Series: Journal of Physics: Conf. Series **1060** (2018) 012045 doi:10.1088/1742-6596/1060/1/012045

Table 3 Specific time section of using mobile phone

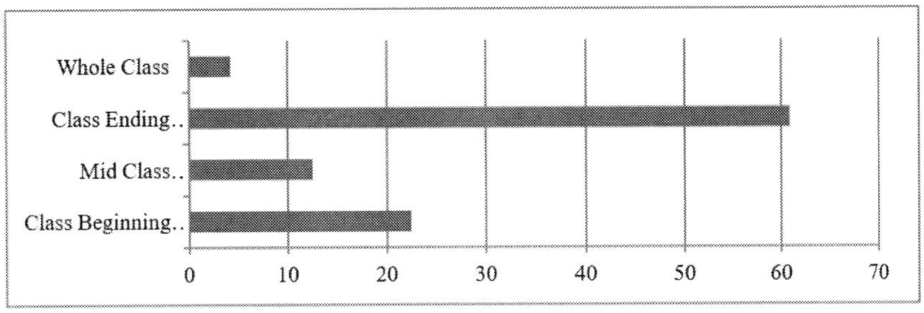

As shown in Table 3, the class is divided into 3 sections to better understand in which section students use mobile phones more: class beginning (about the first 10 minutes); mid class (the middle part of teaching from around the 20th to the 35th minutes); class ending (the last 10 minutes). Results illustrate that 22.47% will begin to use mobile phones at the very beginning of a class. In the mid-class section, the figure drops to12.47% but suddenly climbs to 60.86% at the class ending section. Still 4.19% of the students use their mobile phones in the whole class.

3.1.3 Time spent on mobile phones in class

Table 4 Time spent on mobile phone in class

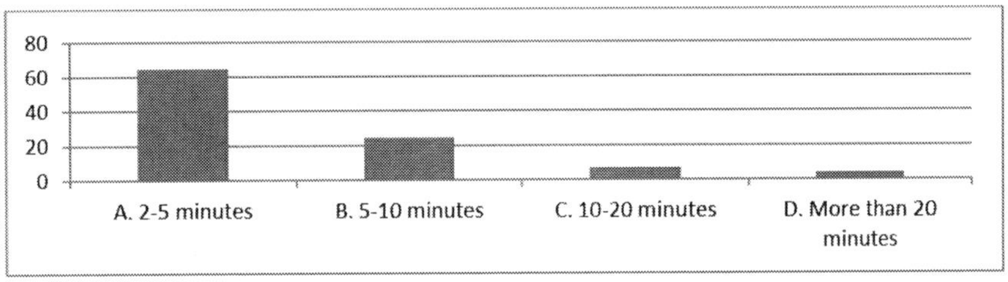

According to the survey and Table 4, 64.74% are carefully controlling the time of using mobile phone between 2 to 5 minutes. 24.3% of the students will use mobile phone 5 to 10 minutes in class. 6.88% will spend 10 to 20 minutes while 4.09% will spend more than 20 minutes. The time of using mobile phones goes up; the percentage of students goes down.

3.1.4 Students' mentation without mobile phones

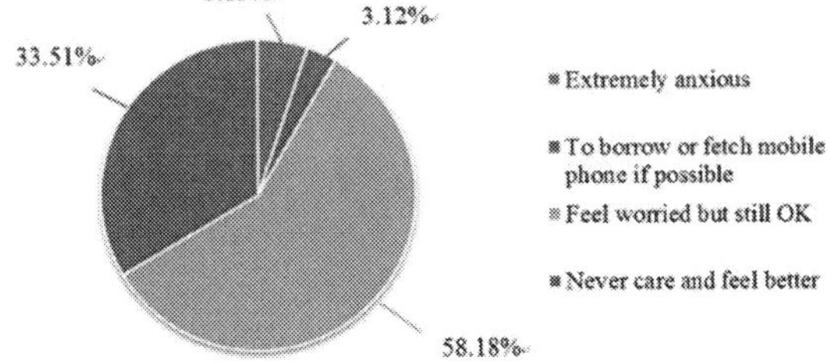

Figure 1 Students' mentation without mobile phones

Finding from Figure 1, if there is no mobile phone, 58.18% of the students feel worried but are still able to continue class study. 33.51% never care and even feel better to learn without mobile phones. 3.12% will try to borrow or fetch their phones if possible. A small group, about 5.19% will be extremely anxious and unable to study.

3.2 Selection of the curriculum: In what subject class students use mobile phone more

Table 5 Selection of the curriculum

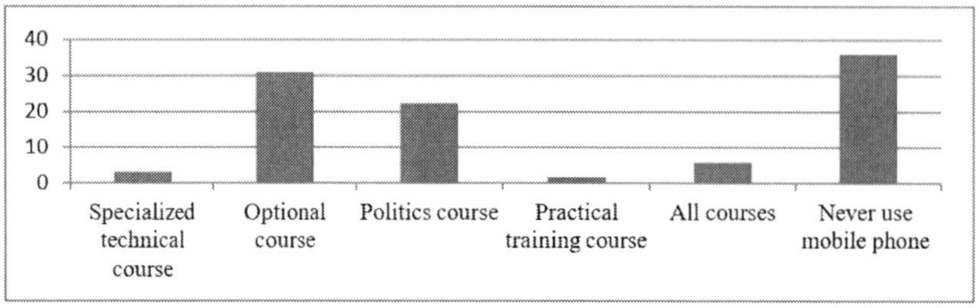

The curriculum generally contains specialized technical course, optional course, politics course and practical training course. Based on results in Table 5, there are 3.12% and 1.72% using mobile phone in specialized technical course and practical training course. Bigger number comes to optional course and politics course, which are 30.79% and 22.37% respectively. About 36% of the students never use mobile phone in all courses while 5.81% use mobile phone in all courses.

3.3 Subjective understanding

3.3.1 How students estimate the behavior of using mobile phone in class

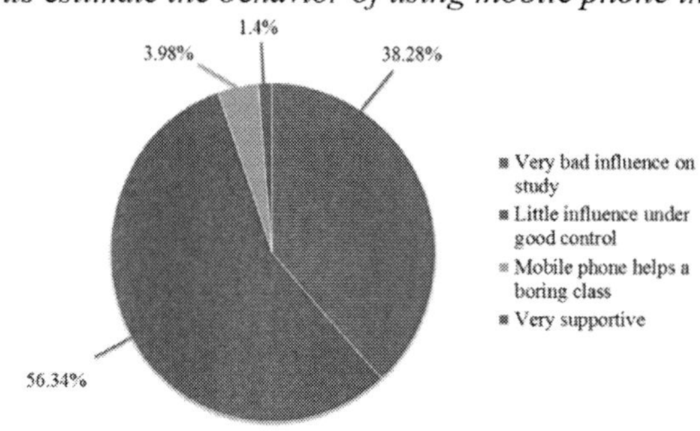

Figure 2 How students estimate the behavior

From Figure 2, 38.28% of the students hold a very negative opinion on how the behavior will affect study: using mobile phone in class will badly influence study. More than half (56.34%) of the students think under good control, using mobile phone little influence the learning. 3.98% of the student say mobile phone helps a boring class and they would prefer using mobile phone for fun. Only 1.4% proposes their support to this behavior.

3.3.2 How students evaluate the influence of using mobile phone in class

Table 6 How students evaluate the influence

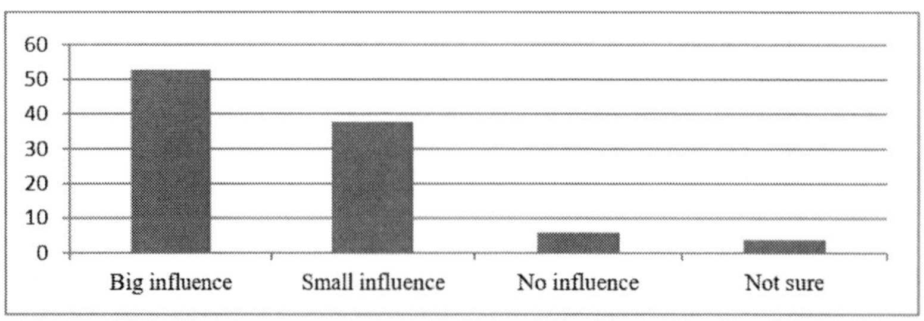

In Table 6, 52.69% agree it has very big influence on study and 37.63% show the similar opinion that it has certain but small influence. 5.91% deny the influence while 3.76% are not sure.

3.4 Reason and purpose

3.4.1 The reason why students use mobile phone in class

Table 7 The reasons

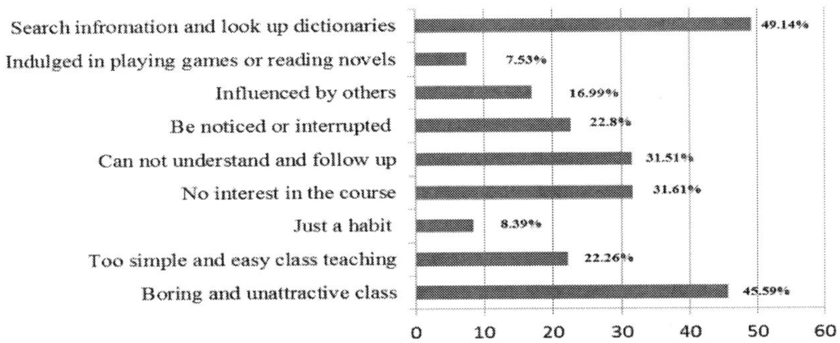

The reasons why student use mobile phone in class are various in Table 7. 45.59% find the class boring and unattractive therefore starting using mobile phone for fun. 31.61% have no interest in the course and content of teaching.31.51% can't understand the teaching but on the contrary 22.26% feel the class is too simple for them. Other causes are also important factors that can't be overlooked. 22.8% will begin using mobile phone once noticed or interrupted (for example a coming message or a phone call). 16.99% will be influenced by other students. 7.53% of the students indulge themselves in playing games or reading novels via mobile phone, and 8.39% just have the habit of using and playing mobile phone in class. What's more, 49.14% use mobile phone to search information and look up dictionaries.

Table 8 Further analyse of the reasons

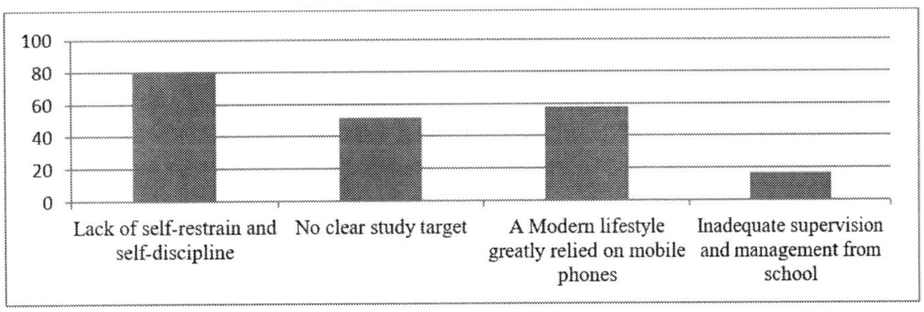

Table 8 tries to further analyse the reasons, 80.43% attribute to the lack of self-restrain and self-discipline of college students. 51.56% don't have clear study target. 57.96% reckon a modern lifestyle greatly relied on mobile phones and 16.77% call for more supervision and management from school and teachers.

3.4.2 What do students use mobile phone to do in class.

Table 9 Purposes of using mobile phone

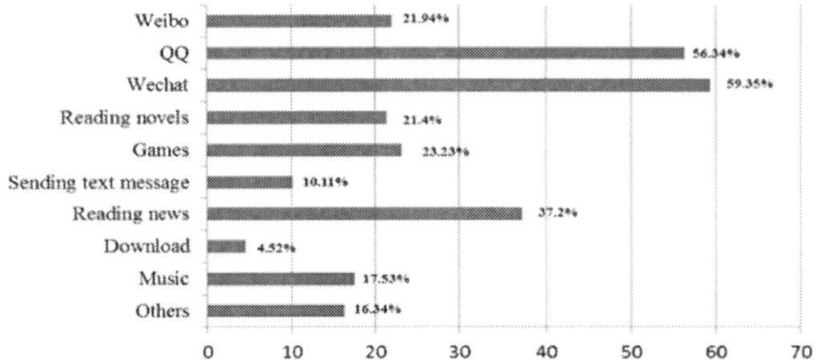

As shown in Table 9, in general, students use their mobile phones for fun and social communication. Social communication tools such as Wechat, QQ, Weibo, Text message respectively contribute 59.35%, 56.34%, 21.94%, 10.11%. On entertainment, 37.2% read news, 23.23% play games, 21.4% read novels and 17.53% enjoy music.

3.5 Suggestions and solutions

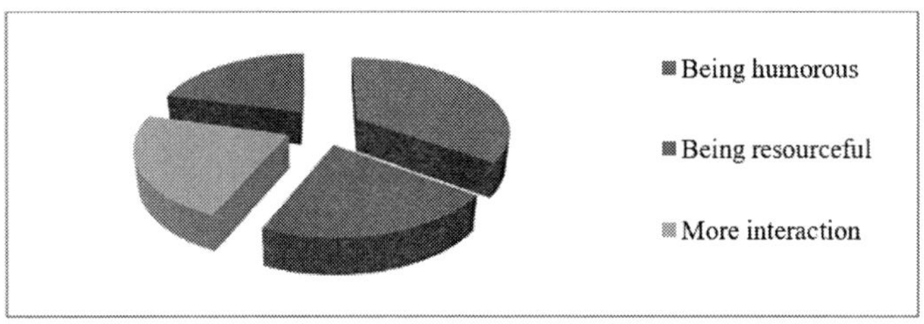

Figure 3 Advices to teachers

As shown in Figure 3, when asking students to give advices to teachers, 83.01% believe a humorous teacher will help students quit playing mobile phones. 53.87% agree more and better interaction between teacher and students will be helpful. 51.61% expect the teachers to be resourceful. 49.35% hope the teachers can enrich their teaching content and involve more related information into classroom teaching.

Table 10 More suggestions

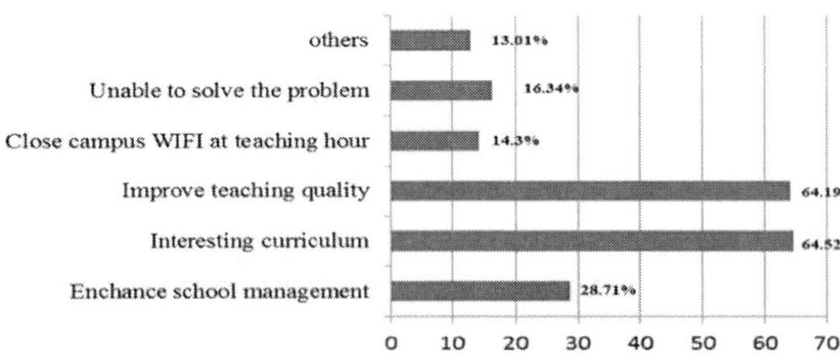

Students also proposed more suggestions as listed in Table 10. 28.71% think the school should enhance management. 64.52% raise the question of adjusting and improving curriculum design. 64.19% highlight the importance of improving teaching quality. 14.3% even suggest closing campus WIFI at teaching hour. 16.34% are pessimistic, saying it is inevitable to stop using mobile phone in class.

4. Suggestion

4.1 Enhance school management, optimize curriculum design

The results of the survey shows the students share a similar view on using mobile phone in class. 56.34% of the students accept using mobile phone in class once under good management and control. 52.96% think using mobile phone would greatly affect class study while 37.63% of them reckon a negative but relatively small influence. That is to say, up to 90.3% of the students agree this behavior would affect class study. Combing these figures, we can find out the fact clearly that using mobile phone in class is widely accepted by students, while students also understand the negative influence it brings up. Therefore, we have to respect the reality and inevitable trend that mobile terminals have already coexisted with classroom teaching. The traditional management and practice such as taking away students' mobile phones or prohibiting bringing mobile phones to class is outdated. [3] We should take the advantage of the students' physiological cognition that "using mobile phone affects class study indeed", follow their needs and instruct them to use mobile phone in class in a more appropriate way.

Students' feedback on curriculum design is also worth discussing. According to the survey, 30.79% choose to use mobile phone in optional courses. In a similar research conducted by Yang Yun [4], the same result was found out that optional courses were students' first choice to use mobile phones. 31.61% state that a boring class will drive the students moving to their mobile phones. 64.52% suggest optimizing curriculum design will greatly help to solve the problem. A conclusion could be draw that there is an urgent call for optimizing curriculum design and among all courses; optional course enjoys the highest priority to be reviewed and optimized.

4.2 Innovate classroom management, co-build class agreement

There is actually a misunderstanding on mobile phones in class among teachers. Due to inadequate support from big data and in-depth survey, teachers always have a judgment that students can't handle without mobile phone and they are not aware of its negative influence at all. Most teachers feel helpless and incapable in class. In fact, the results of the survey surprise us. Without mobile phones, 58.18% are worried but feel all right to continue study. 33.51% never care and even feel better. Altogether, 91.69% of the students could attend a class without mobile phone. Other causes and interruptions are also important factors. 22.8% will use mobile phone once noticed or interrupted by a coming message or a phone call and 16.99% will be affected by classmates. 83.01% will stop using mobile phone if the teacher will be humorous and 53.87% suggest more interaction between teacher and students would be helpful. 14.3% even propose to close WIFI connection at class hours.

From the statistics above, the co-existence of mobile phone and classroom teaching is actually achievable. Teachers should be more optimistic and tolerant to upgrade their classroom management. By investigating the needs of the students, teachers should have more communication and interaction with students to encourage positive class engagement, finally to enrich their learning experience. Meanwhile, innovative classroom management is requested. Wang Bing believed that an excellent class administrator could solve problems before the problems came out. The smartest way was to settle down classroom rules and regulations with students, and use the internet when necessary and needed. [5] A class agreement based on mutual acknowledgement by students and teachers could be set up and observed. This agreement helps to regulate the use of mobile phones in class and contribute to harmonious class teaching.

4.3 Rebuild class structure, redefine mobile terminals

In which section of a specific class would students use their mobile phones more? The answer from the survey is really inspiring. In the core section of a class – mid class section, only 12.47% will choose to use their mobile phones. At class beginning, 22.47% will use their phones while as many as 60.86% will use phones at class ending section. Students reveal a consistency on time selection: mobile phones little influence the core teaching section, but dramatically affect class beginning section and ending section, which are considered to be the non-core section. This fact offers the possibility of rebuilding our class structure.

Liu Wei suggested that Teachers could assign tasks to students before class. After self-learning in their spare time, students attended class with problems and questions they encountered. Teachers could focus classroom teaching on problem shooting. [6] More importantly, educational environment based on Internet Plus makes it possible to rebuild our class structure. The lead-in section of a traditional class could be relocated and inverted with the wide application of MOOC, Micro-lecture, SPOC, Video Open Online Course, online courses and so on. With the internet, students could finish the lead-in part at home ahead of time. Teachers could be concentrated in a face to face class to teach and explain core knowledge. In the Internet Plus module, the roles and functions of mobile terminal are redefined. Rather than an entertaining tool, with internet mobile phone could be used as a tool for searching, commenting and discussing, [7] and could coordinate classroom teaching.

In the age of Internet Plus, college classrooms are gradually evolving into smart classrooms. Featured by "sharing, melting and interacting", smart classroom is an educational information environment which encourages both teachers and students to use mobile devices and WIFI in class to get connected to campus network and internet, therefore obtaining education resources. [8] College teachers should follow the steps of technology innovation and embrace social progresses. With integrated mobile technology, internet and data platform, we could make great education innovation.

Reference

[1] Wang Junxiu, Yang Yiyin. Annual Report on Society Mentality of China 2015 [M]. Beijing: Social Sciences Academic Press. 2015:257.
[2] Chen Huaizhi. Impact and Enlightenment of Intelligent Mobile Phones to College Classroom Teaching [J]. Journal of Xichang College (Social Science Edition), 2014.26(4): 148-150.
[3] Dong Zhenhua. The Influence of Mobile Phones in Vocational College Classroom and Coping Strategies [J]. Journal of Huinan Vocational & Technical College, 2015.1(15): 58-61.
[4] Yang Yun, Zhang Qian, Liu He, Pu Xiangran, Liu Ying. Investigation and analysis on Mobile Phone Usage in Class of College Students [J]. Suevey of Education, 2014.3(7): 84-86.
[5] Wang Bing. The Impact of Mobile Internet on Large Class Teaching in University [J]. Journal of Beijing University of Posts and Telecommunications (Social Science Edition), 2010(5): 7-9.
[6] Liu Wei. Exploration and Practice in the Classroom Teaching of Higher Vocational Education in the Intelligent Mobile Phone and "Internet Plus" [J]. Education Teaching Forum, 2017(39): 159-160.
[7] Guo Kaiying. How Does Mobile Device Influence Teaching in age of Information? – Taking Intelligent Mobile Phone as an Example [J]. Education Modernization, 2016(21): 4-6.
[8] Hu Qingtai, Zheng Kai, Lin Nanhui. The Development and Transformation of Education Informatization: From Digital Campus to Smart Campus [J]. China Educational Technology, 2014(1): 35-39.

IOP Conf. Series: Journal of Physics: Conf. Series **1060** (2018) 012046 doi:10.1088/1742-6596/1060/1/012046

The Design of Adaptive User Interface Based on the Grey Relational Grade

Fangzheng Li[1],Wei WANG[1],Jue QU[1] and Huating Han[1]

[1]Air and Missile Defense College ,Air Force Engineering University, Xi'an 710051

Corresponding author Wei Wang,email:913059960@qq.com

Abstract. In order to improve the efficiency of human-computer interaction and reduce operating load, a new design method of adaptive GUI(graphical user interface) was proposed in this paper. Firstly, The common GUI was analysed and abstracted, and the five basic composition styles of GUI: Trees, Tabs, Forms, Lists, Filters were obtained, as the basis of interface partion. Then the functional object (large area)activity and area attention degree were defined. According to Shannon's information theory, the amount of information that the user needs to obtain from a area within some time is obtained. Therefore the activity of functional object(large area) was determined.The attention degree of area was determined by eye movement tracking experiment datas. Eye-movement factors F_i (fixation count), T_i(Dwell time) and V_i (revisits) were selected as characteristic parameter.In accordance with the principle of equal importance weighted average, the area attention degree are obtained. The Grey relational grade of the functional objects (large area)activity and area attention degree was obtained through Grey relational analysis as the trigger condition for interface change.

1.Introduction

Although automation is becoming more and more important in control system, human operators still play an irreplaceable role in many aspects such as aviation flight operation, weapon equipment operation, manufacturing operation system monitoring and operation,surgery and so on. As the functions of all kinds of control systems are becoming more and more powerful and the control is becoming more and more complex, higher requirements have been put forward to human operators. User interface is the main medium of human-machine interaction. Whether user interface design is good or bad, whether it adapts to user interaction characteristics will directly affect the realization of system function. The traditional graphical user interface uses fixed structure to provide users with interactive information, and does not consider user interaction habits. It does not consider the connection of functional objects in the interface, which seriously restricts the development of users and system functions.

Adaptive interface interaction through monitoring the user's intention, operation level, style preferences and other characteristics of information, dynamically change the interaction style and layout form for different users, different tasks and provides adaptive and personalized information services[1], which greatly improves the efficiency of man-machine interaction, and reduces the operating load.

Many scholars at home and abroad have done a lot of research on adaptive interface design. Rothrock et al. (2002), Stephanidis et al.(1997), Viano et al.(2000), Norcio and Stanley (1989), Haas and Hettinger (2001)[2-4] discussed adaptive interfaces in a number of backgrounds. The adaptive interface is to automatically adjust the information processing mechanism and behavior to adapt to the current task target and the ability of the user. This process is accompanied by monitoring user status, system tasks, and current status requirements.

In-Jee Song, Sung-Bae Cho[5] used Bayesian network and behavior network,and an adaptive user interface based on environment and user status is designed in the family environment. Talia Lavie, Joachim Meyer[6]analyzed the advantages and disadvantages of the adaptive interface, set up four different adaptive levels, different task modes, familiar situations and unfamiliar situations, and considered the advantages and disadvantages of the adaptive interface, he thinks that the adaptive interface is not beneficial under any conditions, it depends on many factors, including the frequency of task execution, the age of users, the difficulty of tasks, and the degree of users' participation in tasks.

Zhejiang Sci-Tech University Liezhong Ge team[7-9] studyed the adaptive bubble cursor, fixed, the rules and characteristics of fixed, adaptive and adaptable, three different mobile phone mail list and adaptive mouse pointing task efficiency, which showed that the adaptive interaction is better than the traditional fixed interaction.Shandong University, Chenglei Yang and Xu Yin [10] to study the adaptive interface with good touch experience size, improveed the touch experience of Android application; Dalian University of Technology Weidong Li, Zuoxin Zhu[11] studied the adaptive displays of the power system operation information, according to the user model and the different interactive scenes, such as normal, emergency, they select the appropriate information display technology and theme display interface.

About the research on adaptive user interface layout, only the Realization mechanism of the adaptive user interface in virtual Home Furnishing customization system[12], proposed by Yinting Fan.He put forward an adaptive prediction method of the activity based on history interaction sequence and adaptive distribution mechanism,which magnified the object of the user's attention, and the object that is not concerned is reduced,but the user's operating habits and personalities are not considered. Guohua Zhan[13] make the functional object activity and region interest degree match in the Web interface according to the interactive habits of the user. But the adaptive layout of interface is overall, it is easy to cause the confusion, increasing the load of users, and in the calculation of the interest degree of region, only considering the region information density, and there is no specific consideration of users' eye attention in interface region.

Based on the above considerations, this paper according to the characteristic of the graphical user interface and provides the basis for the division of the interfacial region; Shannon theorem was determined the amount of information in the interactive process required from the user interface region and function object access, namely the function object activity; through tracking eye movement, overall considering the number of fixation the residence time, and the revisit times determine users concern of a certain area.

Finally, grey relational analysis was used to determine the activity of functional objects and the area attention degree. Trigger condition of adaptive layout was determined by the change of the gray correlation degree. Firstly, the positions of large areas are adjusted, then positions of the functional objects in the large area are adjusted. The adjustment of functional objects is only in its large area, thus ensuring the stability of the user interface and improving the interaction efficiency.

2.Brief introduction to the User Interface Graphic Theory & grey correlation analysis

In the first generation of information systems, the user interface is character interface. With the appearance of graphical interface, the user interface has been rapid development, the first graphical user interface is created by Xerox Corporation in Palo Alto Research Center. New form of user interface such as Microsoft's Windows 10, apple computer, apple OSX mobile phone IOS, and Android. These interfaces have concepts such as windows, icons, menus, pointers, and so on, which are defined by Palo Alto.

2.1.Basic behavior of the user interface

IOP Conf. Series: Journal of Physics: Conf. Series **1060** (2018) 012046 doi:10.1088/1742-6596/1060/1/012046

The behavior of the user interface determines how the data is managed and stored, and the basic behavior of the user is as follows:

CRUD: Its meaning is Create, Read, Update,and Delete, which represent the four basic operations of user in the process of interaction.For example, in the mobile phone address book,creating is creating a new contact; Updating refers to updating mobile phone address book; and deleting refers to deleting the information of the contacts that are not used.

2.2.The basic components of the user interface

The basic user interface components are obtained through abstracting the common user interfaces, such as hotels, finance, accounting, Microsoft windows system, apple OS system, etc, [14]. As shown in Figure 1, an instance of the user interface abstraction.The basic components of the graphical user interface include:

Form: the form is generally in a certain window, consisting of a series of control functions, and usually associates with the CRUD operation.

List: there is an additional list around the Form, which allows users to select and access to responding data in the list, and the detailed information of the corresponding data will be displayed in the Forms. Lists and Forms are usually associated with four basic operations (CRUD) in the same window.

Filter: the main implementation of Filter is filtering functions and data.

Tree: generally in the side of the graphical user interface, the interface structure can be quickly mastered through a tree, and the Tree can facilitates the user to find the desired functional objects.

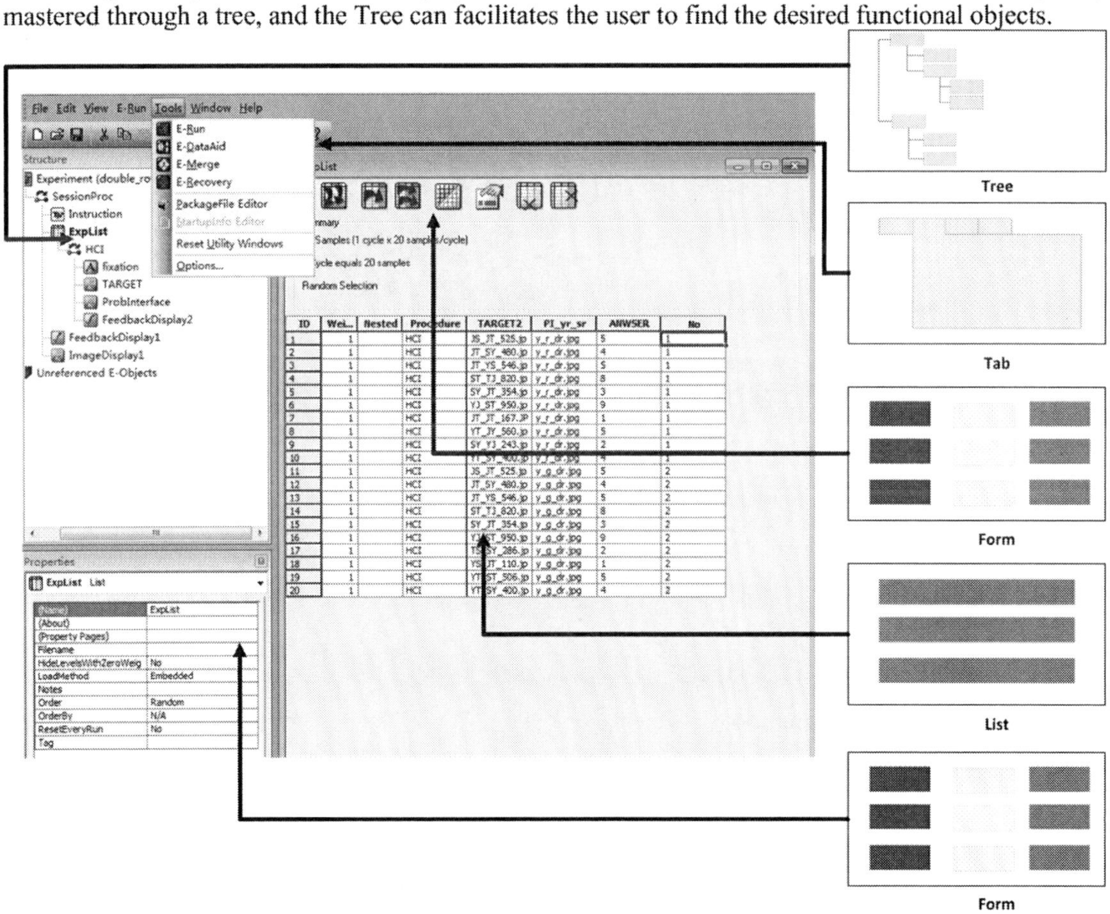

Figure.1 The common graphical interface abstraction

The basic components of the user interface is shown in Figure 2.

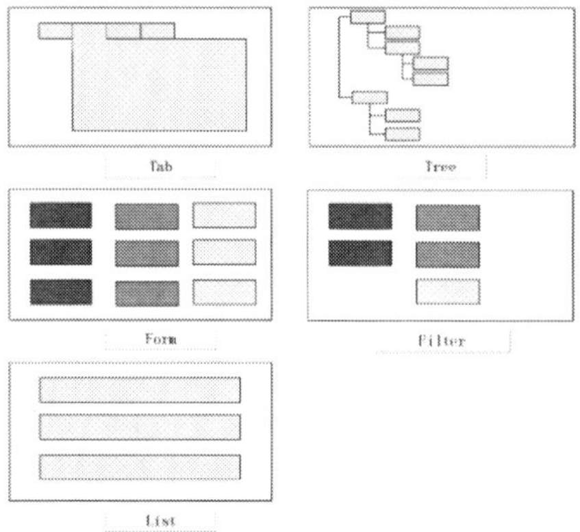

Figure.2 The five basic composition style of the user interface

Common graphical user interfaces can generally be abstracted into Figure 2 five basic components or combinations of one or several basic components. Based on the basic components partition, the graphical user interface partition can retain the basic structure of the interface and keep the spatial and logical connections between the functional objects. Of course, the above abstract components can not include all the elements of graphical user interface,and still need to partition the interface reasonably according to the actual composition of the interface.

2.3.Brief introduction to grey correlation analysis

The basic idea of grey relational analysis is to judge whether the relationship is close according to the similarity of several shapes of the sequence curves. The closer the curve is, the greater the degree of correlation[15].

Set the system behavior sequence

$$\begin{aligned} X_0 &= \left(x_0(1), x_0(2), \cdots, x_0(n)\right) \\ X_1 &= \left(x_1(1), x_1(2), \cdots, x_1(n)\right) \\ &\cdots\cdots \\ X_m &= \left(x_m(1), x_m(2), \cdots, x_m(n)\right) \end{aligned} \tag{1}$$

It is sequences of related factors.

Given real number

$$\varphi\left(x_0(k), x_i(k)\right) = \frac{\min\limits_i \min\limits_k \left|x_0(k) - x_i(k)\right| + \xi \max\limits_i \max\limits_k \left|x_0(k) - x_i(k)\right|}{\left|x_0(k) - x_i(k)\right| + \xi \max\limits_i \max\limits_k \left|x_0(k) - x_i(k)\right|} \tag{2}$$

$i = 0, 1, \cdots, m; k = 1, 2, \cdots, n$

$$\varphi\left(X_0(k), X_i(k)\right) = \frac{1}{n}\sum_{k=1}^{n} \varphi\left(x_0(k), x_i(k)\right) \tag{3}$$

$\xi \in (0,1)$ Called the resolution coefficient, $\varphi\left(X_0(k), X_i(k)\right)$ is the Grey relational degree between X_0 and X_i.

The calculation steps of grey correlation degree are as follows:

①Calculating the initial value (or mean value, or the interval value etc.) for each sequence;

$X_i' = \left(x_i'(1), x_i'(2), \cdots, x_i'(n)\right), i = 0, 1, \cdots, m$;

②Calculating the difference sequence;

$$\Delta_i(k)=\left|x_0^{'}(k)-x_i^{'}(k)\right|, i=1,2,\cdots,m \tag{4}$$

$$\Delta_i=\left(\Delta_i(1),\Delta_i(2),\cdots,\Delta_i(n)\right) \tag{5}$$

③Calculating the polar difference most and least;

$$M=\max_i\max_k\Delta_i(k), m=\min_i\min_k\Delta_i(k) \tag{6}$$

④Calculating the correlation coefficient;

$$\varphi_{0i}(k)=\frac{m+\xi M}{\Delta_i(k)+\xi M}, \xi\in(0,1) \tag{7}$$

⑤Calculating the correlation degree;

$$\varphi_{0i}=\frac{1}{n}\sum_{k=1}^{n}\varphi_{0i}(k), i=1,2,\cdots,m \tag{8}$$

3.Adaptive user interface

The user interface, as the medium of human interaction with the system, plays an important role in the process of communication computer. The requirements for the interface are different because of the people's habits, interests, the degree of perception, and the degree of familiarity with the system operation. This provides a higher requirement for the user interface, which is able to adapt to the characteristics of different users.

To achieve this, the first thing to do is to dig out the user's personalized information. Interface information mining can be divided into three main categories[16]: structure mining, to find useful knowledge from the link of the representation structure; content mining, to extract useful information and knowledge from the interface content; Usage Mining, to mining the user's access pattern from the log of each user clicking on the situation. These three mining methods play an important role in adaptive interface. This paper mainly uses eye tracking equipment and user click log, to mining user interaction habits and azimuth perception.

The user interface is generally composed of a plurality of basic components, each component of interface contains multiple functional objects, and each functional object contains subfunction. The active degree of each functional object is different, and there is a certain connection between the various functional objects.When users use functional objects on the interface, there are several situations and the worst case is to spend a lot of time searching for functional objects that they will use. The goal of this adaptive interface is to dynamically adjust the layout of functional objects based on tracking and analyzing user interaction habits, so that users can find corresponding functional objects in the most habitual location and improve interaction efficiency. For this reason, the interface designer must know the degree of activity of the functional object and the user's attention to the functional object area, and judge whether the two match. If not, then adjust the location of the function object. Therefore, there are the following definitions:

Definition 1. The activity of functional objects (large areas): in an interactive period, the user needs the amount of information obtained from the function object (large area).

Definition 2. Area attention degree: the attention which users pay attention to the corresponding interface areas, mainly based on the user's eye movement data in this area.

The goal of the adaptive interface is to achieve the overall adaptation of the two and reduce the user's search load.

Based on the basic components of the user interface, the interface is divided into large areas. Because the large area of the user interface is relatively independent, it does not consider the transformation of the location of the function objects in the large area, and maintains the basic stability of the interface. The oversize of the interface change will reduce the user's trust in the interface, and increase the user's operating load. In this paper, a "hierarchical adaptive method" is used to determine whether the large area is adaptive or not, First. Then we can determine the layout of the basic components of the interface based on user interaction habits, and further determine the location of the functional objects in the corresponding large areas.

Due to the interrelation between functional objects, including functional objects in parts of the same area and different regions, we must consider the relationship between the contexts of operations in the process of adaptation.

3.1.Determination of attention degree of large area

It is assumed that the basic elements that contain information such as graphics, symbols and words in the user interface are independent of each other and are calculated according to the average amount of information. Formula (12) can be simplified to large area

The number of basic units Forms, Lists, Tabs, Trees, Filters, etc. is N in the user interface,and the corresponding interface is divided into N regions. F_i (fixation count), T_i(Dwell time) and V_i (revisits) etc. in the corresponding region are recorded through eye movement tracking experiment. Because the units of the data are different, the original data are processed without Dimensionalization.

$$F_i^* = \frac{F_i}{F_{\max}}, T_i^* = \frac{T_i}{T_{\max}}, V_i^* = \frac{V_i}{V_{\max}} \tag{9}$$

In accordance with the principle of equal importance weighted average, The area attention degree are obtained:

$$I_i = \frac{1}{3}\left(F_i^* + T_i^* + V_i^*\right) \tag{10}$$

3.2 Determination of the activity of large area

According to Shannon's information theory [17], for an attribute in a set of data, its attribute domain is $\{a_1, a_2, \cdots, a_n\}$, and the corresponding probability distribution is $\{p(a_1), p(a_2), \cdots, p(a_n)\}$,The following matrix can be obtained:

$$(X,P) = \begin{pmatrix} a_1, a_2, \cdots, a_n \\ p(a_1), p(a_2), \cdots, p(a_n) \end{pmatrix} \tag{11}$$

The average amount of information for each symbol of the source, that is the entropy of information

$$h(x) = \sum_{i=1}^{n} p(a_i) \log_2 \frac{1}{p(a_i)} \tag{12}$$

It is assumed that the basic elements that contain information such as graphics, symbols and words in the user interface are independent of each other and are calculated according to the average amount of information. Formula (13) can be simplified to:

$$h(x) = \log_2 \frac{1}{p(a_i)} \tag{13}$$

The total amount of information contained in the area i is

$$H_i(x) = \sum h_i(x) \tag{14}$$

then normalization processing is carried out:

$$H_i^*(x) = \frac{H_i(x)}{H_{\max}(x)} \tag{15}$$

The use probability of area i in a monitoring time period is P_i. Assuming that the user's operation in large area is the same as a whole, then the amount of information that the user needs to obtain from the area i within this period is

$$\tilde{H}_i^*(x) = P_i H_i^*(x) \tag{16}$$

Two behavior sequences are got:

$$I = (I_1, I_2, \cdots, I_N)$$
$$\tilde{H}^* = (\tilde{H}_1^*, \tilde{H}_2^*, \cdots, \tilde{H}_N^*) \tag{17}$$

Grey correlation analysis is used to determine the two sequences'initial grey correlation that is $\varphi_{t_0}(I(k), \tilde{H}^*(k))$, after t_α monitoring time,the grey correlation degree changed into $\varphi_{t_\alpha}(I(k), \tilde{H}^*(k))$, if $\varphi_{t_\alpha}(I(k), \tilde{H}^*(k)) < \varphi_{t_0}(I(k), \tilde{H}^*(k))$, The active components of interface is adjusted to the area of great concern, or no adjustment of the interface. At the same time, the initial grey correlation degree of the next interface adjustment becomes $\varphi_{t_\alpha}(I(k), \tilde{H}^*(k))$.

3.3 Determination of area attention degree for the functional object

The partitions of the user interface are A,B,C, $\cdots$, and the functional objects that each area contains are $A_1, A_2, \cdots, A_\lambda, B_1, B_2, \cdots, B_\theta, C_1, C_2, \cdots, C_\delta, \cdots$. In order not to add to the user's operating load, the location of the function object is adaptive only in the corresponding partition.

Because the area of the functional object is smaller, the error is larger if 4.1 is used to determine the attention of the large area.

Take the area A, for example, to establish a rectangular coordinate system in the area A, as shown in Figure 3.

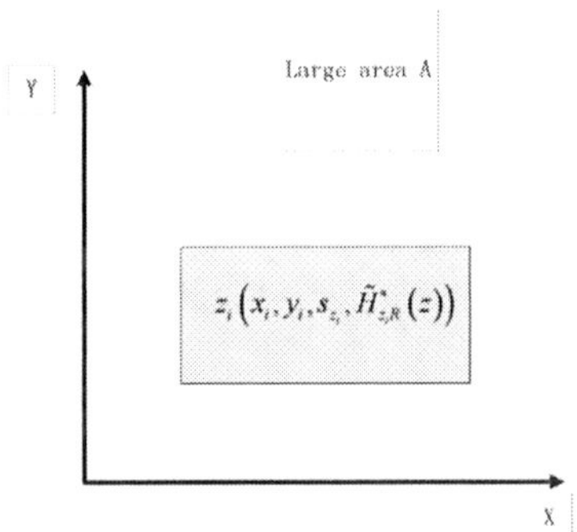

Figure.3 Schematic diagram of the right angle coordinate system in A large area

The subregion where each function object is located is $z_i(x_i, y_i, s_{z_i}, \tilde{H}_{z_iR}^*(z))$, x_i, y_i represent the horizontal and longitudinal coordinates of the area center, s_{z_i} is Area area, $\tilde{H}_{z_iR}^*(z)$ is the amount of modified information that needs to be obtained for the user to operate on this functional object.

The area attention degree determined by 4.1 is $I_{zi} = \frac{1}{3}(F_{zi}^* + T_{zi}^* + V_{zi}^*)$. The area of the large area is large, and the influence of the area factor on user attention is not significant.The area of the functional object is small, and the difference between the areas will affect the attention of the functional object area. It is therefore necessary to further modify area attention of the functional objects.as follows:

$$I_{z_iR} = \frac{1}{3}\eta(F_{zi}^* + T_{zi}^* + V_{zi}^*) \tag{18}$$

$\eta \in (0,1)$ is the area coefficient, $\eta = \frac{s_{\min}}{s_{z_i}}$, The larger the area, the smaller the η.

IOP Conf. Series: Journal of Physics: Conf. Series **1060** (2018) 012046 doi:10.1088/1742-6596/1060/1/012046

3.4 Determination of function object activity

User interface is U , and all the functional objects that the user interface contains are $U_1,U_2,\cdots,U_m$. The vector consisting of a functional object is $U=\left[U_1,U_2,\cdots,U_m\right]^T$. Through the tracking and analysis of functional objects' usage, the relationship between functional objects can be obtained.

$$U=\begin{bmatrix} u_{11} & u_{12} & \cdots & u_{1m} \\ u_{21} & u_{22} & \cdots & u_{2m} \\ \vdots & \vdots & \ddots & \vdots \\ u_{m1} & u_{m2} & \cdots & u_{mm} \end{bmatrix} \tag{19}$$

u_{ij} represents the usage count of the functional object U_j , which is used after the functional object U_i . $u_i=\sum_j u_{ij}$ is the total usage count of U_i . $u=\sum_i\sum_j u_{ij}$ is the number of use of all functional objects. Thus, the use probability of the function object U_i is

$$p_{U_i}=\frac{u_i}{u} \tag{20}$$

By the 4.2 information theory, the amount of information that the user needs to get from the function object in a given time period is

$$\tilde{H}^*_{U_i}(U)=P_{U_i}H^*_{U_i}(U) \tag{21}$$

$H^*_{U_i}(U)$ is the amount of information processed by normalization. Because the user operation of each functional object is different, the user operation of the functional object directly affects the amount of information transmitted from the functional object. Amend the form (21) and get

$$\tilde{H}^*_{U_iR}(U)=\mu P_{U_i}H^*_{U_i}(U) \tag{22}$$

μ is the corresponding operating coefficient. Finally, two association sequence of the functional object is got.

$$\begin{gathered} I_{zR}=\left(I_{z_1R},I_{z_2R},\cdots,I_{z_\lambda R}\right) \\ \tilde{H}^*_{AR}(U)=\left(\tilde{H}^*_{A_1R}(U),\tilde{H}^*_{A_2R}(U),\cdots\tilde{H}^*_{A_\lambda R}(U)\right) \end{gathered} \tag{23}$$

According to the calculation step by grey correlation degree, the initial grey relational degree of I_{zR} and $\tilde{H}^*_{AR}(U)$ was got, $\varphi_{t_0}\left(I_{zR},\tilde{H}^*_{AR}(U)\right)$; After t_α monitoring time, gray correlation degree change into $\varphi_{t_\alpha}\left(I_{zR},\tilde{H}^*_{AR}(U)\right)$. If $\varphi_{t_\alpha}\left(I_{zR},\tilde{H}^*_{AR}(U)\right)<\varphi_{t_0}\left(I_{zR},\tilde{H}^*_{AR}(U)\right)$, we adjusts the function objects with large activity to the area of more attention. If $\varphi_{t_\alpha}\left(I_{zR},\tilde{H}^*_{AR}(U)\right)>\varphi_{t_0}\left(I_{zR},\tilde{H}^*_{AR}(U)\right)$, there is no adjustment. At the same time, the grey correlation degree of the next interface adjustment becomes $\varphi_{t_\alpha}\left(I_{zR},\tilde{H}^*_{AR}(U)\right)$. The interface adapts to the optimal through the adaptive of the large area activity and the attention degree, the function object activity and the area attention of the region.

4 Conclusion

A method of adaptive layout is proposed in this paper. The feasibility and effectiveness of the method need further experimental verification.

Reference

[1] Kobsa A . Generic user modeling systems [J].User Modeling and User Adapted Interaction, 2001, 11(1 /2) :49-63.
[2] Rothrock, L., Koubek, R., Fuchs, F., Haas, M., Salvendy, G., 2002. Review and reappraisal of adaptive interfaces: toward biologically inspired paradigms. Theoretical Issues in Ergonomics Science 3 (1), 47–84.
[3] Shneiderman, B., 1992. Designing the User Interface: Strategies for Effective Human-computer Interaction. Addison-Wesley Longman Publishing Co., Inc,Boston, MA, USA.
[4] Stephanidis, C., Karagiannidis, C., Koumpis, A., 1997. Decision Making in IntelligentUser Interfaces. Paper Presented at the Proceedings of the 2nd International Conference on Intelligent User Interfaces
[5] In-Jee Song, Sung-Bae Cho. Bayesian and behavior networks for context-adaptive user interface in a ubiquitous home environment[J]. Expert Systems with Applications,40(2013):1827-1838.
[6] Talia Lavie, Joachim Meyer .Benefits and costs of adaptive user interfaces[J]. Int. J. Human-Computer Studies 68 (2010) 508–524.
[7] Xinqin Jin.Study of Adaptive Bubble Cursor[D].Hangzhou:Zhejiang Sci-Tech University,2015.
[8] Lu Zheng. Ergonomic Study on Static, Adaptive and Adaptable in Mobile Directory[D]. Hangzhou:Zhejiang Sci-Tech University,2011.
[9] Hua Liu.Pointing task performance study on adaptive-based mouse[D]. Hangzhou:Zhejiang Sci-Tech University,2014.
[10] Xu Yin,Yalin YU,Yuan WANG,et al.A Method to Generate Adaptive User Interface with Touch –Friendly Size.[J]. Journal of Computer—Aided Design&Computer Graphics,2014,26(4):662-668.
[11] Zuoxin Zhu.Preliminary research on adaptive display of power system operating information[D].Dalian:Dalian University of Technology,2013.
[12] Yinting Fan,Dongxing Teng,Gongzheng Wang,et a1 .Implementation Mechanism of Adaptive User Interface for Furniture Layout Customizing System[J]. Journal of Computer-Aided Design& Computer Graphics,2011,23(4):705-712.
[13] Guohua Zhan,Wei Wei.A Web Adaptive Interface Design[J].Journal of Hangzhou Normal University(Natural Science Edition) ,2011,10(2):163-167.
[14] Paulo Roberto Lumertz , Leila Ribeiro , Lucio Mauro Duarte. User interfaces metamodel based on graphs[J]. Journal of Visual Languages and Computing,32(2016)1-34.
[15] Linghao Kong.Study on Substation-Area Protection of Fuzzy Cluster Based on Fault Degree and Gray Correlation Degree[D].Beijing:North China Electric Power University,2014.
[16] Bing Liu.Web data mining[M]. Beijing:Tsinghua University Press,2009:5-9.
[17] Long Zhang.Research on feature selection and classification algorithms based on information theory[D].Chongqing:Southwest Agriculural University,2005.

Biometric Identification Technology and Development Trend of Physiological Characteristics

ZHAO Feng[1, 2]

[1] Key laboratory of Evidence-Identfying in Universites of Shandong, Shandong University of Political Science and Law, Jinan 250014, China;
[2] Judicial identification center of Shandong University of Political Science and Law, Jinan 250014, China.

E-mail: zhaof0531@126.com

Abstract. Biometric recognition is an interdisciplinary and comprehensive technology that combines computer technology with optics, acoustics, sensors and biostatistical technology closely, with the help of physiological and behavioral characteristics common to human beings to authenticate the identity of the individual. Physical characteristics are born with innate characteristics, while behavioral characteristics are habitual characteristics formed by long-term accumulation of behavior, most of them have acquired characteristics. Usually, fingerprints, hand, faces, iris, finger vein, DNA, palmprints and retinas are classified as human physiological characteristics.

1. Introduction

At present, there are several common recognition technologies in the field of biometrics: fingerprint identification, iris recognition, thermogram recognition, palmprint recognition, retinal recognition, DNA recognition, signature recognition, hand recognition, face recognition and speech recognition, etc.Face, fingerprint, finger vein, iris, DNA, palmprint, thermogram,etc[1]. Are classified as physiological features. These recognition techniques have been widely used in finance, public security organs, attendance, medicine and health departments that need individual identification.In the public security organs, biometric identification system is needed to compare the fingerprints left by criminals at the scene of the crime with the fingerprint database that has been established in the public security organs to determine the individuals to whom the fingerprints belong and the locking of the suspects[2]. In IT, the use of individual biometrics to replace traditional passwords, prevent others from illegally performing, and ensure the security of personal computers. In the financial trading industry, by encrypting individual biometric information into chips, more secure transactions can be ensured.

2. Common recognition technologies

2.1. Fingerprint recognition.

Fingerprint identification technology is to use image acquisition device to obtain fingerprint image[3], and then analyze the global and local characteristic points of fingerprint such as valley, kurtosis, bifurcation point, bifurcation point and so on by computer program software[4]. Fingerprint identification is a kind of long history research and mature identification technology[5], which can be used to identify individuals reliably by fingerprint identification.[6].

With the development of pattern recognition and computer technology, automatic fingerprint recognition has been developing more and more[7].The first to invest in automatic fingerprint recognition technology is the United States[8], and other developed countries are also engaged in research. Because in recent years the fingerprint collection device[9], the PC's price performance ratio increased significantly, fingerprint identification as a kind of identity authentication technology already has a more practical software and hardware products, including fingerprint collection device, recognition software and corresponding software development kit, etc[10].

2.2. Iris recognition

Iris recognition technology is based on individual iris diversity and lifetime invariance to authenticate the identity of the individual. The iris, located inside the pupil of the eye, is a ring, and each iris has unique structural features that make it difficult to change its structure through surgery[11]. At present, there is no Iris repetition in the world, even if the iris of the same person's left and right eyes are very different, even though the iris information of all mankind is registered in a database. The occurrence of repetition is minimal. Although the overall shape of the iris is genetically determined, many of its details are influenced by environmental factors (For example: the condition of the embryo).Therefore, it is almost impossible to reproduce the iris through natural means. In addition, the eye is the most sensitive parts of the body, it is very difficult and dangerous to modify the iris structure through surgery, it also became the iris "optical fingerprint" as a person and a favorable factors. Therefore, the iris recognition technology used in individual identification will achieve a very good effect, but the cost of iris acquisition equipment is too high to accept for people [12].

2.3. Retinal recognition

An optical method is used to obtain the vascular images of the retina on the surface of the human eyeball, which can be used as the basis for individual identification. This method is called retinal recognition technology. If the retina on the surface of the eyeball had not been damaged, the blood vessels distributed on it for biometric use would not have changed from the age of three. Similar to iris recognition technology, it is recognized as the most reliable biometric recognition technology at present. Considering that laser irradiation on the back of the eyeball can capture the features used for recognition, it is difficult to recognize the retina in practical application. The retina technology has the advantages that it is basically a life - invariant biological characteristic, meanwhile, it is in a concealed position, is difficult to be forged , and the acquired process belongs to a non - contact type , so that the retina technology is clean and sanitary , The same shortcomings: the high cost of the retinal image collection, the immature technology, may bring health damage to the human body.

2.4. Face recognition

Face recognition is based on the analysis of the features of the face and the relative position of the eyes, nose and mouth. Standard video and thermal Imaging has become two kinds of techniques for obtaining face Images [13]. The standard video is captured by camera to obtain faces images, and the thermal imaging is the formation of facial images through a hot wire generated by blood in the facial capillaries [14].

2.5. Palmprint recognition

Human palmprint is also unique and stable. By studying and analyzing palmprint midline, dot and texture characteristics, we can do individual identification. Palmprint recognition plays an important role in biometrics research. At present, the two methods of obtaining palmprint image are mainly offline palmprint and on-line palmprint. The off-line palmprint method is mainly to press the ink-coated palm on the white paper, leave the figure of the palm on the white paper, and scan it to obtain the digital palm image. The online palmprint method is acquired automatically through a special palmprint acquisition device, and the images collected are more ideal.

2.6. Hand recognition

The technique of hand shape recognition is based on the structure drawing of the human hand, and the individual identification is carried out by analyzing the information contained in the finger structure diagram. The results of scientists' experiments show that the shape of the hand is stable for a long time, and the hand shapes of any two people are different, and the hand images are easy to obtain. So it has become the research direction of biometric recognition field.

2.7 Finger vein recognition

Finger vein recognition technology was developed by Hitachi in the medical field and gradually applied to personal identification. It is one of the world's cutting-edge of authentication technology with high precision, high speed, because it is the use of the individual's internal biometric identity authentication technology, because of its high security and industry attention. This technology started early in foreign countries, the end of the 20th century, the study of finger vein authentication technology began in the middle of the Hitachi institute, Japan's Hitachi has been using finger vein image recognition technology[15].

The vein recognition technology is a venous image through infrared transmission of the fingers, and the identification and identification of the venous images are performed. Finger vein recognition technology is widely used in Japan, but it also attracts other countries. Southeast Asian countries such as Singapore have also gradually adopted Hitachi's finger vein access system, and the office of the New York branch of the U.S. central Treasury Department has adopted a mobile phleboid device. It is believed that with the continuous improvement of the security level requirements, there will be a large potential demand market for venous identification. The finger vein plays an important role in the biometric identification family, and has been recognized internationally because of its high security features[16].

2.8. DNA recognition

There are differences in the structure of DNA in different human cells, so it is used as a biological feature for individual identification. DNA in human body has unique and eternal characteristics in the whole human being. With high accuracy and authority. DNA mode is the same in every cell and tissue of the human body, which determines that its authentication effect is much better than other biometric identification technology. At present, it is used by public security organs to identify criminals. DNA recognition have the problem of people's acceptance, meanwhile DNA acquisition must be in the related laboratories, Therefore, it is difficult to achieve real-time performance for it takes too long.

3. The Development trend of Biometrics in the Future

Biometrics is the most convenient and safe identification technology. Biometric technology has a very broad development. It can be applied to almost every aspect of People's Daily lives.

It is rated as the top ten new technology that can change the future world. With the increasing of the society's requirement for identification, the importance of biometrics will become more and more important. It is mainly reflected in the following aspects:

3.1 Multi-feature information fusion

At present, the research of biometric recognition technology is still in the process of continuous development, how to further improve the accuracy and robustness of system identification is a technical problem in this research field. The above biometric identification technology has a specific scope in practical application, so it has great limitations[17]. Statistics show that up to now, there is no single biometric recognition technology that can achieve a perfect and error-free state. Therefore, in order to improve the recognition performance of biological systems, a new research direction has emerged in the field of biometric recognition.

The combination of multiple individual authentication information, including a variety of biometric features, has become a feasible solution. for example, The fusion of fingerprint and finger vein. Before fusion, The matching scores of fingerprint identification system and venous recognition system are normalized. By comparing and analyzing different normalized models, found. the suitable model. By the improved matching fusion algorithm, the optimal or suboptimal allocation weight of different populations is obtained by adaptive algorithm. The fusion of the two biological characteristics of fingerprint and finger vein in the matching layer is realized, and the recognition rate of the system is improved effectively[18].

3.2 Real-time requirements
Biometric recognition system requires a lot of numerical processing, including image preprocessing, feature extraction, recognition and classification, etc. Real-time and fast computing ability is the key to meet the specific performance requirements of the system. There are two main ways to implement the algorithm: one is to improve the algorithm at the software level, to improve the efficiency of the algorithm, and the other is to implement feature recognition on the hardware level.

3.3 Combination of Biometrics and traditional identity Authentication
The intelligent IC card uses its own storage function and computing function to store the personal biometric information in the smart card and realize the function of offline authentication in the inquiry field, which not only improves the efficiency of authentication, but also improves the efficiency of authentication. And it saves the cost of online query. Therefore, the combination of biometric identification technology and IC card for traditional identity authentication has a promising future. The Technical level of Biometrics Research in China at present is still a certain distance compared with the developed countries in Europe. But the pace of research almost keep pace with countries in Europe.

In particular, some achievements have been achieved in the fields of fingerprint, iris, face and palmprint recognition in recent years. At the same time, there are a lot of technical problems in this field. This needs to include China and other countries in the world to continuously explore solutions to promote biometric identification technology to a higher level.

Fund projects: Scientific Research Project of Shandong University of political Science and Law (2017Z08B); Sichuan Provincial Center for the study of Medical Health and the Rule of Law (YF17-Q19); Supported by Program for Young Innovative Research Team in Shandong University of Political Science and Law.

References
[1] A Jain, A Ross, S Prabhakar. Fingerprint Matching Using Minutiae and Texture Features[C].*Proc.of International Conference on Image Processing* (ICIP), 2001:282-285.
[2] Y. Wang, T. Liu and J. Jiang. A rapid iris localization algorithm based on image sampling. *Opt-Electronic Engineering,* 2008, 35(9), Page(s):122-126.
[3] Li Xiuyan, Miao C hangyun, Liu Tiegen, Yuan Chenhu. Research on Personal Identify Verification Based on Hand Vein Iris and Fingerprint. *IEEE Computer Science and Society International Symposium*, 2011, Page(s): 16-19.
[4] Dinesh P.Mital, Earn Khwang Teoh, An Automated Matching Technique for Fingerprint Identification, *IEEE Trans*, 1997
[5] P.Baldi and Y.Chauvin, Neural Networks for Fingerprint Recognition, *Neural Computation*, 1993, 5(3):402-418.
[6] B.Moayer, K.S. Fu, An Application of Stochastic Languages to Fingerprint Pattern Recognition, *Pattern Recognition*, 1976, vol. 8, 173~179
[7] G.T.Candela, P.J.Grother, C.I.Watson, R.A.Wilkinson, and C.L.Wilson, PCASYS:A Pattern-Level Classification Automation System for Fingerprints, 1995, *NIST Tech.Report NISTTR* 5647
[8] M.Chong, T.Ngee, L.Jun, and R.Gay, Geometric Framework for Fingerprint Image Classification, *Pattern Recognition*, 1997, 30(9):1475-1488.

[9] Andrew Senior, A Hidden Markov Model Fingerprint Classifier, FBM T.J.Watson Research center, P.O.Box 704, *Yorktown Heights*, NY 10598-0218.USA.

[10] D.Isenor and S.Zaky,Fingerprint Identification using Graph Matching, *Pattern Recognition,*1986, 19:113-122.

[11] Eric Sung, "Towards non-cooperative iris recognition systems", Proceedings of the 7th International Conference on Control, *Automation, Robotics and Vision*, 2002

[12] Boles W, Boashah B, A human identification technique using images of the iris and wavelet transform. *IEEE Trans. on Signal Processing*, 1998, 46(4):1185~1188

[13] M. K, H. Ueki and s. Ume. A new method for the identification of individuals by using of vein Pattern of a finger[J].*Proceedings of the 5th Symposium on Measurement*, 2002:9-12.

[14] Kejun Wang, Yan Zhang, Zhi Yuan. Hand Vein Recognition Based on Multi Supplemental Features of Multi-C lassifier Fusion Decision[J].*2006 IEEE International Conference on Mechatronics and Automation.*2006:1790-1795.

[15] Lv X G, Zhou J, Zhang C S. A Novel Algorithm for Rotated Human Face Detection [J]. *Proc. of IEEE Conference on Computer Vision and Pattern Recognition, Hilton Head Island, South Carolina, USA,* 2000:760-764.

[16] Wang J G, Tan T N. A New Face Detection Method Based on Shape Information [J]. *Pattern Recognition Letters,* 2000, 21(6-7):463-471.

[17] Yuanning Liu, Senmiao Yuan, Xiaodong Zhu:A Time-frequency Field Fingerprint Enhancement Technology and Three-order Spline Curve Fitting Matching Algorithm Research.*IEEE Instrumentation and Measurement Technology Conference*, IMtc2003, USA,May20,2003.

[18] Liu Yuanning, Yang Xiaoguang, Yuan Shenmiao, Zhu Xiaodong, Xu Tao: A practical fast point pattern matching algorithm Based on matching pair support and registration condition. *Discrete and Continuous Dynamical Systems*. 2004.1

An adaptive image denoising method based on Deep Rectified Denoising Auto-Encoder

Hongqiang MA, Shiping MA, Yuelei XU, Mingming ZHU

Graduate School, Air Force Engineering University, Xian, China

E-mail: 18049025189@163.com

Abstract. Stacked Sparse Denoising Auto-Encoder(SSDA) has been successfully applied to image denoising, which is superior to the most of the traditional image denoising algorithms. However, the algorithm has low training convergence speed and poor universality. To address these limitations, we present an adaptive image denoising method based on Deep Rectified Denoising Auto-Encoder. To reduce training difficulty and speed up convergence, the rectified linear units (ReLu) is used as activation function of the network, and batch normalization(BN) is used to normalizes the input data of the middle layer in network. Different from the previous image denoising model, the proposed model is trained to learn the mapping relationship between noisy image and noise by residual learning. To overcome the problem that the new model cannot effectively deal with noise not seen in training, we perform adaptive training of multiple channels on the improved model to obtain the optimal channel weights and jointly output the denoised image. The experimental results show that the proposed algorithm can not only outperform SSDA in the convergence speed, but also adaptively remove the noise that is not seen in training.

1. Introduction

Image denoising is a classical yet still active topic in low level vision since it is an indispensable step in many practical applications [1], such as image segmentation, target tracking and recognition, etc. However, during the process of formation and transmission, images are often disturbed by different levels of noises, which degrade image quality. Therefore, it is of great theoretical and practical value to study the image denoising algorithms.

At present, there are mainly three kinds of image denoising algorithms. One is method based on wavelet decomposition or dictionary learning. Wavelet decomposition is to transform the image signal into wavelet domain for multi-layers decomposition where they can be more easily separated from the noise, such as BLS-GSM [2]. The dictionary-based method is to sparsely represent the noisy image on an overcomplete atom library so as to remove unnecessary information in the image and achieve denoising [3], such as KSVD. Another one is methods based on global image statistics or other image properties, such as self-similarity. BM3D is generally considered one of the state-of-the-art image denoising algorithms. It uses a collaborative form of Wiener filter for high dimensional block of patches by grouping similar 2D blocks into a 3D data array [4]. The last one is methods based on deep learning. In recent years, deep learning has received extensive attention from researchers and has gradually become an upsurge of internet and artificial intelligence [5]. It also provides new ideas for image denoising. Harmeling et al. [6] trains the a multi plain perceptron to learn the mapping relationship between noisy images and clean images, and its denoising performance is superior to most traditional denoising algorithms. Later, SSDA is successfully used to remove noise from corrupted images [7-8], especially when the noise intensity is large, the denoising performance is better than that of KSVD and BM3D, PSNR is higher and image blurring has been further improved.

The image denoising method based on deep learning has improved both visual effects and objective evaluation metrics compared with traditional methods [9]. However, deep learning is to learn feature expression from massive data and fit complex nonlinear functions. Therefore, the network is very deep, the parameter is huge, and the gradient dispersion phenomenon is serious. It takes a long time to adjust the parameters and often depends on the prior knowledge. Secondly, this method usually only adds a single type noise to train, and it can effectively remove the noise that appears in the training. If we want to remove other noise types, we have to spend a lot of time to retrain.

In view of this, we make the following three improvements to SSDA. 1) To prevent gradient dispersion, we adopt ReLu function instead of the sigmoid function as the activation function of hidden layers in SSDA, and the sparse constraint term is removed. 2) we use BN and residual learning to facilitate the adjustment of the parameters and speed up the convergence of the network. 3) We perform multi-channel training on the improved model (Deep Rectified Denoising Auto-Encoder, DRDA) to achieve adaptive image denoising and improve the robustness of the model to different types of noise.

2. Stacked Sparse Denoising Auto-Encoder(SSDA)

Vincent et al. proposed a denoising auto-encoder(DAE) in 2008. DAE is a 3 layers neural network, composed of input layer, hidden layer and output layer. The algorithm flow of DAE is shown in Figure 1. The original input data $\boldsymbol{x}$ first adds noise to get corrupted version , then through the coding function s_f to get the feature representation $\boldsymbol{h}$ of the input, and last through the decoding function s_g maps $\boldsymbol{h}$ to the output layer to obtain the reconstructed input data $\boldsymbol{y}$.

The encoding and decoding process of the DAE network is respectively $\boldsymbol{h}=s_f(\boldsymbol{w}\tilde{\boldsymbol{x}}+\boldsymbol{b})$, $\boldsymbol{y}=s_g(\boldsymbol{w}'\boldsymbol{h}+\boldsymbol{b}')$ .Where, s_f and s_g is a nonlinear activation function(for which sigmoid function $s(z)=(1+e(-z))^{-1}$ is often used). $\boldsymbol{w}$ and $\boldsymbol{w}'$ is respectively the encoding and decoding weights, $\boldsymbol{b}$ and $\boldsymbol{b}'$ is respectively the encoding and decoding bias. $\theta=\{\boldsymbol{w},\boldsymbol{w}',\boldsymbol{b},\boldsymbol{b}'\}$ is the parameter of the model DAE. Given training data $D=\{x^{(i)}\}_{i=1}^{N}$, the DAE is trained by backpropagation to minimize the reconstruction loss given by

$$L_{\mathrm{DAE}}(\theta)=\frac{1}{N}\sum_{i=1}^{N}\left\|y^{(i)}-x^{(i)}\right\|_2^2+\frac{\lambda}{2}(\|w\|_{\mathrm{F}}^2+\|w'\|_{\mathrm{F}}^2) \tag{1}$$

Where, λ is the weight of decay term, which can reduce the magnitude of the weight and prevent overfitting.

Later, researchers studied the information processing mechanism of human brain, and found that when images enter into human brain from vision, only a small number of neurons are stimulated, most of the neurons are still in a state of inhibition [10]. Inspired by this, Bengio adds an extra penalty term KL to the reconstruction loss function, which enables the activation of neurons in the hidden layer to satisfy a certain sparsity, that is, the activation value of most neurons is 0. The loss function of the Sparse Denoising Auto-Encoder(SDA) is as follows

$$L_{\mathrm{SDA}}(\theta)=\frac{1}{N}\sum_{i=1}^{N}\left\|y^{(i)}-x^{(i)}\right\|_2^2+\beta\sum_{j=1}^{k}KL(\hat{\rho}_j\|\rho)+\frac{\lambda}{2}(\|w\|_{\mathrm{F}}^2+\|w'\|_{\mathrm{F}}^2) \tag{2}$$

Where, $KL(\hat{\rho}_j \| \rho) = \rho \log \frac{\rho}{\hat{\rho}_j} + (1-\rho)\log\frac{1-\rho}{1-\hat{\rho}_j}$, ρ is a sparsity parameter, β controls the weight of the sparsity penalty term. $\hat{\rho}_j = \frac{1}{N}\sum_{i=1}^{N} \boldsymbol{h}_j(\tilde{x}^{(i)})$ is the average activation value of hidden unit j-th (averaged over the training set).

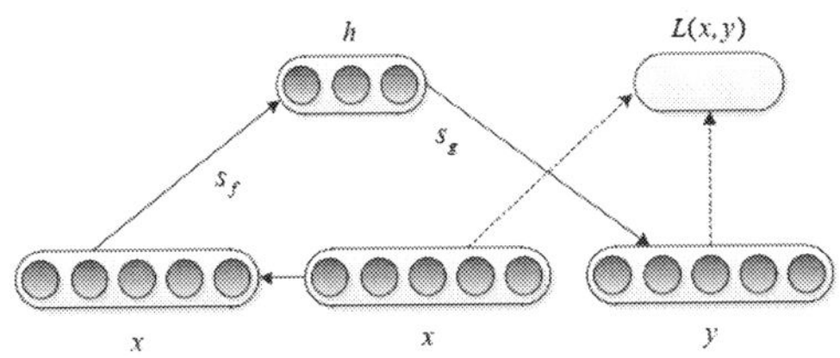

Figure 1. The algorithm flow of DAE

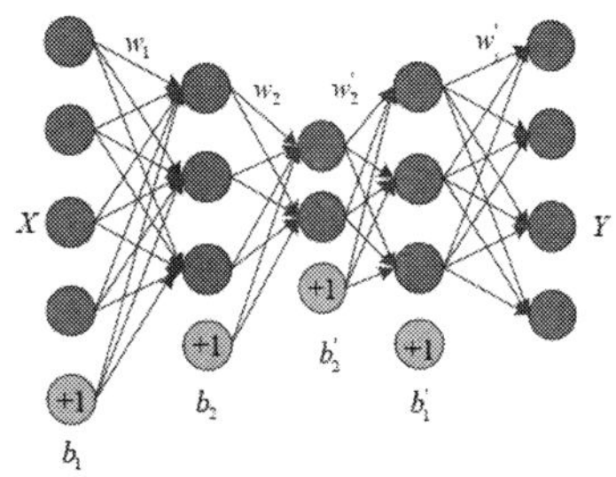

Figure 2. The network diagram of SSDA

SSDA is a deep neural network stacked by a series of SDAs. Between each SDA, the activation value of the previous SDA's hidden layer is used as the input of the next SDA. Figure 2 shows an SSDA network structure with two hidden layers.

A good way to obtain good parameters for SSDA is to use Hinton's unsupervised greedy layer-wise training method [11], which can relieve the phenomenon of dispersion to a certain extent. That is to say, each SDA is separately trained first and then the optimal weights of the network are obtained. Then, these weights are taken as the initial values of the SSDA network weights. Finally, the Back Propagation (BP) algorithm is used to fine-tune the whole network until the optimal parameters of the whole network are obtained. The loss function of SSDA in the fine-tune phase is as follows:

$$L_{\mathrm{SSDA}}(\theta) = \frac{1}{N}\sum_{i=1}^{N}\left\| \boldsymbol{y}^{(i)} - \boldsymbol{x}^{(i)} \right\|_2^2 + \frac{\lambda}{2}\sum_{l=1}^{l}(\left\| \boldsymbol{w}^{(l)} \right\|_F^2 + \left\| \boldsymbol{w}'^{(l)} \right\|_F^2) \tag{3}$$

Where, l is the number of stacked SDA. The sparsity-inducing term is not needed for this step because the sparsity was already incorporated in the pre-trained SDAs.

3. Adaptive image denoising

3.1 Deep Rectified Denoising Auto-Encoder (DRDA)

3.1.1 Activation function selection. The main reason for the gradient dispersion is still the residual, and the residual expression is as follows. The sigmoid function is used in SSDA. Since the derivative value f' of the sigmoid function is always less than 1 and the residual propagates backward from the output layer to the bottom layer. After each layer, the residual is multiplied by f'.Thus, after the residual is passed through many layers, the deep neural network cannot update network weights because of the low residual of the bottom.

$$\boldsymbol{\delta}^{(n_l)} = -(\boldsymbol{y} - \boldsymbol{a}^{(n_l)}) \bullet f'(\boldsymbol{z}^{(n_l)}) \tag{4}$$

$$\boldsymbol{\delta}^{(l)} = -((\boldsymbol{W}^{(l)}))^{\mathrm{T}}\boldsymbol{\delta}^{(l+1)} \bullet f'(\boldsymbol{z}^{(l)}) \tag{5}$$

$$\nabla \boldsymbol{W}^{(l)} = \boldsymbol{\delta}^{(l+1)}(\boldsymbol{a}^{(l)})^{\mathrm{T}} \tag{6}$$

Where, $\boldsymbol{\delta}^{(n_l)}$ is the residual of output layer, $\boldsymbol{a}^{(n_l)}$ is the actual output of output layer, $\boldsymbol{y}$ is the expected output of output layer, $\boldsymbol{\delta}^{(l)}$ is the hidden layer residual, $\boldsymbol{z}$ is the input of activation function, and $\boldsymbol{a}^{(l)}$ is the output of hidden layer. $\nabla \boldsymbol{W}$ represents the gradient of the loss function to the network weights.

The function expression of ReLu is $\max(0, x)$, whose derivative is 1. Therefore, the gradient can flow well in the backpropagation, which can effectively relieve the gradient dispersion. Compared with the sigmoid function, ReLu has 3 main characteristics: unilateral inhibition, relatively wide exciting boundary and sparsity activation. After initialization of network weights by experience rule, almost half of the neurons in the sigmoid function are activated, which is not consistent with neuroscience research. However. ReLu can sparse negative values and adjust dynamically, which is closer to the activation model of biology [12].

3.1.2 Residual learning and batch normalization. The goal of image denoising is to recover a clean image $\boldsymbol{x}$ from a noisy observation $\tilde{\boldsymbol{x}}$ which follows an image degradation model $\tilde{\boldsymbol{x}} = \boldsymbol{x} + \boldsymbol{n}$. Rather than directly outputing the denoised image $\hat{\boldsymbol{x}}$, in this paper, the algorithm we proposed is designed to predict the residual image $\hat{\boldsymbol{n}}$, i.e., the difference between the noisy observation and the latent clean image. We can find a mapping f that minimizes the cost function:

$$\arg\min_{f} \| f(\tilde{x}) - \boldsymbol{n} \|_2^2 \tag{7}$$

Therefore, we use residual learning, the input is a noisy image, and the output is the extracted noise. The denoised image can be obtained by formula $\boldsymbol{x} = \tilde{\boldsymbol{x}} - f(\tilde{\boldsymbol{x}})$.

The purpose of the BN algorithm is to batch and normalize the input data of each intermediate layer to reduce the influence of changes in the distribution of data in the middle of the network on the training of neural network parameters [13], thereby improving the stability of the neural network. Supposed that the input data collection of an intermediate layer of the network is $\{x_1, \cdots, x_n\}$, and the algorithm process is as follows:

$$\mu = \frac{1}{m}\sum_{i=1}^{m} x_i \tag{8}$$

$$\sigma = \frac{1}{m}\sum_{i=1}^{m} (x_i - \mu)^2 \tag{9}$$

Where, μ and σ is respectively the mean and variance, and then the data are normalized to data $\hat{x}_i$ with a mean of 0 and a variance of 1. ε is a constant to avoid disagreement when the variance is zero.

$$\hat{x}_i = \frac{x_i - \mu}{\sqrt{\sigma^2 + \varepsilon}} \tag{10}$$

In order to avoid damage to the feature distribution due to normalization of the data, it is necessary to restore the original feature distribution by reconstruction shift.

$$z_i = \mathrm{BN}_{\alpha,\beta}(\hat{x}_i) = \alpha\hat{x}_i + \beta \tag{11}$$

α and β are reconstruction parameters introduced by the reconstruction shift, which can obtain through network training.

According to [14], stochastic gradient descent algorithm is simple and efficient for deep network training, but it needs artificial choice of parameters. The choice of parameters is crucial to the training results so that a lot of time is wasted on the parameters adjustment. The use of BN greatly reduces the difficulty of network training and does not need to adjust the parameters deliberately to achieve a better performance.

In summary, we first propose a DRDA network. This model uses ReLu as an activation function and removes the sparse constraints in the SSDA loss function, using direct training method. At the same time, the network introduces the BN algorithm to normalize the middle layer data, uses the residual learning to learn the mapping between the noisy image and the noise, and then obtains the denoised image by $\boldsymbol{x} = \tilde{\boldsymbol{x}} - f(\tilde{\boldsymbol{x}})$. The network model DRDA is shown in Figure 3.

The loss function of DRDA is shown as follows

$$L_{\text{DRDA}}(\theta) = \frac{1}{N}\sum_{i=1}^{N}\left\|\boldsymbol{y}^{(i)} - (\tilde{\boldsymbol{x}}^{(i)} - \boldsymbol{x}^{(i)})\right\|_2^2 + \frac{\lambda}{2}\sum_{l=1}^{l}\left(\left\|\boldsymbol{w}^{(l)}\right\|_F^2 + \left\|\boldsymbol{w}'^{(l)}\right\|_F^2\right) \tag{12}$$

3.2 Adaptive Deep Rectified Denoising Auto-Encoder(ADRDA)

The image denoising methods using deep learning have achieved good results in recent years. However, this kind of method is to learn the mapping between the noisy image and the clean image by adding a single type of noise (such as Gaussian white noise), which is only effective for the same noise in the test phase. In view of this, we train adaptively the improved network model—DRDA to remove the noise that is not trained. The universality and robustness of the algorithm have been improved.

ADRDA is the linear combination of several DRDAs, or channels, each trained on a single type of noise can learn the optimized features generated by the activation of the DRDA's hidden layers. Then we use the all features as inputs to a neural network-based regression component, referred to here as the weight prediction module. this module then uses these features to compute the optimal channel weights used to linearly combine the column outputs into a weighted average. The network model is shown in Figure 4.

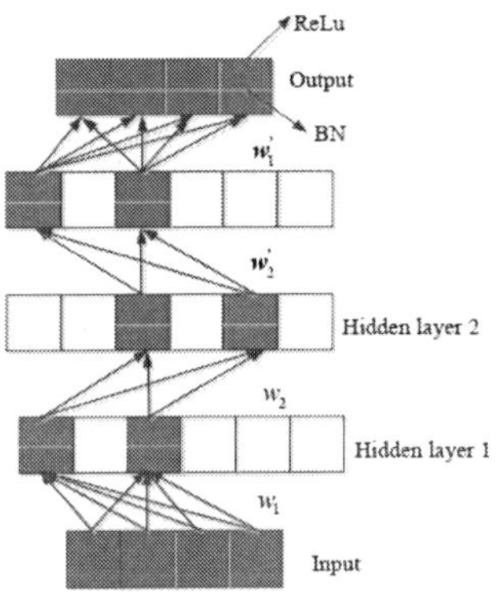

Figure 3. The network diagram of DRDA

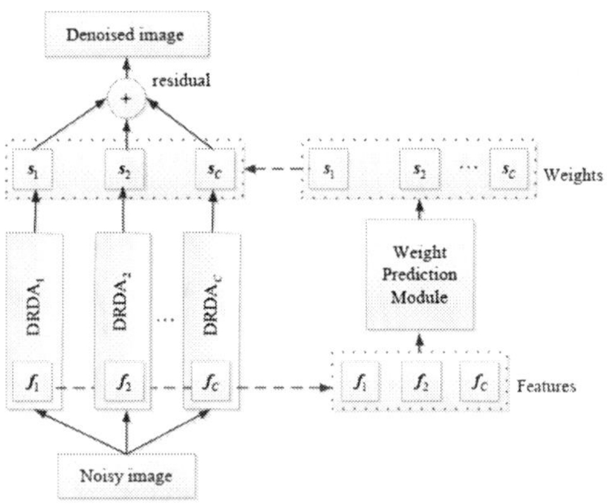

Figure 4. The network model of ADRDA

The ADRDA has three training phases: training the multi-channel DRDAs, finding ideal channel weights and training weight prediction model. The multi-channel DRDAs are trained as discussed in Section 3.1, with each DRDA provided a noisy training set, corrupted by a single noise type along with the original versions of those images as a target set.

3.2.1 Finding ideal channel weights. Once the DRDAs are trained, we make full use of the feature learning capability of the DRDA network and construct a new feature training set extracted from the hidden layers of the DRDAs. Specifically, for DRDA column c, the activations of hidden layers $[\boldsymbol{h}^{(1)};\cdots;\boldsymbol{h}^{(i)}]$ are concatenated into a feature vector f_c, and $f_1, f_2, \cdots f_C$ are concatenated to form the entire feature training set $\phi = [f_1;\cdots;f_C]$,where, i represents the number of stacked DRDA and C is the number of channels.

In addition, for each input image, each channel produces an output. $\hat{\boldsymbol{Y}} = [y_1,\cdots,y_C]$ represents all channel outputs, $\boldsymbol{n} = [\boldsymbol{n}_1,\cdots \boldsymbol{n}_C]$ is the expected noise of the corresponding channel. In order to find the channel weight of the ADRDA, the following loss function is minimized by:

$$\arg\min \frac{1}{2}\left\|\hat{\boldsymbol{Y}}\boldsymbol{s}-\boldsymbol{n}\right\|^2 \tag{13}$$

$$0 \le s_c \le 1, \forall c \tag{14}$$

$$1-\delta \le \sum_{c=1}^{C} s_c \le 1+\delta \tag{15}$$

Here, s is the vector of channel weights, and s_c represents the weight of the channel c. The constraint in Eq. [15] helps to avoid degenerate cases where some weights are too large or too small due to the sum of weights. Although making the weights sum exactly to one is more consistent with human intuition, we found that the performance slightly improved when given some flexibility. In this paper, δ=0.04 is used.

3.2.2 Training weight prediction model. The last stage is the training of the weight prediction model. Since a radial basis function(RBF) network can approximate any nonlinear function, it can handle the hard analytic regularity in the system and has good generalization ability [15]. Therefore, we choose an RBF network as the weight prediction module that is trained to take the feature training set as input and fit the ideal channel weight calculated by 3.2.1.

After the above three phases, the ADRDA training has been completed. A noisy image $\tilde{\boldsymbol{x}}$ is supplied as input to each of the channel, which together produce the output $\hat{\boldsymbol{Y}}$,which is an approximation of noise. The feature vector ϕ is created from the activation of the hidden layers of each DRDA and fed into the weight prediction module, which then computes the predicted column weights $\boldsymbol{s}^*$ according to the deep features of the data. The final denoised image of the proposed model is $\tilde{\boldsymbol{x}}-\hat{\boldsymbol{Y}}\boldsymbol{s}^*$.

4. Results and discussion

4.1 Experimental data

The experimental data in this paper are taken from natural image data sets commonly used in image denoising. we follow [1] to use 400 images of size 180×180 for training, called Train400. Because deep neural network training requires massive data samples, to enrich the training data and improve the generalization ability of the model, we use the data augmentation method to amplify the original database. That is, for Train 400, 90°, 180° rotation and 270° rotation and flip is performed respectively, and a total of 1600 images are used as the final training samples. We use BSD68(a dataset containing 68 natural images from Berkeley segmentation dataset) as validation set and Set12(a dataset containing 12 natural images with rich details and textures) as test set.

4.2 Analysis of convergence

In order to prove the validity of the model DRDA, we first compare DRDA and SSDA. In this paper, we set the number of layers in SSDA and DRDA to be 4. 8000 small image patches of size 8×8 are selected as the input of the network. The number of hidden neurons in the experiment was 40, λ=10^{-4} , β=10^{-2} , ρ=0.01 . The experiment was trained by the method of momentum gradient descent. The batch size is 1000, epoch is set to 10, momentum is [0.9 0.95 0.99] ,which is updated every 3 epochs. The initial value of learning rate is set to 0.1, and it is reduced by 10 times after 3 epochs.

We record the training curves of two models in Figure 5. We can see that DRDA can better fit the sample data than the SSDA network. First, both the training set and the Validation set, it can achieve a very low mean square error (MSE) which the SSDA network may not be able to achieve, and the number of iterations required is less than SSDA. The MSE reflects the difference between the denoised image and the clean image, which indirectly verifies that the denoised images using DRDA retain more information of the original images and has better denoising performance. Second, the mean square error curves of the two networks for train set and validation set are basically the same, which explains the rationality of the hyperparameters selection and the network generalization ability.

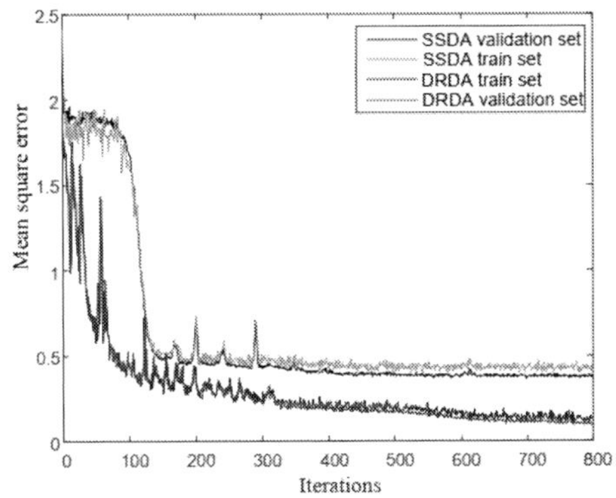

Figure 5. The convergence effect comparison

4.3 Adaptive image denoising

In order to better train the ADRDA model, we add Gaussian noise, speckle noise, and salt & pepper noise to the training samples, and each type of noise has 6 levels. In total, 18 channels of DRDA need to be trained. Among them, the variance of Gaussian noise is respectively 0.02,0.06, 0.10, 0.14, 0.18, 0.22. The density of speckle noise is respectively 0.05, 0.10, 0.15, 0.20, 0.25, 0.30. The density of the salt & pepper noise is the same as the speckle noise's. At the same time, to facilitate the experimental comparison and analysis, we respectively use the same type of noise images with all noise levels as training sample to train model. Here, we respectively refer to these as G-DRDA, S&P-DRDA, S-DRDA. MC-DRDA refers to training 18 channels, but the weight of each channel is equal. When testing the performance of the ADRDA model, Gaussian noise, speckle noise, salt and pepper noise, Poisson noise, and uniform noise is respectively added to the data set Set12. Each type of noise has two degrees, which are as follows: G1: 0.01, G2: 0.07, S1:0.1, S2:0.15, S&P1:0.1, S&P2:0.2, P1:0.5, P2:1.5, U1:0.2, U2:0.4.

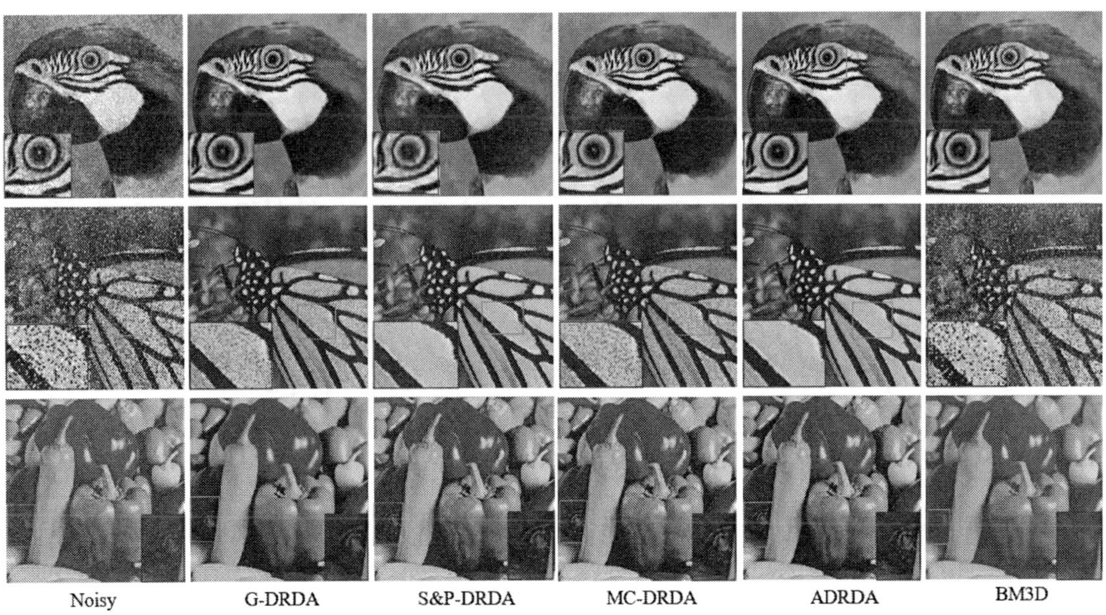

Figure 6. The comparison of denoising performance of different algorithms

In the subjective evaluation, we select three typical images in the data set Set12 as an example to demonstrate the denoising effect of the proposed algorithm. Additive Gaussian white noise, salt and pepper noise, and Poisson noise are added from top to bottom in Figure 6. It can be clearly seen that G-DRDA and BM3D have a good performance of removing Gaussian noise. The detailed features of denoised image such as edge and texture are well preserved, but they are not ideal to remove salt &pepper noise and still remain a lot of noise. We can also see that they can remove Poisson noise, but they will make the non-noise pixels of the image too smooth and the details lost. Although S&P- DRDA can remove these three types of noise, the denoised image remains to be improved. Compared with other algorithms, the algorithm ADRDA is effective for all three types of noise. The denoised image has clear outlines and the texture information remains more complete.

Table 1. The average PSNR/SSIM of the experimental results of different algorithms on the Set12 data set.

Noisy Type	G-DRDA	S&P- DRDA	S-DRDA	MC-DRDA	ADRDA	BM3D
G1	27.63/0.8156	25.39/0.7504	25.41/0.7561	27.44/0.8075	29.87/**0.8517**	**29.88**/0.8486
G2	25.39/0.7233	23.13/0.6194	22.78/0.6003	25.38/0.7171	**25.44/0.7321**	25.39/0.7232
S&P1	28.27/0.8406	34.90/0.9851	23.9/0.6795	26.62/0.8002	**36.87/0.9863**	19.50/0.4326
S&P2	25.75/0.7998	31.66/0.9661	22.92/0.6284	23.79/0.7347	**33.66/0.9711**	13.71/0.1646
S1	23.87/0.6851	24.45/0.6971	**28.58/0.8411**	24.91/0.7029	28.22/0.8396	19.40/0.4743
S2	23.6/0.6596	23.85/0.6644	27.75/**0.8147**	24.24/0.6713	**27.77**/0.8136	16.64/0.3769
P1	23.79/0.6845	24.44/0.7420	23.90/0.6877	24.26/0.6720	**28.83/0.8479**	27.13/0.8289
P2	23.4/0.6520	23.77/0.6561	22.92/0.6310	23.40/0.6463	**25.63/0.7517**	20.61/0.7454
U1	23.65/0.6770	24.11/0.6738	**25.05**/0.6739	24.14/**0.6815**	**25.05**/0.6669	16.98/0.3098
U2	19.28/0.6368	19.41/0.6366	20.21/0.6500	19.88/0.6443	**21.10/0.6610**	14.13/0.3036
Average	24.46/0.7210	25.51/0.7391	24.80/0.7014	24.35/0.7083	**28.25/0.8122**	20.34/0.5208

In objective evaluation, PSNR and SSIM are used to evaluate the quality of denoised image. The PSNR reflects the degree of noise reduction of the image, and the SSIM reflects the integrity of the structure information. Table 1 shows the denoising performance of different algorithms for different noise types. We can conclude that G-DRDA, S&P- DRDA and S-DRDA algorithm have higher PSNR and SSIM when removing the corresponding types of noise, while the PSNR and SSIM are lower for other noise types. The algorithm ADRDA is slightly worse than S-DRDA in removing speckle noise. However, its PSNR and SSIM are all higher than other methods, whether the types of noise seen in training (such as Gaussian, salt and pepper noise), or the noise that does not participate in training (such as Poisson and uniform noise). At the same time, in order to verify the effectiveness of the proposed algorithm, compared with BM3D that is considered as one of the state-of-the-art image denoising methods, we can find that BM3D is only effective for Gaussian noise, while the two objective values for other noise types are lower than ADRDA.

In summary, the algorithm ADRDA has a better denoising performance, which can be improved both in subjective visual and objective evaluation indicators. In addition, the algorithm overcomes the disadvantages of most denoising algorithms, and it can also deal with different types of noise within a certain range without acquiring the prior knowledge of test images.

5. Conclusion and future work

In this paper, we present an adaptive image denoising method based on Deep Rectified Denoising Auto-Encoder. The network uses ReLu as an activation function to effectively alleviate the gradient dispersion. In addition, the joint use of BN and residual learning greatly reduces the difficulty of network training and improves the training speed and image denoising performance. In order to overcome the problem that the new model DRDA can only effectively remove the single type of noise seen in the training, we take advantage of feature learning ability of the model to carry out multi-channel adaptive training. The experimental results show that ADRDA algorithm is better than other methods in terms of subjective visual and objective evaluation indicators, and that can adaptively handle different noise within a certain range. It has certain reference significance and practical value. In the future, we will train the convolution neural network(CNN) adaptively to further improve the image denoising performance.

References

[1] Zhang K, Chen Y, Chen Y, et al. Beyond a Gaussian Denoiser: Residual Learning of Deep CNN for Image Denoising[J]. IEEE Transactions on Image Processing, 2016, 26(7):3142.

[2] Portilla J, Strela V, Wainwright M J, et al. Image denoising using scale mixtures of Gaussians in the wavelet domain[J]. IEEE Transactions on Image Processing, 2003, 12(11): 1338.

[3] Yan X, Yang B, Zhang W, et al. An Improved Denoising Algorithm of Feather and Down Image Based on KSVD[C]// International Conference on Information Technology in Medicine and Education. 2016 pp 419-423.

[4] Dabov K, Foi A, Katkovnik V, et al. Image restoration by sparse 3D transform-domain collaborative filtering[J]. 2008, 6812(8): 681207.

[5] Yu K, Jia L, Chen Y Q, et al. Deep learning: yesterday, today and tomorrow[J]. Journal of Computer Research and Development, 2013, **50(9)** pp 1799-1804.

[6] Harmeling S, Schuler C J, Burger H C. Image denoising: Can plain neural networks compete with BM3D?[C]// Computer Vision and Pattern Recognition. IEEE, 2012 pp 2392-2399.

[7] Xie J Y, Xu L L, Chen E H, et al. Image de-noising and inpainting with deep neural networks[C]// Proceedings of the 25th International Conference on neural information processing systems. New York, USA: ACM Press, 2012 pp 341-349.

[8] Li H M. Deep Learning for Image Denoising[J]. International Journal of Signal Processing Image Processing and Pattern Recognition, 2014, **7(3)** pp 171-180.

[9] Zhang Y G, Yi B S, Wu C Y, et al. Low-dose CT image denoising method based on convolutional neural network[J]. Acta Optica Sinica, 2018, 38(4): 0410003.

[10] Xu J, Xiang L, Liu Q, et al. Stacked Sparse Autoencoder (SSAE) for Nuclei Detection on Breast Cancer Histopathology Images[J]. IEEE Transactions on Medical Imaging, 2016, **35(1)** pp 119-130.

[11] Hinton G E, Osindero S, The Y W. A fast learning algorithm for deep belief nets[J]. Neural Computation, 2006, **18(7)** pp 1527 -54.

[12] Glorot X, Bordes A, Bengio Y, et al. Deep Sparse Rectifier Neural Networks[C]// International Conference on Artificial Intelligence and Statistics. 2012 pp 315-323.

[13] Zhu W, Qu J Y, Wu R B. Straight convolutional neural networks algorithm based on batch normalization for image classification[J]. Journal of Computer-Aided Design & Computer Graphics, 2017, **29(9)** pp 1650-1657.

[14] Ioffe S, Szegedy C. Batch Normalization: Accelerating Deep Network Training by Reducing Internal Covariate Shift[J]. 2015 pp 448-456.

[15] Gai J F, Zhao G R, Song C. Model predictive control based on linearization and neural network approach[J]. Systems engineering and electronics, **2015(2)** pp 394-399.

IOP Conf. Series: Journal of Physics: Conf. Series **1060** (2018) 012049 doi:10.1088/1742-6596/1060/1/012049

LDA Extension via Oblique Projection

B J Yang[1], W G Shang[2] and Y J Wu[2,*]

[1] School of Mathematic Sciences, Guizhou Normal University, Guiyang, P.R. China
[2] School of Information Engineering, Guangdong Medical University, Dongguan, P.R. China

[*]E-mail: wuyingjiang@163.com

Abstract. The classical linear discriminant analysis (LDA) was previously modified by orthogonal projection into null space LDA (N_LDA) and direct LDA (D_LDA) for solving small sample size (SSS) problem. In this paper, the author proposes an extension of LDA by oblique projection, wherein N_LDA and D_LDA are included as special cases, to reduce discriminative information loss resulted from single N_LDA or D_LDA. The effectiveness of the proposed algorithm is tested by image forensics and face recognition.

1. Introduction

Linear discriminant analysis (LDA) is one of the most popular means in pattern recognition [1]. The objective of LDA is to find a projection W that maximizes the ratio of the between-class scatter matrix, S_b, against the within-class scatter matrix, S_w, i.e., via the following Fisher criterion:

$$J(W_{opt}) = \arg \max_{W} \frac{|WS_bW^{\mathrm{T}}|}{|WS_wW^{\mathrm{T}}|}. \tag{1}$$

One of the major drawbacks of LDA is the so-called small sample size (SSS) problem encountered in many practical applications involving high dimensional data [1]. Many techniques have been proposed to solve this problem, and Ref. [2] may be consulted in detail. Among them, the most notable approaches are null space LDA (N_LDA) and direct LDA (D_LDA) [2, 3], which are based on the following modified Fisher criterion [4]:

$$WS_wW^{\mathrm{T}} = 0 \quad and \quad WS_bW^{\mathrm{T}} \neq 0. \tag{2}$$

Since N_LDA and D_LDA are based on orthogonal projection, and generally the two conditions in equation (2) cannot be simultaneously satisfied, the processes are usually fulfilled in two steps: in N_LDA the null space of S_w is found first and then that of S_b is discarded, while in D_LDA the reverse order is taken. In this manner, much discriminated information may be lost.

In this paper, an extension of LDA is proposed by oblique projection, and it is called OP_LDA, for reduction of possible loss of discriminative information in N_LDA or D_LDA. Experimental results on image forensics and face recognition show satisfactory performance of the proposed method.

The rest of this paper is organized as follows. An overview of N_LDA and D_LDA is briefly introduced in Section 2. Section 3 describes OP_LDA and discusses the relationship among N_LDA, D_LDA and OP_LDA. In Section 4, we evaluate the performance of OP_LDA and compare its performance with LDA, N_LDA, D_LDA, and 2-dimensional LDA (2D_LDA) [5]. Finally, conclusions are given in Section 5.

2. N_LDA and D_LDA

2.1. N_LDA

In N_LNA, the discriminative information can be found in the null space of Sw by two steps [2]:

Step 1: Find the null space of S_w, Null(S_w).

Step 2: Let $S_b'=P_{\mathrm{Null}(S_w)}(S_b)$, and then

$$W_1 = P_{\mathrm{Col}(S_b')}, \tag{3}$$

where $P_{\mathrm{Null}(S_w)}(S_b)$ is a matrix obtained by orthogonal projecting the column vectors of S_b onto Null(S_w), Col(S_b') is the column space of S_b', $P_{\mathrm{Col}(Sb')}$ is an orthogonal projection operator onto Col(S_b').

Figure 1 depicts a simple example of this algorithm, where v is an arbitrary vector in the input space, and v' is a new vector obtained from v by N_LDA.

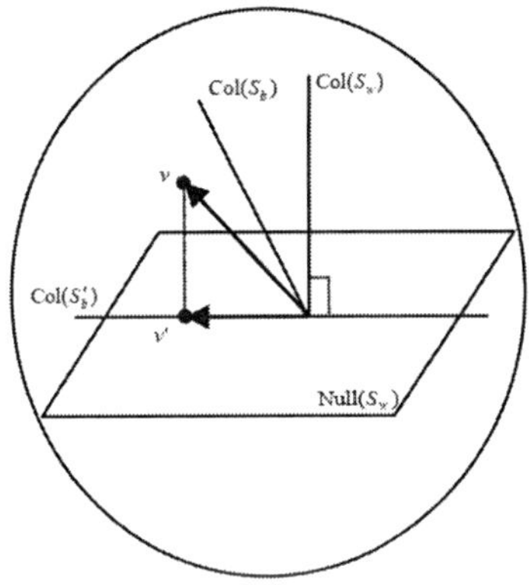

Figure 1. A simple example of N_LDA.

2.2. D_LDA

In D_LNA, the discriminative information can be found in the column space of S_b by two steps [3]:

Step 1: Find the column space of S_b.

Step 2: Let $S_w'=P_{\mathrm{Col}(Sb)}(S_w)$, and then

$$W_2 = P_{\mathrm{Null}|_{\mathrm{Col}(S_b)}(S_w')}, \tag{4}$$

where $P_{\mathrm{Col}(Sb)}(S_w)$ is a matrix obtained by orthogonal projecting the column vectors of S_w onto Col(S_b), $\mathrm{Null}|_{\mathrm{Col}(Sb)}(S_w')$ is the null space of S_w' in Col(S_b), $P_{\mathrm{Null}|_{\mathrm{Col}(Sb)}(Sw')}$ is an orthogonal projection operator onto $\mathrm{Null}|_{\mathrm{Col}(Sb)}(S_w')$.

Since rank(S_b), the rank of matrix S_b, is usually smaller than rank(S_w), and in this case Col(S_w')=Col(S_b) and $\mathrm{Null}|_{\mathrm{Col}(Sb)}(S_w')=\{0\}$, the solution of D_LDA is given by

$$W_2 = P_{\mathrm{Col}(S_b)}. \tag{5}$$

Figure 2 depicts a simple example of this algorithm, where v is an arbitrary vector in the input space, and v' is a new vector obtained from v by D_LDA.

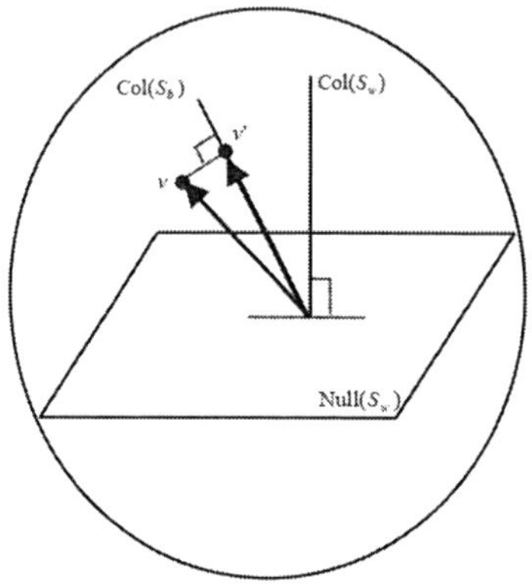

Figure 2. A simple example of D_LDA.

3. Proposed method

3.1. OP_LDA

We present a new LDA algorithm based on oblique projection, named OP_LDA. The objective of OP_LDA is to find a projection operator W that satisfies equation (6), which also is special circumstance of equation (2).

$$WS_w = 0 \quad and \quad WS_b \neq 0. \tag{6}$$

The key idea of our algorithm is that W is an oblique projection operator, whereas the traditional methods take it as orthogonal projection. A particular solution of equation (6) is given by

$$W_3 = P_{\mathrm{Col}(S_b)|\mathrm{Col}(S_w)}, \tag{7}$$

where $P_{\mathrm{Col}(Sb)|\mathrm{Col}(Sw)}$ is an oblique projection operator onto Col(S_b) along Col(S_w), and W_3 can be obtained from Col(S_w) and Col(S_b), Ref. [6] may be consulted in detail.

Figure 3 depicts a simple example of this algorithm, where v is an arbitrary vector in the input space, and v' is a new vector obtained from v by OP_LDA.

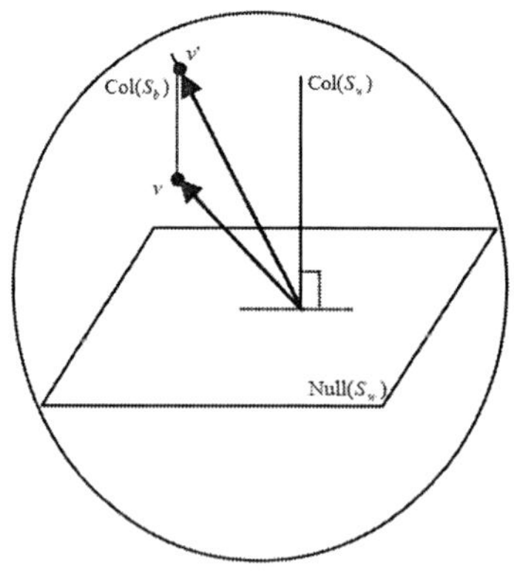

Figure 3. A simple example of OP_LDA.

3.2. Discussion

As shown in Figure 1, Figure 2 and Figure 3, there are three means to obtain v': a. discarding the component of Col(S_w), b. keeping that of Col(S_b), and c. discarding that of Col(S_w) and keeping that of Col(S_b) simultaneously. So it is clear that OP_LDA including N_LDA and D_LDA as special cases when Col(S_w) and Col(S_b) are orthogonal.

The traditional methods emphasize S_w or S_b respectively, and our method emphasizes S_w and S_b simultaneously. For discarding Col(S_w) in N_LDA, S_b is converted into S_b' by projecting onto Null(S_w). Since there are differences between S_b and S_b', discriminated information may be lost by adopting S_b'. The similar drawback may appear in D_LDA. Therefore discriminative information loss in N_LDA or D_LDA may be reduced by adopting S_w and S_b directly in OP_LDA.

3.3. Analysis

The following condition needs to be satisfied in using $P_{\mathrm{Col}(Sb)|\mathrm{Col}(Sw)}$: the direct sum of Col($S_b$) and Col($S_w$) must be equal to the input space [6]. Since the direct sum is the sum of two disjoint spaces, it includes two meanings: the sum of Col(S_b) and Col(S_w) must be equal to the input space, and Col(S_b) and Col(S_w) must be disjoint, i.e., $\mathrm{Col}(S_w)\cap\mathrm{Col}(S_b)=\{0\}$.

Since the sum of Col(S_b) and Col(S_w) is a subspace of the input space, $P_{\mathrm{Col}(Sb)|\mathrm{Col}(Sw)}$ is invalid for vectors which do not belong to the sum space. Therefore we project vectors in this case onto Col(S_t) first, where S_t denotes the total scatter matrix. Since $\mathrm{Null}(S_w)\cap\mathrm{Null}(S_b)=\mathrm{Null}(S_t)$ has been proven in Ref. [7], and S_w, S_b and S_t are symmetric matrixes, it is clear that Col(S_t) is the sum of Col(S_b) and Col(S_w).

If $\mathrm{Col}(S_w)\cap\mathrm{Col}(S_b)\neq\{0\}$, $P_{\mathrm{Col}(Sb)|\mathrm{Col}(Sw)}$ does not exist. This is a more complicated problem. In this paper we consider a simple scheme as follow:

$$W_3=\begin{cases}P_{\mathrm{Col}(S_b)|\mathrm{Col}(S_w)}P_{\mathrm{Col}(S_t)},\mathrm{Col}(S_w)\mid\ \mathrm{Col}(S_w)=\{0\}\\ P_{\mathrm{Col}(S_b)|P_{\mathrm{Col}(S_w)}\mathrm{Null}(S_b)}P_{\mathrm{Col}(S_t)},\mathit{otherwise}.\end{cases}\tag{8}$$

where $P_{\mathrm{Col}(St)}$ is a orthogonal projection operator onto Col(S_t), $P_{\mathrm{Col}(Sw)}$Null(S_b) is a space obtained by orthogonal projecting the vectors of Null(S_b) onto Col(S_w), and $P_{\mathrm{Col}(Sb)|P_{\mathrm{Col}(Sw)}\mathrm{Null}(Sb)}$ is an oblique projection operator onto Col(S_b) along $P_{\mathrm{Col}(Sw)}$Null(S_b).

Figure 4 depicts a simple example of our method, where v is an arbitrary vector in the input space, and v' is a new vector obtained from v by equation (8). It is clear that we want v' to lie in Col(S_b) and discard the partial components of Col(S_w).

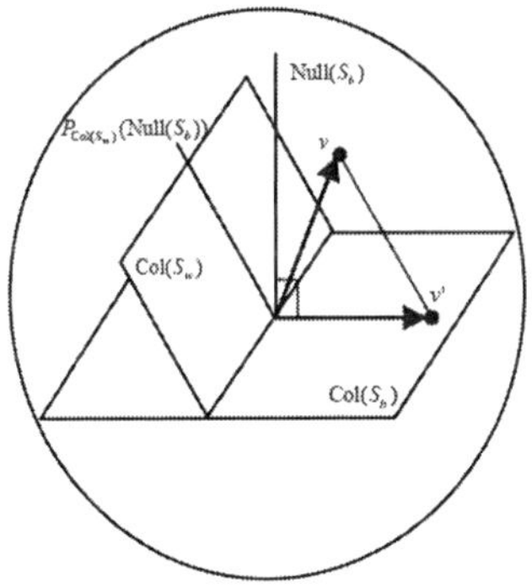

Figure 4. A simple example of proposed method for $\mathrm{Col}(S_w)\cap\mathrm{Col}(S_b)\neq\{0\}$.

4. Experimental results

4.1. Experimental setup

We illustrate the efficacy of the proposed method on image forensics and face recognition. As application-based development, some parameters and schemes have been empirically determined in our implementation.

For finding the null space and the column space of an arbitrary symmetric matrix, M, we perform first the singular value decomposition of M as $M=U\sum V^{\mathrm{T}}$, and then let $\mathrm{Col}(M)=\mathrm{span}\{u_1,\ldots,u_r\}$ and

Null(M)=span$\{u_{r+1},\ldots,u_n\}$, where $U=[u_1,\ldots,u_r,u_{r+1},\ldots,u_n]$, $u_1,\ldots,u_r$ correspond to the nonzero eigenvalues, and $u_{r+1},\ldots,u_n$ correspond to the zero eigenvalues [8].

In order to conform whether Col(S_b) and Col(S_w) are disjoint, we consider a simple scheme via the following dimensional formula [9]: if rank(S_t)=rank(S_b)+rank(S_w), then Col(S_w)∩Col(S_b)=$\{0\}$, otherwise Col(S_w)∩Col(S_b) $\neq\{0\}$.

To evaluate the performance of the proposed method, the training samples are randomly selected about half of the image dataset to train the classifier, and the remaining images are used in testing.

4.2. Image forensics

OP_LDA is tested by blurring detection in image forensics. The dataset and feature extraction are same as Ref. [10], wherein details can be found. The dataset consists of 183 authentic images and 183 forged images, and three types of feature vectors respectively consisted of the singular values of gray image matrix, correlation coefficients for double blurring operation, and image quality metrics are extracted and then fused for forgery detection.

In comparison, Table 1 lists the detection rates of the following four different detection schemes: feature fusion obtained respectively by LDA, N_LDA, D_LDA, and OP_LDA, respectively plus Euclidean distance (ED) classifier. In this table, "--" means that LDA is ineffective, due to the SSS problem.

Table 1. Detection rate vs. different schemes in image forensics.

Scheme	Detection rate (%)
LDA+ED	--
N_LDA+ED	86.81
D_LDA+ED	60.55
OP_LDA+ED	86.81

Since image forensics is in fact a binary classification problem, and in this case Null(S_w) is lost in D_LDA, D_LDA performs poor. While Col(S_b) is kept in N_LDA , so N_LDA obtained the same result as OP_LDA in this experiment.

4.3. Face recognition

OP_LDA is tested by face recognition using three face datasets: YALE, ORL, and PIE, details can be found in Ref. [5]. The YALE dataset contains 165 face images of 15 persons. The image size is 100×100. We subsample the images down to a size of 25×25. The ORL dataset contains 400 face images of 40 persons. The image size is 92×112. We subsample the images down to a size of 23×28. The PIE dataset is a subset of the CMU_PIE face image dataset. It contains 6615 face images of 63 persons. The image size is 486×640. We subsample the images down to a size of 98×128. Table 2 lists the recognition rates of the following five different schemes: feature extraction obtained respectively by LDA, N_LDA, D_LDA, OP_LDA, and 2D_LDA, respectively plus ED classifier.

Table 2. Recognition rate vs. different schemes in face recognition.

Scheme \ Database	YALE	ORL	PIE
LDA+ED	--	--	--
N_LDA+ED	92.93	93.6	100
D_LDA+ED	87.87	89.1	96.39
OP_LDA+ED	93.07	93.75	98.59
2D_LDA+ED	92.53	93.4	99.99

The results demonstrate that our proposed method can achieve comparable performance to 2D_LDA in [5]. Compared to the prior arts in [2, 3], OP_LDA performs better than N_LDA and D_LDA in using YALE and ORL face datasets. Due to the SSS problem, LDA is ineffective in these experiments.

5. Experimental results

An extension of LDA by oblique projection for solving SSS problem is proposed in this paper. Instead of using orthogonal projection, oblique projection is adopted to reduce possible discriminative information loss in N_LDA or D_LDA, two popular modified versions of the traditional LDA. Experimental results on image forensics and face recognition showed satisfactory performance of the proposed method. OP_LDA may be converted into a nonlinear version by kernel method, which is one of our future tasks.

Acknowledgments

This work was supported in part by the Doctor Start-up Fund of Guizhou Normal University (2016)

References

[1] Fukunaga K 1990 *Introduction to Statistical Pattern Recognition 2th ed.* (New York: Academic Press) chapter 10

[2] Chen L F, Liao H Y M, Ko M T, Lin J C and Yu G J 2000 A new LDA-based face recognition system which can solve the small sample size problem *Pattern Recognition* **33(10)**, 1713–26

[3] Yu H and Yang J 2001 A direct LDA algorithm for high-dimensional data: with application to face recognition *Pattern Recognition* **34(10)** 2067–70

[4] Liu K, Cheng Y Q and Yang J Y 1993 Algebraic feature extraction for image recognition based on an optimal discriminant criterion *Pattern Recognition* **26(6)** 903–911

[5] Ye J P, Janardan R and Li Q 2004 Two-dimensional linear discriminant analysis *Advances in Neural Information Processing Systems* **17** 1569–76

[6] Liu B Y 2012 Incremental learning of oblique projection-based kernel discriminator: Theory and algorithm *Sciencepaper Online*

[7] Yang J, Zhang D and Yang J Y 2003 A generalised K-L expansion method which can deal with small sample size and high-dimensional problems *Pattern Analysis and Applications* **6(1)** 47–54

[8] Zhang X D 2004 *Matrix Analysis and Applications* (Beijing: Tsinghua University Press) pp 609–612

[9] Li D C 1995 On Theorems of Dimensions and Bases of Subspace's Intersections *Journal of Southwest Nationalities College, Natural Science Edition* **21(2)** 224–226

[10] Yang B J and Liu B Y 2013 Image forgery detection via multifeature classification *Journal of Pattern Recognition and Image Processing* **4(4)** 413–421

A Clustering-Based Fast and Stable Routing Protocol for Vehicular Ad Hoc Networks

P F Zhao[1,2,3], K Liu[1,2,3], Y Zhang[1,2,3], T Zhang[1,2,3] and F Liu[1,2,3]

[1] School of Electronics and Information Engineering, Beihang Univ., Beijing, China
[2] Beijing Key Lab for Network-Based Cooperative ATM, Beijing, China
[3] Beijing Laboratory for General Aviation Technology, Beijing, China

zpf_wqt@163.com

Abstract. For the sake of improving network performances, i.e. end-to-end delay and packet drop ratio (PDR), in vehicular ad hoc networks (VANET), we propose a clustering-based fast and stable routing (CFSR) protocol. The CFSR protocol introduces a link quality assessment mechanism that evaluates the link quality on road segments between intersections and adopts the end-to-end delay as an evaluation metrics to assign a weight to the link corresponding to each road segment. The delay-sensitive application can directly use this weight as the link quality indicator to directly establish the routing path in the sub-zone by using Dijkstra algorithm. At the same time, the concept of local coordinator (LC) is introduced into the routing protocol to help construct the routing path conveniently and quickly. In addition, it can also avoid the local optimum problem caused by single-step decision and reduce the network transmission delay. In this paper, the Simulation of Urban Mobility (SUMO) and Network Simulation 3 (NS3) are used to simulate this protocol. Simulation results reveal that CFSR protocol can obtain lower end-to-end delay and PDR than those of the AODV and DSDV protocols.

1. Introduction

Vehicular ad hoc network (VANET) basically is a specific form of mobile ad hoc network (MANET) applied in modern transportation system, serving as providing communication among the adjacent vehicles and the nearby fixed road-side equipment. VANET includes two communication approaches, Vehicles-to-Vehicles (V2V) and Vehicles-to-Infrastructure (V2I). Some applications of VANET focus on improve the driving safety via reminding drivers of avoiding dangerous transportation events, such as car crashes, extreme weather and so on, while others pay attention to making trips more joyful and relaxing. For instance, passengers can acquire real-time information about traffic statistics so as to select a faster route to save time. In addition, business services of malls and restaurants can be accessed through road side units (RSUs) [1].

There is a rising demand for a lower end-to-end delay (E2ED) for real-time non-safety applications. That is to say, it is crucial to choose a routing protocol with minimum E2ED to handle this issue. Most existing routing protocols select the routing paths by using greedy-based algorithms which are likely to account for local maximum problem, leading to higher E2ED [2][3]. For example, most position-based routing protocols, such as GPSR and GSR, select the shortest distance path between source and destination, while GyTAR and A-STAR choose roads which are well connected. As said before, these protocols that use greedy algorithm and single-step decision are prone to the local maximum problem which occurs when no other connected roads are closer to the destination than the current one. Although this issue can be settled down by carry-and-forward mechanism, it incurs longer delivery delay.

Giving preference to the global network topology is our motivation to design a clustering-based fast and stable routing (CFSR) protocol, which can avoid the local optimal problem. To achieve this goal, the CFSR protocol introduces a link quality assessment mechanism that evaluates the link quality on road between intersection and uses the end-to-end delay as an evaluation indicator to assign a weight to the link corresponding to each road, as designed in [4]. The delay-sensitive application can directly use this weight as the link quality indicator to directly establish the routing path in the sub-zone by using Dijkstra algorithm. In this way, we can achieve the goal of reducing the E2ED.

2. Network Model

This model aims at a typical city environment and assumes that it merely consists of road segments and intersections as shown in figure 1. For this network model, the article adopts a hierarchical structure to implement network division and management. Specifically, the road segment can be divided into several stationary clusters, meaning that the vehicles on the specific road segment belong to relative clusters [5]. Clusters can communicate with each other through cluster heads; and communication between vehicles of different road segments can be relayed through the gateways node selected at the intersections. Here are some assumptions and explanations for the typical scenario.

- The hierarchical cluster structure is shown in figure 2 and can be divided into three parts. The underlying network consists of cluster members (CMs) within each road segment. The middle-level network consists of the cluster heads (CHs) of each link and the gateways of the crossroads. The high-level network consists of local coordinators (LCs) selected by each sub-zone topology.
- Rij is defined as the road between intersections i and j. At the same time, the hierarchical network can be mapped into a graph G(V, E), where V is a set of network nodes v and E is a set of edges e between nodes. In addition, V' is a path set, and if this set exists a path ei, ea, eb, … , en, ej between the vehicle nodes i and j, then Rij is called as "connected " state.
- Assume that nodes in the network are all assigned a state: cluster head (CH), cluster member (CM), gateway node (GW) or local coordinator (LC).
- Each node has a unique ID, equipped with GPS devices.

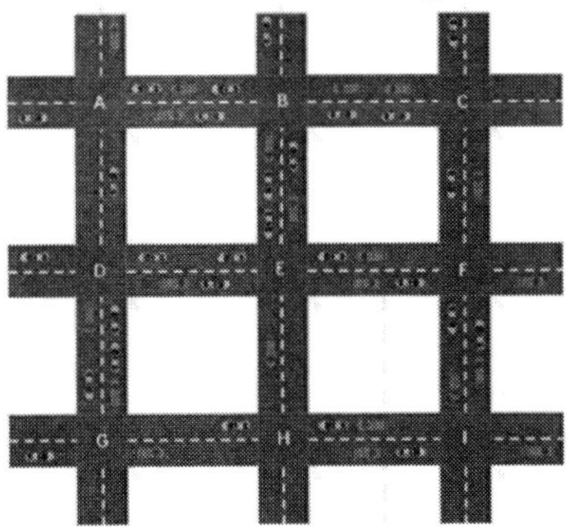

Figure 1. Network model.

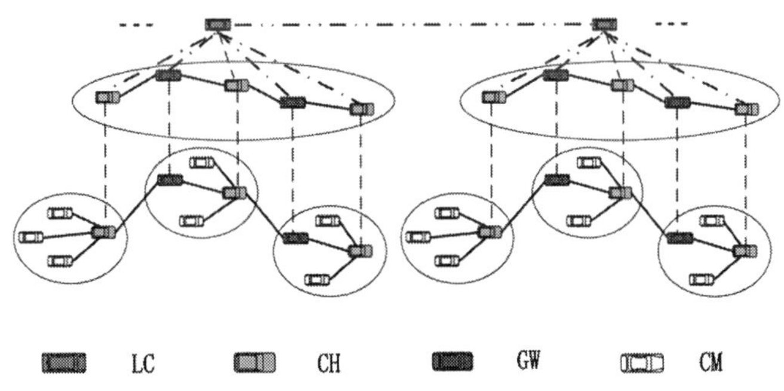

Figure 2. Hierarchical clustering structure.

3. Clustering mechanism

The road segment is designed as a bi-direction road, and each is divided into multiple cluster. Moreover, the diameter of clusters equals half of the transmission range of a standard vehicle. As shown in figure 3, due to the particularity of the lateral extension of the road, the length of a cluster is much larger than its width, which results in that the cluster shape on the road can be approximated as a rectangle.

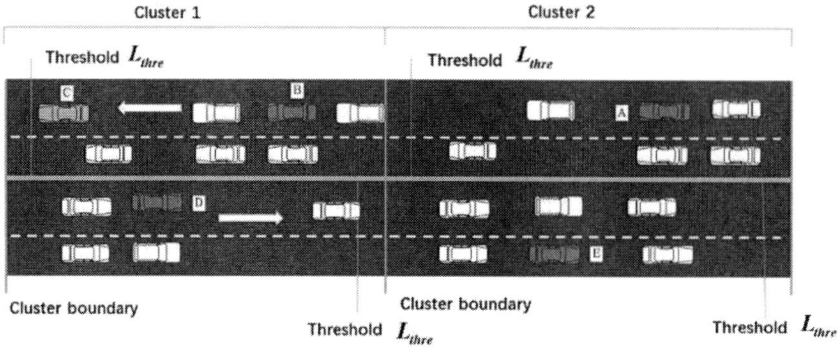

Figure 3. Cluster structure for road segment.

3.1. Clustering head selection process

After the stationary cluster division ends, the cluster head selection process begins. Vehicles within a cluster interacts information <ID, x, y, v, d, b> - where ID is the identification of each vehicle, <x, y> is the European coordinates of the vehicle, v is the speed, d is the direction, and b is the vehicle node flag to indicate the status of the vehicle node (LC: 11, GW: 10, CH: 01, CM: 00). Based on these interactions, this paper selects two factors, velocity and location, as the cluster head selection criteria to calculate stability of nodes [6].

3.1.1. Velocity factor. This performance of speed is crucial in VANET. It results in high dynamics of the cluster structure and very fast change of the cluster topology. If the node with strong mobility is selected as the cluster head, it will cause the instability of the communication link and the cluster structure. The optimal result makes us choose a node as the vehicle node with smallest speed differences between neighbor vehicles.

Supposing that the CH selection mechanism takes the instantaneous speed v_i for each vehicle, where v_u is the average of all vehicle nodes at time t for calculating the speed weighting factor $P_v(i)$ for each vehicle.

$$P_v(i) = \frac{v(i) - v_\mu}{v_\mu}, \forall i \in \phi_n \tag{1}$$

For a vehicle node within a cluster, its velocity weighting factor can be expressed as a vector in the following formula:

$$P_v(CID) = [P_v(1), P_v(2), P_v(3), ..., P_v(n)] \tag{2}$$

These values can be normalized by scaling them between zero and one as:

$$P_{v_{norm}}(i) - \frac{P_v(i) - \min(P_v(CID))}{\max(P_v(CID)) - \min(P_v(CID))}, \forall i \in \phi_n \tag{3}$$

The vehicle with the least $P_{v_{norm}}(i)$ value has the highest speed priority.

3.1.2. Location factor. When the position of the cluster head is close to the start cluster boundary position, the cluster head will stay for a longer time to prevent frequent replacement of cluster heads, which can improve the stability of the cluster. Therefore, we also consider using position weight factor as one of metrics.

The location factor is defined as the distance from current position to stationary cluster boundary, the formula is as follows:

$$P_l(i) = \frac{L(i)}{R}, \forall i \in \phi_n, 0 \le L(i) \le R \tag{4}$$

For a vehicle node within a cluster, its location weight factor can be expressed as a vector in the following formula:

$$P_l(CID) = [P_l(1), P_l(2), P_l(3), ..., P_l(n)] \tag{5}$$

These values can be also normalized by scaling them between zero and one as:

$$P_{l_{norm}}(i) = \frac{P_l(i) - \min(P_l(CID))}{\max(P_l(CID)) - \min(P_l(CID))}, \forall i \in \phi_n \tag{6}$$

The vehicle with the least $P_{l_{norm}}$ value has the highest location priority.

3.1.3. Integrated stability factor. For one cluster, the integrated stability of vehicles can be obtained by:

$$S(i) = \alpha P_{l_{norm}}(i) + (1\text{-}\alpha) P_{v_{norm}}(i),\ \alpha \in (0,1) \tag{7}$$

The former part of the formula evaluates the impact of distance between vehicles on cluster stability, because the greater the distance from the boundary of the vehicle, the more frequently the cluster heads are replaced, which is not conducive to cluster stability. The latter part mainly reflects the impact of vehicle speed on vehicle stability, because when the vehicle speed is small, the topology of the network changes slowly and the service lifetime is relatively long. On the other hand, if the vehicle speed is relatively fast, the performance of the network life will deteriorate. In general, the smaller the value of the stability S, the more stable the cluster.

3.2. Construction of clusters

Assuming that vehicle nodes on the road are equipped with GPS, information of the vehicle such as speed, position, and direction can be obtained from the respective on-board GPS. Therefore, at any time, each vehicle node can confirm which cluster it belongs to, based on the location information. The ID of the vehicle is generated randomly and is not repeated. And we set the initial state of all nodes to the UNDIFIED state. Cluster construction flow chart shown in figure 4.

3.3. Intersection gateway selection

The division of the segmented road clusters and the CH selection are described above. At this time, the communication on the segmented roads is basically established. However, when the data transmission crosses a road segment to an intersection, the communication should be transmitted without interruption. Therefore, this paper selects gateway node at the intersection to perform relay transmission of information. Assuming that the vehicle nodes will obtain the turn direction (turn left, turn right, go straight) itself through the turn signal when each vehicle crosses the intersection, and use the reactive location service (RLS) to obtain its current location. Then, the node located in the intersection area will calculate its own residence time according to formula (8). The criteria for selecting the gateway node at the intersection is to make the vehicle node as long as possible to ensure the relay of information. For instance, the gateway node is more inclined to select the vehicle node as a gateway node that spends the longest time passing through the intersection. The selected gateway node will announce its own status and establish a connection with the neighboring cluster head.

As shown in figure 5, for the calculation of the dwell time of intersections at the intersection, the following are specifically divided into three categories: straight, left, and right.

$$t_i = \begin{cases} \dfrac{W - d_{i-A}}{v_a} & ,0 \le d_{i-A} \le W, \quad straight \\ \dfrac{3W - 2d_{i-A}}{2v_a} & ,0 \le d_{i-A} \le \dfrac{3W}{2}, \quad left \\ \dfrac{W - 2d_{i-A}}{2v_a} & ,0 \le d_{i-A} \le \dfrac{W}{2}, \quad right \end{cases} \tag{8}$$

Vehicle nodes determine whether they are located in the intersection area according to their own position information, if located at the intersection, calculate and interact with their respective dwell time according to the current direction of rotation, the final selected gateway node requires the longest residence time. After the gateway node is selected successfully, it sends a HEAD frame to establish a connection with the cluster head nodes in the neighboring area.

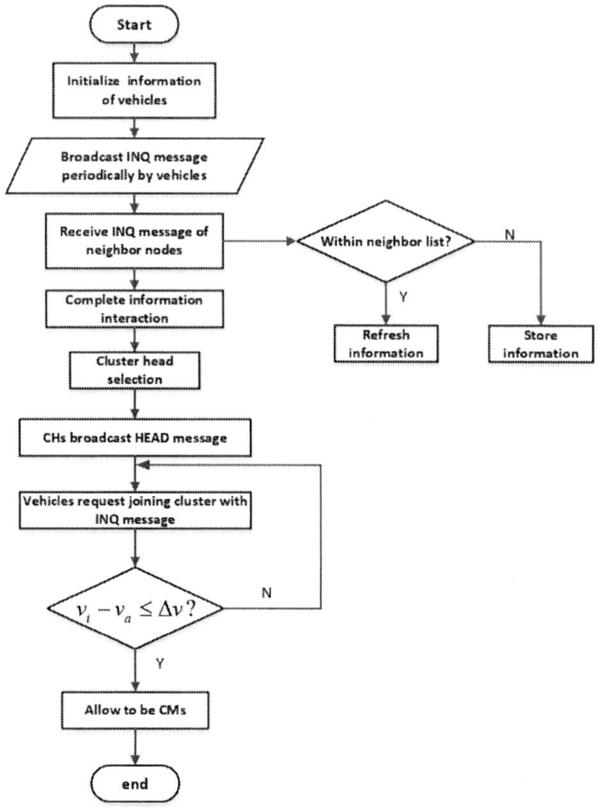

Figure 4. Flow chart of cluster construction.

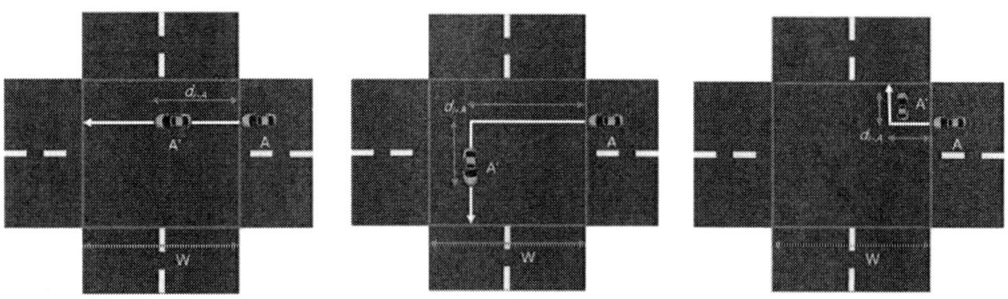

Figure 5. Intersection gateway selection.

4. CFSR protocol

Based on completion of the cluster structure, we start to establish the routing path for the data transmission. To achieve this goal, the process of assessing link quality and assigning weight to each link is necessary. Then, the weight can be directly used as the link quality metric to establish an optimal routing path in every sub-zones.

4.1. Link quality assessment

The quality of each road link will be evaluated and the weight w_R will be assigned according to the end-to-end delay of each road segment.

$$w_{R_{ij}} = \begin{cases} d_{R_i} & , \quad d_{rap} \le T_s \\ d_{R_i} + d_e & , \quad T_s < d_{rap} < T_{\max} \\ \infty & , \quad d_{rap} \ge T_{\max} \end{cases} \tag{9}$$

where d_R is the transmission delay of a Link Assessment Packet (LAP), d_e is the extra delay when a LAP encounters a disconnected path and caused by the carry-and-forward strategy, and d_{rap} ($d_{rap} = t_r - Timestamp$) is the propagation delay which is related with the received time and timestamp of LAP. T_s ($T_s = \sum_{i=1}^{h_{\max}} t_{ts}$) is the sum of transmission delays of road segments along the path, and T_{max} ($T_{\max} = \frac{L}{v_{av}}$) is maximum tolerable transmission delay.

d_R represents the average delay for transmitting a new data packet over a road segment. It is defined as below.

$$d_{R_i} = \sum_{i=1}^{n} d_{C_i} \tag{10}$$

d_C consists of two parts: T_q: representing the queuing delay and T_{ts}: denoting the transmission delay.

$$d_{C_i} = E[t_q + t_{ts}] \tag{11}$$

Actually, T_q and T_{ts} are correlated. The queuing delay of a packet corresponds to the service time of those packets queued ahead of it, which indicates that when a new packet arrives to a queue containing k packets, d_C can be defined as below.

$$d_{C_i} = (k+1)E[t_{ts}] \tag{12}$$

In addition, the extra delay d_e can be interpreted as the product of additional delay and the disconnected probability.

$$d_e = d_{dis-i} \times \rho_{dis-i} \tag{13}$$

d_{dis-i} is the additional delay to incur when choosing carry-and-forward. To compute it, we divide L into n sub-segments of length R, meaning n clusters as shown in figure 6. We assume that the cluster is connected if it contains at least one vehicle. Thus, the total additional delay of m disconnected sub-segments is expressed as below.

$$d_{dis-i} = \frac{m * R}{v_i}, m \le n \tag{14}$$

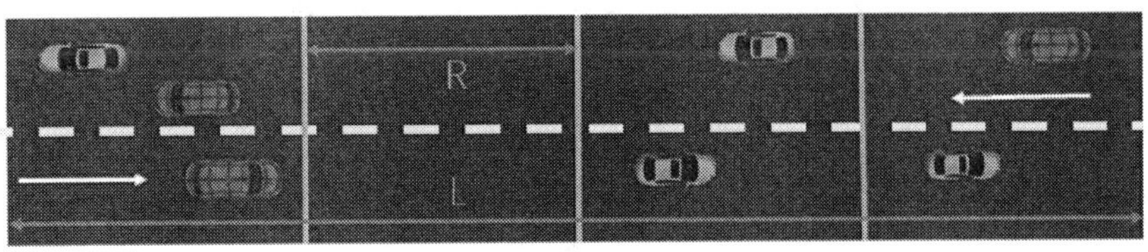

Figure 6. Extra delay caused by carry-and-forward.

ρ_{dis-i} represents the probability of carry-and-forward. We use Poisson process as the model of the vehicle arrivals to road segments. Hence, we obtain:

$$\rho_{dis-i} = C_n^m \times \rho_{null-i}{}^m \times (1 - \rho_{null-i})^{n-m} \tag{15}$$

$$\rho_{null-i} = \exp^{(-\frac{\lambda_1}{n})} \tag{16}$$

Based on all above, we can assign weights to road segments as below.

$$w_{R_{ij}} = \begin{cases} d_{R_i} & , \quad d_{rap} \le T_s \\ d_{R_i} + d_{dis-i} \times \rho_{dis-i} & , \quad T_s \le d_{rap} \le T_{\max} \\ \infty & , \quad d_{rap} \ge T_{\max} \end{cases} \tag{17}$$

4.2. Local coordinator selection

The method to solve the path local optimum problem is to understand the global topology structure and weight information of the entire network, but it is extremely difficult given the considerable size of modern city. Therefore, we divide the urban network topology into sub-zones and establish a local coordinator in every one of them through information exchange. Considering this, we set the size of sub-zone to 3*3 blocks for the sake of limiting LAPs flooding and complexity of algorithm.

The local coordinator is defined as the specific gateway node that connects the most intersection gateway nodes in their respective scopes. In doing so, on the one hand, the efficiency of route discovery is high. On the other hand, it also avoids the problems of excessive information storms and maintenance overhead of the LAPs.

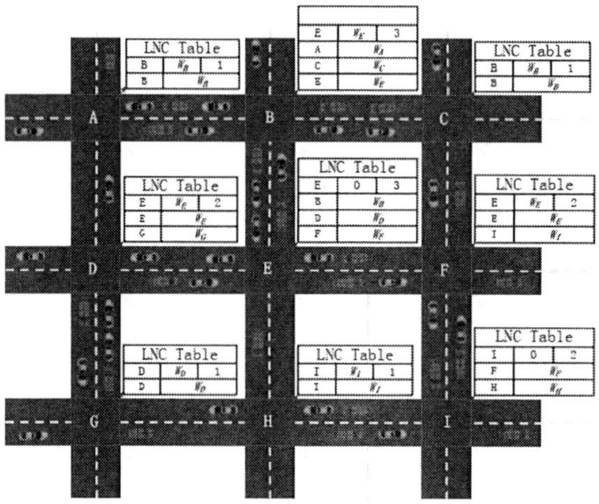

(a)

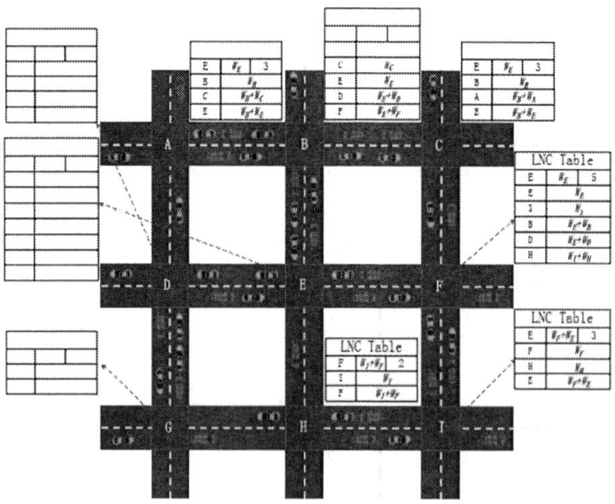

(b)

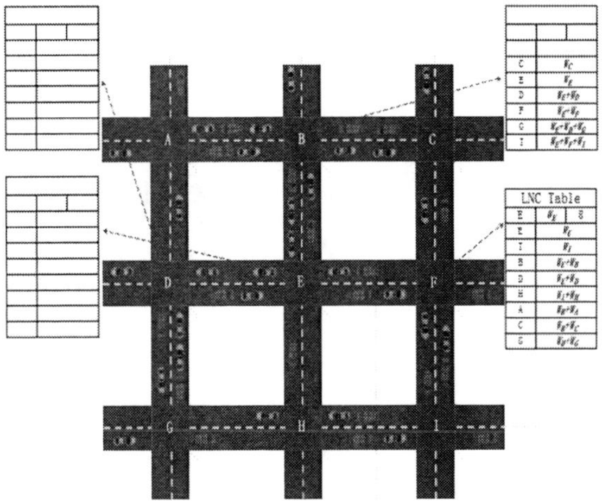

(c)

Figure 7. Local coordinators selection process.

As shown in figure 7, in the initialization phase, each gateway node in the local-area topology sets itself as LC, and sets the number n of other gateway nodes connected to itself to zero. After the first round of LAP information interaction, each gateway node updates its neighboring link information and n value according to its own received LAP condition and compares it with other gateway nodes, mainly comparing the number of its connecting gateway nodes, n determines the coordinator after the first round of comparison. For example, in figure 7(a), the gateway nodes A, G, and H specify B, D, and I as their coordinators, and B, D and F designate E as their own coordinator. For A, B, they have the same value of n, then select the node with the lower total weight as the coordinator. In the second round of interactions, each gateway node embeds its own routing table in the LAP (options field) before broadcasting. Such interaction process continues until the local area network generates a common coordinator. As shown in figure 7(c), the gateway node E can reach any gateway node within three hops, and therefore the gateway node E is selected as the local coordinator.

5. Performance evaluation

5.1. Simulation environment

This paper uses the SUMO [7] and NS3 [8] simulation platform to build a typical 3*3 urban road simulation scenario as shown in figure 8 and 9 to verify the performance of the CFSR protocol which compares with DSDV [9] and AODV [10]. It assumes that there is an intersection, two-direction carriageway, four lanes in each direction, including left-turn lanes, straight lanes, and right-turn lanes. The length of each road segment is 1000m, and a total of 9 crossroads are set.

Table 1. Parameter settings.

Parameters	Value
Numbers of vehicles	500
Average speed of vehicles (km/h)	40
Communication range (m)	50 ~ 350
Simulation time (s)	300

The simulation parameters are set out in Table 1 above. In the table, the total number of vehicles in the 3*3 sub-zone network is 500, which is generated by different nodes. The average speed of the urban vehicles is set to 40 km/h, assuming that the communication range of each vehicle is 50-350 m. The simulation time is set to 300s.

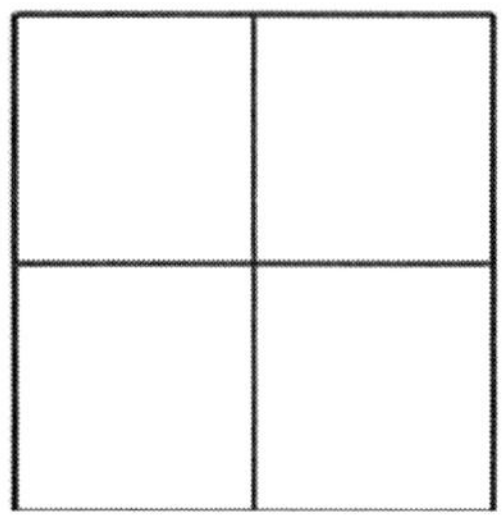

Figure 8. Urban road simulation scenario.

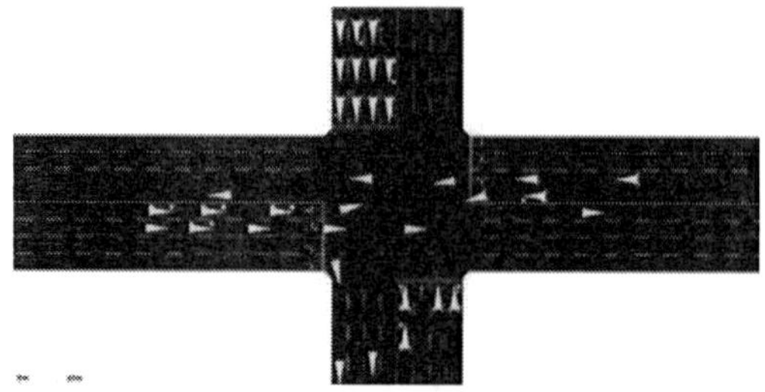

Figure 9. Urban road simulation scenario (SUMO).

5.2. *Performance metrics*

We propose two performance metrics to evaluate the CFSR protocol, including end-to-end delay (D_c) and packet drop ratio (p_d). D_c is defined as the average duration of a packet from source to destination. p_d is the ratio of discarded packets to total existing packets. The relevant formulas are as follows.

$$D_c = \frac{\sum_{i=1}^{N_{success}} D_i}{N_{success}}, \qquad p_d = \frac{N_{failure}}{N_{failure} + N_{success}}. \tag{18}$$

where $N_{success}$ is the number of packets that are sent successfully, $N_{failure}$ is the number of dropped packets, and D_i is the delay of the ith packet.

5.3. *Simulation results*

In the section, we compare CFSR protocol with AODV and DSDV, and simulation results are as follows.

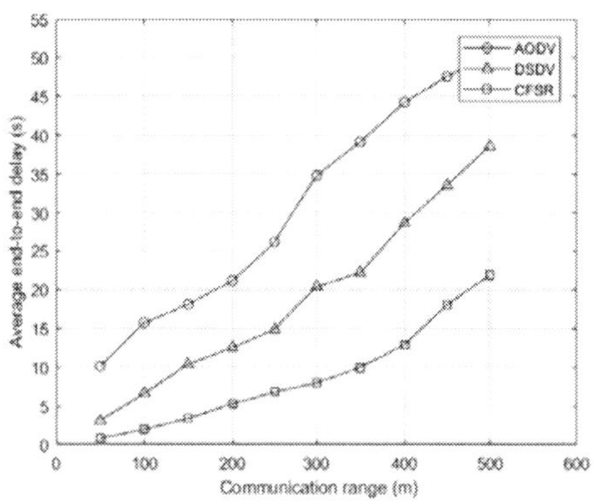

Figure 10. End-to-end delay.

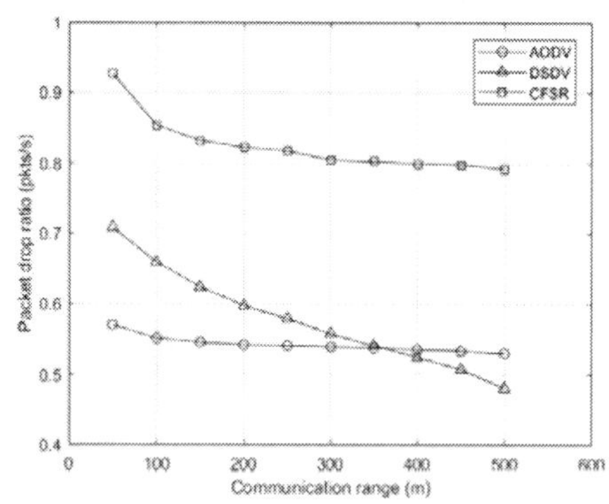

Figure 11. Packet dropping rate.

The change of E2ED relative to the communication range is shown in figure 10. We have observed that the average E2ED of all protocols increases as the communication range becomes larger. Since both AODV and DSDV use hop-by-hop forwarding, they can all lead to data congestion and thus increase E2ED. Since the CFSR uses a clustering structure, the data forwarding only passes through the cluster head and the gateway node, so the efficiency is higher and a lower E2ED is provided.

The relationship between the packet delivery rate and the maximum communication range of the node is shown in figure 11. We observe that the PDR for all protocols decreases as the range of communication. The PDR of AODV and DSDV is poor because the cost of route maintenance becomes large for large-scale networks and the high dynamics of the vehicle, the efficiency becomes very low, the link detection is not timely, and thus many packets are lost. The CFSR does not need to maintain every node. It only needs to maintain the relevant information of the cluster head and gateway nodes. In the route discovery phase, the CFSR can quickly query the local coordinator related routes to establish the routing path. The efficiency is high, and the link is reliable, so PDR is higher.

6. Conclusion

In this paper, we propose the CFSR protocol based on the clustering structure, introduce the link quality assessment, the selection process of the local coordinator, and establishes and updates the routing based on the weight distribution. In the sub-zone, we use Dijkstra algorithm to build the routing path. On the one hand, the optimal path of the routing is selected in the case of ensuring acceptable maintenance overhead, and on the other hand, sub-zone division is used to solve the local optimum problem and maintenance overhead of LAPs proliferation. Finally, performance comparison of the AODV, DSDV and CFSR protocols is made by using SUMO and NS3. Simulation results show that CFSR has lower E2ED and higher PDR than the other protocols.

Acknowledgments

This work was supported in part by the National Key R&D Program of China under Grant No. 2016YFB1200100, NSFC under Grant Nos. 61231013, 61271195 and 61301083, National Science Foundation for Innovative Research Groups of China under Grant No. 61521091.

References

[1] Zeadally S, Hunt R, Chen Y S, et al. Vehicular ad hoc networks (VANETS): status, results, and challenges[J]. Telecommunication Systems, 2012, 50(4): 217-241
[2] Cheng J J, Cheng J L, Zhou M C, et al. Routing in Internet of Vehicles: A Review[J]. IEEE Transactions on Intelligent Transportation Systems, 2015, 16(5): 2339-2352
[3] Sharef B T, Alsaqour R A, Ismail M. Vehicular communication ad hoc routing protocols: A survey[J]. Journal of Network & Computer Applications, 2014, 40(1): 363-396
[4] Togou M A, Hafid A, Khoukhi L. SCRP: Stable CDS-Based Routing Protocol for Urban Vehicular Ad Hoc Networks[J]. IEEE Transactions on Intelligent Transportation Systems, 2016, 17(5): 1298-1307
[5] Abuashour A, Kadoch M. Performance Improvement of Cluster-Based Routing Protocol in VANET[J]. IEEE Access, 2017: 15355-15371
[6] Nasr M M M, Wang Z G, Shen L F. VANET Clustering Based Routing Protocol Suitable for Deserts[J]. Sensors, 2016, 16(4): 1-23
[7] SUMO: Simulation of Urban Mobility, [Online]. Available: https://www.dlr.de/ts/sumo/en/
[8] The Network Simulator: ns3. [Online]. Available: https://www.nsnam.org/
[9] Perkins C E, Bhagwat P. Highly dynamic Destination-Sequenced Distance-Vector routing (DSDV) for mobile computers[C]// Conference on Communications Architectures, Protocols and Applications. ACM, 1994: 234-244
[10] Perkins, BeldingRoyer. Ad hoc On-Demand Distance Vector (AODV) Routing[J]. Internet-Draft, draft-ietf-manet-aodv-13.txt, 2003, 6(7):90

IOP Conf. Series: Journal of Physics: Conf. Series **1060** (2018) 012051 doi:10.1088/1742-6596/1060/1/012051

Task Scheduling Method Based on Feasible Task Execution Sequence and Greedy Strategy

Jieyong Zhang[1,2,3], Songhua Huang[1,3], Peng Sun[2] and Guanyu Chen[2,*]

[1] Science and Technology on Information Systems Engineering Laboratory, Nanjing Jiangsu, China
[2] College of Information and Navigation, Air Force Engineering University, Fenghao East Road NO.1, Xi'an, Shanxi, China
[3]Collaborative Innovation Center for Novel Software and Industrialization, Nanjing

[*]ChenPTN@163.com

Abstract. In the battlefield environment, pre-war mission plan needs to be made adaptive adjustments in the course of the war. The adjustment of the mission plan is a difficult problem in the field of command and control. Aiming at the task planning adjustment problem in the command and control organization, a task planning adjustment model is established to get the minimum mission completion time as the objective function and to complete the precision as the constraint condition. This paper proposes a task planning adjustment method based on feasible task execution sequence and greedy algorithm (GA), and gives the design idea and detailed steps of the task planning adjustment method. Finally, combined with the simulation of joint landing operations, the simulation experiment verifies the feasibility and effectiveness of the method in dealing with emergencies.

1. Introduction

The method of getting a reasonable and efficient task-platform relation has become a hot issue in the field of command and control [1]. Many scholars study mission planning problem and have achieved some valuable results.

Levchuk proposed a multidimensional dynamic list scheduling (MDLS) algorithm combined with PWE to solve the allocated problem [2]; Yang proposed a multi-priority list dynamic scheduling (MPLDS) algorithm based on the MDLS algorithm [3]. Zhang proposed a task-platform resource assigning method based on dynamic list scheduling and genetic algorithm [4].

The battlefield environment is complex and full of various uncertainties [5]-[7]. When the mission and platform are changed, the task-platform relation planned before change should be adjusted in time to make sure the mission goal will come true successfully.

At present, few studies are made on dynamic planning adjusting during the war. In this paper, we establish a mathematical model of task plan adjustment. Then, we proposes a task planning adjustment method based on feasible task execution sequence and greedy algorithm.

2. Task planning adjustment problem modelling

2.1. Basic concepts

The key entities in military organizations are defined as follows [1].

Definition 1 Task: A mission is the goal of a battle, mission is made up of multiple tasks. The completion of a task needs the support of one or more platforms.

Definition 2 Resource: A resource is a measurable physical or virtual entity used in the processing of tasks, and each task is specified by resource requirements in each functional resource category.

Definition 3 Platform: A platform has specified resource capabilities that can be assigned to execute tasks. The relationship matrix of task and platform is $M(t)^{\text{T-P}} = (m(t)_{ij}^{\text{T-P}})_{N(t)\times PN(t)}$.

2.2. *Variable definition*

1) at t time, if emergency event happened, the set of unprocessed task is $S_{\text{T}}^{\text{wait}}(t')$. The set of platform is $S_{\text{P}}(t')$.Otherwise, it is $S_{T}^{\text{wait}}(t)$ and $S_{\text{P}}(t)$ dividedly.

2) $R_i = [R_{i1},\dots,R_{il},\dots,R_{iL}]$: resource requirement vector, where R_{iL} is the number of units of resource type $l(l=1,\dots,L)$ required for successful processing of task i. $r_j = [r_{j1},\dots,r_{jl},\dots,r_{jL}]$: resource capacity vector, where r_{jl} is the number of units of resource type l possessed by platform type j.

3) initial task execution sequence is $TS = [\mathrm{T}_{s_1}, \mathrm{T}_{s_2},\dots,\mathrm{T}_{s_{NT}}]$, NT is the number of task.

4) BEFORE (i) is the set of direct predecessor task set of task i.

5) BACK (i) is the set of direct successor of task i.

6) the platform allocation of task i is $y_i = [y_{i1},\dots,y_{il},\dots,y_{iL}]^T$. Where y_{il} indicates how many platforms of type j are allocated to task i.

7) m_{ij} : assignment variables, if P_j is assigned to process T_i, then m_{ij}=1 , otherwise m_{ij}=0 .

8) MIT: mission completion time, mission completion time is the time when last task is completed. It can be obtained by the equation (1). $t_{s,i}$ is the start time of task I, $t_{p,i}$ is the processing time of task i.

$$MIT = \max_{i=1,2,\cdots,N}(t_{s,i} + t_{p,i}) \tag{1}$$

2.3. *Objective function*

With the constraint condition above, the plan whose completion time is the shortest is the best plan. Thus, the minimized mission completion time is taken as the objective function.

$$\min MIT \tag{2}$$

2.4. *Constraints*

1) Accuracy

To define the accuracy well, we adapt the geometric mean as our accuracy metric. Formally.

$$Acc(i) = \left(\prod_{l\in\gamma(i)} \min\left(1, \frac{\sum_{j=1}^{J} r_{jl} y_{ij}}{R_{il}}\right)\right)^{\frac{1}{|\gamma(i)|}} \tag{3}$$

Here, $Acc(i)$ denotes the accuracy of task i, $\gamma(i)$ denotes the set of resource categories required by task i, $|\gamma(i)|$ is the cardinality of $\gamma(i)$.The $Acc(i)$ is 0, if missing any required resource.

$$Accuracy = \sum_{i=1}^{NT} w_i Acc(i) \tag{4}$$

w_i is the weight of task i. $Acc(i)$ and *Accuracy* should greater than thresholdγ and δ respectively.

2) New added task allocation order

Battlefield situation changes at t time. The newly added task is T_{add}. After inserting T_{add} into $TS_{wait} = [\mathrm{T}_{S_1}, \mathrm{T}_{S_2},\dots,\mathrm{T}_{S_k},\dots,\mathrm{T}_{S_K}]$, the new task sequence is $TS' = [\mathrm{T}_{S_1},\dots,\mathrm{T}_{S_{add}},\dots,\mathrm{T}_{S_{K+1}}]$.In this sequence, the order of T_{add} 's direct predecessor task is in front of T_{add} 's order, the order of T_{add} 's direct predecessor is behind T_{add} 's order.

$$BEFORE(T_{add}) \subseteq \{T_{s_1},\ldots,T_{s_{b-1}},T_{add}\}$$
$$BACK(T_{add}) \subseteq \{T_{add},T_{s_{b+1}},\ldots,T_{s_{K+1}}\} \tag{5}$$

3) The stability of adjusting

The set of damaged platform is $S_{\mathrm{P}}^{\mathrm{broken}} = S_{\mathrm{P}}(t) - S_{\mathrm{P}}(t) \cap S_{\mathrm{P}}(t')$ .The set of impacted task is $S_{\mathrm{T}}^{\mathrm{impact}} = \{\mathrm{T}_{gi} \mid \mathrm{P}_{h_j} \in S_{\mathrm{P}}^{\mathrm{broken}}, m(t)_{ij}^{\mathrm{T\text{-}P}} = 1\} \cap S_{\mathrm{T}}(t')$ .The tasks whose $Acc(i)$ above threshold belongs to $S_{\mathrm{T}}^{\mathrm{keep}}(t') = S_{\mathrm{T}}^{\mathrm{wait}}(t') - S_{\mathrm{T}}^{\mathrm{add}} - S_{\mathrm{T}}^{\mathrm{impact}}$ won't be reassigned platform.

2.5. Problem modelling

Combined with the objective function and various constraints, the mathematical model for the adjustment problem is as follows.

$$\begin{aligned}
&\min\ MIT \\
&s.t. \\
&\begin{cases}
Acc(i) \geq \gamma \\
(i = 1, 2, \cdots, NT) \\
Accuracy \geq \delta \\
BEFORE(\mathrm{T}_{add}) \subseteq \{\mathrm{T}_{s_1},\ldots,\mathrm{T}_{s_{b-1}},\mathrm{T}_{add}\} \\
BACK(\mathrm{T}_{add}) \subseteq \{\mathrm{T}_{add},\mathrm{T}_{s_{b+1}},\ldots,\mathrm{T}_{s_{K+1}}\} \\
m(t')_{ij}^{\mathrm{T\text{-}P}} = m(t)_{kl}^{\mathrm{T\text{-}P}} \\
(\mathrm{T}_{g_i} \in S_{\mathrm{T}}^{\mathrm{keep}}(t'), \mathrm{T}_{g_i} = \mathrm{T}_{e_k}, \\
\mathrm{P}_{h_j} \in S_{\mathrm{P}}(t') \cap S_{\mathrm{P}}(t), \mathrm{P}_{h_j} = \mathrm{P}_{f_l})
\end{cases}
\end{aligned} \tag{6}$$

3. Solving method based on feasible task execution sequence and greedy strategy

The insertion position of new task that meets the constraint may be one or more. It means TS' may be one or more. The set of TS' is $Sequence = \{TS_1', TS_2', \ldots, TS_{SQ}'\}$, SQ is the total number of sequence. Using traversal method to find out all feasible TS' , if new task appear.

The task that need to be reallocated is $S_{\mathrm{T}}^{assign} = \mathrm{T}_{add} \cup S_{\mathrm{T}}^{\mathrm{impact}} = \{\mathrm{T}_{assign_1}, \mathrm{T}_{assign_2}, \ldots, \mathrm{T}_{assign_{as}}, \ldots, \mathrm{T}_{assign_{AS}}\}$. The platform set assigned to T_{assign} is S_P^{as} . M_{BEST} is scheduling scheme corresponding to each TS' . O is the set of scheduling scheme $O = \{M_{BEST_1}, M_{BEST_2}, \ldots, M_{BEST_V}\}$. The specific steps are as follows.

Step 1: Let v = 1.

Step 2: Select the task sequence TS_v' from *Sequence*.

Step3: Assign all of platform to task in S_{T}^{assign} , when the task being executed. It means $\forall \mathrm{T}_{assign_{as}} \in S_{\mathrm{T}}^{assign}, S_P^{\mathrm{T}_{assign_{as}}} = S_{\mathrm{P}}(t')$. This is initial scheduling plan M_0 , calculate the mission completion time MIT_0, $as = 0$.

Step 4: If $as \leqslant AS$, select $\mathrm{T}_{assign_{as}}$ from S_{T}^{assign} to delete redundant platform, as = as +1; Otherwise, go to step 7.

Step 5: If as=1, then $M_0 = M_{best}$, $MIT_0 = MIT_{best}$. Let $M_{temp} = M_{best}$, $MIT_{temp} = MIT_{best}$, delete one redundant platform from S_P^{as} . The principle of deleting platform is delete this platform will reduce mission complete time most. After the redundant platform is deleted, the accuracy of $T_{assign_{as}}$ should fulfill $Acc \geq \gamma$. The scheduling plan is M_{best}, the complete time is MIT_{best}.

Step 6: If deleting any platform in S_P^{as}, the accuracy of $T_{assign_{as}}$ will can't meet $Acc \geq \gamma$, or the mission complete time does not get less, then stop deleting, return to step 4. If the stop condition is not fulfilled, then return to step 5.

Step 7: The optimal solution recorded in the allocation solution set O; v = v + 1, if v≥SQ, output the mission distribution strategy with the M_{best} from the set O; Otherwise, return to step 2.

4. Case analysis

This paper simulates the landing operation in [8]. In this case, the operational mission was dismantled into 18 sub-tasks, and the relationship between the tasks is shown in figure 1.

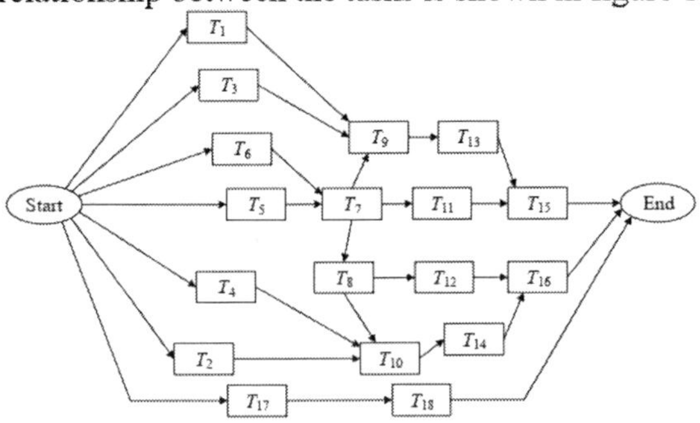

Figure 1. Task's relationship

The MDLS algorithm is applied to pre-war task planning [9]-10], and WL algorithm is selected as priority function [2], and the following task assignment plan is obtained according to the information of task and platform in [8], as shown in figure 2. The execution accuracy of each task is 1.

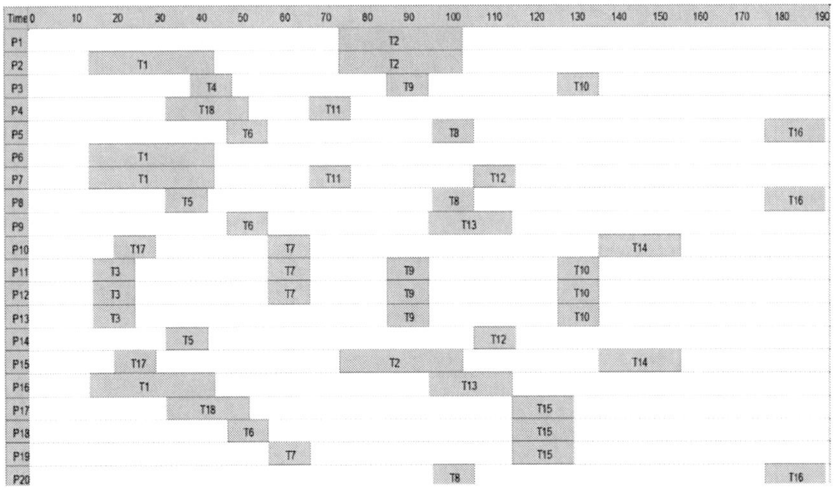

Figure 2. Task-platform relationship without uncertain factors

Under the initial operational conditions, the task execution sequence is TS=[T_5,T_6,T_1,T_3,T_4,T_{17},T_{18}, T_2,T_7,T_8,T_9,T_{11},T_{13},T_{10},T_{12},T_{15},T_{14},T_{16}], and the mission completion time is 192.9529.

4.1. Simulation Analysis of Task Added Event

When the mission is carried out to t=60, the battlefield situation changed and the operational mission was adjusted. A new increase is task T_{19}, and T_{19} is inserted between T_{14} and T_{16} in Figure 2. The location of T_{19} is (5, 95), the vector of resource requirement is [0,0,0,0,0,8,0,6], and the processing time of task is 20.

It can be seen from figure 2 that when t = 60, T_1, T_3, T_4, T_5, T_6, T_{17} and T_{18} have already been executed. In figure 2, TS_{wait} =[T_7,T_8,T_9,T_{11},T_{13},T_{10},T_{12},T_{15},T_{14},T_{16}]. After new task T_{19} is added to TS_{wait}, TS_{wait} changes to TS' =[T_7,T_8,T_9,T_{11},T_{13},T_{10},T_{12},T_{15},T_{14}, T_{19},T_{16}].

Using the adjustment method above to assign platform to T_{19}, the task-platform relationship after adjusting is shown in figure 3. The thresholds γ and δ are both set to 1.

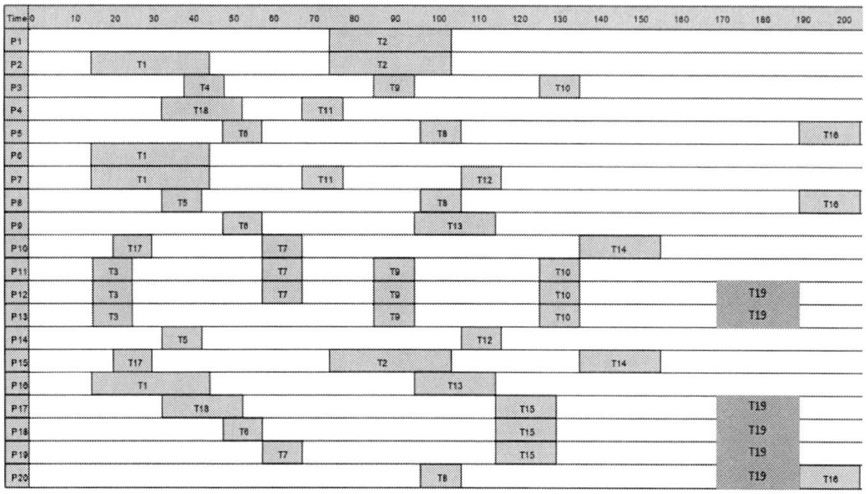

Figure 3. Task-platform relationship after adjusting (add task)

In figure 3, P_{12}, P_{13}, P_{17}, P_{18}, P_{19} and P_{20} are assigned to T_{19}. The allocated platform resources meet the requirements of T_{19}. The platforms to which other tasks are assigned are the same as the original allocation.

After new task is added, the completion time of the mission is increased from 192.9529 to 206.1240, the completion time increases by 6 seconds. The processing time of newly added task is 20. The adjustment algorithm solves the problem of task added effectively.

4.2. *Simulation Analysis of Platform Damage Event*

When the mission is carried out to t=60, platform P_7 is damaged. T_{11} and T_{12} whose accuracy is lower than the threshold should be reassigned platform with γ and δ both are set to 1.

From figure 2, before the platform P_7 is damaged, the platform assignment of tasks T_{11} and T_{12} is shown in table 1.After platform P_7 is destroyed, the mission plan adjustment is carried out, and the platforms reassigned for T_{11} and T_{12} are shown in table 2.

Table 1. Before platform broken

Task	Assigned platform
T_{11}	P_7 ,P_4
T_{12}	P_{14} ,P_7

Table 2. Algorithm run time comparison

Task	Assigned platform
T_{11}	P_1 ,P_2 ,P_3 ,P_4
T_{12}	P_1 ,P_2 ,P_3 ,P_4

The adjusted Gantt chart is shown in figure 4 under the condition that the platform-related constraints are satisfied. The mission completion time changes to 206.2094 after adjustment. Compared with the initial mission completion time, the new mission completion time increased by 14, which meets the requirements, and solving the task plan adjustment problem after the platform loss.

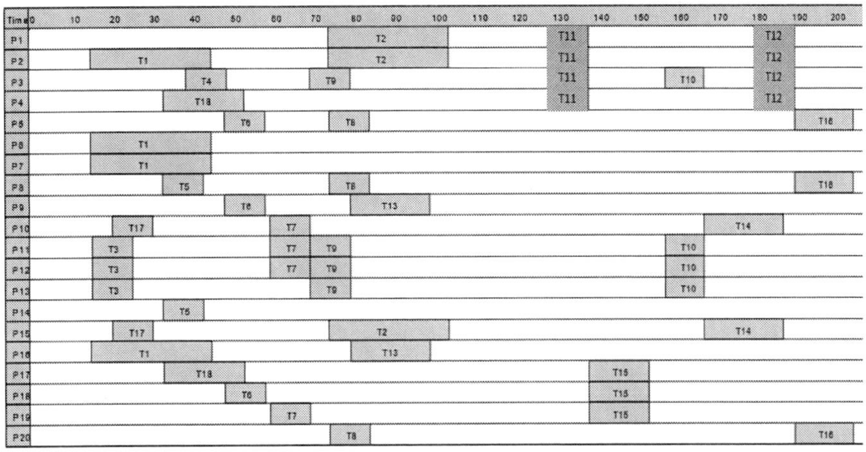

Figure 4. Task-platform relationship after adjusting (platform damage)

5. Conclusion

This paper studies the adjustment of mission plan and establishes the mission plan adjustment model with objective of minimizing mission completion time. A solution to the model is proposed based on the feasible task execution sequence and greedy algorithm. At last, the experimental simulation of landing operation is carried out. The mission plan is adaptively adjusted after battlefield environment changed. The experimental results demonstrate the effectiveness of the method for mission planning adjustment.

Acknowledgments

Supported by Natural Science Foundation (Grant No. 61573017) and Natural Science Foundation of Shanxi Province (Grant No. 2017JM6062).

References

[1] Xie B,Lin H. 2013 Generation time analysis of operation resource planning Command Control & Simulation vol 35 no 6 pp 21-25.

[2] Levchuk G M, Levchuk Y N, Luo J, et al. 2002 Normative design of organizations - part I: Mission planning*IEEE Trans. on Systems, Man, and Cybernetics* vol 32 no 3 pp 346-359.

[3] Yang D S,Zang W M,Lu Y L.2006 Research on Mathematical description and solving algorithms of tasks scheduling for campaign *Systems Engineering-Theory & Practice* vol 1 pp 26-34.

[4] Zhang J Y,Yao P Y,Zhou X X,Wang X. 2012 Approach to operation task and platform resource matching based on DLS and GA *Systems Engineering and Electronics* vol 34 no 5 pp 947-954.

[5] Liu H X. 2010 Design and adjust methodology of command and control organization Changsha: National University of Defense Technology.

[6] Xiu B X. 2006 Design methodology of C2 organizational structure and its analysis of robustness and adaptivity Changsha: National University of Defense Technology.

[7] Zhang J Y,Yao P Y,LI F. 2012 Task-platform relation design model and its algorithm under completion time constraint *Systems Engineering and Electronics* vol 34 no 8 pp 1621-1629.

[8] Yu F, Tu F, Pappipati K R. 2008 Integration of a holonic organizational control architecture and multiobjective evolutionary algorithm for flexible distributed scheduling IEEE Transactions on Systems, Man, and Cybernetics vol 38 no 5 pp 1001-1017.

[9] Zhou X X,Yao P Y,Zhang J Y,Wang X. 2012 Platform resource scheduling method based on DLS and ACO *Computer Science* vol 39 no 6 pp 98-103.

[10] Lu Y L,Yang D S,Liu Z,Dai C H. 2006 Research on algorithm of resource allocating in joint operation *Fire Control and Command Control* vol 2 pp 12-16.

Research on the Application of Cluster Analysis in Criminal Community Detection

Tian Junjing[1,*]

[1]The Chinese People's Armed Police Forces Academy, Lang fang, Hebei, 065000, China

[*]a88241@163.com

Abstract. With the development of China's economy and technology, the mobility of people enhances rapidly, the trend of crime towards to community and organization. Criminal behaviour is a kind of social behaviour. Many malignant cases, especially those that are highly harmful to the lives and property of the citizens and have bad social influence, are not executed by a single person, but need teamwork to complete. At present, the city's population base is large, the police force is insufficient, and it is difficult to expedite the investigation of suspicious personnel by conventional methods in a short time. This paper establishes the social network model according to the local telephone call records, and then uses the cluster analysis algorithm looking for people with close links to the criminal suspect to identify suspected criminal accomplices. The cluster analysis algorithm proposed in this paper can help the public security organs filter suspicious people rapidly, narrow the scope of possible criminal gangs, and improve the efficiency of detected cases.

1. Introduction

Criminal social network is not separated but consists of a series of associations. Community detection technology in social network is an important method for mining criminal gangs, and the quality of community detection technology depends mainly on clustering algorithms. At present, there have been a large number of researches on clustering analysis in academic circles. However, these algorithms are mostly too theoretical and have not been specifically optimized in combination with actual application scenarios. In particular, there are only a handful of optimizations for police crimes [1]. Moreover, the public security department has limited manpower in criminal investigation, and it takes a lot of time to investigate criminal gangs related to criminal suspects. Therefore, it is very necessary to use the cluster analysis algorithm to detect the suspect's society, determine close suspects, reduce the inspection targets, and provide intelligent decision support for the staff in the criminal investigation department.

2. Clustering algorithms in social network

It is of great significance to grasp the network structure of criminal gangs for the investigation of crimes and the prevention of crimes in criminal investigation work. Therefore, a social network analysis (SNA) method can be used to build a communication network, and then a cluster analysis algorithm can be used to determine the degree of closeness between the community members, to identify associated gangs according to specific suspects, and provide the public security department with a list of suspicious criminal gangs to help them effectively prevent and fight crimes [2].

2.1. Community Division in Social Network

The social network can be represented as SN = (V, E), V is a set of nodes that are non-empty, E is the edge set connecting the two nodes in the network. Social networks are divided into undirected networks and directed networks. If any two node pairs (i, j) and (j, i) correspond to the same edge, the network is called an undirected network, otherwise it is directed network [3]. In an undirected network, if each edge has a corresponding weight, the network can be called a weighted network; otherwise it is called an un-weighted network. In this paper, communication network that the nodes communicate with each other belongs to undirected network. The two connected nodes in the network have their call duration and number of calls. Therefore, the communication network is a weighted undirected network.

Community division is an important technology in the current social network analysis. It can extract some closely-associated communities from the social network. The result of this extraction is that the nodes in the community are closely connected and the nodes between different communities are loosely connected. Based on this characteristic, those who are related to known criminals in the community can be classified as key monitoring objects in the criminal investigation. Thereby, Community division can greatly reduce the burden of surveillance, and quickly discover potential criminal partners. The methods of community division in social network mainly include the division methods based on graph theory and the division methods based on hierarchical clustering algorithm[4][5]. The representative methods based on graph theory mainly include the spectral bisection method, Kernighan-Lin algorithm and so on. The algorithms based on hierarchical clustering include decomposed hierarchical clustering algorithms and agglomerative hierarchical clustering algorithms.

2.2. The Process of Cluster Analysis

The process of cluster analysis is shown in Figure 1. Generally, it includes the following steps:

(1) Data Preprocessing

In the data preprocessing stage, the validity of the data is mainly checked to identify whether the data is null or error data and so on. In some cases, it is also necessary to perform operations such as summation and averaging in the data preprocessing stage. In order to find the frequently contacted community in this paper, the duration of calls and the number of calls between two nodes that are two people communicating with each other in a social network should be summed up [6].

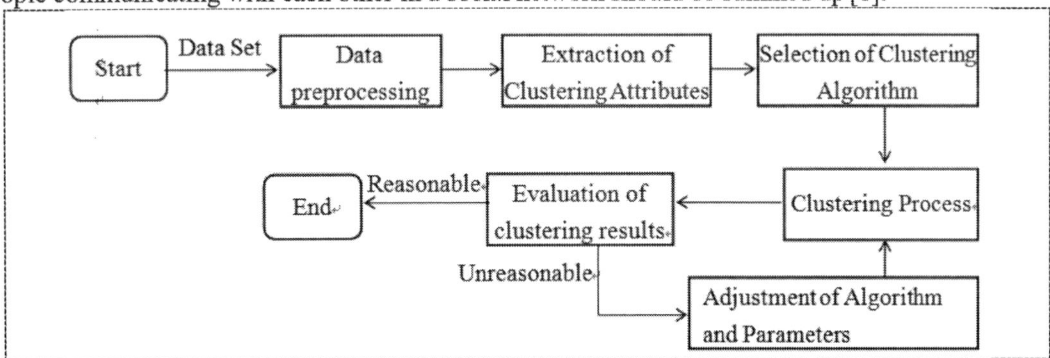

Figure 1. The process of cluster analysis

(2) Extraction of Clustering Attributes

The data in the data set may contain many attributes, such as name, gender, call duration, and so on. Some of these attributes are continuous attribute, and some are binary discrete attribute. Choosing the attributes of the data is a key issue in the process of cluster analysis. The selected attributes should be sufficiently representative of the data sample in the data set. Different attributes need to adopt different node similarity measurement methods. This paper selects two attributes that can reflect the density of the two nodes, which are the duration of the calls and the number of calls.

(3) Selection of Clustering Algorithm

Clustering algorithms include hierarchical clustering algorithms, partitioned clustering algorithms, grid-based clustering algorithms and many other algorithms [7]. According to the application and data set, this paper chooses the nearest neighbor hierarchical clustering algorithm.

(4) Clustering Process

In the clustering process, similarity is calculated for the nodes according to the distance between the two nodes in the social network. According to the input parameters and the selected clustering method, the nodes are clustered.

(5) Evaluation of Clustering Result

The clustering result is evaluated to determine the rationality in the evaluation process. If the clustering result is reasonable, the entire clustering process is ended. If the clustering result is not reasonable, the clustering algorithm and parameters should be adjusted, and the clustering process should be repeated until a reasonable clustering result is obtained.

3. Clustering Analysis Based on Call Records

3.1. Construction of Social Network

When the public security department conducts a case investigation, the relationship strength and relationship model between the parties can be revealed by the telephone records analysis. In the data analysis and extraction stage in this paper, call records of the users are analysed, key information is extracted, and the information such as the calling telephone number, the called telephone number, and the call duration are obtained, and a sum operation is performed on the call duration and the number of calls. Call duration and the number of calls reflect the strength of the relationship between the two contacts. Substituting the duration of the calls and the number of calls into the formula can obtain the weight of the edge between the two nodes. Each person is abstracted to a node. The connection between two people is abstracted to the edge in the network, thus a weighted undirected graph is formed.

3.2. Detection and Division of Criminal Associations

This paper constructs a weighted and undirected graph which shows the calling network of the users. Each edge in the network needs to be assigned a weight value. The weight value is inseparable from the call situation of the two users. When determining the weight value, not only the duration of two users' calls but also the number of calls made by two users must be taken into consideration. Therefore, the weight value of the edge is defined as WE (i,j).The value of WE(i,j) can be calculated by Equation(1).

$$WE(i,j)=M*t(i,j)+N*c(i,j) \quad (1)$$

t(i,j) is the weight value of the call duration between user i and user j, c(i,j) is the weight value of the number of calls between user i and user j. M is the influence factor of the call duration and is used to determine the influence of the call duration on the edge weight. N is the influence factor of the call number and used to determine the influence of the number of calls on the edge weight. M+N=1, $M \in [0,1]$, $N \in [0,1]$.

The value of t (i,j) can be calculated by Equation (2).

$$t(i,j) = \frac{t_{ij}}{t_i + t_j} \quad (2)$$

t_{ij} represents the call duration between user i and user j, t_i represents the total call duration between user i and everyone, t_j represents the total call duration between user j and everyone. If the call duration between user i and user j is a large proportion, it means that the relationship between them is relatively close.

The value of c(i,j) can be calculated by Equation(3).

$$c(i,j) = \frac{c_{ij}}{c_i + c_j} \quad (3)$$

IOP Conf. Series: Journal of Physics: Conf. Series **1060** (2018) 012052 doi:10.1088/1742-6596/1060/1/012052

c_{ij} represents the number of calls between user i and user j, c_i represents the total number of calls between user i and everyone, c_j represents the total number of calls between user j and everyone. If the number of calls between user i and user j is a large proportion, it means that the relationship between them is relatively close.

In the process of clustering analysis, the similarity between node i and node j is related to the distance between the two nodes. The closer the distance between them, the higher the similarity is. The distance D (i,j) between node i and node j is related to the edge weight. The calculation of the distance between them should consider the following three situations:

(1)If the two nodes are directly connected, the distance D (i,j) between them is calculated according to Equation(4).

$$D(i,j) = \frac{1}{1+WE(i,j)} \tag{4}$$

(2) If the two nodes are not directly connected but connected through other nodes, the distance between the two nodes is the shortest path between them. In this case, the distance between the two nodes is calculated by Equation (5).

$$D(i,j)=D(i,k)+D(k,p)+\cdots+D(o,j) \tag{5}$$

D (i,k), D(k,p), D(o,j) can be calculated according to Equation(4),node k, node p, and node o are the intermediate nodes on the shortest path between node i and node j.

(3) If the two nodes are not connected by any path, the distance between them is $D(i,j)=+\infty$.

3.3. *Cluster Analysis Based on Node Distance*

During the process of community detection, the suspect nodes are assigned by the user, which are used as the initial cluster centers, and then the distances between the other nodes in the social network and these cluster centers are calculated. These nodes are assigned to the community where the nearest cluster center is located. A threshold can be set to control the size of the community. If the threshold is not set, the size of the community is determined by the algorithm.

The process of cluster analysis is as follows:

Input parameters: an weighted and undirected social network G=<V, E>, where V is a set of nodes in the network and E is a set of edges in the network;

Output result: M communities;

Follow these steps for cluster analysis:

Step1: Initialize the network and use the M ($M \geqslant 1$) nodes set by the user as the initial clustering center(s).

Step2: Calculate the distances between two nodes in the social network according to Equation (4) and Equation (5), and store the results in an n-dimensional distance matrix.

Step3: Calculate the distances between each node and M cluster centers according to the n-dimensional distance matrix, then assign the nodes into the community represented by the nearest cluster center, and delete these nodes in the node set. If the distance between a node and two cluster centers is equal, a modularity function is used to determine the node belongs to which community.

Step4: If the node set is empty, the algorithm ends. If the node set is not empty go back to step3 to continue.

4. Presentation of the Output Result

In the criminal network, different roles have their own contact characteristics. Contacting members or gangs have higher frequency of contact before committing the crime. The links between closely-connected members are generally characterized by high frequency, long average talk time, and contact at dinner time or at night.

4.1. *Call data set*

First of all, the call data set should be pre-processed. The total call duration of two users and the total number of calls should be calculated. The edge WE (S, T) and distance D (S, T) also should be calculated according to Equation (1)-Equation (5). When constructing the social network and detecting the communities, clustering analysis is executed based on the distance between two nodes. The call data set after pre-processing is shown in Table 1.

Table1. Call data set

S	138****2321	132****8890	......	158****3421	139****4580
T	183****7659	137****0909	......	139****4578	158****4664
S_ Duration	1056432	423098	......	1276580	265780
T_ Duration	876534	456522	......	956412	42383
Tol_Duration	324784	200514	......	78902	33454
S_ Count	1235	432	......	2876	7890
T_ Count	879	321	......	1983	341
Tol_Count	544	243	......	106	246
M	0.5	0.5	......	0.5	0.5
N	0.5	0.5	......	0.5	0.5
WE (S, T)	0.213	0.276	......	0.065	0.139
D (S, T)	0.824	0.784		0.939	0.878

4.2. Analysis of Output Result

After the Clustering analysis, the clustering result is shown in Figure 2.

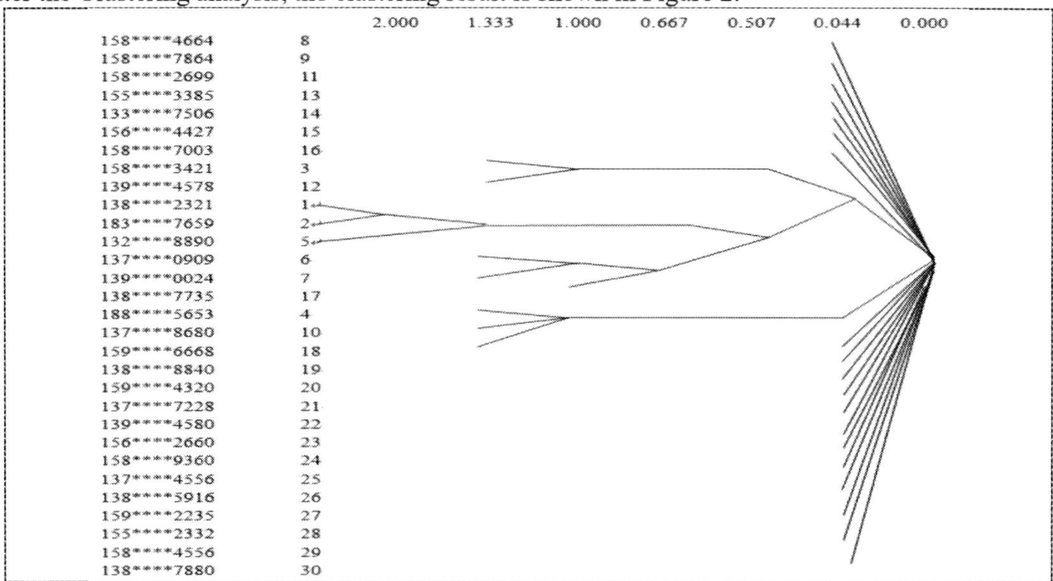

Figure 2. Dataset clustering result

The communities in the social network are detected and the divisions are shown in Figure 3. According to the division result, the social network should be divided into two clusters that are represented by two colors. Each cluster represents a community. The centrality of the nodes is different in each community. The shape of the nodes with a large centrality is relatively small, and the shape of the nodes with a small centrality is relatively large. A node with a large centrality indicates that it is of greater importance in social network and it belongs to core nodes in the community.

According to the result of community division in Figure 3, the closeness of these nodes is exhibited. If it is determined that a certain mobile phone number belongs to a criminal suspect, then according to the clustering result, the contact person related to him can be found. And, according to the weight values of the edges in the social network, the closeness of the contact and the suspect can be judged so as to judge possible criminal gangs.

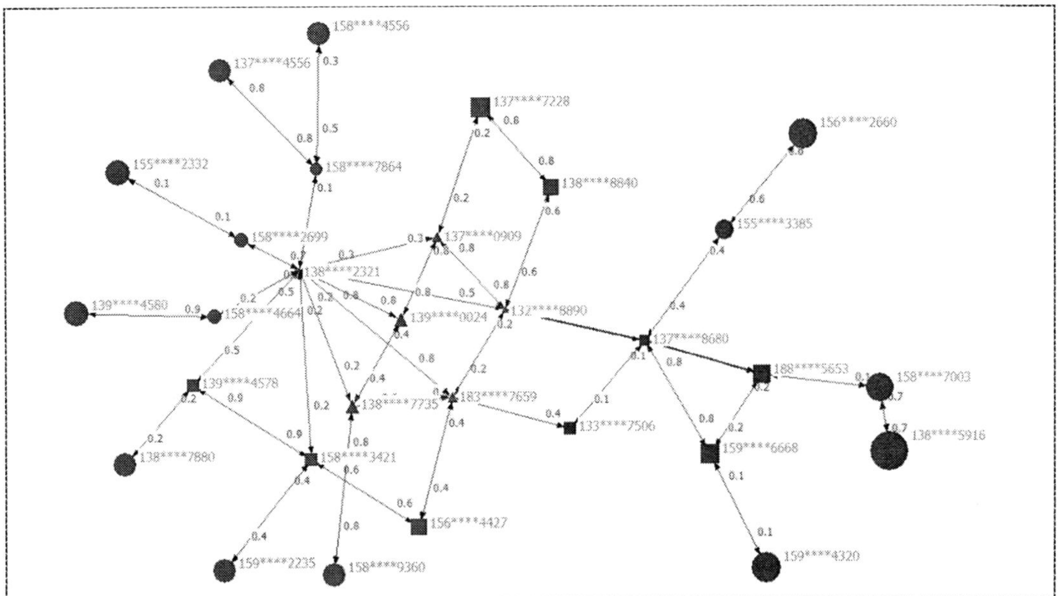

Figure 3. Community division result

5. Conclusion

Sometimes people are not concerned about the structure of the entire network in real life, but only need to know the structure of part of the network. That is, people only need to know the situation of the group you care about in the network. Based on the characteristics of communication data, this paper uses the communication records of the criminal suspects to establish social networks. Based on the shortest distance core mining method, the nearest neighbour hierarchical clustering method is used to detect communities and identify possible criminal gangs automatically. Clustering analysis can facilitate criminal investigation personnel of the Public Security Bureau to conduct targeted investigations of suspicious individuals and speed up the detection of cases. In addition, the core members of possible criminal gangs can also be automatically identified. Although the core member does not necessarily have to be a leader within a possible gang, his influence in the gang is very large, and he(or she) is likely to be a key figure in the gang.

References

[1] Wu Shaozhong 2016.Potential Terrorist Gang Discovery Techniques in Counterterrorism Intelligence Based on Cluster Analysis.*Police Technology*. **6** 18-21

[2] PanFang 2009. Criminal Network Analysis Based on Fuzzy Hierarchical Clustering. *Journal of Southwest China Normal University(Natural Science Edition)*. **34** 210-214

[3] Li Liang 2008 *The Recognition System of Crime Syndicate Based on Social Network Analysis*(Shanghai: School of Software Shanghai Jiaotong University) pp 6-9

[4] Wang Wei 2015 *Research on Social Network Community Division Method Based on Clustering Analysis*(Nanjing:Nanjing University of Posts and Telecommunications) pp 17-25

[5] Shi Taibin 2017 *Research onCommunity Detection Algorithm and Community Profile Based on User Call Record*(Zhejiang:Zhejiang University) pp 4-5

[6] Ao Wenjing 2013 *Study of mining social network based on call records*(South China University of Technology) pp 7-11

[7] Zhang Yuchao 2015 *The Research on Cluster Analysis for Target Group of the Police*(Jinan:Shandong University) pp 5-14

Research on Incremental Learning of SVM Based on Robustness

Li luying[1], Zhou liling[1] and Tong xiaojun[1,*]

[1] College of Math and Computer, Wuhan Textile University, Wuhan 430200, China

[*]1907672151@qq.com

Abstract. The following problems exist in the process of incremental learning of support vector machines. If the number of support vectors will increase with the increase of the increment sample, the time of training will become longer and longer; if the vector that has no effect on the hyperplane of the classification is abandoned, this part of the vector may become a support vector in the subsequent training. This will have an impact on the effect of the classification. In this paper, a support vector machine incremental learning algorithm based on fuzzy C mean clustering and central density is used to determine support vector set by fuzzy C mean clustering, and the selected non support vector sets are obtained by using the ratio of center density to determine non support vectors. The effect of non support vector on the robustness of support vector incremental learning is studied by comparing the classification efficiency of the four kinds of classification, and the conclusion of this paper is finally obtained.

1. Introduction

Nowadays, the rapid development of social information is mainly reflected in the information development caused by the progress of network technology, which puts forward new problems to explore the traditional knowledge. In today's society, the information content is huge and the speed of information updating is getting faster and faster. This leads to the information development and the technological progress cannot be synchronized. Therefore, how to keep the synchronization of the two is very important. In order to solve this problem, incremental learning is needed. By incremental learning, technological innovation can help people sort out information and make information more intelligent [1].

The reason for the support vector machine learning is to find out the support vector in all the solutions of the training set, so as to obtain the ideal state of the classification hyperplane. Many factors will affect the final result when looking for support vector. [2] For example, the number of samples will affect the time spent on the final finding of support vector machine. In order to reduce the scope of the search, we lock the existence range of support vector to the boundary vector according to the law. This method can improve the speed and performance of the algorithm.

In the introduction of the previous part, we learned about the learning strategy of the vector machine. [3] Based on this part of the learning, we can find that the support vector is more important in this incremental learning algorithm, and it may have a direct or indirect effect on the final result. In the process, different strategies are adopted, and the resulting incremental learning algorithm has different performance. In the past, the most suitable support vector set is obtained by repeated use of samples. Only when there is no element in the prediction classification error is the ideal state of the incremental learning algorithm. As for KKT, this factor also plays a significant role in the incremental process. [4]

2. Kernel fuzzy c mean clustering algorithm

The fuzzy C mean clustering algorithm can initialize the cluster center first, and initialize the partition matrix formed by the membership function value first, then seek the value of each cluster center, then redetermine the partition matrix and circulate in turn. [5]

Usually, before using the fuzzy C mean clustering algorithm, first of all, we must determine the appropriate clustering criteria. In all the clustering criteria, the most commonly used criterion is the minimum sum of square errors, which is related to the predefined class.[6] According to the clustering criterion, we can get the objective function and the concept of a fuzzy division of the objective function. First, the distance between the data sample points and the cluster centers is calculated. Then the average weighting method of the membership function is used to express the distance between the data sample points and the cluster centers.[7]

The principles of the algorithm are described below.

Set the existence of a set of samples $T = \{x_1, x_2 \dots x_n\}$,Mapping T to the sample spaceΦ,the goal function is:

$$\begin{aligned} J = (U,V) &= \sum_{i=1}^{C}\sum_{j=1}^{N} U_{ij}^{m} \Phi(x_j) - \Phi(v_i)^2 \\ &= \sum_{i=1}^{C}\sum_{j=1}^{N} U_{ij}^{m} \left[K(x_j,x_j) - 2K(x_j,v_i) + K(v_i,v_i)\right] \end{aligned} \tag{1}$$

In the above formu U is C×N membership degree matrix;V is cluster center matrix; $K(x, y)$ is kernel function.$m \in [0, \infty)$ indicate the fuzzy weighting index,

The Lagrange multiplier method can be obtained U_{ij} and v_i

$$u_{ij} = \frac{\left[\Phi(x_j) - \Phi(v_i)^2\right]^{1/1-m}}{\sum_{k=1}^{C}\left[\Phi(x_j) - \Phi(v_k)^2\right]^{1/1-m}} \quad , \quad v_i = \frac{\sum_{j=1}^{N}(U_{ij})^m x_j}{\sum_{j=1}^{N}(U_{ij})^m} \tag{2}$$

When the kernel function is selected as the Gauss kernel function $K(x, x') = \exp\left(\frac{-\|x-x'\|^2}{\sigma^2}\right)$, it can be obtained

$$K(x_i, x_i) = \exp\left(\frac{-x_i - x_i^{2}}{\sigma^2}\right) = 1 \tag{3}$$

so,

$$\begin{aligned} \left\|\Phi(x_j) - \Phi(v_i)\right\|^2 &= K(x_j,x_j) - 2K(x_j,v_i) + K(v_i,v_i) \\ &= 2\left(1 - \left(K(x_j,v_i)\right)\right) \end{aligned} \tag{4}$$

By using the fuzzy C mean clustering algorithm, the linear separable rate of SVM model is improved, and the degree of difference between categories is further expanded. In this case, when mapped to high dimensional feature space, the sample will be linearly clustered.[8] In the C mean clustering algorithm, the membership degree matrix is used to reflect the sample points belonging to the category. The degree of correlation is different from that of different samples.

3. Select the center density ratio algorithm of non support vector

Under the condition that the training set is linearly separable, for the positive class sample vector group$\{x_1, x_2, \dots, x_n\}$,vector groups of positive class samples is: [5]

$$m_+ = \frac{1}{n_+}\sum_{i=1}^{n_+} x_i \tag{5}$$

In the same way, for the negative class sample vector group $\{x_1, x_2, \ldots, x_n\}$,vector groups of negative class samples: [5]

$$m_- = \frac{1}{n_-}\sum_{i=1}^{n_-} x_i \tag{6}$$

The ratio of the number of samples less than θ (θ >0) to the total number of samples in a class of sample sets is called the central density of such samples, that is

$$\rho_+ = \frac{M\left(\left\{x_i \mid d\left(x_i, m_+\right) < \theta\right\}\right)}{n_+} \tag{7}$$

$$\rho_- = \frac{M\left(\left\{x_i \mid d\left(x_i, m_-\right) < \theta\right\}\right)}{n_-} \tag{8}$$

$M(A)$ is the number of elements in a set A,θ is parameter, the actual operation can make θ two 1/10 of the center of the sample set, which can be adjusted according to the actual situation. [9]

For the proper value of θ,the greater the value of the central density $\rho_+(\rho_-)$, the more dense the class sample is, the closer the classifier interval boundary is to the center of the class sample and the more non support vectors near the interval boundary, so the two classes of non support vectors are controlled respectively according to the value of the value of $\rho_+(\rho_-)$.

4. Support vector machine incremental learning algorithm based on fuzzy c means clustering

Based on the above analysis, a new incremental learning algorithm for support vector machines is proposed. The idea of this algorithm is: first, the fuzzy C means clustering method is used to find the boundary vector set, and the set is trained to get the initial classifier. [10]The support vector set is S_V at this time, and the classifier is found in the initial sample. The sample of the corresponding condition is selected by the center density ratio method, the non support vector set is T_V, and then the support vector which is close to the support vector of the initial sample is found in the new sample. The sample is recorded as S_V', and the sample which is contrary to the classifier's corresponding condition is found, and is recorded as T_V', and the samples without the violation of the condition are incorporated into the initial training. Set, these samples are likely to be transformed into support vectors in the next training. Take S_V, S_V' and the new classifier training set. The classifier is used to determine whether $T_V \cup T_V'$ is related to the classification hyperplane. If relevant, the non support vector set will also be added to training.[11]

The algorithm process is described as follows:

step1. Premise: suppose U is the initial sample set, $U_i, (i = 1,2, \ldots, n)$ for the new sample set;

step2. Goal: find the classifier based on $U \cup U_i$.

step3. calculates C clustering centers for each sample point in U, and extracts the boundary set of the initial sample set accordingly;

step4. U_i is trained to get the classifier, the corresponding support vector is S_V, and the non support vector set is T_V.

step5. Checks whether the incremental process continues. If not, the end of algorithm is the final classifier;

step6. To judge whether there is a condition that violates the classifier's condition in the new sample setU_i, if there is no algorithm, the algorithm is ended as the final classifier.

step7. The support vector that does not violate the classifier condition is S_V', which violates the condition of the classifier. The non support vector set is T_V'.

step8. Order $U_i = S_V \cup S_V' \cup T_V \cup T_V'$.

In the algorithm, the support vector, which is used as the boundary vector, is used instead of the initial sample for training. This method is used to deal with the initial samples, which can not only make the training speed faster, but also do not affect the accuracy of the classification. [12]The KKT condition is used to determine whether it is necessary to retrain and study in order to deal with the new samples. If the new sample needs to be re studied, the non support vector in a band area outside the classification interval of the original classifier, together with the original support vector set, will be added to the sample which is contrary to the KKT condition in the new sample. This training can give up more useless information, retain useful information, reduce training time, and improve classification accuracy. [13]

5. Experimental results and analysis

In this paper, the fuzzy C mean clustering and the incremental learning algorithm based on the center density support vector machine (SVM) can achieve better results in the process of use. Therefore, the effectiveness of the algorithm is compared by experiments and the experimental analysis is given. [14]

The experimental data come from the handwritten numeral classification data set in UCI database. The traditional SVM incremental learning means incremental learning algorithm based on KKT condition and support vector machine, the number of samples in the sample set is 800, initial sample set is 400,incremental sample set is 100,The kernel function is the Gauss kernel,$\delta = 0.16, C = 1000$,in Table ,the time and accuracy of four algorithms in different steps are listed in detail.

1) support vector machine constructed by original support vector and incremental vector, written i-SVM.
2) support vector machine based on original support vector, retention selected non support vector and incremental support vector, written sn-SVM.
3) support vector selection, retention selected non support vector and incremental support vector, and incremental selection of non support vector machines are selected, written s-SVM.
4) select support vector machine and incremental support vector to build classifier, written si-SVM.
5) The results of the experiment are shown in table 1.

Table 1. A comparison table for experimental results of handwritten numeral classification data sets

Sample set type	Algorithm	Training time(S)	Training precision(%)
Initial sample set	i-SVM	18.26	83.98
	sn-SVM	20.77	83.12
	s-SVM	17.35	83.65
	si-SVM	15.26	82.80
New sample set 1	i-SVM	33.52	82.84
	sn-SVM	37.47	81.67
	s-SVM	28.63	83.75
	si-SVM	20.15	82.99
New sample set 2	i-SVM	47.11	85.47
	sn-SVM	28.34	84.59
	s-SVM	25.27	86.83
	si-SVM	22.64	82.72
New sample set 3	i-SVM	64.75	84.55
	sn-SVM	36.65	85.74
	s-SVM	38.44	86.91
	si-SVM	32.67	82.72

New sample set 4	i-SVM	69.37	82.08
	sn-SVM	27.96	85.25
	s-SVM	29,43	87.16
	si-SVM	27.49	83.09

From the experimental data, we can see that if the non support vectors selected from the original sample are retained and the non support vectors in the newly added samples will have some influence on the classification efficiency of the algorithm, [15] the accuracy of the classification is greatly improved. Then it can be shown that the non support vector is robust to the incremental learning algorithm.[16]

6. Summary

In this paper, the support vector preselection algorithm based on fuzzy C mean clustering is applied, and the non support vector preselection method based on the ratio of center density is applied. The support vector machine model is established by these two algorithms, and the commonly used data sets are classified. The experiments have achieved more ideal results.

References

[1] Zhang D Q, Chert S C. Kernel-based clustering incorporating spatial constraints for image segmentation[C]. Xi'an: Proceedings of the Second International Conference On Machine Learning and Cybernetics, 2003. 11.
[2] Du Chengjin, Sun Dawen, Patrick Jachman, et a1. Development of a hybrid image processing algorithm for automatic evaluation of intramuscular fat in beef M. 10ngissimusdorsi[f1. Meat Science. 2008,80(4):1231-1237.
[3] Tang Chenglong, Wang Shigang, Xu Wei. An improved fuzzy clustering algorithm based on data weighting strategy. Journal of electronics and information technology, 2010, 32(6):1277-1283.
[4] Wu Xueqian, Liao Yitao, Fan Yuxia, Cheng Fang. Pork longissimus segmentation technology based on KFCM and improved watershed algorithm. Journal of agricultural machinery, 2010, 41(1):172-176.
[5] Xu Xin Gong. Support vector machine incremental learning research [D]. Xi'an Electronic and Science University.2012
[6] V. N. Vapnik, S. Golowich, A. Smola. Support vector method for function approximation, regression estimation and signal processing [J]. Advances in Neural Information Processing Systems, 1997, 9:281-287.
[7] J. T. Jeng. Hybrid approach of selecting hyper-parameters of support vector machine for regression [J].IEEE Transactions on Systems, Man, and Cybernetics, Part B: Cybernetics, 2006, 36(3): 699-709.
[8] G. A. Carpenter, M. A. Rubin, W. W. Streilein. ARTMAP-FD: familiarity discrimination applied to radar target recognition [C]. International Conference on Neural Networks, 1997, 3: 1459–1464.
[9] D. M. J. Tax, R. P. W. Duin. Outlier detection using classifier instability [J]. Advances in Pattern Recognition, 1998, 1451: 593–601
[10] R. S. Guh, F. Zorriassatine, J. D. T. Tannock, C. ÓBrien. On-line control chart pattern detection and discrimination: a neural network approach [J]. Artificial Intelligence in Engineering, 1999, 13:413-425.
[11] P. Du, J. Peng, T. Terlaky. Self-adaptive support vector machines:modelling and experiments [J].Computational Management Science, 2009, 6(1): 41-51.
[12] S. C. Sun, D. Huang. A novel robust smooth support vector machine [J]. Materials and Energy Saving,2001, 148: 1438-1441.
[13] K. N. Wang, W. X. Zhu, P. Zhong. Robust support vector regression with generalized loss

function and applications. Neural Processing Letters, 2014; 41(1): 89-106.
[14] Chen Liping fuzzy C- means clustering research [D]. Yanshan University,2009
[15] Qin Yuping, Li Xiangna, Wang Xiukun and so on. Research on incremental learning algorithm based on hypersphere support vector machine. Computer Science, 2008. 35(8): p. 116-118
[16] Ju Xu, Wang Hao, et al. Based on concentric hyperspherical segmentation, support vector pre extraction method. Computer engineering and application.2006,31.55-56

IOP Conf. Series: Journal of Physics: Conf. Series **1060** (2018) 012054 doi:10.1088/1742-6596/1060/1/012054

Air-Attack Weapon Identification Model of Weighted Navie Bayes Based on SOA

CHEN Xi-cheng[1, *], FAN Bing-bing[1], YANG Wen-jia[1], TIAN Gui-long[1] and WU Kai[1]

[1]Air and Missile Defense College, Air Force Engineering University Xi'an, Shaanxi, 710051, China

[*]1139899855@qq.com

Abstract. Traditional Navie Bayes algorithm exists the issues of low inefficiency for the Air-attack Weapon Identification. Inorder to solve this problem, Air-attack Weapon Identification model of Weighted Navie Bayes Based on Seeker Optimization Algorithm is proposed. Firstly, the model reduces the dimension of the data sam-ples using rough set theory. Secondly, Seeker Optimization Algorithm searches the best attribute weights of Weighted Naïve Bayes. Finally, Navie Bayes classifier is structured with the best attribute weights to complete detection. The combination of the two algorithms can not only solve the feature redundancy problem of the traditional Navie Bayes algorithm, but also can optimize the strong independence between features. Through the experiments, prove that using this model for Air-attack Weapon Identification in air defense combat identification

1. Introduction

The modern air warfare, with air weapons combat style more diversified and air density increasing, the amount of information required to deal with the air also increases, so more and more important for recognition of air raid weapons effectively and accurately, help the commanders to make correct decisions, improve the air defense weapon system combat capability [1-2]. Study on the recognition of air strike weapons, currently is the most used BP neural network, BP neural network in classification requires a large number of training samples, and the slow convergence speed, easy to fall into local optimum, thus the training effect is not stable[3]; literature[4] constructs the recognition model of particle group average clustering, but the model is easy to appear larger convergence oscillation, the convergence speed is not fast;Paper [5] is established based on object recognition model Naive Bayesian and target recognition model based on probabilistic neural network, and through the comparison of probabilistic neural network and Naive Bayesian method in the accuracy of aerial target identification, found Naive Bayesian model better recognition effect.

Naive Bayes algorithm is widely applied, and has the characteristics of simple structure, fast computation speed and good classification effect [6], however, the classifier can get the optimal solution [7] only when it satisfies the class independent condition. In practical applications, there is a certain correlation between attributes and attributes. Therefore, the independence assumption of naive Bayes algorithm cannot be satisfied. In order to break through the shortcomings of Naive Bayesian algorithm, weighted Harry et al [8-9] of Naive Bayesian algorithm, the algorithm based on the attributes of the contribution of the classification of the different weights, so as to optimize the independence of the property, and get good results. Literature [10] uses the information gain as a weight to improve the naive Bayes weight, but it does not solve the problem of inaccurate classification caused by the large correlation between attributes. Document [11] first introduces rough set to attribute reduction to the data set of naive Bias model, and obtains the simplest attribute set. Then it sets the weight of the condition by maximizing the logarithmic conditional likelihood estimation, and proposes a weighted rough naive Bias model. According to the distribution of feature items between classes and classes, combined with the correlation degree between feature items, document [12] calculates weights, which effectively improves the classification effect of naive Bayes. The above improvement method has certain subjective factors for the determination of the weight value, and it is difficult to get the optimal weight.

Based on the technology of air raid weapons reduction attribute set of rough set, and then use the weighted population Naive Bayesian algorithm attribute weights search algorithm for optimization calculation, the optimal weights according to the Bayesian recognition model of air raid weapons, thereby eliminating the human factors influence on the weight given, simulation results show that the recognition effect is better than the the model of other models.

2. Basic theories and related models

2.1. rough set theory

Rough set theory is first proposed by Pawlak[13]-[14]., is a new mathematical method for data dealing with imprecise, uncertain and incomplete problem, knowledge reduction method of rough set theory, is retained in the basic knowledge and basis to ensure constant ability of classification on the object attributes and eliminate repeated and redundant the value of knowledge and realize the compression and refining, the attributes of air raid weapons set reduction based on discernibility matrix.

Definition 1: Decision tables $IS=(U,A,V,F)$, $U=\{x_1,x_2\cdots x_n\}$ as the non empty finite set of objects, also known as the universe of discourse. $A=RUD$ is the non empty finite set of conditional attributes, $D=\{d\}$ is the set of decision attributes, V represents the range of information functions, and V_j is a_j information function, f represents the information function of IS , and f_j is the a_j attribute information function.

Definition 2: $a_k(x_i)$ is the value of the sample x_i on the attribute of a_k The resolution matrix of the system is defined as:

$$m_{ij}^{*}=\begin{cases} a_k\in A\ ,\ a_k(x_i)\neq a_k(x_j)\cap D(x_i)\neq D(x_j) \\ \varnothing, D(x_i)=D(x_j), i,j=1,2,...,n \end{cases} \quad (1)$$

The D kernel of C is the sum of all the single elements in the matrix, that is:

$$M(S)=[\cup m_{ij}^{*}]_{n\times n} \quad (2)$$

Definition 3: The resolution function of the decision table is defined as follows :

$$f_{M(S)}(a_1,a_2,...,a_m)=\cap\{\cup m_{ij}^{*},1\leq j<i\leq n, m_{ij}^{*}\neq\varphi\} \quad (3)$$

The reduction steps are as follows:

1) the decision table IS is converted to the resolution matrix m_{ij}^{*} .

2) the resolution function $f_{M(S)}$ is obtained according to the resolution matrix.

3) for each conjunct minimal disjunctive form in function $f_{M(S)}$.

4) the output results of attribute reduction, and that each conjunctive minimal disjunctive form contains attributes.

IOP Conf. Series: Journal of Physics: Conf. Series **1060** (2018) 012054 doi:10.1088/1742-6596/1060/1/012054

2.2. *Weighted Naive Bayes model*

Definition 4: From the decision table IS, A is the attribute VARIABLE A1, and the n-dimensional attribute sample X is used to express the measured value. Given the m class C. For an unknown sample X, when and only then P, it is divided into the class, that is, the maximum:

$$p(C_i \mid X) = \frac{p(X \mid C_i)p(C_i)}{p(X)} \tag{4}$$

Because denominator is constant for all classes, it only needs to solve the maximum value of $p(C_1) = p(C_2) = \cdots = p(C_m)$. If the prior probability of class is unknown, it can be P1 and P2. s_i is the number of training samples in class C_i, and s is the total number of training samples.

Definition 5: The simple Bias classification model is the [8] in the class that divides the sample into the maximum posterior probability, and the model formula is:

$$C(X) = \arg\max_{c_i \in C} P(x_j \mid C_i) \tag{5}$$

Definition 6: The attribute A variable is given to the weight value W, and the formula of the weighted naive Bayes model [8]:

2.3. *Seeker Optimization Algorithm*

Seeker Optimization Algorithm (SOA), from human long life social experience through scientific analysis of human intelligent behavior in search activity, the intelligent search behavior is divided into egoism, altruism, pre dynamic behavior, uncertainty reasoning behavior, and the behavior modeling for computing search direction and step. In the iteration of the algorithm, the search direction and the step length are constantly updated to update the searcher's position, so as to get a better solution.

The algorithm flow is as follows:

Step 1: $t \to 0$.

Step 2: initialization, s initial positions are randomly generated in the feasible solution domain:

$$\left\{ \vec{x}_i(t) \mid \vec{x}_i(t) = (x_{i1}, x_{i2} \mathrm{L}, x_{iM}) \right\}$$

Among, $i=1, 2, 3, \mathrm{L}, s;\ t=0$

Step 3: evaluate, calculate the fitness value for each position.

Step 4: Step4: calculate the search direction for each searcher i in the one dimension j Search direction $d_{ij}(t)$ and the step length $a_{ij}(t)$.

Step 5: update the location of the searcher according to the formula.

Step 6: $t = t + 1$.

Step 7: if the algorithm stops the condition, stop the search; otherwise, turn to Step3.

3. Model architecture

3.1. *model structure*

This paper proposes a weighted naive Bayes air attack weapon recognition model based on crowd search algorithm. It can provide reliable information for tactical actions of next target threat assessment and target assignment by accurately and rapidly identifying air attack weapons. This model is divided into two stages, the data preprocessing stage and the crowd search algorithm search the optimal feature attribute weight phase.

In the data preprocessing stage, because the characteristics and attributes of air raid weapons are more, and include continuous and discrete data, in order to use rough set theory to process data, we need to discretize the original data. Then the rough set technique is used to reduce the feature set of the sample, and the optimal set is obtained. According to the formula (2), the posterior probabilities and priori probabilities of the attributes of the reduced attribute concentration are calculated. In the stage of search for optimal weight, the position of the searcher is generated randomly, and the position of each searcher represents a weight value. The weights are taken into the type (5) for identification and detection, each of which corresponds to a recognition accuracy. The weight of the highest recognition accuracy is the global optimal value.

The best location of individual history is compared with the best location of the group's history to update its position, search step length and direction. By constantly updating the location of the searcher to get better solutions until the iteration value is set, the iteration is over, and the optimal weight is saved. According to the optimal weight obtained, the Bias classifier is constructed to complete the air attack weapon recognition.

3.2 SOA algorithm verification

This paper uses the Sphere test function to test the optimization performance of the SOA and PSO algorithms.

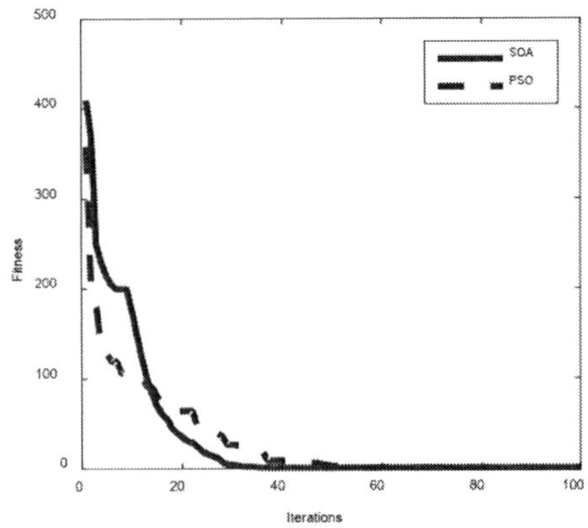

Figure 1. Comparison of the variation trend of the fitness value of SOA and PSO algorithm

The fitness curve of Figure1 shows that the SOA algorithm converges faster than the PSO algorithm. According to the figure1, it can be seen that the convergence accuracy of the SOA algorithm is higher and the robustness is better. It is proved that the optimization performance of the SOA algorithm is better than that of the PSO algorithm. Therefore, the use of SOA algorithm to optimize the weight of attributes is feasible and effective.

4. Experimental verification

According to the activity rules of air attack weapons, they can be divided into 5 categories: ballistic missiles, heavy weapons (including bombers, fighters, and fighter bombers, etc.), light weapons (including air to surface missiles, anti radiation missiles and cruise missiles, etc.), armed helicopters and bait. The feature attribute parameters are divided as shown in Table 1.

Table 1. air raid weapons after discretizatio

Flight height(H)m	>27000,27000-150,<150
Target discovery distance (R)km	>250,<250
Flight speed(V)m/s	>1800,1800-400, 400-200, <200
Route features (Y)	Equal high flying, climbing or subduction, subduction, bifurcation
Target attack angle(C) ℃	0-30, 30-60,60-90
Target interference ability (Y)	Strong, medium, weak, no
The incoming weapons （D）	Ballistic missiles, heavy weapons, light weapons, armed helicopters, bait

The characteristic parameters of the discrete air attack weapons are shown in Table 2.

Table 2. characteristic parameters of air raid weapons after discretization

Flight height(H)m	1,2,3
Target discovery distance (R)km	0,1
Flight speed(V)m/s	1,2,3,4
Route features (Y)	1,2,3,4
Target attack angle(C)°C	1,2,3
Target interference ability (Y)	1,2,3,4
The incoming weapons (D)	1,2,3,4,5

The attribute reduction is carried out according to the formula (3), and the attributes of the incoming target are calculated, and the attributes of R, Y and L are redundant.The set of feature attributes after reduction is shown in Table 3.

Table 3. sample data table after reduction

U HVCD	U HVCD
1 1121	9 2335
2 1131	10 1432
3 2312	11 2332
4 2433	12 1533
5 3434	13 3232
6 2223	...
7 2312	249 1435
8 3414	250 2235

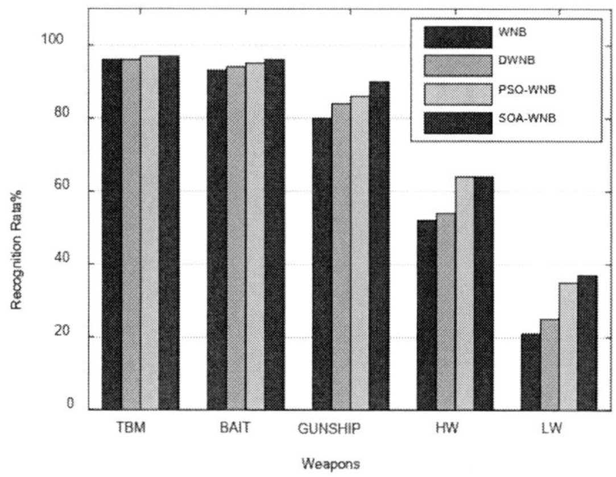

Figure 2. Algorithm result contrast diagram

Figure 2 is the comparison chart of recognition and detection results based on attribute weighted naive Bias classification algorithm (WNB), improved weighted naive Bias classification algorithm (DWNB) and PSO-WNB algorithm in document [11] based on [10]. The recognition rate of the four algorithms for tactical ballistic missiles, bait and armed helicopters is generally high, but the recognition rate of two incoming weapons for heavy weapons and light weapons is relatively low. The recognition accuracy of DWNB model is higher than that of WNB model, because DWNB removes redundant attributes from the samples to be classified by rough set technology, avoiding the interference of redundant attributes to classification. The recognition effect of the SOA-WNB model is better than that of DWNB, because on the basis of eliminating redundant attributes, the weights found using SOA are better than those given by DWNB. After testing and testing, the average detection rate of the SOA-WNB model constructed in this paper is 94.08%, and the average detection rate of the model built by PSO-WNB is 92.15%, which validates the validity of the SOA-WNB model. The simulation experiments show that the model is not only effective, but also superior to other models.

5. Conclusion

This paper presents a search algorithm based on population weighted Naive Bayesian air raid weapons recognition model, model firstly by using rough set theory attribute reduction of the sample set, to solve the problem of air combat weapon of the redundant attributes, attribute weights and then use the population search algorithm optimization weighted naive Bayes Juliu algorithm, thus solving the insufficiency of traditional Naive Bayesian attributes the assumption of independence, according to the identification of the optimal weights obtained improved Naive Bayesian classifier and the air attack weapons. It is proved by the experiment that the recognition effect of this model is better than that of other classification models, and the target of improving the recognition rate of air attack weapons is achieved. In the experiment, the most important influence on the accuracy of recognition is the attributes of weapons. In this paper, only 6 attributes of air raid weapons are involved. In order to improve the accuracy of recognition, we need to further explore other attributes of air raid weapons.

References

[1] Tait, P. (2005). *Introduction to Radar Target Recognition. Introduction to radar target recognition*. Institution of Electrical Engineers.
[2] Anagnostopoulos G C. SVM-based Target Recognition from Synthetic A perture Radar Using Target Region Outline Descriptors [J]. NonlinerAnalysis, 2009, 71(12):2934-2939.
[3] Qu, F., Zhang, M., Li, X., Zhang, J., & Liu, J. (2017). Parameters optimization of improved air-assisted orchard sprayer based on rsm and rbf neural network. *International Agricultural Engineering Journal, 26*(3), 64-74.
[4] Cheng, X. (2016). Multi-target tracking algorithm based on improved simplified particle swarm optimization. *Journal of Chongqing Normal University, 765-767*(02), 486-488.
[5] Wang, G., Liu, G., Yu, N., & Liu, X. (2017). An intelligent algorithm for infrared target recognition. *IEEE International Conference on Information and Automation* (pp.1818-1821). IEEE.
[6] Taniar, D., Taniar, D., & Rusu, L. I. (2009). *Strategic Advancements in Utilizing Data Mining and Warehousing Technologies: New Concepts and Developments*. Information Science Reference - Imprint of: IGI Publishing.
[7] Wu Jia, Cai Zhihua. Attribute weighting via differential evolution algo-rithm for attribute weighted naive Bayes (WNB) [J]. Journal of Computational Information Systems, 2011, 7(5):1672-1679.
[8] HarryZ, Sheng S L.Learning Weighted Naïve Bayes with accurate ranking[C]//Proceedings of the 4th IEEE International Conference on Data Mining (ICDM 04), Brighton, UK 2004: 567-570.
[9] Ueda N, Tanaka Y, Fujino A. Robust Naive Bayes combination of mul-tiple classifications [M] / /The Impact of Applications on Mathema-tics. Berlin: Springer, 2014:141-155.
[10] Taheri S, Yearwood J, Mammadov Metal. Attribute weighted Naive Bayes classifier using a local optimization [J] Neural Computingand Applications, 2014, 24(5):995-1002.
[11] Chen, L., Yang, X., University, H. P., & Polytechnic, H. Q. (2015). Most preferred strategy of pda library book selection based on weighted rough algorithm. *Bulletin of Science & Technology*.
[12] Pawlak Z.Rough sets[J].International Journal of Parallel Programming, 1982, 11(5):341-356
[13] Dai Chaohua, Zhu Yunfang and Chen Weirong. Seeker optimiza-tion algorithm [C]. n: Proc. 2006 Inter Conf. ComputationalIntelligence and Security, Guangzhou, China, IEEE Press, 2006, 1:225 -229.
[14] Wang, X., & Sun, X. (2017). An Improved Weighted Naive Bayesian Classification Algorithm Based on Multivariable Linear Regression Model. International Symposium on Computational Intelligence and Design (pp.219-222). IEEE.

IOP Conf. Series: Journal of Physics: Conf. Series **1060** (2018) 012055 doi:10.1088/1742-6596/1060/1/012055

A Method for Adaptive Adjustment of Command and Control Structure

Guanyu Chen[1,*], Peng Sun[1], Jieyong Zhang[1] and Yu Tao[2]

[1] College of Information and Navigation Air Force Engineering University, Fenghao East Road NO.1, Xi'an, Shanxi, China

[2] Academy of Information and Telecommunication, National University of Defense Technology, Wuhan Hubei, China

[*]ChenPTN@163.com

Abstract. The adaptive adjustment of command and control organizational structure is a hot issue in the field of agile command and control. First of all, the basic components of command and control organization are described. For the emergent events with changes in the action plan, the optimization constraints and optimization objectives of the command and control organizational structure are analysed, an optimization model for the adaptive adjustment of the command and control organizational structure is constructed, and a model solving method based on the discrete firefly algorithm is designed. The specific flow of the discrete firefly algorithm is adopted. Finally, the simulation model of the adaptive organization adjustment model and its solution algorithm are verified and compared by concrete examples. The simulation results show the feasibility and efficiency of the solution method.

1. Introduction

In the face of a complex and changeable battlefield environment, Command and Control (C2) organization structure should be adapted to changes in the battlefield environment [1].

The current research on adjustment of C2 organization structure has achieved some results. The literature [2] considers parameters such as response time, average backup number and load balance, and uses colored Petri nets to study the adaptive adjustment of C2 organization structure; Levchuk proposes an adjustment method based on three-phase idea [3]; literature [4] proposes an adaptive design method for organizational structure based on granularity calculation. During current research, algorithm time-consuming can be improved to meet the rapidly changing battlefield environment.

This paper presents a method for measuring the performance of C2 organization structure. An adaptive structure adjustment model is proposed for the change of the action plan, and a solving method based on the discrete firefly algorithm is proposed. Finally, the validity of the algorithm is verified by case simulation.

2. The entity of the command and control organization

There are three kinds of entities in C2 organization [4], namely, Task (T), Platform (P) and decision entity (Decision-Maker, DM). These three types of entities are the elements of the C2 organization.

Definition 1 Task: A task is an activity that requires a set of relevant resources to be processed. The set of all tasks in organization is recorded as $S^T=\{T_1,T_2,...,T_{NT}\}$ and NT is the total number of task.

Definition 2 Platform: A platform is a physical entity with specified resource capabilities that can be used to process tasks. The set of all platforms is recorded as $S^P = \{P_1, P_2, ..., P_{NP}\}$ and NP is the total number of platform in the organization.

Definition 3 DM: DMs are intelligent entities who utilize knowledge and wisdom of the planning domain to select an appropriate plan among a range of alternative scenarios. Tactical Decision-makers (TDMs) that control the platform to perform specific combat missions. The set of all TDM is recorded as $S_{TDM} = \{TDM_1, TDM_2, ..., TDM_{ND}\}$ and ND is the total number of TDM in the organization.

3. Modelling of adaptive adjustment of C2 organization structure

3.1. *Variables definition*

The various relationships in the C2 organization are defined as follows.

1) Command and control relationship between TDM and platform $R_{\text{TDM}-\text{P}}$

$R_{\text{TDM}-\text{P}}$ is represented by a matrix $M^{\text{TDM}-\text{P}} = (m_{ij}^{\text{TDM}-\text{P}})_{ND\times NP}$.

$$m_{ij}^{\text{TDM}-\text{P}} = \begin{cases} 1 & TDM_i \text{ control } P_j \\ 0 & \text{otherswise} \end{cases} \tag{1}$$

2) Cooperative relationship between tactical decision-maker $R_{\text{TDM-TDM}}$

$R_{\text{TDM-TDM}}$ is represented by a matrix $M^{\text{TDM-TDM}} = (m_{ij}^{\text{TDM-TDM}})_{ND\times ND}$.

$$m_{ij}^{\text{TDM}-\text{TDM}} = \begin{cases} 1 & TDM_i \text{ command } TDM_j \\ 0 & \text{otherwise} \end{cases} \tag{2}$$

3) The processing relationship between platform and task $R_{\text{T}-\text{P}}$

$R_{\text{T}-\text{P}}$ is represented by a matrix $M^{\text{T-P}} = (m_{ij}^{\text{T}-\text{P}})_{NT\times NP}$.

$$m_{ij}^{\text{T}-\text{P}} = \begin{cases} 1 & P_j \text{ is assigned to process } T_i \\ 0 & \text{otherwise} \end{cases} \tag{3}$$

4) Relationship between TDM and task $R_{\text{TDM}-\text{T}}$

When M^{T-P} and $M^{\text{TDM-P}}$ are determined, $M^{\text{TDM-T}} = (m_{ij}^{\text{TDM-T}})_{ND\times NT}$ can be determined by equation (4).

$$m_{ij}^{\text{TDM-T}} = \begin{cases} 1 & \text{if } m_{ik}^{\text{TDM-P}} = 1, m_{jk}^{\text{T-P}} = 1, 1 \le k \le NP \\ 0 & \text{otherwise} \end{cases} \tag{4}$$

3.2. *The optimization objective of C2 structure adjustment*

The performance of the C2 organization structure can be measured by the workload level of the TDM within the organization. The workload is composed of the command load and the cooperative load [4].

3.2.1. *Definition of workload.*

As for $\forall TDM_i \in S_{\text{TDM}}$, when TDM_i performs its responsible task T_j, the set of platforms controlled by TDM_i is recorded as $S_{\text{TDM}_i\text{-T}_j}^{\text{P}} = \{P_k \,|\, m_{ik}^{\text{TDM-P}} = 1, m_{jk}^{\text{T-P}} = 1, 1 \le \text{k} \le NP\}$. When this task is performed, the set of TDM that have a collaboration relationship with TDM_i is $S_{\text{TDM}_i\text{-T}_j}^{\text{TDM}} = \left\{\text{TDM}_k \,\middle|\, m_{kj}^{\text{TDM-T}} = 1, k \ne i, 1 \le k \le ND\right\}$, so the workload of executing T_j is as follows.

$$\text{W}_{\text{TDM}_i}^{T_j} = \omega_{C2} \cdot \left\| S_{\text{TDM}_i\text{-T}_j}^{\text{P}} \right\| + \omega_{\text{TC}} \cdot \left\| S_{\text{TDM}_i\text{-T}_j}^{\text{TDM}} \right\| \tag{5}$$

Here, ω_{C2} is the command load factor, ω_{TC} is the cooperative load factor, $\left\| S_{\text{TDM}_i\text{-T}_j}^{\text{P}} \right\|$ is the number of elements in the set $S_{\text{TDM}_i\text{-T}_j}^{\text{P}}$, and $\left\| S_{\text{TDM}_i\text{-T}_j}^{\text{TDM}} \right\|$ is the number of elements in the set $S_{\text{TDM}_i\text{-T}_j}^{\text{TDM}}$.

The load that is defined in the whole operation cycle of is as follows.

$$\mathrm{W}^{\mathrm{T}}_{\mathrm{TDM}_i} = \sum_{T_j \in S^{\mathrm{T}}_{\mathrm{TDM}_i}} W^{T_j}_{\mathrm{TDM}_i} + W^{t}_{\mathrm{TDM}_i} \tag{6}$$

Here, $W^{t}_{\mathrm{TDM}_i}$ refers to the load that TDM_i command task before time t.

3.2.2. The root mean square of workload. The RMS (Root Mean Square, RMS) load of all in the organization is as follows.

$$\begin{aligned} \mathrm{RMS}^{\mathrm{T}}_{\mathrm{TDM}} &= \sqrt{\frac{1}{ND}\sum_{i=1}^{ND}\left(W^{\mathrm{T}}_{\mathrm{TDM}_i}\right)^2} \\ &= \sqrt{\left(W^{\mathrm{T}}_{\mathrm{TDM}}\right)^2 + \frac{ND-1}{ND} D^{\mathrm{T}}_{TDM}} \end{aligned} \tag{7}$$

$W^{\mathrm{T}}_{\mathrm{TDM}} = \frac{1}{ND}\sum_{i=1}^{ND} W^{\mathrm{T}}_{\mathrm{TDM}_i}$ is the average load in the organization; D^{T}_{TDM} is the variance of the full load.

The smaller the $\mathrm{RMS}^{\mathrm{T}}_{\mathrm{TDM}}$, the lower the mean and variance of the work load of TDM, the more reasonable the design of the organizational structure is. Therefore, minimum $\mathrm{RMS}^{\mathrm{T}}_{\mathrm{TDM}}$ can be used as an objective function of adaptive adjustment.

3.3. The optimization objective of C2 structure adjustment

After mission plan is changed, the C2 organization structure is adjusted to improve the performance of the organization. The C2 relationship between TDM and platform before adjusting is $R^{1}_{\mathrm{TDM-P}}$, and the matrix $M_1^{\mathrm{TDM-P}} = (m1_{ij}^{\mathrm{TDM-P}})_{ND\times NP}$ .After adjusting the relationship is $R^{2}_{\mathrm{TDM-P}}$, $M_2^{\mathrm{TDM-P}} = (m2_{ij}^{\mathrm{TDM-P}})_{ND\times NP}$.

When the action plan is changed, the model of adjustment of organizational structure is as follows.

$$\min \mathrm{RMS}^{\mathrm{T}}_{\mathrm{TDM}}$$
$$\text{s.t.}\begin{cases} NP^{\mathrm{Trans}} = \frac{1}{2}\sum_{i=1}^{ND}\sum_{j=1}^{NP}\left|m2_{ij}^{\mathrm{TDM-P}} - m1_{ij}^{\mathrm{TDM-P}}\right| \le \sigma \\ \sum_{i=1}^{ND} m2_{ij}^{\mathrm{TDM-P}} = 1, j = 1,2,\mathrm{K}\ ,NP \\ \sum_{j=1}^{NP} m2_{ij}^{TDM-P} \ge 1, i = 1,2,\mathrm{K}\ ,ND \end{cases} \tag{8}$$

In equation (8), the objective function is $\min \mathrm{RMS}^{\mathrm{T}}_{\mathrm{TDM}}$, the purpose of adjustment is to get a better C2 structure. The smaller the $\mathrm{RMS}^{\mathrm{T}}_{\mathrm{TDM}}$, the better the performance of the corresponding C2 structure.

σ is the maximum adjustment cost that an organization can afford. The first constraint indicates that the adjustment cost must be within the scope of the organization; the second constraint indicates that each platform in the organization can be controlled by only one TDM. The third constraint represents each TDM control at least one platform.

4. Model solving method based on discrete firefly algorithm

4.1. Discrete firefly algorithm

In 2008, Yang, a scholar of University of Cambridge, proposed the Firefly Algorithm (FA) [5]that simulates the firefly population behaviour. The discrete form of firefly algorithm is applied in task scheduling, assembly sequence planning and traveling salesman [6]-[10] combination optimization problems. In the process of the adaptive adjustment problem of C2 organization structure, several key elements of discrete firefly algorithm are: firefly position, objective function, distance and movement.

4.1.1. Firefly location initialization and coding. Combined with command and control relation matrix adjusted in this paper, integer coding is selected in the coding method.

Each firefly represents a $M^{\text{TDM-P}}$. For any firefly i, an n-dimensional vector $X_i = (x_{i,1}, x_{i,2}, \ldots, x_{i,k}, \ldots, x_{i,n})$ is used to represent the position of firefly i, and n is determined by total number of platform in organization $x_{i,k} \in [1, \text{ND}]$, $x_{i,k}$ indicates that P_k is under the command and control of $TDM_{x_{i,k}}$. Based on this coding rule, for the $M^{\text{TDM-P}}$ shown below. After encoded, it is $X_i = (1,3,2,4,2,1,3,1)$.

$$M^{\text{TDM-P}} = \begin{bmatrix} 1 & 0 & 0 & 0 & 0 & 1 & 0 & 1 \\ 0 & 0 & 1 & 0 & 1 & 0 & 0 & 0 \\ 0 & 1 & 0 & 0 & 0 & 0 & 1 & 0 \\ 0 & 0 & 0 & 1 & 0 & 0 & 0 & 0 \end{bmatrix} \rightarrow X_i = (1,3,2,4,2,1,3,1) \tag{9}$$

4.1.2. Objective function. In order to find the solution with the smallest solution, each firefly i decoded matrix has the corresponding RMS value. The smaller the RMS value, the better the corresponding solution of the firefly. Therefore, brightness is defined as follows. If $NP^{\text{Trans}} \le \sigma$ is not satisfied, the brightness value is 0, a solution satisfying the condition is randomly generated.

$$I_{0i} = \begin{cases} \dfrac{1}{RMS_i} & NP^{\text{Trans}} \le \sigma \\ 0 & NP^{\text{Trans}} > \sigma \end{cases} \tag{10}$$

4.1.3. The distance between fireflies. The distance between two fireflies reflects the degree of differentiation of two schemes. Therefore, in this paper, the distance between firefly i and firefly j in discrete space is calculated using equation (11).

$$r_{ij} = \left\| X_i \oplus X_j \right\| = \sqrt{\sum_{k=1}^{n} \left(X_{i,k} \oplus X_{j,k} \right)^2} \tag{11}$$

4.1.4. The movement of the firefly. Based on the optimization and adjustment features of the $M^{\text{TDM-P}}$, the following equation for updating the position of the firefly is constructed

$$X_i' = X_i \oplus \beta_0 e^{-\gamma r_{ij}^2} \left(X_j \ominus X_i \right) \oplus \alpha \left(\text{rand} - \frac{1}{2} \right) \tag{12}$$

4.2. The steps of the discrete firefly algorithm

The implementation steps of discrete firefly algorithm are as follows.

Step 1: Set parameters of discrete firefly algorithm, initialize the firefly population, the maximum number of iterations, and the light intensity absorption coefficient γ, the maximum degree of attraction β_0.

Step 2: Follow the integer encoding method, the maximum brightness of each firefly is calculated by equation (10).

Step 3: Calculate the distance between each firefly by equation (11). Further, the relative brightness I and the attractiveness β are calculated according to equations.

Step 4: Firefly moves according to equation (12). The brightest firefly searches for the nearest optimal solution and moves toward to it.

Step 5: After moving, calculate the brightness of all the fireflies and rank the brightness. Record the brightness value of the firefly with the highest brightness and the corresponding position. The number of iterations increases by 1.

Step 6: If the maximum iteration number $Max\text{cycle}$ is reached, end the algorithm and output the optimal solution saved in step 5, otherwise go to step 3.

The optimal solution is decoded, and the resulting $M^{\text{TDM-P}}$ is the best C2 relationship.

5. Results and analysis of the simulation

5.1. Simulation case

This paper selects the joint operations scenario in [11] as an experimental case, verifies the performance of the discrete firefly algorithm to solve adaptive adjustment problem of C2 structure. The experiment was carried out in the windows 7 system environment using Matlab 2014a for simulation analysis. The initial task-platform relationship and the TDM-platform relationship are shown in Tables 1 and 2, respectively.

Table 1. Relationship of task and platform before adjusting

Task	Platform	Task	Platform
T_1	P_2,P_6,P_7,P_{16}	T_{10}	P_6,P_{10}
T_2	P_1,P_2,P_{15},P_{19}	T_{11}	P_7,P_{10},P_{14}
T_3	P_{11},P_{12},P_{13}	T_{12}	P_4,P_7,P_{12}
T_4	P_3	T_{13}	P_8,P_{11},P_{16}
T_5	P_8,P_{14}	T_{14}	P_3,P_9,P_{16}
T_6	P_5,P_9,P_{18}	T_{15}	P_5,P_{13},P_{17},P_{20}
T_7	P_{10},P_{11},P_{12},P_{13},P_{20}	T_{16}	P_4,P_8,P_{18},P_{19}
T_8	P_3,P_5,P_9,P_{18}	T_{17}	P_{10},P_{15}
T_9	P_6,P_8	T_{18}	P_4,P_{17}

Table 2. Relationship of tactical decision-maker and platform before adjusting

TDM	Platform
TDM_1	P_1,P_{10},P_{11},P_{14},P_{15}
TDM_2	P_2,P_6,P_8,P_{16},P_{19}
TDM_3	P_7,P_{12},P_{13},P_{20}
TDM_4	P_3,P_4,P_5,P_9,P_{17},P_{18}

5.2. Simulation experiment

In order to verify the validity of this method, 20 simulation experiments are carried out and compare with other methods. Comparisons are made with Greedy Search Algorithm (GSA) [4] and Discrete Artificial Bee colony algorithm (DABC) [12], which can also be used to solve the model of this paper. The experimental results are shown in figure1 and figure 2.

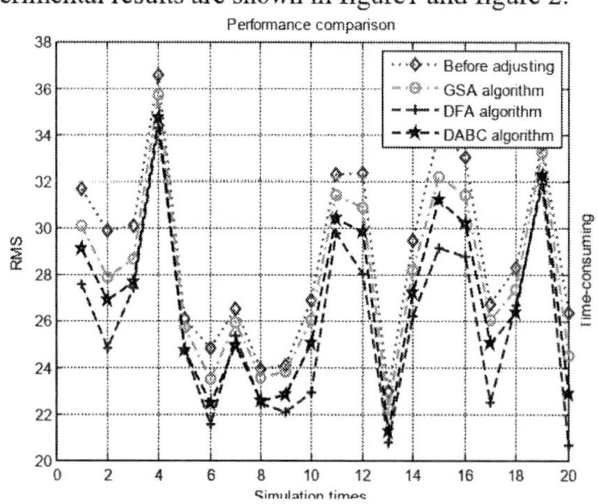

Figure 1. Performance comparison

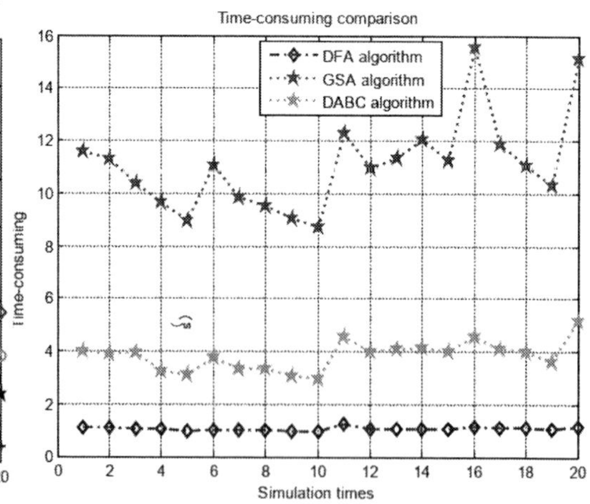

Figure 2. Time-consuming comparison

5.3. Simulation analysis

In figure 1, the RMS values after the adjustment by DFA algorithm, GSA algorithm and DABC algorithm are all lower than RMS value before the adjustment indicates that these three methods can solve this problem. The DFA algorithm has the largest reduction in the RMS value, which shows that the DFA algorithm is better than the DABC and GSA algorithm in solving the adaptive adjustment ofC2 organization structure, and the effect of the DABC and GSA algorithms is close. In the time consuming of algorithm, the time consuming of DFA algorithm is the lowest, and the time consumption of GSA is the highest, which shows that DFA algorithm is the most efficient.

In summary, using the DFA algorithm can better reduce the RMS value of the load, and has a great advantage in the algorithm time-consuming. It can meet the requirement of the timeliness of the adjustment of C2 structure in the battlefield environment.

6. Conclusion
This paper studies adaptive adjustment of C2 structure. Based on the discrete firefly algorithm, the adaptive adjustment method is proposed. The algorithm is simulated and compared with GSA and DABC algorithm. The results verify the feasibility and efficiency of the discrete firefly algorithm used in C2 structure adjustment.

7. Acknowledgments
Supported by Natural Science Foundation (Grant No. 61773396) ,Natural Science Foundation (Grant No. 61573017) and Natural Science Foundation of Shanxi Province (Grant No. 2017JM6062).

References
[1] Hollenbeck J R, Moon H, Ellis A P, et al. 2002 Structural contingency theory and individual difference: examination of external and internal person team fit. *Journal of Applied Psychology* vol 87 no 3 pp 599–606.
[2] Perdu D M, Levis A H.1998 Adaptation as a Morphing Process: A Methodology for the Design and Evaluation of Adaptive Command and Control Teams. *Computational & Mathematical Organization Theory*, vol 4 no1pp 5-41.
[3] Levchuk G M, Levchuk Y N, Meirina C, et al. 2004 Normative design of organizations part III: Modeling congruent, robust, and adaptive organizations. *IEEE Trans. on Systems, Man, and Cybernetics* vol 34 no 3 pp 337-350.
[4] Sun Y, Yao P Y, Li M H, et al. 2016 Adaptive adjusting method of command and control structure of army organization. *Systems Engineering and Electronics* vol 38 no 9 pp 2086-2092.
[5] Yang X S. 2008 Nature-inspired metaheuristic algorithms. London: Luniver Press, 2008:83-96.
[6] Xie J. 2015 Research on ETL scheduling of data warehouse based on discrete firefly algorithm. Shanghai: East China University of Science and Technology.
[7] Liu C P, Ye C M. 2014 A diecrete firefly algorithm for minimizing the makespan in the no-idle permutation flow shops. *Journal of System & Management* vol 23 no 5 pp 723-727.
[8] ZENG B. 2013 Research on assembly sequence planning based on discrete firefly algorithm. Xiangtan: Xiangtan University.
[9] Liu Z M, Zhang R, Zhang G M, et al. 2016 Research on near infrared wavelength optimization method based on discrete firefly algorithm. *Spectroscopy and Spectral Analysis* vol 36 no 12 pp 3931-3936.
[10] Xu B Z, Ren Q H, Meng Q W,et al. 2018 Research on consecutive mean excision algorithm in wavelet domain communication system[J]. Journal of Air Force Engineering University(Natural Science Edition) vol 19 no 5 pp 32-39.
[11] Yu F, Tu F, Pappipati K R. 2006 Novel congruent organizational design methodology using group technology and a nested genetic algorithm. *IEEE Transactions on Systems, Man, and Cybernetics* vol 36 no 1 pp 5-18.
[12] Huo F C, Du Y, Liu Y. 2016 Artificial bee colony algorithm and its application. *Journal of Jilin University (Information Science Edition)* no 12 pp 3931-3936.

IOP Conf. Series: Journal of Physics: Conf. Series **1060** (2018) 012056 doi:10.1088/1742-6596/1060/1/012056

Research on the Operation Mode Switching Strategy of Wind Power-Photovoltaic Micro Grid

Chi ZHANG

Qianhu College of Nanchang University, Nanchang, Jiangxi province, China

573559614@qq.com

Abstract. Wind energy and solar energy are complementary in time. The power generation system composed of wind and solar energy can make full use of the clean energy for wind and solar energy and have high reliability. It is very important to smooth transition between grid-connected mode and the island mode. The micro grid of hybrid power generation system based on wind power is established in DigSILENT/Power Factory. The improved control strategy, P-Q control and the control strategy based on droop characteristics, have been proposed to be applied in the transition state simulation study mode based on the micro grid. The results show that this advanced control strategy can ensure the power balance and reasonable power distribution, and can maintain the voltage and frequency stability of the connection bus during the operation mode switching.

1. Introduction

The micro grid is a system with loads and power. In the micro grid, power electronics are used to change energy and provide the necessary controls. Compared with the external grid, the micro grid is a single controllable unit, it can meet the demand of the customer on the quality, reliability and safety of electric power [1].

When the power grid fails, the micro grid can be operated in the isolated mode, and the reliability and safety of the power supply will be improved. The island operation relies on the smooth transition between the grid-connected and the island mode. Therefore, it is necessary to study the control strategy in the process of micro-grid switching.

2. Micro grid parallel operation and island operation

2.1. The method of Micro grid parallel operation and island operation

The micro grid can be switched between parallel operation and the island operation. Grid mode means that there is a connection between the micro grid and the big power grid, and the load flows between them. The island model means that when the grid fails, the micro grid is isolated from the big grid, and distributed generation systems provide power to the load. The island operation improved the reliability of the power supply [2].

The existence of isolated islands is very dangerous: firstly, the voltage and frequency may be unstable and may be dangerous to the safe operation of power equipment because the isolated island operation area is not controlled by the power grid. Secondly, when the island is in operation, the staff cannot judge whether the equipment is charged or not, thus it is dangerous to the life safety of the staff. Finally, when the micro grid is connected to the main grid, the main grid will be affected.

2.2. *The provisions of the large power grid on the island operation*

Due to the great harm of the isolated island operation, UL1741 and IEEE 929-2000 require that once distributed generation is in an island operation, distributed generation needs to disconnect with the large power grid. The island effect can guarantee the normal operation of the power grid, but with the increase of distributed generation capacity, there are some problems with the island effect. For example, distributed power generation can't be fully used, reducing the reliability of distributed power networks, and so on. It is well known that faults are unavoidable in power systems. When a fault occurs, the distributed generation is disconnected from the power grid, it is difficult to guarantee the reliability of power supply and the advantages of distributed generation.

Based on the above analysis, when failure occurs, switching off the power supply will no longer be a practical or reliable solution. Ieee1547-2008 proposed a new standard for the operation of the planned islands and the unplanned island operation [3]. The planned island operation means that a small grid or part of the grid, composed of a load and a distributed generation system, is separated from the power system and according to the previous plan. The planned island operation can maintain the stable operation of micro grid and improve the reliability of power supply. The unexpected island operation state refers to the distributed generation system independent of the load power supply. The operating state is contingent and unstable. This method allows the island to operate as planned, prevent accidental island operation.

3. Control strategy for the micro grid operation and the parallel grid operation

3.1. *Traditional control strategies*

Currently, master-slave control strategy and peer control strategy are applied in the micro grid. Master-slave control strategy refers to the micro grid all distributed power supply in parallel mode adopted in the P-Q control strategy, but in the isolated operation mode, in order to maintain the stability of voltage and frequency, one or more distributed power uses V/f control strategy based on droop characteristics. The peer control strategy refers to the use of droop control in the distributed power supply in the micro grid, thus ensuring a reasonable power distribution.

The P-Q control refers to the maximum power output of a distributed power source or to specify a specific power generation according to the actual situation. The principle of P-Q control is shown in figure 1, when the frequency of the micro grid is 50Hz and the bus voltage is rated, the distributed generation system runs at point B and the output power is P_{ref} and Q_{REF} [4]. If the frequency of connecting bus voltage and micro grid increases or decreases, the working point moves from B to A or C to maintain power output at point P_{ref} and Q_{REF}. The purpose of P-Q control design is the maximum use of renewable energy, which is suitable for intermittent energy. P-Q control is relatively easy, but it can't keep the voltage and frequency stably [5].

P-Q control follows the reference current by adjusting the active current (ID) and the unreactive current (IQ).

$$\begin{bmatrix} U_a \\ U_b \\ U_c \end{bmatrix} = \begin{bmatrix} U_m \cos(\omega t) \\ U_m \cos(\omega t - \frac{2\pi}{3}) \\ U_m \cos(\omega t + \frac{2\pi}{3}) \end{bmatrix} \tag{1}$$

U is network voltage, U_m is the amplitude of phase voltage. Do a change in U:

$$\begin{bmatrix} U_d \\ U_q \end{bmatrix} = T_{abc-dq} \begin{bmatrix} U_a \\ U_b \\ U_c \end{bmatrix} = \begin{bmatrix} U_m \\ 0 \end{bmatrix} \tag{2}$$

In the d_q coordinate system, decoupling the d axis and the q axis, U_d is constant, U_q is 0. Then the active power is set to the P_{ref} and reactive power set to Q_{ref}. The reference current of the inverter is expressed as follows:

$$\begin{aligned} \dot{i}_{dref} &= \frac{P_{ref}}{U_d} \\ \dot{i}_{qref} &= \frac{Q_{ref}}{U_d} \end{aligned} \tag{3}$$

In summary, the research on control strategy has shifted from the output power to the current research.

Figure 2 shows the principle of droop control: When distributed generation is connected in parallel to the micro grid, the relationship between the real-time output power and the frequency of the micro grid is linear, and the relationship between the reactive power output and the connection bus voltage is the same.

When the output power of distributed generation increases, the operating point moves from A to B. When the bus voltage and micro grid frequency change slightly, the power output should be adjusted reasonably.

$$\Delta P = k_{\frac{P}{f}}(f_0 - f_1) \tag{4}$$

$$\Delta Q = k_{\frac{Q}{V}}(V_0 - V_1) \tag{5}$$

Droop control is a kind of control mode that does not require the communication between power supplies. It is suitable for the control of distributed power inverter.

3.2. Improved control strategy

The single control strategy cannot meet the operation demand of micro grid. In the model based on wind power and photovoltaic hybrid system, the photovoltaic power generation system is intermittent and the control goal is to maximize the use of renewable energy, so P-Q control is applicable to the photovoltaic system [6]. The output characteristic of the wind power system containing accumulator is relatively stable, so the droop control is suitable for wind power system. When operating on the micro-grid island, it is used to ensure reasonable power distribution and to maintain the frequency of the voltage and micro-grid connected to the bus line to remain stable.

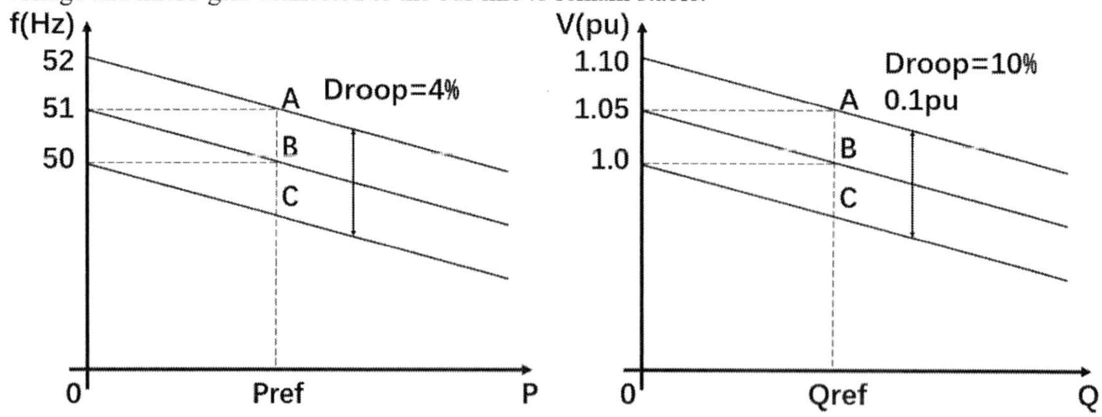

Figure 1. Principle of P-Q control

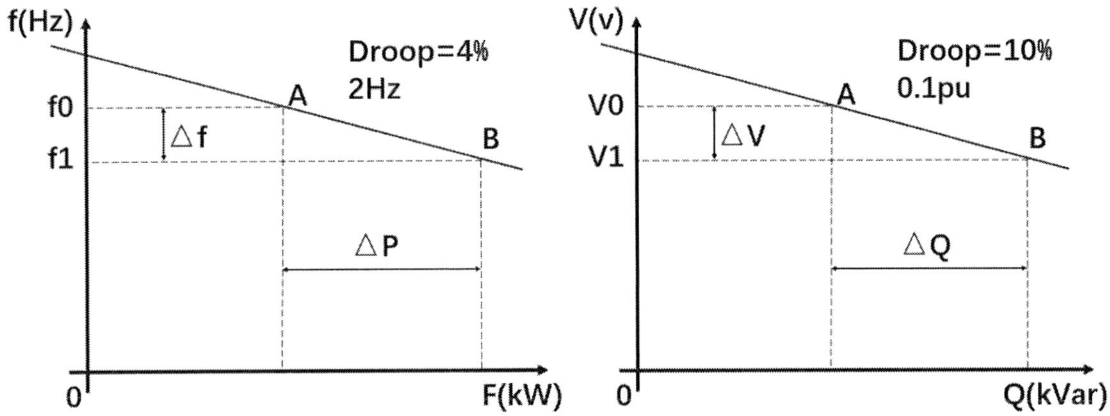

Figure 2. Principle of droop control

4. Model and simulation

Figure 3 shows the model of wind and PV hybrid power system. In this model, the voltage of the micro grid is 0.4kV, which is connected with the power grid of 10kV. The micro grid consists of a 0.045MW wind powered generator and 10 pieces of 0.01MW photovoltaic array. The wind turbine is composed of the synchronous generator, the converter, the battery and the control system. Photovoltaic generation system consists of the dc voltage source, the inverter and the control system. Design capacity of the photovoltaic system is 100kW active, 0kVar reactive power. Wind power output is 10kW active power, 10kVar reactive power. The power of load 1, 2 and 3 is respectively 5kVar, 10kW; 0kW, 5kVar; 10kW, 10kVar. Build a hybrid power system in DIgSILENT/PowerFactory.

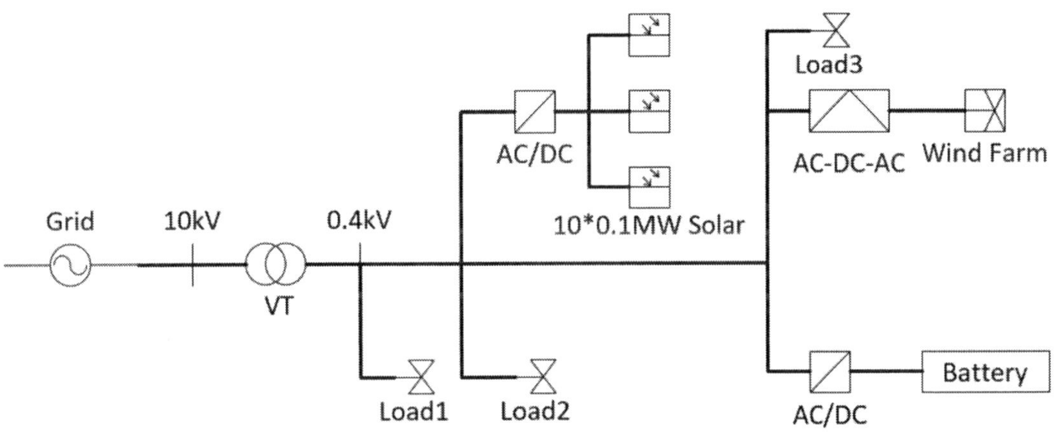

Figure 3. Model of wind and PV hybrid power system

When a transient fault occurs in the power grid, the micro grid is separated from the main power grid. When the fault is removed, the micro grid is connected to the main power grid. The simulation time of the simulated micro grid switch in this paper is 1 second. When the power grid fails 0.2 s, the micro grid is isolated from the distributed network. Then, at 0.5s, the fault clearing micro grid is reconnected to the main grid.

IOP Conf. Series: Journal of Physics: Conf. Series **1060** (2018) 012056 doi:10.1088/1742-6596/1060/1/012056

4.1. Output power

Figure 4 shows the output power of wind power system and photovoltaic power generation system respectively. The wind power generation system adopts the droop control strategy, and the power output of the wind power unit is 10kW, which was originally 10kvar. At 0.2s, the grid connection mode is transformed into an isolated island, while the power of the micro grid is unbalanced. The output power of the wind turbine is changed to ensure the stability of the micro grid, and its power output changes to 5KW, -5kvar. At 0.5s, the micro grid is reconnected to the main grid, and the output power of the wind power system will become stable after a slight change. Photovoltaic power generation adopts PQ control strategy, the output of active power and reactive power output of photovoltaic power generation system is 100kw, 0kVar. At 0.2s, the micro grid is transformed from the grid-connected mode to the isolated island mode, and the output power of the photovoltaic system is slightly fluctuated. After adjustment, the power output will be stable. After 0.5s, the micro grid is reconnected to the main grid, and the output power of the photovoltaic system is still 100kW, 0kvar.

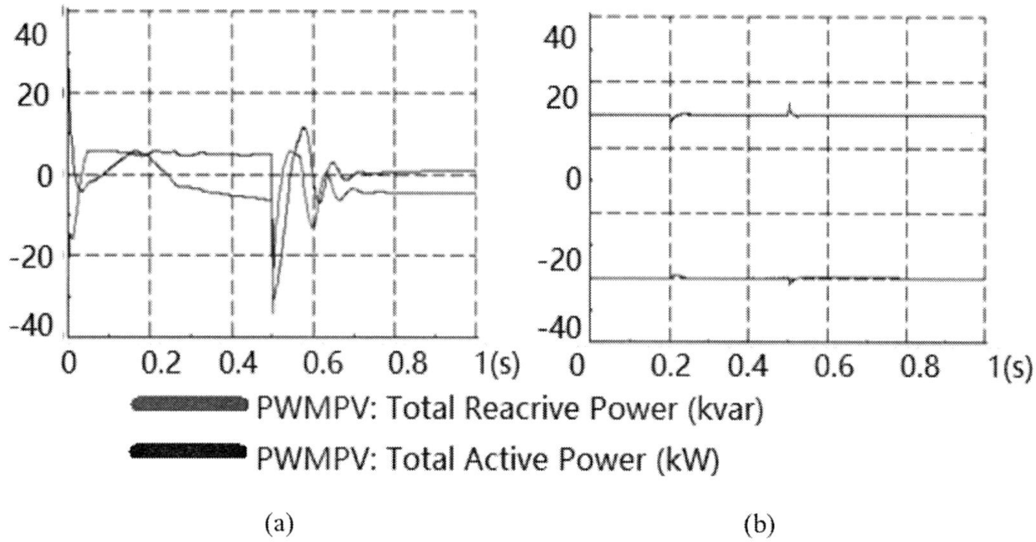

Figure 4. (a) Output of wind power; (b) Output of photovoltaic.

4.2. Frequency

As shown in figure 5 (a), in parallel operation, its frequency is kept at 50 Hz. At 0.2s, the micro grid has the grid-connected mode to the isolated island mode, and the micro grid quickly adjusts the frequency to maintain the power balance. At 0.5s, the micro grid is reconnected to the main grid, and the frequency of the micro grid becomes 50Hz.

4.3. Voltage of connecting bus

Figure 5 (b) shows the bus voltage of the micro grid connection. Before 0.2s, the micro grid was connected to the main network, and the voltage of the connecting bus is 0.4kV. At 0.2s, the micro grid is transformed from the grid-connected mode to the island mode. Due to the droop control, the change of the connecting bus voltage is 0.4003kV, and the change range is less than 10%. At 0.5s, the micro grid is reconnected to the main power grid, and the connected bus voltage is changed to 0.4kV.

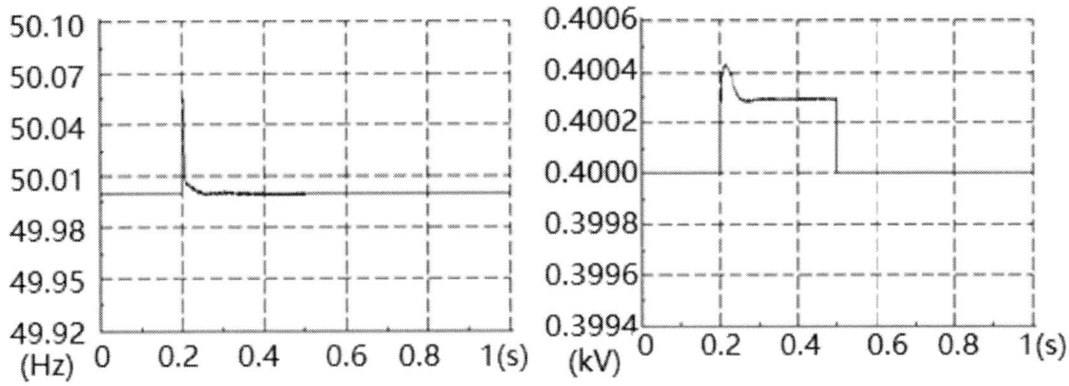

(a) (b)

Figure 5. (a) Frequency; (b) Voltage of connecting bus.

5. Conclusion

The simulation study of wind power and photovoltaic hybrid system shows that the improved control strategy model conforms to P-F and Q-V control characteristics. The simulation results show that the control strategy combined with P-Q control and droop control can guarantee the power balance and reasonable power distribution. The strategy can be further applied to practical application.

References

[1] A. Linder and R. Kennel. Model predictive control for electrical drives. Proc. IEEE PESC, Recife, Brazil, Jun. 12–16, 2005, pp. 1793–1799.

[2] Miranda H, Cortes P, Yuz J I, et al. Predictive torque control of induction machines based on state-space models [J]. IEEE Transactions on Industrial Electronics, 2009, 56 (6): 1916-1924.

[3] R. H. Lasseter and P. Piagi. Micro Grid: a conceptual solution[C]. Power Electronics Specialists Conference, IEEE 35th Annual, Germany, 2004, 6: 4285-4290.

[4] J. Rodriguez,J. Pontt, C. Silva,et al. Predictive current control of a voltage source inverter. Proc.35th IEEE PESC, Jun. 2004, vol. 3, pp. 2192–2196.

[5] J. Rodríguez, J. Pontt, C. Silva,et al. Predictive current control of a voltage source inverter. IEEE Trans. Ind. Electron., Feb. 2007,vol.54, no.1, pp.495–503.

[6] Yang Z, Zhang G, Wen Y, et al. The design and simulation of the 3MW direct drive wind turbine's controller to decrease vibration of the tower[C]// Automation. IEEE, 2017:52-55.

IOP Conf. Series: Journal of Physics: Conf. Series **1060** (2018) 012057 doi:10.1088/1742-6596/1060/1/012057

Multi-Instance Multi-Label with Application to High Resolution Remote Sensing Images

DJELLALI Mohamed[1,*]

[1]College of computer science and technology, Nanjing University of Aeronautics and Astronautics, Nanjing 210016 china

[*]mohameddjellali@nuaa.edu.cn

Abstract. In this paper, we seek to solve the problem of remote sensing image classification using Multi Instance Multi Label Learning (MIML) framework, where each image contain multiple regions (instances) corresponds to multiple objects (labels). MIML framework is more efficient than traditional learning framework for complicated objects such as satellite images. Existing MIML algorithms such as MIMLboost, MIMLsvm and MIMLfast have been found useful in scene classification and can provide multiple labels for each instance in complicated objects. The proposed approach was performed in two steps: In the first step, we process the dataset using segmentation and feature extraction. Images are ambiguous because they contain numerous objects so we consider that each image is a bag and various blocks in the image are instances. In the second step, we apply MIML to classify each block of images. According to the experimental results, the proposed method outperforms the state-of-the-art methods.

1. Introduction

Recently, MIML attracts much attention in machine learning community. The MIML formalization is learning the model function: $2^X \rightarrow 2^Y$ for dataset $\{(X_1, Y_1), \ldots, (X_l, Y_l)\}$, where $X_i \subseteq X$ is set of instances. $\{x_{i1}, x_{i2}, \ldots, x_{in}\}, Y_i \subseteq Y$ is set of associated labels $\{y_{i1}, y_{i2}, \ldots, y_{im}\}$ such that n and m are sizes of X_i and Y_i respectively .$\{(X_1, Y_1), \ldots, (X_l, Y_l)\}, X_i \subseteq X$ is set of instances$\{x_{i1}, x_{i2}, \ldots, x_{in}\}, Y_i \subseteq Y$ With the development of satellite technology, the plentiful information provided in high resolution images totally changes the perspective of satellite image classification. HR image as one individual object with only a single semantic label. Actually, with more information the HR image contain numerous semantic meaning and multi-labels emerged from image segmented regions instead of the entire image. Feng [1] introduced an algorithm of multi-instance semi-supervised learning with hierarchical sparse representation. Liu and al [2] presented the key instance detection task and propose a new framework as solution based on the relationship between instances. T. Gärtner [3] used multi-instance bags after extraction of feature vector in order to prepare multi-instance kernel function. In order to predict one label of multi instance bags, MIML framework was proposed by Zhou et al. [3, 4], which has advanced toward becoming another exploration area. MIML framework provides a useful formulation for HR satellite images. In MIML HR images are described as bags, each bag contains many instances associated to segmented regions of the original image. Labelling a bag as negative required that all its instances are negative. We defined a bag as positive if at least one from its instances is positive. The rest of this paper is organized as follows: In section 2 we introduce the related works for both remote sensing image classification and other application of MIML framework.

In section 3 we discuss the experimental results with preparation of dataset and evaluation criteria. The experiments and result analysis are given in section 4. Section 5 offers our conclusion.

2. Related work

2.1 MIML

Multi Label Learning (MLL) [5, 6], can be used in classification problems, where each sample can be mapped to various classes at the same time. The main function of MLL is to learn the classifier: $H_{MLL}: X \rightarrow 2^Y$. In a set of MLL training set: $\{(x_i, Y_i), 1 \leq i \leq n\}$ x_i is the i-th instance and $Y_i$is the label set of x_i. In [7], Chaudhuri B proposed semi-supervised graph-theoretic method related to the approach of multi-label image retrieval problem. This approach diffuses the labels in the training images using a label correlation propagation technique to the unlabelled image in order to map them with multiple labels. In order to recuperate the images out of archives with the same classes of regions as the image in question. MLL has been applied in many real world fields, such as medical, banking, text classification and ecological studies [8, 9]. Numerous methods are proposed in order to solve the task of MLL for instance: ML k-Nearest neighbour, ML decision tree and ranking SVM are proposed in [10].

Multi Instance Learning (MIL) model was formalized by Dietterich et al, in which they propose algorithm for dry activity prediction [11]. The main function of MIL is: $L_{MIL}: 2^x \rightarrow \{-1, +1\}$ From a set of MIL training set $\{(X_i, Y_i), 1 \leq i \leq n\}$; $x_i is\ a\ bag = \{x_1^i, x_2^i, \dots, x_{ni}^i\}\ and\ y_i \in \{-1, +1\}$ is the label of the bag x_i. Jaume amores[12], present comparative studies about multi-label solutions of MIL.

Zhou et al [13, 14] introduce two MIML algorithms, MIMLboost and MIMLsvm algorithm respectively and their application in dataset scene classification. Li et al [15] proposed a Key Instances Sharing among Related labels (KISAR) algorithm which can find key-instance activating several labels in a bag. In [16] they propose the RankLossSIM approach to optimize with label ranking loss .To effectively deal with large dataset and minimize time complexity of MIML, MIMLfast algorithm is proposed with an approximated ranking-loss by using stochastic gradient descent (SGD).

2.2 Remote sensing Images

Remote sensing data is about the object or phenomenon through devises sensor without contact the objects physically. The use of satellite sensors is to detect and classify ground objects for the interest of numerous fields: including geography, geology, military, intelligence and humanitarian application.

In fact remote sensing is referred to satellite or aircraft sensor devises, according to the process of emission/reception of signals, remote sensing is divided into active (signal is emitted by satellite and reflection is used in the sensor) and passive (reflection of sunlight is detected by sensor).

2.3 Feature extraction methods

- Discrete Cosine Transform (DCT) is applied for each block after subdividing the image into regions of 8x8 pixels. The precision of classifiers on DCT features extraction is better than statistical and texture feature extraction techniques. The main idea of DCT is that the intensity of an image is often high in the low frequency elements. Generally the DCT only uses the coefficient with the largest values to reconstruct the original image [17].
- Local Binary Pattern (LBP) corresponds the functions of the descriptor, i.e. a local neighbourhood is started from gray scale of middle pixels into a binary Pattern [18].
- Gabor texture features is efficient for analysing remote sensed imagery [19]. They were shown to surpass other texture features in the images classification. Gabor method uses a set of scales and orientations filters to extract the texture features [20].

  We summary the general process of object detection and mention the step of previous feature extraction methods in Figure 1.

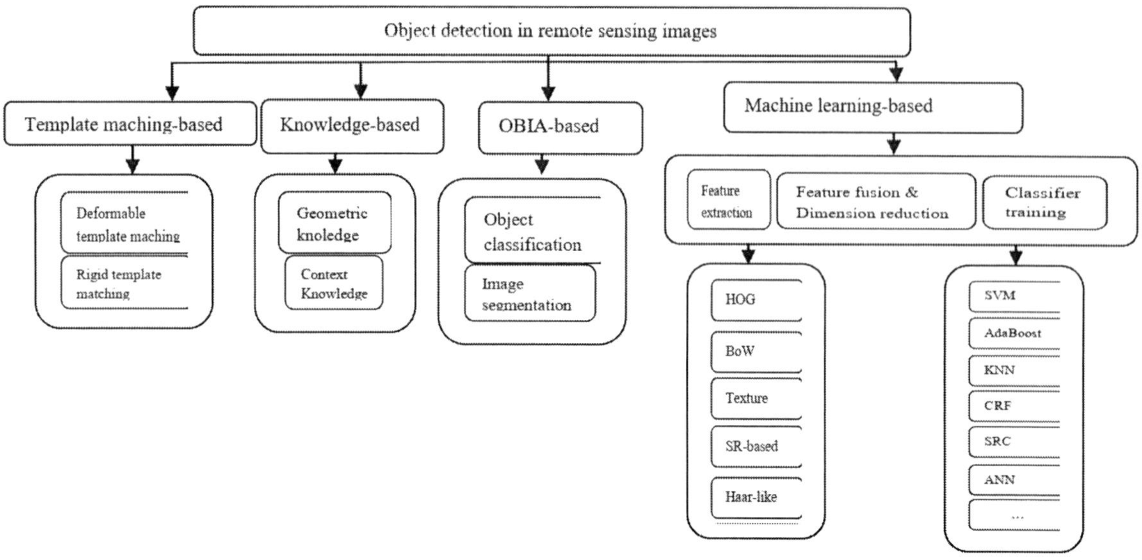

Figure 1. Object detection methods in remote sensing images.

3. Proposed method

Remote sensing images are complicated where each sample can correspond to multi-instances and multi-labels at the same time. For example, a satellite image contain several region, each one can be associated to an instance. Simultaneously the image can have the part with buildings and trees. In this paper we apply MIML into remote sensing images. When we learn the model by numerous bags; each bag has several instances and several labels as well.

MIML is the continuation of classical supervised learning (MIL and MLL). In MIL[21] an object presented by several instances is attached to one class label. The training dataset used under the form $\{(I_1, l_1), (I_2, l_2), \ldots, (I_n, l_n)\}$, where$I_1 \subseteq I$ is set of instances$\{i_{1,}i_2, \ldots, i_{nj}\}$ is and $n_j$is number of instances, $l_j \in \{-1, +1\}$is the label of I_j. In MLL [22] an object presented by one instance is attached with numerous class labels. The training dataset used under the form$\{(i_1, L_1), (i_2, L_2), \ldots, (i_n, L_n)\}$, where $i_k \in I$ is an instance and L_i is an array of labels in class label L.

MIML framework provides the mapping between objects instances and corresponding labels as show in figure 2. The classification on label l is defined by the function f: $f_l(X) = W_l^T W_0 X$, W_0 is a m×d matrix, W_l is m-dimensional weight vector corresponding to label l, d and m are the sizes of the feature space and the shared space successively. The prediction of instance x on label l as:

$$f_X(X) = max_{k=1..K} f_{l,k}(X) = max_{k=1..K} W_{l,k}^T W_0 X \quad (1)$$

According to Dietterich, a bag is positive equivalently it contains an minimum one positive instance, the prediction of a bag X on label l is defined by the following function: $f_l(X) = max_{x \in X} f_l(X)$. The basics parameters of gradient descent are as follow:

$$W_0^{t+1} = W_0^t - \gamma_t S_{\bar{Y},v}(W_{\bar{y},\bar{k}}^t \bar{X}^T - W_{y,k}^t X^T) \quad (2)$$

$$W_{y,k}^{t+1} = W_{y,k}^t + \gamma_t S_{\bar{Y},v} W_0^t X \quad (3)$$

$$W_{\bar{y},\bar{k}}^{t+1} = W_{\bar{y},\bar{k}}^t - \gamma_t S_{\bar{Y},v} W_0^t \bar{X} \quad (4)$$

Such that $\gamma_t$is the step size of SGD, $W_{y,k}$, $W_{\bar{y},\bar{k}}$ and all column are standardised to Norm L2.

IOP Conf. Series: Journal of Physics: Conf. Series **1060** (2018) 012057 doi:10.1088/1742-6596/1060/1/012057

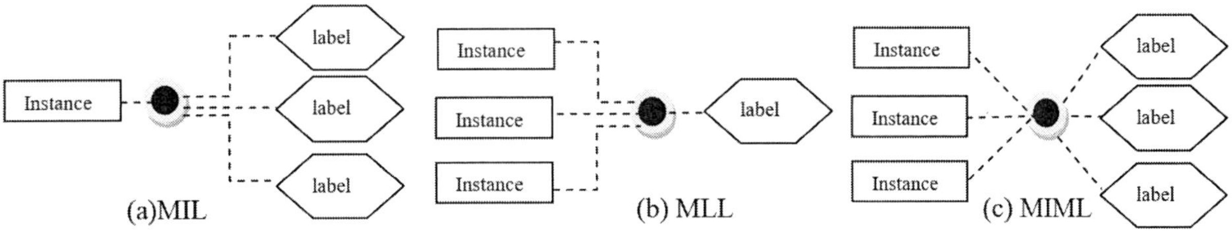

Figure 2. Four different learning frameworks

Algorithm

1: **INPUT:** training images, parameters m, C, K and t

2: **TRAIN:** initialization w_0 and $w_{l,k}$(l=1….L, k=1….K)

3: **repeat:** randomly sample a bag X and one of its relevant label y

4: select key instance and sub-concept by $(x,k) = \arg max\ max_{x\in X, k\in\{1....K\}} f_{y,k}(x)$

5: **for** $i = 1{:}\,|\overline{y}|$; 6: sample an irrelevant label $\overline{y}$ from $\overline{Y}$

7: select key instance and sub-concept by $(\overline{x},\overline{k}) = \arg max\ \ ax_{x\in X,\overline{k}\in\{1....K\}} f_{\overline{y},\overline{k}}(x)$

8: **if** $f_{\overline{y}}$(x) >f_y(x)-1; 9: v = i

10 : update w_0 , $w_{y,k} and w_{\overline{y},\overline{k}}$ as Eqs. 4 to 9, and perform normalization; 11: break

12 : **until** stop criterion reached

13: **OUTPUT**: Relevant labels set for the test bag X test is: $\{l|1 + f_l(x_{test}) > f_{\hat{y}}(x_{test})\}$

Figure 3 show the scenario of image classification and the illustration according to selected approaches (MIL, MLL, MIML). MLL Principe associate the labels (building, trees and water) to the input image by founding the class objects in the image and fill the matrix instance-label by 1(-1) if it exist (not).MIL approach link multi sub images with one label if and only if at least one class object exist in the numerous sub images. Finally MIML provide the opportunity to label the image linking each sub image with its conform label and this is the developed approach to classify the image with complicated objects.

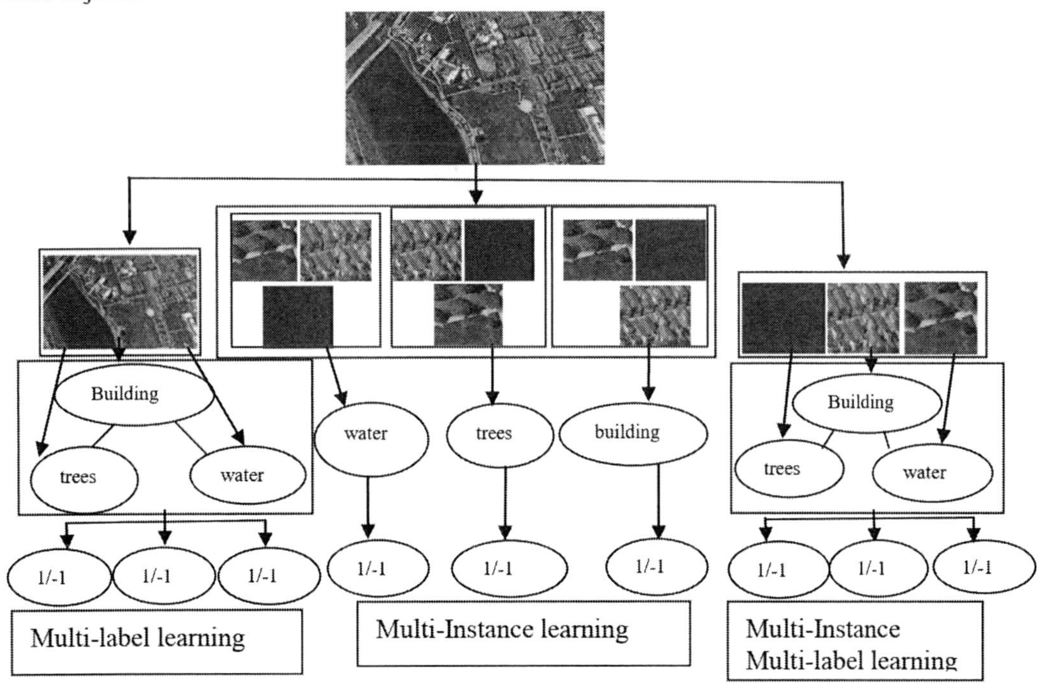

Figure 3. Explicative examples clarify the difference between MIL, MLL and MIML.

4. Experiment

4.1 Data collection

The dataset we used contains 2000 high resolution images affiliated to the classes: ground, water, buildings, trees and roads. We collected those images from Google earth in the area of California USA, more than 90% belong to multiple classes in the same image. We obtained 2000 sub-images after segmentation of original images. Totally we use 1800 samples for training and 200 for test, for each sample we divide it into four regions and extract the features using dct features. This project required to have labelled dataset in order to train the model and evaluate its performance. The images are resized to be a uniform scale for the object extraction.

4.2 Multi label Performance criteria

Multi-label classification problem needs special criteria for evaluation different from traditional signal label classification that uses just accuracy as performance evaluation criteria [23, 24]. In the experiment part we employ five criteria Hamming loss (h.l), One-error (o.e), Coverage (co), Ranking loss (r.l) and Average precision (a.p) to check the performance of the proposed MIML classification method. These criteria are defined as follows:

- **Hamming loss:** calculate how frequently an instance-label pair is wrong classified.

$$HL(h) = \frac{1}{p}\sum_{i=1}^{p}\frac{1}{|y|}|h(X_i)\Delta Y_i| \quad (5)$$

Δ is the difference between two sets .The less value, the improved achievement .

- **One-error:** calculate how frequently the object is mislabelled. The ideal performance is OE(h) = 0; the less value of OE(h) ,the improved achievement of h.

$$OE(h) = \frac{1}{p}\sum_{i=1}^{p}[\![[\mathrm{argmax}_{y\in Y} h(X_i, y)] \notin Y_i]\!] \quad (6)$$

- **Coverage:** calculate the required value, on the normal, to check the array of labels and include the corresponding labels of the object. It is approximately linked to the accuracy at the level of perfect recall. The less value of coverage (h), the improved achievement of h.

$$coverage(x) = \frac{1}{p}\sum_{i=1}^{p} \max_{y\in Y_i} rank^h(X_i, y)\text{-}1 \quad (7)$$

Ranking loss: Calculate the medium portion of labels that are not correctly requested.

$$\text{Rank Loss}(h) = \frac{1}{p}\sum_{i=1}^{p}\frac{1}{|y_i||\bar{y_i}|}|\{(y_1, y_2)|\ h(X_i, y_1) \le f(X_i, y_2), (y_1, y_2) \in Y_i \times \bar{Y_i}\}| \quad (8)$$

Where the $\bar{y_i}$de notes the supplementary set of $Y_i$in C. The performance is ideal when RankLoss (f) =0, The less value, the improved achievement.

- **Average Precision:** Calculate the medium portion of labels classified to a certain label y $\in$ Y_i which actually is in Y_i.

$$\text{AverPrec}(h, D_u) = \frac{1}{p}\sum_{i=1}^{p}\frac{1}{|Y_i|}\sum_{y\in Y_i}\frac{|\{y'|rank^h(X_i,y')\le rank^h(X_i,y), y'\in Y_i\}|}{rank^h(X_i,y)} \quad (9)$$

4.3 Efficiency Comparison

Table 1 describes the experimental results of many feature extraction methods in order to compare the evaluation performance.

Table 1. Experimental results (mean±std) on satellite image Dataset: With the corresponding criteria h.l, o.e , c.o and r.l ,the smaller value the better performance and the inverse for a.p criteria.

	h.l	o.e	co	r.l	a.p
Lbp	.272±.041	.146±0.01	2.842±.041	.024±.006	.843±.002
Dct	.368±.014	.177±.008	3.181±.019	.291±.008	.792±.005
Lbp+dct	.402±.074	.239±.311	2.94±.041	.0328±.121	.793±.082
gabor	.469±.065	.409±.203	3.267±.234	.471±.107	.702±.068

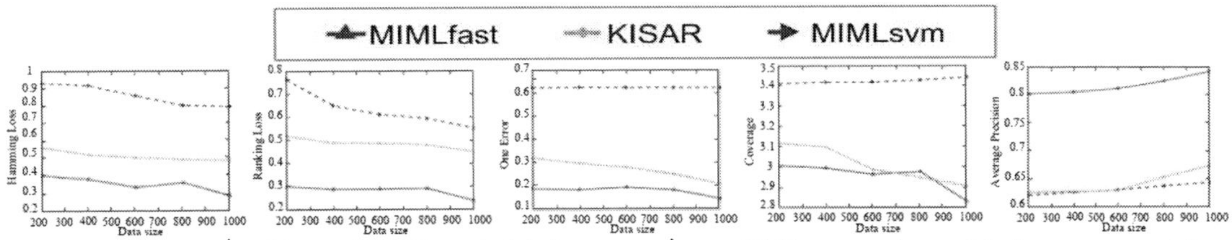

(a)Hamming loss↓ (b) Ranking Loss↓ (c) One error↓ (d)Coverage↓ (e) Average precision↑

Figure 4. Performance criteria (h.l,o.e,r.l,co,a.p) versus size of dataset according to three following methods MIMLfast, KISAR and MIMLsvm .

It is essential to evaluate the effectiveness of the model trained with MIML approach, for this purpose we calculate the five MIML criteria as shown in Figure 4. Finally we conclude that MIMLfast give more accurate result than two other methods (MIMLsvm, KISAR) especially with LBP feature extraction method.

5. Conclusion

MIML approach is the appropriate for Remote sensing object recognition thus it is created for learning the complicated objects. MIML can be comprehended by transformation MIML problem into its corresponding in classical supervised learning using MIML or MLL as intermediate. Although, this transformation may miss information in ignore the relation between labels. When taking in consideration the relation between labels in common area and finding sub-ideas for complex labels. Comparison results demonstrate the efficiency of proposed method in terms of classification accuracy. Future work will cover the optimization of hamming loss function by implementing active learning in the process of training the model.

References

[1] S. H. Feng, W. H. Xiong, B. Li, C. Y. Lang, and X. K. Huang,"Hierarchical sparse representation based multi-instance semisupervised learning with application to image categorization," Signal Processing, vol. 94, pp. 595–607, 2014.

[2] G. Liu, J. Wu, and Z. H. Zhou, "Key instance detection in Multi-instance learning," in Proceedings of the 4th Asian Conference on Machine Learning, SIN, pp. 253–268, 2012.

[3] T. Gärtner, P. A. Flach, A. Kowalczyk, and A. J. Smola, "Multi-instance kernels," in Proceedings of the 19th International Conference on Machine Learning, Sydney, AU, pp. 179–186, 2002.

[4] Z. H. Zhou and M. L. Zhang, "Multi-instance multi-label learning with application to scene classification," Advances in Neural Information Processing Systems 19, Cambridge, MA: MIT Press, pp. 1609–1616,2007.

[5] A. McCallum, "Multi-label text cassification by EM," AAAI'99 Workshop on Text Learning., 1999.

[6] R. E. Schapire and Y. Singer, "BOOSTEXTER: a boostingbasedsystem for text categorization," Machine Learning,vol. 39, no. 2/3, pp. 135–168, 2000.

[7] Chaudhuri B, Demir B, Chaudhuri S, et al. Multilabel Remote Sensing Image Retrieval Using a Semisupervised Graph-Theoretic Method[J]. IEEE Transactions on Geoscience & Remote Sensing, 2017, PP(99):1-15.

[8] Kang F, Jin R, Sukthankar R. Correlated Label Propagation with Application to Multi-label Learning[C]// IEEE Computer Society Conference on Computer Vision and Pattern Recognition. IEEE Computer Society, 2006:1719-1726.

[9] Guo L, Jin B, Yu R, et al. Multi-label classification methods for green computing and application for mobile medical recommendations[J]. IEEE Access, 2016, 4:3201-3209.

[10] Zhang M L, Zhou Z H. A Review on Multi-Label Learning Algorithms[J]. IEEE Transactions on Knowledge & Data Engineering, 2014, 26(8):1819-1837.

[11] T. G. Dietterich, R. H. Lathrop, T. Lozano-Perez, Solving the multiple-instance problem with axis-parallel rectangles, Artificial Intelligence 89 (1997) 31–71.
[12] Amores J. Multiple instance classification: Review, taxonomy and comparative study[J]. Artificial Intelligence, 2013, 201(4):81-105.
[13] Z. H. Zhou and M. L. Zhang, "Multi-instance multi-label learning with application to scene classification," Advances in Neural Information Processing Systems 19, Cambridge, MA: MIT Press, pp. 1609–1616, 2007.
[14] Z. H. Zhou, M. L. Zhang, S. J. Huang, and Y. F. Li, "Multi-instance multi-label learning," Artificial Intelligence, vol. 176, pp. 2291–2320,2012.
[15] Y. F. Li, J. H. Hu, Y. Jiang, and Z. H. Zhou, "Towards discovering what patterns trigger what labels," in Proceedings of the 26th AAAI Conference on Artificial Intelligence, Toronto, Ontario, CA, pp. 1012–1018, 2012.
[16] F. Briggs, X. Z. Fern, and R. Raich, "Rank-loss support instance machines for MIML instance annotation," in Proceedings of the 18th ACM SIGKDD international conference on Knowledge discovery and data mining, Beijing, CHN, pp. 534–542, 2012.
[17] Al-Shehhi S, Al-Muhairi M, Abuoeida D, et al. Robust and Fragile Watermarking Scheme Based on DCT and Hash Function for Color Satellite Images[C]// International Conference on Developments in Esystems Engineering. IEEE, 2013:253-258.
[18] T. Ojala, M. Pietikainen, and T. Maenpaa, "MultiresolutionGray-Scale and Rotation Invariant Texture Classification with Local Binary Patterns", IEEE Transactionon Pattern Analysis and Machine Intelligence – PAMI, vol. 24, no. 7, 2002, pp. 971-987 .
[19] S. Newsam, L. Wang, S. Bhagavathy, and B. S. Manju-nath, "Using texture to analyze and manage large collec-tions of remote sensed image and video data,"Journalof Applied Optics: Information Processing, vol. 43, no. 2, pp. 210–217, 2004.
[20] B. S. Manjunath and W. Y. Ma, "Texture features forbrowsing and retrieval of image data,".IEEE Trans.OnPattern Analysis and Machine Intelligence, vol. 18, no.8, pp. 837–842, 1996
[21] S. Andrews, I. Tsochantaridis, and T. Hofmann, "Support vectormachines for multiple-instance learning," in The Neural Information Processing Systems (NIPS), 2002, pp. 561–568.
[22] Kong X, Ng M K, Zhou Z H. Transductive Multilabel Learning via Label Set Propagation[J]. IEEE Transactions on Knowledge & Data Engineering, 2013, 25(3):704-719.
[23] Zhang M L, Zhou Z H. Multilabel Neural Networks with Applications to Functional Genomics and Text Categorization [J]. IEEE Transactions on Knowledge & Data Engineering, 2006, 18(10):1338 1351.
[24] Zhou Z H, Zhang M L, Huang S J, et al. Multi-Instance Multi-Label Learning[J]. Artificial Intelligence, 2008, 176(1):2291-2320.

New Application of Children's Sign Language Enlightenment Based on Data Glove

Ruchang Xu[1], Linbo Zhai[1,*], Jingyu Li[1], Jiaxiu Jiang[1], Weicheng Zhai[1] and Huayan Yu[1]

[1] School of Information Science and Engineering, Shandong Normal University, Jinan, China

Email: *110940323@qq.com

Abstract. In order to enhance the learning interest of deaf-mute children as beginners in sign language and promote the balance of educational reform, we have developed a new application which based on data gloves, to be the enlightenment for children learning sign language. The application mainly includes three parts: data acquisition and conversion, wireless communication, and cartoon interactive application. The application, taking *STM32* development board as the control core, is integrated with multiple sensor function modules including *Wi-Fi* wireless communication, three-axis gyroscope, bending resistor and others, therefore, it can realize the gesture recognition and data transmission. After repeated testing, the system has high accuracy in gesture recognition, not only data transmission is reliable but also the application response timely. Deaf-mute children can use the cartoonish application to get sign language examination questions and to answer by wearing data gloves, and then the system will give a star rating comprehensively based on whether the answer is correct, on whether the cue is clicked in answering question and on other aspects, so as to realize the interaction with the *PC* program and let deaf-mute children contact and learn sign language in games. The design concept of the application has received favorable response from Qingdao school for the deaf, which has a certain practical and promoted value.

1. Introduction

Language is the most important communication tool for human beings, which is closely related to thinking and is the direct manifestation of thought, so is the sign language, which is the first language of the deaf-mutes, a tool for communication and their thought [1]. Research suggests that 0-12 years old is the best period of language learning and there is plenty of nice teaching software assisting sound children study, while the way for deaf-mute children to learn sign language is relatively single. Usually teachers demonstrate personally in class and children learn sign language picture books on their own after class. Although hearing and speech disorder aggravates the difficulty of mental and intellectual development of deaf-mute children, it has a positive impact on their development of visual perception so that their visual advantage is very obvious. In the related tests, the observation ability of deaf-mute children is more sensitive and their visual motion ability is stronger compared to healthy children[2].Besides, deaf-mute children are more competent than those normal children in terms of identifying and distinguishing important features of different objects according to their vision, and the visual memory of color and pattern. Therefore, in addition to using language communication, increasing the stimulus of nonverbal materials in the process of education for deaf-mute children can give play to the advantages of visual memory to get more information [3], which is also our idea and original intention of designing the new application of children's sign language enlightenment based on data glove. The proposed application of this article is a kind of Internet of things application. It puts forward the new application of the enlightening sign language teaching according with children's psychology, interest and motivation. Compared with traditional teaching method of sign language, this application is more enlightening for children who have just begun to access sign language, help them understand sign language and stimulate their interest in learning sign language.

The first part of the article describes the overall design and implementation of the application. It mainly introduces function of the application, design and implementation of the model. The second part gives an overview of centre control, communication module and data acquisition module. The third part introduces design ideas of system software, testing methods of hardware and software, and makes an example analysis to some debugging results. The fourth part summarizes design ideas and design method of the application.

2. The overall design and realization of application

2.1. The function profile of application

By wireless communication, sensor technology, embedded system and other significant technology, we develop a children's sign language enlightenment application that accords with children's curiosity and has prominent visual communication effect. Its main functions are as follows:

- Application realize gestural recognition through detecting tortuosity of the bending resistor and changes of palm displacement.
- Contrast whether gesture is accord with the requirements of application and judge whether gestural answer is right.
- Based on the comprehensive evaluation of the answer situation, the application will give star rating to realize Inspired teaching.
- Sign language teaching mode which can through passing levels stimulates the enthusiasm of deaf-mute children to learn sign language.

2.2. The design of application module and realization of function.

The design of application integrates *STM32F103VET6*, *Wi-Fi* module and other digital sensor inspection module in order to realize above function, which are shown in the figure 1. We use *Visual Studio* developing Interactive cartoon application and realizing sign language enlightenment of deaf-mute children.

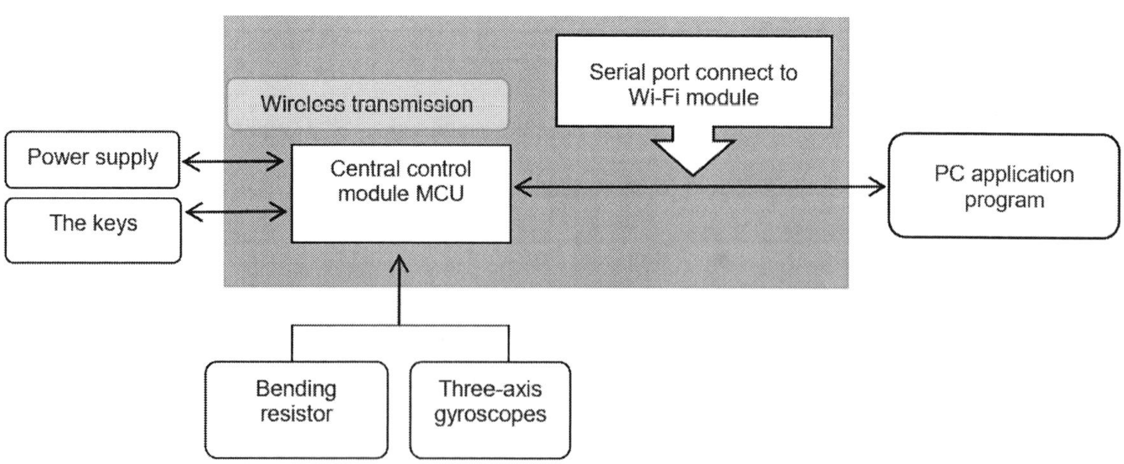

Figure 1. System Architecture

The bending resistor takes charge of collecting tortuosity of figures and its resistance decreases with the increase of tortuosity. Three-axis gyroscopes takes charge of collecting changes of palm displacement and then the recognition of gestures can be realized by the above two modules. Considering sign language is a dynamic language and children are more active, so application integrates serial port connect to *Wi-Fi* module to realize wireless transmission of data, which provides more comfortable use experience for deaf-mute children. In addition, wireless transmission provides conditions for applications to be ported to mobile devices such as mobile phones and tablets [4], so the use of APP can make sign language learning more convenient and fast.

2.3. Instruction of application module

2.3.1. Central control module and Communication module. Application selects *STM32F103VET6* which is high performance and low power consumption as Central control module of SCM selects *Wi-Fi* module to realize communication with *PC*.

The *Wi-Fi* module of application is *ESP8266* series chip. With its high integration, low power consumption and ease of use, the chip has gained recognition in the Internet of things field and it is the best choice to realize remote control which is based on *Wi-Fi* [5]. *ESP8266* supports soft AP mode, Station mode and soft AP + Station coexistence mode. Application selects soft AP mode which regards *ESP8266* as a *Wi-Fi* hotspot.

The fundamental process of *ESP8266* realize wireless control: Firstly, *PC* joins up *ESP8266*'s hotspot by *Wi-Fi* and builds Socket links through the application. Secondly, the instructions of the starting data detection were sent to the *ESP8266* chip via *Wi-Fi*. The chip converts the *Wi-Fi* signal into a serial signal and sends it to the *MCU* via serial communication. Thirdly, the *MCU* begins processing the acquired instruction and data. Finally, *MCU* sends the data back to the application in real time via the *ESP8266* chip, so application realize the data interconnection between *STM32* development board and *PC*. The integral control process is shown in the figure 2.

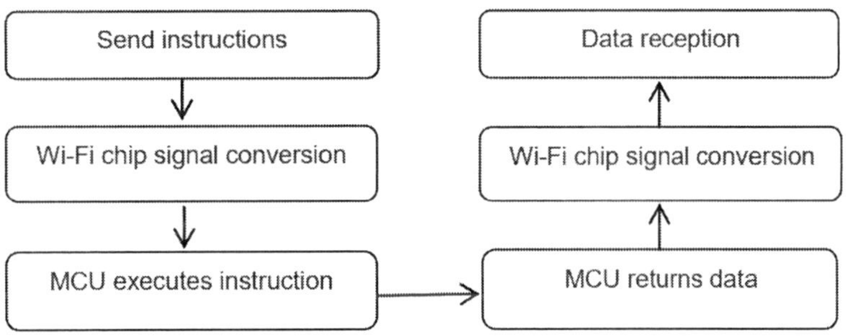

Figure 2. System Communication Frame

2.3.2. Data acquisition module. Application select bending resistor as acquisition module of figure condition and output of bending resistor as analog signal. So *MCU* need use *ADC* to collect data. The change of palm displacement is detected by the three-axis gyroscope *MPU6050*. The module's interior integrates *AD*, so we simply use the protocol configuration to read the data. Selection of system front-end data acquisition module is shown in the table 1. The data glove is shown in the figure 3.

Table 1. Selection of Data Acquisition Sensor Module

Module	Mode l
Bending resistance module	RFP
Three-axis gyroscopes module	MPU6050

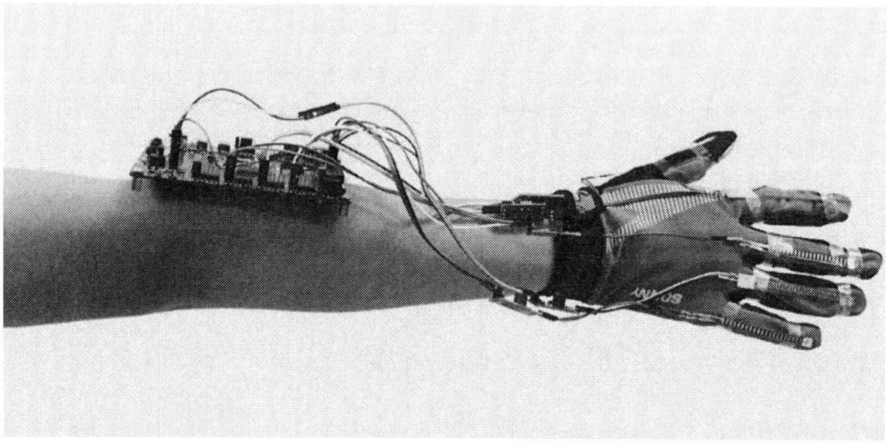

Figure 3. Data glove

3. Software design of application

The application mainly uses *Keil uVision5*, computer and *DAP* simulator as the development tool.

Programming on *STM32* development board with *Keil uVision5* and programming on Bending resistance, three-axis gyroscope according to the agreement.

We adopted *C#* language in cartoon interactive application. Using the controls provided by *Visual Studio* makes application forms to sets up interactive interface which appeals to children's psychology. Through *Wi-Fi* to receiving the information coming from hardware and achieve the data connection. And then judge gestures to make further instructions.

4. Application test

The application test divided into *STM32* development board test part and *PC* application test part.

For the *STM32* development board test, testing the design of the power is the first step, then testing whether it can successfully start *Wi-Fi*, *ADC* module is the second step when providing power supply for development board. Thirdly, testing whether bending resistance and three-axis gyroscope can successfully collect data and return to *PC* through *Wi-Fi*.

For the *PC* application test, start with the development board started correctly. Get sign language examination questions from *PC* first and then wear data glove made up of bending resistance and three-axis gyroscope to answer the question. The glove will get the bending combination of each finger in order to achieve the capture of sign language action. *PC* receiving data through *Wi-Fi* and judge the answer right or not, if right, you can enter the next level, or you can click point to see the correct answer.

5. Summary

The article outlines a design method of new application of children's sign language enlightenment based on Internet of things, expounds the design idea of the application and the realization way of key functional modules. The application integrates the sensor technology, embedded technology and wireless communication technology and other important technologies. Tests have proven that the application which according with children's psychology, interest, motivation and other characteristics can achieve an effective gesture recognition, and give star rating to children's answer. This achieves the desired design effect. With the development of technology, intellectualization and modernization are the trends of future society. Currently, sign language teaching limited to traditional class sign language teaching. The development of the application improves this situation. This application with preferable interactive performance is easy to use and can enhance the learning interest of deaf-mute child. Moreover, it makes sign language teaching emission its brilliance. Not only bring multiple reading experience but also bring stimulate of fresh vision to children.[6] The 19th session of national congress of the communist party of China pointed out that do a good job of special education. Efforts will be made to give every child access to fair and quality education. This consistent with our design philosophy and meets the development trend of society which beneficial to product promotion.

Acknowledgment
The work is supported by Key Research and Development Program of Shandong Province, China (No. 2017GGX10142).

References

[1] Cao Z J 2007 Some Experiences of Sign Language Application in Deaf Children Teaching *Basic Education Theory Research Achievements*
[2] Li Y Q 2017 Study of Sign Language Design with Visual Communication Concept *Yunnan Art Institute*
[3] Wang F and Hu X J 2012 A Comparative Study of Visual Memory Ability between Hearing impaired Children and Normal *Children Chinese special education*
[4] Gao Z L 2014 Design and Implementation of Online Education APP Based on Mobile Terminal *Dalian institute of Technology*
[5] Cao Z M, Chen N S, Ma Q, Wu L and Wu J 2017 Design of Wireless Control Circuit Based on ESP8266 *Industrial control computer*
[6] Liu H Y 2013 Visual Shape Design and Enlightenment Education of Children with Young Children *Qiqihar University*

Decision Support for Complicated Irrigation System: A Case Study of Lower Pak Phanang River Basin

Natapon Kaewthong[1] and Pakorn Ditthakit[2,*]

[1]Graduate Student, School of Engineering and Resources, Walailak University.
[2]Assistant professor, School of Engineering and Resources, Walailak University.

[*]Correspondence: dpakorn@hotmail.com

Abstract. Decision support for complicated irrigation system is presented in this article. The optimization mathematical model was developed mainly based on water balance concept for both command areas and coupled river and canal delivery system to suitably determine planted area for each plant species under constraints of available water supplies and limited areas to meet the maximization of economic value for the lower Pak Phanang river basin, Southern Thailand. Due to very complex problem-solving type with many decision variables of 410 and constraints of 861, an open source Excel add-in, widely known as Open Solver, was applied to solve nonlinear mixed-integer problems. The developed model was calibrated by comparison of the required inflow quantities to satisfy the calculated agricultural water demand in the study and the observed runoff quantities at X.105 runoff station, upstream of study area, during dry season (Jan to Apr). Land use data in 2012 obtained from the Land Development Department of Thailand (LDD) was interpreted to determine agricultural water demand. It was found that there was satisfying agreement between both considered values with agricultural water requirement quantities of 21.54 MCM and observed runoff quantities of 20.98 MCM. We hence applied the developed model with 8 scenarios for guidance for irrigation management in the study area.

1. Introduction

As rapidly growing urbanization, more and more water resources are required as a fundamental manufacture factor. However, many regions have been facing water shortage problem due to the effects of climate change, land use change, mismanagement, no enough water supply and other uncontrollable factors [1-5]. With limited area of developing large water infrastructure projects such as dam or reservoir and current environmental concerns, development of small water resources projects or rainwater harvesting project have been widely acceptable in water resources management [6, 7]. In addition to water supply management as mentioned previously, water demand management like optimal cropping pattern determination in irrigated area is commonly applied [8]. For those suitable managements and maximization of possible benefit, mathematical model development for water resources planning and management is an important decision support tool for irrigation engineer. It is a coupled model of water balance analysis model and optimization model [9, 10].

The area of the lower Pak Panang river basin may be claimed as one of the most complex irrigated areas in Thailand. In spite of having Huay Nam Sai reservoir located the upper Pak Panang river basin, there is no sufficient amount of water for agricultural activities in the lower Pak Panang river basin. Additionally, because of flatted area with nearly-zero bed slope of river and most irrigated area having level higher than water surface level in river and canals, the pumping irrigation system was used in this area. The river and canals are inevitably used as major water storage, especially during dry season. We might call this storage system as river and canal storage system. The irrigated pumping stations were built along with Pak Phanang River and tributary canals by Royal Irrigation Department (RID). Water in river or canals can flow arbitrarily depending on hydraulic gradient direction due to pumping by farmer group. At present, only gentle agreement was assigned for pumping, that is, all farmer groups can independently pump the water used for their agriculture until water level in Pak Phanang river reaching to +0.3 m MSL.

Therefore, the model for water resource planning and management considered an important tool in reducing future conflicts in the Lower Pak Panang Basin. This model uses the principle of the highest economic value to be derived from the management of water resources under limited resources. The results of this research will use as a model for practical implementation in the future for other watersheds experiencing similar problems. To manage water shortage problem for agricultural activities during dry season, this article presents determination of suitable planted area for each plant species under constraints of available water supplies and limited areas to meet the maximization of economic value.

2. Literature

At present, several mathematical modellings based on optimization techniques have been being developed to support decision-making in irrigation management [11-14] and water resources development [15, 16]. Although linear equations are widely used in solving management-type problems, mixed-integer nonlinear programming (MINLP) are more suitable for complex problem-solving type as considering in this research work. Nonlinear programming (NLP) have been applying for solving in water resource management problems, such as reservoir operation and surface water management [17]. MINLP was used for identifying optimal locations of pressure reducing valves in water distribution systems [18], optimal storm sewer system design [19] and water resource management [20, 21]. An open source Excel add-in, widely known as Open Solver developed by Mason [22, 23] in the Department of Engineering Science at the University of Auckland, was selected for this research work since it has been successfully applied for many fields of application, such as chemical engineering [24], airline business [25], and complex university timetabling problem [26].

3. Methodology

3.1. Water balance and water flow network

To comprehend the complicated irrigation system in lower Pak Phanang river basin, the schematic diagram was intentionally delineated under consultant of irrigation officers from Royal Irrigation Department (RID) and local stakeholders as shown in figure 1. It shows components of the system including main river (Pak Phanang river), canals, command area, locations of pumping station, agricultural water demand in each command area, water flow direction, return flow, and water storage or community pond in each command area. Main Canal (MC) command area is the area which irrigation water is obtained from pumping water from river or canals into gravitational irrigation canal system, whilst Main Drainage (MD) command area which irrigation water is obtained from pumping water from river or canals into drainage canal system (water flowing with adverse slope).Water balance analysis was firstly considered for each command areas and then for the whole system with coupled river and canal delivery network.

Figure 2 shows more details for a coupled model of water balance analysis and optimization model. Water supply in the study area are obtained mainly from river storage and canal storage and some from surface runoff. The community ponds in each command area are alternative water sources and could be more considered in present and near future since no enough water source meet more growing demand. Water surface runoff was evaluated by the SCS model, which transfer rainfall to runoff. The water in river storage and canal storage is delivered to the command areas by gravitational irrigation

IOP Conf. Series: Journal of Physics: Conf. Series **1060** (2018) 012059 doi:10.1088/1742-6596/1060/1/012059

canal system and some is pumped to the command areas. Land use data in 2012 obtained from the Land Development Department of Thailand (LDD) was interpreted to determine agricultural water demand based on Hargreaves equation and estimation of water storage (community pond) in each command area. The data of electricity charge and crop price were used to evaluate benefit for the lower Pak Phanang river basin.

Tables 1 and 2 show water storage volume in the river, main canals, and community ponds in each command area, and the amount of water flowing into the study area. The amount of water in these water storage were assumed to be full since rainy season or flooding event has just encountered in December. These values were assigned as initial values for the simulation run. As presented previously, mixed-integer nonlinear optimization mathematical model was developed for solving complex problem-solving type with many decision variables of 410 and constraints of 861. To simplify the model and relaxzation for a large optimization problem, one time period of 4 months (Jan to Apr) in dry season was considered herein. The details of the objective equation and the constraint equation are discussed in the next section.

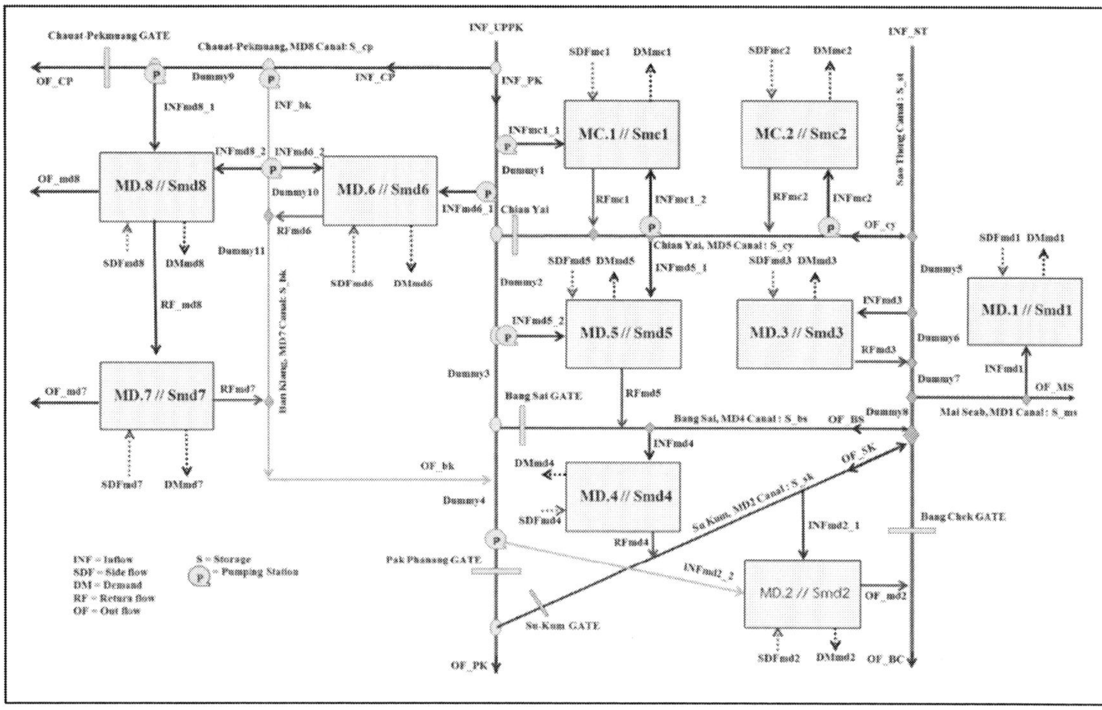

Figure 1. Schematic diagram of lower Pak Phanang river basin

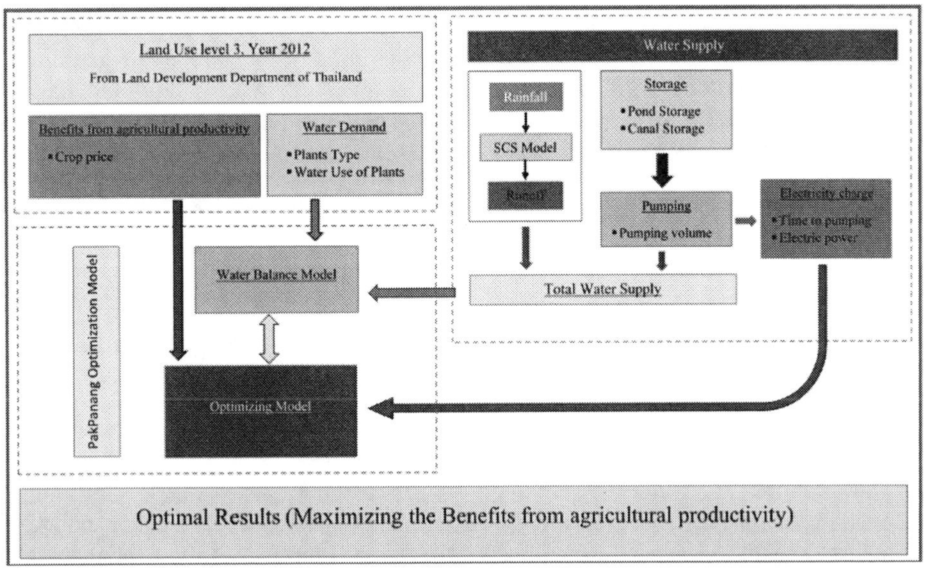

Figure 2. A coupled model of water balance analysis and optimization model

Table 1 Water storage volume in the river and main canals and the amount of water flowing into the study area (MCM)

Inflow from Upper Pakpanang River	Inflow from Upper SaoThong Canal	Pakpanang River storage	Chauat Prakmuang Canal storage	Ban Klang Canal storage	Chian Yai Canal storage	Bang Sai Canal storage	Sukhum Canal storage	Maisiab Canal storage
20.79	4	25	5	3.2	3	2.3	2	0.5

Table 2 Water storage volume in community pond (MCM)

water storage in area									
MC1	MC2	MD1	MD2	MD3	MD4	MD5	MD6	MD7	MD8
0.50	0.71	0.94	0.34	0.50	0.69	0.22	0.13	0.23	0.14

3.2. Objective function and constraints

3.2.1. Objective function. The objective function was developed for maximizing economic value for the lower Pak Phanang river basin. The decision variables include planted area for each plant species in each command area, time taken for pumping water from river or irrigation canals at each pumping station, water flow direction in river and canals. In this work, we only considered in dry season (January to April) due to being the period of time having limited water sources. The objective function can be written as follows.

$$Max\ Z = \sum (\mathrm{Ac,m} * \mathrm{Pc}) - \sum (\mathrm{Oi}) \tag{1}$$

$Max\ Z$ = net benefit obtained from agricultural crops (baht)
$A_{c,m}$ = planted area for each plant species (c) in each command area (m) (m^2)
P_c = selling price of agricultural crop for each plant species (c) (baht/ m^2)
O_i = pumping costs for electric irrigation pumps depending on time taken for pumping water at each pumping station (i) (baht)

3.2.2. Constraints. Constraint equations can be implemented by introducing parametric constraints to determine the relationship of nonlinear inequalities to the limits of the objective equation generated by the variant of the problem. The plant species involving in this study were paddy field, palm oil, rubber, grapefruit, orchard, vegetable, ruzy, and biennial crop. The considered constraints in the study can be alphabetically explained as follows.

- planted area in each command area ≤ total area of each command area
- total agricultural water demand ≤ amount of available water storage in river and canals
- water allocated to each command area ≤ available pump capacity at each command area
- sum of water inflow at junction node = sum of water outflow at junction node
- some water flow direction in river or canals must be assigned to be corresponding the natural water flow direction of study

4. Model Calibration

To calibrate the model, planted areas and time taken for pumping water in each command area as decision variable were determined. The planted areas obtained from simulation run were not more than those derived from interpretation of land use data in 2012 given by Land Development Department of Thailand (LDD). Such areas then were used for calculation of agricultural water demand. The exact time taken for pumping water in each command area was ambiguous due to no official record. However, we obtained those values, which were not exceed 20 hrs per day, from interviewing irrigation officers. The comparison between the required inflow quantities to satisfy the calculated agricultural water demand in the study and observed runoff quantities at X.105 runoff station, upstream of study area, during dry season (Jan to Apr) was undertaken. . It was found that there was satisfying agreement between both considered values with agricultural water requirement quantities of 21.54 MCM and observed runoff quantities of 20.98 MCM.

5. Scenarios Analysis for water management in Lower Pak Panang basin

Two main scenarios analysis were considered herein, depending on variable of inflow from upper Pak Phanang river, that is, it was one of decision variables as defined in scenario#01 or it was constant variable of 20.79 MCM as defined in scenario#02. Each sub-scenarios under main scenarios were established with altering cases of planted area variable as well as pump control variable as shown in table 3. The detained information can be explicitly explained as follows.

Table 3 Scenarios Analysis

Scenarios	Inflow from upper Pak Phanang river	Planted Area				Pump Control		community pond
		land use in 2012	conditionaly decreased	Increased abandoned areas of paddy fields and orchards	new planted area allocation	realistic in-situ	allowing additional	
1.1	○	●				○		
1.2	○	●					○	
1.3	○	●		○			○	
1.4	○				○	○		
2.1	●		○			○		
2.2	●		○				○	
2.3	●	●		○		○		
2.4	●		○			○		○

● = fixed value ○ = decision valiable

Scenario#01: The aim of this scenario was to determine amount of required inflow from the upper Pak Phanang river under various consideration of planted area and pump control variables with maximizing net benefit of area.

- 1.1 Using land use data in 2012 and pump control with realistic in-situ, no more than 20 hrs a day
- 1.2 Using land use data in 2012 and allowing additional pump control
- 1.3 Using land use data in 2012, allowing to increase planted areas in abandoned areas of paddy fields and orchards and allowing additional pump control
- 1.4 New planted area allocation under total planted areas obtained land use data in 2012, but not allowing to increase planted areas in abandoned areas of paddy fields and orchards, and pump control with realistic in-situ, no more than 20 hrs a day

Scenario#02: The aim of this scenario was to analyse net-benefit maximization of area under recorded data of inflow from the Upper Pak Phanang river.

- 2.1 Allowing to reduce planted areas based on land use data in 2012 and pump control with realistic in-situ, no more than 20 hrs a day
- 2.2 Allowing to reduce planted areas based on land use data in 2012 and allowing additional pump control
- 2.3 New planted area allocation, the considered total area including total planted areas obtained land use data in 2012 and planted areas in abandoned areas of paddy fields and orchards, and pump control with realistic in-situ
- 2.4 Allowing to reduce planted areas based on land use data in 2012, pump control with realistic in-situ, and allowing community pond based on 10% of the planted areas in each command area.

6. Results and Discussion

The various study results for two main scenarios are presented in table 4. It gave the information concerning water inflow for upper basin, economic value or net benefit, agricultural water demand, and planted areas for each plant species, i.e. paddy filed, palm oil, para rubber, grapefruit or pomelo, orchard, vegetable, biennial plant. They can be explained and discussed as follows.

Table 4. The study results obtained from applying MINLP for complicated irrigation system

Case study	Water inflow from Upper basin	Economic value	Water Demand	Area MC1-MD8 (hectare)								
				paddy field	palm	rubber	grapefruit	orchard	vegetable	ruzy	biennial	Total
Land Use 2012				39,272	12,101	1,374	1,096	4,568	2,281	5,839	242	66,772.33
Case 1.1	-	-	-	-	-	-	-	-	-	-	-	-
Case 1.2	209,307,690	1,677,667,863	326,574,029	39,272	12,101	1,374	1,096	4,568	2,281	5,839	242	66,772.33
Case 1.3	216,845,570	1,820,084,433	334,771,298	39,272	12,101	1,380	1,096	4,602	4,162	6,586	267	69,466.17
Case 1.4	21,006,507	1,005,378,896	134,416,843	881	513	4,122	2,705	6,027	6,843	17,516	578	39,184.15
Case 2.1	20,790,000	983,164,644	138,075,252	14,842	1,434	1,374	1,096	4,568	2,281	5,839	242	31,675.07
Case 2.2	20,790,000	984,763,058	138,108,252	14,842	1,434	1,374	1,096	4,568	2,281	5,839	242	31,675.07
Case 2.3	20,790,000	1,004,589,819	138,371,994	843	513	4,122	2,705	6,027	6,843	17,516	578	39,146.18
Case 2.4	20,790,000	1,701,978,223	326,574,032	39,272	12,101	1,374	1,096	4,568	2,281	5,839	242	66,772.33

Case 1.1 indicated that it was no feasible solution found due to insufficiently existing pump capacity for agricultural water demand. Although amounts of water inflow from upper basin were plentiful, they were not be able to be used to meet water requirement. It was obviously shown as in case1.2 that allowing additional pump control could conquer the limitations as found in case 1.1. In this scenario, it required agricultural water demand of 326.57 MCM, which was from upper basin 209.31 MCM and water storage in area 117.26 MCM, for the agricultural area of 66,772.32 hectares. Undoubtedly, the maximum net benefit of 1.82 thousand billion baht (more than 42.41 million

baht as compared to case 1.2) was found in case 1.3 due to allowing to increase planted areas in abandoned areas of paddy fields and orchards and allowing additional pump control. It required agricultural water demand of 334.77 MCM, which was from upper basin 216.85 MCM and water storage in area 117.92 MCM, for the agricultural area of 69,466.17 hectares. Case study 1.4 showed altering of planted area under pump capacity limitation. This proposed scenario was to test the reduction of agricultural water demand by increasing the planted area of low water demand plant (para rubber and ruzy) and decreasing the planted area of high water demand plant (paddy filed and palm oil). As shown in table 4, minimum agricultural water demand of 134.42 MCM was found. In addition, it required water from the upper basin only 21.01 MCM with the economic return of 1.01 thousand billion baht.

The results of cases 2.1 and 2.2 were very similar, but not the same values, for planted areas of each plant species, but it could not be shown in table 4 due having difference in the 5th digits decimal. It can be explained that reduction measure of planted areas were more effective than measure of additional pump control under the limitation of amount of water supply form the upper basin. The total planted area of approximately 31,675.07 hectares with the economic return between 983.16 and 984.73 million baht for cases 2.1 and 2.2, respectively. It was found that cases 2.3 and 1.4 was similar concept for irrigation management, that is, increasing the planted area of low water demand plant (para rubber, grapefruit, orchard, vetgetable, and ruzy) with high income and decreasing the planted area of high water demand plant (paddy filed and palm oil). The total planted area of approximately 39,146.18 hectares with the economic return of 1.00 thousand billion baht. Both total planted area and the economic return were more than case 2.2 with 7,471.11 hectares and 19.83 million baht. For the final case study of 2.4 with allowing community pond based on 10% of the planted areas in each command area, the maximum economic return of 1.70 thousand billion baht and the maximum total planted area of 66,772.32 hectares , when considering only case 2, was obtained. Moreover, it indicated that when water supply sources were plentiful, growing rice gave the maximum benefit for the Lower Pak Phanang River Basin.

7. Conclusions

This study illustrated the application of mixed integer nonlinear programming as decision support tool for complicated irrigation management. The Lower Pak Phanang River Basin, located Nakhon Si Thammarat province, Southern part of Thailand was used as a study area with limited water supply sources during dry season. To solve the water shortage problem, we proposed water demand management by determination of suitably plated area for each plant species. The abandon areas of paddy field and orchard were additionally considered to increase planted areas for more obtained net benefit. Not only were water supply sources limited, but also the operations of pump control for each command areas at present were restricted and obstructed to get more net benefit. The additional pump control, which is water supply management, then was proposed to release this restriction. Finally, the community pond, which applies the concept of rainfall harvesting, supplementally undertaken in this study. All above scenarios analysis could be used as guidance for irrigation management of the Lower Pak Phanang River Basin.

References

[1] Erica DeNicola, Omar S. Aburizaiza, Azhar Siddique, Haider Khwaja, and David O. Carpenter, 2015. Climate Change and Water Scarcity: The Case of Saudi Arabia, Annals of Global Health (2015), VOL.81, NO.3, MayeJune 2015: 342– 353.

[2] Anil Kumar Misra 2014. Climate change and challenges of water and food security, International Journal of Sustainable Built Environment (2014), 3, 153–165.

[3] J. A. Duran-Encalada, A. Paucar-Caceres, E. R. Bandala, and G. H. Wright 2017. The impact of global climate change on water quantity and quality: A system dynamics approach to the US–Mexican transborder region, European Journal of Operational Research 256 (2017) 567–581.

[4] Jawad Taleb Al-Bakri, Mohammad Salahat, Ayman Suleiman, Marwan Suifan, Mohammad R. Hamdan, Saeb Khresat and Tarek Kandakji 2017. Impact of Climate and Land Use Changes on Water and Food Security in Jordan: Implications for Transcending "The Tragedy of the Commons", Sustainability (2013), 5, 724-748.

[5] Sujay S. Kaushal, Arthur J. Gold, and Paul M. Mayer 2017. Land Use, Climate, and Water Resources-Global Stages of Interaction, Water 2017, 9, 815.

[6] Md. Tariqul Islam, Md. Mohabbat Ullah, M. G. Mostofa Amin, and Sahadat Hossain 2016. Rainwater harvesting potential for farming system development in a hilly watershed of Bangladesh, Appl Water Sci (2017) 7:2523–2532.

[7] N. Areerachakul 2013. Overviews of Rainwater Harvesting and Utilization in Thailand: Bangsaiy Municipality, International Journal of Environmental and Ecological Engineering (2013), Vol7, No:7.

[8] Ehsan NeamatollahiJ, Vafabakhshi M.R, Jahansuz, F.Sharifzadeh 2017. Agricultural Optimal Cropping Pattern Determination Based on Fuzzy System, Fuzzy Information and Engineering (2017), 9(4), 479-491.

[9] Min Goo Kang, A.M.ASCE; and Seung Woo Park 2014. Combined Simulation-Optimization Model for Assessing Irrigation Water Supply Capacities of Reservoirs, Journal of Irrigation and Drainage Engineering (2014) ,Volume 140 Issue 5 - May 2014.

[10] Wafa Difallah, Khelifa Benahmed, Belkacem Draoui, and Fateh Bounaama 2017. Linear Optimization Model for Efficient Use of Irrigation Water, International Journal of Agronomy (2017), Volume 2017, Article ID 5353648.

[11] Laxmi Narayan Sethi, D. Nagesh Kumar, Sudhindra Nath Panda and Bimal Chandra Mal 2002. Optimal Crop Planning and Conjunctive Use of Water Resources in a Coastal River Basin, Water Resources Management (2002), 16, 145–169.

[12] Laxmi Narayan Sethi, Sudhindra N. Panda and Manoj K. Nayak 2006. Optimal crop planning and water resources allocation in a coastal groundwater basin, Orissa, India, agricultural water management 83 (2006), 209–220.

[13] Takeshi Itoh, Hiroaki Ishii and Teruaki Nanseki 2003. A model of crop planning under uncertaintyin agricultural management, Int. J. Production Economics 81–82 (2003) 555–558.

[14] Angel Galan-Martin, Carlos Pozo, Gonzalo Guillen-Gosalbez, Assumpcio Anton Vallejo and Laureano Jimenez Esteller 2015. Multi-stage linear programming model for optimizing cropping plan decisions under the new Common Agricultural Policy, Land Use Policy 48 (2015), 515–524.

[15] Zhanqi Wang, Jun Yang, Xiangzheng Deng and Xi Lan 2015. Optimal Water Resources Allocation under the Constraint of Land Use in the Heihe River Basin of China, Sustainability (2015), 7, 1558-1575.

[16] M. W. Rose grant, C. Ringler, D. C. McKinney, X. Cai, A. Keller and G. Dono so 2000. Integrated economic-hydrologic water modeling at the basin scale: the Maipo river basin, Agricultural Economics 24 (2000) 33-46.

[17] Anongrit Kangrang, and Cetthaphan 2013. Optimal Reservoir Rule Curve Considering Ant Colony Optimization with Simulation Model, Journal of Applied Sciences (2013), 13(1): 154-160.

[18] Pham Duc Dai, and Pu Li 2014. Optimal Localization of Pressure Reducing Valves in Water Distribution Systems by a Reformulation Approach, Water Resour Manage (2014) 28:3057–3074.

[19] Omer Karovic, and Larry W. Mays 2014. Sewer System Design Using Simulated Annealing in Excel, Water Resour Manage (2014) 28:4551–4565.

[20] POFF, N.L…et al 2016. Sustainable water management under future uncertainty with eco-engineering decision scaling, Nature Climate Change (2016), 6(1),pp.25-34.

[21] Ajay Singh 2013. Irrigation Planning and Management Through Optimization Modelling, Water Resour Manage (2014) 28:1–14.

[22] Andrew J Mason 2010. OpenSolver: Open Source Optimisation for Excel, Proceedings of the 45th Annual Conference of the ORSNZ, November 2010, pp.181-190.

[23] Andrew J Mason 2011. OpenSolver: An Open Source Add-in to Solver Linear and Integer Programmes in Excel, Operations Research Proceeding (2011), pp.401-406.

[24] Jorge R. Paloschi 2001. Improving robustness using homotopy as an open solver in dynamic simulation package, Computer Aided Chemical Engineering (2001), Volume 9, Pages 255-260.

[25] Zachary Cesaretti, Matthew Reyers, Reiko Tominaga, and Samantha Zimmerman 2017. Airline Schedule Planning: Optimization Approaches, Analytics Now 2015-2016 Edition, Pages 2-10.

[26] R.A.Runga Prabodanie 2016. An Interger Programming Model for a Complex University Timetabling Problem: A Case Study, Industrial Engineering & Management Systems, Vol 16, No 1, March 2017, pp 141-153.

IOP Conf. Series: Journal of Physics: Conf. Series **1060** (2018) 012060 doi:10.1088/1742-6596/1060/1/012060

A Satellite Navigation Signal Acquisition and Interference Suppression Method in a Complex Electromagnetic Environment

Jie GU[1,*], Kai-chuang JIANG[1], Jia-tao LING[1], Ming-li Zhang[1] and Jia-qing QU[1]

[1]Shanghai Radio Equipment Research Institute, 203 Li-ping Rd, Shanghai, China

Email:[*]gujiecmzx@163.com

Abstract. Under a complex electromagnetic environment, satellite navigation signals are affected by interference and the receiver cannot effectively locate them. A satellite navigation signal acquisition and interference suppression method is proposed. Prior to signal acquisition synchronization, the power inversion method is used to cancel interference, and then multiple half-bit information is used for incoherent accumulation to effectively capture satellite signals, and after satellite signals are captured, a parallel multi-beam method is used to beamform different satellites separately. The simulation results show that this method can not only suppress the interference signal, but also produce a large gain for the satellite signal, with its performance superior to the conventional methods.

1. Introduction

Since the distance between the navigation satellite and the ground is far, and the signal power at the end of the receiver is lower than the noise, the receiver is easily disturbed by the complex electromagnetic environment. Therefore, anti-jamming technology continues to develop, and the adaptive filter algorithm based on array antennas is the most common and most effective measure.

The power inversion (PI) method commonly used in GPS receivers can achieve a good anti-interference effect without the need to know the Direction of Arrival (DOA) of the desired signal[1]; in addition, reference[2,3] also studies the suppression methods of specific interferences, such as pulse interference, FM interference, etc. In the calculation process, to avoid matrix inversion, the rank reduction method using multi-stage Wiener filter has been widely used[4]; since GPS is a code division multiple access (CDMA) system, blind adaptive beamforming algorithm using PN code has also become a direction of research[5]. The reference[6] uses a multi-beam antenna to point the main beam toward the satellite, but the DOA of the desired signal must be known, so it is seldom used.

This paper proposes a signal acquisition and parallel multi-beam interference suppression method in a complex electromagnetic environment, which can achieve the best reception performance through high-sensitivity acquisition and main beam gain while cancelling interference.

2. Signal Acquisition in Complex Electromagnetic Environments

2.1. Interference power minimization

In a complex electromagnetic environment, the power of the interfering signal is high, and the direction is unknown. The GPS signal is from multiple satellites and cannot be accurately tracked.

Therefore, the direction is unknown. However, the power of the GPS signal to reach the ground signal is much lower than the natural noise. Thus, before the signal acquisition synchronization, the power minimization can be used as a calculation criterion, that is, the PI method[1] is used, and derived as follows:

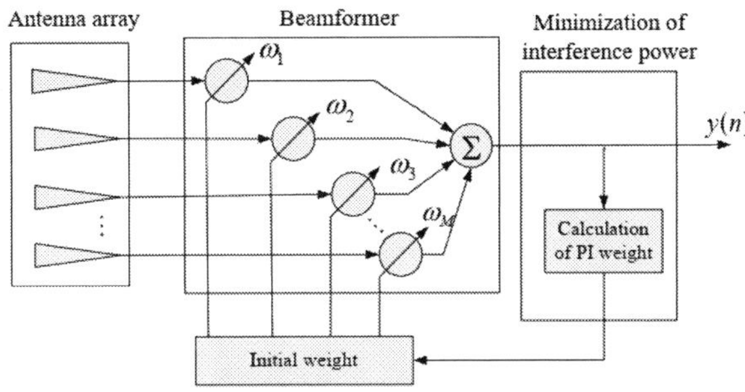

Figure 1. The structure of the PI method

As shown in the array structure of Figure 1, the received signal is written in vector form:

$$x(n)=\left[x_1(n),x_2(n),\cdots,x_M(n)\right]^{\mathrm{T}} \tag{1}$$

The weighted output of the array is

$$y(n)=w^{\mathrm{H}}x(n) \tag{2}$$

The output power can be expressed as

$$P_{out}=E\left|y(n)\right|^2=w^{\mathrm{H}}E\left(x(n)x^{\mathrm{H}}(n)\right)w \tag{3}$$

The constraints are as follows:

$$\begin{aligned}&Minimize\ P_{out}=E\{\left|y(n)\right|^2\}\\&w^H a\left(r_j^{gps}\right)=1\end{aligned} \tag{4}$$

Where, w is for the antenna array weighting coefficient, $a\left(r_j^{gps}\right)$ for the jth GPS signal, and H for the conjugate transpose.

After construction of a Lagrangian extremal function, the equation is solved to get the optimal weight:

$$w_{\mathrm{opt}}=\frac{R^{-1}s}{s^{\mathrm{H}}R^{-1}s},R=E\left(x(n)x^{\mathrm{H}}(n)\right) \tag{5}$$

In(5), the solution of the optimal weight requires matrix inversion. To avoid inversion, the initial weight $w(0)$ can be set so that the weight is iterated along the direction in which P_{out} reduction is fastest. The recurrence formula is

$$\begin{cases} w(0)=(1,0,0,\ldots)^T \\ s=(1,0,0,\ldots)^T \\ w(n+1)=w(n)-2\mu\left[I-\frac{ss^T}{s^Ts}\right]x^*(n)x^T(n)w(n) \end{cases} \tag{6}$$

Where, $w(0)$ is for the initial weight, s for the constraint condition, x for the input vector, μ for the iteration step size, and $w(n+1)$ for the updated weight value.

2.2. High sensitivity signal acquisition

The GPS signal uses a GOLD code with a length of 1023, a chip rate of 1.023 Mcps, and a data rate of 20 bps. Figure 2 shows the data model of the signal.

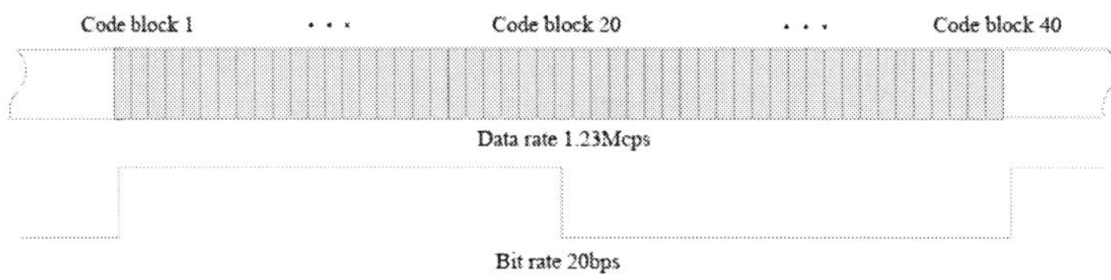

Figure 2. Data model of the signal

As shown in Figure 2, to improve the acquisition sensitivity, complete correlation processing can be performed on the entire bit length data to obtain a 43 dB spreading gain. However, before the signal acquisition, the bit flipping position is unknown. If the whole bit data is used for correlation, the bit flipping causes the correlation peak to be cancelled and the signal cannot be successfully captured.

To solve this problem, techniques such as half-bit data matching filtering and incoherent accumulation (HDMF-IA) are used, as shown in Figure 3.

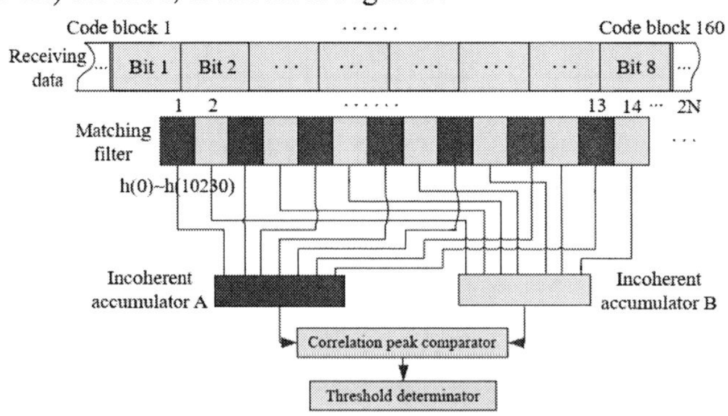

Figure 3. The structure of the HDMF-IA method

The length of each segment of the matching filter is designed to be half of the data bit period, so that the correlation result of one of the adjacent two segment matching filters is not affected by the bit flipping. The length of the matching filter is related to the data bit period, which creates conditions for incoherent accumulation.

The power-minimized I and Q baseband signals, i.e. $S_{b,I}$ and $S_{b,Q}$ are sent to a matching filter for processing. Each segment has a length of 10,230 chips. Assume that the cumulative number of segments is N, each segment performs related calculations.

$$Z_{\mathrm{I},k}(n)=\sum_{i=0}^{10230} S_{\mathrm{b,I}}(n-i)\times h(-i), k=1\sim 2N \tag{7}$$

$$Z_{\mathrm{Q},k}(n)=\sum_{i=0}^{10230} S_{\mathrm{b,Q}}(n-i)\times h(-i), k=1\sim 2N \tag{8}$$

Where, $Z_{\mathrm{I},k}(n)$ and $Z_{\mathrm{Q},k}(n)$ represent the matching results of the k-th segment of I and Q baseband data respectively.

After incoherent accumulation of matching filter results where k is an odd segment, we can obtain:

$$G_{\mathrm{o}}(n)=\sum_{k}\left(Z_{k,\mathrm{I}}(n)+Z_{k,\mathrm{Q}}(n)\right) \tag{9}$$

After incoherent accumulation of matching filter results where k is an even segment, we can obtain:

$$G_{\mathrm{e}}(n)=\sum_{k}\left(Z_{k,\mathrm{I}}(n)+Z_{k,\mathrm{Q}}(n)\right) \tag{10}$$

Where, $G_{\mathrm{o}}(n)$ and $G_{\mathrm{e}}(n)$ are the incoherent accumulated value of the matching filtering results of the odd and even segments respectively. Through peak search and real-time comparison of the two correlation results, the correlation peak value of a large value is compared with the capture threshold, and if the correlation peak value exceeds the threshold value, the satellite signal is considered to have been successfully captured.

3. Parallel multi beam (PMB) interference suppression

The PI method and high-sensitivity signal acquisition method ensure the effective capture of GPS signals, but in the tracking stage, if the PI method is still used, and when the interference is high, the gain of the useful signal direction cannot be satisfied, resulting in decline of reception performance.

Therefore, multiple parallel beams can be used to obtain adaptive weights for multiple satellites and perform beamforming independently. Using the despreading and re-spreading method[5], a main beam is formed at the direction of a single desired receiving satellite through the beamformer, and nulling is formed at the interference of other directions. Figure 4 shows the structure of the PMB method.

The number of antennas of the receiver is M, and each satellite corresponds to an adaptive beamformer. By adaptively processing the weights, the method minimizes the square-error between the array output and the expected response. The cost function is defined as

$$D(w_i)=\sum_{k=1}^{K}\left|w_i^{\mathrm{H}}X(k)-r_i(k)\right|^2 \tag{11}$$

Where, K is the size of the data block. For the GPS synchronizing with spreading code and data bit, to obtain the best spread-spectrum processing gain, K is taken as sampling data within one bit period.

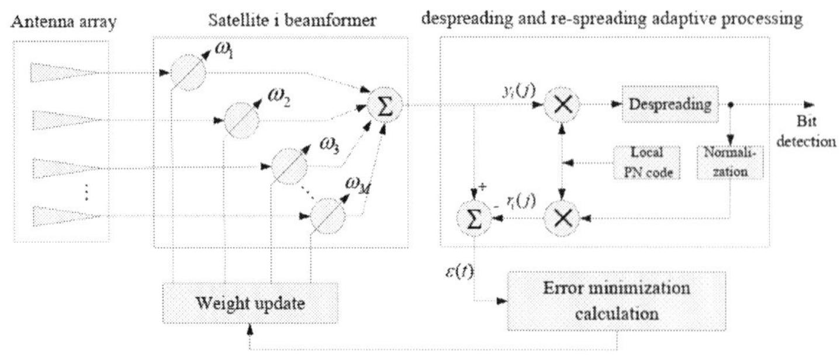

Figure 4. The structure of the PMB method

After the satellite signal is captured and steadily tracked, the data bits of the i-th satellite are detected. If the n-th data bit is detected correctly, the signal waveform of the i-th satellite transmitted within the time interval [(n-1)Tb, nTb] can be considered to be obtainable through respread of the detected data bits $\hat{b}_i$ using the c_i code of the i-th satellite, i.e.

$$r_i(t)=\hat{b}_i c_i,\quad (\text{n-1})T_b \le t \le nT_b \tag{12}$$

The i-th beamformer uses this respread signal to adaptively process the weight vector of the i-th satellite. Since the c_i code repeats once every bit period, c_i and $r_i(t)$ have the same period Tb. For iteration at each step, a different input data block may also be used to adaptively process the weight vector. Let

$$X(j)=\left[x(1+jK),x(2+jK),\cdots,x((l+1)K)\right],\quad j=0,1,\cdots,J \tag{13}$$

Where, J is the number of iterative steps required for the convergence of the algorithm; K is the number of data samples per bit, and all data samples in one bit period are used in the adaptive process. The output data of beamforming corresponding to the i-th satellite is as follows:

$$y_i(j)=\left[w_i^H(j)X(j)\right]^{\mathrm{T}}=\left[y_i(1+jK),y_i(2+jK),\cdots,y_i((j+1)K)\right]^{\mathrm{T}} \tag{14}$$

$\hat{b}_{il}$ is the estimate of the j-th data bit of the signal transmitted from the i-th satellite, and the respread signal can be written as

$$r_i(j)=\hat{b}_{ij}\,c_i^{\mathrm{T}} \tag{15}$$

$r_i(j)$ is used as the reference value for calculation of the weight, and $y_i(j)$ is the output data corresponding to the weight w_i at iteration j. The weights are adaptively updated to minimize the cost function in equation (11).

$$w_i(j+1)=\left[X(j)X^{\mathrm{H}}(j)\right]^{-1}X(j)r_i^*(j) \tag{16}$$

In(16), the method uses matrix inversion, since the PI method has already calculated the initial weight, to reduce the amount of calculation, an iterative method can be used instead of the matrix inversion. The steps are as follows:

$$\begin{cases} w_i(j+1) = w_i(j) + \mu X(j) E_i(j+1)^H \\ E_i(j) = r_i(j) - y_i(j) \end{cases} \tag{17}$$

Where, μ is for the step size; $X(j)$ for the array receiving vector of the current bit; $r_i(j)$ for the local reference signal; $E_i(j)$ for the weight vector correction error vector of current bit; $y_i(j)$ for the current bit data output.

4. Simulation results

To verify the effectiveness of the method, the simulation uses a 4-element uniform linear array with half-wavelength spacing. Assume that a singleton interference is located at -60° and the interference-to-noise ratio (INR) is 60dB; an equal bandwidth DSSS-BPSK modulation signal is located at 50° and the INR is 50dB. The desired GPS satellite 1(Sat.1) is located at 18° and the signal to noise ratio (SNR) is -30 dB; the desired GPS satellite 2(Sat.2) is located at -25° and the SNR is -27 dB.

The frequency spectrum before the interference cancellation is shown in Figure 5, and the frequency spectrum after cancellation is shown in Figure 6. The PI method can suppress the narrowband interference and the broadband interference well.

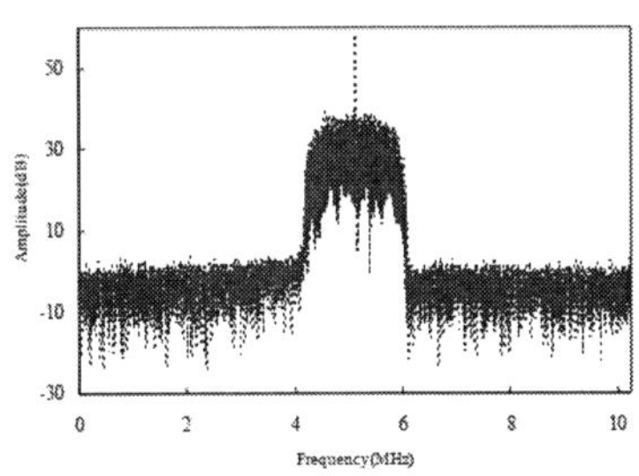

Figure 5. The frequency spectrum before PI method

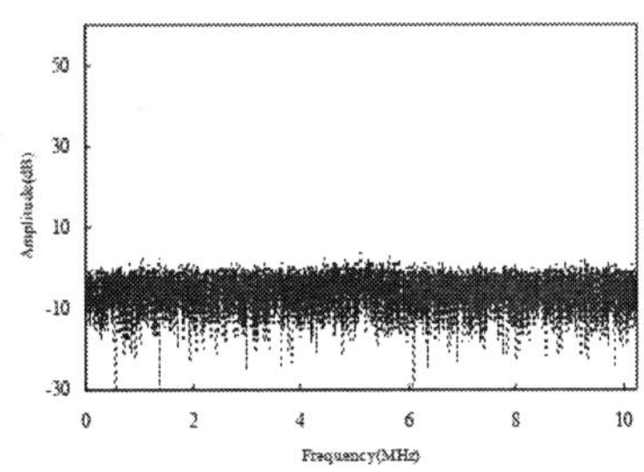

Figure 6. The frequency spectrum after PI method

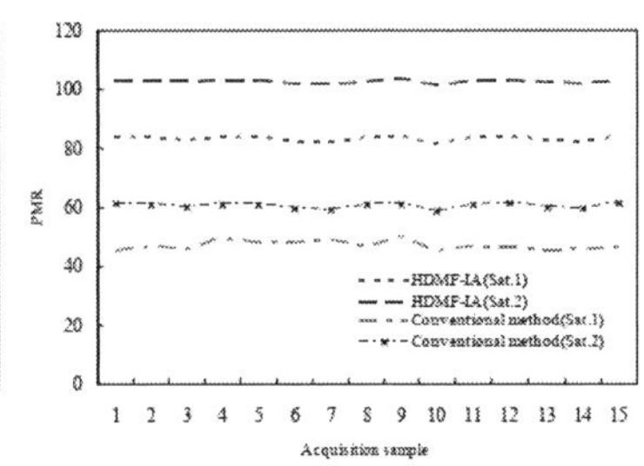

Figure 7. The PMR of the correlation

The GPS signal acquisition uses a matching filter and takes N=8 segments for incoherent accumulation, i.e. the HDMF-IA method. The matching filter produces correlation peaks when the received signal is aligned with the local PN code, and approximates noise when not aligned. Therefore, the peak to mean ratio (PMR) of the correlation value is an important basis for capture. The result is shown in Figure 7.

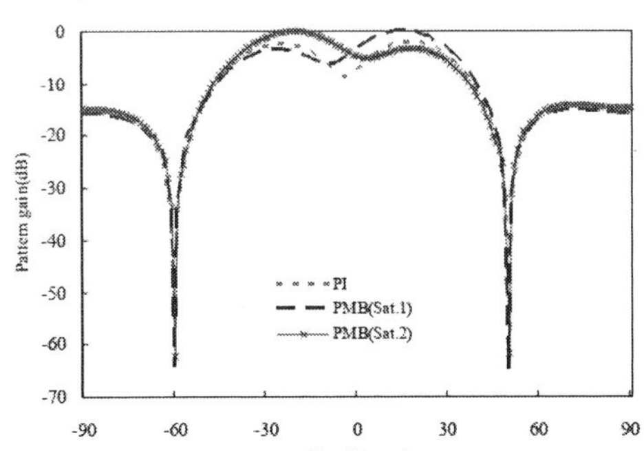

Figure 8. Beam and nulls pattern gain

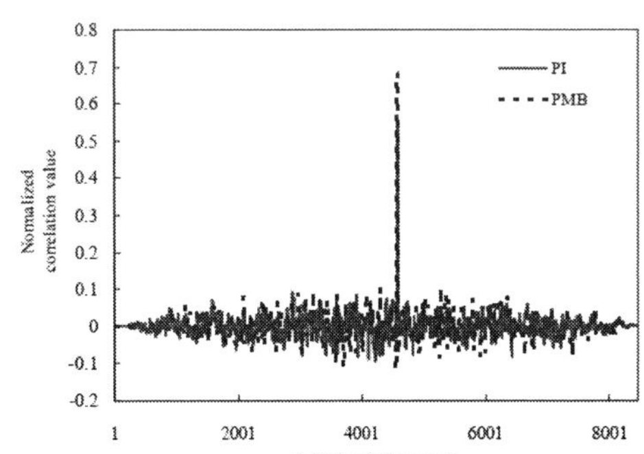

Figure 9. Normalized correlation value

After the satellite signal is captured, different satellites are beamformed using a parallel multibeam method, with the directional pattern as shown in Figure 8. Compared with the PI method, the PMB method not only does not affect the nulling of the interference direction, but also can obtain greater gain in the desired direction. Figure 9 is comparison of correlation peaks before and after use of the method, showing that the method suppresses narrow-band and wide-band interference and also improves the SNR of satellite signals.

5. Conclusion

This paper presents a satellite navigation signal acquisition and interference suppression method under complex electromagnetic environment. Before the satellite signal is captured, the PI method is used to cancel the interference, and then the half-bit incoherent signal acquisition method is used to increase the acquisition sensitivity. After satellite signals are captured, a different beamformer is used for each satellite, and the adaptive weights are respectively obtained through the despreading and respreading method, so that a main beam can be formed at the direction of the desired receiving satellite and nulling formed at other directions of interference.

The simulation results show that the proposed method can effectively implement interference suppression and signal acquisition for multiple satellites in the airspace under the condition that the desired satellite and interference direction are unknown, and the performance is superior to conventional methods.

Acknowledgement

Project funded by the Shanghai Municipal Science and Technology Commission, No.16DZ1120901

References

[1] Schwegman C W and Compton R T 1972 Power inversion in a two-element adaptive array *AD0758690* (Ohio State, USA: Ohio State Univ Columbus Electroscience Lab) pp 1-40
[2] Fante R L and Vaccaro J J 2000 Wideband cancellation of interference in a GPS receiver array *IEEE Trans. on Aerospace and Electronic System* 36(2) pp 549-564
[3] Amin M G Zhao L and Lindsey A R 2004 Subspace array processing for the suppression of FM jamming in GPS receivers *IEEE Trans. on Aerospace and Electronic System* 40(1) pp 80-92
[4] Werner S With M and Koivunen V 2007 Householder multistage Wiener filter for space-time navigation receivers *IEEE Trans. on Aerospace and Electronic System* 43(3) pp 975-988
[5] SANG Huai-Sheng LI Zheng-Rong and YONG Shao-Wei 2002 The blind adaptive beamforming algorithms using PN code *Journal of China Institute of Communications* 23(11) pp 103-108
[6] Miura Ryu Tanaka Toyohisa and Chiba Isamu 1997 Beamforming experiment with a DBF multibeam antenna in a mobile satellite environment *IEEE Transactions on Antennas and Propagation* 45(4) pp 707-714

IOP Conf. Series: Journal of Physics: Conf. Series **1060** (2018) 012061 doi:10.1088/1742-6596/1060/1/012061

Ship License Number Layout Normalization Based on Regional Texts Fine Localization

Baolong Liu[1, 2], Sanyuan Zhang[1], Zhenjie Hong[2] and Xiuzi Ye[2,*]

[1]College of Computer Science and Technology, Zhejiang University, Hangzhou 310027, China
[2]College of Mathematics, Physics and Electronic Information Engineering, Wenzhou University, Wenzhou 325035, China

[*]Yexiuzi@wzu.edu.cn

Abstract. Ship license numbers (SLNs) are frequently used to identify ships. An automatic SLNs recognition system makes it convenient and fast to recognize SLNs. A significant prerequisite of many characters recognition algorithms is that characters need to be recognized should be segmented as single ones or at least be listed in one horizontal line. However, SLNs from different ships may have different layouts and the characters of SLNs are often written in more than one lines. In order to recognize these multi-layout SLNs, it is momentous to normalize then into one line. In this paper, we present a SLN layout normalization framework based on regional texts fine localization. For an input SLN image, a series of pre-processing operations are first conducted to enhance the quality of the input SLN image. Then, a single shot text detector with regional attention is used to fine detect characters in that SLN. The normalization algorithms are finally proceeded to normalize the input SLN image based on the detected text bounding boxes and the layout priors of SLNs. At last, the proposed framework is tested on 3216 SLN images, the proposed framework is proved to be effective with a normalization rate of 87.38%.

1. Introduction

What are ship license numbers? Simply speaking, SLNs are ships names. Every ship has a name. Ship names are not randomly given, they are generally given by following the ship naming conventions issued by the local official authorities. Therefore, it is more reasonable to call them “license”.

Similar with a car license plates recognition system, an automatic ship license numbers recognition system can be designed to identify ships by recognizing the characters in SLNs. The recognition pipeline of SLNs are similar with car license plates’, it is mainly comprised of image acquisition, license location, tilt correction, character segmentation and character recognition.

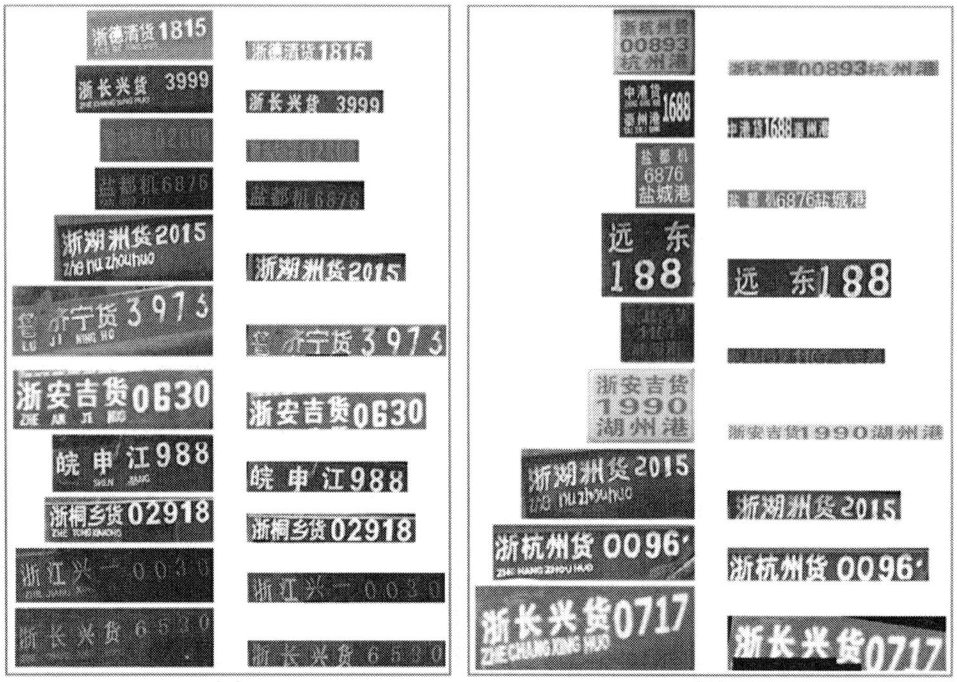

(a) (b)

Figure 1. Some sample normalization results of SLN images (note that some images may be resized for the convenience of display). (a) Left: input SLNs. Right: Normalization results. (b) Left: input SLNs. Right: Normalization results.

Several papers concerning SLNs recognition have been published [1-3]. [1] used a transferred deep CNN model in combination with prior features to localize multi-style ship license numbers in nature scenes. Liu et al. [2] presented an effective coarse-to-fine approach for locating various SLNs in the wild. [3] proposed a horizontal tilt correction method for SLNs.

A significant prerequisite of many license (or generally characters) recognition algorithms is that characters need to be recognized should be segmented as single characters. Even for the most advanced deep convolutional neural network-based characters recognition algorithms, e.g., CRNN [4], the characters need to be recognized also should be listed in one horizontal line. However, as shown in Figure 1, the styles of SLNs are multiple. Characters contained by these SLNs may be written in different lines with different angles. To boost the accuracy of SLNs recognition, in this paper, we present a SLNs layout normalization framework based on regional texts fine localization. The main objective is to normalize those SLNs that are of multiple styles into one nearly horizontal line.

SLNs from different countries or different ships may have different layouts. In this paper, the SLNs from Chinese cargo ships are used as experimental samples to demonstrate the pipeline of the proposed framework. SLNs of Chinese cargo ships from one same ship may have different layouts. It is thus reasonable to use SLNs of Chinese cargo ships as experimental samples.

The core difficulties of SLNs layout normalization mainly lie in two aspects: 1) Characters contained in the same SLN may be different in script, angle, size, aspect ratio and stroke width, and these characters, especially Pinyin, are of low resolution, at the same time. 2) Characters of SLNs may be written in different lines. The most difficult thing is that, some characters in an SLN are written in two line, while others are written in one line. Therefore, it is hard for image processing-based methods, e.g., MSER or Connected-Component analysis, to distinguish these cluttered characters out. In this paper, for an input SLN image, a series of pre-processing operations are first conducted to enhance the quality of the input SLN image. Then, a single shot text detector with regional attention (SSTD) [5] is used to fine detect characters in that SLN. The normalization algorithms are finally proceeded to normalize the input SLN image based on the detected text bounding boxes and the layout priors of SLNs. Figure 2 shows the procedures of the proposed framework.

The proposed framework is experimented on 3216 collected SLN images. The experimental results in Sec. 3 indicate that the proposed approach can effectively normalize SLNs with a normalization rate of 87.38%.

2. The proposed framework

For an input SLN image with cluttered characters contained, the objective of the proposed framework is to normalize the characters need to be recognized into an approximately horizontal line and discard the interference characters, e.g., Pinyin, the phonetic transcriptions of Chinese characters.

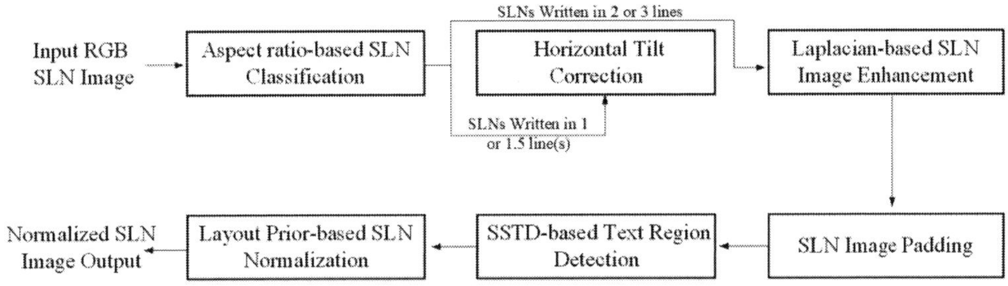

Figure 2. The processing procedures of the proposed framework.

SLNs from Chinese cargo ships can be classified into three categories according to the line number, namely, one-line SLNs, one-and-a-half-line SLNs and SLNs comprised of multiple (greater or equal to two) lines. One-and-a-half-line SLNs mean that some characters contained in one SLN are written in two line, while the others are written in one line.

Figure 2 shows the processing procedures of the proposed framework. It mainly consists of six steps. The technical details of each step are described as follows:

1) Aspect ratio-based SLN classification. SLNs from Chinese cargo ships generally contains at least 5 characters. In this situation, one-line and one-and-a-half-line SLNs are often have greater aspect ratios than multiple-line SLNs. In our experiments, if the aspect ratio of an input SLN image is greater than 2.0, this SLN will be classified as a multiple-line SLN, otherwise it will be classified as a one-line/one-and-a-half-line SLN. The reason of conducting aspect ratio-based SLN classification is that the horizontal tilt correction method [3] we used can only handle one-line or one-and-a-half-line SLNs.

2) Horizontal tilt correction. The horizontal tilt SLNs correction method presented in [3] is used to correct the input SLN because the SSTD-based text region detection method [5] can only detect axis-oriented horizontal text lines.

3) Laplacian-based SLN image enhancement. Ships, especially cargo ships, usually have large volume, while SLNs contained by them have very small sizes and low resolutions by comparison [2]. It is hard even for the state-of-the-art convolutional neural network (CNN)-based text detector, e.g., SSTD [5], to directly detect these small characters out. Therefore, a Laplacian-based enhancement algorithm is used to enhance the input SLN images.

4) SLN image padding. Padding is often used in deep convolutional neural networks to preserve the size and features of some convolutional layers. As is well known, the fine information captured by each conv layer of a CNN becomes coarser with its receptive field size increasing [1]. To boost the detection accuracy, the input SLN image is padded in advance. Assume that w and h represent the width and height of the input SLN image, respectively, the padding size of width and height are w/2 and h/2. The black rectangle in Figure 3(a) is the padding result of the SLN image in Figure 3(a) (red rectangle).

5) SSTD-based regional texts fine detection. SSTD [5] is a single-shot text detector that directly

outputs word-level bounding boxes in a natural image. It uses an attention mechanism which roughly identifies text regions via an automatically learned attentional map, which is helpful to substantially suppresses background interference and produce accurate inference of words, especially for the words at extremely small sizes. SLNs usually are of very small and low resolutions, it is thus reasonable to use SSTD to detect the characters, especially Pinyin, and then discarded them.

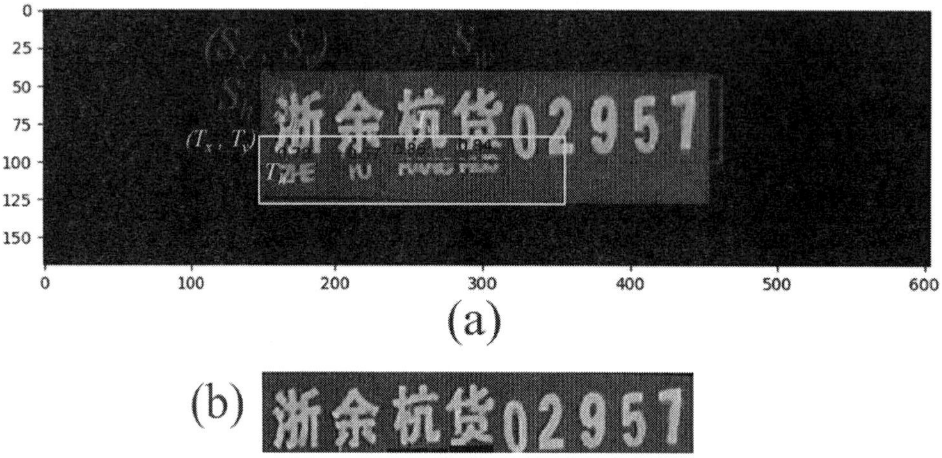

Figure 3. Layout prior-based SLN normalization.

6) Layout prior-based SLN normalization. The normalization procedures of one-line, one-and-a-half-line and multiple-line SLNs will be described in this part, respectively. (1) one-line and one-and-a-half-line SLNs. Denoting rectangle $S(S_x, S_y, S_w, S_h)$ as the input SLN image (red rectangle in Figure 3(a), for example), where S_x and S_y are the top-left coordinates of S, S_w and S_h are the width and height of S, respectively. $D(D_x, D_y, D_w, D_h)$ denotes the rectangle with the greatest width from the text rectangles detected by SSTD (green rectangles in Figure 3(a)). $T(T_x, T_y, T_w, T_h)$ represents the bottom-left rectangle of S. First of all, the bottom-left rectangle is defined as $T'(S_x, S_y + 0.5 * S_h, 0.67 * S_w, 0.5 * S_h)$, based on the layout prior of one-and-a-half-line SLNs, the Pinyin characters would appear inside this area very likely. Thus, the Pinyin characters (green rectangles inside of T' in Figure 3(a)) will be masked with black masks to reduce noises. Then, D is adjusted as $D'(S_x, D_y, S_w, D_h)$. D' is treated as the final normalization result (Figure 3(b)). (2) Multiple-line SLNs. Multiple-line SLNs will be normalized into one line by chaining the detected bounding boxes one by one from top to bottom. Figure 4 shows the detail processing procedures of some SLN instances of the proposed framework.

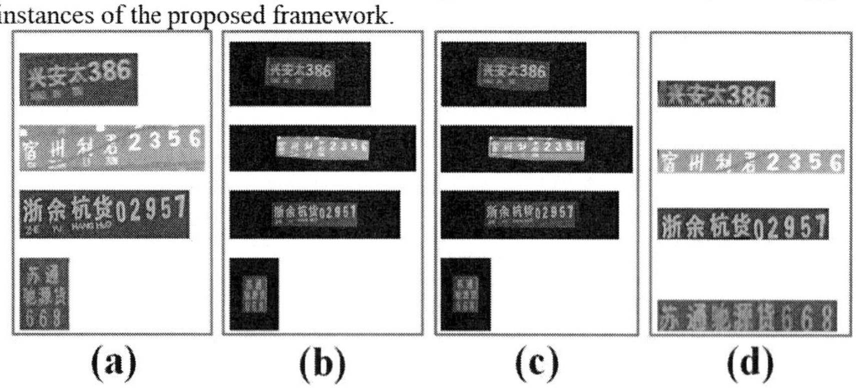

Figure 4. SLNs layout normalization (note that some images may be resized for the convenience of display). (a) The input SLN images. (b) SLN images after tilt correction, enhancement and padding. (c) Texts fine detection results. (d) The normalization results.

3. Experimental results

3.1 Testing dataset

The effectiveness of the proposed framework was tested on 3216 SLN images. These SLN images were collected in the practical application environment of the proposed framework — Beijing Hangzhou the Grand Canal (Hangzhou). The cameras were set up along the river, and photographed SLNs from thousands of ships over day and night in different seasons. The background of them are complex, illuminations and characters are various (This SLNs recognition dataset will be released publicly in the near feature). Therefore, these testing SLN images are challenging enough. Figure 1 shows some testing SLN images.

3.2 Experiment results

The input SLNs are considered successfully corrected if the cluttering Pinyin characters of one-and-a-half-line SLNs are successfully cropped, and all the characters contained in those SLNs are observed to be in one fairly horizontal line. Otherwise, they will be treated as failures. The quantitative experiment results are listed in Table 1. Table 1 tells that the proposed framework achieves a normalization rate of 87.38%, suggesting the proposed approach has good SLNs normalization ability. One can also recognize the impressive SLNs normalization performances of the proposed framework from Figure 1.

Table 1. SLNs normalization results of the proposed framework

Number of SLN images	Number of valid normalizations	Number of failures	Normalization rate
3216	2810	406	87.38%

4. Conclusion

An effective layout normalization framework of SLNs have been proposed in this paper. The proposed framework was tested on 3216 collected SLN images, as the first work concerning the layout normalization of SLNs, the proposed framework is proved to be effective with a layout normalization rate of 87.38% and is very helpful to the characters recognition algorithms of SLNs.

Acknowledgments
This work was supported partly by National Natural Science Foundation of China (61272304 and 61772374), and Zhejiang Provincial Natural Science Foundation of China (LY16F020023 and LQ18F030010).

References

[1] B. Liu, X. Lyu, C. Li, S. Zhang, Z. Hong, and X. Ye, “Using transferred deep model in combination with prior features to localize multi-style ship license numbers in nature scenes,” in *IEEE International Conference on TOOLS with Artificial Intelligence*, 2018, pp. 506–510.
[2] B. Liu, J. Sheng, J. Dun, S. Zhang, Z. Hong, and X. Ye, “Locating various ship license numbers in the wild: An effective approach,” *IEEE Intelligent Transportation Systems Magazine*, vol. 9, no. 4, pp.102–117, 2017.
[3] B. Liu, S. Zhang, Z. Hong, and X. Ye, “A Horizontal Tilt Correction Method for Ship License Numbers Recognition,” in *IOP Journal of Physics: Conf. Series,* vol. 976, 012013, 2018.
[4] B. Shi, X. Bai and C. Yao, "An End-to-End Trainable Neural Network for Image-Based Sequence Recognition and Its Application to Scene Text Recognition," in *IEEE Transactions on Pattern Analysis and Machine Intelligence*, vol. 39, no. 11, pp. 2298-2304, Nov. 1 2017.
[5] P. He, W. Huang, T. He, Q. Zhu, Y. Qiao, and X. Li, “Single Shot Text Detector with Regional Attention,” in *IEEE International Conference on Computer Vision*, 2017, pp. 3066-3074.

Research on Moving Target Detection System Based on Video Surveillance

Linglong Tan[1, *], Wenju Zhang[1] and Yehui Chen[1]

[1]Electronic communication engineering college, Anhui Xinhua University, Hefei, 230088, China

[*]tanlinglong@axhu.edu.cn

Abstract. In order to quickly and accurately obtain useful information needed by people from massive video information, a structured, hierarchical video retrieval platform is constructed from the perspectives of reliability, efficiency, universality, and ease of use. In the monitoring of pedestrian flow information, multi-scale analysis and fractal texture analysis methods are used. First, two-level wavelet decomposition is performed on the crowd image, and the original image is decomposed into multiple sub-graphs. Then the box model is used to represent the texture model. Then, the support vector machine is used to classify the fractal dimension. In vehicle image detection, a hidden Markov model is used to identify vehicle models. First, observation sequence O and model $\lambda = (A, B, \pi)$ are used to calculate the observation sequence $P(O \mid \lambda)$ produced by the model. Then, the Viterbi algorithm is used to find the best state sequence. Finally, the Baum-Welch algorithm is used to train the model. Judging the abnormal situation of the vehicle through the output of the model, the video surveillance system can monitor the people and vehicles.

1. Introduction

Computer vision is the study of video captured by surveillance cameras for target detection, tracking, analysis, and recognition. It also considers the temporal and spatial relationships between people and objects in video information, and understands the implications of various events under visual information——Video understanding [1,2].Traditional video surveillance mainly relies on manual monitoring and recording, but manual search of data is very inefficient. With the increasing application of surveillance video, the amount of video information is very large. According to statistics, after the operator monitors the screen for more than 10 minutes, 90% of video information will be missed, and it is easy to miss suspicious characters and abnormal events. This makes the monitoring system lose its original meaning and cannot play a role in quickly handling and reducing the loss of security risks[3,4,5].

2. System design

The first paragraph after a heading is not indented (Bodytext style). Based on digital image processing, artificial intelligence and database technology, this paper quickly and accurately extracts the feature information of moving objects such as people and cars from the large-scale video surveillance database, and stores these feature information in a database for security personnel to quickly query. Unattended intelligent monitoring of roads, plazas, and confidential departments[6,7].

The system is designed from its four perspectives: reliability, efficiency, universality, and ease of use to achieve: (1) reliability. The design of the completed moving target detection system has an adaptive threshold adjustment function. The system is suitable for 80% of surveillance video scenes and can accurately capture small targets. (2) High-efficiency, rapid processing of collected video, and generation of condensed video, rapid positioning and retrieval of targets, with an average concentration ratio of 20-40 times. (3) Versatility: The video surveillance system supports all major video formats and can be customized and developed. (4)Easy-to-use, standardized case management process is adopted to achieve a quick grasp of the operation flow without basic 10 minutes.

To efficiently and accurately realize the investigation of sports targets (people and vehicles), and provide efficient data support for all types of case detection. This article analyses the video surveillance images from the characteristics of human flow analysis, vehicle characteristics analysis and abnormal target detection.

3. System implementation

This article uses the basic principles and techniques of digital image processing technology, computer vision technology, simulation and other related disciplines. Based on the research method of "mechanism analysis-system modeling-simulation research-algorithm development", the key feature information extraction algorithm of video motion target based on digital image processing technology is proposed. The technical route is shown in Figure 1.

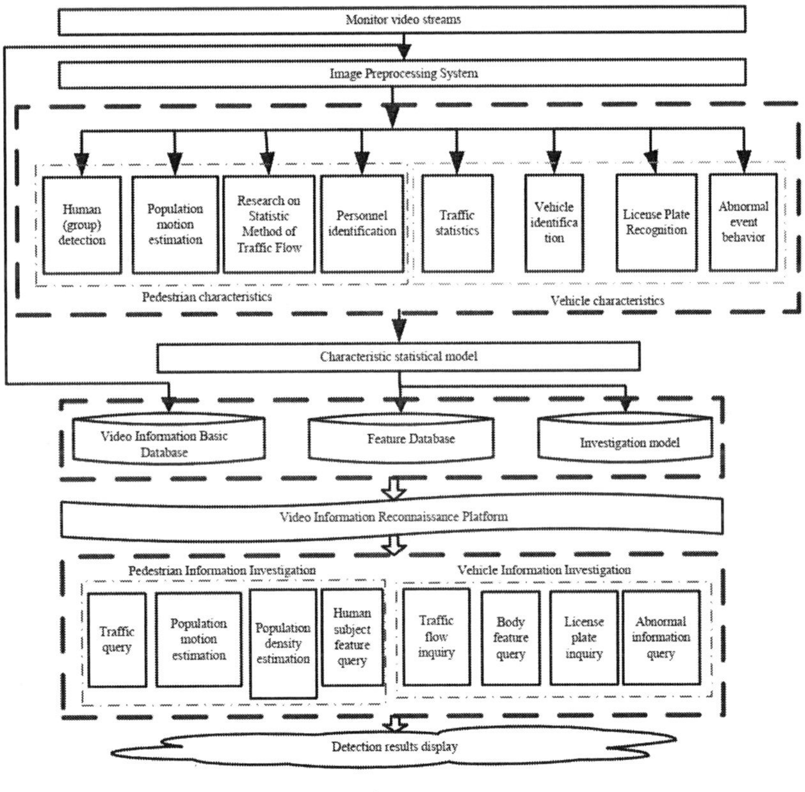

Figure 1. Technical route

By analyzing the characteristics of dynamic video images, starting from the principle of computer vision, video image processing technology is adopted to realize the recognition of the characteristics of pedestrians and vehicles in video sequences. Analyze the characteristics of video feature information data, use database technology to build a basic information database for storing the original massive video and dynamic video feature database of storage vehicle and human subject features. Analyze the monitoring and retrieval requirements and the characteristics of the data model supported by the database, and establish a structured and hierarchical retrieval model. The B/S architecture is used to build a video retrieval platform, and a search platform is used to quickly and effectively screen video clips that meet the conditions.

The specific research plan is:

(1) Automated population density estimation method based on multi-scale analysis and fractal texture analysis

After capturing the image from the video acquisition system, the background of the crowd is masked by the template and the region of interest (ROI) of the crowd is calculated. If the ROI is lower than 40%, the threshold value method is used to obtain the binary image of the crowd $C(i, j)$, that is:

$$C(i,j) = \begin{cases} 0 & D(i,j) \le T \\ 255 & D(i,j) > T \end{cases} \quad (1)$$

Among them, the value of T is used to determine whether the current pixel is a crowd.

If the ROI is higher than 40%, considering that the position of the video capture device is fixed, the crowds of the collected images are projected onto the camera at different scales, and the person near the camera is larger than the far person, thus introducing multi-scale analysis. . In order to use multi-scale analysis to estimate population density, a two-level wavelet packet decomposition is performed on the crowd image, and the two-level wavelet packet decomposition is illustrated in Figure 2.

After dividing the original image into 20 sub-band images by two-level wavelet packet decomposition, the box-counting dimension is calculated by Equation (2), and the box-counting dimension is used as the fractal dimension to represent the texture pattern.

$$\dim_{\beta} F = \lim_{\delta \to 0} \frac{\ln N(F,\delta)}{\ln(1/\delta)} \quad (2)$$

Due to the limitation of the training volume of the crowd image, the support vector machine was used to classify the fractal dimension.

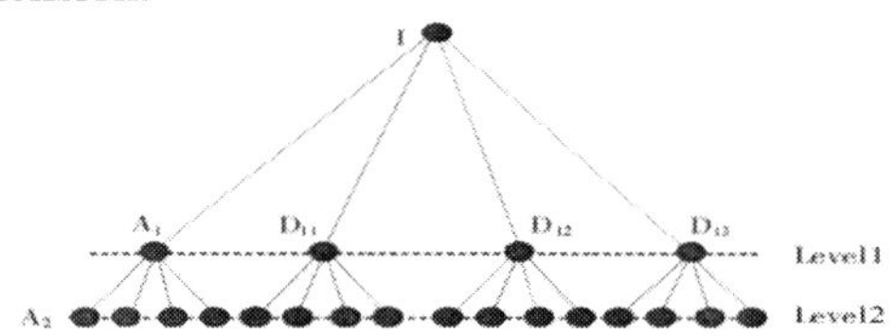

Figure 2. Two-level wavelet packet decomposition tree

(2)Estimation of crowd movement based on the block matching method (SDS) of the small diamond template

The two small diamond templates designed in Figure. 4 are all composed of 5 points and their distribution has obvious bias. A horizontal diamond template is used to search from the center of the search window. Five points of the matching template are calculated, and the average absolute error (MAD) value is calculated by Equation (3). If the MAD minimum point is the center point of the template, the search of the current block is nearly completed. The matching error between the current center point and the two empty points is calculated and compared, and the position where the current block is located is the one with the smallest error. If the MAD minimum point appears at the two distal points of the template, the horizontal point diamond is used to calculate the matching point error with the current minimum point as the center; if the MAD minimum point appears at the two near ends of the template, Then use the smallest point found in the previous step as the center, and use the vertical diamond template to calculate the matching point error until the minimum MAD appears on the center point of the template. The search process is shown in Figure 3.

$$MAD(i,j) = \frac{1}{MN}\sum_{m=1}^{M}\sum_{n=1}^{N} | f_k(m,n) - f_{k-1}(m+i,n+j) | \quad -w \le i,j \le w \quad (3)$$

(3) Identification of Person Subject Characteristics Based on HOG Method

The color space standardization method using the square root is first used for the collected image, this method can effectively suppress the influence of noise, and then calculate the gradient of the image. Image gradient horizontal differential operator $dx = I(x+1,y) - I(x-1,y)$ and vertical differential operator $dy = I(x,y+1) - I(x,y-1)$, then the gradient is:

$$G(x,y) = \sqrt{dx^2 + dy^2} \tag{4}$$

The gradient direction at any point in the image is: $\alpha(x,y) = \arctan(\frac{dy}{dx})$.

Next, the gradient values in the space and direction are statistically calculated, and the gradient direction histogram is constructed by using weighted voting (using the amplitude to represent the weights). Using the histogram average distribution of the omnidirectional gradient value, if the number of directions is less than 9, the number of added can significantly improve the classification effect, if the number of directions is greater than 9, the number of changes has no significant effect on the effect.

The actual acquired image is affected by external light, background and other factors, which makes the gradient intensity change range larger, so it needs to further standardize the gradient intensity, use formula (5) to do the standardization.

$$L2-norm, v \leftarrow v \Big/ \sqrt{\|v\|_2^2 + \varepsilon^2} \tag{5}$$

v is an unnormalized eigenvector, $\| v \|_k$ is the k order norm of v, and ε is a very small constant.

At this point, the HOG feature vector of the image area can be obtained. Taking the 64×128 image as an example, it is divided into image blocks of 16×16. When the image blocks are allowed to overlap, the step size is divided into 8 pixels to obtain $7 \times 15 = 105$ image blocks, and each block is divided into four units that do not overlap. Divide $[0°, 180°]$ into 9 directional gradient histograms so that each block is represented by a $9 \times 4 = 36$-dimensional vector.

(4) Vehicle Recognition Based on Hidden Markov Model (HMM)

The Hidden Markov Model is a bi-directional stochastic process. It is used to count the characteristics of the short-time stationary features and each short-term stationary to the next short-term stationary. Using this model for vehicle identification requires solving three key issues:

①Identify, by observing sequence o and model $\lambda = (A,B,\pi)$, calculating the probability $P(O \mid \lambda)$ of the observed sequence produced by the model. To solve this problem, a forward-backward algorithm is introduced. The forward probability is defined as Equation (6), and the backward probability is defined as Equation (7).

$$a_t(i) = P(o_1 o_2 .. o_t, q_t = i \mid \lambda) \tag{6}$$

$$\beta_t(i) = P(o_{t+1} o_{t+2} .. o_t, q_t = i \mid \lambda) \tag{7}$$

Then we get $P(O \mid \lambda)$ from the forward and backward probability:

$$P(O \mid \lambda) = \sum_{i=1}^{N} a_t(i) \tag{8}$$

② Finding the best state sequence, from observation sequence o to model λ, how to choose the best sequence $q = \{q_1, q_2, \cdots, q_T\}$ to get the best o. The best criterion is Viterbi algorithm:

$$\delta_t(i) = \max P[q_1 q_2 .. q_{t-1}, q_t = i, o_1, o_2, \ldots o_t \mid \lambda] \tag{9}$$

Selecting state i at state t moment, the probability of model λ moving along $\{q_1, q_2 \cdots, q_t\}$ to produce observation sequence $\{o_1, o_2, \ldots, o_t\}$.

③ Model training, correcting (A,B,π) by observing the sequence o, makes $P(O \mid \lambda)$ maximal. This project plans to use the Baum-Welch algorithm to train the model. By selecting an appropriate parameter re-evaluation formula, such as Equation (10), Equation (11), Equation (12), the new model λ is estimated from the current model λ' to satisfy $P(O \mid \lambda') \le P(O \mid \lambda)$. When the process is continuously repeated to bring its parameters into a converged state, the model is a maximum likelihood model.

$$\bar{\pi}_i = \frac{P(O, q_0 = i \mid \lambda)}{P(O, \lambda)} = \frac{a_0(i)\beta_0(i)}{\sum_{j=1}^{N} a_T(j)} \tag{10}$$

$$\overline{a}_{ij} = \frac{\sum_{t=1}^{T} P(O, q_{t-1} = i, q_t = j \mid \lambda)}{\sum_{t=1}^{T} P(O, q_{t-1} = i, \lambda)} = \frac{\sum_{t=1}^{T} a_{t-1}(i) a_{ij} b_j(o_t) \beta_t(j)}{\sum_{t=1}^{T} a_{t-1}(i) \beta_{t-1}(i)} \quad (11)$$

$$\overline{b}_i(k) = \frac{\sum_{t=1}^{T} P(O, q_t = i \mid \lambda) \delta(o_t, v_k)}{\sum_{t=1}^{T} P(O, q_t = i, \lambda)} = \frac{\sum_{t=1}^{T} a_t(i) \beta_t(i) \delta(o_t, v_k)}{\sum_{t=1}^{T} a_t(i) \beta_t(i)} \quad (12)$$

(5) The Measurement of Abnormal Behavior of Vehicles Based on the Pascal Distance Measure Method

On the basis of the detection and tracking of the moving vehicle in the front, through the effective learning of the target trajectory information, the motion law of the moving vehicle is obtained, thereby explaining the behavior of the moving vehicle. First, the normal vehicle behavior is modeled and trained, and then the deviation from the abnormal vehicle behavior and normal vehicle behavior is used to judge whether the behavior is normal. Due to the limitations of the sample size, this paper selects the Papsi metric method with superior processing time and performance to detect abnormal vehicles. Using the Bhattacharyya distance to classify the features, under the assumption of normal distribution, the upper bound ε_{ij} of the minimum error probability of the Bayesian error is calculated by Equation(13), and the Bahrain distance is measured by Equation (14). The effective distance between two adjacent frames is expressed as a geometric mean. The clustering effect of this method is sensitive, and the inter-category variation with frame-to-frame variation is significant, which is suitable for abnormal behavior detection of sports objects.

$$\varepsilon_{ij} = \sqrt{p(\phi_i) p(\phi_j)} e^{-\beta} \quad (13)$$

$$\sum_r (p, q) = \frac{1}{s-1} [\sum_{r=1}^{s} \Psi_r(p) \Psi_r(q) - \frac{1}{s} \sum_{r=1}^{s} \Psi_r(p) \sum_{r=1}^{s} \Psi_r(q)] \quad (14)$$

Formula (14) is conceptualized to obtain the Pap distance between all classes as in Equation (15)

$$u_i - u_j = \begin{bmatrix} \frac{1}{m} \sum_{i=1}^{n} x_i - \frac{1}{n} \sum_{j=1}^{n} x_j \\ \frac{1}{m} \sum_{i=1}^{n} y_i - \frac{1}{n} \sum_{j=1}^{n} y_j \\ \frac{1}{m} \sum_{i=1}^{n} \delta_i - \frac{1}{n} \sum_{j=1}^{n} \delta_j \\ \frac{1}{m} \sum_{i=1}^{n} a_i - \frac{1}{n} \sum_{j=1}^{n} a_j \end{bmatrix} \quad (15)$$

Then use the geometric mean to represent the effective distance G_β of the adjacent frames (the algorithm is shown in Table 1). In the traffic scene, if you encounter an abnormal situation will affect the position of the image direction, resulting in β and G_β values become larger .

Table 1. Effective Distance Algorithm

Calculate effective distance G_β algorithm:
//F: Total number of clusters in frame f
//S: Total number of clusters in frame f+1
// C_m : counter in frame f
// C_n : counter in frame f+1
$C_m \leftarrow 1 : C_n \leftarrow 1$
While $C_m \le F$
While $C_N \le S$
$c_n = c_n + 1$
End
Calculate the C_m and C_N values to get β and save as Ω_{cm} :
$\Omega_{cm} = [\prod_{i=1}^{c_n} \beta_i]^{\frac{1}{c_n}} = \exp[\sum_{i=1}^{c_n} \log_e \beta_i]$
$F = F - 1$
$c_n = 1; c_m = c_m + 1$
End

Set the predefined threshold Γ_β to partition the motion distance of each frame and determine whether the behavior is normal.

(6) Design and Implementation of Video Information Retrieval Platform

The retrieval platform is to provide users with a human-computer interaction interface and complete video retrieval. This article plans to adopt the B/S (browser/server) architecture. The video retrieval platform includes the main interface (Main.aspx), the query result display interface (VQResult.aspx) and the result playback interface (play.aspx). Use C# language to build the platform and write the 3 main asp.net form files of the platform. The design ideas are shown in Figure 4.

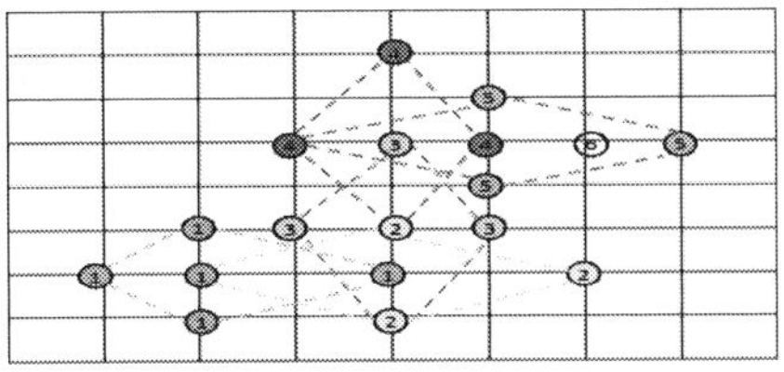

Figure 3. SDS search algorithm

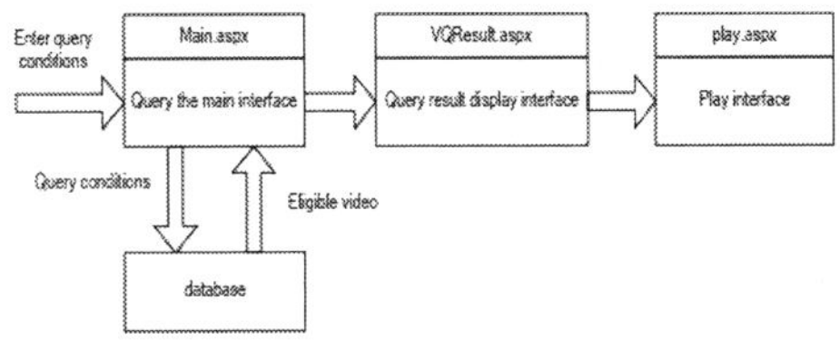

Figure 4. Search platform design ideas

4. Conclusion

In view of the large amount of surveillance video information, the complexity, dynamics, and uncertainty of the moving target, as well as the need for rapid retrieval of massive objects, the paper focuses on the research of moving target monitoring technology for massive amounts of video information. Based on the analysis of the analysis methods of the flow characteristics and vehicle characteristics in the surveillance video, the reliability and effectiveness indicators of the video surveillance technology are analyzed, the detection and understanding methods of the abnormal behavior of the video surveillance target are explored, and the target abnormal behavior detection model is developed. Considering data security and system maintenance costs, a structured and hierarchical video retrieval model is constructed. Integrate monitoring purposes and monitoring environment in intelligent video surveillance, provide theoretical basis, data support and visual management for the security department's mission planning and resource allocation decisions, provide a reliable reference for public safety prevention, and ensure the stable and stable social life.

Acknowledgments

This work was supported by the Scientific Research Project of Anhui Xinhua College No. 2017zr003, Anhui Natural Science Research Project No.KJ2017A628 and the domestic outstanding young backbone talents in universities of visiting program No. gxgnfx2018070.

References

[1] Jianfang Dou,Jianxun Li. Modeling the background and detecting moving objects based on Sift flow[J]. Optik -International Journal for Light and Electron Optics,2014,125(1).

[2] Constance S. Royden,Daniel Parsons,Joshua Travatello. The effect of monocular depth cues on the detection of moving objects by moving observers[J]. Vision Research,2016,124.

[3] Seungwon Lee,Nahyun Kim,Kyungwon Jeong,Kyungju Park,Joonki Paik. Moving object detection using unstable camera for video surveillance systems[J]. Optik - International Journal for Light and Electron Optics,2015,126(20).

[4] Qing Ye,Rentao Gu,Yuefeng Ji. Human detection based on motion object extraction and head–shoulder feature[J]. Optik - International Journal for Light and Electron Optics,2013,124(19).

[5] Constance S. Royden,Daniel Parsons,Joshua Travatello. The effect of monocular depth cues on the detection of moving objects by moving observers[J]. Vision Research,2016,124.

[6] Royden Constance S,Parsons Daniel,Travatello Joshua. The effect of monocular depth cues on the detection of moving objects by moving observers.[J]. Vision research,2016,124.

[7] Shuai Yuan,Guo Yun Zhang,Jian Hui Wu,Long Yuan Guo. Study of Motion Detection Technology Based on Davinci Platform[J]. Advanced Materials Research,2013,2450(712).

Research on Vehicle Detection Algorithms in Surveillance Video Images

Linglong Tan[1,*], Wenju Zhang[1] and Yehui Chen[1]

[1]Electronic communication engineering college, Anhui Xinhua University, Hefei, 230088, China

[*]tanlinglong@axhu.edu.cn

Abstract. Intelligent traffic video surveillance system, including vehicle inspection, vehicle tracking and license plate recognition module. The video vehicle detection in this paper is divided into four steps: video image pre-processing, background modeling and updating, foreground detection and adhesion vehicle area segmentation. After the collected video sequence is subjected to mean-value down-sampling and gray-scale conversion preprocessing, the background model is built using a mixture of Gaussian models. A sliding window is set on the time axis, and the grayscale video frame image in the sliding window is median-filtered to output a background gradient map. By merging the results of the spatial background difference between the color space and the grayscale gradient, the foreground image detected by the vehicle is obtained. Finally, the background difference is obtained in the gradient domain to obtain the foreground image to achieve vehicle detection.

1. Introduction

With the development and progress of today's society, the monitoring of traffic scenes has also become a subject of concern for a person. It is common in social life and people's daily lives to monitor traffic scenes by connecting cameras to computers. However, today's science and technology are developed. It is not only inappropriate for people to analyze videos, but also to spend a lot of manpower and material resources. Due to the development of technology, the use of computers to process video images has become an increasingly common method. Its universal application scope and future commercial value have drawn great attention from academic and business circles at home and abroad.

At present, our monitoring of traffic monitoring sites is mainly achieved through computer connection monitoring systems. Based on computer and image processing technology, people have established an intelligent traffic monitoring and management system, which basically does not require human control in order to achieve the goal of saving manpower and material resources. Through the analysis of video images, the positioning and identification of vehicles can be achieved. While performing routine management tasks, it can also take the most safe and effective measures when abnormal conditions occur, thus providing a more advanced and feasible monitoring program. In order to solve these problems, the supervising department used video cameras to take pictures in some places, and then used manual analysis and management.

2. Vehicle Detection Algorithm Research

2.1. Image Preprocessing

Due to the huge difference in imaging features of the scene images, for example, the probability of discrimination is very low, the sizes are different, and the illumination is uneven. These features affect the positioning of moving objects, image cutting to character recognition and other processes. Before inputting the text image under the premise of the same scene into each module, the video image must undergo a necessary pre-processing process, which will help to improve the accuracy of positioning and recognition.

(1) Expansion and corrosion

Inflation [1,2] refers to the process of merging all the background points in contact with an object into the object and expanding the boundary line to the outside. Use this process to fill the space in the object.

Assume that A and B are sets in Z^2, ϕ is an empty set, $\bar{B}$ is an image of B, B performs an expansion operation on A, and it is denoted as $A \oplus B$, $\oplus$ is an expansion operator and expands to:

$$A \oplus B = \{x \mid [(\bar{B})_x \cap A] \neq \phi\} \tag{1}$$

The expansion process shown in Equation (1) is that B first maps the far point and then translates x. The expansion operation performed by B on A is that $\overline{B}$ is translated by all X and A has at least one non-zero common element, and the definition can be changed to:

$$A \oplus B = \{x \mid [(\bar{B})_x \cap A] \not\subset \phi\} \tag{2}$$

Corrosion [3, 4] is a process that eliminates boundary points and shrinks the boundaries to the inside. With this operation, small, non-meaningful objects can be eliminated.

Assume that A and B are collections in Z^2. A is corrupted by B, and it is written as $A\Theta B$. It is defined as:

$$A\Theta B = \{x \mid (B)_x \subseteq A\} \tag{3}$$

That is to say, the result of A being corrupted by B is the set of points x that all include A after B is translated by x.

(2) Filter enhancement of images

The purpose of the panning and sliding operation is to eliminate or minimize noise interference as much as possible and improve the quality of the image. If the additive noise is randomly distributed, the use of the average or weighted average of the neighborhoods can effectively suppress noise interference.

If S is the neighborhood of pixel (x_0, y_0) (including (x_0, y_0)), where (x, y) represents the element in S, $f(x, y)$ represents the gray value of point (x, y), and $a(x, y)$ represents the weight of each point, smoothing (x_0, y_0) can be expressed as:

$$f'(x_0, y_0) = \frac{1}{\sum_{(x,y)\in S} a(x, y)} \left[\sum_{(x,y)\in S} a(x, y) f(x, y)\right] \tag{4}$$

In general, the weights are symmetrical with respect to the middle point. For the following 3*3-size templates, their weights are equal:

$$T = \frac{1}{5}\begin{bmatrix} 0 & 1 & 0 \\ 1 & 1 & 1 \\ 0 & 1 & 0 \end{bmatrix} \tag{5}$$

The corresponding function expression for this template is:

$$f'(x, y) = \frac{1}{5}[f(x, y-1) + f(x-1, y) + f(x, y) + f(x+1, y) + f(x, y+1)] \tag{6}$$

Can be expressed as:

$$f'(x, y) = \frac{1}{5}[1 \times f(x, y-1) + 1 \times f(x-1, y) + \mathrm{L} + 1 \times f(x, y+1)] = T \times f \tag{7}$$

In other words, the neighborhood operation can be obtained by the convolution of the neighborhood and the template, which is also very convenient for calculation.

2.2. Detection Algorithm for Moving Vehicles

(1) Inter-frame difference method

The inter-frame difference method [5, 6, 7] is a method for obtaining a target (vehicle) surface in motion by performing a differential operation on a moving target between two adjacent frames in a video surveillance image. This shows that if the difference between each frame is relatively large, then it can explain the existence of the moving target; conversely, if the difference between each frame is relatively small, then it can indicate that the moving target does not exist. . The difference here mainly refers to the continuity and independence between images and images. The principle of this calculation method is shown in Figure 1. That is, the extracted image is first subjected to image enhancement processing, ie, filtering, to remove various noise disturbances in the image, and then the image is subjected to binary processing and differential processing. In order to detect the goal in motion.

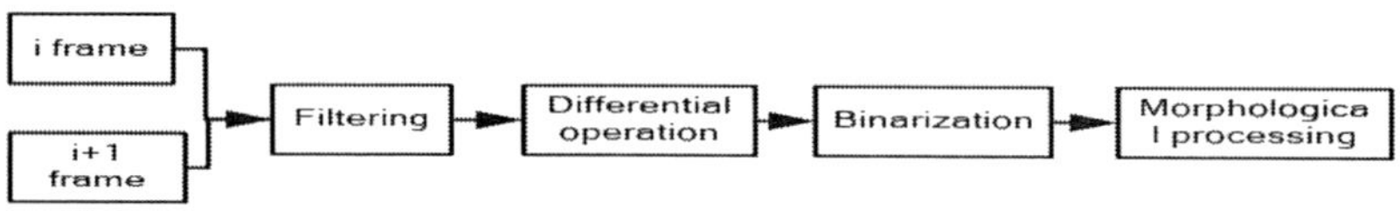

Figure 1. Schematic diagram of difference between frames

The inter-frame difference method has two frame methods and three frame methods. The two-frame method is to process dynamic objects, and the binary and differential images must be processed first. The two-frame difference method works as follows:

The original image is denoted as $f_i(x, y)$. Here, i=1, 2, 3...,,n, which represents the number of image frames, denotes the image after the difference and the binarized image as $D_k(x, y)$ and $T_k(x, y)$, respectively. The two-frame difference method can be expressed as:

$$D_k(x, y) = |f_{k+1}(x, y) - f_k(x, y)| \tag{8}$$

$$T_k(x, y) = \begin{cases} 1, & D_k(x, y) \geq T \\ 0, & D_k(x, y) \leq T \end{cases} \tag{9}$$

Although this method captures the target in motion, it has a significant effect, but its requirements are also more demanding. It requires the target to be in a state of motion, and its brightness changes must be very obvious. If the target is stationary, it is very easy to catch the target information. This will make the moving target unable to guarantee its continuity.

The three-frame difference algorithm is used in three-frame images. This method is used if two consecutive frames are overlapped in the detection target. Its specific operations are as follows: First, each frame image is processed with various grayscale processing methods to obtain a corresponding grayscale image; then, the processed image is binarized to obtain a corresponding binary image;

Finally, the target region is extracted on the binarized image. The expression is as follows:

$$D_{k1} = (k-1, k)\begin{cases} 1, & if|f_k(x, y) - f_{k-1}(x, y)| \geq T \\ 0, & otherwise \end{cases} \tag{10}$$

$$D_{k1} = (k, k+1)\begin{cases} 1, & if|f_k(x, y) - f_{k-1}(x, y)| \geq T \\ 0, & otherwise \end{cases} \tag{11}$$

$$D_{k1} = (k, k+1)\begin{cases} 1, & if|(D_{k1}(k-1, k) \wedge D_{k2}(k, k+1) = 1| \\ 0, & otherwise \end{cases} \tag{12}$$

In the formula, $f_{k-1}(x, y)$, $f_k(x, y)$, and $f_{k+1}(x, y)$ respectively represent consecutive three-frame images, $D_{k1}(k-1, k)$ and $D_{k2}(k, k+1)$ represent grayscale images of three consecutive frames of images, and T is the binarization threshold. In the formula, K and D represent the intersection of two frames of images, ie coincident portions, which represent the moving target.

(2) Background Difference Method

The background difference method [8, 9] is a method for testing a moving object by comparing the current frame with a background reference model in a video surveillance image, and it relies on a background modeling technique.

In the target detection of the motion based on the background difference method, the accuracy of the model image and the simulated background image will have a great influence on the test result. However, any moving object detection algorithm must meet the requirements of any image scene processing. However, it is sometimes difficult to achieve background modeling because of large motion scenes, unpredictability, external environment interference, and different noise. . Therefore, the research in this paper is based on the background difference algorithm in the background. Let (m, n) be the plane coordinates of the two-dimensional digital image. The binary mathematics based on the background subtraction is described as:

$$D_i(m, n) = |I_i(m, n) - B_i(m, n)| \tag{13}$$

$$M_i(m, n) = \begin{cases} 1, & D_i \geq T \\ 0, & D_i \leq T \end{cases} \tag{14}$$

$I_i(m, n)$ represents the grayscale image after the current frame transition in the video surveillance image, $B_i(m, n)$ represents the monitored background grayscale image of the current frame in the video surveillance image, $M_i(m, n)$ represents the subtracted binarization result, and T represents the grayscale image threshold.

3. vehicle detection algorithm simulation

3.1. Realization of the Difference Between Frames

Use mmreader() function to read video files, and nframes = get(videoObj, 'NumberOfFrames') function is to get the number of video file frames. After reading the video file, use currentFrame = read(videoObj, k) to read any frame video image, and then use function grayFrame = rgb2gray(currentFrame) to grayscale it to obtain a grayscale image. Subtract the two adjacent images to obtain the result of the difference between the frames.

The following are the effect diagrams of the 19th and 20th frames, 79th and 80th frames, and 139th and 140th frames of the selected video, respectively, as shown in Figure 2.

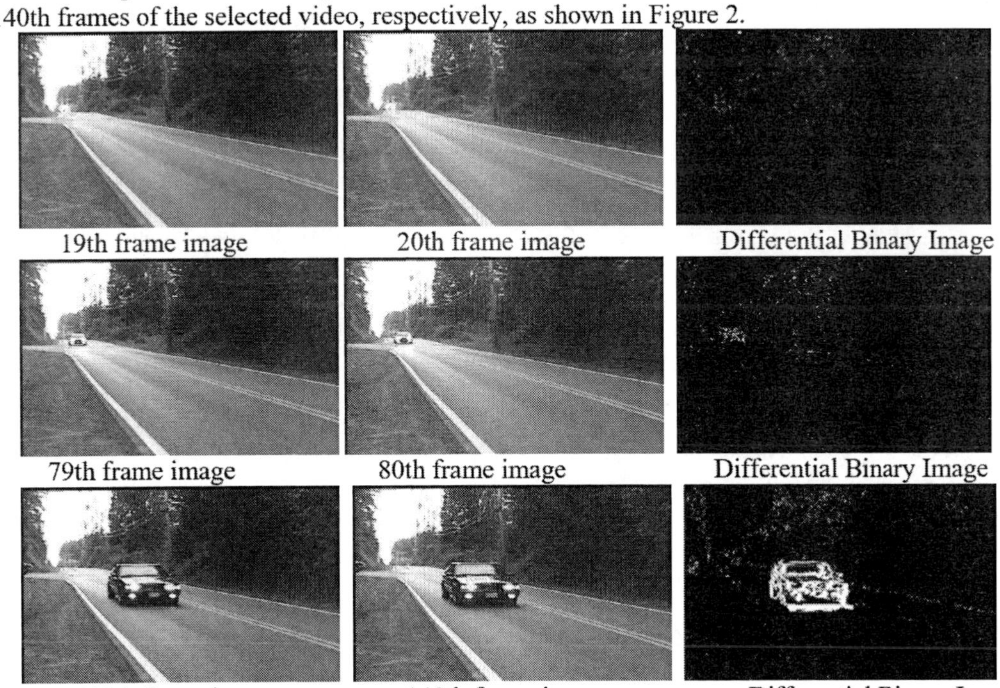

Figure 2. Experimental results of inter-frame difference

3.2. Background Difference Method Implementation

The background difference method is to obtain the background, and then perform the difference operation on the background. Read the video file, use m=rgb2gry(m) to grayscale the read video file [10], obtain the grayscale image and then perform the differential operation on the adjacent two frames before and after to get the differential effect, as shown in Figure 3. For the aspects of operation, a vehicle detection interactive interface is designed in the MATLAB GUI, and the running effect is shown in Figure 4.

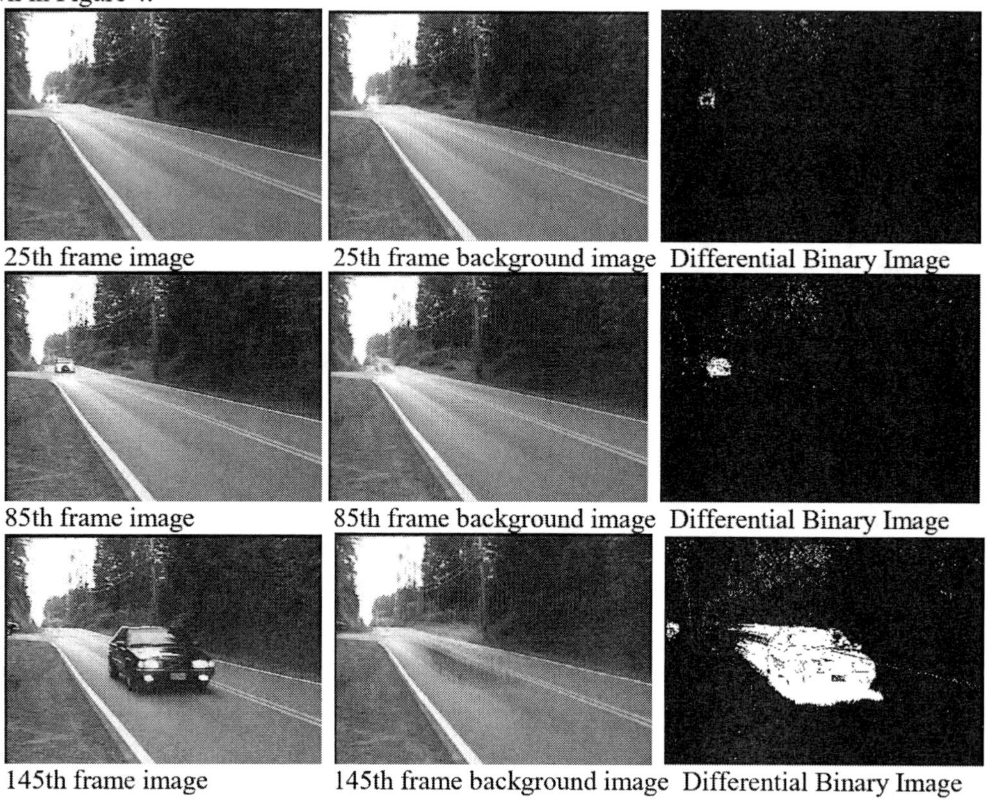

Figure 3. Moving target detection image with background difference method

4. Conclusion

The vehicle video detection in this paper is processed from video image preprocessing, background modeling and updating, foreground detection, and segmentation of stuck vehicle areas. The collected video image is subjected to mean sampling and grayscale conversion, and the extracted image is subjected to image filter processing to remove various noise interferences in the image. Then use the binarization method and differential processing method to detect the moving target in the image. In the design, the inter-frame difference method, background difference method and edge detection method were analyzed. Among them, the inter-frame difference method and the background difference method were implemented by subtraction. However, they are also different. The interframe difference method is a method for obtaining the target contour in motion by performing a differential operation on the moving target between two adjacent frames in the video surveillance image. The background difference rule is a comparison of the current frame and the background reference model in the video surveillance image.

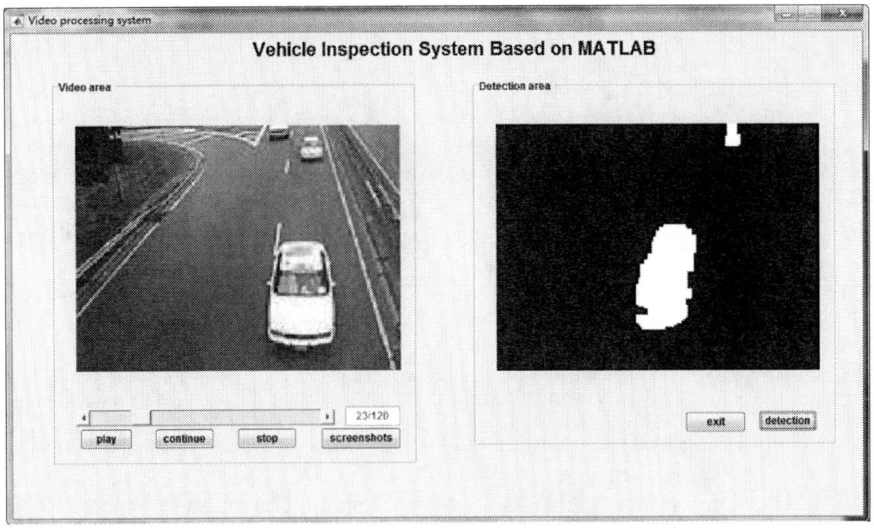

Figure 4. vehicle detection interface

Acknowledgments
This work was supported by the Scientific Research Project of Anhui Xinhua College No. 2017zr003, Anhui Natural Science Research Project No.KJ2017A628 and the domestic outstanding young backbone talents in universities of visiting program No. gxgnfx2018070.

References

[1] Fan Yao,Chu Yanli. Study of a new vehicle detection algorithm based on linear CCD images[J]. Optik - International Journal for Light and Electron Optics,2015,126(24).
[2] Bao Xu,Li Haijian,Xu Dongwei,Jia Limin,Ran Bin,Rong Jian. Traffic Vehicle Counting in Jam Flow Conditions Using Low-Cost and Energy-Efficient Wireless Magnetic Sensors.[J]. Sensors (Basel, Switzerland),2016,16(11).
[3] Meng Xiaoli,Wang Heng,Liu Bingbing. A Robust Vehicle Localization Approach Based on GNSS/IMU/DMI/LiDAR Sensor Fusion for Autonomous Vehicles.[J]. Sensors (Basel, Switzerland),2017,17(9).
[4] Fan Yao,Chu Yanli. Study of a new vehicle detection algorithm based on linear CCD images[J]. Optik - International Journal for Light and Electron Optics,2015,126(24).
[5] Xiaorong Gao,Yingyong Wu,Kai Yang,Jinlong Li. Vehicle bottom anomaly detection algorithm based on SIFT[J]. Optik - International Journal for Light and Electron Optics,2015,126(23).
[6] Zhonghua Zhang,Xuecai Yu,Feng You,George Siedel,Wenqiang He,Lifang Yang. A Front Vehicle Detection Algorithm for Intelligent Vehicle Based on Improved Gabor Filter and SVM[J]. Recent Patents on Computer Science,2015,8(1).
[7] Hong He,Rui Zheng,Zhi Hong Zhang. Vehicle Detector Based on the AMR Sensor[J]. Applied Mechanics and Materials,2014,3082(543).
[8] Y H Lin,R Bai,Z H Qian. The design method and research status of vehicle detection system based on geomagnetic detection principle[J]. Journal of Physics: Conference Series,2018,986(1).
[9] Zi Yang,Lilian S.C. Pun-Cheng. Vehicle detection in intelligent transportation systems and its applications under varying environments: A review[J]. Image and Vision Computing,2018,69.
[10] Yingfeng Cai,Ze Liu,Xiaoqiang Sun,Long Chen,Hai Wang,Yong Zhang,Mi Zhang. Vehicle Detection Based on Deep Dual-Vehicle Deformable Part Models[J]. Journal of Sensors,2017,2017.

Ship License Numbers Recognition Using Deep Neural Networks

Baolong Liu[1], Shuangqing Wu[2,*], Sanyuan Zhang[1], Zhenjie Hong[3] and Xiuzi Ye[3]

[1]College of Computer Science and Technology, Zhejiang University, Hangzhou 310027, China
[2]Ningbo Institute of Technology, Zhejiang University, Ningbo 315100, China
[3]College of Mathematics, Physics and Electronic Information Engineering, Wenzhou University, Wenzhou 325035, China

[*]wsqing1999@163.com

Abstract. Ships are important waterway transportation vehicles. Ship license numbers can be used to identify ships due to their uniqueness. However, rare attentions have been paid to the domain of SLNs recognition. In this paper, a SLNs recognition framework is proposed to recognize multi-style SLNs. First of all, the styles of SLNs are normalized. In this phrase, the characters of SLNs that are written in multiple lines are normalized into one nearly horizontal line. Then, a convolutional recurrent neural network is trained to recognize these characters. A spatial transformer network is configured to rectify distorted SLN images. A sufficient and proper training dataset is a basic prerequisite for a deep neural network to play out its advantages. Therefore, a synthetic SLNs dataset which contains more than 6 million synthetic SLN images were generated to train the convolutional recurrent neural network. This synthetic SLNs dataset was generated based on the image backgrounds, character fonts, character types and text-line angels of the collected real-world SLN images. Finally, the phonetic transcriptions are eliminated. In the experimental period, the Levenshtein distance is configured to estimate the recognition efficiency. Experimental results on 1846 collected real-world SLNs proved the good SLNs recognition ability of the proposed framework.

1. Introduction

With the needs of security surveillance continually growing, many image processing, machine learning and deep learning technics have been applied to the domain of car license plates location and recognition [1], and impressive progresses on car license plates recognition have been achieved also.

Waterway traffic serves as a powerful assurance for the development of economic and commerce. Ship is the most important vehicle in waterway transportation. In some country, e.g., China, ships that are hoped to work legally in public waters need to be registered on local official authorities with a unique name in advance, these unique ship names are also referred as ship license numbers (SLNs). Therefore, SLNs can be treated as unique identities of ships. However, rare attentions have been paid to the domain of SLNs recognition.

One related research area of SLNs recognition is car license plates recognition [1]. Many applications which based on car license plates recognition technics, e.g., parking charge, vehicle entrance control and transportation surveillance, are already a part of everyday life. Even though, there exists no universal method that can recognize different car license plates from different countries,

IOP Conf. Series: Journal of Physics: Conf. Series **1060** (2018) 012064 doi:10.1088/1742-6596/1060/1/012064

most car license plates recognition algorithms can only recognize car license plates that are of specific styles. However, SLNs from different ships have various styles [2]. It is thus difficult to use existing car license plates recognition algorithms to proceed SLNs recognition.

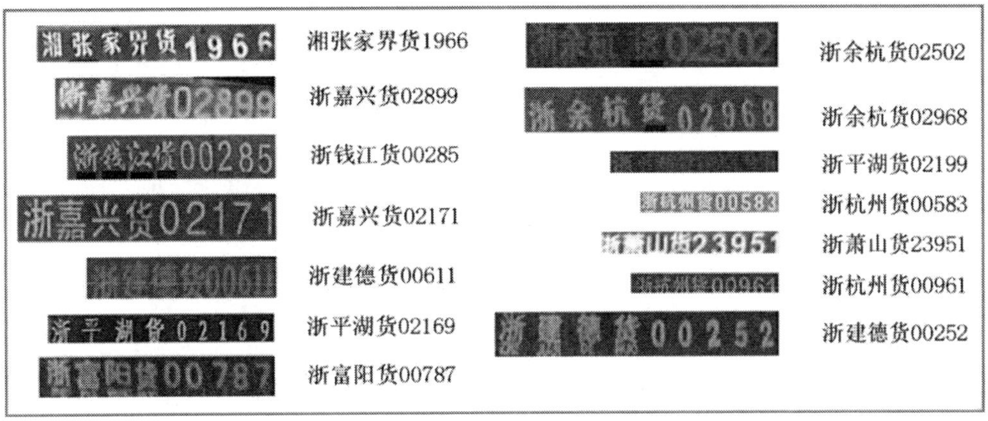

Figure 1. Some 0-Levenshtein-distance recognition results of the proposed approach. Left: the input SLN images. Right: the corresponding recognition results.

In [3], a transferred deep CNN model in combination with prior features is used to localize multi-style ship license numbers. An effective coarse-to-fine approach for locating various SLNs in the wild is proposed by Liu et al. [2]. A horizontal tilt correction method for SLNs recognition is presented in [4].

Recently, deep neural networks-based methods have achieved great progresses on many tasks, e.g. object detection [5] and character recognition [6]. An end-to-end trainable neural network, which named as Convolutional Recurrent Neural Network (CRNN), is presented in [6], it integrates feature extraction, sequence modelling and transcription into a unified framework, and can accomplish character sequence recognition task. However, on the one hand, CRNN can only recognize English alphabet so far. On the other hand, for CRNN, the characters need to be recognized should be nearly listed in one line, hence, CRNN is unable to read characters that are written in multiple lines. It is hard to directly use CRNN to recognize SLNs because multilingual characters may be found in SLNs. For example, SLNs of Chinese cargo ships contain not only English alphabet but also Chinese characters. Moreover, characters of SLNs may be also written in more than one lines.

To this end, CRNN is improved in this paper to recognize multi-style SLNs. First of all, SLNs that have different styles are normalized, characters that are written in different lines are normalized into one nearly horizontal line. Then, CRNN is improved to recognize these normalized characters. Specifically, for an input SLN image, the CNN features are computed by the CNN module, a spatial transformer network [7] is inserted after the CNN to rectify distorted SLN images, finally a RNN with long short-term memory (LSTM) mechanism is used to recognize the character sequences contained by the input SLN image. A sufficient and proper dataset is a basic prerequisite for deep neural network to play out its advantages. A synthetic SLNs dataset is thus produced. The dataset contains more than 6 million SLNs images, in which about 5.46 million are used as the training data and the remaining are used as validation data. Several influence factors including SLN image backgrounds, character fonts, character types and text-line angels are taken into consideration when making these synthetic SLN images. Meanwhile, a real-world SLNs dataset is also collected. The networks is firstly trained on the synthetic SLNs dataset, then fine turned on the collected real-world SLNs dataset. Testing experiments on 1846 real-world SLN images is conducted to validate the recognition efficiency of the proposed approach. Figure 1 shows some recognition results of the proposed approach.

2. The proposed SLNs recognition approach

Figure 2 shows the SLNs recognition pipeline of the proposed approach. It consists six main parts.

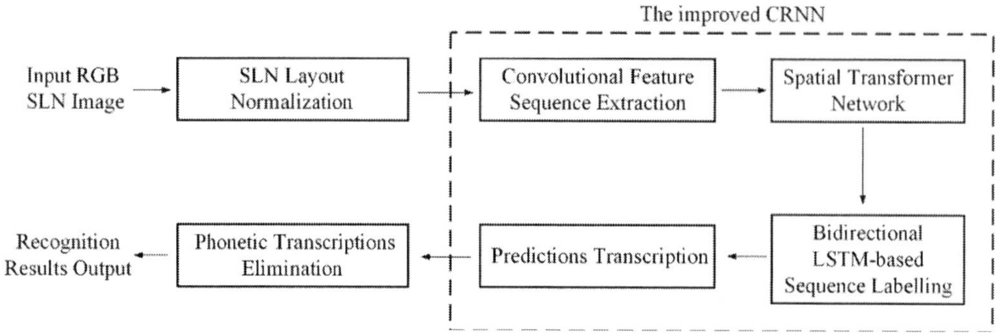

Figure 2. The SLNs recognition pipeline of the proposed approach.

The technical details are described as follows:

(1) SLN layout normalization. Characters that are written in two or more lines are normalized into one line before they are fed to CRNN to recognize. For one-line SLNs, some phonetic transcriptions which comprised by English alphabet are removed, then our horizontal tilt SLN correction method presented in [4] is used to correct the input SLN image; For multiple-lines SLNs, the text region detection method proposed in [8] is fist used to fine localize the texts contained in the SLN image, multiple-line SLNs then are normalized into one line by chaining the detected bounding boxes one by one from top to bottom.

(2) Convolutional feature sequence extraction. The input SLN is resized into 32*256 (height*width) grayscale image. Then, the grayscale image is fed to the convolutional feature sequence extraction module to extract sequential features of the input SLN image.

(3) Spatial transformer network. The spatial transformer networks presented in [7] is inserted after the CNN layers to rectify distorted SLN images. STN is kind of powerful neural network architecture that can achieve spatial invariance by automatically rectifying the input images before they are fed to a normal neural network. STN is end-to-end differential and can be applied to the existing network architectures without extra supervision.

(4) Bidirectional LSTM-based sequence labelling. The extracted convolutional feature sequences are fed to bidirectional LSTM layers to predict a label distribution. Long-Short-Term-Memory (LSTM) [9] is a kind of RNN unit that can alleviate the vanishing gradient problem of RNN [6].

(5) Predictions transcription. The predictions resulted by LSTM are converted into a label sequence in this phrase.

(6) Phonetic transcription elimination. The phonetic transcriptions (also referred as Pinyin), appear in SLNs frequently. Phonetic transcriptions are composed by English alphabet and need to be discarded since they are just some pronunciations.

The network architectures of step 2, 4 and 5 are kept same with that in CRNN, please refer [6] for more details.

3. Experimental results

3.1 Dataset

(1) Synthetic SLNs dataset. No large SLNs training dataset is published so far, a synthetic dataset is generated in this paper to handle the problem of SLNs training data insufficient. These synthetic SLNs data were generated based on the image backgrounds, character fonts, character types and text-line angels of the collected real-world SLN images. Fifty images that have different RGB value are used as the basic background images. Twenty one kinds of Chinese fonts that may appear in real-world SLN images are used as the basic character fonts. Characters that appeared in real-world SLNs are used to generate corpuses, the character appearance frequencies were also taken into consideration. Random noises were added. More than 6 million synthetic SLN images were generated, in which 5,468,671 synthetic SLN images were used as the training data, the remaining were used as the validation data.

Each synthetic SLN image has a ground truth label. Figure 3 demonstrates some generated synthetic SLNs.

(2) Real-world SLNs dataset. 10,975 SLNs images with their reverse images (totally 21,950) are collected as the training dataset. Another 1846 SLNs images are used as testing instances. These SLN images were collected in the practical application environment of the proposed framework — Beijing Hangzhou the Grand Canal (Hangzhou).

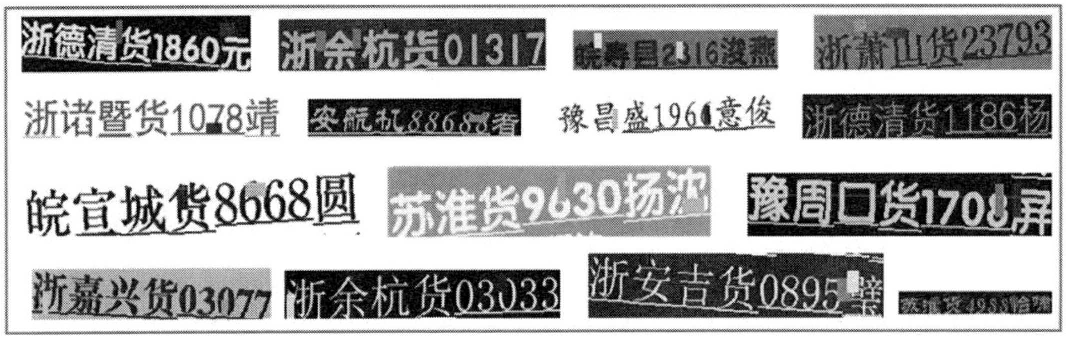

Figure 3. Some samples of the generated synthetic SLNs.

3.2 Implement details

Experiments were carried out on a workstation with a 3.50 GHz Intel(R) Core(R) i7-5930 CPU, 64GB RAM and an NVIDIA(R) GeForce GTX 1080 GPU. All images were scaled to 256* 32 (width*height) in order to accelerate the training process. Networks were implemented using PyTorch, and trained with RMSProp. The networks were firstly trained on synthetic dataset for 160k iterations (learning rate was set to $10^{-2)}$, then fine-tuned on the real-world SLNs dataset for 10k iterations (learning rate of 10^{-7}).

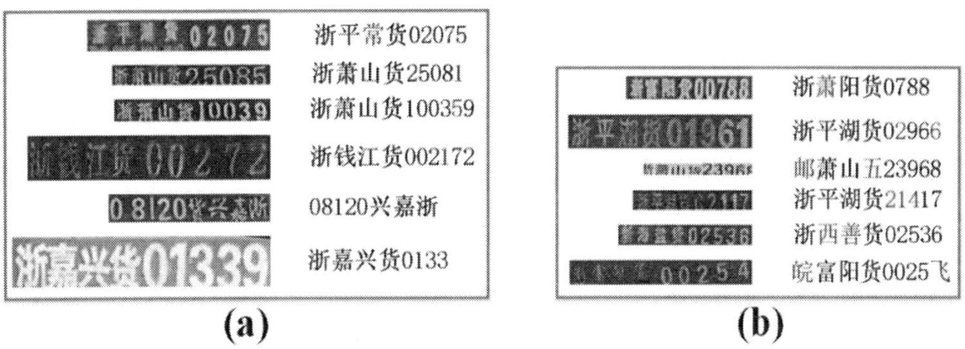

Figure 4. Some recognition results of (a) 1-Levenshtein-distance and (b) 2-Levenshtein-distance. Characters with red color were missed or mistakenly recognized.

3.3 *Experimental results*

For an input SLN image, the corresponding recognition output is a sequence of characters. The Levenshtein distance (Edit distance) between the recognized characters and the ground truth labels are computed to evaluate the recognition accuracy. The smaller the Levenshtein distance is, the more approximate the recognized outputs and the ground truth label are. The Levenshtein distance between a recognized sequence and a ground truth label is equal to 0 represents that the output sequence is totally the same with the corresponding ground truth.

In our experiments, recognition accuracies are computed under the conditions of different Levenshtein distance thresholds. For example, the Levenshtein distance threshold is set to 1 means that, those character sequences whose Levenshtein distances are less than or equal to 1 are regarded as corrections. 1460 in Table 1 means that there are 1460 resulted sequences whose Levenshtein distances are 0.

Table 1. Recognition results of SLNs

Levenshtein distances	0	1	2	Greater than 2
Number of recognized instances	1460	204	118	64
Overall recognition accuracy	79.09%	90.14%	96.53%	100%

Table 1 demonstrates the overall recognition accuracies of the proposed approach on 1846 real-world SLNs images under the Levenshtein distance threshold of 0, 1 and 2. Table 1 suggests that the proposed approach achieved a SLNs recognition accuracy of 79.09% under the condition of 0 Levenshtein distance, a recognition accuracy of 90.14% under the condition of 1 Levenshtein distance and a recognition accuracy of 96.53% under the condition of 2 Levenshtein distance. Table 1 proves the effectiveness of the proposed SLNs recognition framework. Figure 1 shows some 0-Levenshtein-distance recognition results. From Figure 1 one can recognizes the good recognition ability of the proposed approach. Figure 4 shows some 1-Levenshtein-distance and 2-Levenshtein-distance recognition results, from which one can see some characters of SLNs were missed or mistakenly recognized (characters with red color), the reason is that these characters are in lower resolutions or are interfered.

4. Conclusions

In this paper, a SLNs recognition framework have been proposed to recognize multi-style SLNs. CRNN is improved to recognize SLNs. A synthetic SLNs dataset which contains more than 6 million synthetic SLN images were generated. Experimental results on 1846 collected real-world SLNs proves the good SLNs recognition efficiency of the proposed framework.

Acknowledgments

This work was supported partly by National Natural Science Foundation of China (61272304 and 61772374), National Key Research Project "Cloud Computing & Big Data" of China (2017YFB1002600), Ningbo Natural Science Foundation (2017A610108), and Zhejiang Provincial Natural Science Foundation (LY16F020023, LQ18F030010 and LQ17F030002).

References

[1] S. Du, M. Ibrahim, M. Shehata, and W. Badawy, "Automatic license plate recognition (alpr): A state-of-the-art review," *IEEE Trans. Circuits Syst. Video Technol.*, vol. 23, no. 2, pp. 311–325, 2013.

[2] B. Liu, J. Sheng, J. Dun, S. Zhang, Z. Hong, and X. Ye, "Locating various ship license numbers in the wild: An effective approach," *IEEE Intelligent Transportation Systems Magazine*, vol. 9, no. 4, pp.102–117, 2017.

[3] B. Liu, X. Lyu, C. Li, S. Zhang, Z. Hong, and X. Ye, "Using transferred deep model in combination with prior features to localize multi-style ship license numbers in nature scenes," in *IEEE International Conference on TOOLS with Artificial Intelligence*, 2018, pp. 506–510.

[4] B. Liu, S. Zhang, Z. Hong, and X. Ye, "A Horizontal Tilt Correction Method for Ship License Numbers Recognition," in *Journal of Physics: Conf. Series*, vol. 976, pp. 012013, 2018.

[5] Liu, W., Anguelov, D., Erhan, D., Szegedy, C., Reed, S., & Fu, C. Y., et al. "SSD: Single Shot MultiBox Detector," in *European Conference on Computer Vision,* 2016, pp. 21-37.

[6] B. Shi, X. Bai and C. Yao, "An End-to-End Trainable Neural Network for Image-Based Sequence Recognition and Its Application to Scene Text Recognition," in *IEEE Transactions on Pattern Analysis and Machine Intelligence*, vol. 39, no. 11, pp. 2298- 2304, Nov. 1 2017.

[7] Jaderberg M, Simonyan K, Zisserman A, "Spatial transformer networks," in *Advances in Neural Information Processing Systems*. 2015: 2017-2025.
[8] P. He, W. Huang, T. He, Q. Zhu, Y. Qiao, and X. Li, "Single Shot Text Detector with Regional Attention," in *IEEE International Conference on Computer Vision*, 2017, pp. 3066-3074.
[9] S. Hochreiter and J. Schmidhuber. Long short-term memory. *Neural Computation*, 9(8):1735–1780, 1997. 3

IOP Conf. Series: Journal of Physics: Conf. Series **1060** (2018) 012065 doi:10.1088/1742-6596/1060/1/012065

Research on Probability of Detection of Submarine Time Limit Based on Continuous Time within Finite Region

ZHANG Chi*, CHEN Jianyong and TANG Shuai

Address: Navy Submarine Academy. Jinshui Road No. 1, Qingdao, Shandong, China, 266199

*E-mail address: troy100@qq.com

Abstract. The sonar detection of underwater submarines in a certain sea area was modeled by using the sonar in the sea area. The problem of continuous detection time of underwater large random targets was studied based on the fixed detection points. According to the characteristics of the target's motion, this paper puts forward the research method of detecting the probability of finding the time limit of the random moving target by fixed point detection, establishes the random constant velocity target distribution model under the target initial position distribution, uses this model to carry on the duration . The effective probing time and the redundant probing time exist in the probing and continuous probing times. The probabilistic trend of the probing probability under the probing limit time decreases with the increase of the distance between the probing point and the distribution center. The effective probing time length increases with the probing point The distance from the center generally increases, but there is no linear rule.

1. Introduction

In the study of the fixed detection point on the continuous detection of underwater large random targets, the probability model of continuous detection of fixed-point detection was studied in reference [1]-[3], and the researches on the initial distribution under the condition of a circular normal distribution, the time-varying discovery probability and the variation characteristics of integral domain are studied[4][5]. Modeling method for the problem of continuous Detection in a finite area of a moving Target based on Hellman and Foraker[6]-[9]. On the basis of finding probability calculation model in the fixed point finite region of moving target, a random constant velocity target distribution model is established under the initial position distribution of the target, and the model is used in the region. The duration detection is carried out at a detection point in the domain. The duration detection time tends to infinity. A method to calculate the time limit detection probability is proposed, and the limited detection time length included in the duration detection time is analyzed. In addition, the integral domain tends to be a definite finite field, and the integral time of the detection function on the characteristic trace is also limited at points outside the distribution center.

2. Time limit probability

Suppose for a simply connected convex domain Ω ,have $b(\boldsymbol{x},t)>0$,if $\boldsymbol{x}\in\Omega$; have $b(\boldsymbol{x},t)=0$,Then it's called a probe region $\boldsymbol{x}\notin\Omega$. According to the probability model of continuous detection discovery in reference [1], The initial distribution density function of the target position is $\rho[x(t_0),t_0]$, From the time of continuous detection t_0 to the moment of discovery time t, the expression of the discovery probability of continuous detection is as follows:

$$P(t)=\int_{\omega}\rho[x(t_0),t_0][1-\mathrm{e}^{-\int_{t_0}^{t}b(x(s),s)\mathrm{d}s}]\mathrm{d}x(t_0) \tag{1}$$

Where $\omega=\{x(t_0):\int_{t_0}^{t}b[x(s),s]\mathrm{d}s>0\}$, Ω is the Collection of starting points for intersecting feature traces. When $t\to\infty$, ω with an index $\mathrm{e}^{-\int_{t_0}^{t}b(x(s),s)\mathrm{d}s}$ may be changed. A deterministic moving object with velocity mean motion $\bar{v}(x,t)$ in integral domain at the time t_0, During the time $t_0\sim t$, one moment in the exploration zone at least.

With the initial distribution center as the origin, a coordinate system is established. For the diffusion random moving target, The characteristic trace equation $x(t)=G[x(t_0),t]$ obtained from the differential system $\frac{\mathrm{d}x}{\mathrm{d}t}=\bar{v}(x,t)$ and the initial point $x(t_0)$ [10]. If $t>t_0$, then $|x(t)|=|G[x(t_0),t]|>|x(t_0)|$ can be launched, The characteristic trace from $x(t_0)$ always radiates outward away from the origin of the coordinates. As shown in Figure 1 and Figure 2. When $(0,0)\in\Omega$,have $\omega=\Omega$; When $(0,0)\notin\Omega$, have $\Omega\subset\omega$,And for all of the point $x_{\Omega}\in\Omega, x_{\omega-\Omega}\in\omega-\Omega$ on the Characteristic trace,have $|x_{\Omega}|>|x_{\omega-\Omega}|$. So, when $t\to\infty$,the $(0,0)\in\omega$,and ω is a finite field.

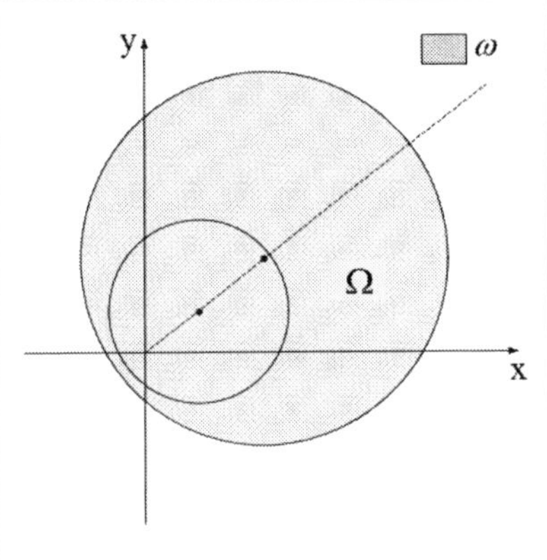

Figure 1. If $(0,0)\in\Omega$, Schematic diagram of ω.

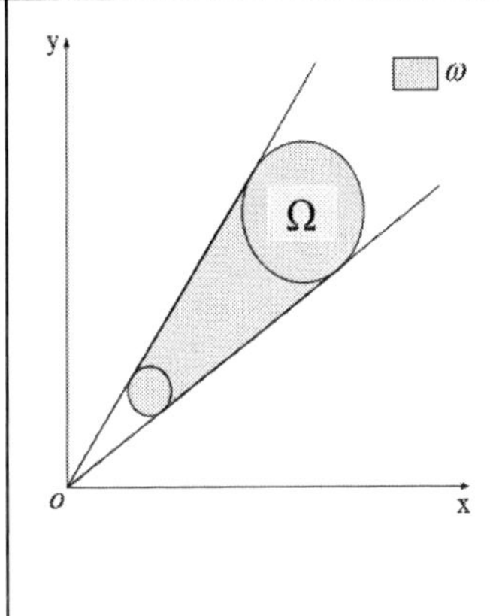

Figure 2. If $(0,0)\notin\Omega$, Schematic diagram of ω.

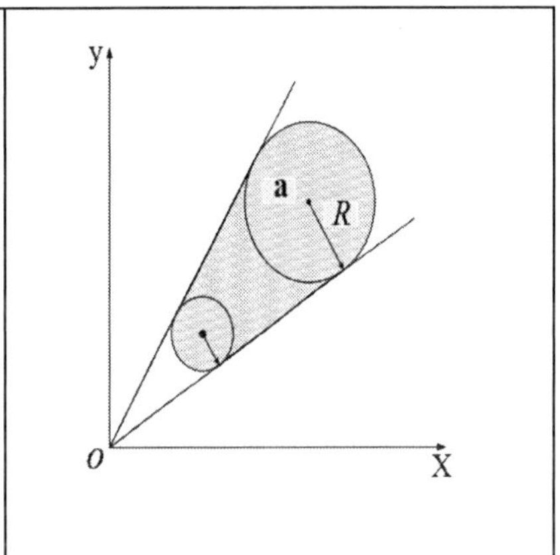

Figure.3 Detection probabilistic integral domain without distribution center ω.

Assuming initial position distribution and velocity distribution are both circular normal distribution [11], where:

$$\rho(x_0)=\frac{1}{2\pi\sigma^2}e^{-\frac{x_0\cdot x_0^{\mathrm{T}}}{2\sigma^2}},\ w(v,t_0)=\frac{1}{2\pi\mu^2}e^{-\frac{v\cdot v^{\mathrm{T}}}{2\mu^2}}. \tag{2}$$

Set time-independent detection rate function: $b(x,t;\mathbf{a})=\begin{cases}B, & |x-\mathbf{a}|\le R\\ 0, & |x-\mathbf{a}|>R\end{cases}$. where B is a constant, The radius of the circular detection field (Ω) is R, where centre point is located at $\mathbf{a}$.

Any time $t>t_0$,The expected value of the spatial conditional distribution of velocity is: $\bar{v}(x)=\frac{\sigma_2^2(t-t_0)^2}{\sigma_1^2+\sigma_2^2(t-t_0)^2}\cdot\frac{x}{t-t_0}$, characteristic line equation is $x(t)=\frac{\sqrt{\sigma_1^2+\sigma_2^2(t-t_0)^2}}{\sigma_1}x_0$.

For circumstances $(0,0)\in\Omega$, Discovery probability integral domain $\omega=\{\boldsymbol{x}_0:|\boldsymbol{x}_0-\mathbf{a}|\le R\}$, according to characteristic line equation is: $\left|\mathbf{a}-\frac{\sqrt{\sigma^2+\mu^2(t_1-t_0)^2}}{\sigma}\boldsymbol{x}_0\right|=R$, The only positive solution can be obtained: $t_1-t_0=\mathrm{F}_1(\boldsymbol{x}_0,\mathbf{a},R)$.

Time limit found probability is:

$$P_\infty=\int_\Omega\rho(\boldsymbol{x}_0)[1-\mathrm{e}^{-\int_{t_0}^{t_1}b(x(s),s)\mathrm{d}s}]\mathrm{d}\boldsymbol{x}_0=\int_\Omega\rho(\boldsymbol{x}_0)[1-\mathrm{e}^{-BF_1(\boldsymbol{x}_0)}]\mathrm{d}\boldsymbol{x}_0 \tag{3}$$

For circumstances $(0,0)\notin\Omega$, As is shown in Figure 3, shadow region is the domain of probabilistic integral of discovery ω. When $t\to\infty$, ω extends along the fan toward the origin to include the origin.

When $|\boldsymbol{x}_0-\mathbf{a}|\le R$, then $\boldsymbol{x}_0\in\Omega$, There is also a unique positive solution: $t_1-t_0=\mathrm{F}_1(\boldsymbol{x}_0,\mathbf{a},R)$.

When $\boldsymbol{x}_0\in\omega-\Omega$, then $x_0\in\omega$ and $|\boldsymbol{x}_0-\mathbf{a}|>R$, with $\left|\mathbf{a}-\frac{\sqrt{\sigma^2+\mu^2(t_T-t_0)^2}}{\sigma}x_0\right|=R$, then Obtainable two positive solutions of t_T-t_0 is : $\mathrm{F}_M(\boldsymbol{x}_0,\mathbf{a},R)>\mathrm{F}_m(\boldsymbol{x}_0,\mathbf{a},R)$.

Take $t_3-t_2=\mathrm{F}_M(\boldsymbol{x}_0,\mathbf{a},R)-\mathrm{F}_m(\boldsymbol{x}_0,\mathbf{a},R)$, Then the discovery probability of time limit is:

$$P_\infty=\int_\Omega\rho(\boldsymbol{x}_0)[1-\mathrm{e}^{-BF_1(x_0,\mathbf{a},R)}]\mathrm{d}\boldsymbol{x}_0+\int_{\omega-\Omega}\rho(\boldsymbol{x}_0)[1-\mathrm{e}^{-B[\mathrm{F}_M(x_0,\mathbf{a},R)-\mathrm{F}_m(x_0,\mathbf{a},R)]}]\mathrm{d}\boldsymbol{x}_0 \tag{4}$$

The analysis of the solution process of the concrete problem is given by using the concrete expression.

First of all, let the calculation and analysis of the probability of time limit be divided into two cases, and take the limit of time $t\to\infty$ [12].

First case:If $\omega_u=\Omega$, origin in the detection domain, where point $(0,0)\in\Omega$. The detection probability is:

$$\begin{aligned}P_\infty&=\int_\omega\rho(\boldsymbol{x}_0)[1-\mathrm{e}^{-\int_{t_0}^{\infty}b(x(s),s)\mathrm{d}s}]\mathrm{d}\boldsymbol{x}_0=\int_\Omega\rho(\boldsymbol{x}_0)[1-\mathrm{e}^{-\int_{t_0}^{\infty}b(x(s),s)\mathrm{d}s}]\mathrm{d}\boldsymbol{x}_0\\&=\int_\Omega\rho(\boldsymbol{x}_0)[1-\mathrm{e}^{-\int_{t_0}^{t_1}b(x(s),s)\mathrm{d}s}]\mathrm{d}\boldsymbol{x}_0\end{aligned} \tag{5}$$

Because of $1-\mathrm{e}^{-\int_{t_0}^{\infty}b(x(s),s)\mathrm{d}s}<1$, probe $P_\infty<\int_\Omega\rho(\boldsymbol{x}_0)\mathrm{d}\boldsymbol{x}_0$, it indicates that an infinitely long probe in an area containing a distribution center has a probability of finding that is less than the probability that the target exists in the probe domain at the initial time.

$$\begin{aligned}P_\infty&=\int_\Omega\rho(\boldsymbol{x}_0)[1-\mathrm{e}^{-\left[\int_{t_0}^{t_1}B\mathrm{d}s+\int_{t1}^{\infty}0\mathrm{d}s\right]}]\mathrm{d}\boldsymbol{x}(t_0)\\&=\int_\Omega\rho(\boldsymbol{x}_0)[1-\mathrm{e}^{-B(t_1-t_0)}]\mathrm{d}\boldsymbol{x}(t_0)\end{aligned} \tag{6}$$

Where t_1 is the time of the characteristic line leaves Ω, the one point $\boldsymbol{x}_0$ is on it .

To calculate the probability of detection of limit time, we first need to find out the intersection point x_1 of the point along the feature curve x_0 with the detection area at time t_0, the circle equation of the joint detection area and the characteristic equation of the line is:

$$\begin{cases}(x-a)^2+(y-b)^2=R^2\\ y=\dfrac{y_0}{x_0}x\end{cases} \tag{7}$$

Get the intersection point x_1, x_2:

$$x_{1,2} = \frac{ax_0 + by_0 \pm \sqrt{(x_0{}^2 + y_0{}^2)R^2 - (ay_0 - bx_0)^2}}{x_0{}^2 + y_0{}^2} x_0 \tag{8}$$

Because in the trace equation:

$x(t) = \frac{\sqrt{\sigma_1{}^2 + \sigma_2{}^2(t-t_0)^2}}{\sigma_1} x(t_0)$, $\frac{\sqrt{\sigma_1{}^2 + \sigma_2{}^2(t-t_0)^2}}{\sigma_1} \geq 1$, Positive and negative solutions for coefficients in eq. (8), put Negative root, Preserve only the positive root:

$$\begin{cases} x_1 = \dfrac{ax_0 + by_0 + \sqrt{(x_0{}^2 + y_0{}^2)R^2 - (ay_0 - bx_0)^2}}{x_0{}^2 + y_0{}^2} x_0 \\ y_1 = \dfrac{ax_0 + by_0 + \sqrt{(x_0{}^2 + y_0{}^2)R^2 - (ay_0 - bx_0)^2}}{x_0{}^2 + y_0{}^2} y_0 \end{cases} \tag{9}$$

Make $\gamma(x_0, y_0) = \frac{ax_0 + by_0 + \sqrt{(x_0{}^2 + y_0{}^2)R^2 - (ay_0 - bx_0)^2}}{x_0{}^2 + y_0{}^2}$, By this definitions eq. (9) can be rewritten in the following form:

$$\gamma(x_0, y_0) = \frac{\sqrt{\sigma_1{}^2 + \sigma_2{}^2(t-t_0)^2}}{\sigma_1} \tag{10}$$

The result is: $t_1 - t_0 = \frac{\sigma_1\sqrt{\gamma^2(x_0, y_0) - 1}}{\sigma_2}$,

$$P_\infty = \iint_\Omega \rho(x_0, y_0)[1 - \mathrm{e}^{-\frac{B\sigma_1\sqrt{\gamma^2(x_0,y_0)-1}}{\sigma_2}}]dx_0 dy_0 \tag{11}$$

The second case: $w_u \neq \Omega$, The origin is detecting beyond the border, $(0,0) \notin \Omega$, The detection probability at the limit time is:

$$\begin{aligned} P_\infty &= \int_\omega \rho(\boldsymbol{x}_0)[1 - \mathrm{e}^{-\int_{t_0}^{\infty} b(x(s),s)\mathrm{d}s}]\mathrm{d}\,\boldsymbol{x}_0 \\ &= \int_\Omega \rho(\boldsymbol{x}_0)[1 - \mathrm{e}^{-\left[\int_{t_0}^{t_1} b(x(s),s)\mathrm{d}s + \int_{t1}^{\infty} 0\mathrm{d}s\right]}]\mathrm{d}\,\boldsymbol{x}_0 + \int_{\omega-\Omega} \rho(\boldsymbol{x}_0)[1 - \mathrm{e}^{-\left[\int_{t_0}^{t_2} 0\mathrm{d}s + \int_{t_2}^{t_3} b(x(s),s)\mathrm{d}s + \int_{t_3}^{\infty} 0\mathrm{d}s\right]}]\mathrm{d}\,\boldsymbol{x}_0 \\ &= \int_\Omega \rho(\boldsymbol{x}_0)[1 - \mathrm{e}^{-\int_{t_0}^{t_1} b(x(s),s)\mathrm{d}s}]\mathrm{d}\,\boldsymbol{x}_0 + \int_{\omega-\Omega} \rho(\boldsymbol{x}_0)[1 - \mathrm{e}^{-\int_{t_2}^{t_3} b(x(s),s)\mathrm{d}s}]\mathrm{d}\,\boldsymbol{x}_0 \\ &= P_{\infty\Omega} + P_{\infty(\omega_k - \Omega)} \end{aligned} \tag{12}$$

Where the time t_1 in $P_{\infty\Omega}$ account form is the same way as in the first case.

In the eq. (12), t_1 is the time of the characteristic line leave Ω,the one point $\boldsymbol{x}_0$ is in Ω.t_2 is the time of the characteristic line get into Ω, which one point $\boldsymbol{x}_0$ is in $\omega - \Omega$. t_3.is the time of the characteristic line leave Ω,which one point $\boldsymbol{x}_0$ is in $\omega - \Omega$. From the characteristic line equation and the boundary equation of Ω, we can know, t_1, t_2 and t_3 are functions of $\boldsymbol{x}(t_0)$.

3. Example analysis

Through the above discussion of the time limit probabilities of continuous detection of random moving targets, the proper detection function, simplified probabilistic model and target distribution model of submarine are selected and analyzed[13].

The following gives specific values to explore the detection of continuous detection of the time limit probabilities of changes with the detection of location.

Example 1:Hypothesized normal distribution of circles $\sigma_1 = 8nm$, $\sigma_2 = 8kn$ .Probe radius $R = 6nm$, Detection rate constant $B = \frac{1}{6s}$. Let the distance from the center of the probe to the origin of the coordinates $|\mathbf{a}|$ change from 0 to 50 n-mile, Figure 4 shows the time limit discovery probability for continuous detection.

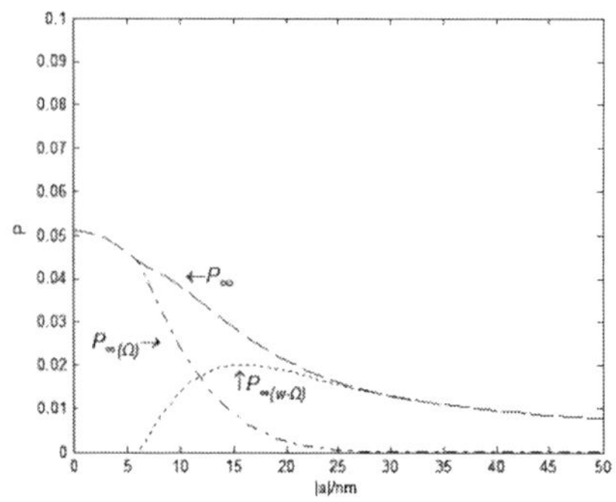

Figure. 4 The relationship between the discovery probability of time limit and the location of the probe point.

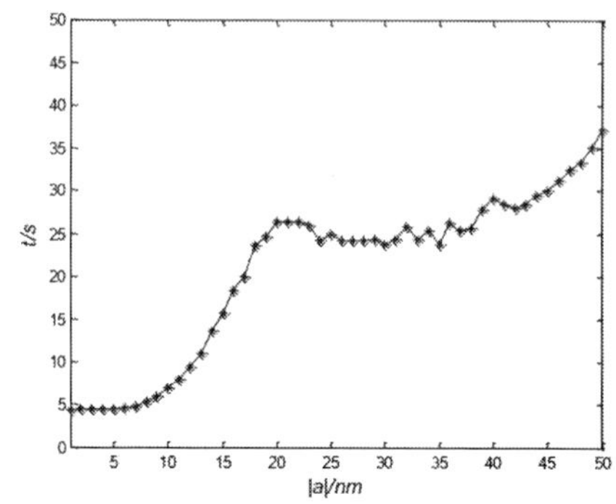

Figure. 5 The detection time of 0.95 times the limit probability.

$P_{\infty(\Omega)}$ is the calculation result of the first integral eq. (12) shown in Figure 4, Represents the time limit probability that a target in the probe domain was discovered at the time of detection, which is the result of the second integral formula in the eq. (12), Represents the time limit probability that a target outside the detection area enters and is discovered at the time of detection. Because the moving target is a diffusive distribution. When the detection region includes a diffusion center. That is, the distance between the center of the circle and the center of diffusion is not great-er than $6nm$,which can be saw from Figure 4. discovery probability does not exist the probe $P_{\infty(w-\Omega)}$. With the increase of $|\mathbf{a}|$, $P_{\infty(\Omega)}$ Continuous decline, And there is a maximum in $P_{\infty(\omega-\Omega)}$. The proportion of $P_{\infty(\omega-\Omega)}$ components gradually increased in time limit discovery probability even when $|\mathbf{a}|$ is coming bigger. Contribution of $P_{\infty\Omega}$ in the probability of time limit discovery may be passed.

Example2:Study on the relationship between the effective detection time length and the distance between the probe point and the center. Make $t_0 = 0$, defined proportional coefficient [14] $\eta(t, \mathbf{a}) = \frac{P(t, \mathbf{a})}{P_\infty(\mathbf{a})}$. For a given value,time t is a function of $|\mathbf{a}|$.

The calculation condition is the same as that of the example 1. Calculating the probability of continuous detection of time varying $P(t, \mathbf{a})$. Set ratio coefficient value η =0.95, $|\mathbf{a}|$ change from 0 to $50n\text{-}mile$, the calculated time values are shown in figure 5. The result of the calculation clearly shows, Detection time for a fixed ratio of discovery probability reaching the near time limit,when the detection domain comprises a distribution center or near the distribution center , almost invariable and when the probe is away from the center of the distribution, the detection time increases significantly with the increase of $|\mathbf{a}|$.

It can be seen from the results of the example that the probability of finding the time limit of continuous detection decreases with the increase of the distance between the detection point and the distribution center, and the integral value in the non-detection region shows a peak value of probability. When the detection limit of 95% of the time is reached, the detection time tends to increase with the increase of the distance between the detection point and the distribution center, which $|a|$ varies from about 20 to 40*n-mile* fluctuations in detection time, The fluctuation of the detection time is due to the change of the time limit probability and the non-synchronization of the temporal change of the discovery probability with the change of $|a|$.

4. Conclusions

The probability of detecting the time limit of random moving targets by fixed point detection is discussed. And the method of numerical integration is used to calculate the example. The research results show that there are effective detection time and redundant detection time in the continuous detection time. The simulation results of the discovery probability under the limit time show that the detection probability of the continuous detection increases with the distance between the detection point and the distribution center And decreases, in which the integral value in the non-probe domain shows a probability peak. From Case 2, it can be concluded that the effective detection time tends to increase as the distance between the detection point and the center increases, but there is no linear rule.

References

[1] CHEN Jianyong and ZHANG Chi 2015 *Search Equation of Moving Target with Random Constant Velocity and the Probability of Persistence Detecting (Journal of Naval Aeronautical and Astronautical University* vol 30*)* chapter 6 pp 522-525
[2] CHEN Jianyong,QU Xiaohui and WANG Jian. 2014 *Probability increment with detection duration in limited space for random moving targets (Systems Engineering and Electronics* vol 36*)* chapter 6 pp 1039-1043
[3] CHEN Jianyong, SHAN Zhichao and ZHANG Chi 2015 *Appr-oximate Probability Model of Area Persistence Detection for A Random Moving Target (Journal of Naval Aeronautical and Astronautical University* vol 30*)*.chapter 3 pp 201-205
[4] ZHANG Chi and CHEN Jianyong 2016 *Optimize Algorithm of the Maximum Probability of Detection Points in Limited Detection Area (Journal of Sichuan Ordnance* vol 37*)* chapter 4 pp 118-120
[5] CHEN Jianyong,WANG Jian and SHAN Zhichao 2013 *Optimal search algorithm for constant velocity random moving target detecting in discrete time (Systems Engineering and Electronics* vol 35*)* chapter 8 pp 1627-1630
[6] Hellman O 1970 *On the Effect of a Search upon the Probability Distribution of a Target whose Motion is a Diffusion Process (The Annals of Mathematical Statistics* vol 41*)* pp 1717- 1724
[7] Hellman O. 1971 *Optimal Search for a Randomly Moving Object in a Special Cases (Journal of Applied Probability* vol 8*)* pp 606-611
[8] Hellman O. 1972 *On the Optimal Search for a Randomly Moving Target (SIAM Journal of Applied Mathematics* vol 22*)* chapter 4 pp 545-552
[9] Foraker III J C 2011 *Optimal Search for Moving Targets in Continuous Time and Space Using Consistent Approximations*(Monterey California: Naval Postgraduate School)
[10] CHEN Jianyong,SHAN Zhichao and WANG Jian 2014 *Proba-bility Optimality of Area Persistence Detection for a Random Moving Target (Fire Control and Command Control* vol 39*)* chapter 6 pp 9-12
[11] CHEN jianyong,LENG jiang and YU chuan-jian 2004 *The Detection Probabilities Of on-call Helicopter Antisubmarine Search Using Dipping Sonar (Journal of Naval Aeronautical and Astronautical University* vol 19*)* chapter 5 pp 559-561
[12] XIE kexin,HAN jian and LIN you-man 2004 *Optimal method* (Tianjin:Tianjin University Press) pp 94-96
[13] XIAO bin and XU hongfei 1998 *Solution and Convergence Analysis of Search Optimal Configuration (Fire Control & Command Control* vol 23*)* chapter 4 pp 36-39
[14] LI Chagming and YANG Jian *Analysis of quasi-optimal incremental search efficiency (Fire Control & Command Control* vol 25*)* chapter 4 pp 49-51

Dynamic Relationship between Chinese FDI, Agricultural and Economic Growth in West African: An Application of the Pool Mean Group Model

Kankou Hadia Fofana[1,*], Enjun Xia[2] and Mamadou Bado Traore[3]

[1]PhD student, School of Management and Economics Beijing Institute of Technology
[2]PhD Professor School of Management and Economics Beijing Institute of technology
[3]PhD student, Food Technology Beijing Forestry University

*fofanakankouhadia@yahoo.fr

Abstract. This study is an attempt to test the long run relationship between Chinese foreign direct investment (FDI), agricultural and economic growth in host countries is known to have an important role in economic literature suffering from unemployment problems, food security and lack of technological progress. This paper examines this issue for West Africa by applying Pool Mean Group (PMG) and panel-Granger causality Models over the period of 2003 to 2015.The Pool Mean Group tests suggest cointegration between China FDI, economic growth, domestic investment and land agricultural used The article indicates that Chinese FDI, domestic investment and land agricultural spur economic growth contrary to some studies, which found that China FDI does not cause economic growth. The results also show that there is no significant Panel-VECM Granger causality from China FDI to economic growth, from economic growth to Chinese FDI, from agricultural to economic growth and from economic growth to agricultural. This implies that the increment in Chinese FDI inflows would definitely lead to increasing the economic growth, domestic investment and agricultural land in West Africa.

1. Introduction

The most important issue in this economic research is how to accelerate West Africa's economic growth; since this growth can have many impacts on its development Indeed, the agricultural sector and FDI inflows have been widely recognized as very important factors in the economic growth process. The agricultural has played a crucial role for raises revenues, generates employment opportunities, the achievement of national objectives as regards food security, regional equilibrium, and social cohesion. The[1]observed that the important effect of agricultural development not only enhances the agricultural sector but also triggers overall economic growth. Recently, it is generating about 35% of the total GDP and 65% of the employment opportunities of the West African.The impact of Agricultural on economic growth can be positive and significant mainly due to the accumulation of employment as much as to earn export earnings and malnutrition. Inward Chinese FDI is of crucial importance for the reorganization of agricultural sector in the host country by offer of funds. Moreover, China FDI can increase African's export capacity and augment its foreign exchange earnings. The researchers of the majority of past empirical studies dealt with both Chinese FDI inflows and trade interaction on economic growth[2] or the relationship between China FDI and economic growth [3] and the relationship between agricultural and economic growth[4].Completely these studies concluded that both Chinese FDI inflows and agricultural promote economic growth. However, these studies argue that the impact of China FDI by multinational enterprises (MNEs) on growth is complex in particular in many developing countries. The growth enhancing effects from FDI inflows and agriculture vary from country to country and over time. For some countries, agricultural sector and FDI can even affect economic growth negatively[5].The study is also relevant because West Africa is an interesting case study for other Africa countries which have applied incentive programs in order to attract China FDI.

The rest of the paper is structured as follows; Section 2 presents a theoretical and empirical literature review. Section 3 describes the model and the data, while Section 4 deals the methodology and the empirical analysis of the results. Conclusion and accompanying recommendations are presented in the final section (section 5).

2. Review of the literature

Over the past decade, a number of studies have found how Chinese FDI inflows can contribute to the growth of a host country's economy[2],[3]. In general, these studies show that the impact of China FDI by trade, investments and aid on growth is complex.

Regarding the empirical economic literature existing studying China FDI-led growth have found that china FDI promotion can greatly benefit host countries through the introduction of new technologies and skills, new jobs, surging domestic competition, foreign exchange earnings, and expanding access to international marketing networks.[2]found that china FDI may affect growth positively due to concentration of sectors in the Manufacturing, Building & Construction and General Trade of the economy of Ghana for the past five years. Relying on a sampling technique of 30 persons regressions for testing the impact of china FDI on per capita GDP in Kenya,[6] found that China FDI promotes economic growth by through human capital development, employment, and capital supply.By estimating a panel data model that allows for study the Chinese FDI flows to Africa between 2003 and 2008,[7] found that the causal relationship between African GDP and Chinese FDI was bi-directional, while uni-directional relationships were established between Chinese FDI and African infrastructure and corruption, respectively. They justified these results by the domestic investment, market size, agricultural potential and oil are important and significant determinants of Chinese FDI.For example, by using time series data and Johansen and Juselius multivariate Cointegration and Granger causality model,[8]evaluated the effect of China FDI, trade and its effects on agricultural sector development in Nigeria for 1980-2009. The article indicates that both foreign direct investment and exports spur economic growth contrary to some studies, which found that FDI does not cause economic growth. The article found that Chinese FDI and trade spur Nigeria' agricultural sector, which found that Labor Force and China FDI inflow into Nigeria, has no causal relationship with agricultural output. Also the result shows that Labor Force and China foreign direct investment inflow into Nigeria has no causal relationship with agricultural output.[9]studied the effect of Nigeria's industry process on economic growth by investigating a vector-regressive model relationship between industry, China FDI and economic growth during the period 1988–2008 on an Annual time series data. They found that there is evidence that more Chinese investments are highly needed in the Nigerian textile industry to make it internationally competitive.[10] investigated the quantitative effect of Chinese FDI on economic growth in Africa using a panel data model with fixed effects for a sample of 20 African countries from 2003 to 2012. They showed that china FDI increases Africa's gross domestic product (GDP) growth. Nevertheless, Granger Causality results in Africa suggested that unidirectional causality between GDP growth and China FDI. Over the past decade, studies examined the impacts of agricultural sector and China FDI on economic growth within the same model has also delivered ambiguous results.[11]found that the impact of China investment on economic growth is positive while there is an increase of FDI through manufacturing and service sector in Kenya. This result implies that China FDI appears to bring about close to a one-for-one increase loans to Kenya for hospital and schools construction in less developed areas and the impact of loans fall below those of the other flows.

3. Data sources and description of variables

3.1 Data sources and Model construction

The main objective of the present paper is to investigate the annual time series data on economic growth, Chinese FDI, agricultural sector, labor and capital stock covering the 2003 – 2015 period have been used in this study. The data of all variable was used and collected from World Development Indicator (WDI)(http://www.worldbank.org),except China foreign direct investment in West Africa (CFDI) variable was accumulated from ASE (www.pairault.fr/sinaf/index).

$$lnCFDI_t = \beta_1 + \delta_2 lnCFDI_{t-1} + \delta_3 lnY_{t-1} + \delta_4 lnAgri_{t-1} + \delta_5 lnLab_{t-1} + \delta_6 lnDI_{t-1} + \delta_7 lnLAND_{t-1} + \sum_{i=1}^{p} a_1 lnY_{t-1} + \sum_{i=1}^{p} \alpha_2 lnCFDI_{t-1} + \sum_{i=1}^{p} \alpha_3 lnAgri_{t-1} + \sum_{i=1}^{p} \alpha_4 lnLab_{t-1} + \sum_{i=1}^{p} \alpha_5 lnDI_{t-1} + \sum_{i=1}^{p} \alpha_6 lnLAND_{t-1} + \varepsilon_t \quad (1)$$

Economic growth is measured by the increase of real GDP per capita in each successive time period. The proxy real GDP per capita is noted by (Y.) It is expressed in constant 2000 dollar America (us). FDI is the value of real gross foreign direct investment inflows to GDP ratio. China FDI is the prime source of human capital and new technology to acquire a lasting management interest in an enterprise operating in an economy other than that of the investor. Agricultural sector (Agic) is the total sum of major crops, minor crops, livestock, fishing and forestry divided by GDP; Lab is measured as the volume of the total labour force; capital investment (DI) is measured by the real value of gross fixed capital formation (GFCF constant 2000 us) For the construction of the CFDI, conventional neo-classical production function was used to test the impact of FDI on growth, but added foreign direct investment as an additional variable because FDI is the prime source of human capital and new technology.

3.2. Methodology specification

Several steps of the methodology were used, to examine the relationship between Chinese FDI, agricultural sector and economic growth in West Africa. First, we use the conventional augmented Fisher Augmented Dickey–Fuller (ADF) and the Phillips–Perron test following[12](ADF), and[13][to detect the level of stationarity either at I(0) or I(1) therefore is an rule for ARDL bounds test. In econometric analysis, stationarity tests are effectuated to determine the stationarity of variables. Second, we investigate the long-run relationships and short-run dynamic interactions among the variables. After, we apply the autoregressive distributed lag (ARDL) cointegration technique, which is used to capture cointegrating non-stationary time series. For so doing, the ARDL modelling approach was used for first by[14] and extended by[15]. However, ARDL model has advantages over the conventional methods of [16, 17].

4. Results and discussion

Table1 presents the descriptive statistics results of the variables used in the current paper. According to the Jarque– Bera test, these statistics reveal that all the series such as (agricultural sector, Chinese FDI, labor, capital investment, exchange rate and economic growth) have normal distributions.

4.1. Stationary and co-integration tests

Table 2 shows the results of Fisher ADF and P.P unit root tests indicate that all the variables set are integrated of I(0) or I(1) (see Table 2). The results clearly indicated that the H0 of unit root tests are rejected at 1%, 5% and 10% significance levels. Therefore, if all the variables are stationary at I (0), I (1) or both, the bounds testing approach would be used and are the major advantages of using the ARDL estimators. Generally, none of the data series are I (2) or above. Table 3 shows the results pointed out there are co-integration relationships between the variables in West Africa. The result shows that the H0 of no co-integration among all variables in the models are rejected at 5%

significance level, evidencing a variables strong correlation in long run. of sections to divide the text of the paper is optional and left as a decision for the author. Where the author wishes to divide the paper into sections the formatting shown in table 2 should be used.

Table 1 Descriptive statistics and correlation matrix results

Variable	$LAND_t$	Lab_t	$CFDI_t$	DI_t	Y_t	$AGRI_t$
Mean	1.4367	1.6523	6.4745	21.5824	1.2834	1.2925
Median	1.4464	1.6888	6.5824	2.2477	1.3000	1.4424
Maximum	1.7626	1.9000	7.7423	390.35	1.9999	2.0110
Minimum	0.8849	1.2545	5.2333	-195.94	0.0000	-1.6923
Std. dev.	0.1926	1.1656	0.5435	58.83	0.3145	0.5723
$LAND_t$	1000					
Lab_t	0.5507	1000				
$CFDI_t$	0.3214	0.5812	1000			
DI_t	- 0.2567	-0.1003	-0.0370	1000		
Y_t	-0.1204	-0.0758	-0.0698	0.2542	1000	
$AGRI_t$	0.4621	0.3035	0.1251	-0.2260	-0.5871	1000

Table 2: Pedroni and Kao panel cointegration tests

Test	Intercept	Intercept and trend	None	Kao
$Y_t, K_t, Lab_t, FDI_t, EXP_t, EXR_t,$				
Panel–v	-3.2645	-4.3734	-3.7220	0.0002
Panel–rho	4.2432	5.9467	3.6689	
Panel–PP	-3.5023	-5.4898	-5.6024	
Panel–ADF	-2.6034	-2.1332	-7.0110	
Group–rho	5.1723	6.3432	4.5445	
Group–PP	-14.5243	-17.64 43	13.2723	
Group–ADF	-8.5034	-8.0222	-8.9220	

Note: *, **, *** indicate 10%, 5% and 1% level of significance.

4.2. Long-run relationship analyses

Table 4 presents the estimated coefficients of the long-run relationship report that there is a positive relationship between China FDI, economic growth, agricultural land and domestic investment at 1% significance level, which means that an increase of Chinese FDI inflows would definitely lead to increasing the economic growth, agricultural land and domestic investment as well in the West Africa economy. In the endogenous growth model theory reported FDI inflow has positive spillover effects on the scarce economy through capital accumulation, technological advance and management skills. .[18, 19]. Moreover, the results is similar with the recent empirical studies, for example ,[20]for Vietnam; [21]for Nigeria; which confirms the FDI-led growth hypothesis. It shows a significance long-run relationship among the variables at 1% and 10% levels. On the contrary, the relationship between China FDI and real agricultural GDP is negative at 1% significance level. Even though there is a widespread belief that FDI to positively impact agriculture for the host country, our empirical results fail to confirm this belief for the case of West Africa .Furthermore, several studies argued that FDI to positively impact agriculture for the host country[22], while[23]show a negative relationship between FDI and agricultural sector. Thus, the Low FDI is found to be responsible for the negative economic implications in countries with very fast growing population[24].Therefore, the results recorded that the relationship between economic growth and the labor force is positive at 1% significance level, means that the rising of economic growth affects labor force positively. These results are consistent with the findings of[25].Furthermore, the results recorded corrected relationships between domestic investment and real agricultural GDP at 1% significance level. Also, these results are consistent with the findings by [24, 26],who hypothesized a similar significant relationship through the effects of domestic investment variables on real agricultural GDP.

Table 3. Long-run ARLD result

Variable	Coefficient	SE	T-ratio	Prob
Constant	29.8324	6.2619	4.7641	0.000
$AGRI_t$	-0.905	0.858	-1.0544	0.294
lab_t	-12.26	1.580	-7.761	0.000
$LAND_t$	9.425	1.884	5.001	0.000
DI_t	1.724	0.379	4.548	0.000
Y_t	4.724	0.935	4.566	0.000
Diagnostic test statistics (p-values)				
$\chi 2$ (normality)			0.2234	
$\chi 2$ (heteroscedasticity)			0.8134	
$\chi 2$ (functional form)			0.1902	
$\chi 2$ (serial correlation)			0.6345	

4.3. Short-run and ECMt − 1 analyses

After the long run estimates, the short-run ECMt − 1 model in (4) is estimated and the results are presented in Table 4. These results indicated that at 1% significance levels, the agricultural land variables, real GDP per capita and capital are positively associated with Chinese FDIt model. In addition, the coefficients of ECMt − 1 are significant with appropriate signs for all models implies a relative speed of achieving the long-run equilibrium. Specifically, Chinese FDI, real GDP per capita and agricultural models reported the highest ECMt − 1 coefficient in absolute value among other models with−56%, −25% and−70% respectively. This implies that these models are corrected from the short-run towards the long-run equilibrium by 56%, 25% and 70%. Also, this means that the long-run would be shortly corrected back by 12 months for the China FDI, agricultural and real GDP per capita models.

Table 4. Short-run and ECMt – 1 ARLD result

Variable	Coefficient	SE	T-ratio	Prob
$\Delta AGRI_t$	-2.2160	3.2487	-0.066	0.9471
ΔLab_t	69.6389	42.009	1.657	0.100
$\Delta LAND_t$	86.004	83.894	1.025	0.308
ΔDI_t	-1.784	7.594	0.378	0.705
ΔY_t	2..876	38.3333	-0.4128	0.6802
ECM_{t-1}	-0.5693	0.1121	-5.074	0.0000

4.4. Granger causality analyses

The results further indicated that there is no significant Granger causality from Chinese FDI to economic growth, from economic growth to China FDI, from agriculture to economic growth and from economic growth to agriculture. However, these results are consistent with the findings of[27]for Tunisia;[28]for China;[29]for Ghana.

5. Conclusion and Policy Implications

The current study examines the dynamic causal relationship among the series of economic growth, Chinese foreign direct investment, agricultural , land agricultural used, labor, and capital investment in west Africa for the period of 2003– 2015. The Pool Mean Group (PMG) test suggested that there is cointegration among the variables specified in the model when FDI is the dependent variable. In addition, the associated equilibrium corrections are significant with appropriate signs for all models confirming the existence of long-run relationships. The equilibrium correction is fairly fast and is restored by the first quarter of the year. The results indicate that Land agricultural, capital investment and economic growth are important in promoting foreign direct investment in West Africa in the long run. Finally, the Panel-VECM causality indicates that there is no significant Granger causality from china FDI to economic growth or from economic growth to china FDI in the short run. The study detected that there is no significant Granger causality from Chinese FDI to economic growth, from economic growth to China FDI, from agriculture to economic growth and from economic growth to agriculture. The topic merits special attention due to the possible interrelations among the series with implications for economic growth.

Even though there is a wrong impression about Chinese FDI can generate positive productivity externalities for the host country, our results confirm the China FDI-led growth this belief for the case of West Africa. Chinese FDI and Domestic capital investment is the main driver of agricultural growth in West Africa. From the results, a policy can generate important implications and recommendations for policymakers in West Africa. It has to improve its attraction of China FDI through a series of more ambitious structural policies (regional integration such as activating the ECOWAS, WAEMU Union, reforming its educational and financial system, and developing its infrastructure).

The results current of this paper could be very interesting for many developing countries to learn the lesson that the attraction of china FDI is important to promote economic growth, but not sufficient. One may wonder about the effectiveness of china FDI to promote economic growth, which depends on the level of FDI, its nature, the sectors of investment, etc. The positive FDI effect which is consistent with past studies, confirms the agricultural sector Chinese FDI dominance which does generate direct growth impacts on the wider economy. Attracting export-oriented China FDI into the food industry and mining sector of the economy of West Africa is of paramount importance. Unfortunately, the major handicap here is the availability of data, which are still lacking.

Acknowledgments
The authors would like to thank the China Scholarship Council (CSC) for their technical and financial supports of this research. Hospitality of the Department of Management and Economics is gratefully acknowledged.

References
[1] World Bank, "Annual Report, World Bank, Washington, DC and New York, NY," 2009.
[2] D. Tang and K. B. Gyasi, "China – Africa Foreign Trade Policies: the Impact of China's Foreign Direct Investment (FDI) Flow on Employment of Ghana," *Energy Procedia,* vol. 16, Part A, pp. 553-557, // 2012.
[3] A. G. Ross, "An empirical analysis of Chinese outward foreign direct investment in Africa," *Journal of Chinese Economic and Foreign Trade Studies,* vol. 8, pp. 4-19, 2015.
[4] H. Eddine Chebbi, "Agriculture and economic growth in Tunisia," *China Agricultural Economic Review,* vol. 2, pp. 63-78, 2010.
[5] S. K. Frimpong, "Research on relationship between China and Ghana: Trade and foreign direct investment (FDI)," *Journal of Economics and Sustainable Development,* vol. 3, pp. 51-61, 2012.
[6] R. N. Gathaiya, J. Kinyua, P. Machuki, and V. Keraro, "The Impact of Foreign Direct Investments by Chinese Companies in Kenya," *International Journal of Management & Business Studies,* vol. 4, pp. 9-17, 2014.
[7] C. Claassen, E. Loots, and H. Bezuidenhout, "Chinese Foreign Direct Investment in Africa," *African Journal of Business Management,* vol. 6, pp. 11583-11597, 2012.
[8] O.-O. Izuchukwu, H. Huiping, A. S. Abubakar, and E. A. Olufemi, "FDI Trade and Its Effects on Agricultural Development in Nigeria: Evidence From Time Series Analysis," *International Journal of Management Science and Business Administration,* vol. 1, pp. 28-40, 2014.
[9] M. A. Eneji, I. J. Onyinye, D. N. Kennedy, and S. L. Rong, "Impact of foreign trade and investment on Nigeria's textile industry: The case of China," *Journal of African Studies and Development,* vol. 4, p. 130, 2012.
[10] I. Doku, J. Akuma, and J. Owusuafriyie, "Effect of Chinese foreign direct investment on economic growth in Africa," *Journal of Chinese Economic & Foreign Trade Studies,* pp. 00-00, 2017.

[11] K. M. B. Ombaba, P. A. Arogo, P. Bii, and L. Ongeri, "A Study on the Impact of China's Investments in Africa: The Case of Kenya," *Journal of Emerging Trends in Economics and Management Sciences,* vol. 3, p. 529, 2012.
[12] P. C. Phillips and P. Perron, "Testing for a unit root in time series regression," *Biometrika,* pp. 335-346, 1988.
[13] D. A. Dickey and W. A. Fuller, "Likelihood ratio statistics for autoregressive time series with a unit root," *Econometrica: Journal of the Econometric Society,* pp. 1057-1072, 1981.
[14] D. F. Hendry, *Dynamic econometrics*: Oxford University Press on Demand, 1995.
[15] M. H. Pesaran, Y. Shin, and R. J. Smith, "Bounds testing approaches to the analysis of level relationships," *Journal of applied econometrics,* vol. 16, pp. 289-326, 2001.
[16] C. W. J. Granger, "Some recent development in a concept of causality," *Journal of Econometrics,* vol. 39, pp. 199-211, 1988/09/01 1988.
[17] S. Johansen and K. Juselius, "Maximum likelihood estimation and inference on cointegration—with applications to the demand for money," *Oxford Bulletin of Economics and statistics,* vol. 52, pp. 169-210, 1990.
[18] E. Borensztein, J. De Gregorio, and J.-W. Lee, "How does foreign direct investment affect economic growth?," *Journal of international Economics,* vol. 45, pp. 115-135, 1998.
[19] L. D. Mello, "Foreign direct investment-led growth: evidence from time series and panel data," *Studies in Economics,* vol. 51, pp. 133-151, 1999.
[20] S. Anwar and L. P. Nguyen, "Foreign direct investment and trade: The case of Vietnam," *Research in International Business and Finance,* vol. 25, pp. 39-52, 1// 2011.
[21] B. B. Oloyede, "IMPACT OF FOREIGN DIRECT INVESTMENT ON AGRICULTURAL SECTOR DEVELOPMENT IN NIGERIA,(1981-2012)," *Kuwait Chapter of the Arabian Journal of Business and Management Review,* vol. 3, p. 14, 2014.
[22] S. Chaudhuri and D. Banerjee, "FDI in agricultural land, welfare and unemployment in a developing economy," *Research in Economics,* vol. 64, pp. 229-239, 2010.
[23] S. Pilgrim and S. Iyare, "Foreign direct investment (FDI) and the global food crisis. A study of the Windward Islands' agricultural sector," *Economics Bulletin,* vol. 3, pp. 1-15, 2008.
[24] E. Ogbanje, O. Okwu, and S. Saror, "An Analysis of foreign direct investment in Nigeria: The fate of Nigeria's agricultural sector," *Production Agriculture and Technology Journal,* 2010.
[25] O. Tasseven, *The Relationship Between Economic Development and Female Labor Force Participation Rate: A Panel Data Analysis*: Springer International Publishing, 2017.
[26] J. G. Djokoto, Y. Francis, Srofenyoh, and K. Gidiglo, "Domestic and foreign direct investment in Ghanaian agriculture," *Agricultural Finance Review,* vol. 74, pp. 427 - 440, 2014.
[27] M. Belloumi, "The relationship between trade, FDI and economic growth in Tunisia: An application of the autoregressive distributed lag model," *Economic Systems,* vol. 38, pp. 269-287, 2014.
[28] M. S. Z. L. Miao, "Foreign direct investment and agriculture security empirical analysis based on the crowding-out effect on domestic investment and employment," *Journal of International Trade,* vol. 4, p. 12, 2012.
[29] J. G. Djokoto, "OPENNESS AND AGRICULTURAL PERFORMANCE IN GHANA " *Journal of Science and Technology,* vol. 33, pp. 24-36 2013.

Modeling and Simulation Method of Flight Trajectory for Ship-To-Air Missile Based on Vector Rotation

LIU Songtao*, WANG Longtao and WEI Baoyan

Dept. of Information System, Dalian Naval Academy, Dalian 116018, China

*navylst@163.com

Abstract. Aiming at trajectory modeling for ship-to-air missile, a novel modeling method based on vector rotation is proposed, which can be used under different conditions of ship-to-air missile launching scheme. Compared with traditional Euler modeling method and quaternion modeling method, its advantage lies in simple principle and extensive practicality. The implementation method is adopted as mixed programming of C language and Matlab language, which combines both simplicity of Matlab developing environment to display 3D graph and efficiency of C language. Addition, using dynamic link library to implement trajectory calculation is easy to extend. The simulation results verify the effectiveness and practicality of the proposed method.

1. Introduction

Trajectory modeling and simulation plays an important role in the field of ballistic design and simulation. Currently, there are three main methods of trajectory modeling: (1) Based on the mathematical model of missile system, the missile motion equations are established for ballistic calculation. The complete mathematical model of the missile system includes: the target motion model; the projectile body model (kinetic model of centroid motion, kinematic model of centroid motion, kinetic model of motion around the center of mass, kinematics model of motion around the center of mass); control relation model (guidance loop model and attitude control loop model) and relative kinematic model. The established missile motion equations are nonlinear differential equations with variable coefficients, which are computationally intensive and complex; (2) Ignore the control relation, retain the missile kinematics equation and adopt a simplified dynamic model[1]. The dynamic equation is constrained by such conditions as acceleration and overload, that is, dynamic response is determined by maneuvering performance limits. The velocity of centroid motion is obtained by simplifying the dynamic equation, while the position and attitude of the missile are solved by the kinematics equation; (3) According to the characteristics of each stage of missile flight, a universal solution model of missile position is established. The method can be used to simulate the trajectory's certain stage of a missile, the trajectory in different situations and the trajectory of different types of missiles by setting the parameters of each stage of missile flight. In this paper, the third method is used to model the ship-to-air missile trajectory according to the missile model, the specific missile parameters, ballistic characteristics and guiding law.

In the process of trajectory modeling, the literature [2] only simulates the trajectory in the two-dimensional plane, which could not reflect the real situation of the missile flight. The proportional guidance method is usually used in the homing guidance of the ship (ground)-to-air missile. When calculating the kill zone based on the trajectory simulation, the proportional guidance is decomposed into a vertical direction and a horizontal direction in the literature [3]. In the prediction model of the aerial target's hit point, the literature [4] also decomposes the missile motions into vertical plane or horizontal plane. In order to realize the three-dimensional trajectory, the literature [5] does not adopt the thought of decomposition, but establishes the simulation model of three-dimensional trajectory based on the vector idea. However, the description of its concrete implementation is not clear.

In the traditional trajectory modeling method, the attitude angle of the missile is described by the Euler angles, which is not suitable for the trajectory modeling of vertically-launched missiles. The quaternion method can be used to solve this problem[6]. After trajectory modeling, the numerical integration method is usually adopted, including Euler method, Runge-Kutta method and Adams method. The integration step size may be a definite step or a variable step. Similar to Euler method, the literature [7] gives the difference equation of proportional guidance method when calculating the simulation trajectory, for seeking the discrete three-dimensional coordinates of the missile at intervals. The realization of the solution and display after trajectory modeling is usually implemented by Matlab language programming, which is not engineering and real-time.

To overcome the above shortcomings, a new modeling method of flight trajectory for ship-to-air missile based on vector rotation is put forward for the trajectory simulation of ship-to-air missile with different types of launching schemes. Compared with Euler method and Quaternion method, this method is more practical and understandable.

2. Simulation condition

2.1. Simulation Target Analysis

The Russian SA-N-6 vertically launched ship-to-air missile weapon system is taken as an example to conduct trajectory modeling. The characteristics of the missile is maximum speed of 1800 m/s, the average speed of 1100 m/s, high trajectory, maximum distance of 150 km, with 85° quasi vertical cold launch mode. After the launcher ejects the missile for 1.5s, the missile main engine ignites and begins to control the missile turn. When the desired course angle and trajectory inclination are reached, the turn phase ends. At the same time, intermediate guidance begins with semi-active command. After the missile seeker captures the target signal, the intermediate command guidance is converted into the terminal guidance [8].

According to the above description, the trajectory of the missile is divided into four phases: the free trajectory phase (launch phase), the initial guidance phase (the turn phase), the intermediate guidance phase and the terminal guidance phase. Because both of them are guided by proportional guidance, the intermediate guidance and terminal guidance can be regarded as a phase in the trajectory simulation, which is the guidance phase. Therefore, this paper divides the trajectory into three phases to conduct modeling and simulation respectively, that is, the free trajectory phase, the turn phase and the guidance phase.

2.2. Trajectory Simulation Hypothesis

The missile's complete flight control system, projectile body dynamics and kinematics model are very complex, but we are mainly concerned with the change of the missile's center of mass. Therefore, the missile is regarded as a maneuverable particle, and the kinematics and dynamical equations of the missile's center of mass motion are studied to determine the coordinate position of the center of mass in each moment. The dynamics equations are simplified and analyzed from an overload point of view, and we assumed that the gravitational acceleration does not vary with altitude.

In the modeling and simulation, the geodetic coordinate system is adopted, and the initial position of the missiles at the launch time of the ship-to-air missile is taken as the origin of coordinates.

2.3. Target Motion Model

Suppose the target to do uniform linear motion. If the coordinate of the current target is (X_t^t, Y_t^t, Z_t^t), the flying speed is V_t, the course angle is ψ_t, and the velocity vector inclination is θ_t, then the target

IOP Conf. Series: Journal of Physics: Conf. Series **1060** (2018) 012067 doi:10.1088/1742-6596/1060/1/012067

coordinate at the next time is

$$X_t^{t+\Delta t} = X_t^t + V_t I_{x\_t}^t \Delta t \quad , \quad Y_t^{t+\Delta t} = Y_t^t + V_t I_{y\_t}^t \Delta t \quad , \quad Z_t^{t+\Delta t} = Z_t^t + V_t I_{z\_t}^t \Delta t \tag{1}$$

Where the flight velocity vector direction of the target is

$$I_{x\_t}^t = \cos\theta_t \cos\psi_t \, , \; I_{y\_t}^t = \sin\theta_t \, , \; I_{z\_t}^t = \cos\theta_t \sin\psi_t \tag{2}$$

3. Modeling of flight trajectory for ship-to-air missile based on vector rotation

3.1. The Free Trajectory Phase

(1) The velocity vector's direction of the missile remains unchanged, given by

$$I_{wkx\_m}^t = \cos\theta_{wk\_m}^t \cdot \cos\psi_{wk\_m}^t \quad , \quad I_{wky\_m}^t = \sin\theta_{wk\_m}^t \quad , \quad I_{wkz\_m}^t = \cos\theta_{wk\_m}^t \cdot \sin\psi_{wk\_m}^t \tag{3}$$

Where $\theta_{wk\_m}^t$ is the ballistic inclination of the missile launching time, $\psi_{wk\_m}^t$ is the ballistic deflection of the missile launching time.

(2) The coordinates of the missile at the next moment can be expressed by

$$X_m^{t+\Delta t} = X_m^t + V_m^t \cdot I_{wkx\_m}^t \Delta t \quad , \quad Y_m^{t+\Delta t} = Y_m^t + V_m^t \cdot I_{wky\_m}^t \cdot \Delta t \quad , \quad Z_m^{t+\Delta t} = Z_m^t + V_m^t \cdot I_{wkz\_m}^t \cdot \Delta t \tag{4}$$

Where V_m^t is the velocity of the missile, which could be obtained by fitting and sampling the maximum velocity and average velocity.

(3) The end of the free trajectory phase is controlled by the time variable t, i.e., $0 < t \le 1.5s$.

3.2. The Turn Phase

(1) The velocity vector direction $(I_{zwx\_m}^t, I_{zwy\_m}^t, I_{zwz\_m}^t)$ of the missile at the beginning is the velocity vector direction at the end of the free trajectory phase.

(2) The change rate of velocity vector is calculated by

$$\dot{\theta}_{zw\_m}^t = k_v \cdot V_m^t \quad , \quad k_v = k_c \cdot \frac{n_{zw} \cdot g}{V_{\max}^2} \tag{5}$$

W n_{zw} is the available overload of the missile; k_c is the adjustment coefficient; g is the acceleration of gravity.

(3) The change direction of the velocity vector in the turn is determined by the cross product direction of the direction of the missile velocity vector at the beginning of the turn phase and the missile velocity vector determined by the initial bookbinding parameters, that is

$$I_{zdx\_m}^t = \cos\theta_{zd\_m}^t \cdot \cos\psi_{zd\_m}^t \quad , \quad I_{zdy\_m}^t = \sin\theta_{zd\_m}^t \quad , \quad I_{zdz\_m}^t = \cos\theta_{zd\_m}^t \cdot \sin\psi_{zd\_m}^t \tag{6}$$

Where $\theta_{zd\_m}^t$ is the trajectory inclination and $\psi_{zd\_m}^t$ is the trajectory declination binding at the end of the turn.

(4) The velocity vector of the missile at the next moment is

$$(I_{zwx\_m}^{t+\Delta t}, I_{zwy\_m}^{t+\Delta t}, I_{zwz\_m}^{t+\Delta t}) = R \cdot (I_{zwx\_m}^t, I_{zwy\_m}^t, I_{zwz\_m}^t) \tag{7}$$

Where the rotation matrix R is determined by the change rate and change direction of the velocity vector in the turn phase.

(5) Missile coordinates at the next moment are

$$X_m^{t+\Delta t} = X_m^t + V_m^t \cdot I_{zwx\_m}^t \cdot \Delta t \quad , \quad Y_m^{t+\Delta t} = Y_m^t + V_m^t \cdot I_{zwy\_m}^t \cdot \Delta t \quad , \quad Z_m^{t+\Delta t} = Z_m^t + V_m^t \cdot I_{zwz\_m}^t \cdot \Delta t \tag{8}$$

(6) Suppose the angle between the velocity vector's direction of the missile $(I_{zwx\_m}^t, I_{zwy\_m}^t, I_{zwz\_m}^t)$ and the missile velocity vector's direction determined by the initial bookbinding parameters

$(I^t_{zdx\_m}, I^t_{zdy\_m}, I^t_{zdz\_m})$ is α^t_m. When α^t_m meets the following conditions, the turn phase end

$$\alpha^t_m = \arccos(I^t_{zwx\_m} \cdot I^t_{zdx\_m} + I^t_{zwy\_m} \cdot I^t_{zdy\_m} + I^t_{zwz\_m} \cdot I^t_{zdz\_m}) \tag{9}$$

When α^t_m is less than or equal to α^*, stop calculation, where α^* is the precision requirement that control the end of turn.

3.3. The Guidance Phase

The guidance phase adopts the proportional guidance law, that is, the angular velocity of the missile velocity vector $\dot{\varphi}_m$ is proportional to the angular velocity of the target $\dot{q}_m$, which is as follows

$$\frac{d\varphi_m}{dt} = K_n \frac{dq_m}{dt} \tag{10}$$

Where K_n is the proportional guidance coefficient. In the proportional guidance, the relative movement of missile and airplane is shown in Figure 1.

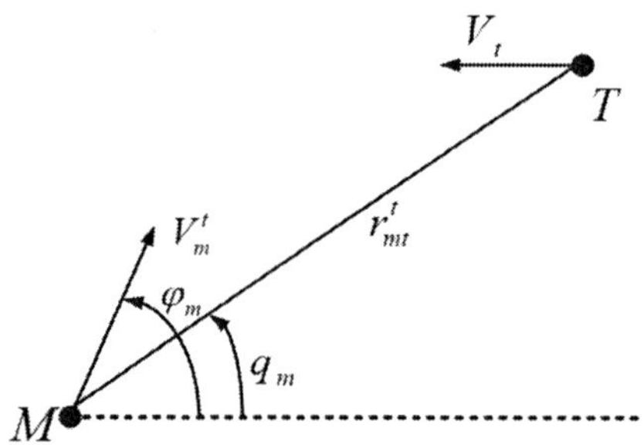

Figure 1. The relative movement of missile and airplane under the proportional guidance law

According to the proportional guidance law, the solution of the trajectory model of the guidance phase is summarized:

(1) At the beginning, the angle between the target-missile line of sight direction and the missile velocity direction is

$$\beta^t_{tm\_v_m} = \arccos(I^t_{x\_tm} \cdot I^t_{dyx\_m} + I^t_{y\_tm} \cdot I^t_{dyy\_m} + I^t_{z\_tm} \cdot I^t_{dyz\_m}) \tag{11}$$

Where the target-missile line of sight direction is

$$I^{t'}_{x\_tm} = X^t_t - X^t_m, \quad I^{t'}_{y\_tm} = Y^t_t - Y^t_m, \quad I^{t'}_{z\_tm} = Z^t_t - Z^t_m \tag{12}$$

After normalization of Eq.(12), the following equation is obtained.

$$I^t_{x\_tm} = \frac{I^{t'}_{x\_tm}}{\sqrt{I^{t'\,2}_{x\_tm} + I^{t'\,2}_{y\_tm} + I^{t'\,2}_{z\_tm}}}, \quad I^t_{y\_tm} = \frac{I^{t'}_{y\_tm}}{\sqrt{I^{t'\,2}_{x\_tm} + I^{t'\,2}_{y\_tm} + I^{t'\,2}_{z\_tm}}},$$

$$I^t_{z\_tm} = \frac{I^{t'}_{z\_tm}}{\sqrt{I^{t'\,2}_{x\_tm} + I^{t'\,2}_{y\_tm} + I^{t'\,2}_{z\_tm}}} \tag{13}$$

The velocity vector direction of the missile $(I^t_{dyx\_m}, I^t_{dyy\_m}, I^t_{dyz\_m})$ at the beginning is the velocity vector direction at the end of the turn phase.

(2) At the beginning, the angle between the target-missile line of sight direction and the target velocity direction is

$$\beta^t_{tm\_v_t} = \arccos(I^t_{x\_tm} \cdot I^t_{x\_t} + I^t_{y\_tm} \cdot I^t_{y\_t} + I^t_{z\_tm} \cdot I^t_{z\_t}) \tag{14}$$

Where $(I^t_{x\_t}, I^t_{y\_t}, I^t_{z\_t})$ is the target velocity vector direction.

(3) The angular velocity of target-missile line of sight direction at the beginning is

$$\dot{q}_m^t = \frac{\dot{r}_{mt}^t}{r_{mt}^t} \tag{15}$$

Where the missile-target distance is $r_{mt}^t = \sqrt{(X_t^t - X_m^t)^2 + (Y_t^t - Y_m^t)^2 + (Z_t^t - Z_m^t)^2}$ and the missile-target distance rate is $\dot{r}_{mt}^t = V_t \cdot \sin\beta_{tm\_v_t}^t - V_m^t \cdot \sin\beta_{tm\_v_m}^t$.

(4) The change rate of the missile velocity vector direction at the beginning is

$$\dot{\varphi}_m^t = K_n \dot{q}_m^t \tag{16}$$

(5) The change direction of the velocity vector at the beginning is determined by the cross product direction of the direction of the missile velocity vector and the missile velocity vector at the beginning.

(6) The velocity vector of the missile at the next moment is the same as step 4 in the turn phase.

(7) Missile coordinates at the next moment are the same as step 5 in the turn phase.

(8)The ending condition of the guidance phase is $r_{mt}^{t+\Delta t} \le r_{mt}^t$ and $r_{mt}^{t+\Delta t} \le R$, where $r_{mt}^{t+\Delta t} = r_{mt}^t + \dot{r}_{mt}^t \Delta t$, R is the ship-to-air missile's kill radius.

4. Simulation implementation method

According to the needs of the project, C and Matlab languages are used to realize the ballistic solution and display. The solution part is realized by dynamic link library in VC6.0 integrated development environment, and the display part is three-dimensionally demonstrated in Matlab7.1 environment. The main advantage of this method is the combination of the convenience of 3D display in Matlab environment and the high efficiency of C language program.

In the trajectory modeling process, when the velocity vector is rotated from the current position to another location, two key parameters need to be known: rotation angle and rotation direction. The solution of these two parameters has been given in the trajectory modeling section. In the case of these two parameters known, the following rotation matrix can be used to realize the vector rotation:

$$R = \begin{bmatrix} x^2+(1-x^2)\cos\alpha & xy(1-\cos\alpha)-z\sin\alpha & xz(1-\cos\alpha)+y\sin\alpha \\ yx(1-\cos\alpha)+z\sin\alpha & y^2+(1-y^2)\cos\alpha & yz(1-\cos\alpha)-x\sin\alpha \\ zx(1-\cos\alpha)-y\sin\alpha & zy(1-\cos\alpha)+x\sin\alpha & z^2+(1-z^2)\cos\alpha \end{bmatrix} \tag{17}$$

Where (x, y, z) is the rotation direction of normalization, and α is the angle of rotation. The rotation matrix multiplies the velocity vector right to get the new velocity vector.

5. Simulation result

During the simulation, at the beginning of launching, the ship-to-air missile trajectory angle is 85°, the ballistic deflection angle is 10°. The turn phase adjustment coefficient is $k_c = 1.6$, the missile overload is $K_n = 1.6$, the guidance phase proportion coefficient is $K_n = 1.6$, and the simulation step is 0.1s.

Let the target parameters at the missile launch time as follows: target direction $B_t = 45^\circ$, the target slant range $R_t = 100km$, the target height $H_t = 10km$, the target flying straight and level at a constant speed , the target speed $V_t = 500m/s$. Figure 2(a) and (b) are the ballistic trajectories of the ship-to-air missiles at the target course $C_t = 180^\circ$ and $C_t = 270^\circ$ respectively.

The following conclusions can be obtained through the simulation results: (1) In Figure 2(a), the shortest distance between the missile and the target is about 26m. In Figure 2(b), the shortest distance between the missile and the target is about 41m. If the kill radius is 50m, the ship-to-air missile can effectively destroy the target; (2) It can realize the combat simulation of ship-to-air missile and the aircraft target in different battlefield environment by changing the target or the missile parameters; (3) In this paper, the trajectory simulation of SA-N-16 ship-to-air missiles is carried out. But the simulation ideas and steps are universal, which can be used in ballistic simulation of other ship-to-air missile weapon systems and can also be widely applied to ballistic simulation of anti-ship missile weapon systems.

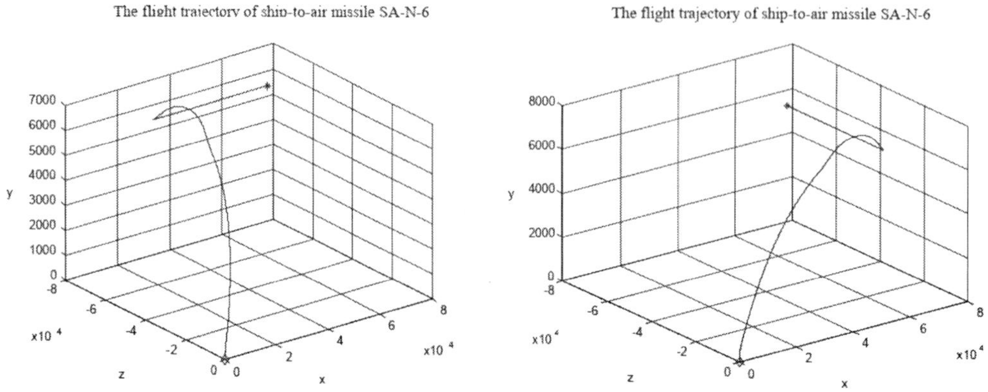

(a) Course 180° (b) Course 270°

Figure 2. The simulation result of flight trajectory of ship-to-air missile SA-N-6

6. Conclusion

In this paper, a new method of ship-to-air missile trajectory modeling is proposed based on vector rotation, which can be effectively applied to different types of ship-to-air missile trajectory simulation. The method of mixed programming of C language and Matlab language is adopted in the concrete realization of ballistic calculation and display. It takes advantage of the convenience of 3D display in Matlab environment and the efficiency of C language design program, and uses dynamic linking Library technology for ballistic calculations with application scalability. The simulation results verify the effectiveness and practicability of the proposed method. In the ballistic modeling process, some reasonable and simplified assumptions are made. Given the known aerodynamic forces, aerodynamic moments, thrusts, standard atmospheric parameters, mass changes, etc., more realistic ballistic trajectories can be obtained, coupled with environmental considerations including wind, flow, atmospheric media and the electromagnetic environment, etc., as well as the impact of the target's mobility, more real countermeasure scene can be simulated.

References

[1] Ku H., Kang F. J., Hang L., et al. 2006. Modeling and simulation technology of typical anti-ship missile motion. *Journal of System Simulation*, 18, 8 (Aug. 2006), 2067-2069.

[2] Li P., Hu Y.. 2010. Modeling and Simulation of typical Motion of Anti-ship Missile. *Computer Simulation*, 27, 5(May. 2010), 46-48.

[3] Zhao J. J., Li W. B., Sang D. Y., et al. 2010. Killing Zone calculation based on trajectory Simulation. *Ordnance Industry Automation*, 29, 6(Jun. 2010), 8-10.

[4] Wang J., Zhou L., Lei H. M.. 2009. Prediction model and algorithm on hit point of ground to air missile and aerial target. *Journal of System Simulation*, 21, 1(Jan. 2009), 80-83.

[5] Ma Q. D., Ma L. T., Zhong Z. T.. 2008. Research on 3-Dimensional trajectory simulation of proportional guidance ship-to-air missile. *Missiles and Space Vehicles*, 3, (Mar. 2008), 15-18.

[6] Fang Y., Sun S. C., Liu F. 2010. Research on trajectory simulation of thermal vertical launching ship-to-air missile. *Ship Electronic Engineering*, 30, 11(Nov. 2010), 110-113.

[7] Gao S.. 2003. Simulation on missile ideal trajectory under proportional guidance. *Computer Engineering and Design*, 24, 8(Aug. 2003), 66-68.

[8] Zhang Y. L.. 1997. *Naval Ship-to-Air Missile Weapon Manual*. Weapons Industry Press, Beijing.

Firepower Allocation of Continuous Multiple Incoming Targets

Yao Liu[1,*], Z Y Zhang[2], J R He[1] and Z C Huang[1]

[1]Department of Graduate Management, Space Engineering University, Beijing, 101416, China
[2]Department of Space Safety Research Center, Space Engineering University, Beijing, 101416, China

[*]liuyao936@126.com

Abstract. Aiming at the problem of firepower allocation under continuous multi-target attack during the course of anti-ballistic missile operations, the time factor is added to the performance function of the evaluation scheme index. A dynamic firepower allocation model under continuous multi-target attack is established .The efficiency changes of the same weapon target matching scheme at different times are fully considered. The problem of target priority is optimized by using the method of generalized difference and rescheduling to solve the dynamic firepower Allocation scheme, which can determine the matching scheme of the target and weapons and the launching time of the interceptor. Thus, make the firepower allocation plan more time-efficient, closer to actual combat. The validity of this method is verified by a specific example. The research results have certain reference value for the application of mid-stage missile defence.

1. Introduction

The purpose of firepower allocation is to determine which positions and which time an interceptor should be reached in to respond to the incoming missile targets in one or more times of anti-missile operations, in order to achieve operational effectiveness with the minimum number of weapons. It can be seen that the firepower allocation includes the matching of weapons targets and the timing of interceptor launching.

At present, the content of firepower allocation focuses on the issue of intercepting missiles allocated to missiles [1-3], and belongs to the Weapon Target Assignment (WTA) issue. The launch timing of an interceptor is determined by the particular operational scenario. Literature [4] studied the firepower allocation in air defence and determined the required number of weapons. The weapon assigned to the target was determined based on the shortest combat time. However, the article did not conduct an in-depth study on the shooting opportunity but conducted a qualitative analysis and did not give an analytical mathematical model. In the paper[5], the concept of "pre-emptive damage" and "firing advancement factor" was used to allocate anti-aircraft firepower to the combined guns. Weapons that meet the expected probability of destruction were assigned to each incoming target, and an appropriate timing of warfare was planned. In the repeatedly attack, the ABM decision-making plan should be adjusted according to real-time operational situations, which increasing the difficulty of choosing the best solution.

In this paper, the time factor is added to the performance function of the evaluation scheme index, a dynamic model of fire Allocation scheme is established, and the model is solved by generalized differential method and rescheduling method. Because time elements are added to the model, the fire allocation scheme is timelier and more practical combat.

2. Analysis of Continuous Multi - target Missile Defence Problem

The text of your paper should be formatted as follows: In actual combat operations, the missile strike is often uncertain. During the entire anti-missile battle, at most one battlefield for each incoming missile is generated for each combat period corresponding thereto. The missile strike is attacking the ground assets with multiple missiles in a very short period of time or a single incoming missile random, offensive position and time are not sure. The firepower allocation plan in actual combat should not only determine the match strategy of the weapon target, but also determine the launch time of the interceptor against the target.

The generalized differential method can solve the problem that the total efficiency of the firepower allocation scheme and the effect of time factor on the quality of the solution when the number of interceptors in firepower allocation scheme is gradually increased. The re-scheduling algorithm can maximize the improvement under the condition of shortage of combat resources Combat effectiveness, increase the use of resources.

2.1. Firepower Allocation based on improved generalized difference method

Theorem[6]:The number of incoming missile targets is n. The optimal solution for intercepting all targets using m interceptors can be written as $(m_1, m_2, \cdots m_n)$, $m_1 + m_2 + \cdots + m_n = m$, m_i $(i = 1, 2 \cdots n)$ indicates the number of interceptors assigned to the target i . Then when the number of interceptors increases by 1, there exists the optimal solution k $(m_1^k, m_2^k, \cdots m_n^k)$ in the $m+1$ optimal interception schemes when the number of interceptors is equal $0, 1 \cdots m$, so that exists the target k in the scheme that makes the final scheme $(m_1^k, m_2^k, \cdots m_{k-1}^k, (m-k+1), \cdots m_n^k)$ the optimal scheme, when the number of interceptors is $m+1$.

The principle of generalized differential method is adding an interceptor and find out the optimal firepower Allocation plan when the number of interceptors is $m+1$, based on the optimal firepower Allocation of allocating m interceptors to n targets.

Suppose the attack target set is N_q in the period q , When using the improved generalized differential method to solve the fire Allocation, the combat time is taken as a factor to evaluate the quality of the Allocation effect[7].The remaining interception time of the missile is the time from the earliest moment when the missile intercepts within the interception time window to the latest moment of the interception window from the moment when the information of the incoming missile is transmitted from the early warning system to the accusation system, which characterizes the urgency of anti-missile operations. If the current moment is before the earliest moment of the interception window, the remaining interception time is the entire interception window time period.

Using adaptive optimization criteria, combined with the remaining interception time and interception probability, the definition of the total performance function is:

$$E = \sum_{j=1}^{n} \varepsilon \cdot p_j + \mu \cdot t_j \tag{1}$$

ε , μ are the weight coefficients. p_j is the intercept probability of target j , t_j is the time cost function, the formula is:

$$t_j = \frac{t_z^j - t_f}{\Delta t_j} \tag{2}$$

t_z^j is the last Interceptor launch time on the target j , t_f is the moment when the last interceptor launched on the target j in the Allocation plan, Δt_j is the length of launch time window of target j .

When the objective function incorporates the time factor, the higher the interception time is, the higher the quality of the prior Allocation plan is. On the contrary, the closer the interception time is to the deadline, the lower the quality of the Allocation plan is.

Steps of generalized differential method are:

1) First, find the number of basic interceptors that satisfy the probability of destroying damage. The minimum number of interception shells per target is $(m_1, m_2, \cdots m_n)$, Find a target to make the number of the interceptors assigned to the target is m_0 ,the optimization plan is $(0, 0, \cdots m_0 \cdots 0)$,when the Launch time is $(t_1, t_2, \cdots t_{m0})$ .The set target is allocated according to the expected damage value, the expected damage probability for each incoming missile target is p_j^* in the period of time t_q ,so that the minimum projectiles m_0 can be solved:

$$m_0 = \min(\log_{(1-p_i)}(1-p^*)),\ j = 1, 2\cdots n \tag{3}$$

2) Add an interceptor on m_0 .There are two cases, in the first case, each target of the current time is assigned to $m_0 + 1$ interceptors, and the intercept performance is calculated and the maximum value is $\sum_1$. In the second case, on the basis of (1), an interceptor is added to one of the targets to calculate the intercept effectiveness, and the maximum value is $\sum_2$ .To compare $\sum_1$ and $\sum_2$, the corresponding solution for the largest is the optimal solution for $m_0 + 1$ interceptors.
3) Add two interceptors on m_0 .There are three cases, in the first case, each target of the current time is assigned to $m_0 + 2$ interceptors, and the intercept performance is calculated and the maximum value is $\sum_1$. In the second case, on the basis of (1), two interceptors is added to one of the targets to calculate the intercept effectiveness, and the maximum value is $\sum_2$. In the third case, on the basis of $m_0 + 1$, find the optimal solution according to Step 2,and the maximum value is $\sum_3$ .To compare $\sum_1$, $\sum_2$ and $\sum_3$, the corresponding solution for the largest is the optimal solution for $m_0 + 2$ interceptors.

Similarly when m interceptors are added on m_0 , it needs to be divided into $m + 1$ steps.

According to the traditional generalized difference method, when the firepower Allocation is solved, only the number of intercepting bombs is considered, and the type of interceptor bombs is not distinguished. For incoming missile targets, intercepting bombs of different launching positions can be distinguished as different types. Therefore, the use of generalized differential method assigned to the target interceptor, it should also consider the choice of launch position. In each step, there are two levels. The first level is the decision plan $(m_1, m_2, \cdots m_n)$ of m interceptors assigned to n incoming targets. The second level is making m_j interceptors assigned to the target j . The allocation plan assigned to s positions is $(m_1^j, m_2^j, \cdots m_s^j)$. According to the generalized differential method, the estimated initial consumption of interceptors of the target j is m_j , and the m_j interceptors are allocated in s positions to find the optimal allocation scheme.

2.2. *Firepower allocation based on rescheduling*

Establish a fire allocation model of continuous multiple incoming objects. Rescheduling[8]refers to when occur new incoming target, updates the program according to the current situation. When the previous decision-making program has not yet been implemented and can't meet the optimal decision-making under the current situation, it is necessary to release the previous decision-making resources to release the occupied resources and regenerate an optimized solution to the current situation.

The division of the time period follows the target incoming process, dividing the entire operation time process into several time periods according to the events, and the time required to generate the solution is ΔT. Suppose that make the moment T_0^q when the first target is discovered as the starting time of phase q , ΔT before the moment is the end of the previous period. In this stage the incoming missile collection is N_q .If a target is not successfully intercepted in a phase, and the target's remaining interception time meets the constraint of the second interception, the target is added as a new target to the target of the next phase to redistribute the allocation scheme. Suppose the current moment is t, then at this moment belonging to $q(t)$, the target of the incoming missile collection is $N(t)$. The current collection of weapons resources $M(t)$ is solved by:

$$M(t)=M(0)-\sum_{q=1}^{q(t)-1} d(t^q) \tag{4}$$

$N_{ap}(q)$ represent the set of new targets that appear during the period q. $N_{die}(q)$ represents the set of targets that have vanished during the period q ,including targets that were successfully intercepted and those that were not successfully intercepted but could not be intercepted for the second time. Then the target set $N(t)$ within the current time as:

$$N(t)=N(0)+\sum_{q=1}^{q(t)-1} N_{ap}(q)-\sum_{q=1}^{q(t)-1} N_{die}(q) \tag{5}$$

There are b_i launchers at the i position, so the maximum number of interceptor launches that can be fired at the same position is b_i. For the same target, in order to prevent interference with each other, the interception launch interval of each position on the same target is Δt_i, the adjacent minimum firing interval for each launch frame on the same target is Δt_d.The position i of the interceptor's maximum launch volume in the period q is:

$$M_i^q=\left(1+\frac{\Delta t_i}{\Delta t_d}\right)b_i \tag{6}$$

The number of missiles launched during the period q should satisfy:

$$\sum_{j=1}^{m}\sum_{i=1}^{n} x_{ij}^q \le \sum_{i=1}^{n} M_i^q \tag{7}$$

x_{ij}^q represents the position i assigned to target j during period q. The matching decision value $x_{ij}(t)$ of weapon and target at any time meet the 0-1 constraints over time, that is $x_{ij}(t)=\{0,1\}$. Thus it can be seen that the entire matrix of fire allocation decision-making also changes over time, indicating whether weapons i are assigned to the target j at time t [9], in the course of the anti-missile operations, the same allocation plan was put in place at different times to represent different interception strategies.

Applying a combination of on-line scheduling and re-scheduling to avoid the phenomenon of "short-sightedness" in the allocation plan due to the appearance of a random target, and to make the second interception of the target that is not successfully intercepted in the first interception solution[10].

3. Steps of continuous multi-target fire allocation

Text Steps of continuous multi-target incoming missile defense firepower allocation are as follows:

1) First, the target set N_1 of the first phase is used to estimate the initial consumption of the interceptor of the current phase. For the type of incoming target, the probability of the expected damage to the target is determined. By combining the hit probability of the missile with different positions ,the initial consumption N_0 of the current time is obtained:

$$M_0=\sum_{j=1}^{n}\min(\log_{(1-p_{ij})}(1-p^*)),\quad i=1,2\cdots m \tag{8}$$

2) The generalized differential method is used to assign M_0 interceptors to the current wave target, and the optimal firepower allocation scheme is selected according to the interception efficiency.
3) Calculate the probability of damage to each incoming target, that is $[p_1,p_2,\cdots p_n]$.
4) If $p_j \ge p^*, j=1,2\cdots n$, indicates that the desired damage value can be achieved under the current energy consumption, and the amount of ammunition will be reduced by one. Repeat

steps(2)and(3)to find the optimal solution and the probability of damage to each target for the number of consumption is $M_0 - 1$ .When $p_j < p^*, j = 1,2 \cdots n$ exists, the corresponding allocation scheme for M_0 is the final solution. If $p_j \geq p^*, j = 1,2 \cdots n$ always meet, reduce the amount of ammunition by 1 again. Repeat steps(2)and(3)until $p_j < p^*, j = 1,2 \cdots n$ exists.

5) If $p_j < p^*, j = 1,2 \cdots n$, indicates that the desired damage value can't be achieved under the current energy consumption, and the amount of ammunition will be increased by one. Repeat steps(2)and(3)to find the optimal solution and the probability of damage to each target for the number of consumption is $M_0 + 1$ .When $p_j \geq p^*, j = 1,2 \cdots n$ exists, the corresponding allocation scheme for $M_0 + 1$ is the final solution. If $p_j < p^*, j = 1,2 \cdots n$ always meet, increase the amount of ammunition by 1 again. Repeat steps(2)and(3)until the probability of intercepting any target meet $p_j \geq p^*, j = 1,2 \cdots n$.
6) After the current missiles and firepower allocation plan are finalized, all interceptor missile launches will be ready for launching and waiting for the launching moment to come.
7) When the new goal appears, there are three situations. First, follow steps (1) - (5) to calculate the allocation plan for the new target. In the first case, the current number of interceptors meets the requirements for a new target allocation scenario, so continue to implement the program. In the second case, the current number of interception shells does not meet the requirements for a new target allocation scheme. If the previous time of fire allocation plan has been executed or the rescheduling has not been implemented, the existing quantity of combat resources will be used as the amount of consumption, and the project will continue to be implemented according to steps (1) and (2). In the third case, the current number of interception missiles can't meet the requirements of the new target allocation plan. However, part of the plan for the previous time of missiles has not yet been implemented, and then the firepower allocation is carried out according to the re-scheduling to generate a new allocation plan.
8) Proceed to step (7) when a new target appears, until the entire battle is over.

4. Instance simulation analysis

To defend against a target as the background of the anti-missile operation, assume that there are four interception positions around the point, its location is shown in Table 1.

Table 1. Intercept position coordinates.

position	Coordinates in the geocentric coordinate system.(km)		
A	(-4583.21	1616.73	4130.31)
B	(-4676.21	1633.02	4018.12)
C	(-4633.64	1746.35	4019.85)

Assuming the incoming missile targets landed near the ground. The initial time of the simulated scenario is 0.The parameters are as shown in Table 2:

Table 2. The first incoming missile information.

Target	Discovery time/s	The position in geocentric coordinate system.(km)	The velocity in geocentric coordinate system(km/s)
T_{01}	340	(-4373.71 4243.42 3928.88)	(-1.81 -2.71 1.72)
T_{02}	259	(-6109.63 1020.23 3051.44)	(0.41 1.64 2.99)
T_{03}	302	(-4521.52 -1365.51 5222.33)	(-2.39 3.75 0.69)

T_{04}	138	(-4859.91 -372.4 4565.07)	(-1.64 3.40 1.05)

On the first incoming missiles, consider the moment of the missile as the starting moment of the interception window to judge. Four interception positions can be obtained for each incoming missile interception window. Using a certain type of interceptor to intercept the incoming targets, the launch time window of the target in the first period is as shown in Table 3:

Table 3. Launch time window of interceptor.

Position	T_{01}^{1}	T_{02}^{1}	T_{03}^{1}	T_{04}^{1}
A	[675 1047]	[367 759]	[563 938]	[240 646]
B	[680 1055]	[415 656]	[579 867]	[256 613]
C	[716 994]	[398 703]	[620 683]	[209 621]

Assuming there are three interceptor launchers, the number of interceptors available at each position at the initial moment expressed as $Q = [3 \quad 4 \quad 3]$. The average intercept probability of interceptor missiles on incoming missiles at different positions is shown in table 4:

Table 4. The intercept probability for the first incoming missiles.

Position	T_{01}^{1}	T_{02}^{1}	T_{03}^{1}	T_{04}^{1}
A	0.81	0.79	0.81	0.71
B	0.82	0.64	0.87	0.79
C	0.69	0.72	0.86	0.85

Suppose the threat value of current incoming missiles is $c = [0.217 \quad 0.185 \quad 0.344 \; 0.253]$.

Suppose that for the incoming missile target, the expected probability P^* of damage for each target is 0.95. The minimum interval launch time of the same position on the same target is $\Delta T = 20s$. Then according to the formula (8) the required number of missiles can be obtained. The target's initial consumption of interceptor is $M_0 = 8$. The number m_0 of minimum consumption of interceptor for each target is 2.

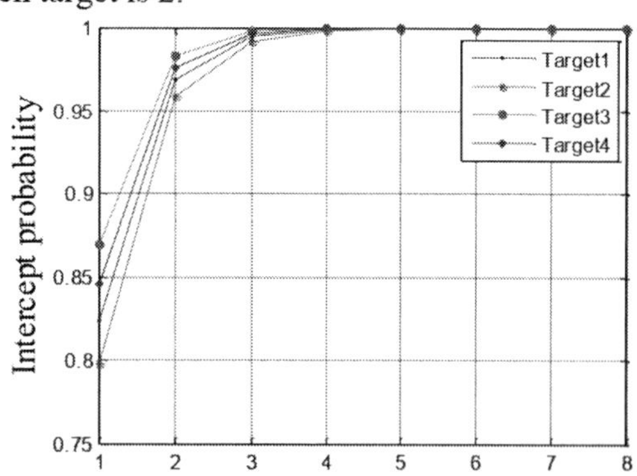

Figure 1. Interception probability on different number of interceptors for single target

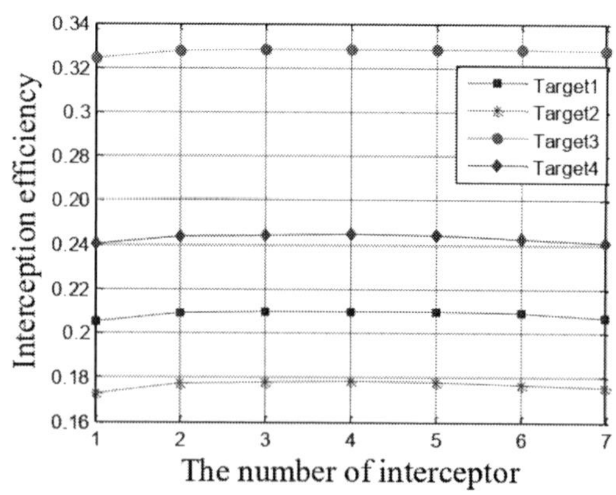

Figure 2. Interception efficiency on different number of interceptors for single target.

Figure 1 shows the change in the probability that the target is successfully intercepted according to the number of interceptors increase from 1 to 8 in the current scenario. The results show that when the number of interceptors allocated to each target is 2, the expected damage probability can be met. Figure 2 shows the trend of the maximum interception efficiency when the number of interceptors assigned to a single target is 1 to 8. The results show that for the 4 incoming missile targets in the current time, the interception performance basically does not change while the number of intercepting missiles assigned to each target is 2 or more. We can also see that there is a downward trend in the increase again, reflecting that the probability of damage does not increase any more, and the quality of the program decreases due to the time factor, when the increase of interceptors to a special target.

Table 5 shows that the allocation of interceptors in different positions, while the number of interceptors assigned to each target to meet the maximum interception performance change from 1 to 8. According to the generalized differential method, it is allocated according to the target level, and the minimum amount of ammunition for each target is 2. According to step 1, the optimal allocation scheme is $(0,0,2,0)$ for the firing of 2 interceptors as shown in Figure 2. The corresponding interceptor allocation scheme is $(0,2,0)$ as shown in table5.

Table 5. Allocation against different target.

Number	Target1			Target2			Target 3			Target 4		
	Position 1	Position 2	Position 3	Position 1	Position 2	Position 3	Position 1	Position 2	Position 3	Position 1	Position 2	Position 3
2	0	2	0	2	0	0	0	2	0	0	0	2
3	1	2	0	3	0	0	0	3	0	0	0	3
4	2	2	0	3	1	0	1	3	0	1	1	2
5	3	2	0	3	1	1	1	3	1	1	1	3
6	3	2	1	3	1	2	2	3	1	2	1	3
7	3	3	1	4	1	2	3	3	1	2	1	4
8	4	3	1	4	2	2	4	3	1	2	2	4

The effect of the program for each step of increasing the amount of interceptors from 2 to 8 is as shown in Table 6:

Table 6. The effect of each step allocation scheme of generalized differential method.

Interceptors number	The optimal interception efficiency in single step of generalized differential method						
	$\sum_1$	$\sum_2$	$\sum_3$	$\sum_4$	$\sum_5$	$\sum_6$	$\sum_7$
2	0.3242						
3	0.3278	0.6485					
4	0.3283	0.7737	0.6716				
5	0.3283	0.8769	0.7957	0.6752			
6	0.3284	0.9000	0.8978	0.7995	0.6755		
7	0.3284	0.9220	0.9035	0.9021	0.8003	0.6755	
8	0.3275	0.9430	0.9259	0.9039	0.9028	0.8003	0.6755

Figure 3 shows the comprehensive interception efficiency of the 4 incoming missile targets according to the generalized difference method and the total number of interceptor from 2 to 8. When the number of interceptor is 8, the comprehensive interception effectiveness is 0.9429.

IOP Conf. Series: Journal of Physics: Conf. Series **1060** (2018) 012068 doi:10.1088/1742-6596/1060/1/012068

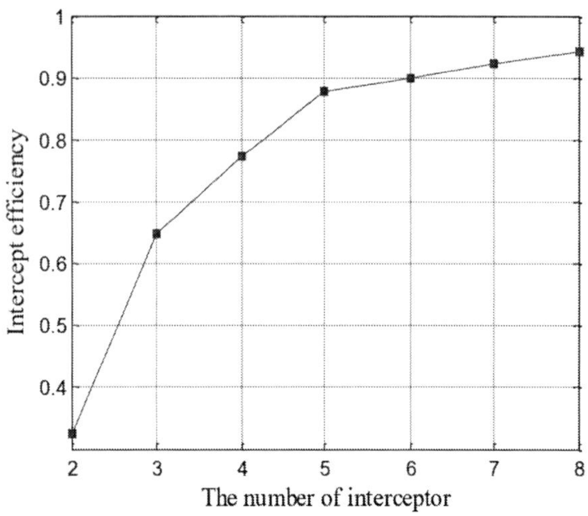

Figure 3. The interception efficiency of different number of interceptors in the optimal allocation strategy.

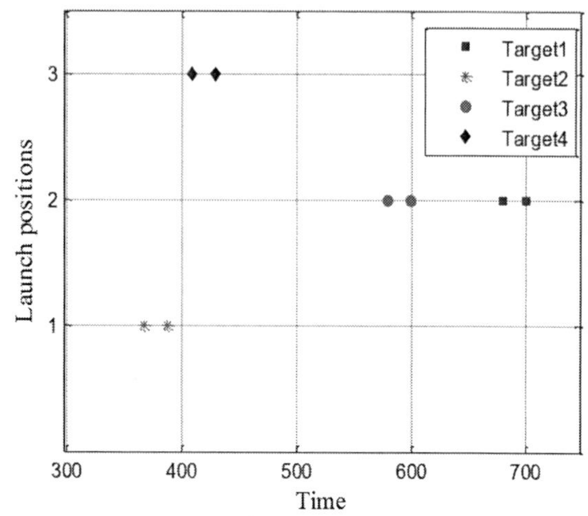

Figure 4. Launch time of interceptor.

Corresponding to Figure 3, the allocation of interceptors for each particular number of interceptors optimal allocation strategy are shown in the Table 7. If $p_j > p^*, j = 1,2\cdots n$, reduce the number of interceptors by one to count according to the generalized differential method to set the desired damage probability for each target. When the number of interceptors is 8, the probability of damage is solved according to the optimal strategy, the result is $(0.9689, 0.9589, 0.9829, 0.9763)$.

Table 7. The optimal allocation scheme for different number of interceptors.

Number	Target1			Target2			Target 3			Target 4		
	Positio n1	Positio n 2	Positio n 3	Position 1	Positio n 2	Positio n 3	Positio n 1	Positio n 2	Positio n 3	Positio n 1	Positio n 2	Positio n 3
2	0	0	0	0	0	0	0	2	0	0	0	0
3	0	0	0	0	0	0	0	2	0	0	0	1
4	0	1	0	0	0	0	0	2	0	0	0	1
5	0	1	0	1	0	0	0	2	0	0	0	1
6	0	1	0	1	0	0	0	2	0	0	0	2
7	0	2	0	1	0	0	0	2	0	0	0	2
8	0	2	0	2	0	0	0	2	0	0	0	2

When the number of interceptors is 7, the probability of damage is solved according to the optimal strategy, the result is $(0.9689, 0.7973, 0.9829, 0.9763)$. We can see that $p_2 < p^*$ exists, so the corresponding scheme is the final allocation scheme when the number of interceptors is 8.

Figure 4 shows the launching times of each position interceptor under the optimal allocation scheme. Position A assigns two interceptors to target 2,and the launching times are [367s, 387s]; Position B assigns two interceptors to target 3 , and the launching times are [579s,599s]; Position B assigns two interceptors to target 1 , and the launching times are [680s,700s]; Position C assigns two interceptors to target 4 , and the launching times are [409s,429s].

The target of the first phase has been assigned according to the expected damage probability and the maximum interception efficiency. When the first phase of the program has not yet been fully implemented, the new target appears, the new target information are shown in Table 8.

Table 8. The second incoming missile information.

Target	Discovery time/s	The position in geocentric coordinate system.(km)	The velocity in geocentric coordinate system(km/s)
T^1_{05}	536	(-5480.41 4417.52 2266.81)	(-1.85 -0.61 3.27)
T^1_{06}	572	(-5661.01 373.31 4573.15)	(-0.45 2.25 0.96)

The parameter information of the new target is shown in Table 9:

Table 9. Launch window and intercept probability in period two.

Position	Launch time window		Intercept probability	
	T^1_{03}	T^1_{04}	T^1_{03}	T^1_{04}
A	[1226,1545]	[888,1068]	0.76	0.83
B	[1271,1484]	[911,1033]	0.70	0.84
C	[1280,1427]	[884,1120]	0.62	0.75

The number of the threat of target 5 and target 6 are $c=[0.3020\ \ 0.4972]$. It can be seen from Figure 4 that when the scene time is 572s, the interceptors against T_{01} and T_{03} at position B has not been launched. The minimum consumption of interceptors against T_{05} and T_{05} is 4 according to expectation interception probability of the targets. The rank of the threat at this time is $c_6 > c_3 > c_5 > c_1$ .At this moment, the number set of interceptor remaining in all positions are $M_2=(1,0,1)$, so that the weapons resources allocated to T_{01} and T_{03} will be released ,and the number set of released weapons resources are $M_2=(1,4,1)$. So the second set targets period are $N_2=(T_{01},T_{03},T_{05},T_{06})$,then using the generalized differential method to allocate the interceptors of N_2 again.

Figure 5 is a comparison diagram of the interception efficiency of the target before and after rescheduling with the generalized differential method. It is shown that the rescheduling method is better than the non-rescheduling method in the case of resource shortage.

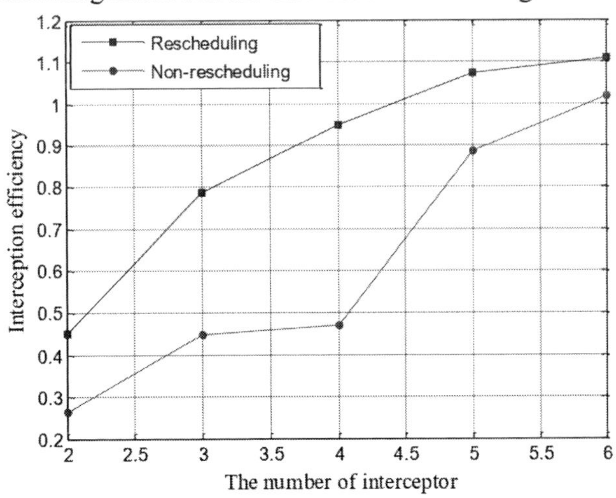

Figure 5. The interception performance after e-scheduling

The allocation scheme and launch time after e-scheduling are as shown in Table 10:

Table 10. The allocation scheme after e-scheduling.

Position	Allocation scheme				Launch time			
	T_{01}	T_{02}	T_{03}	T_{04}	T_{01}	T_{02}	T_{03}	T_{04}
A			1				1226	
B	1	1	1	1	680	579	1271	911
C		1				620		

5. Conclusions

In this paper, the time factor is added to the performance function of the evaluation scheme index, a dynamic firepower allocation scheme model under continuous multi-target attack is established, the dynamic fire allocation scheme is solved by the generalized differential method and the rescheduling method. The differential method can solve the effect of the firepower allocation scheme on the quality of solution when the number of interceptor in the firepower system is gradually increased, and compares the operational effectiveness before and after heavy scheduling. It shows that the re-scheduling algorithm can be used in a situation that the combat resources is of shortage, in this way the operational effectiveness is maximized and the use of resources is increased.

Acknowledgments

Authors gratefully acknowledge Z Y Zhang, J R He and Z C Huang, for provide help to the establishment of simulation model and for providing valuable insights on the solution.

Reference

[1] O Ma, J Han 2015 Research on air defence missile target assignment algorithm *Int. Conf. Intelligent Systems Research and Mechatronics Engineering*

[2] M Z Lee 2009 Constrained weapon–target assignment: Enhanced very large scale neighborhood search algorithm IEEE Transactions on Systems, Man, and Cybernetics - Part A: Systems and Humans, 40(1):198-204

[3] O Karasakal , N E Ozdemirel and L Kandiller 2011 Anti-ship missile defense for a naval task group Naval Research Logistics, 58(3):304–321

[4] H Liu , Z H He 2003 Air defense command automation system principle Xi'an: Xi'an University of Electronic Science and Technology Press

[5] Y X Wang 2008Air defence fire allocation model and algorithm under advanced damage criterion Nanjing University of Science and Technology

[6] J S Shu, Y X Li and H Hao 2013 Examples of missile operations Northwestern Polytechnical University Press

[7] L Wu , F X Lu and P F Jia 2008 Meta - Level Control of Improved Genetic Algorithm in Dynamic Weapon Target Assignment Journal of Tsinghua University (Science and Technology), 48 (s2): 68-71

[8] M Yamamoto, S Y Nof 1985 Scheduling/rescheduing in the manufacturing operating system environment International Journal of Production Research,23(4):705-722

[9] H P Cai ,J X Liu and Y G Chen ,rt al .Survey of the research on dynamic weapon-target assignment problem Journal of Systems Engineering and Electronics,17(3):559-565

[10] J F Luo, C Zhu and Z Liu, et al 2014 Online scheduling of warship air defence and firepower under successive multi-targets Systems Engineering -Theory & Practice, 34 (11): 2929-2937.

An Approach to Space-Debris Optical Image Simulation Considering the Streak and Saturated Star-Background

Yan Ouyang*, Tingxin Xu, Xiaobin Huang, Li Lu, ShuWen Wang and Ting Gao

Wuhan Early Warning Academy, Wuhan, China

*oyy_01@163.com

Abstract. In present, space-object optical image simulation study is very important to the area of space technology, because it can support the development of space-object detection algorithm before the optical platforms are launched to the space and supplement the information of unusual cases which is absent in the real optical images. Although there're many simulation methods focus on different space-object and space-surveillance platforms, this paper mainly proposed an optical image simulation approach focus on the small field-of-view and high detection sensitivity spatial imaging system for detecting space-debris. The proposed simulation method focuses on two key features streak and saturated star-background in the space-debris optical images which caused by high angular speed of space-debris and bright star respectively. Firstly, for the purpose of simulating the saturated star-background, this paper introduces an accurate star radiation calculating model by using bolometric magnitude to replace common visual magnitude. Secondly, in order to simulate the streak, an energy accumulation based linear approximation method is proposed which is more sufficient than the original gauss distribution model when simulate the streak caused by fast moving space-debris. Finally, the experiment results show the efficiency of proposed method.

1. Introduction

For the purpose of collision avoidance and fragmentation analysis, space-debris is one of the most important space-objects need to be survey. In present, space-based optical systems are the main instruments to capture the information of space-debris, for example, the Space-based Visible (SBV) sensor [1-2] and the European Space Situational Awareness System (ESSAS)[3]. It is necessary to develop space-debris optical image simulation technology to help implement a space-based optical system when it is not ready to launch into the space, because space-debris optical simulating image can not only provide enough experiment data to help the engineers to design information process algorithm, but also can simulate the special situations on the images like valid streak and stray light which are unusually happened during the operating period on orbit.

Although the image simulating technology of space-based sensors has recently been a tremendous development, there still remain many problems, especially in the area of space-debris optical image simulation. In fact, the space-debris optical images can not only be influenced by the parameter of optical system, CCD/CMOS noises, exposure time, coordinate transformation, etc., but also be affected by the angular velocity of space-debris. Generally, space-debris move fast and lightless, it can only be detected when shined by solar-light in space. Thereby, space-debris represents as a streak constructed by sequential faint point target in the optical image and the space-based visible surveillance system need high detection sensitivity to find space-debris[4]. However, bight star will be saturated in those optical images captured by space-based optical systems with high detection sensitivity. In summary, the streak and saturated star background simulation are the key problems need to solve.

As mentioned above, there're two important problems of simulating space-debris optical images need to solve. One is how to simulate saturated stars which will commonly appear in the space-debris optical images. Another is how to simulate the streak caused by high velocity of space-debris. Therefore, the main purpose of this paper is trying to find a way to solve those two problems. There're two contributions of this issue. Firstly, this paper theoretically derives the energy calculating equations of saturated stars in CCD/CMOS image plane by using bolometric magnitude instead of visual magnitude and conducts the approach of seeking table to find the spread size of saturated stars. Secondly, an Energy Accumulation based Linear Approximation method (EALA) is proposed to simulate the streak of rapid moving space-debris in optical images.

2. Saturated star background simulation process

In general, if a star is saturated, it means that the radiation from the star to the image plane of CMOS/CCD is larger than the full well capacity (FWC) of the CMOS/CCD. Although a method had been proposed in [5] to calculate electrons generated by star radiation on the image plane, the computing process directly use star visual magnitude to calculate the star radiation which will lead to the inaccurate results. Because the visual magnitude of stars is measured in visible spectrum, the accurate estimation results of star radiation should be on the basis of the bolometric magnitude which is measured in full-wave band. Thereby, the first part of this section proposes an accurate star radiation calculating model based on bolometric magnitude.

2.1. Star Radiation Estimation Model

For the purpose of using bolometric magnitude to replace visual magnitude during star radiation calculating, a correction coefficient BC is needed to modify the difference between them. BC can be represented as equation (1).

$$BC = m_{bol} - m_{vis} \tag{1}$$

Where m_{bol} represents bolometric magnitude, m_{vis} represents visual magnitude.

The BC can be computed due to the effective temperature on the surface of the star.

$$BC(T_{eff}) = -42.54 + 10 \cdot \log(T_{eff}) + (29000K / T_{eff}) \tag{2}$$

Where T_{eff} represents the effective temperature on the surface of the star.

The integral radiation energy density outside the stellar atmosphere is shown in equation (3) and (4):

$$S_{m_{bol}} = 2.512^{-m_{bol}} S_0 \tag{3}$$

$$S_{0-\infty} = 2.512^{-(m_{vis}+BC)} S_0 \tag{4}$$

Where $S_0 = 2.48 \times 10^{-12}$ W/cm^2.

Equation (4) gives the star radiation in full-wave band, but the optical sensors can't receive the whole energy of stars. So the calculation of star radiation in a small bandwidth should be considered. Equation (5) shows the solutions as below.

$$M_\lambda(\lambda, m_{vis}, T_{eff}) = 2.512^{-(m_{vis}+BC)} \cdot S_0 \cdot \frac{M(\lambda, T_{eff})}{M_s} \tag{5}$$

$M(\lambda, T_{eff})$ represents the star radiation in a specific wave-band when Stellar emission rate ε is set to 1.

$$M(\lambda, T_{eff}) = \frac{2\pi hc^2}{\lambda^5} \frac{1}{\left(e^{hc/\lambda KT_{eff}} - 1\right)} \tag{6}$$

Where K is Boltzmann constant, h is Planck's constant, c is speed of light.

M_s represents the star radiation in all wave-band.

$$M_s = \int_0^\infty M(\lambda, T_{eff}) d\lambda = \int_0^\infty \frac{2\pi hc^2}{\lambda^5} \frac{1}{\left(e^{hc/\lambda KT_{eff}} - 1\right)} d\lambda = \sigma T_{eff}^4 \tag{7}$$

Therefore, the atmospheric external radiation of stars among a wave range can be calculated as below:

$$M = \int_{\lambda_1}^{\lambda_2} M_\lambda d\lambda \tag{8}$$

As mentioned above, the single star flux in a specific integration time on the imaging detector can be calculated as below:

$$N_e = \int_{\lambda_1}^{\lambda_2} \tau_0 \cdot M_\lambda \cdot A \cdot QE \cdot t \cdot d\lambda \tag{9}$$

Where N_e represents the electron numbers on the imaging detector, A represents the pupil area, t represents the integration time, QE represents the quantum efficiency, τ_0 represents the transmissivity of optical systems.

2.2. *Star energy distribution model*

After star electrons being calculated, the star energy distribution model needs to be determined. Generally, in ideal imaging condition, star energy diffusion model is subjected to two-dimensional gauss distribution.

$$f(x, y) = \frac{N_e}{\sqrt{2\pi}\sigma} \exp\left(-\frac{(x - x_0)^2 + (y - y_0)^2}{2\sigma^2}\right) \tag{10}$$

Where (x_0, y_0) represents the center coordinate value of gauss function, σ is the size of gauss function.

For space-debris detection visible sensor, bright star will be saturated. As described in many researches, under ideal imaging condition, the pixel size of saturated star will become much bigger than unsaturated star[4]. Therefore, how to confirm the pixel size value is the key point in saturated star simulation. This section proposes a theoretically simple but efficient and fast method based on table-checking to quickly acquire the pixel size value of saturated star in image plane. The process diagram is shown as below. The main contents in the comparison table are the relationships between the value of pixel size and σ. The process diagram is shown in Figure 1.

IOP Conf. Series: Journal of Physics: Conf. Series **1060** (2018) 012069 doi:10.1088/1742-6596/1060/1/012069

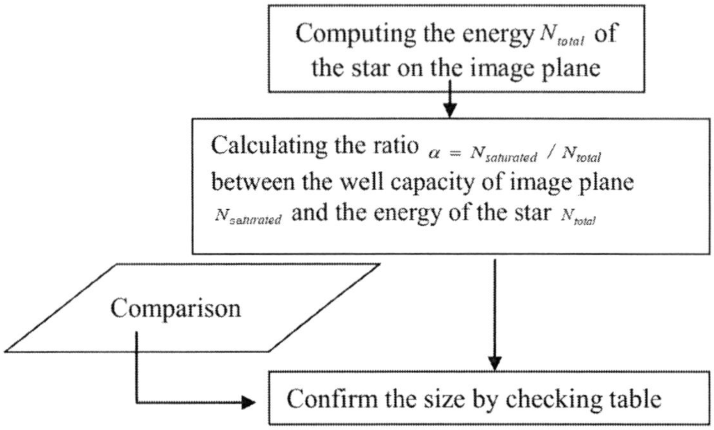

Figure 1. The process of confirming the size of saturated star on the image plane

3. Streak Simulation Method

In the image plane of space-debris visible sensor, fast moving space-objects will generate a line of streak which is the key feature to separate space-object from star background. Therefore, the streak simulation is the most important aspect need to be concern.

This paper proposes an energy accumulation based linear approximation method which is built up on the basis of that the space-debris is moving at a constant speed. Firstly, we assume that the angular moving speed of space-debris on the image plane is *V*, the velocity components in *X* and *Y* coordinate direction are V_x and V_y respectively, so among a specific integration time *t*, the moving pixels of space-debris on the image plane are M_X and M_Y.

$$M_X = V_X \cdot t / P \tag{11}$$

$$M_Y = V_Y \cdot t / P \tag{12}$$

Where, P represents the viewing angle of single pixel on the image plane.

The process of proposed method is shown as below:

Step 1: Build a local image of streak as shown in Figure 2.

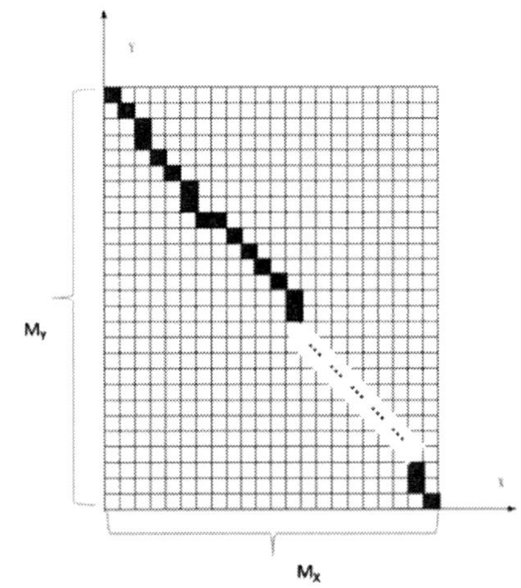

Figure 2 Local image of moving target streak

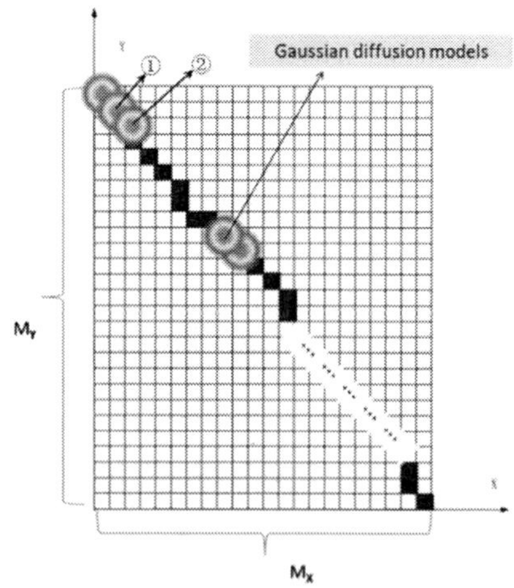

Figure 3 The diagram of a serial overlapped Gaussian diffusion models

In Figure 2, each square represents a pixel, black square represents space-debris reside in the pixel in a very short time Δt in which the space-debris can be seen as stationary object.

Step 2: The energy distribution model described in section 2 is used to calculate the electron number N of space-debris on the image plane in Δt. If N is larger than the full well capacity of image sensor, we can check the ccomparison table to determine the value of σ. Otherwise, a normal gauss distribution model can be used to simulate the stationary object in the black square. Therefore, a serial overlapped Gaussian diffusion models can be generated as shown in Figure 3. And Figure 4 shows the overlapped energy region between ① and ② which are the stationary points in a streak.

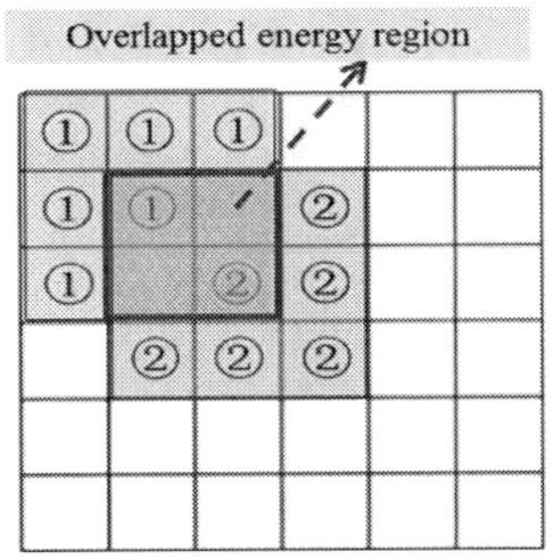

Figure 4 Diagram of energy overlapped region

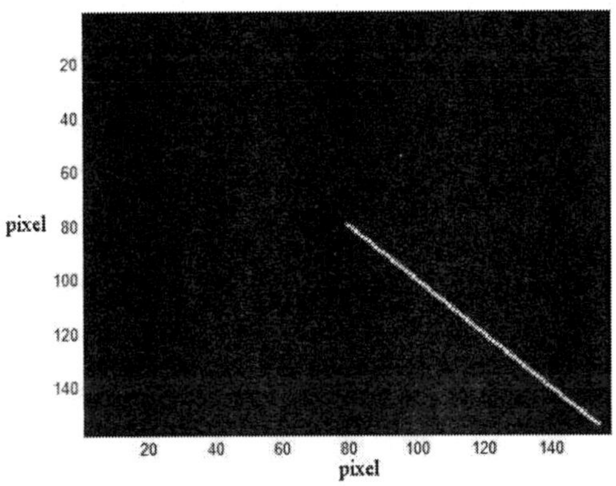

Figure 5 Diagram of simulated streak of space-debris with high velocity

Step 3: By accumulating the energy overlapped region of neighbouring black squares, the simulated streak can be generated as shown in Figure 5.

4. Experiment

This section introduces a detailed implementation of space-debris optical image simulating experiment based on a specific CMOS imaging sensor CIS2521F. Firstly,The parameters of optical system is described in Table 1 and the paremeters of CMOS imaging sensor can be found in[6].

Table 1 Parameters of optical system

Parameter of optical system		
1	Optical transmissivity	0.85
2	Lens caliber	A (cm)
3	Integration time	T (s)
4	Field of View	2°×1.4°

In Table 1, the lens caliber and integration time is adjustable which will directly influence the computing results of star radiation energy.Secondly, according to the parameters in Table 1, we can use the calculation equation (9) in section 2 to compute the electron numbers of different magnitude stars.Thirdly, we use the approach mentioned in [7] to calculate the image noise of CIS2521F. The calculation equation is introduced as below.

$$n_{CIS2521F} = \sqrt{N_S + n_{DCSN}^2 + n_{DCNU}^2 + n_{FPN}^2 + n_{PRNU}^2 + n_{RN}^2} \tag{13}$$

N_s is the electron numbers of star radiation energy on the image plane, n_{DCSN} is the root mean square of dark current noise, n_{DCNU} is the root mean square of Dark current non-uniformity noise, n_{PRNU} is the root mean square of PRNU, n_{FPN} is the root mean square of fixed pattern noise, n_{RN} is the root mean square of read noise. Fourthly, other imaging parameters are introduced as below: (1) the orbit parameters of space-debris are initialized by Satellite Tool Kit (STK) software which is developed by Analytical Graphics Corporation. (2) Four star catalogs are used to simulate the star background such as SKY2000, SAO, Tycho II and UCAC4. The UCAC4 star catalog which includes approximate a hundred million stars, compared to the commonly used star catalogs, indicates more detailed stars between 8th magnitude and 16th magnitude. (3) Space-debris do not give off light of their own, which can only be detected by space-based optical imaging sensor when they are shined by the sunlight. And the proposed method in [8] and [9] can calculate the space-debris energy on the image plane. (4) The stars position on the CMOS image plane can be calculated by using the approach mentioned in [10], while the space-debris position is computed by the aid of coordinate transform theory and SKT software. (5) The Matlab2014 is used as implementing software, and the experiments are conducted on the PC with Intel i5 core and 4G memory.Finally, the simulating experiment results are shown from Figure 6 to Figure 7 which contains streak, star background and image noise. It can be seen that saturated stars have bigger area than unsaturated stars, detected space-debris have clear and continuous streak.

5. Conclusion

This paper conducts the research of space-debris optical image simulation technology and proposes three suggestions for improvement: (1) we use bolometric magnitude instead of visual magnitude to calculate the star radiation, for the reason of acquiring precise electron numbers; (2) we propose a method of seeking table to fast determine the spread area size of saturated star; (3) we introduce an energy accumulation based linear approximation method to simulate the streak on the optical images to avoid error streak. The experiment results show that the proposed image simulation method can imitate the key features of space-debris optical image such as saturated star background and streak.

Figure 6. Simulated image when T=1s, A=15cm

Figure 7. Simulated image when T=1.6s, A=15cm

Acknowledgment

This research was supported by the National Natural Science Foundation of China (Grant No: 61503410)

References

[1] Jayant S 2000 Space-based visible space surveillance performance *Journal of Guidance, Control, and Dynamics*. **23(1)** 153-158

[2] Michael G Curt V B and Jayant S 2000 Space-Based Space Surveillance with the Space-Based

Visible. *Journal of Guidance, Control, and Dynamics* **23(1)** 148- 152

[3] Estrella O Noelia S O Nuria G 2011 Jaime N. & Holger K. Survey-only optical strategies for cataloguing space debris objects in the future European space surveillance system *Advances in Space Research* **48** 535-556

[4] Jenni V Jonne P Tero S Tuomo K and etc. 2016 Streak detection and analysis pipeline for space-debris optical images *Advances in Space Research* **57** 1607-1623

[5] KATAKE A B 2006 *Modeling, Image processing and attitude estimation of high speed star sensors* (Texas : Texas A&M University)

[6] Fairchild Imaging *MAN 0101 CIS2521F Standard and Scientific Package Datashieet_RevE.*

[7] Li J Liu J G Li G Z 2007 *3rd International Symposium on Advanced Optical Manufacturing and Testing Technologies* vol 67235F-1 p 67235F-7

[8] Diao H F Li Z and Ma Z H 2011 *International symposium on photoelectronic detection and imaging* vol 8196 p A1-A9

[9] Sun H Y Yi H Li Y C 2009 *International Conference on Space Information Technology* vol 7651 p c1-c9

[10] Li A J Liu C S and Shen X F 2013 An approach to star map simulation for star sensor considering the effect of image motion *Optics and Photonics Journal* **3** 108-111

Simulation of Second Harmonic Based on Optical Fiber Sensor

QI Hui[1], SUN Chao[1], YANG Piao[1], ZHANG Sheng[1], ZHANG Mingyang[1], MI Tie[2] and DING Jianjun[1,*]

[1]The School of Physics and Information Engineering, Jianghan University, Wuhan 430056
[2]The School of Chemistry and Environment, Jianghan University, Wuhan 430056

*Corresponding author: sycsc115@163.com

Abstract. This paper is based on the Lambert-Beer law and the differential optical absorption method is used to measure the concentration of methane gas. Simulation of the whole experiment is carried out by using MATLAB. Sine waves with periodic changes are simultaneously superimposed on triangular waves are simulating the light source in the experiment. Adding harmonic modulation to eliminate the interference of noise and fundamental components obtain the second harmonic components which carrying the concentration information. Finally, different concentration will bring the second harmonic voltage value. Records are corresponding with concentration of the second harmonic voltage value. Experiments show that the resulting second harmonic signal can meet the gas detection requirements.

1. Introduction

When light passes through the gas, the gas molecules absorbed by light of certain wavelengths. Light passes through the gas and the light intensity attenuates weakly. Although this attenuation is weak, concentration can still be measured by measuring the attenuation of light intensity Beer-Lambert's law describes the phenomenon of light intensity attenuation [1].

Based on the gas spectrum absorption theory, the technique for analyzing the absorbed gas concentration information is based on the synchronous detection using a harmonic voltage signal of a lock-in amplifier having a phase difference of 90° from the target gas modulated signal. By slowly modulating the emission wavelength stabilized at the absorption peak, the first harmonic signal is proportional to the first derivative of the gas absorption line through zero. The second harmonic signal peaks at this point and gives the second derivative profile of the gas absorption line [2].

2. Absorption gas sensor second harmonic

The differential absorption optical path is as follows: a light beam passes through the gas chamber containing the gas to be measured as a measurement signal after the incident light is split into two beams, and the other gas passes through the reference gas chamber (filled with inert gas: nitrogen gas) as contrast signal. Difference between the two signals sent to the lock-in amplifier modulation, measuring their second harmonic voltage values, resulting in gas concentration information. The system diagram is shown in Figure 1.

The advantage of this method is that it can effectively eliminate the interference of the optoelectronic components such as drift and temperature drift on the optical path and the influence of the instability of the output of the optical source on the measurement result as well as the external noise and other uncorrelated signals in the optical path in order to determine the overall optical design of the experimental program.

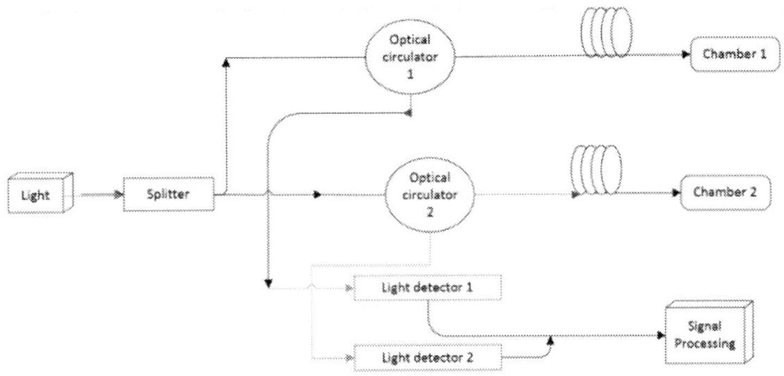

Figure 1. Experimental overall optical path scheme

3. Second harmonic analysis based on lock-in amplification

The laser emits a light wavelength that is always stable at the absorption peak, and the slow modulation signal at the light source drive is canceled, resulting in a transmission signal whose amplitude depends on the gas concentration [3].The spectral parameters of the absorption line determine the intensity of the second harmonic signal, that is, the second harmonic signal contains the gas characteristics. The use of second harmonics to detect gas concentrations has many advantages: the second harmonic signal signal (2f) is sensitive to the spectral shape of the absorption line rather than the absorption value and due to the use of phase-locked phase-sensitive detection, the laser intensity and electronic device. The associated noise is that the intensity modulated signal of the diode laser is the strongest in the 1f signal as the noise falls outside the lock band after passing the filter. Therefore, it can be used to normalize 2f to eliminate noise related to laser signal drift, scattering or beam deviation [4] [5].

In this paper, LIA is used to detect the amplitude of the second harmonic to obtain the gas concentration information. The theoretical analysis shows that the amplitude of the second harmonic has a linear relationship with the gas concentration.

4. Absorption in the optical path of the second harmonic signal system simulation

The process of modeling the system is to use the Beer-Lambert theorem, the modulation scheme of the wavelength modulation spectrum and the lock-in amplification and demodulation scheme by using the mathematical formula combined with the module building model in the simulation program Simulink. The simulation model of methane optical fiber sensor based on wavelength modulation, lock-in amplification and second-harmonic detection demodulation is shown in Figure 2.

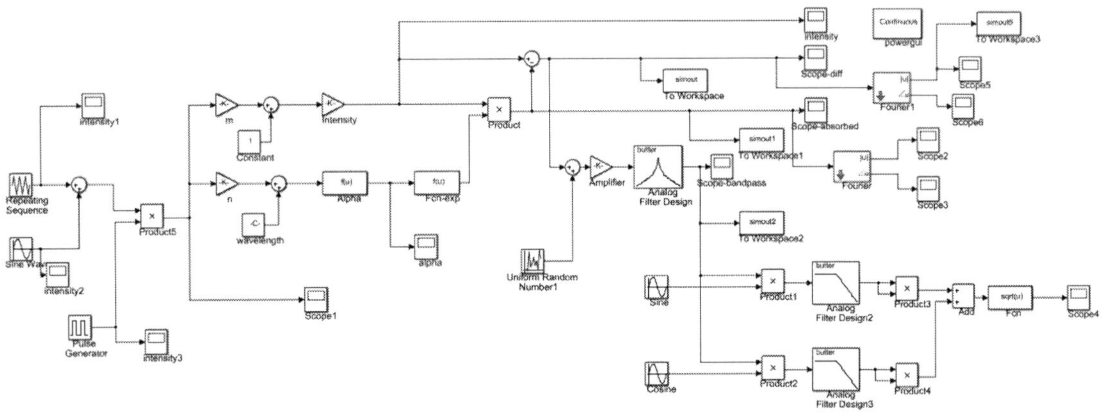

Figure 2. Methane Optical Fiber Sensing Unit System Simulation Model

The parameters of the simulation model are as follows: intensity modulation coefficient m = 0.35, wavelength modulation coefficient n=0.037, gas chamber length L = 2cm, absorption peak wavelength is1653.7nm.

Figure 3 shows the low-frequency triangular wave and high-frequency sine wave superimposed on the modulation signal. We can see that the methane absorption coefficient changes with the periodic changes in the modulation signal. In the experimental simulation, we use the periodic sine wave signal superimposed triangular wave as a system of light. In the simulation of differential absorption, two signals are simulated. According to the Beer-Lambert law, the incident light from the light source enters the optical path of the light to be measured and passes through the gas to be measured and there is a slight absorption.

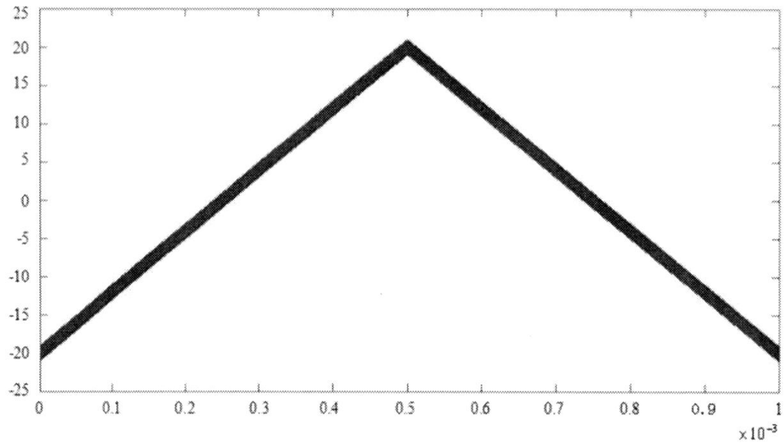

Figure 3. Light source drive waveform

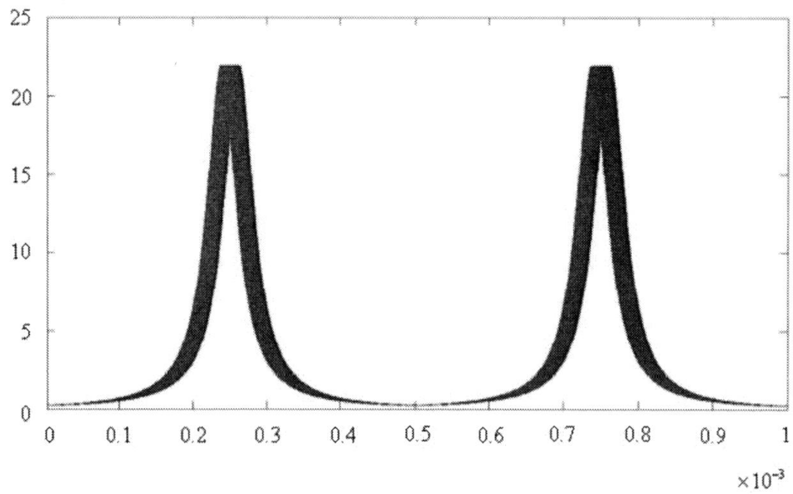

Figure 4. Absorbed waveform

5. System simulation results analysis

Figure 4 shows the change of light intensity of the gas chamber, when the methane gas volume fraction is set to 0.1. Experimental simulation, we use the periodic sine wave signal superimposed triangular wave as a system of light source. In the simulation of differential absorption, the incident light emitted from the light source according to the Beer-Lambert law enters the optical path of the light to be measured and passes through the gas to be measured and there is a slight absorption. After the difference between the reference optical path and the optical path to be measured, the external

IOP Conf. Series: Journal of Physics: Conf. Series **1060** (2018) 012070 doi:10.1088/1742-6596/1060/1/012070

interference signal such as noise is eliminated, but the intensity of the optical signal has also been weakened. In order to ensure that the signal amplitude is not excessive but also to strengthen the detection of the second harmonic signal carrying gas concentration information, we adopt a lock-in amplifier.

Detection of gas chamber light intensity with the modulation signal changes at the same time, and its change cycle and trend are the same as the modulation signal. Detection of light intensity at half the height of the image appeared obvious absorption phenomenon, and showed a cyclical change, the same cycle and modulation signal. Figure.5 shows the second harmonic image of the signal after the detection of the gas cell. It can be seen that the outline of the image is relatively vague because of the introduction of the component of the triangular wave at this frequency.

The peak-to-peak V_{PP} of the second-harmonic signal and the second-harmonic wave obtained by amplifying the signal 10^5 times at different concentrations.

When C1=CC=0.5pp,V_{PP1}=1.2774e-06(v);when C2=CC=1.0ppm, V_{PP2}=2.5549e-06(v);

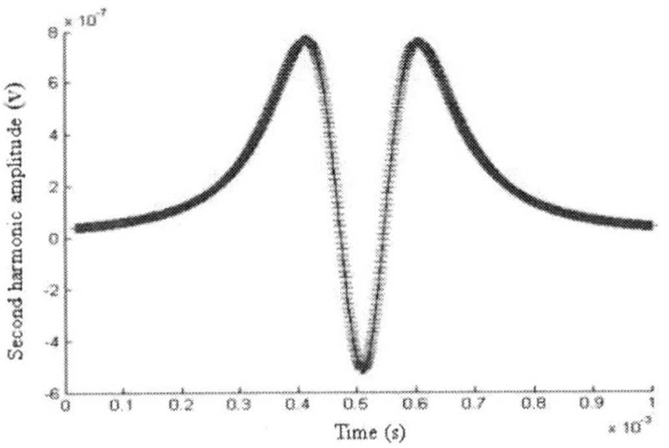

Figure 5. Second harmonic signal at CC=0.5ppm

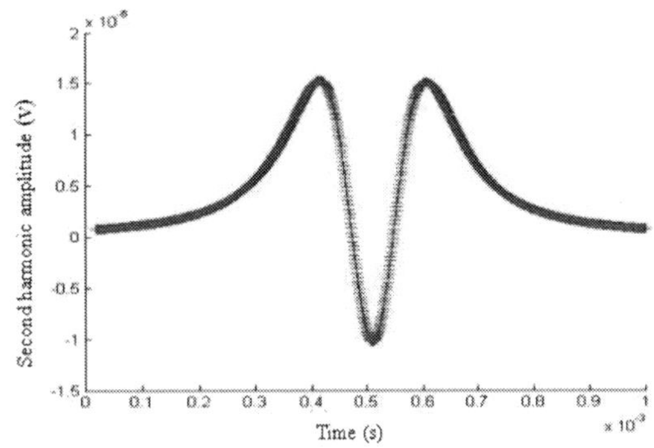

Figure 6. Second harmonic signal at CC=1.0ppm

When C3=CC=1.5ppm, V_{PP3}=3.83234e-06(v);when C4=CC=2.0ppm, V_{PP4}=5.1097e-06(v);

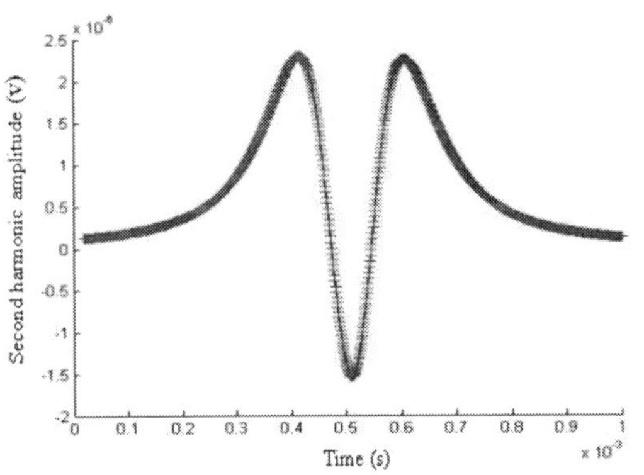

Figure 7. Second harmonic signal at CC=1.5ppm

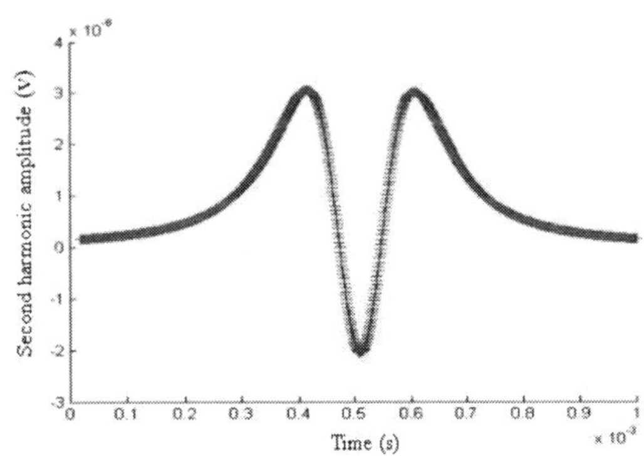

Figure 8. Second harmonic signal at CC=2.0ppm

When C5=CC=2.5ppm, V_{PP5}=6.3871e-06(v).

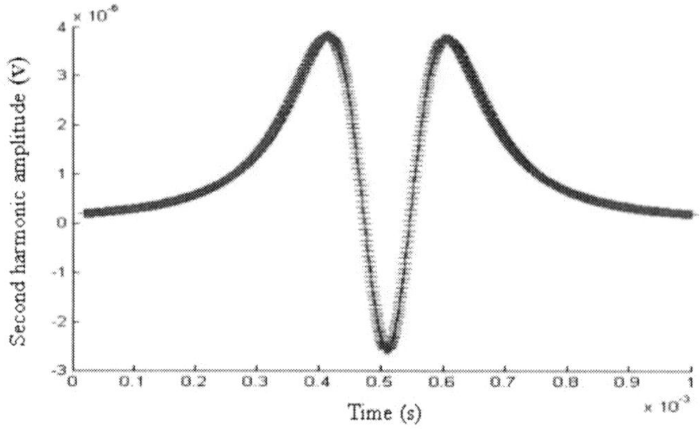

Figure 9. Second harmonic signal at CC=2.5ppm

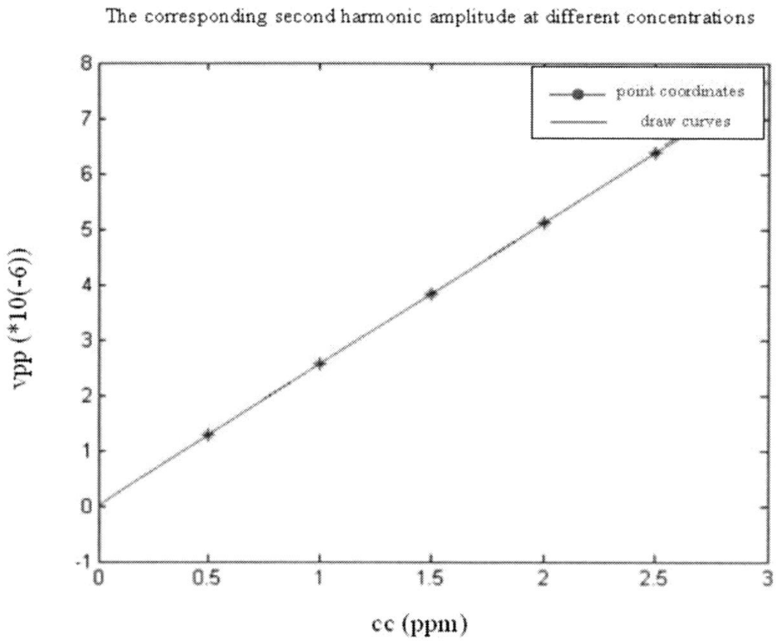

Figure 10. The corresponding second harmonic amplitude at different concentrations

It can be seen from the above five sets of data(Figure.5-Figure.9) that within a certain range, the gas concentration is related to the peak-to-peak value of the second harmonic signal, and is proportional to the ratio. As shown in Figure.10: A linear relationship can be obtained by analyzing the peak-to-peak values of the second harmonics of these concentrations and their corresponding outputs. Using the collected data to fit the linearity between the concentration and the second harmonic voltage, one-to-one correspondence is obtained. Experimental results show that this method of tutor data, good linearity. The reason is: differential absorption through the differential filter out a large number of signal interference signals, and direct absorption of these signals are still there, the measured value of the second harmonic voltage interference. Difference absorption curve obtained as a measure of the concentration of the standard curve. Simulation shows that the linearity is high, so the experimental scheme is reasonable.

6. Conclusion

In this paper, based on the Bill Lambert law, the differential optical absorption method is used to measure the concentration of methane gas. The experimental simulation process uses MATLAB to complete the gas absorption process, and the second harmonic is collected using the method of lock-in amplification. After locking the amplified signal, the larger weakened the higher amplitude of the fundamental signal, the resulting second harmonic signal to meet the gas detection requirements.

Acknowledgement
This work was supported in part by the New Technology Research on Environmental Monitoring and Application Innovation Team(T201623), New Research on Environmental Monitoring Technology(CXY201601) and Hubei Province College Outstanding Young and Middle-aged Science and Technology Innovation Team (T201420) (Corresponding author: DING Jianjun: sycsc115@163.com)

References

[1] Stutz, Jochen, et al. "A new Differential Optical Absorption Spectroscopy instrument to study atmospheric chemistry from a high-altitude unmanned aircraft." Atmospheric Measurement Techniques 10.3 (2017): 1017.
[2] Zhu, F., et al. "Mid-infrared dual frequency comb spectroscopy based on fiber lasers for the detection of methane in ambient air." Laser Physics Letters 12.9 (2015): 095701.
[3] Delahaye, T., et al. "Precise methane absorption measurements in the 1640 nm spectral region for the MERLIN mission." Journal of Geophysical Research: Atmospheres 121.12 (2016): 7360-7370.
[4] Gao, Qiang, et al. "Tunable multi-mode diode laser absorption spectroscopy for methane detection." Sensors and Actuators A: Physical 199 (2013): 106-110.
[5] P. Werle,F. Slemr,K. Maurer,R et al. Near- and mid-infrared laser-optical sensors for gas analysis[J], Optics and Lasers in Engineering 37 (2002) 101–114.

The Test on the Short Period Mode of the Simulator With Time-Frequency Combination

Shanlu Zhao and Guohui Li

Aviation University of Air Force, Changchun China

Email:381355365@qq.com

Abstract: Based on flight simulation data and measured data, the article extracted short cycle time of correlating parameters curve from a certain type of flight simulators. And in the time domain, using the average standardized distance test method and gray correlation analysis method for this type of simulator simulation data. In the frequency domain, the classical spectral estimation method is used for quantitative detection, and the consistency test of flight simulation data and flight parameters is realized. Then the evidence theory is improved and thus the synthesized type simulator short cycle when the correlating parameters and frequency domain the consistency of the test results, verify the longitudinal short-period mode type simulator fidelity is better.

1. Introduction

Flight simulation training is an important means to improve the wingmanship, carring out dangerous training and ensure flight safety.In the verification work of flight simulator, flight quality is a very important verification content, which is closely related to the verisimilitude of flight simulator[1].

At present, the dynamic consistency test of simulation results of flight simulator is mostly verified by single time domain or frequency domain method, and the comprehensive verification method is seldom verified[2].The Thile inequality coefficient method is a widely used dynamic data consistency test method, which has the advantages of simple principle and small calculation, while the TIC coefficient is related to the selection of the origin of coordinates, which can increase and decrease the TIC coefficient by changing the coordinate origin, so that the consistency analysis result can be misjudged[3].However, the average standardized distance test method can analyze the sample data quantitatively, but there is also a case of inaccurate test[4].Grey relational analysis method is a common method for data consistency analysis in time domain without considering the statistical distribution law of sample sequence[5].However, it is only a qualitative verification method, and the analysis results alone have certain problems.The classical spectrum estimation method is mature and fast, but because of the direct addition of "window" to the data, there may be problems such as energy leakage and separation of spectral lines[6].

According to the flight parameters and simulation data collected, six important parameters of the aircraft longitudinal short period mode are extracted.Based on the improved evidence theory, the comprehensive average standardized distance test method, grayscale correlation analysis method and classical spectral estimation method are obtained, and the data consistency test results are obtained.And flight parameters were compared according to the damping ratio and natural frequency of vibration with simulation data, the modal characteristics of further verification, test the longitudinal short-period mode type flight simulator fidelity is better.

2. Correlation theory

2.1. Time domain analysis methods

2.1.1. Average standardized distance test methods. Set $X=\{x(1),x(2),\cdots,x(n)\}^T$ as n entries of sequence $\{x(A)\}$ .Set $Y=\{y(1),y(2),\cdots,y(n)\}^T$ as the simulation sequence. $Y_i=\{y_i(1),y_i(2),\cdots,y_i(n)\}^T$ is the ith sample, $i=1,2,\cdots,m$.

The mean standardized distance between the measured sequence X and the sample mean of the simulation sequence $\bar{Y}$ is $D(X,\bar{Y})$, as shown in equation (1),

$$D(X,\bar{Y})=\sqrt{\frac{1}{n}\sum_{A=1}^{n}\left[\frac{X(A)-\bar{Y}(A)}{S(A)}\right]^2} \tag{1}$$

In the formula, $S=\frac{1}{m-1}\sum_{i=1}^{m}(Y_i-\bar{Y})(Y_i-\bar{Y})^T$, $\bar{Y}=\frac{1}{m}\sum_{i=1}^{m}Y_i$.

If the average standardized distance $D(X,\bar{Y})$ between the sample X and its simulation sample mean $\bar{Y}$ is satisfied with equation (2),

$$D(X,\bar{Y})\geq\sqrt{\frac{m^2-1}{m(m-n)}F_{\alpha,n,m-n}} \tag{2}$$

So, at the significant level α ,the difference between the two sequences can be determined.

2.1.2 Grey relational analysis theory. Define the grey correlation coefficient of X_t to Y_t is,

$$\xi_i(t)=\begin{cases}1, |x_t(t)-y_t(t)|\equiv C\\ \dfrac{\min\limits_{i,t}|x_t(t)-y_t(t)|+\sigma\max\limits_{i,t}|x_t(t)-y_t(t)|}{|x_t(t)-y_t(t)|+\sigma\max\limits_{i,t}|x_t(t)-y_t(t)|}, else\end{cases} \tag{3}$$

In the formula, σ is resolution coefficient, C is constant, when σ is smaller, the resolution is higher. In actual engineering applications, $\sigma\in[0,0.5]$ is usually taken.

The grey correlation between X_t and Y_t is,

$$\gamma_i=\frac{1}{n}\sum_{t=1}^{n}\xi_i(t) \tag{4}$$

The larger the grey relational degree, the higher are there the correlation between the two groups and the higher the consistency of the data.

2.2. Frequency domain analysis methods

The classical spectrum analysis method is used to estimate model validation is to estimate power spectral density of random sequence. Here suppose that x_t and y_t are samples of the output sequence of the simulation model and the samples of the flight parameters. $S_{xx}(\omega)$ and $S_{yy}(\omega)$ are the corresponding self-spectral density functions respectively, and the spectral quotient function of x_t to y_t is defined as

$Q_{xy}(\omega)=S_{xx}(\omega)/S_{yy}(\omega)$.

So for the given significance level α, the $Q_{xy}(\omega)$ confidence interval of the spectrum quotient function $1-\alpha$ can be calculated by the following formula,

$$L(\omega)=\frac{\hat{Q}_{xy}(\omega)}{F_{\alpha/2}(v,v)}\leq Q_{xy}(\omega)\leq\frac{\hat{Q}_{xy}(\omega)}{F_{1-\alpha/2}(v,v)}=U(\omega),0\leq\omega\leq\pi \tag{5}$$

If the confidence interval contains a value of 1, it is considered that the simulation output sequence and flight test results are consistent with the significant level of α.

2.3. Evidence theory

In the evidence theory, Sample space Θ consists of a series of two two repelling each object, and includes current to identify all the objects, namely $\Theta=\{\theta_1,\theta_2,\cdots,\theta_n\}$ .Make Θ domain set as a theory, 2^Θ are all the subsets of Θ collection, if $m:2^\Theta$ as the basic probability distribution function, it meet the formula[7],

$$\sum_{A\in P(\Theta)} m(A)=1 \quad , \quad m(\phi)=0 \tag{6}$$

In the formula,m(A) is called the basic probability assignment of event A, indicating the degree of support for proposition A.The D-S combination rule is,

$$\begin{cases} m(A)=\dfrac{1}{1-k}\displaystyle\sum_{\substack{A_i,B_i\\A_i\cap B_i=A}} m_1(A_i)m_2(B_j) for A\neq\phi \\ m(\phi)=0 \end{cases} \tag{7}$$

In the formule(7),k is the conflict factor, which reflects the degree of conflict between the evidence.

$$k=\sum_{A_i\cap B_i\neq\phi} m_1(A_i)m_2(B_j) \tag{8}$$

And the closer the k value is to 1, the greater the conflict between the evidence.The closer to 0, the smaller the conflict.

2.4. Improved evidence theory

Classic theory of D-S combination contradiction and robustness of the problems, so when the serious conflict between evidence or a proposition of tiny change of basic probability distribution, may cause the results of synthetic radical changes[8].The improvement steps are as follows:

1) The average probability value of the n group of evidence of proposition A_k is assigned to $\bar{m}(A_k)$ to determine the average support degree of all evidence for the proposition A_k.

2) The distance d_i between the probability distribution and the average probability distribution of the ith evidence body is determined one by one. The formula is as follows,

$$d_i=\sum_{k=1}^{m}|m_i(A_k)-\bar{m}(A_k)| \tag{9}$$

3) According to the distance d_i to determine the weight ω_i of each body of evidence, the formula is as follows,

$$\varphi_i=\frac{1}{d_i}, \quad \omega_i=\frac{\varphi_i}{sum(\varphi_i)} \tag{10}$$

4) Identify conflicting evidence. When the weight of the conflicting evidence i $\omega_i < 1/n$ (that is, less than the average weight of each evidence), the modified coefficient is calculated as,

$$discount = \frac{\omega_i}{\max(\omega_i)} \; (discount < 1) \tag{11}$$

3. Verification and analysis

3.1. Time curve extraction

Selecting a group to test the model of the longitudinal characteristics of longitudinal times pulse comparison between the control input validation, the initial height is 5 km, the speed is 0.4 Ma, flying parameters according to the comparision of the data of simulation results as shown in figure 1 , solid line for flight parameters according to the curve, dotted lines for simulation data parameters.

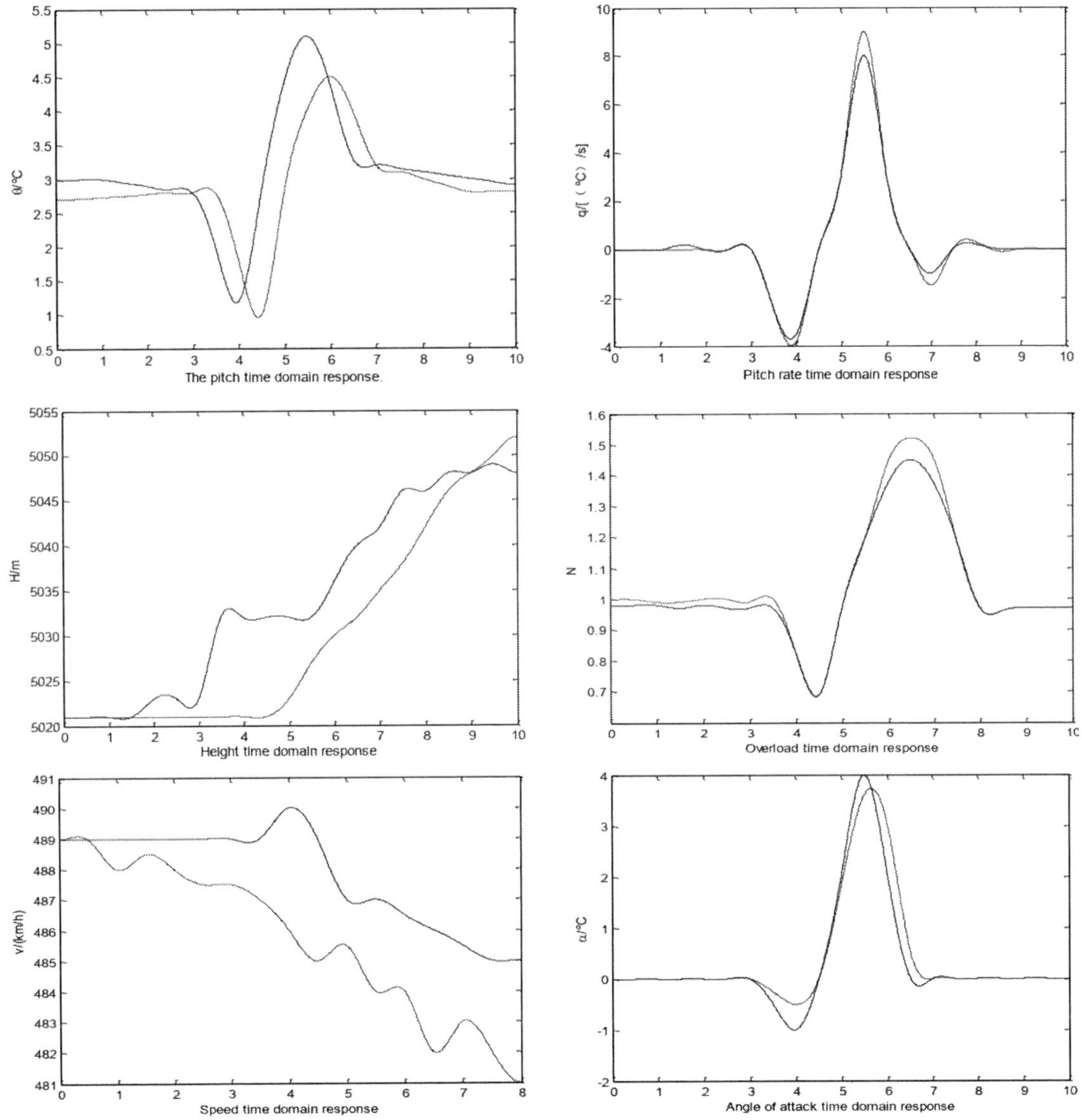

Figure 1. Parameter time domain change curve

3.2. The time domain authentication

3.2.1. Average standardized distance test method validation. Based on the flight parameters and simulation data, the parameters of the short-period mode are tested in the time domain by means of mean standardized distance test.

According to figure 1, the longitudinal short period mode can be completed within 3-8 seconds, so the flight data in 3-8 seconds should be analyzed in consistency.Take the Angle of attack as an example.Under the condition of 5km height and 0.4Ma, a group of flight parameters were obtained with the simulation data of multiple groups, and the average normalized distance between the actual sequence of attack and the simulation sequence of attack Angle was,

$$D(X,\overline{Y}) = \sqrt{\frac{1}{n}\sum_{A=1}^{n}\left[\frac{X(A)-\overline{Y}(A)}{S(A)}\right]^2} = 2.3575 \tag{12}$$

When the significance level is 0.08, its rejection area is $D(X,\overline{Y}) \geq 2.4069$.It can be seen from this that, with the significance level of 0.08, the consistency between the flying parameter sequence and the simulation sequence is very good.

3.2.2. Grey correlation analysis method validation. The flight parameters and simulation data obtained under the condition of height 5km,speed 0.4Ma was taken into equation(3),equation(4), and if taken the resolution coefficient 0.5, the grey correlation is 0.75.

For the consistency result of qualitative method, fuzzy set theory is adopted to define the acceptable criterion.The consistency of data was divided into five levels: very high (1.0~0.9), high (0.9~0.7), general (0.7~0.5), poor (0.5~0.3) and very poor (0.3~0.0).Therefore, it can be seen that the consistency degree of the two sequences is high.

3.3. Frequency domain authentication

The data length is divided into 8 segments and 50% overlap with the sampling frequency for 5km height.In this case, Welch's method is used, and the window function is the hanning window.According to Welch's method, the estimated power spectrum is shown in figure 2,

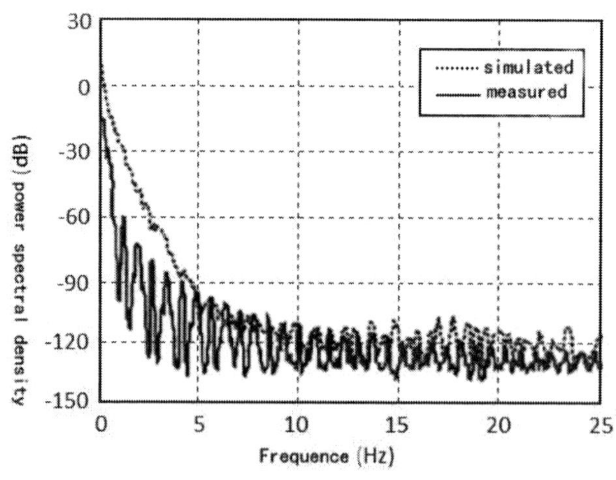

Figure 2. Angle of attack data power spectrum curve

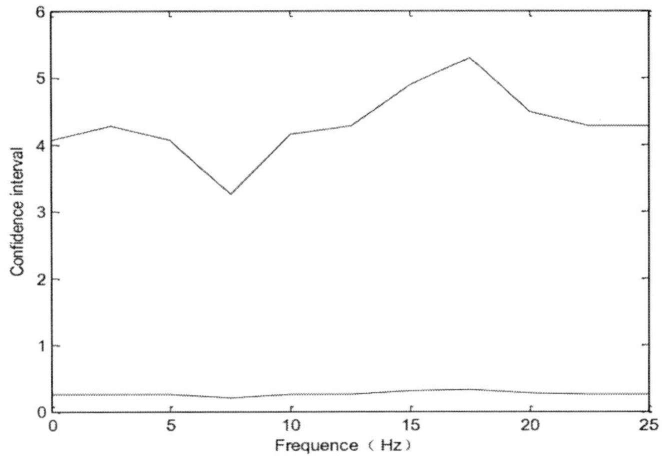

Figure 3. Confidence interval

The test formula is related to α , and take $\alpha = 0.01$ here.The distribution table shows that, $F_{\alpha/2}(15,15) = 4.07$, $F_{1-\alpha/2}(15,15) = 0.25$.

And then you can compute the lower bound at the confidence interval at each frequency point.With the frequency as the x-coordinate and the lower and lower limit of the confidence interval as the vertical coordinate, the confidence belt curve is shown in figure 3.

As can be seen from the diagram, 1 is contained within the confidence upper limit and the confidence limit.Therefore, the power spectrum of the simulation sequence and the measured sequence is consistent with the 99% reliability level.

3.4. Integrated test results

Using the average standardized distance test method, grey correlation analysis method and classical spectrum estimation method,the consistency of the longitudinal short-period damping ratio of the simulator is tested in the time domain and frequency domain.The obtained consistency result is used as the basic probability distribution function of the evidence, and the fusion result is calculated according to the improved d-s method.The basic probability distribution is shown in table 1 below.

Table 1. Basic probability distribution table of evidence

	A	B	C
m_1	0.92	0	0.08
m_2	0	0.75	0.25
m_3	0.005	0.005	0.99

Then, $\bar{m}(A) = \frac{1}{3}(0.92 + 0 + 0.005) = 0.31$, $\bar{m}(B) = 0.25$, $\bar{m}(C) = 0.44$

Plug in formula (9),

$$d_1 = |0.92 - 0.31| + |0 - 0.25| + |0.08 - 0.44| = 1.22, d_2 = 1, d_3 = 1.1$$

$$\varphi_1 = \frac{1}{d_1} = 0.82, \varphi_2 = 1, \varphi_3 = 0.91, \omega_1 = \frac{\varphi_1}{\varphi_1 + \varphi_2 + \varphi_3} = 0.30, \omega_2 = 0.37, \omega_3 = 0.33$$

Since there are three pieces of evidence, the average weight of each evidence is 1/3.So Evidence 1 is calculated to be corrected of the the discount degree.

According to formula(11), $discount = 0.3/0.37 = 0.82$ .Therefore,the dynamic consistency degree of the simulation results at 5km height is 0.83.

In the same way, the consistency degree of the simulation results of the other five parameters such as the pitch Angle and the flight parameters can be obtained, as shown in table 2,

Table 2. Analysis table of parameter consistency degree.

Parameter	Pitching Angle	Pitch rate	Speed	Height	Overload
Consistency	0.82	0.88	0.75	0.71	0.85

It can be seen that the consistency degree of each parameter flight parameter according to the sequence and simulation sequence is higher than 0.7. The short period mode of the simulator is consistent with the actual aircraft well.

4. Conclusion

In this paper, a dynamic consistency test method for simulation data of time domain and frequency domain is presented, and the longitudinal short period mode of flight simulator of a certain aircraft is verified.This paper firstly extracted the short cycle time history curve of modal parameters, and then applied to the average standardized distance test method, grey correlation analysis method and the classic spectrum estimation three methods, time domain and frequency domain of the longitudinal short-period type simulator the inspection of the consistency of related parameters, and the evidence theory is improved and synthesized consistency inspection results, obtained the comprehensive results of dynamic simulation data consistency check type simulator, this type of flight simulator was verified longitudinal short-period mode has good fidelity.

Reference

[1] Jingtao.Modeling and simulation of naval weapon and equipment systems [J].System engineering and electronic technology,2005(10):1767-1770.

[2] Comdale A.Real world data collection for model validation[C]//Proc.of the Winter Simulation Conference,2009:1764-1769.

[3] Fu huimin.Comparison and analysis of simulation results distance test method and TIC method [J]. Journal of aviation dynamics,2009, 24(12):2784-2788.

[4] Fu huimin,Chen jianwei.Method of distance test for dynamic simulation results [J]. Mechanical strength,2007(02):206-211.

[5] Wei Hualiang, Li Zhongwu.Grey correlation analysis and its application in the verification of simulation model of missile system [J].System engineering and electronic technology,1997(02):55-61.

[6] Chen J W.Comparison analysis for validating methods of system simulation models[C]//Proc.of the IEEE Conference on Electric information and Control Engineering,2011:232-235.

[7] Yang F B,Wang X X.Combination method of confliction confidence in D-S evidence the ory[M].Beijing:National Defense Industry Press.2010:66-78.

[8] ZHANG Xibin,ZONG Qun.Wavelet coherence analysis approach to model validation of the hypersonic vehicle model[C] // Proceedings of the 32nd Chinese Control conference,2013.

Study on the Fast Air Heating Method for the Testbed of the Environmental Control System of the Aircraft

Zheng Dai[1,*] and Yi Cui[1]

[1]Shanghai Aircraft Design and Research Institute, Commercial Aircraft Corporation of China Ltd, No. 5188 Jinke Road, Shanghai, China

*daitseng@yeah.net

Abstract. In this paper, the necessity of studying the dynamic performance of the control system was firstly discussed. Through analysing the technical difficulties of the dynamic performance of the thermodynamic testbed, it was found that the most important problem to be solved was to simulate the actual working process of the engine and guarantee the test piece has a sufficiently high supply air temperature to meet the requirements of rapid air temperature rise. Through investigating the method of fast heating, the hot and cold blending method was proposed to realize the rapid response of the temperature. Through the switching of different air supply lines, the process of rapid temperature rising can be achieved.

1. Introduction

With the rapid development of civil aviation, the overall design requirements of large-scale delivery, long range and low fuel consumption have led to the continuous improvement of the performance of aircraft engines. As a result, the range of air temperature for the air-control systems has increased. Advanced Communications, navigation systems and ever-increasing on-board electronics make the heating power on the machine ever-increasing, resulting in increased air flow to the environment control system (ECS). So far much work has been carried out on the ECS. Vargas proposed a method to the thermodynamic optimization of components [1]. Ordonez and Bejan [2] took into account two basic questions in the thermodynamic optimization and analysed the robustness of the optimized models. In addition, the greatly varying ambient conditions undoubtedly change the ram air temperature parameters. As a result, the dynamic performance of the control system is facing a huge challenge [3][4].

Research on the dynamic performance of a control system requires a large number of ground-based laboratory tests to verify. Only in this way can we ensure that the problem is not brought to sky as much as possible. As the core of the environmental control laboratory, the thermodynamic testbed is a universal test platform for performance testing of the aircraft environmental control system. At present, the thermodynamic testbed mainly tests the principle verification and the static performance index, rarely involving the dynamic performance. A dynamic response simulation was performed by Eichler [5], which was to study sensitivity, identify possible problem areas in the design, and act as a tool to evaluate the dynamic effects of any future design changes. However, it did not pay attention to the way to achieve the changes of the external environment. Honeywell thermodynamic testbed can control the gas supply parameters of the thermodynamic test bed in real time through the mathematical model of the engine. The temperature has a certain followability of the engine performance curve in the flight envelope. However, the temperature hysteresis caused by the heat capacity of the test tube cannot simulate the ideal engine working condition.

In this paper, we will discuss the design of the thermodynamic test stand. First, we will discuss the technical difficulties in the design of the thermodynamic test stand, and then find out the design key points suitable for the dynamic character analysis. Through the research, the feasibility of the current measures are discussed, and finally the preliminary design scheme is given based on the dynamic performance requirements of the thermodynamic testbed.

2. Technical difficulties

According to the instantaneous data of the maximum change rate of the air outlet temperature, the rapid rise characteristic of the engine bleed air temperature is plotted. All of these data can be found in aircraft air bleed system function interface control file (FICD). As shown in figure 1, High-pressure bleed air temperature is usually 500 °F ~ 1200 °F and Medium-pressure bleed air temperature range from 200 °F to 700 °F. Furthermore, based on the maximum heating rate and the most severe heating curve facing the aircraft's control system, the maximum inlet temperature rise rate of the high-pressure gas supply meets 235.8 °F / s. The temperature should increase from 580 °F to 1100°F in 5s and stabilize within 10s. For the medium-pressure, the gas supply meets 182.5°F / s and the temperature should increase from 300 °F to 1100 °F in 5s and stabilize within 10s. Therefore, in order to truly simulate the working condition of the engine, the primary problem is that the high and medium pressure air supply lines of the testbed must have a sufficiently high supply air temperature and meet the requirement of rapidly heating for the inlet air of the test piece.

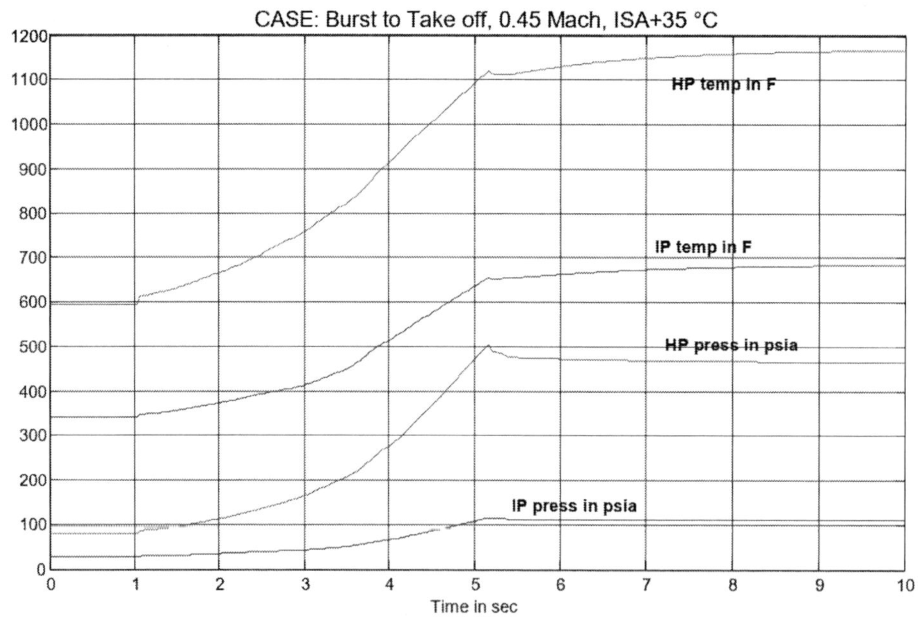

Figure 1. Engine bleed air temperature rapid rise characteristics.

At present, the commonly method of heating air are used: direct heating of electric heater and indirect heat transfer of high temperature gas. As gas combustion temperature control technology is not mature, and the temperature regulation accuracy is not too high, the electric heating has been used to obtain high-temperature heat source. There are two typical high-temperature air electric heaters currently on the market classified into two types for civil and aviation use. They are mainly used for the testbed in static conditions, but useful information can be found from their basic configuration and working principle for the purpose of selecting core components of the thermodynamic testbed heater.

Three-stage air heating was used in civil air heater. The temperature rise step by step and the highest temperature can reach 1500 °F. Aeronautical air heaters are based on the Soviet electric heater, specifically used in aeronautical environmental control thermodynamic testbed. The design maximum temperature of the electric heater is below 1100 °F. Through the comparison of the working principle and the static design indexes of the two kinds of air electric heaters, the aeronautical air heater has small heat capacity, weak thermal inertia and higher temperature rising than the civil heating in the middle and low temperature range when the airflow pressure is the same through the electric heater. In the high temperature range, with the increase of temperature, the air temperature in the isolation chamber is obviously higher than that in the middle and low temperature range. The heat dissipation to the outside increases and the thermal inertia of the structure is smaller, but the temperature rising rate significantly decreased, and the highest temperature of the electric heater outlet air is lower. In contrast, civilian electric heater makes it less heat dissipation due to the use of insulation material, and electric heater outlet air temperature is higher. As dynamic performance is not the focus of designers and users of electric heaters, there is no credible data to support the dynamic performance design of the testbed. Through the time-consuming analysis from normal temperature to maximum temperature, it is impossible to directly use the electric heater to meet the dynamic performance requirement of the testbed. One of the technical difficulties is how to achieve rapid and fast heating demand by using existing electric heaters.

In addition, the change rate of the fastest temperature rise supposed by the thermodynamic testbed should be consistent with the actual situation of the system. In order to achieve such dynamic properties, the thermal inertia and the volume inertia of the rear end of the heater are the key factors. Such inertial cause a large delay in temperature change and make the temperature control system produce some overshoot and fluctuation. Therefore, the originally difficult rapid heating performance become more difficult. In order to illustrate the effect of thermal inertia and volumetric inertia on dynamic changes in supply air temperature, a simulation description is given. The simulation assumes that the duct is stainless steel with an inner diameter of 6 in and a thickness of 0.375 in. the temperature characteristics of duct were compared in different length (10 ft, 20 ft and 30 ft), different initial wall temperature and different supply temperature. As shown in figure 2, the results show that the shorter the ductr, the better the dynamic response of the output airflow temperature when the same supply temperature, the same initial wall temperature and the same air flow in the duct. The higher the supply air temperature, the better the dynamic response of the duct output airflow temperature response. Therefore, in order to guarantee the inlet temperature rising rate, it is helpful to increases the temperature of the outlet air of the electric heater and minimize the length of the duct between the heater and the inlet of the test piece. This study also found that pre-heating of the duct can also speed up the dynamic response of outlet air temperature.

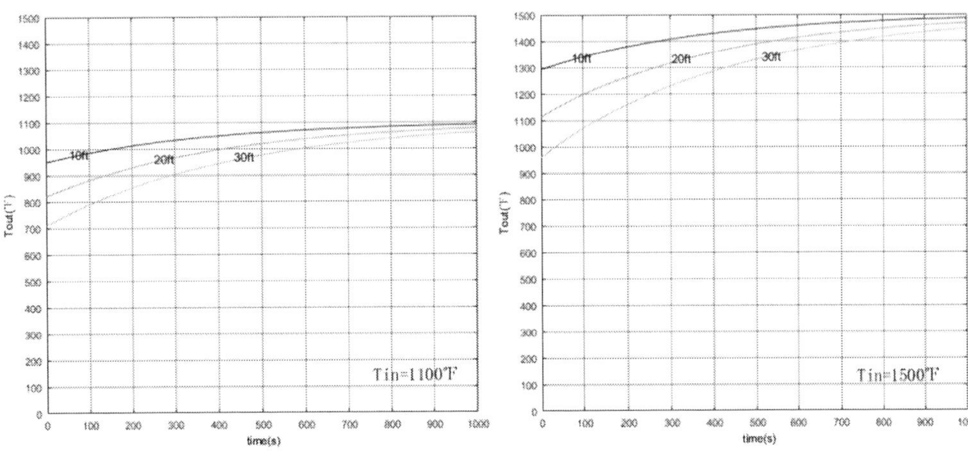

Figure 2. The temperature characteristics of duct

3. Scheme design of thermal dynamic testbed

The dynamic performance test of the environmental control system requires that the inlet air of the pipeline should be heated rapidly. The testbed must overcome the thermal inertia and volumetric inertia. Based on the existing products, the testbed should have the ability of fast air heating and rapid temperature respond through system engineering theory.

3.1. Fast air heating

At present, the fastest temperature rise rate of the electric heater in the low temperature range can only reach 50 °F /s. Due to the increase of the external heat loss in the high temperature range, the temperature rise rate is greatly decreased, which are far from meeting the requirements of the environmental control system demand for dynamic performance testing. On the basis of the static test method, drawing on the system configuration method of aircraft cockpit temperature control, the rapid response of the temperature is achieved by the method of hot and cold gas blending. As shown in figure 3, the specific method is as follows: the test bench adds a bypass in the front of the electric heater, by adjusting heater front valve and bypass valve, one part of the supply air flow through the electric heater and another part flow through the bypass to maintain normal temperature. The two parts were mixed at the rear end of the heater. Through changing the ratio of hot and cold air mixture, we can achieve the purpose of rapid temperature control. During the test, the electric heater kept the outlet air temperature at the maximum temperature that the electric heater could reach. The cold air regulating valve in the test tube was normally closed and the hot air regulating valve was always open. This system overcomes the inherent characteristics of the slow rate of temperature rise of the electric heater and ensures that the gas supply to the test pipe inlet has a faster heating rate.

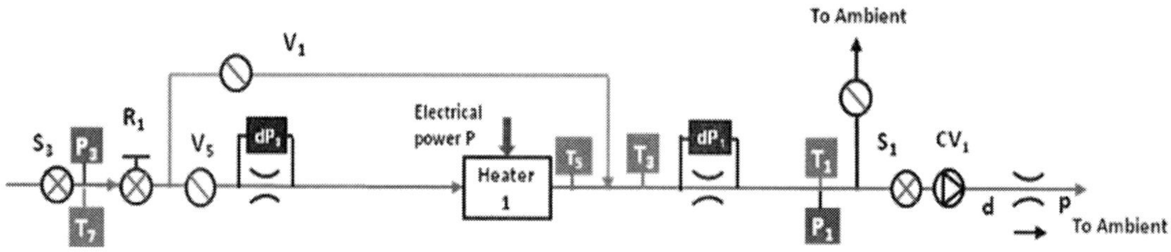

Figure 3. Fast air heating principle diagram.

In order to solve the problem of thermal inertia of the pipeline connecting the outlet of the electric air heater and the control system of the aircraft control system, two aspects need to be taken. On the one hand, under the premise of ensuring the safety of the testbed, the distance between the outlet of the electric heater and the interface of the aircraft loop control test piece should be shortened as much as possible to reduce the inertia and retardation of the system; on the other hand, the outlet air temperature of the electric heater should be increased to satisfy pipe heat capacity requirements.

3.2. Rapid temperature respond

In the take-off phase, the aircraft engine thrust rapidly increased within 10s and high pressure air outlet port was rapid heating. According to the heating curve shown in figure 1, the analysis shows that the temperature change during high and medium pressure switching is similar to the step response, and the testbed can achieve the dynamic performance by switching different air supply lines just as shown in figure 4. During test preparation, the two air supply line of the testbed were used to simulate the initial upstream temperature before and after the temperature change. One air supply line of the testbed has been in the emptying stage before getting testbed control instructions. There was 2-3s delay between valve changes and the response of the step signal slightly slowed down which was closer to the true simulated speed heating process.

IOP Conf. Series: Journal of Physics: Conf. Series **1060** (2018) 012072 doi:10.1088/1742-6596/1060/1/012072

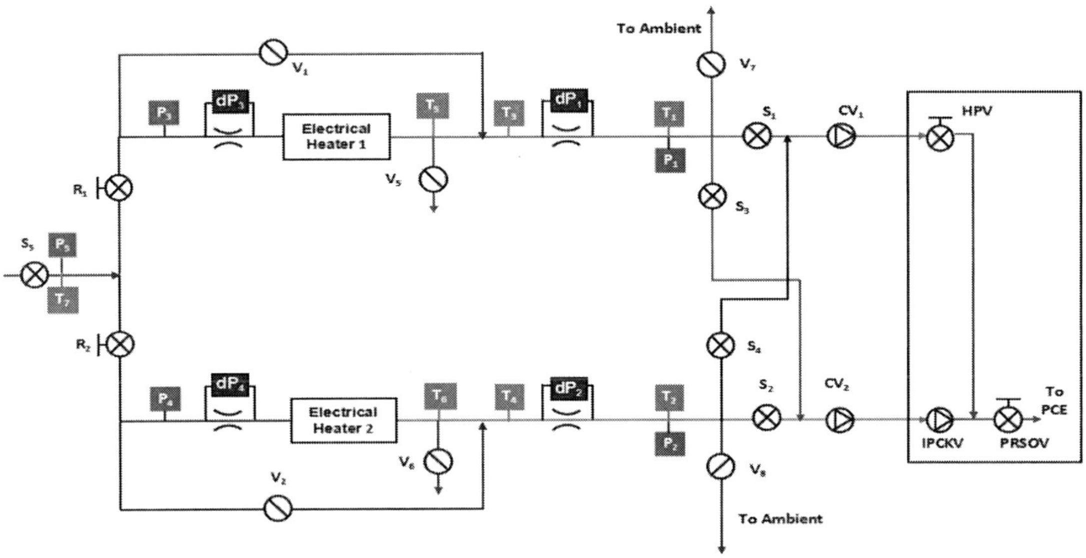

Figure 4. Rapid temperature respond principle diagram.

4. Conclusion

In this paper, the necessity of studying the dynamic performance of the control system is discussed firstly. The design of thermodynamic testbed makes it possible to study this dynamic characteristic in the laboratory. This paper illustrates the technical difficulties of the dynamic performance of the testbed, and the primary problem is that the high and medium pressure air supply lines of the testbed must have a sufficiently high supply air temperature and meet the requirement of rapidly heating for the inlet air of the test piece. Because the current electric heater cannot meet the requirements of the dynamic performance test of the environmental control system at the heating rate, a hot and cold blending method is proposed in this paper to achieve the rapid response of the temperature. Through the switching of different air supply lines, the process of rapid temperature rising can be achieved.

Acknowledgments

Authors wishing to acknowledge assistance or encouragement from colleagues, special work by technical staff or financial support from organizations should do so in an unnumbered Acknowledgments section immediately following the last numbered section of the paper.

References

[1] Vargas Jose V C, Bejan A 2001 Integrative thermodynamic optimization of the environmental control system of an aircraft *INT J HEAT MASS TRAN* **44** 3907

[2] Ordonez JC, Bejan A 2003 Minimum power requirement for environmental control of aircraft *Energy* **12** 1183

[3] Zhao HL, Hou Y, Zhu YF, Chen L, Chen ST 2009 Experimental study on the performance of an aircraft environmental control system *Appl. Therm. Eng.* **29** 3284

[4] J. HE, J. ZHAO. Dynamic Simulation of the Aircraft Environ-mental Control System. *Chinese Journal of Aeronautics*, Vol.**14**, No. 3, 129-133 (2001)

[5] Eichler J 1975 Simulation Study of an Aircraft's Environmental Control System Dynamic Response *J. Aircraft* **12** 757

IOP Conf. Series: Journal of Physics: Conf. Series **1060** (2018) 012073 doi:10.1088/1742-6596/1060/1/012073

Study on the Simulink Model for the Testbed of the Environmental Control System of the Aircraft

Yi Cui*
Shanghai Aircraft Design and Research Institute, Commercial Aircraft Corporation of China Ltd, 5188 Jinke Road, Shanghai, China

***cuiyi@comac.cc**

Abstract. The rapid rise in temperature and pressure and the Switching of high pressure and medium pressure state of the thermal dynamic testbed is realized by the switch between the two channels of gas supply. The dynamic performance depends mainly on the travel time of the switch valve. We simulate the class step change, which basically meet the needs of the dynamic test of the environmental control system. The dynamic rising temperature and pressure of the testbed depends mainly on the mixing of the cold and hot air and the adjustment of the regulating valve. It cannot be directly evaluated that how many dynamic performance indicators can be achieved. The MATLAB/Simulink modelling software is used to build the thermal dynamic test platform system and control model. And we use this dynamic simulation technique to analyse the time of fast response.

1. Introduction

At present, the built thermal dynamic testbed in china mainly serve the development and maintenance of military aircraft. The object of these testbed is the finished product. The purpose is to determine whether the performance parameters of the component meet the static index of the design. Although the development of military aircraft will also verify the function of the system, only is in the middle and later stage of the project development. Its focus is the verification of the principle and the static performance index test. Therefore, the current thermal dynamic testbed has no dynamic performance. In addition, environmental control system of the military aircraft is simpler than the civil aircraft. In the bleed air system of the military aircraft, there is no the high and low pressure switches , no fluctuation in the temperature and pressure of the gas supply. The coupling effect between parameters is not obvious and the problem of dynamic characteristics is not outstanding. Therefore, there is no test requirement for dynamic performance verification. Military aircraft increasingly pursue ultra-manoeuvrability and cause short time fluctuation of air entraining parameters, making the demand for the dynamic characteristics of the military aircraft environmental control system more and more obvious. Chinese users generally agree that dynamic characteristics are an important trend in the future development of thermal dynamic testbed, which will improve the practical value of environmental control system test. The system design of the environmental control of foreign aircraft manufacturers, who providing a large number of products to Boeing and Airbus through years, has accumulated rich experience in the design and test of testbed. But because of their test-bed technology is relatively closed, the technical parameters of the specific testbed are reluctant to be published to the outside world, and the research work is difficult.

IOP Conf. Series: Journal of Physics: Conf. Series **1060** (2018) 012073 doi:10.1088/1742-6596/1060/1/012073

System simulation technology is widely used in the field of aeronautics. In order to support decision analysis and reduce the cost of prototype and parametric testing of large physical prototypes. The environment control system (ECS) simulation is one of the important application directions of aircraft system simulation. Aircraft manufacturers attach great importance to air conditioning system simulation technology research to support system development, shorten the development cycle, improve energy efficiency and reduce costs [1]. In 2007, Boeing and the University of Hamburg in Germany started the loop control system functions (FLECS) project and aim to establish all components of the civil air-control system thermal model used to support the development of civil aircraft environmental control system [2]. Some scholars have also developed steady-state simulation models of air-conditioning systems to compare the advantages and disadvantages of the thermodynamic processes of different system configurations [3]. Müller and Airbus have jointly developed a dynamic simulation model library for supporting the design of aircraft air-conditioning systems [4]. The published literature shows that there are reports of simulation studies of system control [5]. However, the systems studied are mainly used in military aircraft system which have some differences with civil aircraft system. Therefore, the development of the system urgently need to verify and optimize for the aircraft during the development of engineering applications. In this paper, we use dynamic simulation technology to analyse the time of rapid response, and the simulation results can provide reference for subsequent supplier selection.

2. Methods

Matlab/Simulink modelling software is used to build a thermal dynamic testbed system and control model, as shown in Figure 1. So that the dynamic simulation technology is used to analyse the time of rapid response of the system. The data used in the simulation comes from the typical ECS components. Suppose the valves used are all butterfly valves with specific leakage. The parameters of the pipeline are the physical parameters of the stainless steel. The electric heater is a nonlinear model based on the conservation of energy. It is assumed that the time constant of the maximum flow is 3min. All component parameters can be changed according to the actual information provided by the later supplier, which does not affect the overall structure of the model.

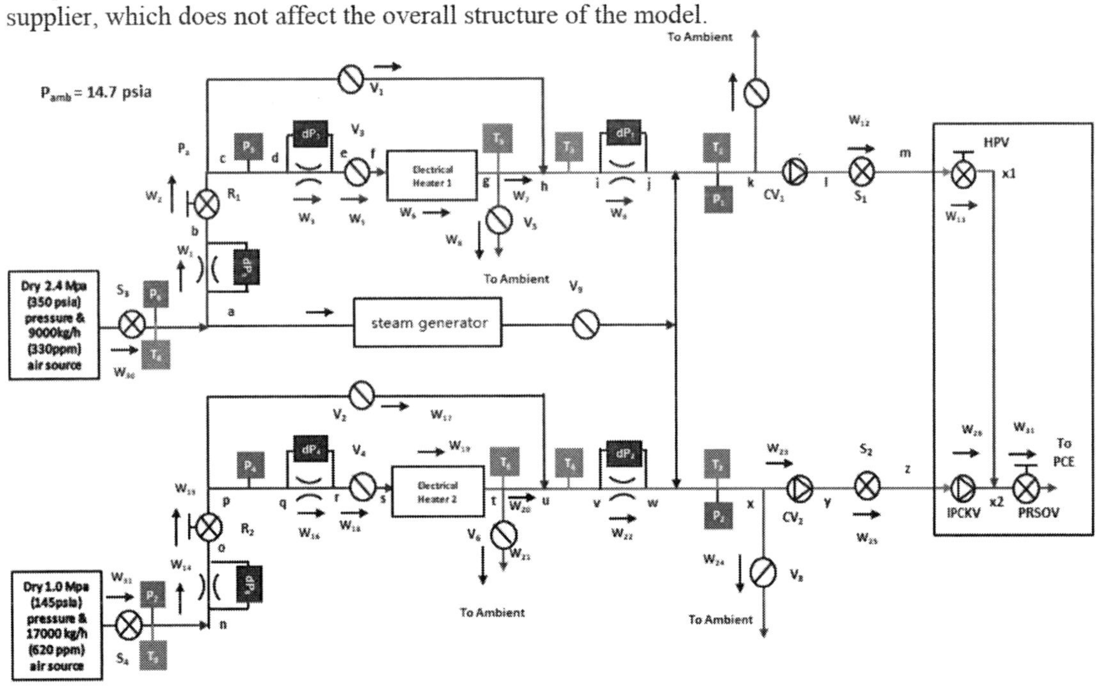

Figure 1. High pressure air supply channel in a thermal dynamic testbed

In the model, the heating of the stainless steel electric heater is based on the principle that the electrical energy of the resistive load is converted into thermal energy and has the heating efficiency of 85%~95%. It is usually controlled by a silicon-controlled power regulator. It is based on the zero-crossing control technology of the cycle, which we usually call solid state relays. The heating voltage is constant, and the voltage is adjusted by the pulse pass signal, so the heater power is changed and the temperature is adjusted. This method is mostly used for the load of the large inertia heater. The mathematical model of the electric heater is as shown by equation 1:

$$M_h C_h \left(dT/dt \right) = W \times C_p \left(T - T_a \right) + \left(Q/1.056 \right) \tag{1}$$

Where $M_h C_h$ is Thermal capacity of electric heater. W is the flow rate of air flowing through the electric heater. Ta is air temperature at the entrance, C_p is air specific heat and Q is the power rate. Factor 1.056 is a power unit conversion factor. Through the Manual Switch in the control model of pulse pass signal, the air temperature setting value at the outlet of the electric heater can be selected.

The dynamic performance of the testbed under the condition of fast rising pressure and temperature is analysed by using the model of the thermal dynamic testbed. It is assumed that the parameters of the high pressure air supply channel at the gas supply outlet of the testbed are P1 and T1 and the medium pressure gas channel is P2 and T2, and the set change curve is shown in Figure 2. P1 has taken into account that the maximum gas supply pressure of the gas source of the thermal dynamic testbed is 350psia. The highest temperature of the T1 setting is lower than the highest temperature of the outlet of the electric heater. The simulation results will show the thermal inertia effect of the pipe.

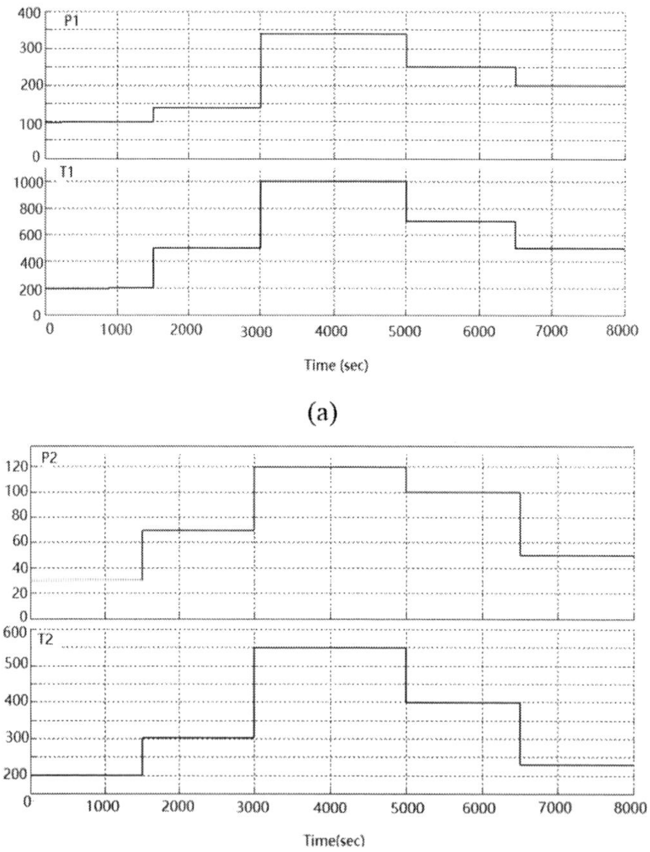

Figure 2. The setting curves: (a) P1 and T1 at the outlet of high pressure air supply channel; (b) P2 and T2 setting curves at the outlet of medium pressure air supply channel

3. Results

In the mixing process of cold and hot air, the air temperature at the outlet of the electric heater maintains a constant value. Because the air temperature at the outlet of the electric heater can reduce the influence of thermal inertia of the rear section pipeline, we compare the two cases of setting the outlet temperature of the electric heater to 1300°F and 1100°F. At 1300°F, the true response curve of T1 first passes through 1000°F at about 3010S, and the response curve of the sensor T1 is at 3060s, as shown in Figure 3; At 1100°F, the real response curve of T1 first passes through 1000 degrees F at about 3020s, and the response curve of the sensor T1 is at 3160s, as shown in Figure 4. There is no obvious difference in the change of pressure in the two cases.

According to the information shown in Figure 3, if a high temperature electric heater is used, the time required for the T1 temperature to rise from 500°F to 1000°F is only about 10s, and the stability time is 60s. The time required for T2 to rise from 300°F to 550°F is only about 5s, and the stability time is 40s, as shown in Figure 5. The overshoot of the test bed is large, and the test part has the risk of 5s over temperature.

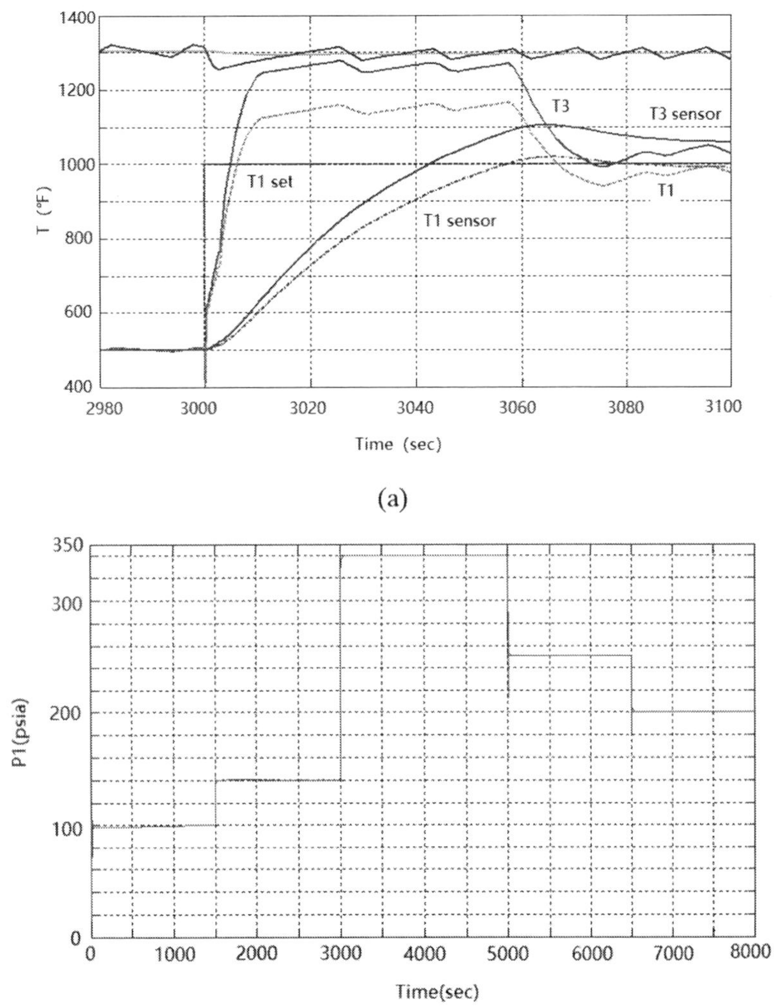

(a)

(b)

Figure 3. Response curves of T1(a) and P1(b) at the outlet of high pressure air supply channel (1300°F of electric heater)

IOP Conf. Series: Journal of Physics: Conf. Series **1060** (2018) 012073 doi:10.1088/1742-6596/1060/1/012073

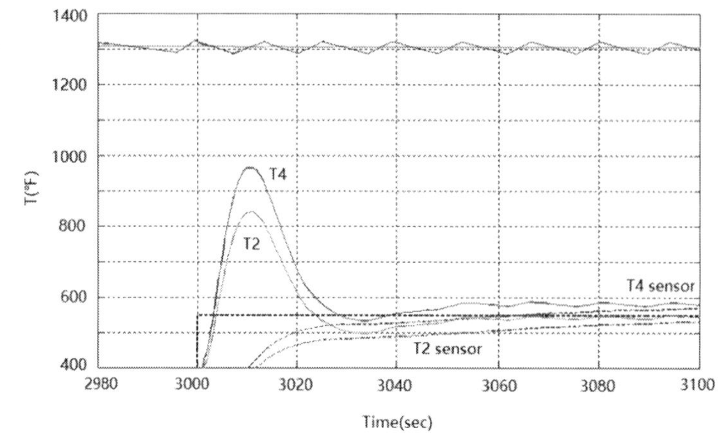

(a)

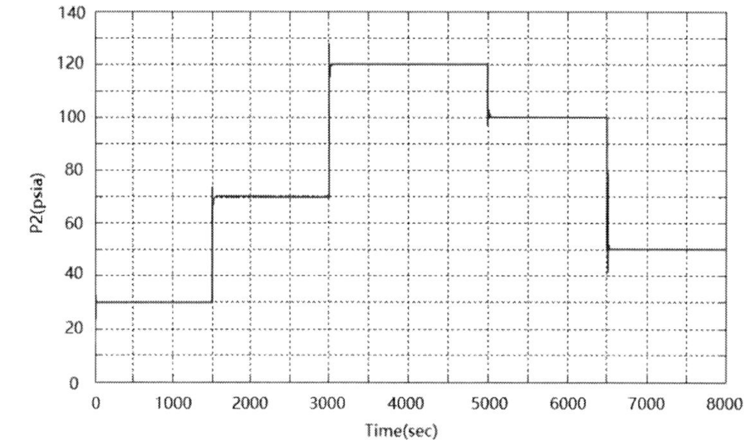

(b)

Figure 4. Response curves of T2(a) and P2(b) at the outlet of high pressure air supply channel (1300°F of electric heater)

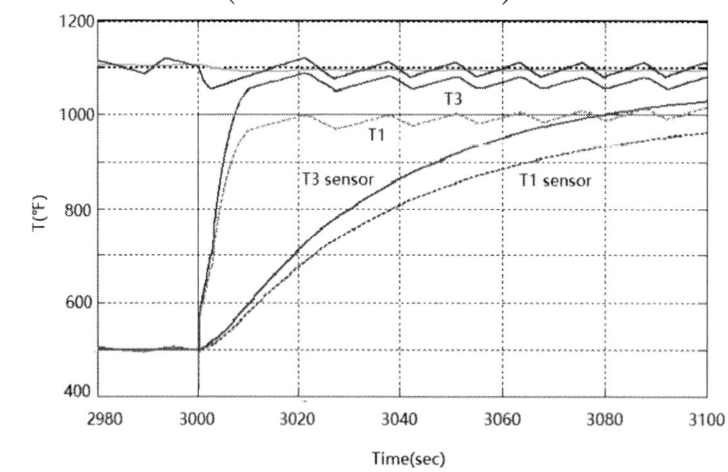

(a)

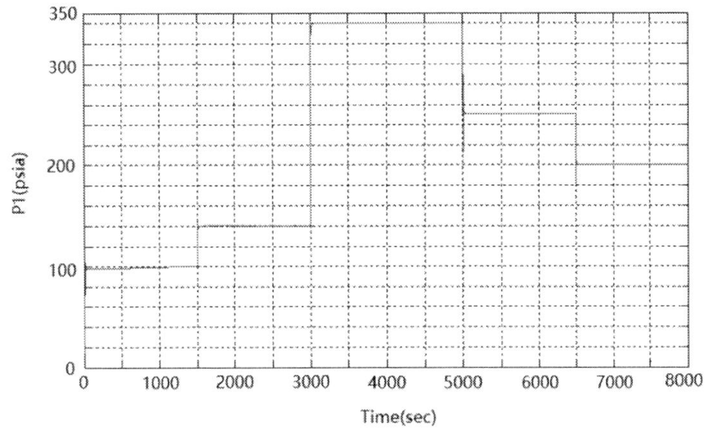

(b)

Figure 5. Response curves of T1(a) and P1(b) at the outlet of high pressure air supply channel (1100°F of electric heater)

4. Conclusion

In this paper, the Simulink model for the testbed of the environmental control system of the aircraft was studied. The results show that the higher the outlet temperature of the electric heater is, the faster the response characteristic of the temperature output of the testbed is. And the temperature overshoot is more severe. The pressure response rate is not affected. The system decoupled temperature and pressure on the structure. This dynamic performance can fully meet the needs of rapid heating up of the environmental control system and overcome the conventional furnace heating mode in the domestic environmental control laboratory. Because of the thermal inertia of the conventional electric furnace and the high thermal inertia of the back-end pipeline, the dynamic performance is completely limited to its real-time heating rate. To achieve the heating up from 350 to 1000 degrees, it is basically needs over 5min to reach a stationary state, which leads to no way to simulate the fast-changing temperature at the entrance of the test.

Acknowledgments

Authors wishing to acknowledge assistance or encouragement from colleagues, special work by technical staff or financial support from organizations should do so in an unnumbered Acknowledgments section immediately following the last numbered section of the paper.

References

[1] Ming ZP, Dong SJ, Wang J 2011 Study on the feasibility and structural optimization of Heat-pipe in abroad Journal of Aerospace Power 26 848-853. (in Chinese)

[2] D 2007 FLECS: functional library of the environment control system—a simulation tool for the support of industrial process Journal of Aerospace Power1st International Workshop on Aircraft System Technologies Hamburg, Germany

[3] Conceição S T, Zaparoli E L, Turcio W H L 2007 Thermodynamic Study of Aircraft Air Conditioning Air Cycle Machine: 3-wheel × 4-wheel SAE Brasil 2007 Congress and Exhibit Brazil

[4] Müller C, Scholz D, Giese T 2007 Dynamic simulation of innovative aircraft air conditioning. The 1st CEAS European Air and Space Conference

[5] Tu Y, Lin GP 2011 Simulation of large-scale cabin temperature control system Journal of Aerospace Power1st International Workshop on Aircraft System Technologies Acta Aeronaut. Astronaut. Sin. 32 49-57. (in Chinese)

IOP Conf. Series: Journal of Physics: Conf. Series **1060** (2018) 012074 doi:10.1088/1742-6596/1060/1/012074

Simulation and Analysis of the Detection Opportunities for the Multi-Azimuth/Multi-Waveband Ultraviolet Imager on Tiangong-02 Spacecraft Based on STK 10

Wang Huiping[1,*], Yan Zhen[1], Jin Shan[1], Rao Jun[1] and Guo Lili[1]

[1]Key Laboratory of Space Utilization, Technology and Engineering Centre for space Utilization, Chinese Academy of Sciences. 100094, Beijing, China

*wanghuiping@csu.ac.cn

Abstract. Aiming at the detection mission of the multi-azimuth/multi-waveband ultraviolet (UV) imager on Tiangong-02 spacecraft, a simulation and analysis method of the detection opportunities based on Satellite Tool Kit 10(STK 10) is introduced. The UV imager has more than one working modes such as limb viewing, nadir viewing, solar occultation viewing. A detection model considering multiple conditions such as orbit, attitude, installation angle, field of view, lighting, etc. is provided by using the Analysis Workbench Tools of STK 10. Calculated results are used in the ground operating control system to schedule the detection plan. By analysis the actual detection images, the simulation method is proved to be effective and credible. This research can provide references for design and system simulation of space-based imaging payload, as well as the flight control task arrangement.

1. Introduction

Space-based atmospheric remote sensing detection is an internationally hotspot at present and plays an important role in the atmospheric environment monitoring and protection. It can be divided into three different detection modes: nadir viewing, solar occultation viewing and limb viewing. Each has its own characteristic and detection conditions. The nadir viewing has lower vertical resolution and cannot satisfy the research requirements for the interaction between atmospheric upper layer and lower layer, and the global change process. The solar occultation viewing has higher vertical resolution but lower space coverage than other two modes, because the detection opportunities are limited to the sunup and sunset. The limb viewing has enough vertical resolution and space coverage, so it finds favour with people increasingly and becomes the main detection mode [1].

At the early stage, the imager developed usually had only one detection mode. Some prime imagers for nadir viewing were Ozone monitoring instrument (OMI) [2] and ultraviolet ozone vertical profile for FY-3A meteorological satellite (SBUVS) [3]. The OMPS developed by the USA was an imager for limb viewing [4]. In recent years, the imagers with more than one detection modes have been developed, for example the UV panoramic sounder for nadir and limb viewing [5,6]. It is a trend that the designs of the imager are becoming more complex.

The multi-azimuth/multi-waveband ultraviolet (UV) imager on Tiangong-02 spacecraft launched in 2016 has three detection modes. The Annular detector and the front detector are integral parts of the imager and have different working modes and detection conditions as shown in the Table 1. The annular detector can detect nadir and limb atmosphere simultaneously and the front detector can detect limb and solar occultation separately. The lighting condition plays an important role in the detection as shown in the Table 1. Some working modes require that the camera must avoid direct sunlight otherwise it will be burned, but some other working modes require that the camera must receive direct sunlight. Because of the complexity of the imager and the multiple detection conditions, the simulation and analysis of detection opportunities is a great challenge for the flight control task arrangement. It is, moreover, important for ensuring the payload security.

Table 1. Working modes and detection conditions

	Detection modes	Field of view	Detection conditions
Annular working mode one	limb/nadir viewing	70°~73° (limb) ±5° (nadir)	The atmosphere in the field of view is in the sunlight; The camera must avoid direct sunlight.
Annular working mode two	limb/nadir viewing	70°~73° (limb) ±5° (nadir)	The atmosphere in the field of view enters the umbra totally; The camera must avoid direct sunlight.
Front working mode one	limb viewing	0.024° × 1.6°	The atmosphere in the field of view is in the sunlight; The camera must avoid direct sunlight.
Front working mode two	limb viewing	0.024° × 1.6°	The atmosphere in the field of view enters the umbra totally; The camera must avoid direct sunlight.
Front working mode three	solar occultation viewing	±7°	The sunlight can get into the field of view through the atmosphere (100km).
Front working mode four	solar occultation viewing	±7°	The sunlight can get into the field of view but higher than the atmosphere (100km).

In this paper, we describe a simulation and analysis method of the detection opportunities based on Satellite Tool Kit 10 (STK 10). Using the older versions such as STK 9, we can only do some simulation and analysis with simple and a small amount of conditions, and rely on other software to get further analytic results. A lot of researches have been made in combined simulation with STK and other software. Cui Chengguang [7] *et al.* introduced a method of passive multi-mode space-based atmospheric sounding spectrometers nadir and limb matching in flight with a combination of MATLAB and STK. Li Yaqiong [8] *et al.* made an application of MATLAB and STK combined simulation for design of spaceborne antenna servo system. By using the Analysis Workbench Tools, which are new and improved in STK 10, we can do some simulation and analysis with complex and a great amount of conditions and get the finial analytic results directly. Calculated results are used in the ground operating control system of the Tiangong-02 spacecraft to schedule the detection plan. By analysis the actual detection images, the simulation method is proved to be effective and credible. This research can provide references for design and system simulation of space-based imaging payload, as well as the flight control task arrangement.

2. Geometry and mathematical model

2.1. Geometry and mathematical model of the annular detector

The annular detector can detect nadir and limb atmosphere simultaneously. The two working modes both require that the camera must avoid direct sunlight. The camera with ±5° field of view for nadir viewing can avoid direct sunlight at any time because of the protection of the Earth. However, the camera with 70°~73° field of view for limb viewing will be exposed to the direct sunlight, when the angle α_0 as shown in the Figure 1 and Figure 2 is between 70° and 73°. To avoid direct sunlight, the equation (1) as below must be satisfied.

$$73° < \alpha_0, or\ , \alpha_0 < 70° \quad (1)$$

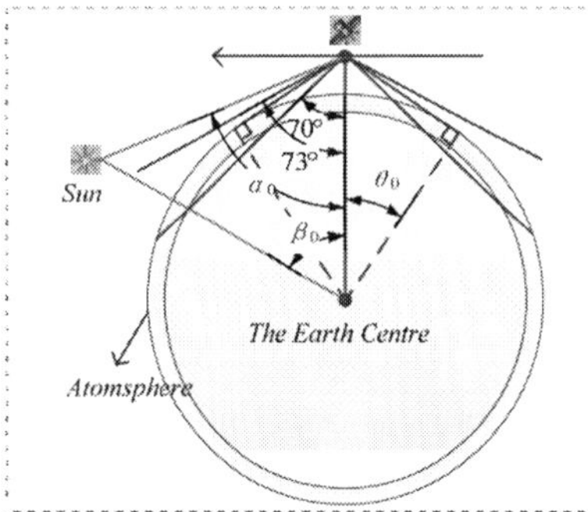

Figure 1. Geometry model for annular working mode one.

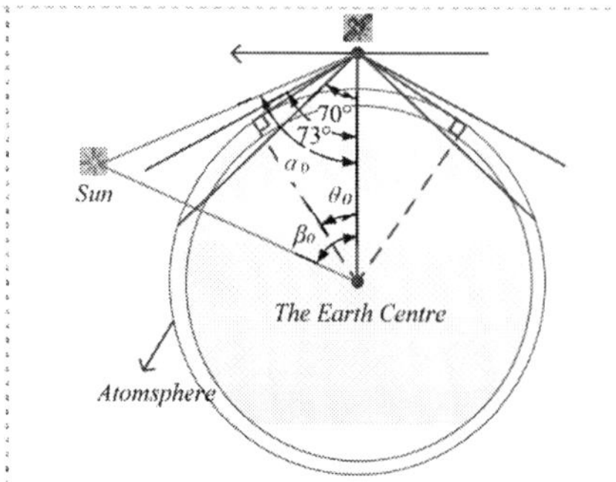

Figure 2. Geometry model for annular working mode two.

As shown in the Figure 1, when the equation (2) is satisfied, the atmosphere in the field of view is in sunlight, both in limb and nadir viewing.

$$\beta_0 + \theta_0 < 90° \quad (2)$$

As shown in the Figure 2, when the equation (3) is satisfied, the atmosphere in the field of view enters the umbra totally, both in limb and nadir viewing.

$$\beta_0 - \theta_0 > 90° \quad (3)$$

The angle θ_0 cannot be calculated exactly, because the Earth is a spheroid and not a sphere, the atmospheric thickness is approximate, and the orbit decays every day. As the field of view is between 70° and 73°, the angle θ_0 must be between 17° and 20° or larger. The angle θ_0 is significant for ensuring the accuracy of the result, so if necessary it can be magnified. We will give the final result in the Section 3.

2.2. Geometry and mathematical model of the front detector

The centroid orbit coordinate system is described in the Figure 3. The original point O_0 is the barycenter of the spacecraft, and O_0Z_0 is in the orbit plane and pointing to the Earth Centre. O_0X_0 is in the orbit plane perpendicular to O_0Z_0 and pointing to the flight direction. The sense of O_0Y_0 is determined as a matter of fixed convention by the right hand rule.The front detector is installed in Y_0Z_0 and pointing to the negative direction of O_0X_0 with a rotation of 18.5° to O_0Z_0. It has two different cameras, one for limb viewing with $0.024° \times 1.6°$ field of view and another for solar occultation viewing with $\pm 7°$ field of view.

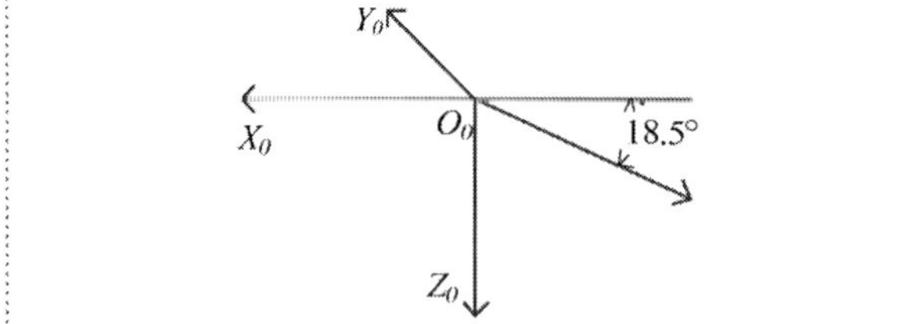

Figure 3. The installation of the front detector

2.2.1. Geometry and mathematical model for limb viewing. The field of view for limb viewing is $0.024° \times 1.6°$. We can magnify the field of view to $\pm 0.8°$. As shown in the Figure 4 and the Figure 5, when the angle α_1 is greater than 0.8°, the camera can avoid direct sunlight. The equation can be expressed as follow.

$$\alpha_1 > 0.8° \quad (4)$$

The geometry and the mathematical model cannot be given exactly because the detection region is irregular. It is a special case when the sun is parallel with the orbit plane in the Figure 4 and Figure 5.

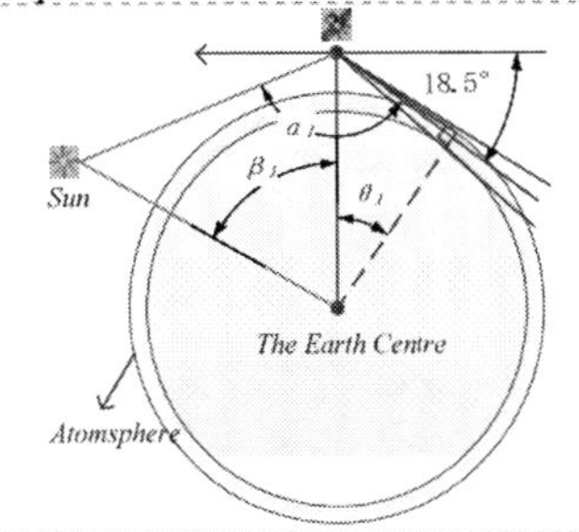

Figure 4. Geometry model for front working modes for limb viewing.

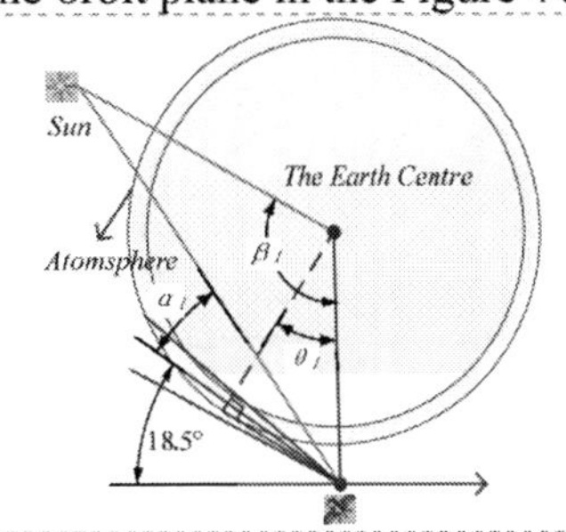

Figure 5. Geometry model for front working modes limb viewing.

As shown in the Figure 4 and Figure 5, when the equation (5) and the equation (6) are both satisfied, the atmosphere in the field of view is in sunlight. The equations can be merged to the equation (2).

$$\beta_1 + \theta_1 < 90° \quad (5)$$

$$\beta_1 - \theta_1 < 90° \quad (6)$$

When the equation (7) and the equation (8) are both satisfied, the atmosphere in the field of view is in sunlight. The equations can be merged to the equation (3).

$$\beta_1 + \theta_1 > 90° \quad (7)$$

$$\beta_1 - \theta_1 > 90° \quad (8)$$

We can get a conclusion that the front detector for limb viewing can take the same model with the annular detector.

2.2.2. Geometry and mathematical model for solar occultation viewing. The two working modes for solar occultation viewing both require that the sunlight can get into the field of view. As shown in the Figure 6, when the angle α_2 is less than 7°, the detection condition can be satisfied. The equation can be expressed as follows.

$$\alpha_2 < 7° \quad (9)$$

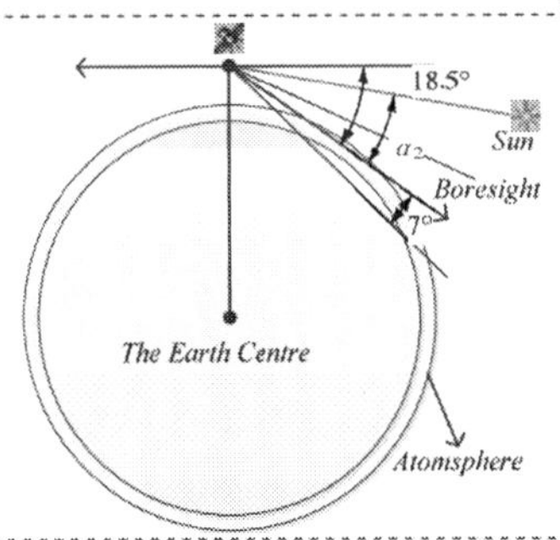

Figure 6. Geometry model for solar occultation viewing.

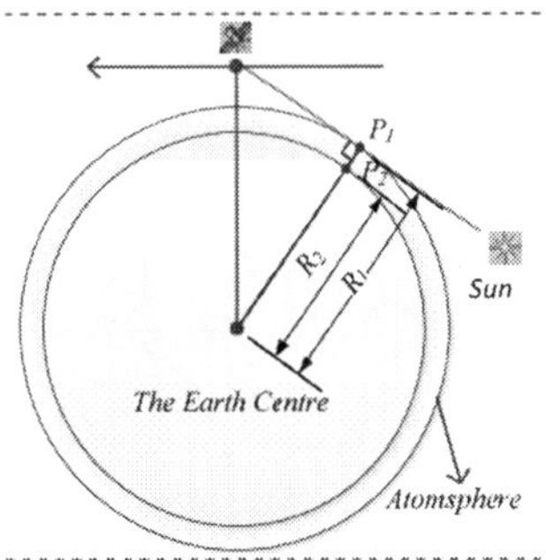

Figure 7. Geometry model for solar occultation viewing.

The detection conditions that the sunlight getting into the field of view through the atmosphere (100km) or higher than the atmosphere (100km) can be described in the Figure 7. When the equation (10) is satisfied, the sunlight can get into the field of view higher than the atmosphere (100km).

$$R_1 - R_2 > 100\text{km} \quad (10)$$

When the equation (11) is satisfied, the sunlight can get into the field of view through the atmosphere (100km).

$$R_1 - R_2 < 100\text{km} \quad (11)$$

IOP Conf. Series: Journal of Physics: Conf. Series **1060** (2018) 012074 doi:10.1088/1742-6596/1060/1/012074

3. Simulation and analysis based on STK 10

In this section, the simulation and analysis for the geometry and mathematical models are given. The STK Analysis Workbench Tools contain three different tools, the Vector Geometry Tool (VGT), the Calculation Tool, and the Time Tool. The vectors and angles described in the geometry and mathematical models can be defined by using VGT, and the equations can be defined by using the Calculation Tool. We can get the results when the equations are satisfied by using the Time Tool.

3.1. Simulation and analysis of the annular detector

The angle α_0 and the angle β_0 can be defined as follows by using VGT.

- Define the vectors V_1, V_2, V_3 and V_4 by using the vector geometry component, and the options are shown in the Table 2.
- Define the angle α_0 by using angle component, select "Between Vectors" for the "Type"option, and from the vector V_1 to the vector V_2.
- Define the angle β_0 by using angle component, select "Between Vectors" for the type option, and from the vector V_3 to the vector V_4.

Table 2. The options for the vectors V_1, V_2, V_3, V_4

	Type	Origin point	Destination point
V_1	Displacement	the spacecraft centre	the earth centre
V_2	Displacement	the spacecraft centre	the sun centre
V_3	Displacement	the earth centre	the spacecraft centre
V_4	Displacement	the earth centre	the sun centre

The equation (1), the equation (2) and the equation (3) can be defined as follows by using the Calculation Tool.

- Add a scalar calculation component named Alpha0, select "Angle" for the "Type" option and select the angle α_0 for the "Input Angle" option.
- Add a scalar calculation component named Beta0, select "Angle" for the "Type" option and select the angle β_0 for the "Input Angle" option.
- Define the conditions Equation1, Equation2, Equation3 by using condition component, and the options are shown in the Table 3.

Table 3. The options for the conditions Equation1, Equation2, Equation3

	Scalar	Operation	Minimum	Maximum
Equation1	Alpha0	Outside Minimum and Maximum	70 deg	73 deg
Equation2	Beta0	Below Maximum	--	111 deg
Equation3	Beta0	Above Minimum	69 deg	--

We can get the results when the equations are satisfied by using the Time Tool, and the options are shown in the Table 4.

Table 4. The options for the intervals.

	Type	condition	the Intervals A	the Intervals B	Operation
Equation1	Satisfaction	Equation1	--	--	--
Equation2	Satisfaction	Equation2	--	--	--
Equation3	Satisfaction	Equation3	--	--	--
Equation1and2	Merged	--	Equation1	Equation2	AND
Equation1and3	Merged	--	Equation1	Equation3	AND

IOP Conf. Series: Journal of Physics: Conf. Series **1060** (2018) 012074 doi:10.1088/1742-6596/1060/1/012074

The intervals calculated form Equation1and2 as shown in the Table 5 are detection opportunities for the annular working mode one. As shown in the Figure 8 and Figure 9, when the angle θ_0 is 21°, the atmosphere in the field of view is in sunlight totally。

The intervals calculated form Equation1and3 as shown in the Table 6 are detection opportunities for the annular working mode two. When the angle θ_0 is 17°, the atmosphere in the field of view is in umbra seemingly, but it may still be lighten by the atmospheric scattering, so we use $\theta_0 = 21°$ to calculate the results as shown in the Figure 10 and Figure 11.

Table 5. Partly detection opportunities for annular working mode one

	Start Time (UTCG)	Stop Time (UTCG)	Duration (sec)
1	1 Mar 2018 18:30:46.881	1 Mar 2018 19:04:56.217	2049.336
2	1 Mar 2018 20:02:34.735	1 Mar 2018 20:36:45.201	2050.466
3	1 Mar 2018 21:34:22.583	1 Mar 2018 22:08:34.267	2051.684
4	1 Mar 2018 23:06:10.457	1 Mar 2018 23:40:23.334	2052.877

Table 6. Partly detection opportunities for annular working mode two

	Start Time (UTCG)	Stop Time (UTCG)	Duration (sec)
1	1 Mar 2018 19:16:39.063	1 Mar 2018 19:50:50.030	2050.967
2	1 Mar 2018 20:48:26.916	1 Mar 2018 21:22:39.020	2052.105
3	1 Mar 2018 22:20:14.844	1 Mar 2018 22:54:28.024	2053.180
4	1 Mar 2018 23:52:02.789	2 Mar 2018 00:26:16.991	2054.202

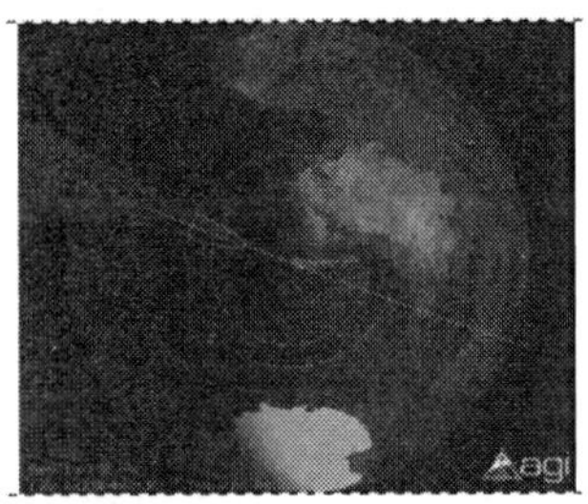

Figure 8. The beginning time of one detection opportunity for annular working mode one when the angle θ_0 is 21°.

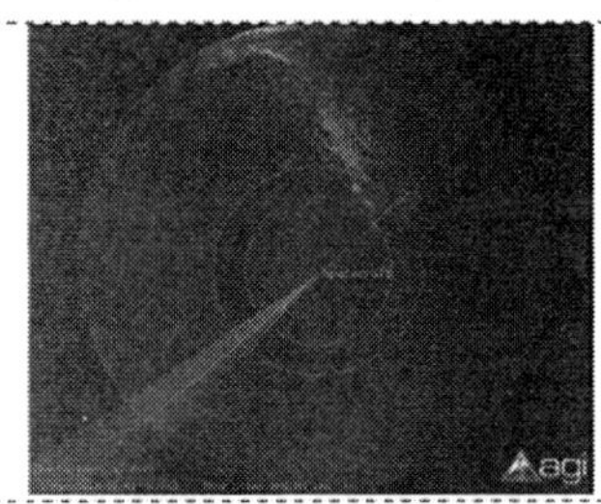

Figure 9. The ending time of one detection opportunity for annular working mode one when the angle θ_0 is 21°.

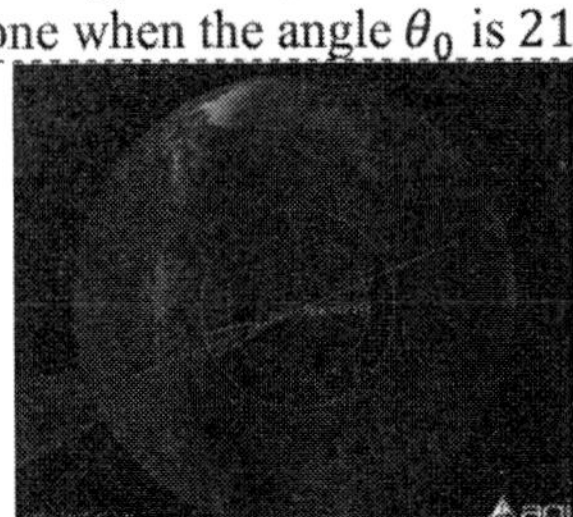

Figure 10. The beginning time of one detection opportunity for annular working mode two when the angle θ_0 is 21°.

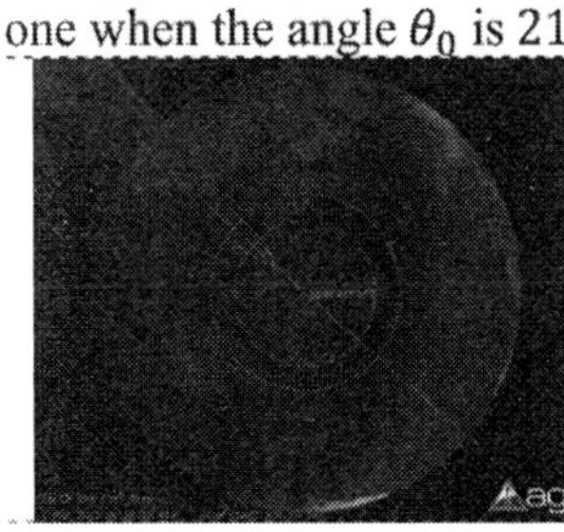

Figure 11. The ending time of one detection opportunity for annular working mode two when the angle θ_0 is 21°.

Every day has 16 detection opportunities and each detection opportunity lasts 1850 seconds to 2150 seconds both for the two annular working modes. We can also use the results for the two front working modes for limb viewing as described in Section 2.2.1.

3.2. Simulation and analysis of the front detector

The angle α_2 can be defined as follows by using VGT.

- Define the angle α_2 by using angle component, select "Between Vectors" for the "Type"option, and from the vector V_2 to the vector Boresight which is represent the optical axis of the camera.

The vectors R_1 and R_2 can be defined as follows by using VGT.

- Define the vectors N_1, N_2 by using the vector geometry component, and the options are shown in the Table 7.
- Add a plane geometry component named Plane1, select "Normal" for the "Type" option, select the vector N_2, the vector V_2 for the "Normal Vector" and "Reference Vector" option, and select the sun centre for the "Reference Point" option.
- Add a point geometry component named P_1, select "Projection" for the "Type" option, select the earth centre for the "Source Point" option and select the plane Plane1 for the "Reference Plane" option.
- Add a point geometry component named P_2, select "Surface" for the "Type" option, select "Earth" for the "Central Body", select the point P_1 for the "Reference Point" option, select "Ellipsoid" for the "Reference Shape Type" option, and select "Detic" for the "Surface Point Type" option.
- Define the vectors R_1 and R_2 by using the vector geometry component, and the options are shown in the Table 8.

Table 7. The options for the vectors N_1,N_2

	Type	Vector A	Vector B
N_1	Cross Product	the vector V_1	the vector V_2
N_2	Cross Product	the vector N_1	the vector V_2

Table 8. The options for the vectors R_1,R_2

	Type	Origin point	Destination point
R_1	Displacement	the earth centre	the point P_1
R_2	Displacement	the earth centre	the point P_2

The equation (9), the equation (10) and the equation (11) can be defined as follows by using the Calculation Tool.

- Define the scalars Alpha2, R_1, R_2, R_1-R_2 by using the condition component, and the options are shown in the Table 9.
- Define the conditions Equation9, Equation10, Equation11 by using the condition component, and the options are shown in the Table 10.

Table 9. The options for the intervals.

	Type	Input Angle	Input Vector	Function	Operation	the x argument	the y argument
Alpha2	Angle	α_2	--	--	--		
R_1	Vector Magnitude	--	R_1	--	--		
R_2	Vector Magnitude	--	R_2	--	--		
R_1-R_2	Function (x, y)	--	--	a*x-b*y	AND	R_1	R_2

Table 10. The options for the conditions Equation9, Equation10, Equation11

	Scalar	Operation	Minimum	Maximum
Equation9	Alpha2	Between Minimum and Maximum	0deg	73deg
Equation10	R_1-R_2	Above Minimum	100km	--
Equation11	R_1-R_2	Between Minimum and Maximum	0km	100km

We can get the results when the equations are satisfied by using the Time Tool, and the options are shown in the Table 11.

Table 11. The options for the intervals.

	Type	condition	the Intervals A	the Intervals B	Operation
Equation9	Satisfaction	Equation9	--	--	--
Equation10	Satisfaction	Equation10	--	--	--
Equation11	Satisfaction	Equation11	--	--	--
Equation9and10	Merged	--	Equation9	Equation10	AND
Equation9and11	Merged	--	Equation9	Equation11	AND

The intervals calculated form Equation9and10 are the detection opportunities as shown in the Table 12. Every day has 16 detection opportunities and the maximum of the duration is 70 seconds. The sunlight can get into the field of view higher than the atmosphere (100km) as shown in the Figure 12 and the Figure 13.The intervals calculated form Equation9and11 are the detection opportunities as shown in the Table 13. Every day has 16 detection opportunities and the maximum of the duration is only 43 seconds. The sunlight can get into the field of view through the atmosphere as shown in the Figure 14, the Figure 15.

Table 12. Partly detection opportunities calculated form Equation9and10.

	Start Time (UTCG)	Stop Time (UTCG)	Duration (sec)
1	7 Mar 2018 15:57:32.940	7 Mar 2018 15:58:42.822	69.882
2	7 Mar 2018 17:29:18.162	7 Mar 2018 17:30:28.299	70.137
3	7 Mar 2018 19:01:03.614	7 Mar 2018 19:02:13.792	70.177
4	7 Mar 2018 20:32:49.311	7 Mar 2018 20:33:59.269	69.958

Table 13. Partly detection opportunities calculated form Equation9and11.

	Start Time (UTCG)	Stop Time (UTCG)	Duration (sec)
1	7 Mar 2018 15:58:42.822	7 Mar 2018 15:59:26.117	43.295
2	7 Mar 2018 17:30:28.299	7 Mar 2018 17:31:11.583	43.284
3	7 Mar 2018 19:02:13.792	7 Mar 2018 19:02:57.073	43.281
4	7 Mar 2018 20:33:59.269	7 Mar 2018 20:34:42.559	43.290

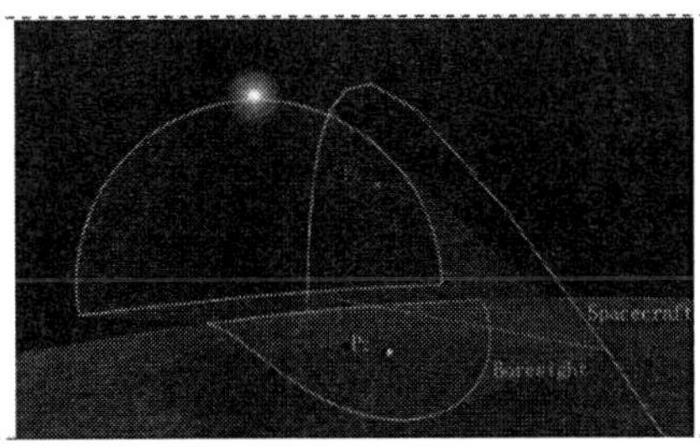

Figure 12. The beginning time of one detection opportunity calculated form Equation9and10.

Figure 13. The ending time of one detection opportunity calculated form Equation9and10.

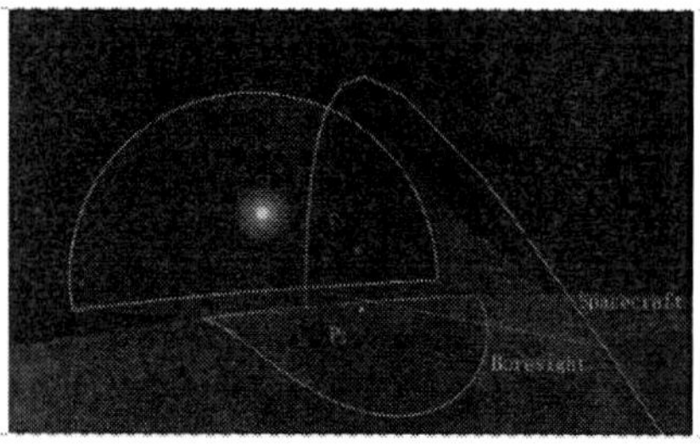

Figure 14. The beginning time of the detection opportunity calculated form Equation9and11.

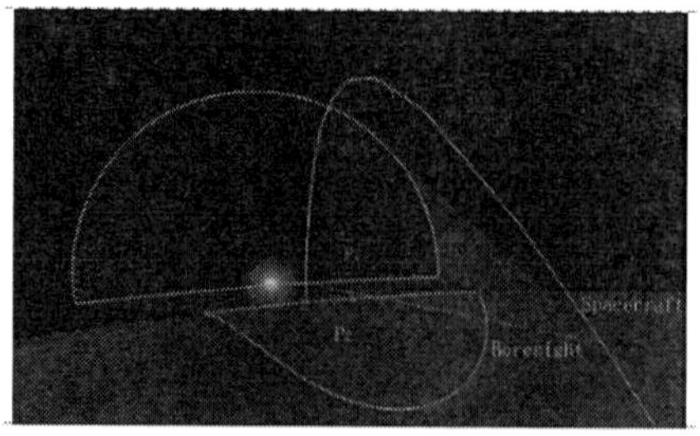

Figure 15. The ending time of the detection opportunity calculated form Equation9and11.

4. Conclusions

In this paper, we provide a simulation and analysis method of the detection opportunities based on Satellite Tool Kit 10 (STK 10). By using the Analysis Workbench Tools, we can do some simulation and analysis with complex and a great amount of conditions and get the finial analytic results directly. We use the calculated results in the ground operating control system of the Tiangong-02 spacecraft to schedule the detection plan. By analysis the actual detection images, we find that the simulation method is effective and credible. This research can provide references for design and system simulation of space-based imaging payload, as well as the flight control task arrangement.

References

[1] Xue Qingsheng. Optical design of spaceborne broadband limb sounder for detecting atmospheric trace gas[J]. Acta Optica Sinica, 2012, 41(6):631-637.

[2] JOHAN D V, GIJSBERTUS H J, VAN D O, *et al.* Ozone monitoring instrument (OMI)[C]. SPIE, 2002, 4480:215-325.

[3] Wang Shurong, Li Futian, Song Kefei. Ultraviolet ozone vertical profile probe for FY-3A meteorological satellite[J]. Acta Optica Sinica, 2009, 29(9):2590-2593.

[4] Glen Jaross, Grace Chen, Mark Kowitt, *et al.* Suomi NPP OMPS limb forfiler initial sensor performance assessment[C]. SPIE, 2012, 8258:825805-1~825805-9.

[5] Yu Xiangyang, Wang Shurong, Huang Yu, *et al.* Development of ultraviolet annular imager for earth limb and nadir sounding simultaneously[J]. Acta Optica Sinica, 2012, 32(12):1211004.

[6] Yang Xiaohu, Wang Shurong, Huang Yu. Research on radiation characteristic of the multi-azimuth/multi-waveband earth limb ultraviolet imager[J]. Acta Optica Sinica, 2014, 34(8): 0801001.

[7] Cui Chengguang, Wang Shurong, Huang Yu, *et al.* Study on the Method of Passive Multi-Mode Space-Based Atmospheric Sounding Spectrometers Nadir and Limb Matching in-Flight. Acta Optica Sinica, 2015, 35(6): 0601001.

[8] Li Yaqiong, Li Yang, Guan Guizhu, *et al.* The application of MATLAB and STK combined simulation for design of spaceborne antenna servo system[J]. Journal of Telemetry, Tracking and Command, 2016, 37(5): 00279.

IOP Conf. Series: Journal of Physics: Conf. Series **1060** (2018) 012075 doi:10.1088/1742-6596/1060/1/012075

Modelling and Forecasting of Commodity Trading Price

Jun Liu[1, 2, 3, 4, *, a], **Yuan Tian**[1, 2, 3, 4, b] **and Qing Yan**[1, 2, 3, 4, c]

[1]Center for Data Science, BUPT, Beijing 100876, China.
[2]School of Information and Communication Engineering, BUPT, Beijing 100876, China.
[3]Beijing Laboratory of Advanced Information Networks, BUPT, Beijing 100876, China.
[4]Beijing Key Laboratory of Network System Architecture and Convergence, BUPT, Beijing 100876, China.

*[a]liujun@bupt.edu.cn, [b]lance6716@bupt.edu.cn, [c]yq@haohandata.com.cn

Abstract. This paper proposes and improves a model for China's cotton reserves trading market, generates a more accurate price level table in contrast to the widely used China Cotton Association(CCA)'s table, and predicts the future trading price with this model. Data used is all 29895 trade records of cotton reserves from May, 2016 to Sept, 2016. In this paper, we firstly give a briefly introduction to the data as well as basic knowledge of cotton, especially the CCA's price level table as the target to improve accuracy. We then present a multiple linear regression with dummy variable model, which can reduce the error on predicting the trading price of current month, but still remains some problems such as the result shows a batch of cotton with better quality may get a lower price. So we finally advance the model with multidimensional isotonic regression and more factors taken into consideration. Our last model could be used to predict the trade price of both current month and next month with approximately 4% mean absolute percentage error(MAPE).

1. Introduction

Commodity trading servers an important role in the national economy and has been widely studied by economics, statistics and many other subjects[1][2][3]. Meanwhile China's commodity market has received increasingly attention and research these years for its unique economic system [4][5]. In this paper, we focus on cotton as a representative of the commodity, study the modelling and forecasting of China's cotton price with the data provided by Information Centre of China National Cotton Reserves Corp(CNCRC). China has sold from its cotton reserves of 12' and 13' and put up for auction, whose trading price is determined of bidding of textile enterprises as well as guided by some policies and plans to some extent. The final trading price is a good indicator of cotton industry. We have studied and proposed models to analysis and find hidden patterns of cotton reserves trading price, which aim to improve policy effectiveness, advance accuracy of cotton index and predict trading price of next month.

2. Data Structure

Data used in this paper was collected from May, 2016 to Sept, 2016 by CNCRC's cotton reserve trading system, with total amount of 29895 cotton reserves transactions. The data could be introduced as three parts as follow.

2.1. Basic cotton knowledge

CNCRC provided cotton index of basic cotton type 3128B (meaning of this code will be interpreted in next section) every week. This index reflects the ups and downs of cotton market and is used as well as the other index to calculate the base price of cotton auction.

CCA also publish a price level table every month to calculate the exact price of cotton of different properties. The price then can be calculated by finding the according difference of price of each properties, adding or subtracting it from the basic type 3128B's price. Table 1 shows part of CCA's price level table.

Table 1. Part of CCA's price level table.

Staple length	Difference of price	Micronaire level	Difference of price
30	600	A	200
29	300	B1	0
28	0	B2	0
27	-400	C1	-600
26	-850	C2	-100

The calculated result should be referenced during the cotton trading as CCA stated.

2.2. Public quality inspection data

Public quality inspection was made before the cotton reserves was listed for sale. It inspects many properties but nine important ones are listed in Table 2. Some of these properties are involved in CCA's price level table and others are concerned by the cotton corporations in actual trading. Others properties beyond those nine, such as production year and production area could be reflected by those nine properties and thus omitted.

Table 2. Cotton properties and values.

Properties	Values and explanations
Colour	'11', '21', '31', '41', '51', '12', '22', '32', '13', '23', '33', '14', '24'. (First character represents brightness and last character represents whiteness. Small value indicates high quality.)
Average micronaire	Decimal and according levels (Value of 3.4-and-lower: C1, 3.5~3.6: B1, 3.7~4.2: A, 4.3~4.9: B2, 5.0-and-higher: C2, A indicate moderate value and high quality.)
Rolling quality	'P1', 'P2', 'P3'. (P1 indicate high quality)
Staple length	Integer from 21 to 33.
Fibre strength	Integer from 21 to 33 and according levels. (Value of 31~33: S1, 29~30: S2, 26~28: S3, 24~25: S4, 21~23: S5.)
Length uniformity	Integer from 75 to 87 and according levels. (Value of 86~87: U1, 83~85: U2, 80~82: U3, 77~79: U4, 75~76: U5.)
Other fibre	'N' for none and 'L' for low.
Heterogeneity	Percentage (Not included in CCA's table)
Moisture regain	Percentage (Not included in CCA's table)

So the basic type 3128B represents for 'colour 31, staple length 28, average micronaire B1 or B2'.

2.3. Trading records

Many information was recorded but trading date, price of 3128B on that day and trading price are merely involved in this paper. Price of 3128B is used for auction base price, the auction is twice a day with over 100 batches of cotton on sale. Bid for any batch of cotton will reset global countdown for every cotton batch, so the final trade price is an equitable and systematic indicator of cotton value.

3. Linear regression model

3.1. Trading records

CCA's price level table was widely used in actual cotton trade these years, but its accuracy and room for improvement remain to be studied. Our first goal is to propose a model which could generate a more accurate price level table. Part of CCA's table (May, 2016) shows as table 1.

Linear regression is a widely used method with a simple form and strong ability to interpret, and can be easily found in most statistic and economic textbooks. So we build our first model on linear regression. Cotton trading data and CCA's table indicate a lot of information which specifies the exact form of linear regression.

We could infer from the price level table that difference price of each cotton property could be calculated independently, thus the correlation among them must be weak. That is a prerequisite for linear regression. We subsequently perform a significance tests of Kendall's Tau-b, the result shows most properties have slightly correlation (absolute value below 0.3) except staple length and length uniformity have a tau value of 0.58. A possible explanation is cotton of high quality will have both long and neat staple. These two properties will be treated carefully in the following work.

Another point is though all properties are order variables, the table shows that difference price between two adjacent level of some properties are not constant. And some properties such as micronaire are complex to find a linear relationship, so it is suitable to use dummy variables for all properties. Standard values of each property are dropped to avoid multi-collinearity, which is known for "dummy variable trap".

We take first seven properties mentioned in table 2 into consideration and calculate the regression for each month using 80% data as training set. The regression has a form of

$$y = \beta_0 + \beta_{colour11} \times x_{colour11} + \beta_{colour21} \times x_{colour21} + \cdots + \beta_{rollingP1} \times x_{rollingP1} + \cdots$$

The coefficient of x is the difference of price in table.

β_0 represents the average trade price of each month, which is similar to China Cotton Index 3128B (CC Index 3128B). The validity of the model can be checked from the relationship between that two variables. Table 3 shows average values of β_0 and CC Index 3128B from May, 2016 to Sept, 2016:

Table 3. Contrast between model and CC Index 3128B.

	May	June	July	Aug	Sept
β_0	12011	12185	15076	14377	14355
CC Index 3128B	12510	12694	14316	14673	14257

Mean absolute percentage error(MAPE) of model and CCA's price level table are listed in Table 4

Table 4. MAPE of CCA's table and model.

	May	June	July	Aug	Sept
CCA's MAPE	4.3%	4.0%	5.6%	4.2%	4.6%
Model's MAPE	2.3%	2.9%	3.9%	2.3%	3.5%

From table 3 and 4 we could draw a conclusion that our linear regression model does improve the accuracy in contrast to CCA's price level table.

3.2. Problems remained

However, there are still some points needed further improvement. Table 5 shows part of the price level table generated by our model.

Table 5. Part of model generated table.

Staple length	Difference of price	Length uniformity	Difference of price
30	342	U1	No data
29	207	U2	187
28	0	U3	0
27	-349	U4	-96.5
26	-567	U5	258

One main problem is this price table doesn't follow the prior knowledge of "higher price better quality", for example, U5 is the worst length uniformity level but gets highest price. This is both a true reflection of the trading record to some extent after we take a drill-down analysis under current assumption, and a common representation of multi-collinearity as we mentioned in section 3.1[6]. There are other regression methods such as lasso regression, ridge regression, to solve this problem[7], but they are not suitable in our situation of dummy variables for order variable.

Another problem is price of a certain property varies a lot among these five months, which may lead to a low acceptance of public and cotton corporations.

After making more research and having a better understanding of cotton trade business, we proposed an improved model using multidimensional isotonic regression.

4. Isotonic regression model

4.1. Improvement to previous model

The improvements we made could be roughly divided into three part: a) using isotonic regression to keep high-quality cotton evaluated high price, b) using CNCRC's 3128B week price to eliminate the error of base price changing within a month, c) taking more properties into model factors.

First improvement is using isotonic regression. It is a widely used regression method of fitting a sequence of variables $\{x_i\}$ into $\{y_i\}$ under some partial ordering constrains, for example $y_i < y_{i+1}$, with a minimum of $\sum_i (y_i - x_i)^2$. Many isotonic regression algorithms have been proposed such as Pool Adjacent Violators Algorithm(PAVA)[8], Least square isotonic regression in two dimensions[9], Merge and Chop Algorithm[10]etc. For our condition, isotonic regression is a constrain to combine the "higher price better quality" prior knowledge with our model. In order to apply isotonic regression, further processing is required to indicate the partial order relationship of properties, which will be introduced in next section.

Next improvement is replacing trade price with difference price between trade price and base price as target variable of regression, in order to eliminate the error caused by market goes up and down within a month. This could account for the above-average MAPE 3.9% of July, because CNCRC's 3128B week price shows there was a sharp ascending in July.

Third improvement is adjustment on taking more cotton properties into model's factors to fit the actual cotton trading business. Levels of properties are split into integer values to make a more accurate prediction, and heterogeneity and moisture regain are used to calculate the net price of cotton which is more concerned by cotton corporations.

4.2. Improvement to previous model

The origin data is firstly calculated the auxiliary variables such as net price or variables used for isotonic regression, then processed through multidimensional isotonic regression with seven properties as input and difference of trading price and base price as output, finally the output prices of each combinations of properties values are decomposed into seven dimensions representing each property to generate the price level table.

The multidimensional isotonic regression algorithm we used is multiisotonic component of Python Scikit-learn package(https://github.com/alexfields/multiisotonic). The interface of this component only accept multidimensional vector X_i as constrains and corresponding y_i as input, returns $y_i^{'}$ with minimum squared distance from y_i, for $y_i^{'} \geq y_j^{'}$ if all components of X_i are greater than or equal to corresponding components of X_j. So colour and micronaire characteristics should be replaced by auxiliary variables since they don't follow an increasing numeric value with increasing quality. As shown in Table 2, first character of colour represents brightness and the last character represents whiteness, and both represent a better quality with lower value. So auxiliary variables for colour could be generated by splitting colour measurements values and negating it. Quality of micronaire follows umbrella order C1 ≤ B1 ≤ A ≥ B2 ≥ C2. Auxiliary variables suitable for that Python package interface shows in Table 6

Table 6. Auxiliary variables for micronaire

Average micronaire	Auxiliary variables
A	(3, 3)
B1	(3, 2)
B2	(2, 3)
C1	(3, 1)
C2	(1, 3)

Then the "higher price better quality" result of cotton property combinations appeared in each month is returned, which need deduplication and decomposition into seven dimensions for generating the table. This could be easily done using multiple linear regression with dummy variables from our first model. Table 7 shows the result generated by isotonic regression model.

Table 7. part of table of isotonic regression model.

Staple length	Difference of price	Length uniformity	Difference of price
30	345	83	349
29	169	82	182
28	0	81	0
27	-258	80	-158
26	-396	79	-345

MAPE of this model as well as CCA's table and our first model lists for convenience in Table 8.

Table 8. MAPE of different tables.

	May	June	July	Aug	Sept
CCA's MAPE	4.3%	4.0%	5.6%	4.2%	4.6%
First model's MAPE	2.3%	2.9%	3.9%	2.3%	3.5%
Second model's MAPE	3.2%	3.1%	3.7%	2.1%	4.0%

Error slightly improved in contrast with our first linear regression model, one possible reason is that there is a trade-off between the accuracy and "higher price better quality" knowledge, which is very similar to the trade-off between empirical error and generalization error. Considering the model will be announced to guide cotton's trade, there is no strong reason to violate the widely accepted rule of "higher price better quality". So it is agreed to us all that isotonic regression model is more acceptable.

5. Prediction of future trade price

Isotonic regression model generates price level tables that have steadier values among months, so it is possible that price level table from previous month could be used to predict the price of next month. That is a valuable application in cotton reserves trade. The accuracy of next-month prediction could be evaluated by the MAPE of difference price between CNCRC's 3128B week price and predicted trading price by the table of every trade cotton. We then apply the price level table of previous month to data of next month to calculate average error of a month. The result shows in Table 9.

Table 9. MAPE of prediction on next month's data.

June	July	Aug	Sept
3.3%	3.5%	3.1%	4.2%

Contrast the prediction result with isotonic regression result of each month, we could find that the error only improves slightly on most month, which indicates that our price level table is general through two adjacent months, and the effect of the prediction is acceptable.

6. Future works

The isotonic regression model generates price level table which keeps a balance between accuracy and "higher price better quality". But there are still some points needed study and improvement.

CCA's price level table has a simple form as well as some limits. Combines of cotton properties, such as staple length and length uniformity, may generate a new auxiliary property such as staple length level, to make use of correlation of current seven properties and advance accuracy.

CCA's monthly table may be advanced to weekly table or a period of floating windows, in order to make a quick response to cotton market changes.

Intercept β_0 of linear regression model and CC Index 3128B may also be used to analysis and predict the cotton market ups and downs. Future cooperation with CNCRC is planning and studies on cotton will be further explored to produce more valuable results.

References

[1] Bazzi S and Blattman C 2014 Economic Shocks and Conflict: Evidence from Commodity Prices *American Economic Journal: Macroeconomics* **6** 1-38.

[2] Brooks C, Prokopczuk M and Wu Y 2013 Commodity futures prices: More evidence on forecast power, risk premia and the theory of storage *The Quarterly Review of Economics and Finance* **53** 73-85.

[3] Hodgson G 2016 Karl Polanyi on economy and society: a critical analysis of core concepts *Review of Social Economy* **75** 1-25.

[4] Hammoudeh S, Nguyen D, Reboredo J and Wen X 2014 Dependence of stock and commodity futures markets in China: Implications for portfolio investment *Emerging Markets Review* **21** 183-200.

[5] Wang X and Zhang C 2014 The impacts of global oil price shocks on China's fundamental industries *Energy Policy* **68** 394-402.

[6] Yoo W, Mayberry R, Bae S, Singh K, He Q P and Lillard Jr J W 2014 A study of effects of multicollinearity in the multivariable analysis *International journal of applied science and technology* **4** 9

[7] Sagaert Y, Aghezzaf E, Kourentzes N and Desmet B 2018 Tactical sales forecasting using a very large set of macroeconomic indicators *European Journal of Operational Research* **264** 558-569.

[8] Leeuw J, Hornik K and Mair P 2009 Isotone Optimization inR: Pool-Adjacent-Violators Algorithm (PAVA) and Active Set Methods *Journal of Statistical Software* **32** 1-24.

[9] Spouge J, Wan H and Wilbur W 2003 Least Squares Isotonic Regression in Two Dimensions *Journal of Optimization Theory and Applications* **117** 585-605.

[10] Hu X and Hansohm J 2008 Merge and chop in the computation for isotonic regressions *Journal of Statistical Planning and Inference* **138** 3099-3106.

IOP Conf. Series: Journal of Physics: Conf. Series **1060** (2018) 012076 doi:10.1088/1742-6596/1060/1/012076

Sliding Mode Observer and Control for Quadrotor Stabilization

Hyukwoo Lee, Hyungkwan Kwon, and Kwanho You*

Department of Electrical and Computer Engineering, Sungkyunkwan University

*The corresponding author's e-mail address : khyou@skku.edu

Abstract. In recent years, the development of unmanned aerial vehicles (UAVs) has brought about brilliant progression in many industrial fields. For UAV motion control, the most fundamental and core factor is the controller design for attitude stability. This paper focuses on the stabilization of a quadrotor. Quadrotor is influenced by external disturbances such as wind gust and meteorological changes. For these reasons, quadrotor has a modeling limitation in precision. In this paper, we propose a control method with the estimated three angles to follow the desired value for quadrotor attitude stabilization. We suggest the sliding mode observer (SMO) based attitude estimation model of a quadrotor. The configured error variable by SMO improves the estimation accuracy of the angles such as roll, pitch, and yaw. Using sliding mode control (SMC), moreover, we obtain the appropriate control input to stabilize the quadrotor attitude with the estimated angle based on SMO. Sliding surface that contains the desired angles is determined to reduce the tracking error. The performance and effectiveness of the proposed system are verified through a simulation result.

1. Introduction

Recently, the interest in UAVs has grown rapidly. The quadrotor is maneuvered or propelled by four motors. The quadrotor is classified as a rotorcraft, as opposed to fixed wing aircraft, because it takes off vertically with four motors. Many researchers have studied on a quadrotor for various purposes, such as video recording, swarm, parcel service, exploration of risk, and vehicle. In diverse studies, the core technology is the attitude control and is to use the proportional-integral-derivative (PID) or SMC as a controller. PID is a widely used control algorithm in controller design. However, PID has a limitation in precision for control. Many researchers have studied on other controller to improve the performance. As a result, SMC has been applied successfully in various design problems [1]. The quadrotor is a system with four inputs, independently and six coordinate outputs. Various control methods have been proposed for quadrotor. Castillo proposed the quadrotor obtained via a Lagrange approach dynamic model and control algorithm based on nested saturation [2], and nonlinear control strategy based on Lyapunov analysis [3]. Bouabdallah proposed nonlinear control techniques to stabilize attitude and altitude by using backstepping and SMC [4]. Davila proposed an observer which estimates only position measurements for mechanical systems[5].

In this paper, we propose the stabilization of quadrotor attitude control with the estimated angle to follow the desired angle. We set up the attitude dynamic model of quadrotor and design the adapted quadrotor attitude dynamics by SMO. We obtain the appropriate input u by SMC with adapted attitude dynamics of a quadrotor by SMO. Input u obtained by SMC helps to follow the desired attitude for the stability of a quadrotor. This paper is composed as follows. We obtain the dynamic model of quadrotor attitude in Section 2. We design the adapted quadrotor dynamic and estimate the angle of a quadrotor by using SMO in Section 3. The input u for stabilization of a quadrotor by SMC and Lyapunov stability is obtained in Section 4 with adapted dynamic designed by SMO. We show simulation results of the proposed method in Section 5. Finally, the conclusion appears in Section 6.

2. Quadrotor Dynamics

The quadrotor can take off and land vertically by four motors that generate the propeller forces. Figure 1 shows the earth fixed frame and rigid body model of a quadrotor. Fig.1 describes two pairs of propellers (Ω_1,Ω_3) and (Ω_2,Ω_4). Two diagonal motors (Ω_1,Ω_3) rotate clockwise, and the others (Ω_2,Ω_4) rotate counterclockwise. Quadrotor can change its direction on the spot by two pairs of motors which rotate in the opposite direction.

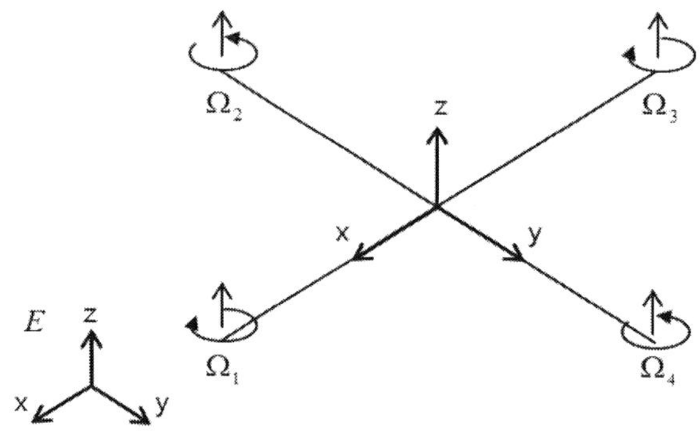

Figure 1. Composition of quadrotor.

We can acquire different position and attitude of the quadrotor by changing the velocity of the four propellers. The quadrotor dynamic model can be presented as follows:

$$\begin{aligned}\ddot{\phi} &= \dot{\theta}\dot{\psi}\left(\frac{I_y - I_z}{I_x}\right) + \dot{\theta}\left(\frac{J_r}{I_x}\right)\Omega + \left(\frac{l}{I_x}\right)u_1 \\ \ddot{\theta} &= \dot{\phi}\dot{\psi}\left(\frac{I_x - I_z}{I_y}\right) + \dot{\phi}\left(\frac{J_r}{I_y}\right)\Omega + \left(\frac{l}{I_y}\right)u_2 \\ \ddot{\psi} &= \dot{\phi}\dot{\theta}\left(\frac{I_x - I_y}{I_z}\right) + \left(\frac{l}{I_z}\right)u_3\end{aligned} \tag{1}$$

Where the angles (ϕ,θ,ψ) obtained from the rotation of quadrotor's body frame represent roll, pitch, and yaw, respectively. $I_{x,y,z}$ is the body inertia, J_r is the rotor inertia, and l is the lever length. The control inputs are represented by u_1, u_2 and u_3, respectively for roll, pitch, and yaw and these are given as:

$$\begin{aligned}u_1 &= t\left(\Omega_4^2 - \Omega_2^2\right) \\ u_2 &= t\left(\Omega_3^2 - \Omega_1^2\right) \\ u_3 &= d\left(\Omega_2^2 + \Omega_4^2 - \Omega_1^2 - \Omega_3^2\right) \\ \Omega &= \Omega_2 + \Omega_4 - \Omega_1 - \Omega_3\end{aligned} \tag{2}$$

Where Ω_i is the i-th motor rotational velocity, t is the thrust factor, and d is the drag factor. The developed model in equation (1) can be rewritten in a state-space form: $\dot{\mathbf{x}} = \mathbf{f}(\mathbf{x}, \mathbf{u})$ and $\mathbf{x} = (x_1 \cdots x_6)^T$ is the state vector of the system. State vector is composed with $x_1 = \phi$, $x_2 = \dot{\phi}$, $x_3 = \theta$, $x_4 = \dot{\theta}$, $x_5 = \psi$, and $x_6 = \dot{\psi}$.

$$c_1 = \frac{(I_y - I_z)}{I_x},\ c_2 = \frac{J_r}{I_x},\ c_3 = \frac{(I_z - I_x)}{I_y},\ c_4 = \frac{J_r}{I_y},\ c_5 = \frac{(I_x - I_y)}{I_z},$$
$$c_6 = \frac{l}{I_x},\ c_7 = \frac{l}{I_x},\ c_8 = \frac{l}{I_x} \tag{3}$$

From equations (1) and (3), we obtain the attitude dynamic of the quadrotor:

$$\dot{\mathbf{x}} = \mathbf{f}(\mathbf{x}, \mathbf{u}) = \begin{pmatrix} x_2 \\ x_4 x_6 c_1 + x_4 c_2 \Omega + c_6 u_1 \\ x_4 \\ x_2 x_6 c_3 + x_2 c_4 \Omega + c_7 u_2 \\ x_6 \\ x_2 x_4 c_5 + c_8 u_3 \end{pmatrix} \tag{4}$$

3. Adapted Quadrotor Dynamics with SMO

The purpose of SMO is to estimate the unmeasurable states of a quadrotor system based only on the measured outputs and inputs. Sliding mode observer works by minimizing the error between quadrotor dynamic model and the observer model by using a switching function and can force the output estimation error to converge to zero in finite time. In this section, we design the SMO which estimates a quadrotor rotational angle and rotational velocity, and design the adapted quadrotor dynamics.

In Section 3, we define the state-space form with the state variables $\mathbf{S} = [\hat{x}_1\ \hat{x}_2\ s_1\ \hat{x}_3\ \hat{x}_4\ s_2\ \hat{x}_5\ \hat{x}_6\ s_3]^T$. We propose the observer based on the differentiation of the state variables x_1 through x_6 of the quadrotor attitude dynamics as follows:

$$\dot{\mathbf{S}} = \begin{bmatrix} \hat{x}_2 + z_1 \\ \hat{x}_4 \hat{x}_6 c_1 + \hat{x}_4 c_2 \Omega + c_6 u_2 + s_1 + z_2 \\ z_3 \\ \hat{x}_4 + z_4 \\ \hat{x}_2 \hat{x}_6 c_3 + \hat{x}_2 c_4 \Omega + c_7 u_3 + s_2 + z_5 \\ z_6 \\ \hat{x}_6 + z_7 \\ \hat{x}_4 \hat{x}_2 c_5 + c_8 u_4 + s_3 + z_8 \\ z_9 \end{bmatrix} \tag{5}$$

In equation (5), the differentiators (z_i) are proposed as:

IOP Conf. Series: Journal of Physics: Conf. Series **1060** (2018) 012076 doi:10.1088/1742-6596/1060/1/012076

$$
\begin{aligned}
z_1 &= k_1 \left|e_1\right|^{\frac{2}{3}} sat\left(e_1\right), z_2 = k_2 \left|e_1\right|^{\frac{1}{3}} sat\left(e_1\right),\ z_3 = k_3 sat\left(e_1\right) \\
z_4 &= k_4 \left|e_3\right|^{\frac{2}{3}} sat\left(e_3\right), z_5 = k_5 \left|e_3\right|^{\frac{1}{3}} sat\left(e_3\right),\ z_6 = k_6 sat\left(e_3\right) \\
z_7 &= k_7 \left|e_5\right|^{\frac{2}{3}} sat\left(e_5\right), z_8 = k_8 \left|e_5\right|^{\frac{1}{3}} sat\left(e_5\right),\ z_9 = k_9 sat\left(e_5\right)
\end{aligned} \tag{6}
$$

The differentiators (z_i) are used in advanced SMO for the estimation of the system state in the presence of unknown external disturbances. In equation (6), e_i is the error between measurement and estimated value as $e_i = x_i - \hat{x}_i$, $(i = 1, 3, 5)$.

4. Sliding mode control for attitude stabilization

In this section, we use the SMC to obtain the input u_i to stabilize the quadrotor attitude with SMO based on quadrotor attitude dynamics. For the first step of SMC controller design, we set up a sliding surface (σ_1) as following:

$$\sigma_1 = \left(\hat{x}_2 - x_{2d}\right) + \lambda_1\left(\hat{x}_1 - x_{1d}\right) \tag{7}$$

In equation (7), λ_1 is a positive constant. Next we use the Lyapunov function to satisfy the stability condition.

$$V(\sigma_1) = \frac{1}{2}\sigma_1^{\ 2}\ ,\ \dot{V}\left(\sigma_1\right) = \sigma_1\dot{\sigma}_1 < 0 \tag{8}$$

Lyapunov function and sliding surface must satisfy the condition (8). As a result, we derive the time derivative of sliding surface as:

$$
\begin{aligned}
\dot{\sigma}_1 &= -\mu_1 sat(\sigma_1) - \mu_2\sigma_1 \\
&= \left(\dot{\hat{x}}_2 - \dot{x}_{2d}\right) + \lambda_1\left(\left(x_2 - x_{2d}\right) + z_1\right) \\
&= \left[\left(\hat{x}_4\hat{x}_6 c_1 + \hat{x}_4 c_2 \Omega + c_6 u_1 + s_1 + z_2 - \dot{x}_{2d}\right) + \lambda_1\left(\left(\hat{x}_2 - x_{2d}\right) + z_1\right)\right]
\end{aligned} \tag{9}
$$

We can obtain the control input u_1 from equation (9) as follows:

$$u_1 = -\frac{1}{c_6}\left[\hat{x}_4\hat{x}_6 c_1 + \hat{x}_4 c_2 \Omega + s_1 + z_2 - \dot{x}_{2d} + \lambda_1\left(\left(\hat{x}_2 - x_{2d}\right) + z_1\right) + \mu_1 sat\left(\sigma_1\right) + \mu_2\sigma_1\right] \tag{10}$$

In equations (9) and (10), we use the saturation function for stabilizing the SMC.

$$sat(\sigma_1, \alpha) = \begin{cases} \text{sgn}(\sigma_1), & \left|\sigma_1\right| > \alpha, \\ \dfrac{\sigma_1}{\alpha}, & \left|\sigma_1\right| \le \alpha \end{cases} \tag{11}$$

α is a width of sliding surface's (σ_1) boundary. We set up a sliding surface (σ_2, σ_3) as follows:

$$\begin{aligned} \sigma_2 &= \left(\hat{x}_4 - x_{4d}\right) + \lambda_2\left(\hat{x}_3 - x_{3d}\right), \\ \sigma_3 &= \left(\hat{x}_6 - x_{6d}\right) + \lambda_3\left(\hat{x}_5 - x_{5d}\right) \end{aligned} \tag{12}$$

We can obtain inputs u_2 and u_3 following the procedures of u_1.

$$\begin{aligned} u_2 &= -\frac{1}{c_7}\left[\hat{x}_2\hat{x}_6 c_3 + \hat{x}_2 c_4 \Omega + s_2 + z_5 - \dot{x}_{4d} + \lambda_2\left(\left(\hat{x}_4 - x_{4d}\right) + z_4\right) + \mu_3 \operatorname{sgn}\left(\sigma_2\right) + \mu_4 \sigma_2\right], \\ u_3 &= -\frac{1}{c_8}\left[\hat{x}_4\hat{x}_2 c_5 + s_3 + z_8 - \dot{x}_{6d} + \lambda_3\left(\left(\hat{x}_6 - x_{6d}\right) + z_7\right) + \mu_5 \operatorname{sgn}\left(\sigma_3\right) + \mu_6 \sigma_3\right] \end{aligned} \tag{13}$$

5. Simulation

In this section, we prove the adapted attitude dynamics of a quadrotor by SMO and the stabilization of quadrotor attitude with SMC. The whole system stabilizes the quadrotor attitude with estimation data. Figure 2 shows the comparison between the attitude estimate values and the desired angles. In figure 2-(a) shows that estimated and stabilized angle (ϕ) of quadrotor attitude. Fig.2-(b) shows the pitch angle (θ). Pitch angle has a difference between attitude estimate value and the desired angle at initial stage because the initial point of pitch angle is not the same. However, after few seconds SMO estimates the quadrotor attitude correctly and pitch angle is stabilized by SMC.

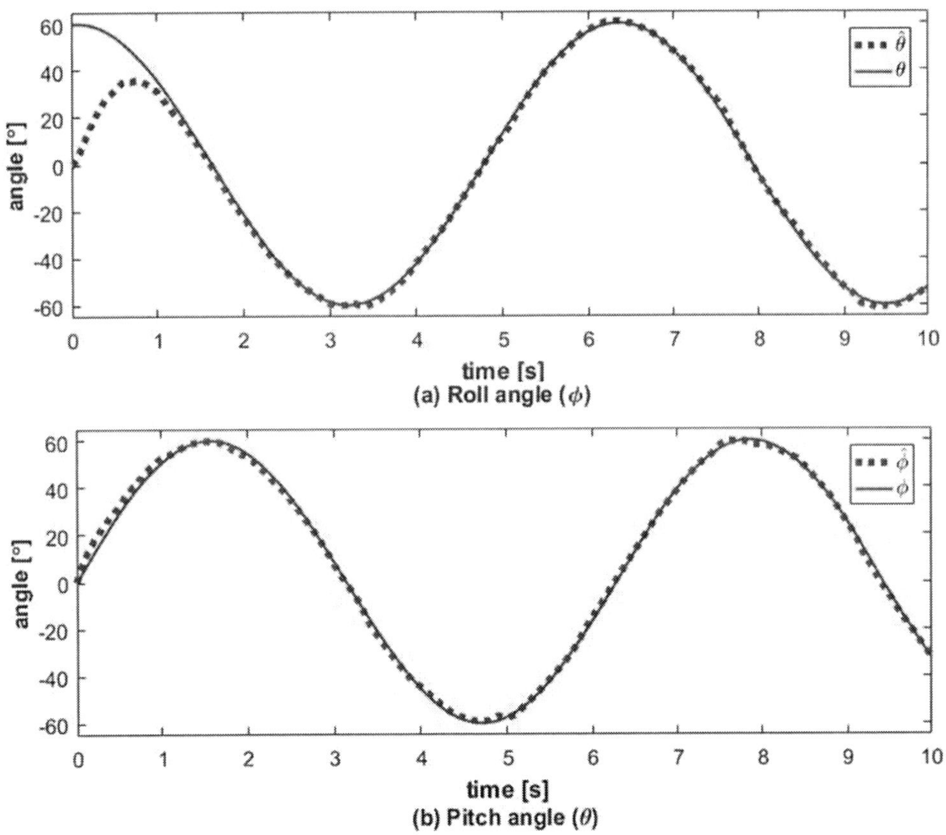

Figure 2. Estimation and stabilization of quadrotor attitude (ϕ, θ).

6. Conclusion

In this paper, we proposed a SMC for stabilization and applied to the quadrotor with SMO. At first, we set up the quadrotor attitude dynamic modeling. Secondly we reconfigured the adapted attitude dynamics of quadrotor, the unmeasured states, and their derivatives through the SMO. Finally we obtained the appropriate input u_i by SMC using the adapted attitude dynamics and estimation data. The estimation data, the differentiator, and the saturation function in SMC and SMO were used to improve the performance in stabilization. We confirmed that SMO estimates the quadrotor attitude and SMC interacts with SMO to stabilize the attitude.

Acknowledgment

This work was supported by the National Research Foundation of Korea (NRF) grant funded by the Korea government (MSIP) (NRF-2016R1A2B4011369).

References

[1] D. Cho, Y. Kato, and D. Spilman, "Sliding mode and classical controllers in magneticlevitation," IEEE Control Systems, vol. 13, pp.42-48, February 1993.
[2] P. Castillo, A. Dzul, and R. Lozano, "Real-time stabilization and tracking of a four-rotor mini-rotorcraft," IEEE Conf., European Control, vol. 12, pp.510-516, September 2003.
[3] P. Castillo, R. Lozano, and A. Dzul, "Stabilization of a mini-rotorcraft having four rotors",IEEE/RSJ Int. Conf., Intelligent Robots and Systems,vol. 3, pp.1543-1548, September 2004.
[4] S. Bouabdallah and R. Siegwart, "Backstepping and sliding-mode techniques applied to an indoor micro quadrotor," IEEE Int. Conf. Robotics and Automation, pp.2247-2252. April 2005.
[5] J. Davila,L. Fridman, and A. Levant,"Second-order sliding-mode observer for mechanical systems," IEEE transactions on automatic control,Vol. 50, pp.1785-1789, November 2005.

AOA Measurement based Localization Using RLS Algorithm under NLOS Environment

Hyukwoo Lee, Kyunghyun Lee, and Kwanho You*

Department of Electrical and Computer Engineering, Sungkyunkwan University

*the corresponding author's email address : khyou@skku.edu

Abstract. In localization problem, the estimation accuracy of a target location is one of the most significant issues. The global positioning system (GPS) that is the most widely used localization method can be disrupted by some disturbances under non-line-of-sight (NLOS) condition such as indoor environment and downtown. In this paper, we suggest a localization algorithm using the angle of arrival (AOA) measurements. AOA measurement based localization is one of the most efficient geolocation methods that use a wireless active signal. Moreover, AOA method has an advantage that only two base stations are required to estimate the target location. In all kinds of localization methods using wireless signal, the NLOS and measurement noise are the significant problems that decrease an estimation accuracy of a target location. In this paper, we suggest a Kalman filter based hypothesis test and a recursive least square scheme to overcome NLOS and measurement noise problem, respectively. Using Kalman filter based hypothesis test, the measurement data set from each base station can be identified whether it contains the NLOS noise or not. Also, recursive least square (RLS) scheme can obtain the precise location of a target with rapid calculation speed when additional measurement data is received from auxiliary base stations. Simulation result confirms the high estimation accuracy and computational speed of our proposed scheme.

1. Introduction

In accordance with the recent improvement of location based service (LBS), localization methods are widely researched and used in many kinds of industrial fields such as internet of things, augmented reality, target tracking, and the mobile application. The AOA scheme using the measurement angle data of signal incidence on an antenna array is one of the typical localization methods using active wireless signal due to its robustness of the environmental disturbance. Although AOA measurements based localization method estimates the relatively exact location of a target compared with GPS, AOA method is insufficient to satisfy the location estimation accuracy that is demanded in recent LBS. AOA method calculates the incidence direction of radio signal by measuring the difference of arrival time at individual elements of the array [1].

However, some noise factors cause that the arrival time of signal propagation at each antenna is delayed and the incidence angle is changed. Especially the NLOS noise that generally occurs in indoor environment is the significant problem of localization. NLOS noise is a radio transmission across a path that is partially obstructed by a physical object such as walls, buildings, and trees. In order to mitigate the NLOS problem, Wang [2] proposed a RSS based localization algorithm to handle the problem of simultaneously localizing the unknown node in LOS/NLOS environments. The proposed algorithm could identify the propagation conditions of the beacon nodes and the measurements which contain the NLOS error are corrected. Silva [3] discussed principles and challenges pertaining to impulse radio ultra wideband ranging and NLOS identification in harsh industrial environments. Xiao [4] has proposed two regressors based on machine learning with propagation models to accurately estimate the transmitter-receiver distances in NLOS conditions.

In this paper, we propose the NLOS data discrimination algorithm based on Kalman filter to recognize whether the measurement data group contains NLOS noise or not. Kalman filter is a recursive estimator in which the estimated state from the previous time step and the current measurement are needed to compute the estimation for the current state. The standard deviation of the distance between a target and each base station that can be obtained from the estimated state using Kalman filter is used to discriminate the NLOS data. Moreover, we suggest the RLS algorithm to estimate a target location precisely. RLS is an iterated optimization method of least square (LS) to minimize the estimation error. The remainder of this paper is organized as follows. The NLOS data is discriminated and eliminated using Kalman filter based hypothesis test in Section 2. In Section 3, the localization model is derived using AOA measurements. Moreover, the recursive estimation algorithm for localization of a target is explained through RLS scheme. The performance of our proposed algorithm is demonstrated through the simulation result in Section 4. Conclusions are presented finally in Section 5.

2. NLOS data discrimination using Kalman filter

In localization problem using wireless signal source, NLOS problem that is occurred by some obstacles on the signal pathway causes the deflection of the signal. Especially in downtown or indoor case, NLOS problem is one of the most significant disturbance factors that cause the discrepancy between the real location and the estimated location of a target [5]. In Section 2, the discrimination algorithm of the NLOS noise included in measurement data at a specific base station is proposed before a localization solution is derived. For the discrimination of NLOS data, the measured distance and velocity data that can be obtained with the measured time of arrival data is divided into some data groups depending on sampling time interval. Each data group is discriminated whether NLOS data or not by the Kalman filter based hypothesis test on the basis of the definite LOS data.

We derive the state equation for the x and y-coordinate of a target independently as follows

$$\mathbf{X}_{k+1} = \mathbf{\Phi X}_k + \mathbf{\Pi \gamma}_k, \quad k = 1, \cdots, N \tag{1}$$

Where

$$\mathbf{\Phi} = \begin{bmatrix} 1 & \Delta n \\ 0 & 1 \end{bmatrix}, \ \mathbf{\Pi} = \begin{bmatrix} 0 \\ \Delta n \end{bmatrix}. \tag{2}$$

State vector of a target $\mathbf{X}_k$ denotes as $[s_{x,k} \ v_{x,k}]^T$ for x-direction or $[s_{y,k} \ v_{y,k}]^T$ for y-direction at the sampling time t_k. $s_{x,k}$ and $v_{x,k}$ are the location and velocity of x-coordinate, respectively. In the state equation (1), γ_k is the driving noise vector with a covariance of $\mathbf{Q} = \sigma_q^2 \mathbf{I}$. The measurement location of a target m_k can be obtained as follows

$$m_k = \mathbf{AX}_k + \rho_k \tag{3}$$

With $\mathbf{A} = [1 \ \ 0]$. ρ_k is the measurement noise value with a covariance of $\rho = \sigma_m^2$.

The iterative process of Kalman filter can be derived using the equations (1) and (3) as follows

$$\begin{aligned} \hat{\mathbf{X}}_k^- &= \mathbf{\Phi}\hat{\mathbf{X}}_{k-1} \\ \mathbf{P}_k^- &= \mathbf{\Phi}\mathbf{P}_{k-1}\mathbf{\Phi}^T + \mathbf{\Pi}\mathbf{Q}\mathbf{\Pi}^T \\ \mathbf{K}_k &= \mathbf{P}_k^-\mathbf{A}^T(\mathbf{A}\mathbf{P}_k^-\mathbf{A}^T + \rho)^{-1} \\ \hat{\mathbf{X}}_k &= \hat{\mathbf{X}}_k^- + \mathbf{K}_k(m_k - \mathbf{A}\hat{\mathbf{X}}_k^-) \\ \mathbf{P}_k &= \mathbf{P}_k^- - \mathbf{K}_k\mathbf{A}\mathbf{P}_k^-. \end{aligned} \tag{4}$$

In equation (4), $\mathbf{P}_k$ means the covariance matrix of a state vector $\mathbf{X}_k$. The updated estimate $\hat{\mathbf{X}}_k$ and covariance $\mathbf{P}_k$ in equation (4) are valid for the optimal Kalman gain which is derived by $\mathbf{K}_k$ in equation (4). Kalman gain gives the relative weight to the measurements and current state estimate. The state of a target $\hat{\mathbf{X}}_k$ at the sampling time t_k can be estimated through this update operation.

We compare the measured distance data with estimated one between a target and each base station for the discrimination of the measurement data group whether under LOS or NLOS condition. The distance between a target and i-th base station can be acquired with the multiplication of the signal propagation speed and TOA data. The measured distance data $d_m(t_k)$ at the sampling time t_k is comprised of real distance $d_i(t_k)$, a measurement error $\Delta d_m(t_k)$, and an NLOS error $\Delta d_{nlos}(t_k)$ as a following equation.

$$d_m(t_k) = d_i(t_k) + \Delta d_m(t_k) + \Delta d_{nlos}(t_k) \tag{5}$$

The estimated location of a target can be derived by using Kalman filter process which is expressed by equation (3). The estimated distance $d_{kalman}(t_k)$ can be calculated with the distance difference between a target and each base station. On the basis of this estimated distance, the standard deviation of measured distance $\hat{\sigma}$ can be formulated as follows

$$\hat{\sigma} = \sqrt{\frac{1}{P}\sum_{k=1}^{P}(d_m(t_k) - d_{Kalman}(t_k))^2}. \tag{6}$$

The parameter P is the number of distance data $d_m(t_k)$ in each partition.

At least one base station is supposed to be located under LOS environment in this paper. Then, the measurement data group which contains the NLOS noise can be discriminated with the following hypothesis test.

$$\begin{aligned} H_1(LOS\ condition)&: \ \hat{\sigma} < \tau\sigma_{LOS} \\ H_2(NLOS\ condition)&: \ \hat{\sigma} \geq \tau\sigma_{LOS} \end{aligned} \tag{7}$$

The parameter τ can be determined depending on a localization environment. The data group of distance measurements $d_m(t_k)$ from each base station can be discriminated whether under LOS or NLOS environment by using hypothesis test of equation (6). Both AOA and TOA data are obtained with the same wireless signal. Therefore, AOA data that is measured at the time interval when the TOA data is determined LOS data is also under LOS environment. Only the discriminated AOA data that is measured under LOS environment is used for localization in Section 3.

3. Recursive estimation of a target location using AOA measurements

In this section, we derive the target localization formula using AOA measurement data under LOS environment. AOA method is one of the most efficient localization approaches. The localization formula can be acquired using the measured angle data of a signal from at least two base stations. When the known location of i-th base station is represented as (x_i, y_i), the measured incidence angle of the propagation signal from the target to the i-th base station, denoted by θ_i, satisfies the following equation.

$$x\sin(\theta_i)-y\cos(\theta_i)=x_i\sin(\theta_i)-y_i\cos(\theta_i) \tag{8}$$

With the unknown target location $\boldsymbol{\varphi}=[x \quad y]^T$. Equation (8) can be rewritten by the matrix form of localization formula as follows

$$\mathbf{Z}\boldsymbol{\varphi}=\boldsymbol{\zeta} \tag{9}$$

Where

$$\mathbf{Z}=\begin{bmatrix}\sin(\theta_1) & -\cos(\theta_1)\\ \vdots & \vdots \\ \sin(\theta_M) & -\cos(\theta_M)\end{bmatrix},\ \boldsymbol{\zeta}=\begin{bmatrix}x_1\sin(\theta_1)-y_1\cos(\theta_1)\\ \vdots \\ x_M\sin(\theta_M)-y_M\cos(\theta_M)\end{bmatrix}. \tag{10}$$

Since the measurement noise causes the refraction of a propagation signal, however, the intersection of AOA lines is not just exact one point in real environment. In equation (11), the matrices $\mathbf{Z}$ and $\boldsymbol{\zeta}$ that are subordinated to the measured AOA data θ_i contain the measurement noise. In order to minimize the localization error that is caused by measurement noise, we suggest RLS algorithm.

For applying RLS algorithm to localization problem, the LS solution needs to be obtained with initial measurement data. When the additional AOA data set is measured at each base station, the solution can be computed recursively with rapid speed using initial solution and RLS technique. The localization formula using n*'s* AOA data and (c+1)-th AOA data can be written based on equation (9) as follows

$$\begin{bmatrix}\overline{\mathbf{Z}}_n \\ \mathbf{Z}_{n+1}\end{bmatrix}\boldsymbol{\varphi}=\begin{bmatrix}\overline{\boldsymbol{\zeta}}_n \\ \boldsymbol{\zeta}_{n+1}\end{bmatrix} \tag{11}$$

Where $\overline{\mathbf{Z}}_n=[\mathbf{Z}_1\ \mathbf{Z}_2\ \cdots\ \mathbf{Z}_n]^T$ and $\overline{\boldsymbol{\zeta}}_n=[\boldsymbol{\zeta}_1\ \boldsymbol{\zeta}_2\ \cdots\ \boldsymbol{\zeta}_n]^T$. $\mathbf{Z}_{n+1}$ and $\boldsymbol{\zeta}_{n+1}$ are $(n+1)$-th AOA parameters. The LS solution of equation (11) is obtained as a following equation.

$$\hat{\boldsymbol{\varphi}}_{n+1}=(\overline{\mathbf{Z}}_{n+1}^T\overline{\mathbf{Z}}_{n+1})^{-1}\overline{\mathbf{Z}}_{n+1}^T\overline{\boldsymbol{\zeta}}_{n+1} \tag{12}$$

With the definition of $\mathbf{H}_{n+1}=(\overline{\mathbf{Z}}_{n+1}^T\overline{\mathbf{Z}}_{n+1})^{-1}$ for a notational simplicity, the equation (12) is rewritten as below

$$\hat{\boldsymbol{\varphi}}_{n+1}=\mathbf{H}_{n+1}[\mathbf{H}_n^{-1}\mathbf{H}_n\overline{\mathbf{Z}}_n^T\overline{\boldsymbol{\zeta}}_n+\mathbf{Z}_{n+1}^T\boldsymbol{\zeta}_{n+1}]. \tag{13}$$

Since the term $\overline{\mathbf{H}}_n\overline{\mathbf{Z}}_n^T\overline{\boldsymbol{\zeta}}_n$ in equation (13) can be represented as $\hat{\boldsymbol{\varphi}}_n$, the equation (13) can be rewritten as follows

$$\hat{\boldsymbol{\varphi}}_{n+1}=\hat{\boldsymbol{\varphi}}_n+\mathbf{H}_{n+1}\mathbf{Z}_{n+1}^T(\mathbf{h}_{n+1}-\mathbf{Z}_{n+1}\hat{\boldsymbol{\varphi}}_n) \tag{14}$$

With $\mathbf{H}_{n+1}=(\overline{\mathbf{Z}}_n^T\overline{\mathbf{Z}}_n+\mathbf{Z}_{n+1}^T\mathbf{Z}_{n+1})^{-1}$. Therefore, the recursive solution $\hat{\boldsymbol{\varphi}}_{n+1}$ can be computed as a following equation.

$$\hat{\boldsymbol{\varphi}}_{n+1}=\hat{\boldsymbol{\varphi}}_n+(\overline{\mathbf{Z}}_n^T\overline{\mathbf{Z}}_n+\mathbf{Z}_{n+1}^T\mathbf{Z}_{n+1})^{-1}\mathbf{Z}_{n+1}^T(\boldsymbol{\zeta}_{n+1}-\mathbf{Z}_{n+1}\hat{\boldsymbol{\varphi}}_n) \tag{15}$$

The location of a target can be computed more rapidly through equation (15) when we receive additional AOA data sets. $\hat{\boldsymbol{\varphi}}_n$ and $\overline{\mathbf{Z}}_n$ are the calculated parameters based on information of $n's$ AOA data in previous n-th iteration. $\mathbf{Z}_{n+1}$ and ζ_{n+1} are the additional measured data. We can estimate the location of a target recursively using previous AOA solution and additionally received information with minimizing the estimation error

4. Simulation results

The effectiveness of our proposed Kalman filter based NLOS discrimination algorithm and RLS scheme based localization using AOA measurements is confirmed with some simulations in Section 4.

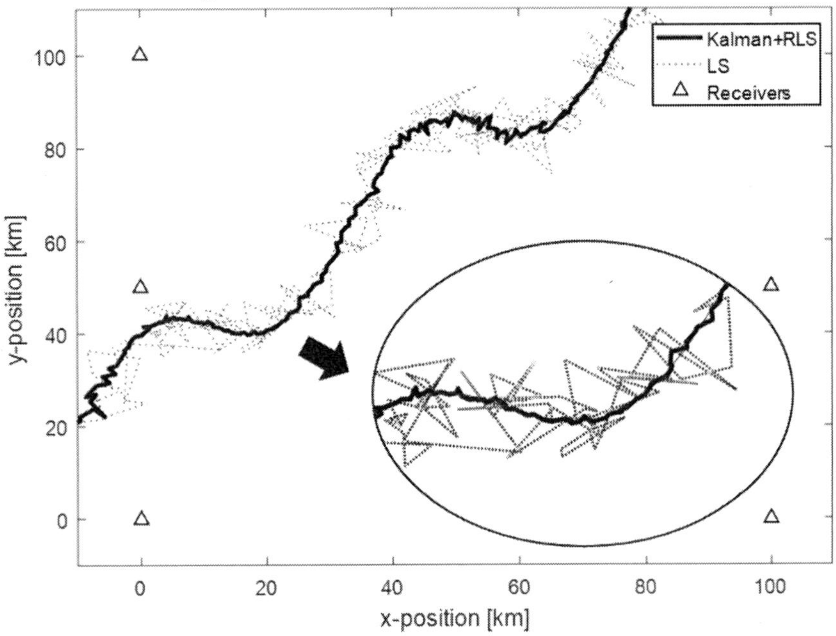

Figure 1. Estimated trajectory of a target.

In our simulation, the wireless propagation signal from a target is supposed to be measured at five base stations of which the locations are (0, 0), (0, 50), (0, 100), (100, 0), and (100, 50) km, respectively. Also, the hypothesis parameter τ is set as 5. The simulated trajectory has 200 time samples with one second interval. In figure 1, the estimated trajectory compares the performance driven by the general LS algorithm and our proposed scheme. The solid line is the estimated trajectory of a target using our proposed scheme. The dotted line is the estimated trajectory through the general LS algorithm. The recursive solution using Kalman filter based NLOS data discrimination algorithm follows the true trajectory more closely than the estimated trajectory using LS algorithm.

5. Conclusions

In this paper, the AOA measurement based localization method under NLOS environment was explained. The NLOS and measurement noises are the typical factors that cause the estimation inaccuracy of a target location. We estimated the state of a target through Kalman filter and applied the hypothesis test to discriminate the NLOS noise contained AOA measurements. Moreover, the recursive solution was obtained using RLS scheme for the estimation error minimization and the fast computational speed. Therefore, the rapidity of calculation through RLS algorithm facilitated to handle more additional measurement data for precise estimation of a target location. We confirmed the effectiveness of the proposed Kalman filter based NLOS data discrimination algorithm and RLS scheme through our simulation result.

Acknowledgment
This work was supported by the National Research Foundation of Korea (NRF) grant funded by the Korea government (MSIP) (NRF-2016R1A2B4011369).

References

[1] J. Bull, "Wireless geolocation," *IEEE Veh. Technol. Mag.*, vol. 4, pp. 45-53, November 2009.
[2] Y. Wang, L. Cheng, G. Han, H. Wu, and B. Jiang, "RSS localization algorithm based on nonline of sight identification for wireless sensor network," *Int. J. Distrib. Sens. Netw.*, vol. 2014, pp.1-8, July 2014.
[3] B. Silva and G. Hancke, "IR-UWB-Based Non-line-of-sight identification in harsh environments: principles and challenges," *IEEE Trans. Ind. Inform.*, vol. 12, pp. 1188-1195, June 2016.
[4] Z. Xiao, H. Wen, A. Markham, N. Trigoni, P. Blunsom, and J. Frolik, "Non-line-of-sight identification and mitigation using received signal strength," *IEEE Trans. Wirel. Commun.*, vol. 14, pp.1689-1702, March 2015.
[5] Y. Qi, H. Kobayashi, and H. Suda, "Analysis of wireless geolocation in a non-line-of-sight environment," *IEEE Trans. Wirel. Commun.*, vol. 5, pp. 672–681, March 2006.

An Integrated Load-balancing Scheduling Algorithm for Nginx-Based Web Application Clusters

Ruoyu Li[1], Yunchun Li[1] and Wei Li[1]

[1]School of Computer Science and Engineering, BeiHang University, Beijing, China

*Corresponding author e-mail: {liruoyu, lych, liw}@buaa.edu.cn

Abstract. In recent years, the LXC (Linux Container)-based Docker technology has attracted great attention due to its lightweight and convenient features. Because of the performance differences among the hosts and the network complexity, it's necessary to improve the resource utilization of WEB application clusters. This paper analyzes the WEB application deployment architecture based on container technology, calculates the host load metrics combined the host's performance indicates and status of the operating containers, including CPU utilization, memory utilization, network utilization, and the proportion of unused memory that has been allocated for the containers, and proposes a Dynamic Weighted Least-Connection Algorithm(DWLC). The experiment shows that, the DWLC algorithm helps the WEB application response 52.6% and 46.4% faster than the ordinary Round-Robin and Least-Connection algorithm.

1. Introduction

As the most popular container technology in recent years, the LXC (Linux Container)-based Docker technology attracted great attention due to its lightweight and convenient features. Docker uses Linux Cgroup for resource allocation and ensures the independence and security of each container resource through Linux Namespace. Docker can run an application on any docker-installed server by packaging its dependent binaries, libraries, configuration files and any other things that the application needed into a lightweight, portable, self-contained container. Compared with virtual machine, Docker can provision performance and user-space isolation with extremely low virtualization overheads [1]. Although Docker uses the host's hardware resources in a shared manner, it can create thousands of containers and improve the overall utilization of system resources.

Since "Build once, run anywhere" is the most appropriate statement about Docker, the development and deployment of applications increasingly tends to depend on Docker containers with its strong scalability and portability. As the vast majority of WEB applications deployed on the cloud platform [2], more and more service providers use the Docker container-based clustering approach for deployment. At the entrance of the cluster, there will always be a load balancer, such as nginx, haproxy, F5, etc. The load balancer minimizes the probability that any particular server will be overwhelmed, optimizes the bandwidth available to each server, facilitates traffic prioritization, provides end-to-end application monitoring, helps protect against malicious activity, and "evenly" distributes the received requests to back-end servers.

However, in the face of container-based WEB clusters, these load balancers can only statically set the load balancing algorithm based on the initial performance of the host, instead of dynamically adjusting based on the real-time monitored host state. Therefore, how to properly monitor the status of the host and the containers on it and form a dynamic adaptive load balancing algorithm, so as to truly achieve load balancing of web requests at the entrance of the cluster, it has great significance in improving system utilization to the current increasing container-based Web application cluster.

The remainder of this paper is organized as follows. We discuss the related work of load balance algorithm in section II. In section III, we design the structure of an integrated load-balancing scheduling algorithm for nginx-based Web application cluster. We introduce the selection of the performance parameters and the algorithm in section IV. Section V shows the experimental results and we conclude our work in the last section.

2. Related Works

Under the comprehensive consideration of the memory usage of the server, the CPU usage of the server and the current network bandwidth, Luo Shuhua et. al. calculated weight of each server, proposed a dynamic weight least-load balancing algorithm [3]. This algorithm realized the dynamic adjustment of server weight, resulting in more stable network operation and better load distribution.

Xu Zongyu proposed an optimized Route Robin load balancing strategy based on WEB server cluster [4]. By calculating and comparing the load values of various servers, this strategy sends incoming WEB requests to the next server node, which is not the most heavily loaded. All this implements a dynamic optimization of the Route Robin algorithm and WEB request allocation, and achieves an even distribution of WEB requests.

Considering features of Docker, various applications' requirements, and available resources in cloud date centers, Guan Xinjie et. al. proposed a communication-efficient resource allocation algorithm and designed a framework for Docker container-based application oriented dynamic resource allocation, to minimize the application deployment cost [5].

Unfortunately, for a Docker container-based WEB application cluster, how to effectively monitor the load status of the host, distribute WEB requests to the back-end service container in an even and balanced manner, thereby balancing the load of the entire system at the entrance of the cluster, there is no better way to deal with it. So we propose a Dynamic Weighted Least-Connection (DWLC) load balancing algorithm based on Nginx.

3. System Structure and Modelling

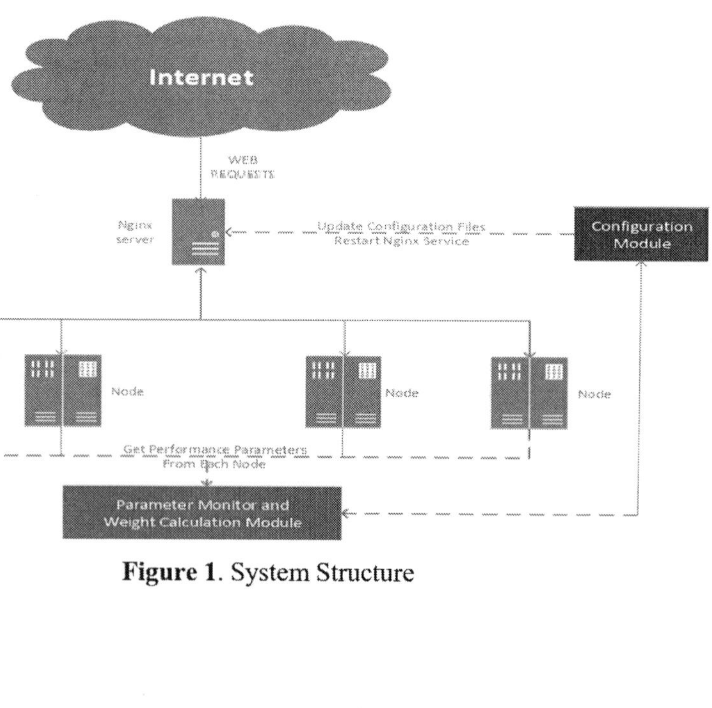

Figure 1. System Structure

Nginx is a software-based load balancing server with eight different load balancing algorithms. According to different requirements, Nginx can reverse proxy multiple web applications at the same time by setting different configuration files. However, the original Least-Connection or Weighted Least-Connection algorithm in Nginx is static and cannot fully reflect the node's overall processing capability during system operation. So we propose a Dynamic Weighted Least-Connection (DWLC) Algorithm combined the host's performance metrics with status of the operating containers.

As shown in figure 1, we establish the parameter acquisition and weight calculation module, as well as the nginx configuration module. The system monitors the performance parameters of each node in real time and calculates the weight of the node according to a certain algorithm. Then the nginx configuration module decides whether to update the weight of the node or restart the nginx service.

4. Performance Parameters and Algorithm

4.1. Selection of Performance Parameters

In a hybrid cloud environment, the load status of the whole docker container-based WEB cluster can not only be calculated by the host's performance indicates, but also the status of the operating containers. So we choose four performance parameters-CPU utilization ratio, Memory utilization ratio, Network utilization ratio and the proportion of used memory that has been allocated for the containers to calculate the load of the node.

We assume that the WEB cluster has n nodes. For each node i ($i \in n$), we get the four performance parameters every t seconds.

The CPU utilization ratio ($CPU\_Ratio_i(t)$) is the ratio of the sum of the cpu time in user mode and system kernel mode to the sum of the cpu time in user mode, kernel mode, and idle mode. It can be calculated using the following formula:

$$\mathrm{CPU}_{\mathrm{Ratio_i(t)}} = \frac{CPU_{usr(t)} + CPU_{kernel(t)}}{CPU_{usr(t)} + CPU_{kernel(t)} + CPU_{free(t)}} \tag{1}$$

The Memory utilization ratio ($MEM\_Ratio_i(t)$) is the non-idle ratio of current host memory. It can be calculated using the following formula:

$$\mathrm{MEM}_{\mathrm{Ratio_i(t)}} = 1 - \frac{MEM_{free(t)}}{MEM_{total}} \tag{2}$$

The Network utilization ratio ($NET\_Ratio_i(t)$) is the ratio of the sum of the bytes received and sent by the node within the time interval t to the network bandwidth(NET_band). When a node join the cluster, first of all, we will use iperf (iperf is a network performance testing tool that can measure the maximum bandwidth between the client and the server) to calculate the network bandwidth (NET_band) between the node and the load balancing server. And then, we calculate the sum of the bytes received and sent by the node within the time interval t. So the network utilization ratio can be calculated using the following formula:

$$\mathrm{NET}_{\mathrm{Ratio_i(t)}} = \frac{NET_{sent(t)} + NET_{rev(t)}}{NET_{band}} \tag{3}$$

The proportion of used memory that has been allocated for the containers (Di(t)). We can use the Docker native command "docker stats" to read the real-time memory information of each container on the node, and then $D_i(t)$ is the ratio of the sum of MEM USAGEs to the sum of MEM LIMITs:

$$\mathrm{D_i(t)} = \frac{\Sigma mem_{usage}}{\Sigma mem_{limit}} \tag{4}$$

4.2. Algorithm

When a node joins the cluster, we will arrange the node an initial weight 100, and then we execute the weight calculation every t seconds.

First of all, we calculate the real-time load $L_i(t)$ of node i every t seconds using the following formula:

$$L_i(t) = a_1 * CPU_{Ratio_i(t)} + a_2 * MEM_{Ratio_i(t)} + a_3 * NET_{Ratio_i(t)} + a_4 * D_i(t) \quad (5)$$

Where $L_i(0)=0$, and $a_1+a_2+a_3+a_4=1$, $0<a_i<1$. Normally, a_i can be set different value due to different kinds of application. Here, We set $a_1 = 0.35, a_2 = 0.35, a_3 = 0.2, a_4 = 0.1$.

Then the current weight $W_i(t)$ of node i is:

$$W_i(t) = 100 - Li(t) \quad (6)$$

In order to avoid the node's weight fluctuation caused by small weight changes, we establish a load variation threshold A. We get the weight of the node i W_i from the nginx configuration file, and then calculate the weight variation λ within the time interval t:

$$\lambda = W_i(t) - W_i \quad (7)$$

If the absolute value of λ is smaller than A, the weight of the node remains, else the weight of the node changes as formula below:

$$W'_i = \begin{cases} W_i(t) & |\lambda| \geq A \\ W_i & |\lambda| < A \end{cases} \quad (8)$$

Here, W_i is the weight of node i in the nginx configuration file, λ is the weight variation, $W_i(t)$ is the current weight of node i and W'_i is the new weight of the node i.

After all this calculation finished, if W'_i is not equal to W_i, we update the nginx configuration file and reload the nginx service.

5. Experiments

Apache Bench (ab) is a single-threaded command line tool for measuring the performance of HTTP web servers. It can simulate multiple users creating a large number of concurrent access threads to stress a URL address [6].

Table 1. the software and hardware of each server

Role	NUM of CPU	MEM	Bandwidth(Iperf)	Operation System	Main Software
Worker	8	4GB	942 Mbits/sec	Ubuntu14.04 LTS	DOCKER1.13
Worker	8	4GB	941 Mbits/sec	Ubuntu14.04 LTS	DOCKER1.13
Worker	4	8GB	23.0 Gbits/sec	Ubuntu14.04 LTS	DOCKER1.13
Worker	4	8GB	18.5 Gbits/sec	Ubuntu14.04 LTS	DOCKER1.13
Worker	4	8GB	21.7 Gbits/sec	Ubuntu14.04 LTS	DOCKER1.13
Nginx Server	4	4GB		Ubuntu14.04 LTS	Nginx1.4.6

We use 1 nginx server and 5 worker servers to conduct the experiment. The software and hardware configuration of the servers are shown in table 1. We deploy the SNS Web application UCenter Home in docker, using the nginx container and the php container. Then we use iperf to measure the bandwidth between the nginx server and each worker server. We use the ab command to simulate 100 users for 10,000 concurrency every time to see the response speed of the web application cluster. In all figures, the abscissa is the proportion of all requests completed, and the ordinate is the response speed (in milliseconds).

First of all, to confirm the threshold of the weight changes, we test the value of the threshold from 1 to 6. As shown in Figure 2, the response time is almost no change when the value of the threshold changes. The experiment result shows that the algorithm is insensitive to the threshold. However, Figure 3 shows that the response speed of the web cluster is relatively faster when the threshold is 5 than others when 95% requests are finished. So we choose 5 as the threshold of the weight change.

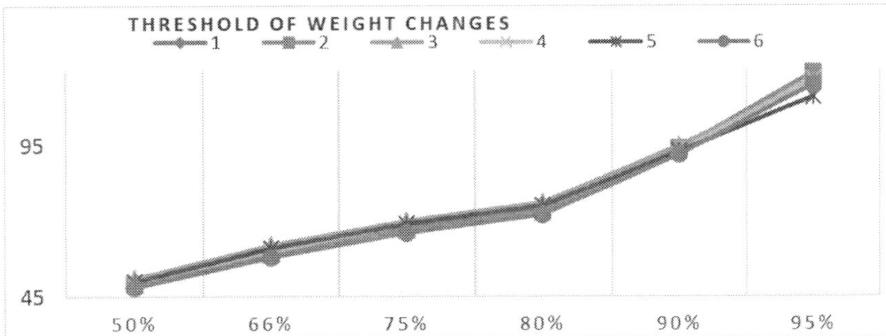

Figure 2. threshold experiment

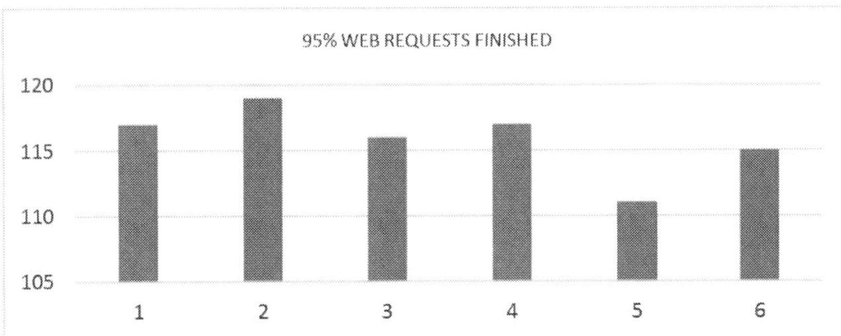

Figure 3. Response speed when 95% finish

Then in order to avoid taking up too much system resources or feeding back the load of the node not in time because of bad time interval choice, we test the time interval t from 1s to 5s. As shown in Figure 4, the response speed of the cluster is the fastest when the time interval is 3s. So we set the time interval 3 seconds.

At last, we set the threshold of weight changes 5 and the time interval 3s, and compare the response speed of Round Robin algorithm, Least-Connection algorithm and our DWLC algorithm. As shown in Figure 5, the DWLC algorithm has obvious advantages and can improve the Web cluster response speed 52.6% and 46.4% faster than the ordinary Round Robin and Least-Connection algorithm when 95% requests are finished.

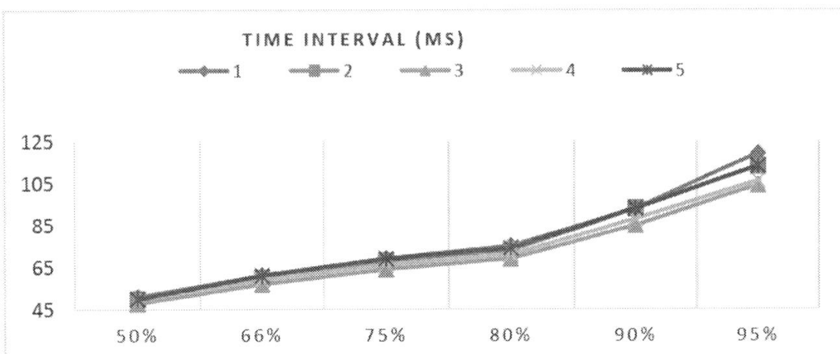

Figure 4. Time interval experiment

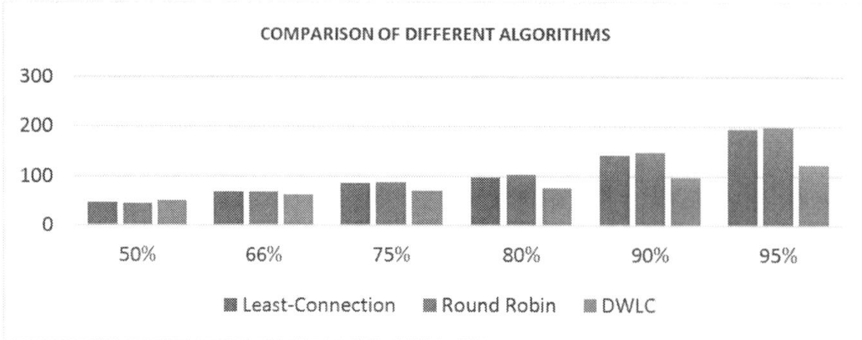

Figure 5. Comparison of different algorithms

6. Conclusion

In this paper, we analyse the WEB application deployment architecture based on docker container, calculate the host load metrics combined the host's performance indicates and status of the operating containers, and propose a Dynamic Weighted Least-Connection algorithm. The experiments show this algorithm can efficiently improve the response speed of the WEB cluster compared to RR algorithm and Least-Connection algorithm.

Acknowledgment

This work is partly supported by the National High-tech R&D Program ("863" Program) of China (No.2015AA01A202) and research project of Beijing Municipal Education Commission.

References

[1] J. García-Galán, P Trinidad , Rana OF, A. Ruiz-Cortés, "Automated configuration support for infrastructure migration to the cloud," Future Generation Computer Systems pp 200-212, Feb-2016 VILAPLANA,JORDI.SLA-aware load balancing in a web-based cloud system over OpenStack[J].Service-Oriented Computing-ICSOC 2013Workshops,2014(9):281-293.

[2] Felter, W., Ferreira, A., Rajamony, R., & Rubio, J. (2015, March). An updated performance comparison of virtual machines and linux containers. In Performance Analysis of Systems and Software (ISPASS), 2015 IEEE International Symposium On (pp. 171-172). IEEE. VILAPLANA,JORDI.SLA-aware load balancing in a web-based cloud system over OpenStack[J].Service-Oriented Computing-ICSOC 2013Workshops,2014(9):281-293.

[3] Ying Wu, Shuhua Luo,Qing Li.An Adaptive Weighted Least-Load Balancing Algorithm Based on Server Cluster[C].Intelligent Human-Machine Systems and Cybernetics(IHMSC), 2013,pp,224-227.

[4] Xu Zongyu, Wang Xingxuan.A Modified Round-robin Load-balancing Algorithm for Cluster-based Web Servers[C].Control Conference(CCC),2014,pp,3580-3584.

[5] Guan Xinjie, Wan Xili.Application Oriented Dynamic Resource Allocation for Data Centers Using Docker Containers. 23 December 2016, IEEE Communications Letters, pp. 504-507.

[6] Ab-Apache HTTP server benchmarking tool.http://httpd.apache.org/docs/2.4/programs/ab.html

An Improved P-DPC for A New Inverter Based on MLD Model

Li Xue-feng, Lei Xiao-ben,Han Jan-ding, Wang Chuan-qi and Zeng Jia-qi

Aeronautics Engineering Institute, Air Force Engineering University, Xi'an 710038, China

519132089@qq.com;2625335209@qq.com;hanjiandinghjd@sina.com;1176058172@qq.com;1258509463@qq.com

Abstract. Building an accurate mathematical model for inverter is the key to achieving a precise control. Contrasted to the traditional switching function model of power electronic circuits, this paper built the mixed logical dynamic(MLD) model for a new inverter, and the MLD model was used as a prediction model, then a predictive direct power control(P-DPC) method was researched for the new inverter. A symmetrical 4+4 voltage vector sequence was employed to obtain constant switching frequency and lower THD of output voltage, the action time of vector sequence was calculated by minimizing objective function. The feasibility and effectiveness were proved by simulations.

1. Introduction

Constraint conditions and nonlinear factors of the control object could be fully considered by model predictive control (MPC), which is also suitable to the multivariable system control, multiple targets can be reached by minimizing objective function, therefore MPC is appropriate for power electronic circuits control [1]. While a challenge we faced with is the problem of mixed integer quadratic programming (MIQP) [2].

There are actually two approaches for solving MIQP, one is solving MIQP problem off-line, the results are stored in formulated table, circuit states are monitored to select the optimal solution by searching table, but this method would occupy computer storage, and an effective searching algorithm is needed[3-5]. Another is using on-line optimization algorithm, for example, Branch and Bound, which needs a large amount of calculation [6].

The document [7-10] researched Finite Control Set Model Predictive Control(FCS—MPC) for circuits, the discrete properties of power electronic circuits were fully used by FCS-MPC, each possible combination of switch states was considered, and the switch state, which minimized objective function was used as circuit control, the complex MIQP problem was avoided, but there still had two problems:

1) Switching frequency was inconstant, which increased the difficulties to design filter;

2) Each switch state should be considered, which was not suitable for more switch states circuits.

P-DPC, which combined direct power control and MPC, had both advantages, documents [11] adopted 3+3 voltage vector sequence to decrease switching loss.

An improved 4+4 voltage vector sequence was selected by this paper to obtain constant switching frequency and lower THD of output voltage, the action time of vector can be got by minimizing objective function.

2. The prediction model of inverter based on MLD model

The inverter shown at figure 1 is improved from [12], switch S1-S4 are increased to balance the voltage of neutral point O. The inverter could be used in high reliable system for its simple control and fine fault tolerant.

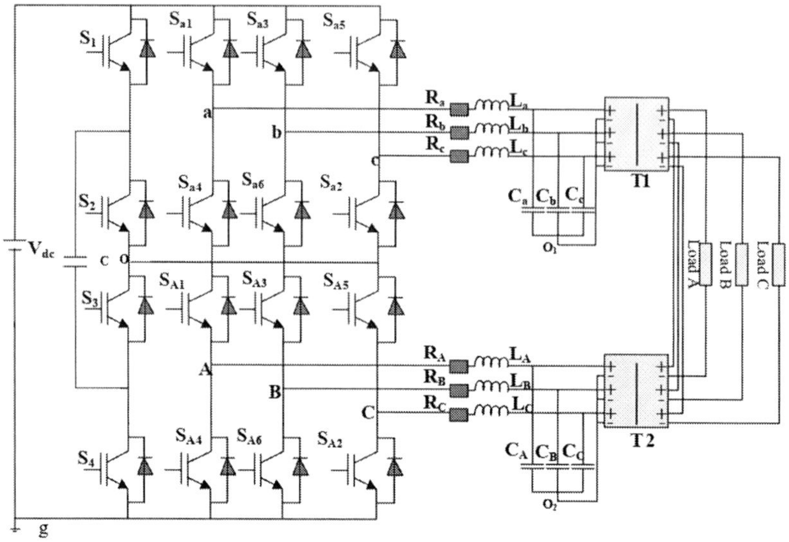

Figure.1 New inverter topology

The switch states of one phrase of inverter is shown in table 1, similarly to the rest two phrases, U_{ao} is voltage between a and g, U_{Ag} is voltage between A and g, $s_1 - s_6$ are control signals of switch $S_{a1} - S_{a6}$, $s_7 - s_{12}$ are control signals of switch $S_{A1} - S_{A6}$. The control of neutral point voltage is achieved by setting switches $S_1 - S_4$ on and off.

Table.1 switch states of one phrase

(s_1, s_4, s_7, s_{10})	(U_{ao}, U_{Ag})
$(1,0,1,0)$	$(V_{dc}, V_{dc}/2)$
$(0,1,0,1)$	$(V_{dc}/2, 0)$
$(0,1,1,0)$	$(V_{dc}/2, V_{dc}/2)$
$(1,0,0,1)$	$(V_{dc}, 0)$

The direction of filter current i_a is shown in Figure 1, for phase a, operating models are shown in (1), when "1" represents "on", "0" represents "off".

$$\begin{aligned} &i_a>0: s_1=0, s_4=1 \Rightarrow u_{ao}=U_{dc}/2; s_1=1, s_4=0 \Rightarrow u_{ao}=U_{dc}; s_1=0, s_4=0 \Rightarrow u_{ao}=U_{dc}/2 \\ &i_a<0: s_1=0, s_4=1 \Rightarrow u_{ao}=U_{dc}/2; s_1=1, s_4=0 \Rightarrow u_{ao}=U_{dc}; s_1=0, s_4=0 \Rightarrow u_{ao}=U_{dc} \end{aligned} \tag{1}$$

Logical operators are introduced [13], "$\wedge$" represents conjunction, "$\vee$" represents disjunction, "$\leftrightarrow$" represents equivalence, "-" represents not, meanwhile logical variable σ_a is also introduced, the discrete event $i_a > 0$ and $i_a < 0$ are respectively shown by $\sigma_a = 1$ and $\sigma_a = 0$, as (2).

$$\begin{cases} [\sigma_a = 1] \leftrightarrow [i_a > 0] \\ [\sigma_a = 0] \leftrightarrow [i_a < 0] \end{cases} \tag{2}$$

The discrete events in (1) are replaced by logical variables, then mixed logical relation of phase a is shown in (3).

$$\begin{aligned}&[s_1=0,s_4=1,\sigma_a=1]\vee[s_1=0,s_4=0,\sigma_a=1]\vee\\&[s_1=0,s_4=1,\sigma_a=0]\leftrightarrow[u_{ag}=U_{dc}/2]\\&[s_1=1,s_4=0,\sigma_a=1]\vee[s_1=0,s_4=0,\sigma_a=0]\vee\\&[s_1=1,s_4=0,\sigma_a=0]\leftrightarrow[u_{ag}=U_{dc}]\end{aligned} \tag{3}$$

(3) is expressed as one equation, and the mixed logical expression of u_{ag} is shown in (4).

$$u_{ao}=U_{dc}[\bar{s}_4(s_1+\bar{s}_1\bar{\sigma}_a)+\frac{1}{2}\bar{s}_4\overline{(s_1+\bar{s}_1\bar{\sigma}_a)}] \tag{4}$$

The expression of phase 'b' and 'c' can be also obtained by the same way

$$\begin{cases}u_{ao}=V_{dc}[\bar{s}_4(s_1+\bar{s}_1\bar{\sigma}_a)+\frac{1}{2}\bar{s}_4\overline{(s_1+\bar{s}_1\bar{\sigma}_a)}]\\u_{bo}=V_{dc}[\bar{s}_6(s_3+\bar{s}_3\bar{\sigma}_b)]+\frac{1}{2}\bar{s}_6\overline{(s_3+\bar{s}_3\bar{\sigma}_b)}]\\u_{co}=V_{dc}[\bar{s}_2(s_5+\bar{s}_5\bar{\sigma}_c)]+\frac{1}{2}\bar{s}_2\overline{(s_5+\bar{s}_5\bar{\sigma}_c)}]\end{cases} \tag{5}$$

If three phase of inverter are balanced, then the MLD expressions of u_{ao_1}, u_{bo_1} and u_{co_1} are shown in (6).

$$\begin{bmatrix}u_{ao_1}\\u_{bo_1}\\u_{co_1}\end{bmatrix}=\frac{V_{dc}}{3}\begin{bmatrix}2&-1&-1\\-1&2&-1\\-1&-1&2\end{bmatrix}\cdot\begin{bmatrix}\left[\bar{s}_4\left(s_1+\bar{s}_1\bar{\sigma}_a\right)\right]+\frac{1}{2}\bar{s}_4\overline{\left(s_1+\bar{s}_1\bar{\sigma}_a\right)}\\\left[\bar{s}_6\left(s_3+\bar{s}_3\bar{\sigma}_b\right)\right]+\frac{1}{2}\bar{s}_6\overline{\left(s_3+\bar{s}_3\bar{\sigma}_b\right)}\\\left[\bar{s}_2\left(s_5+\bar{s}_5\bar{\sigma}_c\right)\right]+\frac{1}{2}\bar{s}_2\overline{\left(s_5+\bar{s}_5\bar{\sigma}_c\right)}\end{bmatrix} \tag{6}$$

Then, (6) is transformed into space vector form, have:

$$u_{o_1}=\frac{2}{3}\left(u_{ao_1}+\alpha u_{bo_1}+\alpha^2u_{co_1}\right) \tag{7}$$

Where: $\alpha=e^{j(2\pi/3)}$.

By the same way, inductance current, i_{af} i_{bf} and i_{cf}, capacitance voltage u_{ac}, u_{bc} and u_{cc}, are all transformed into space vector form, which are shown in (8).

$$\begin{cases}i_f=\frac{2}{3}\left(i_{af}+\alpha i_{bf}+\alpha^2i_{cf}\right)\\u_c=\frac{2}{3}\left(u_{ac}+\alpha u_{bc}+\alpha^2u_{cc}\right)\end{cases} \tag{8}$$

u_{o_1}, u_c, i_f are transformed into α, β, coordinate system, which can obtained $u_{o_{1\alpha}}$, $u_{o_{1\beta}}$, $u_{\alpha c}$, $u_{c\beta}$, $i_{f\alpha}$, $i_{f\beta}$ the active and reactive power are shown in (9)

$$\begin{bmatrix}P\\Q\end{bmatrix}=\begin{bmatrix}u_{c\alpha}&u_{c\beta}\\u_{c\beta}&-u_{c\alpha}\end{bmatrix}\begin{bmatrix}i_{f\alpha}\\i_{f\beta}\end{bmatrix} \tag{9}$$

Then, (9) is derived to (10)

IOP Conf. Series: Journal of Physics: Conf. Series **1060** (2018) 012079 doi:10.1088/1742-6596/1060/1/012079

$$\begin{bmatrix} \frac{dP}{dt} \\ \frac{dQ}{dt} \end{bmatrix} = \begin{bmatrix} u_{c\alpha}\frac{di_{f\alpha}}{dt} + i_{f\alpha}\frac{du_{c\alpha}}{dt} + u_{c\beta}\frac{di_{f\beta}}{dt} + i_{f\beta}\frac{du_{c\beta}}{dt} \\ u_{c\beta}\frac{di_{f\alpha}}{dt} + i_{f\alpha}\frac{du_{c\beta}}{dt} - u_{c\alpha}\frac{di_{f\beta}}{dt} - i_{f\beta}\frac{du_{c\alpha}}{dt} \end{bmatrix} \tag{10}$$

Ignored the filter resistance, Considered:

$$\frac{di_f}{dt} = \frac{1}{L}\left(u_{o_1} - u_c\right) \tag{11}$$

Then the (12) can be obtained by (11)

$$\begin{cases} \frac{di_{f\alpha}}{dt} = \left(u_{o_{1\alpha}} - u_{c\alpha}\right)/L \\ \frac{di_{f\beta}}{dt} = \left(u_{o_{1\beta}} - u_{c\beta}\right)/L \end{cases} \tag{12}$$

The target of this control is to get a standard sinusoidal voltage from the input of transformer. So that $u_{c\alpha}$ and $u_{c\beta}$ can be described as (13)

$$\begin{cases} u_{c\alpha} = U_s \sin \omega t \\ \frac{du_{c\alpha}}{dt} = U_s \omega \cos \omega t = -\omega u_{c\beta}; \\ u_{c\beta} = -U_s \cos \omega t \\ \frac{du_{c\beta}}{dt} = U_s \omega \sin \omega t = \omega u_{c\alpha} \end{cases} \tag{13}$$

Where: U_s is the amplitude which is expected. Substituting (10) with (12) and (13), have:

$$\begin{cases} \frac{dP}{dt} = u_{c\alpha}\left[\left(u_{o_{1\alpha}} - u_{c\alpha}\right)/L + \omega i_{f\beta}\right] + u_{c\beta}\left[\left(u_{o_{1\beta}} - u_{c\beta}\right)/L - \omega i_{f\alpha}\right] \\ \frac{dQ}{dt} = u_{c\alpha}\left[\omega i_{f\alpha} - \left(u_{o_{1\alpha}} - u_{c\alpha}\right)/L\right] + u_{c\beta}\left[\left(u_{o_{1\beta}} - u_{c\beta}\right)/L + \omega i_{f\beta}\right] \end{cases} \tag{14}$$

It is clear that the variation rate of power is related with the filter inductance and switch status from (14).

Under the condition of a certain voltage vector $u_{o_1} = \left[u_{o_{1\alpha}} \quad u_{o_{1\beta}}\right]^T$ and a large switch frequency, the $u_c = \left[u_{c\alpha} \quad u_{c\beta}\right]^T$ can be regarded as a constant. Considering under that condition, the $i_{f\alpha}$ and $i_{f\beta}$ only have little change, and in the period when the voltage vector effect, the variation rate of power can be approximately regarded as a constant.

$$k_{P_i} = \left.\frac{dP}{dt}\right|_{u_{o_1} = u_{o_1 i}} \quad k_{Q_i} = \left.\frac{dQ}{dt}\right|_{u_{o_1} = u_{o_1 i}} \tag{15}$$

Where: $u_{o_1 i}$ is one of the eight space voltage vectors (i=0 1 2……7). The space voltage vectors are shown in figure 2.

IOP Conf. Series: Journal of Physics: Conf. Series **1060** (2018) 012079 doi:10.1088/1742-6596/1060/1/012079

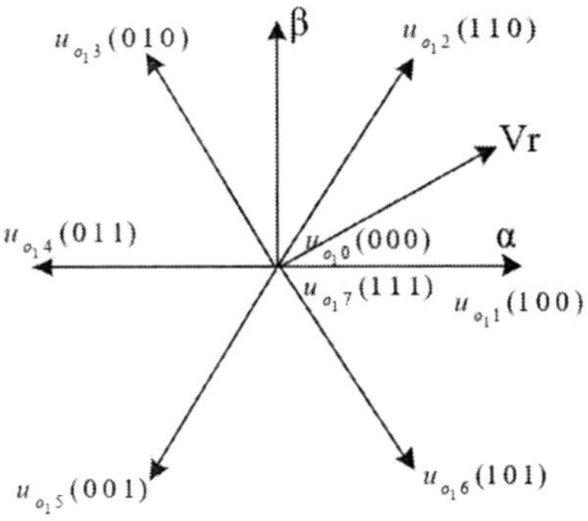

Figure.2 The space voltage vector

Through (15) the track of active and reactive power can be predicted.

$$P_i = P_{i-1} + k_{P_i} \cdot t_m \quad Q_i = Q_{i-1} + k_{Q_i} \cdot t_m \tag{16}$$

Where: t_m is the effecting period of voltage vector $u_{o_1 i}$; P_{i-1}, Q_{i-1} are the power before the vector effecting, and P_i, Q_i are the after.

3. Symmetrical 4+4 voltage vector sequence

3.1. The principle of 4+4 vector sequence

Symmetrical 4+4 vector sequence is improved from 3+3 vector sequence [14], as Figure 2 shown, two voltage vectors which near reference voltage vector are selected as effective vectors, in addition zero vectors $u_{o_1 0}$ and $u_{o_1 7}$ are selected, the sequence is constituted by 4 vectors, and these vectors are used as inverter control.

The sequence is divided into two subsequences, and the first subsequence is completely symmetrical to the second one, the last vector and its action time of the first subsequence are the same as the first vector and its action time of second subsequence, the other vectors of the two subsequences also follow the same symmetrical law.

When the reference voltage vector located in different 6 sections, the 4+4 vector sequences of inverter are shown in table.2.

Table.2 4+4 voltage vector sequence

Section	The first sequence	The second sequence
Section 1	$u_{o_1 0}, u_{o_1 1}, u_{o_1 2}, u_{o_1 7}$	$u_{o_1 7}, u_{o_1 2}, u_{o_1 1}, u_{o_1 0}$
Section 2	$u_{o_1 0}, u_{o_1 2}, u_{o_1 3}, u_{o_1 7}$	$u_{o_1 7}, u_{o_1 3}, u_{o_1 2}, u_{o_1 0}$
Section 3	$u_{o_1 0}, u_{o_1 3}, u_{o_1 4}, u_{o_1 7}$	$u_{o_1 7}, u_{o_1 4}, u_{o_1 3}, u_{o_1 0}$
Section 4	$u_{o_1 0}, u_{o_1 4}, u_{o_1 5}, u_{o_1 7}$	$u_{o_1 7}, u_{o_1 5}, u_{o_1 4}, u_{o_1 0}$
Section 5	$u_{o_1 0}, u_{o_1 5}, u_{o_1 6}, u_{o_1 7}$	$u_{o_1 7}, u_{o_1 6}, u_{o_1 5}, u_{o_1 0}$
Section 6	$u_{o_1 0}, u_{o_1 6}, u_{o_1 1}, u_{o_1 7}$	$u_{o_1 7}, u_{o_1 1}, u_{o_1 6}, u_{o_1 0}$

IOP Conf. Series: Journal of Physics: Conf. Series **1060** (2018) 012079 doi:10.1088/1742-6596/1060/1/012079

3.2. Vector action time computation

If reference voltage vector locates i^{th} section at k moment, adjacent vectors $u_{o_1 0}, u_{o_1 m}, u_{o_1 n}$ and $u_{o_1 7}$ are selected as effective vectors. From (14) and (15) the $k_{P_0}, k_{P_m}, k_{P_n} k_{P_7}$ can be obtained.

$$\begin{aligned} &P_1 = P_k + k_{P_0} \cdot t_0, Q_1 = Q_k + k_{Q_0} \cdot t_0 \\ &P_2 = P_1 + k_{P_m} \cdot 2t_m, Q_2 = Q_1 + k_{Q_m} \cdot 2t_m \\ &P_3 = P_2 + k_{P_n} \cdot 2t_n, Q_3 = Q_2 + k_{Q_n} \cdot 2t_n \\ &P_{k+1} = P_3 + k_{P_7} \cdot t_7, Q_{k+1} = Q_3 + k_{Q_7} \cdot t_7 \\ &T_s = t_0 + 2t_m + 2t_n + t_7 \end{aligned} \tag{17}$$

Where t_0, t_m, t_n, t_7 respectively are the action time when $u_{o_1 0}, u_{o_1 m}, u_{o_1 n}, u_{o_1 7}$, effect. P_k, Q_k are the active power and reactive power at the k moment, and P_{k+1}, Q_{k+1} are the k+1 moment. Ts are the sampling time.

If the action time of $u_{o_1 0}$ is equal to the action time of $u_{o_1 7}$, (17) can be simplified as:

$$\begin{aligned} &P_1 = P_k + k_{P_0} \cdot 2t_0, Q_1 = Q_k + k_{Q_0} \cdot 2t_0 \\ &P_2 = P_1 + k_{P_m} \cdot 2t_m, Q_2 = Q_1 + k_{Q_m} \cdot 2t_m \\ &P_{k+1} = P_2 + k_{P_n} \cdot 2t_n, Q_{k+1} = Q_2 + k_{Q_n} \cdot 2t_n \\ &T_s = 2t_0 + 2t_m + 2t_n \end{aligned} \tag{18}$$

The objective function is selected as (20) in order to compute vector action time.

$$g = (P_{k+1} - P_{k+1}^*)^2 + (Q_{k+1} - Q_{k+1}^*)^2 \tag{19}$$

Where: P_{k+1}^*, Q_{k+1}^* are the reference power at 'k' moment .

To minimize the objective function, the effective vectors action time should satisfy:

$$\frac{\partial g}{\partial t_m} = 0, \quad \frac{\partial g}{\partial t_n} = 0 \tag{20}$$

Thus, combined (20)-(18) the action time of each vector can be obtained, then the inverter can be controlled by these 4 vectors,

4. The structural design for P-DPC

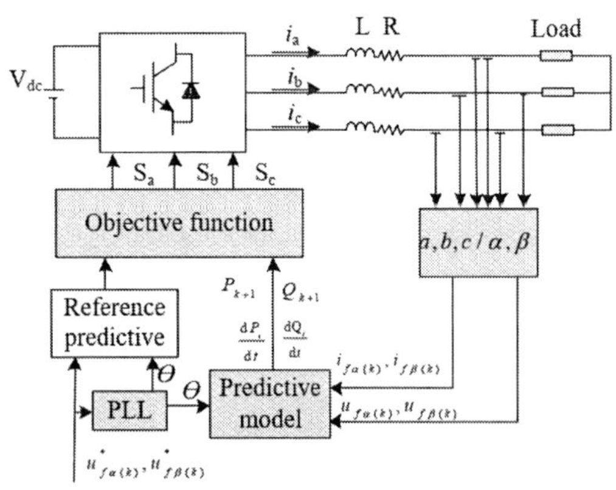

Figure.3 The block diagram of P-DPC

The block diagram of P-DPC is shown in figure 3, which are constituted by sampling module, prediction module, PLL module, reference voltage prediction module, objective function module, the inductance current and capacitor voltage is sampled, and $i_{f\alpha}(k)$, $i_{f\beta}(k)$, $u_{c\alpha}(k)$, $u_{c\beta}(k)$ are obtained by $abc/\alpha\beta$ transformation, the phase angel θ is calculated by PLL module, adjacent two voltage vectors is selected as effective vectors by judging the section of reference voltage vector [15,16].

P_{k+1} and Q_{k+1} at k+1 moment are calculated by prediction module according to the information offered by sampling module and PLL module, each vector action time is calculated by objective function module by (18)-(20), S_a, S_b, S_c are generated as inverter control signals.

IOP Conf. Series: Journal of Physics: Conf. Series **1060** (2018) 012079 doi:10.1088/1742-6596/1060/1/012079

5. Simulation

According to the inverter topology shown in Figure 1, the strategy researched in this paper was proved by MATLAB/Smulink, main simulation parameters were provided as following. $V_{dc} = 270V$, filter inductance L=2mH, filter capacitance C=780mF, rated frequency is 400Hz, sampling period $T_s = 10\mu s$.

The results of P-DPC based on switching function model and 6 voltage sequence were shown in figure 4, THD=5.98%, the output voltage had a dispersive harmonic seen from spectrogram, which was difficult for filter design. figure 5 were results which based on MLD model and 3+3 voltage vector sequence, THD=4.26%, the THD was reduced compared to the results in figure 4, the reason was the setting up of MLD model for inverter, which was more accurate than switching function model, but the harmonic was still dispersive. figure 6 were results based on MLD model and 4+4 voltage vector sequence, THD=2.43%, and the harmonic concentrated on low-frequency, which was easy to clear up.

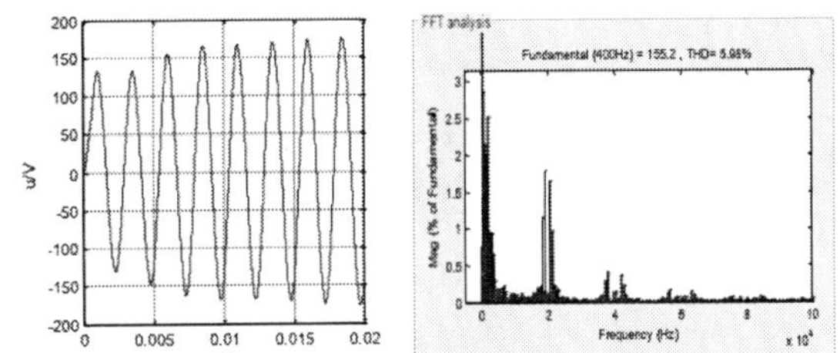

Figue.4 The control result based on switch function model and 3+3 voltage vector sequence

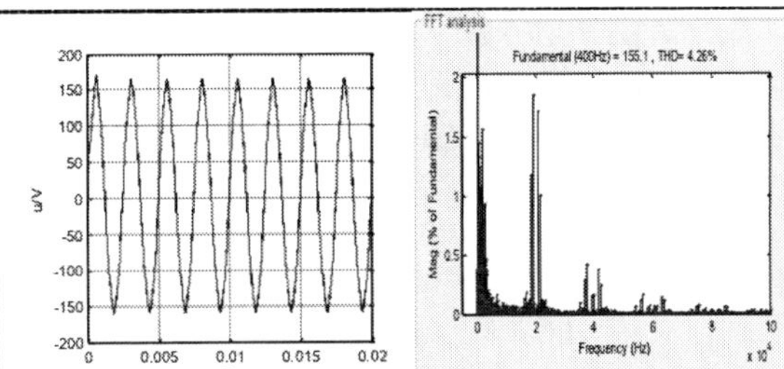

Figure.5 The control result based on MLD model and 3+3 voltage vector sequence

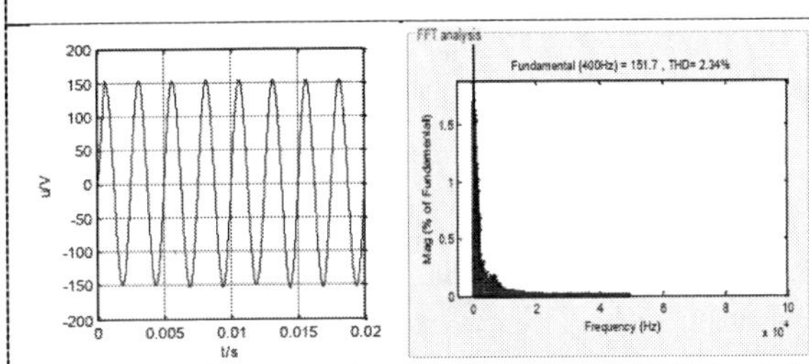

Figure.6 The control result based on MLD model and 4+4 voltage vector sequence

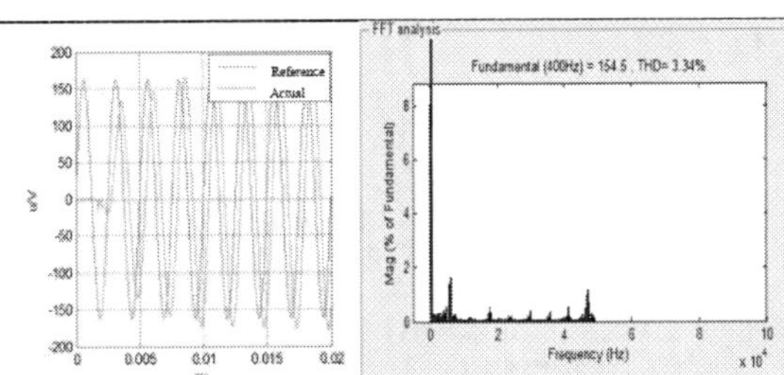

Figure.7 The transient performance of control

The inverter transient properties from no-load to full-load was shown in figure 7, a resistance with 200Ω was installed at 0.0015s, the output voltage can adjust to reference voltage quickly, after full-load, the output voltage THD=3.44%.

6. Conclusions

MLD prediction model was built for a new inverter, then the paper researched P-DPC strategy based on 4+4 voltage vector sequence, the low THD and fixed harmonic of output voltage were obtained, which was easy for filter design, and the favorable steady state and transient properties of the strategy was proved by simulations.

References

[1] P. Cortes, M. P. Kazmierkowski, R. M. Kennel. 2008 *J. IEEE Trans. Ind. Electron,* Predictive control in power electronics and drives, VOL.55, NO.12, pp:4312-4324.

[2] D. Axehill, T. Besselmann, D.M. Rainmondo. 2011 August 28 – September 2, *Conf.on 18th*

IFAC World Congress, Milano:Italy. Suboptimal Explicit Hybrid MPC via Branch and Bound.

[3] Tobias Geyer, Georgios Papafotiou, Roberto Frasca. 2008 *J. IEEE TRANSACTIONS ON POWER ELECTRONICS,* Constrained Optimal Control of Step-Down DC-DC Converter, VOL 55, NO.12, pp:2454-2464,.

[4] Alberto Bemporad, Francesco Borrelli, Manfred Morari. 2002 *J. IEEE TRANSACTIONS ON AUTOMATIC CONTROL,* Model Predictive Control Based on Linear Programming – The Explicit Solution. VOL.47, NO.12, pp:1974-1985,.

[5] J. Kan, J. Qiu, K. Tang, K. Zhu and C. Shao. 2009 *J. International Journal of Applied Electromagnetics and Mechanics*, Modeling and simulation of piezoelectric composite diaphragms for energy harvesting, 30 (2009), pp:95-106.

[6] Daniel Axehill, Manfred Morari. 2010 *Conf.on 49th IEEE conference on decision and control,* Improved Complexity Analysis of Branch and Bound for Hybrid MPC., USA:IEEE, pp:4216-4222.

[7] P. Cortes, G. Ortiz, J. I. Yuz. 2009 *J. Industrial Electronics, IEEE TRANSACTIONS ON INDUSTRIAL ELECTRONICS.* Model Predictive Control of an Inverter with Output LC Filter for UPS Application [J]., VOL.56, NO.6, pp: 1875-1883.

[8] Yalin Zhang, Hua Lin. 2011 May 30-June 3, *Conf. on Simplified 8th International Conference on Power Electronics. The Shilla Jeju, Korea.* Model Predictive Current Control Method of Voltage-source Inverter.

[9] Jose Rodriguez, Jorge Pontt, 2007 *J. IEEE TRANSACTIONS ON INDUSTRIAL ELECTRONICS.* Predictive Current Control of a Voltage Source Inverter. VOL.54,NO1, pp:495-503.

[10] Stefan Almer, Sebastien Mariethoz, Manfred Morari. 2012 *J. IEEE TRANSACTIONS ON CONTROL SYSTEMS TECHNOLOGY.* Sampled Data Model Predictive Control of a Voltage Source Inverter for Reduced Harmonic Distortion, VOL.24, NO.14,pp:89-95.

[11] S. Aurtenechea, M. A. Rodriguez, E. Oyarbide. 2007 *J. IEEE Trans. Ind. Electron,* Predictive control strategy for dc/ac converter based on direct power control. VOL.54, NO.12, pp:1261-1271,.

[12] Li Ning, Li Ying-hui, Lei Hong-li. 2013 *J. Power Electronics*, A Multi-fault Tolerant Inverter Topology. VOL.47, NO.3, pp:106-108.

[13] SREEKUMAR C, VIVEK Agarwal. 2008 J. IEEE Transactions on Industrial Electronics, A Hybrid Control Algorithm for Voltage Regulation in DC-DC Boost converter. VOL.55, NO.6, pp:2530-2538.

[14] Sergio Aurtenechea Larrinaga, Miguel Angel Rodríguez Vidal, Estanis Oyarbide. 2007 *J. IEEE TRANSACTIONS ON INDUSTRIAL ELECTRONICS.* Predictive Control Strategy for DC/AC Converters Based on Direct Power Control, VOL. 54, NO.3, pp:1261-1271,.

[15] Yalin Zhang, Hua Lin. 2011 *Conf. on 8th International Conference on Power Electronics – ECCE Asia May 30-june 3, The Shilla Jeju, Korea.* Simplified Model Predictive Current Control Method of Voltage-Source Inverter.

[16] Junwu Kan, Shuyun Wang, Jianming Wen. 2011 *J. International Journal of Applied Electromagnetics and Mechanics*. Performance of a constant-stress piezoelectric generator for battery-less remote control/switch., 35 (2011) pp:1-13.

An Adaptive Sampling Algorithm and Approximate-Model-Based Optimization Method

J L Tian[1, *] and X Y Tong[2]

[1] School of Astronautics, Beihang University, B921, 37 Xuyuan Road, Haidian District, Beijing100191, China
[2] School of Astronautics, Beihang University, B330, 37 Xuyuan Road, Haidian District, Beijing100191, China

[*]tianjiuling@163.com

Abstract: The approximate model based optimization algorithm can improve the speed of optimization convergence and save the time of calculation, which is very consistent with the development requirements of modern industrial rapid design. In this work, an adaptive and approximate model based optimization method is introduced, which is improved on the basis on HAM. This method uses three different types of approximate models at the same time to the full advantage of the three models' characteristics. Besides, an adaptive sampling algorithm is applied to achieve the self-setting mechanism for new sampling points in the search process. By testing on several standard test functions, the method performed better than GA, PSO and HAM in computational efficiency and accuracy. The method is also applied in the optimization design of rocket engine design to get the maximum thrust. Only after fifty-seven times of simulation calculation, optimal is obtained and the thrust is improved by 2.56%. the method is particularly suitable for design problems involving computation intensive analyses and simulations.

1. Introduction

Today, with the great improvement in the accuracy of computer analyses and simulations, the physical characteristics of the modelled systems can be well predicted and expensive physical experiments for a given design are replaced by computer simulations widely. Besides, these computer simulation softwares are also used for optimization design. Then appropriate search algorithms for design optimization are essential. In recent years, many efficient, capable, and robust optimization methods has been developed. Optimization methods based on approximate model are widely applied to solve time-consuming design optimization problems.

Many approximate models with high efficiency and accuracy are proposed and applied widely, for instance, response surface method (RSM), radial basis function (RBF), multivariate adaptive regression spline (MARS) and Kriging methods [1]. These approximate models have different characteristics and suitable for different kinds of design problems [2]. Then a strategy to choose the most suitable approximate model from several models for an optimization design problem becomes invaluable.

Many strategies proposed to solve the problems are using several approximate models to predict at the same time. Zerpa proposed the concept of combination of approximate models [3]. Combination of approximate models superimposes several approximate models together in the form of weights and the weight coefficient is generally determined by a certain error index. The robustness and accuracy of the combined approximate model are higher than that of a single approximate model. This strategies has been applied on the optimization of helicopter propeller blade noise reduction [4]. Gu introduced a hybrid and adaptive meta-model-based global optimization methods (HAM) and applied in vehicle design [5]. This method incorporates three different approximate models into the search process.

In this work, an approximate model based optimization method is introduced to meet the need of optimization design problems which are characteristics unknown, computation intensive, time-consuming and computer simulation based. Developed on the basis of HAM, the method incorporates three different approximate models and sets new sampling points adaptively in each iteration. The adaptive sampling mechanism increases the accuracy of the approximate models and the efficiency of optimization process. Performance tests on several standard test functions showed considerable improvements over other global optimization methods that are single approximate model based and conventional. A real industrial design problem in rocket engine nozzle is used to demonstrate the feasibility and advantages of the method.

2. Introduced method

In this article, a method for global optimization is introduced that three different approximate models are used for modelling and new sampling points are set adaptively in the search process. In this strategy, three approximate models are constructed on initial sample points. Then, three set of points are calculated using the three constructed approximate models. Finally, several new sample points are choose from the three set of points, and the models are reconstructed.

The use of three approximate models presents a good balance in complexity and performance. Because the three approximate models have different characteristics for different types of optimization design problems, when one of the three models is not suitable for the optimization design problem, the other two approximate models are likely to be suitable for the problem. Response surface method (RSM), radial basis function (RBF) and Kriging methods are chosen in this work, because each of them have different performance characteristics and present a unique capability to cover a specific type of objective functions. Kriging method is better for high order problems; RSM is suitable for lower order problems; and RBF has the advantage of local approximation.

Given a simulation function, $f(\mathrm{x})$, the proposed optimization method contains the following parts:

2.1. Get the initial sampling points

The simple Latin hypercube sampling (LHS) technique is used to obtain design data samples that best represent the objective function. A few points sampled by LHS and several boundary points are used as initial sampling points. The purpose of adding boundary points as initial sampling points is Improve the efficiency of calculation under these circumstances where the optimum points are on the boundary.

2.2. Adaptive sampling algorithm (ASA)

(1) Construct the initial approximate models, $f'(\mathrm{x})$, $g'(\mathrm{x})$ and $h'(\mathrm{x})$, using the three different approximate models with the initial points from $f(\mathrm{x})$.

(2) In design space, *S*, Generate *N* points using LHS method and put these points into group *P*.

(3) Use the three constructed approximate models to calculate the values of the *N* points in *P*, and sort these points in the descending order of the values calculated by the models independently. The obtained three group of points have the same elements, but in different order.

(4) Select a small portion (n=0.01N) of the three group of points independently with the low values associated with each of the three approximate models. Then three group of points are obtained: *F* for $f'(\mathrm{x})$, *G* for $g'(\mathrm{x})$ and *H* for $h'(\mathrm{x})$.

(5) The points in the three groups, *F*, *G* and *H*, are divided into seven new groups again:

$S_1 = F \cap G \cap H$, covers t_1 points that are selected by all three approximate models: $f'(\mathrm{x})$, $g'(\mathrm{x})$ and $h'(\mathrm{x})$.

$S_2 = F \cap G - S_1$, covers t_2 points that are selected by two approximate models: $f'(\mathrm{x})$ and $g'(\mathrm{x})$.

$S_3 = F \cap H - S_1$, covers t_3 points that are selected by two approximate models: $f'(\text{x})$ and $h'(\text{x})$.
$S_4 = H \cap G - S_1$, covers t_4 points that are selected by two approximate models: $h'(\text{x})$ and $g'(\text{x})$.
$S_5 = F - S_2 \cup S_3$, covers t_5 points that are selected only by approximate models: $f'(\text{x})$.
$S_6 = G - S_2 \cup S_4$, covers t_6 points that are selected only by approximate models: $g'(\text{x})$.
$S_7 = H - S_3 \cup S_4$, covers t_7 points that are selected only by approximate models: $h'(\text{x})$.

(6) Select new sample points to reconstruct the approximate models from the seven groups of points, as in equation (1):

$$k_i = int\{7 \times w_i\} \quad (i = 1, \dots, 7) \tag{1}$$

Where i is the number of the group , w_i is the weight of the group, and k_i is the integer value after rounding.

If $t_1 > 0$ and $k_1 = 0$, k_1 is reset to 1. This can avoids the situation that the points in S_1 is ignored after weighted calculation because the number of points in S_1 is too small..

(7) Use the objective function, $f(\text{x})$ to calculate the values of these new chosen points. Combine the new sample points with the former sample points as the original sample points in next loop to reconstruct the three approximate models.

2.3. Weight calculation

The weight of each group is calculated based upon the number of points in the group, as in equation (2):

$$w_i = \frac{t_i \times l_i}{\sum_{i=1}^{k} t_i \times l_i} \quad (i = 1, \dots, 7) \tag{2}$$

Where t_i is the number of the points in the ith group, $l_1 = 10,\ l_2 = l_3 = l_4 = 5,\ l_5 = l_6 = l_7 = 1$.

2.4. Termination criteria

The mean of the ten lowest function values, $\overline{min}$, are set as the terminating parameters of the search process. When the improvement on $\overline{min}$ becomes negligible, the search process terminates. As in equation (3):

$$|\overline{min}_{i+1} - \overline{min}_i| \le \varepsilon \tag{3}$$

$$\overline{min} = \frac{\sum_{j=1}^{10} f_j}{10} \tag{4}$$

Where ε is a small value given by designer, and f_j is the jth smallest function value.

3. Test of the new ASA method on standard test functions

The proposed ASA method has been tested on several well-known test functions, and compared with several widely used global optimization methods, including particle swarm optimization (PSO) and genetic algorithm (GA). Besides, GU's HAM method is also used as comparison, since ASA is improved on the basis of HAM.

The standard test functions are:

(1) Generalized polynomial function(GF)

$$f(x_1, x_2) = \left(2.625 - x_1(1 - x_2^3)\right)^2 + \left(2.25 - x_1(1 - x_2^2)\right)^2 + \left(1.5 - x_1(1 - x_2)\right)^2, \quad x_1, x_2 \in [-2, 2] \tag{5}$$

Its known minimum is 0.5233.

(2) Goldstein and Price function (GP)

$$f(x_1, x_2) = \left(1 + (x_1 + x_2 + 1)^2(19 - 14x_1 + 3x_1^2 - 14x_2 + 6x_1x_2 + 3x_2^2)\right) \times \left((30 + 2x_1 - 3x_2)^2(18 - 32x_1 + 12x_1^2 + 48x_2 - 36x_1x_2 + 27x_2^2)\right) \quad x_1, x_2 \in [-2, 2] \tag{6}$$

Its known minimum is 3.

(3) Griewank functions(GN)

$$f(x) = 1 + \frac{(x_1^2+x_2^2)}{200} - \cos(x_1)\cos\left(\frac{x_2}{\sqrt{2}}\right),\ x_1, x_2 \in [-100, 100] \tag{7}$$

Its known minimum is 0.

(4) Six-hump Camel-Back function (SC)

$$f(x_1, x_2) = 4x_1^2 - 2.1x_1^4 + \frac{1}{3}x_1^6 + x_1x_2 - 4x_2^2 + 4x_2^4, x_1, x_2 \in [-2, 2] \tag{8}$$

Its known minimum is -1.0316.

(5) Leon function

$$f(x_1, x_2) = 100(x_2 - x_1^3)^2 + (x_1 - 1)^2, x_1, x_2 \in [-10, 10] \tag{9}$$

Its known minimum is 0.

In the test, seven points sampled by LHS and four boundary points are used as initial sampling points. And N is set as 10000, then n is 100. For each standard test function, the search process runs 10 times to avoid unrepresentative numerical results. The number of iterations (NIT) and the number of sample points (NSP) are used to measure computation efficiency. The average of all values are illustrated in table 1. The calculated minimum is also shown to illustrate the accuracy of the search.

Table 1. the test data of GA, PSO, HAM and ASA on five standard test functions.

(a) calculated minimum(MIN).

	GA	PSO	HAM	ASA
GF	0.5234	0.5233	0.5467	0.5461
GP	11.2539	3.0036	3.0006	3.0039
GN	0	0	0.0537	0.0306
SC	-1.0316	-1.0316	-1.0310	-1.0311
Leon	0.0379	0.0059	0.0053	0.0055

(b) number of iterations(NIT).

	GA	PSO	HAM	ASA
GF	51.5	276.2	12.8	11.8
GP	103.1	335.7	23.4	19.3
GN	51	382.7	13.1	8
SC	51	412.8	6.7	6.3
Leon	84.4	406.6	21.9	19.1

(c) number of sample points(NSP).

	GA	PSO	HAM	ASA
GF	1030	5524	88.4	87.7
GP	2062	6714	165.1	138.9
GN	1020	7654	72.8	65.1
SC	1020	1729	49.6	51.8
Leon	1688	8132	160.6	132.6

As the results shown in Table 1 (a), the ASA method preserves the accuracy of HAM, which is equivalent to the genetic algorithm. Only on test function GN, ASA perform poor in accuracy, and the reason is that the GN function has hundreds of Local optimum points in the design space.

As the results shown in Table 1 (b) and (c), the ASA method performed well on all functions, with both NEF and NIT lower than GA and PSO. GA and PSO require too many sample points and cost function evaluations. ASA method performed well than HAM on all test functions except SC, because ASA has more initial sample points than HAM which lead to larger NIT.

In summary, the proposed ASA method has improved the calculation efficiency on the premise of keeping the accuracy of the HAM.

4. Application on the nozzle profile optimization

As an important part of the rocket engine, the design of the nozzle has an important influence on the thrust of the rocket engine. The change in the nozzle profile of the rocket engine will affect the thrust, when all the other conditions remain unchanged. In order to obtain larger engine thrust, the ASA method is used to optimize the design of the nozzle of a small thrust rocket engine.

4.1. Simulation model

The physical model used in numerical calculation is a curved nozzle. Considering the symmetry of the physical model, the model in the numerical analysis and calculation is simplified, that is, only half of the model is established, so that the computation cost of the whole simulation can be largely reduced. The FLUENT software based on the finite volume method is used to simulate the flow field of the nozzle. The surface mesh of the rocket engine nozzle is shown in Figure 1.

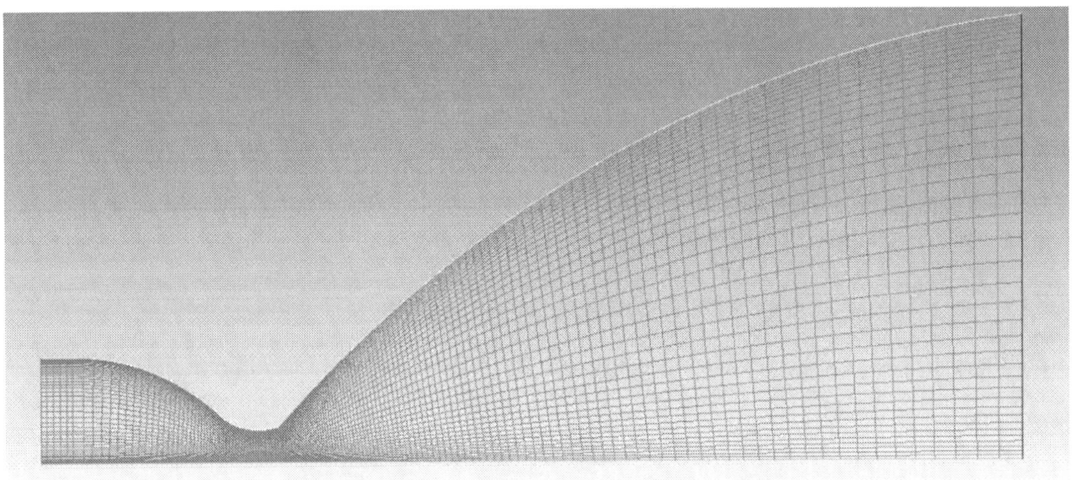

Figure 1. The rocket engine nozzle

In the process of numerical calculation, Nitrogen is set to a gas flow medium inside the nozzle, the inlet and outlet boundary condition are set to the pressure inlet boundary, wall boundary condition is set to non-slip adiabatic wall boundary

4.2. Optimization process

Three key design variables are considered, which are nozzle diffuser angle β_e, ratio of the radius of the contraction section surface to the nozzle throat radius, $k = R_2/R_t$, and ratio of the radius of the throat section surface to the nozzle throat radius, $\rho = R_1/R_t$. Other parameters, such as the nozzle throat radius, inlet radius and outlet radius, remain unchanged. The design objective is to maximize the thrust:

$$\begin{aligned} &\text{Min} \quad (\text{-F}) \\ &\text{s.t.} \quad 1.3 \ll k \ll 2 \\ &\qquad 2 \ll \rho \ll 4 \\ &\qquad 5 \ll \beta_e \ll 10 \end{aligned} \tag{10}$$

Optimization results using the ASA optimization program are obtained as shown in table 2. The optimization process has only seven iterations and fifty-seven times of simulation calculation, taking

550 minutes in total. Compared with the initial design, the optimal design increases 2.56% of the thrust.

Table 2. the comparison between initial design and optimal design

	k	ρ	β_e	F/N
Initial design	1.5	3	7.5	44.252
Optimal design	2.001	2.046	5.194	45.386

5. Conclusions

In this work, an adaptive and approximate model based optimization method is introduced, which is improved on the basis on HAM. The objective of this work is to further improve the efficiency and accuracy of the approximate model based optimization technique and make it applicable to a broad spectrum of design optimization applications.

This method uses three different types of approximate models at the same time, so that it can takes the full advantage of the three models' characteristics and be suitable for different kinds of optimization design problems. Besides, the method applied an adaptive sampling algorithm (ASA) to achieve the self-setting mechanism for new sampling points in the search process.

By testing on several standard test functions, the method preserves the accuracy of HAM, which is equivalent to the genetic algorithm. The method also performed well than GA, PSO and HAM in computational efficiency.

The method has been applied in the optimization design of rocket engine design to get the maximum thrust. Only after seven iterations and fifty-seven times of simulation calculation, optimal is obtained and the thrust is improved by 2.56%, then the feasibility and efficiency of this method have been proved.

Acknowledgements

This article is the National Natural Science Fund Project (approval No. 51306012). Thank you for Prof F Y Xie, Professor G B Cai and Professor H X Wang for their valuable suggestions.

Conflict

The authors declare that there is no conflict of interests regarding the publication of this article.

References

[1] Friedman J H. 1991 Multivariate adaptive regression splines. *J. The Annals of Statistics.* 19(1):1-67.

[2] Jin R, Chen W and Simpson T W. 2001 Comparative studies of metamodelling techniques under multiple modelling criteria. *J. Structural and Multidisciplinary Optimization.*,23(1):1-13.

[3] Zerpa L E, Queipo N V and Pintos S. 2005 An optimization methodology of alkalinesurfactant-polymer flooding process using field scale numerical simulation and multiple surrogates. *J. Journal of Petroleum Science and Engineering.* 47(2):197-208.

[4] Glaz B, Goel T, and Liu L. 2009 Multiple-surrogate approach of helicopter rotor blade vibration reduction. *J. AIAA Journal*, 47(1):271-282.

[5] Gu Jichao. 2011 New Hybrid meta-model-based Global Optimization Methods and applications in Vehicle Design *D.* Hunan University.

Inter Chamber Leakages in Orbit Motors- Estimation by Modeling and Simulation in FEM and Fluent® Environment

D Roy[1], R Maiti[1,*] and P K Das[1]

[1]Indian Institute of Technology, Kharagpur 721302, West Bengal, India.

*rmaiti@mech.iitkgp.ernet.in

Abstract. An analysis is carried out to predict inter-chamber leakage flows through the two transition active contacts, which separate the high pressure chambers from the adjacent low pressure chambers, in the cycloidal class toothed star-ring of ORBIT® motors. Among geometrically form closed contacts which separates chambers from each other, few active contacts support high reaction loads during the process of transmitting torque in such unique LSHT hydrostatic units. Contact deformations in those contacts generate gaps in the other contacts for a partial period of time. High inter-chamber leakages occur through such gaps at transition active contacts. In an attempt to visualize the characteristics of such leakages, which contribute towards the poorer leakage characteristics of such machines, detail mapping i.e., magnitudes, rotational position etc., of such gaps of a commercially available unit, are carried out. For estimating gaps FEM technique is used. The FEM results have very good agreement with the results predicted earlier using trial method. Such gaps are difficult to measure as the contact positions are inaccessible to equip any conventional measuring instrument. The characteristics of leakage flows through such gaps are then established by CFD analysis using Fluent® in Ansys environment.

1. Introduction

The GEROTOR type ROPIMA (**Ro**tary **Pi**ston **Ma**chines) having geometrically form closed chambers, was first proposed by Galloy in 1848 [1] as a rotary steam engine, although, later the same design principle has been adopted to develop a unique class of positive displacement machines like rotary piston combustion (Wankel) engines, lubrication pumps, hydrostatic units (pumps/motors) and compressors. Such ROPIMAs come under the category of positive displacement machines, with fixed axis rotating star and ring [Figure 1(a)] of cycloidal class profiles. Their operations/working principles are more similar to the common and conventional involute toothed internal gear pumps and motors. GEROTOR pumps and motors, the fixed axis units where both star and ring rotates [Figure 1(a)], possess High Speed Low Torque (HSLT) features. It is to be noted that the ring may be with integral lobes or separate rollers as shown in Figures 1(a) and 1(b) respectively. The models shown in Figure 1 are of constant difference modified epitrochoidal star and its envelope as ring. The active portions of the envelope are circular arcs. The orbital rotor (with the same elements) principle was patented in the midst of the 1960s, as reported in a technical bulletin of Danfoss, by Lynn I. Charlson – the founder of Char-Lynn Company, Minnesota, USA. In such a unit, usually the star in epicyclic motion experiences rotary motion about its own axis (connected to the torque shaft) and revolving motion about the fixed ring axis. Due to kinematic advantages such a unit is used as motor only. Such a unit, popularly known as ORBIT ® motor, possesses Low Speed High Torque (LSHT) feature.

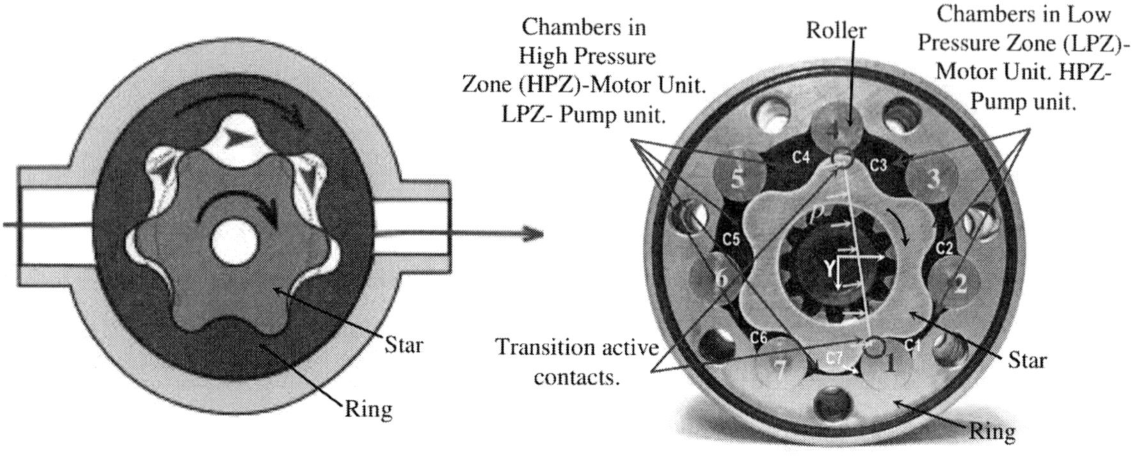

(a) Epitrochoidal Star and Ring with Integral Lobes.

(b) Epitrochoidal Star and Ring with Roller Lobes.

Figure 1. Typical epitrochoidal Star and Ring.

Gamez-Montero and Codina [2] has evaluated the flow characteristic of a trochoidal (modified cycloidal) -gear pump analytically and experimentally to understand the performance of this type gear pump. Coulbourne [3] has shown a method of estimating contact deformations in such machines. Maiti [4-6] has done exhaustive investigations on estimating such gaps. To solve the statically indeterminate problem, as there are multiple contacts, a trial method is proposed. Later an analytical method [7, 8] is also proposed. Gamez-Montero et al. [9] have established a method of estimating contact deformations in trochoidal gear (GEROTOR) pump using FEM. Bonandrini et al. [10] presented a complete and original method to analyze both the epitrochoidal and the hypotrochoidal ROPIMA type HST units. In the present investigation, the gaps at transition contacts are estimated at first for different angular position of torque transmitting shaft. Then the CFD analyses are done to estimate leakage considering the leakage path geometry and pressure profile.

2. Estimation of deformation / gap at active contacts due to fluid pressure

The calculation of deformations and generated gaps at the active contacts with respect to the relative angular positions of star and ring is the main task prior to the leakage analysis. In the present investigation both analyses are done using numerical techniques in Ansys environment. It is assumed that the star and ring are perfectly form closed and there is no manufacturing error. Analyses are carried out with respect to the ORBIT unit, in which form closed star is considered as the floating axis member. The gear coupling in the star transmits torque to the output shaft through cardan shaft. All radial loads supported by the active contacts and not transmitted to the output shaft. Few chambers are in high pressure one side (chambers between contact points 1 to 4 as an instant as shown in Figure 1(b) & Figure 2 of the imaginary plane and in low pressure in other side. It may be considered that the fluid pressure acts as the uniformly distributed load [6] on the plane joining the higher pair contacts at lobe 4 and lobe 1 The net forces [Figure 2] due to deformations at some active contacts, must be balanced with the pressure force. As well as, all moments are also to be balanced. These can be expressed as [4,6]:

$$\mathbf{F_i} = \sum_{n=1}^{Z} \mathbf{F_n} \tag{1}$$

$$|\mathbf{F_i}| = \Delta p_i L b \tag{2}$$

IOP Conf. Series: Journal of Physics: Conf. Series **1060** (2018) 012081 doi:10.1088/1742-6596/1060/1/012081

$$\mathbf{M_i} = \sum_{n=1}^{Z}(\mathbf{r_n} \times \mathbf{F_n}) = \mathbf{r_i} \times \mathbf{F_i} \tag{3}$$

Based on Hertzian contact theory and on simplification the contact deformation can be expressed as:

$$\delta_n = \frac{8F_n}{\pi b}\left[\frac{1-\nu^2}{E}\left(1+\frac{\pi}{4}\right)\right] \tag{4}$$

3. Trial and error method

Deformations (δ_n) are estimated by iterating the infinitesimal displacement of star centre assuming the star is rigid except allowing local contact deformations, and solving equations 1 to 4 till the equilibrium is achieved. Resolving the displacement vectors in normal direction of the lobes and the values of deflection/gap at all contact points are calculated. The values for transition active contact points (in this case 1 & 4) are recorded for further analysis. After the shaft rotation of a cycle i.e. $\xi_o = \pi / Z(Z-1)$, degree in case of an ORBIT unit and $\pi/(Z-1)$ degree in case of a GEROTOR unit (torque transmitting shaft is connected to star unit) the phase changes [4]. For next ξ_o contacts 1 & 5 will be in transition (clockwise rotation for torque shaft).

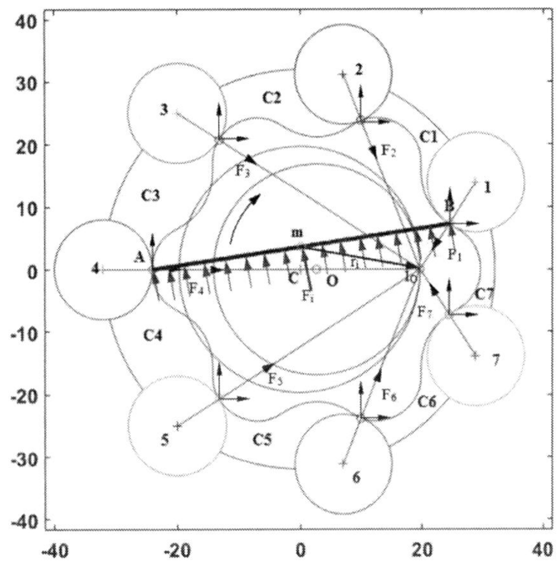

Figure 2. Deformations and Hertzian contact forces at active contacts due to fluid pressure [see also Figure 1(b)].

4. FEM modeling and analysis in ANSYS

The main focus for the geometric modeling of ORBIT motor is on the theoretical profiles of two major components: star and ring. Basically, three types of software have been used these are Matlab, Solid Works and Ansys. SOLID186 element has been chosen for analysis. In this model min. mesh size is 0.00021994 mm. and max. mesh size is 1.09970 mm. So, the number of nodes and elements are 717490 and 166137 respectively. Regarding boundary conditions consider stator part of the orbit motor is rigid and star-roller contacts, stator-roller contacts are frictional (C.O.F- 0.15). The formulation used here is augmented Lagrange and operating pressure is 5 MPa. The deformation analysis is carried out on FEM model varying the shaft rotational angle ξ from 0^0 to 8.57^0 [$2\xi_0=2\pi/Z(Z-1)$] degree (one cycle/phase) in twelve steps.

5. Result and discussion based on FEM model

As revealed from graphs in Figure 3 (a & b), a gap is generated at transition contact point 4 [Figure 1(b) and Figure 2 point A] from the beginning of the cycle/phase i.e., $\xi=0^0$. The contact 4 is separating the chamber C3 and C4, where C4 just started expanding i.e., in HPZ. The chamber C3 is continuing compression phase, i.e., it is in LPZ. Gaps start at the beginning of $\xi=0^0$ and resulting inter-chamber leakage from C4 to C3. In the present case it continues up to $\xi=4.285^0$ .In Figure 3 (a & b), which are very prominently shown the gap or deformation found in trial and error method have good agreement with the deformation and gaps comes from FEM analysis (using Ansys software).

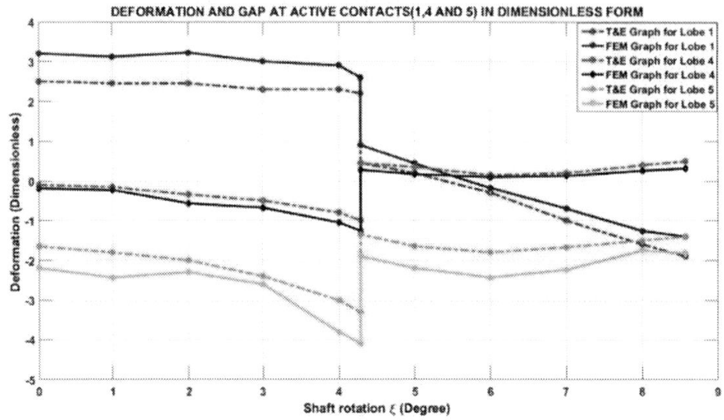

Figure 3(a). Deformation and gap at active contacts in lobe 1, 4, and 5.

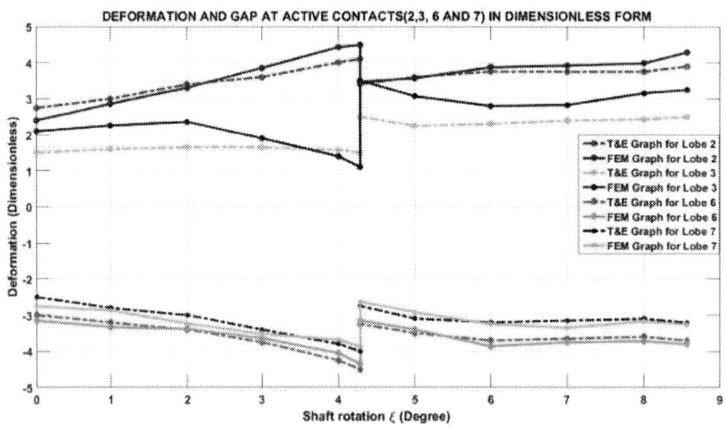

Figure 3(b). Deformation and gap at active contacts in lobe 2, 3, 6 and 7.

6. Analysis of leakage model in Ansys® Fluent

After mapping the gaps through the transition contacts for this model, leakage through the gaps require to be estimated. Patterns of leakage flow will be studied through CFD modeling in Fluent® environment. Basically, three types of software have been used for three main stages of the simulation. These are Matlab, Ansys apdl for modelling and Fluent module in Ansys work bench. 2-D geometric 1st. lobe, 4th. lobe and 5th. lobe contact models are imported in fluent module for meshing. For 1st. lobe min. and max. mesh sizes are 0.002226 mm. and 0.42 mm, corresponding nodes and elements are 2190 and 1850 numbers. Similarly for 4th. lobe min. and max. mesh sizes are 0.000381834 mm. and 0.3 mm, corresponding nodes and elements are 4796 and 4320 and for 5th. lobe min. and max. mesh sizes are 0.00037304 mm. and 0.2 mm , corresponding nodes and elements are 5270 and 5030.

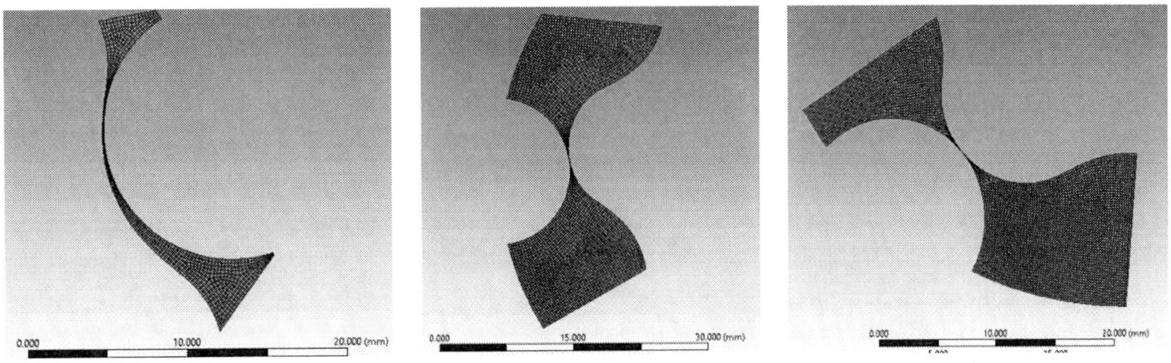

(a): At 1st. lobe leakage path at ξ=8.57^0 **(b):** At 4th. lobe leakage path at ξ=4.285^0 **(c):** At 5th. lobe leakage path at ξ=4.286^0

Figure 4. 2-D meshed model of leakage paths at 1st., 4th. and 5th. lobes in Ansys Work bench.

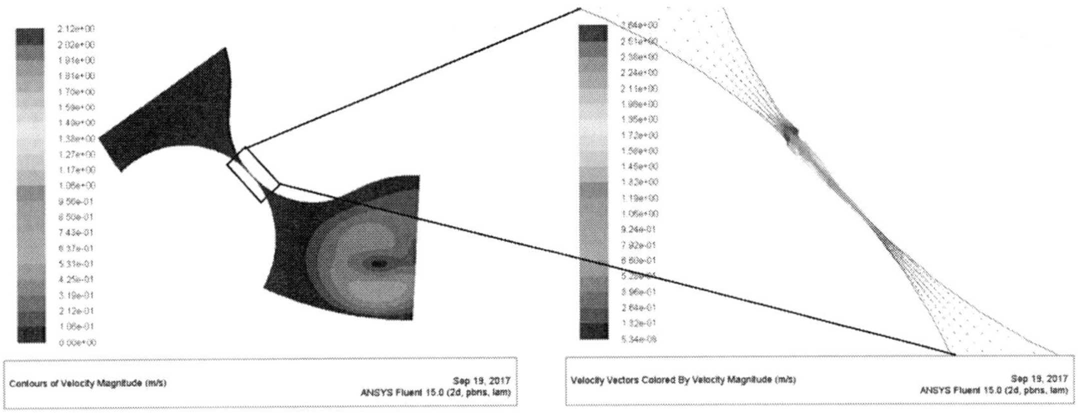

Figure 5. Leakage oil velocity through gap at transition contact at lobe 5 ($\xi = 4.286^0$) at p=5 MPa.

Table 1. Leakage flow rate through gaps at 5 MPa.

ξ	Gap at 1st. lobe (δ_n, mm)	Velocity (v_n, m/s) at 1st. lobe	Flow Rate (10^{-8}) 1st. lobe ($q_{1,n}$, m^3/sec)	Gap at 4th. lobe (δ_n, mm)	Gap at 5th. lobe (δ_n, mm)	Width (b, mm)	Velocity (v_n, m/s)	Flow Rate (10^{-8}) ($q_{1,n}$, m^3/sec) at 4th. and 5th. lobe
0^0				0.00160055		14	0.97	2.17354
1^0				0.00198987		14	1.14	3.175829
2^0				0.00493141		14	1.54	10.63212
3^0				0.00588309		14	1.65	13.58993
4^0				0.00908418		14	2.12	26.96185
4.285^0				0.01090102		14	2.34	35.71173
4.286^0					0.0164380	14	2.55	58.6838
5^0					0.0190335	14	2.64	70.34789
6^0	0.0015572	0.96	2.0929951		0.0210234	14	2.98	87.70957
7^0	0.0060561	2.23	18.9072066		0.0193796	14	2.86	77.59585
8^0	0.0109010	7.72	117.818180		0.0151403	14	2.54	53.83891
8.57^0	0.0121122	15.5	262.835608		0.0157459	14	1.97	43.42723

Figure 4 shows 2-D meshed model for shaft rotation of ξ=4.285^0, 4.286^0 and 8.57^0. Regarding the boundary condition, consider the star wall and ring wall are stationary.

7. Result and discussion based on Fluent® model

After the simulation, the velocity profile for ξ=4.285^0, 4.286^0 and 8.57^0 are shown in Figure 5. The fluent analysis is defined with 2-D, pressure based Navier Stokes, and laminar flow. The instantaneous leakage flow rate through the gap at contact in nth lobe can be estimated as:

$$q_{l,n} = v_n \, (\delta_n \times b) \tag{5}$$

Based on Equstion-5, the leakage flow rate at 5 MPa were calculated and shown in table-1.

8. Conclusion

A method of estimating flow through the narrow gap, generated at transition contacts separating a high pressure chamber from the adjacent low pressure chamber in ORBIT motor, is established through an analysis using Fluent. The generation of gap at transition contacts occurs at certain phases, due to deformations at form closed contacts, while transmitting torque. Such gaps are estimated using FEM in Ansys in the present investigation. Results have good agreement with a trail method developed earlier. The estimated quantity of such leakage flow indicates the relatively poor volumetric efficiencies that are shown in product catalogues.

9. Abbreviations

CFD	:	Computational Fluid Dynamics.
FEM	:	Finite Element Method.
HPZ	:	High Pressure Zone.
LPZ	:	Low pressure Zone.
ROPIMA	:	Rotary Piston Machine.
TEM	:	Trial and Error Method.

References

[1] White G 1988 Epicyclic Gear applied to early Stem Engines. *Mechanism and Machine Theory*, **23(1)** pp 25-37.

[2] Gamez-Montero P J and Codina E 2007 Flow characteristics of a trochoidal-gear pump using bond graphs and experimental measurement *J. Systems and Control Engineering, Proc. IMechE*, **221** pp 331- 346.

[3] Colbourne, J R 1976 Reduction of the Contact Stresses in Internal Gear Pumps *J. Engg. for Industry, Trans ASME* **98(4)** pp 1296-1300.

[4] Maiti R 1993 Torque Characteristics of Epitrochoid Generated Orbital Rotary Piston type Hydraulic Motors *Mechanism and Machine Theory* **28(2)** pp 225-231.

[5] Maiti R 1993 Active Contact Problems in Epitrochoid Generated 'Floating Axis' Orbital Rotary Piston Machines *ASME J. of Engg. for Industry* **115 (3)** pp 337-340.

[6] Maiti R 1991 Active Contact Stresses at Epitrochoid Generated ROTOR-STATOR set of Fixed Axis or Equivalent system 'ROPIMA' type Hydrostatic units *ASME J. of Engg. for Industry* **113(4)** pp 465-473.

[7] Nag A, Basu S and Maiti R 2008 Estimation of Stresses in Components and Gap in Active Contacts of Epitrochoid Generated Floating Axis ROPIMA type Hydrostatic Units- An FEM Approach *In Proceeding of ICHP*-08 Prague pp 129-136.

[8] Nag A 2015 Analysis of Active Contacts in Epitrochoid Generated Form Closed Rotor-Stator set with specific reference to Hydrostatic Motor Performance *PhD Thesis* IIT Kharagpur India.

[9] Gamez-Montero P J, Castilla R, Khamashta M and Codina E 2006 Contact problems of trochoidal-gear pump *Intl. Journal of Mech Sc.* **48** pp 1471-1480.

[10] Bonandrini G, Mimmi G and Rottenbacher C 2009 Theoretical analysis of internal epitrochoidal and hypotrochoidal machines Proceedings of the Institution of Mechanical Engineers, Part C: Journal of Mechanical Engineering Science **223** pp 1469-1480.

Simulation and Analysis of ESB Algorithm Based on Frequency Diverse Array

Xiaoyu Feng[a, *], Junwei Xie[b], Xiao Yang[c] and Dejiang Chen[d]

Air Force Engineering University, Xi'an, Shaanxi, China

[*a]790482771@qq.com; [b]xjw_xjw_123@163.com; [c]15594979691@163.com; [d]1037984332@qq.com

Abstract. The beam of phased array radar has angular resolution and no range resolution. The frequency diverse array adds a frequency increment which is much smaller than the carrier frequency between the array elements, thus forming a beam which is related to angle, range and time. On the basis of analyzing the characteristic of frequency diverse array beam, the mechanism of receiving only the transmitting array element frequency signal is introduced. The eigenspace-based beamforming algorithm is used to simulate the transmitting pattern, the receiving pattern and the dual path pattern, and the simulation results are analyzed. The performance of SINR under the ESB algorithm is simulated and analyzed. The results verify the effectiveness of the application of the ESB algorithm in the frequency diverse array.

1. Introduction

Phased array can efficiently change the beam space pointing, and is widely used in radar, electronic warfare, radio astronomy and airport security and other fields [1]. Phased array can form high gain in the specified direction, to detect/track the weak signal and suppress other aspects of interference signal [2]. However, the limitation of phased array is that the spatial distribution of beam is only related to angle, and has nothing to do with range. That is, the beam has only angle resolution and no range resolution. In addition, the transmitting range dependent beams become more and more urgent in some specific fields, such as range dependent interference suppression, directional communication and range ambiguity suppression, [3]-[5].

In 2006, P. Antonik proposed the concept of frequency diverse array (FDA) radar at the international radar conference, which attracted wide attention from scholars both at home and abroad [6], [7]. Different from phased array, FDA adds an increment which is much smaller than the carrier frequency across the array element, so that the antenna beam spatial distribution will change with time, range and angle [8]. There is a lot of literature on FDA. Zhuang et al. [9] points out that beam scanning characteristics of FDA is related to frequency offset between the array elements. The research status of FDA is summarized, and the structure and beam characteristic of FDA are analyzed, and the application is prospected in [10]-[12]. Basit et al. [13] studies the application of FDA in multiple target localization in the same direction and different ranges. A logarithmically increasing frequency offsets is introduced in [14], and decoupled the range-angle beampattern.

Most of the current research focuses on the structure characteristics of FDA and its application, while the beamforming algorithms of FDA are rarely involved. For phased array, the essence of beamforming is to adjust the weight coefficients of each array element, so that the high gain is aligned to the desired signal direction, and the nulls are aligned to the direction of the interference, so as to maximize the reception of useful signals. For FDA, because the spatial distribution of beam is related to range and angle, we need to get the high gain of the beam to the specific location of the desired signal, and zero notch is aligned to the location of the interference, so as to receive useful signals as far as possible. There are many beamforming algorithms in phased array, such as linear constrained minimum variance (LCMV) beamforming algorithm [15], and Eigenspace-based beamforming (ESB) [16].

In this paper, the ESB algorithm in phased array is introduced into the FDA, and the performance of the algorithm is analyzed. Finally, the effectiveness of the algorithm in the frequency control array is verified by simulation.

2. FDA signal model

The uniform linear array FDA structure is shown in Figure 1.

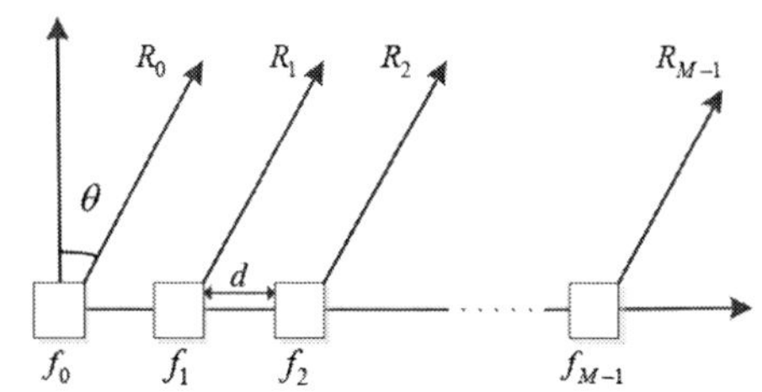

Figure 1. Uniform linear array frequency diverse array.

Under the narrow band condition, the signal of each array element can be expressed as

$$s_n(t) = \exp(\mathrm{j}2\pi f_n t),\ n = 0,1,\ldots,M\text{-}1 \tag{1}$$

Where $f_n = f_0 + n\Delta f, n = 0,1,\ldots,M-1$ is the radiation frequency, f_0 is the carrier frequency, Δf is the frequency increment, and M is the number of array elements.

The signal arriving at a far-field point (R,θ) can be expressed as

$$s_{rn}(t) = \exp\left[\mathrm{j}2\pi f_n\left(t - \frac{R_n}{c}\right)\right] \tag{2}$$

Where $R_n = R - nd\sin\theta$ is the target slant range for the nth element, d is element spacing and c is the speed of light. the array factor at position (R,θ) can be expressed as

$$AF(t;R,\theta) = \sum_{n=0}^{N-1}\frac{1}{R_n}s_{rn}(t) = \sum_{n=0}^{N-1}\frac{1}{R_n}\exp\left\{\mathrm{j}2\pi f_n\left[t - \frac{R}{c} + \frac{nd\sin(\theta)}{c}\right]\right\} \approx \frac{\exp\{\mathrm{j}\phi_0\}}{R}\cdot\frac{\sin\left[N\pi\left(\Delta ft - \frac{\Delta fR}{c} + \frac{df_0\sin(\theta)}{c}\right)\right]}{\sin\left[\pi\left(\Delta ft - \frac{\Delta fR}{c} + \frac{df_0\sin(\theta)}{c}\right)\right]} \tag{3}$$

With ϕ_0 being

$$\phi_0 = 2\pi f_0\left(t - \frac{R}{c}\right) + \pi(N-1)\left(\frac{f_0 d\sin\theta}{c} + \Delta ft - \frac{\Delta fR}{c}\right) \tag{4}$$

When $f_0 = 3\text{GHz}$, $\Delta f = 2\text{KHz}$, $M = 10$, $d = 0.5c/f_0$, $\theta = 20^\circ$, $R = 100\text{km}$, the transmitting beampattern of the ULA-FDA is shown in Figure 2, and Figure 3 shows the transmitting beampattern of the ULA-PAR.

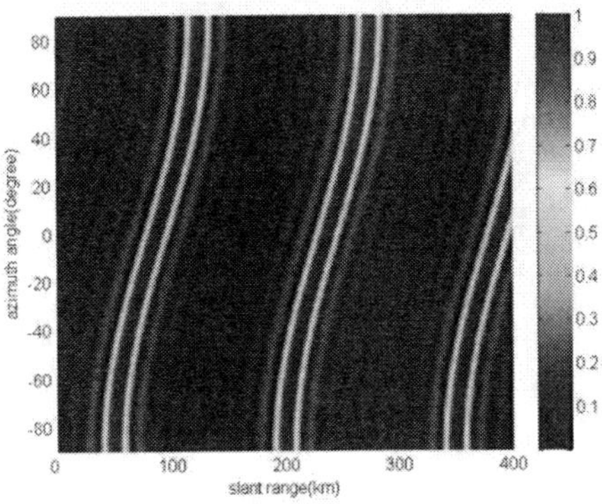

Figure 2. The transmitting beam-pattern of ULA-FDA.

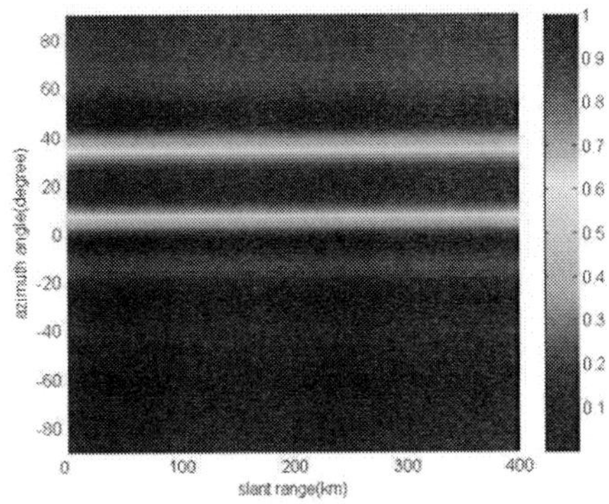

Figure 3. The transmitting beam-pattern of ULA-PAR.

As is obvious from figures 2-3, the beampattern of FDA is range-angle-dependent, and the phased array beampattern has only angle resolution.

3. Receiving signal processing

A steering vector is added to the FDA transmitting antenna.

$$\boldsymbol{w}_t=\left[\alpha_0\left(\hat{R},\hat{\theta}\right),\alpha_1\left(\hat{R},\hat{\theta}\right),\cdots\alpha_{M-1}\left(\hat{R},\hat{\theta}\right)\right]^{\mathrm{T}} \tag{5}$$

Where $\alpha_n\left(\hat{R},\hat{\theta}\right)=\exp\left(-\mathrm{j}2\pi f_n\hat{R}/c\right)$ is the weight coefficient of the nth element. For a fixed target position $\left(\hat{R},\hat{\theta}\right)$ in the far field, the weighted arrival target position of the signal transmitted by the array element n can be expressed as

$$s\left(t;\hat{R},\hat{\theta}\right)=\sum_{n=0}^{M-1}\exp\left\{\mathrm{j}2\pi f_n\left(t-\frac{R-\hat{R}}{c}+\frac{nd\left(\sin\theta-\sin\hat{\theta}\right)}{c}\right)\right\} \tag{6}$$

The echo signal received by the mth array element is

$$r\left(t;\hat{R},\hat{\theta}\right)=\sum_{n=0}^{M-1}\exp\left\{\mathrm{j}2\pi f_n\left(t-\frac{R-\hat{R}}{c}+\frac{nd\left(\sin\theta-\sin\hat{\theta}\right)}{c}-\frac{R_m}{c}\right)\right\}=\sum_{n=0}^{M-1}\exp\left\{\mathrm{j}2\pi f_n\left(t-\frac{2R-\hat{R}}{c}+\frac{nd\left(\sin\theta-\sin\hat{\theta}\right)}{c}+\frac{md\sin\theta}{c}\right)\right\} \tag{7}$$

At this time, the echo signal received by each element contains all the transmitting signals. By adding filters, the mth element only extracts the signal with f_m, and by weighting, the output can be obtained.

$$y\left(t;\hat{R},\hat{\theta}\right)=\sum_{m=0}^{M-1}\boldsymbol{w}_r^{\mathrm{H}}\left(\hat{R},\hat{\theta}\right)r\left(t;\hat{R},\hat{\theta}\right)=\sum_{m=0}^{M-1}\exp\left\{\mathrm{j}2\pi f_m\left(t-\frac{2\left(R-\hat{R}\right)}{c}+\frac{2md\left(\sin\theta-\sin\hat{\theta}\right)}{c}\right)\right\} \tag{8}$$

4. The ESB algorithm

Consider an expected signal and J interference signals in a space. The eigendecomposition of the sample covariance matrix $\hat{\boldsymbol{R}}$ is decomposed.

$$\hat{R} = \sum_{i=1}^{J+1} \lambda_i u_i u_i^{\mathrm{H}} + \sigma_n^2 \sum_{i=J+2}^{M} u_i u_i^{\mathrm{H}} \tag{9}$$

Where $\lambda_1 \geq \lambda_2 \geq \cdots \geq \lambda_{J+1} > \lambda_{J+2} = \cdots = \lambda_M = \sigma_n^2$ are the corresponding M eigenvalues, and the corresponding eigenvector is $u_i, i = 1, 2, \cdots M$.

$$\boldsymbol{D}_s = \mathrm{diag}(\lambda_1, \cdots \lambda_{J+1}), \boldsymbol{D}_n = \mathrm{diag}(\lambda_{J+2}, \cdots \lambda_M) \tag{10}$$

$$\boldsymbol{U}_s = [u_1, u_2, \cdots, u_{J+1}], \boldsymbol{U}_n = [u_{J+2}, \cdots, u_M] \tag{11}$$

The column vectors of $\boldsymbol{U}_s$ constitute the subspace of the signal, and the column vectors of $\boldsymbol{U}_n$ constitute the noise subspace. In the LCMV algorithm, the adaptive weight is

$$w_0 = \frac{\hat{\boldsymbol{R}}^{-1}\boldsymbol{a}(R,\theta)}{\boldsymbol{a}^{\mathrm{H}}(R,\theta)\hat{\boldsymbol{R}}^{-1}\boldsymbol{a}(R,\theta)} \tag{12}$$

In the ideal case, the expected signal is in the signal subspace, then $\boldsymbol{U}_n^{\mathrm{H}}\boldsymbol{a}(R,\theta) = 0$. So the weight vector $\boldsymbol{a}(R,\theta)$ is only the component of the signal subspace, and does not contain the subspace component of the noise. ESB algorithm is based on this principle, the weight vector onto the signal subspace, so as to give in the noise subspace components retained in the signal subspace of the component, i.e.

$$\boldsymbol{w}_p = \boldsymbol{U}_s \boldsymbol{U}_s^{\mathrm{H}} \boldsymbol{w}_0 \tag{13}$$

5. Simulation and analysis

Consider an 10 element uniform linear FDA at the carrier frequency $f_0 = 3\text{GHz}$, with the element spacing being fixed to be $d = 0.5c / f_0$. Simulation have been carried out for $J = 4$, $\Delta f = 4500\text{Hz}$, $\hat{R} = 50\text{km}$, $\hat{\theta} = 20^\circ$, SNR=10dB, INR=10dB. The azimuth of the interference is -60° , 15° , 20° , 25° , respectively. And the corresponding slant range of the interference is 30km, 90km, 65km, 45km in turn. Simulations of transmitting, receiving, and dual path pattern under the ESB algorithm of FDA are shown in Figure 4-6.

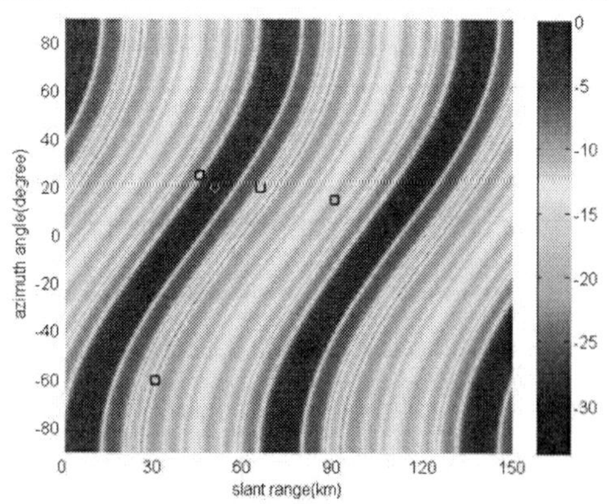

Figure 4. The transmitting beampattern with ESB algorithm.

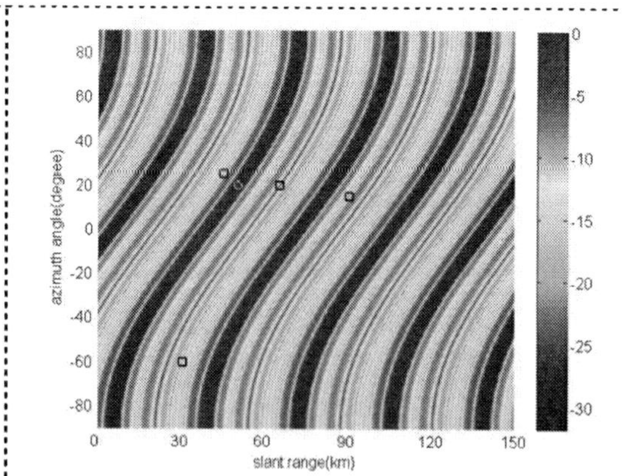

Figure 5. The receiving beampattern with ESB algorithm.

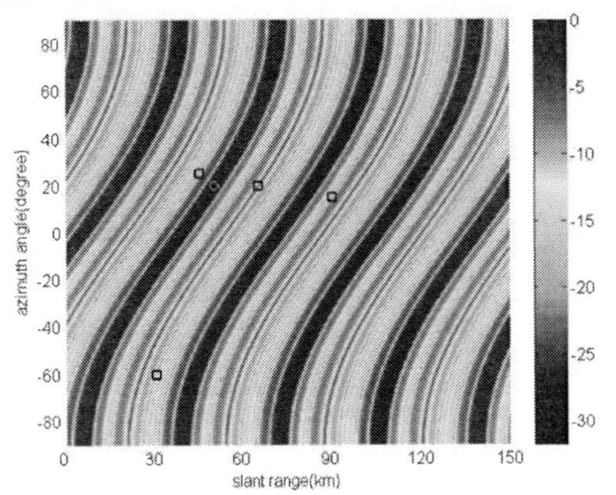

Figure 6. The dual path beampattern with ESB algorithm.

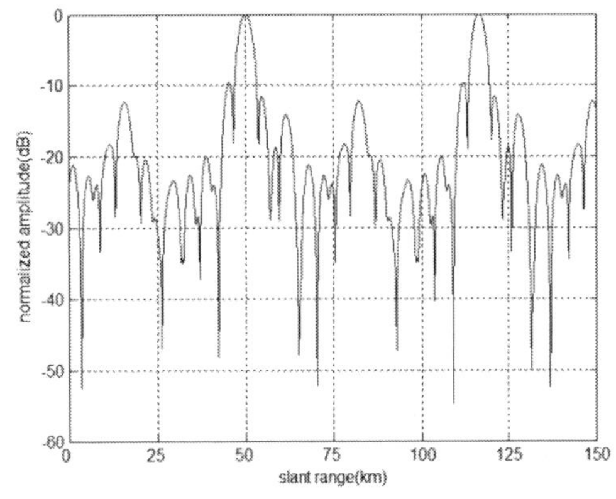

Figure 7. The beampattern of the range dimension with $\hat{\theta}=20^{\circ}$.

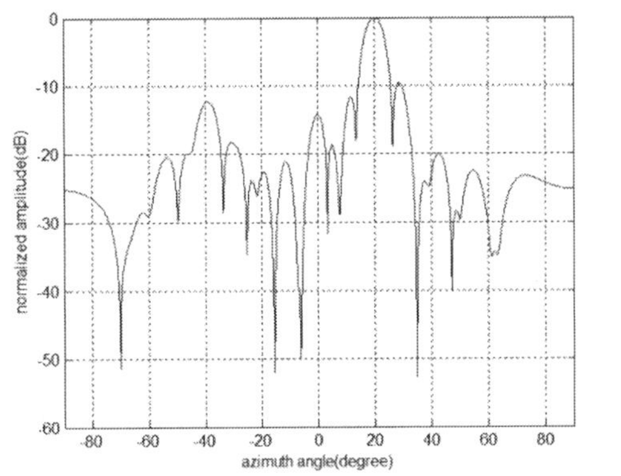

Figure 8. The beampattern of the range dimension with $\hat{R}=50\text{km}$.

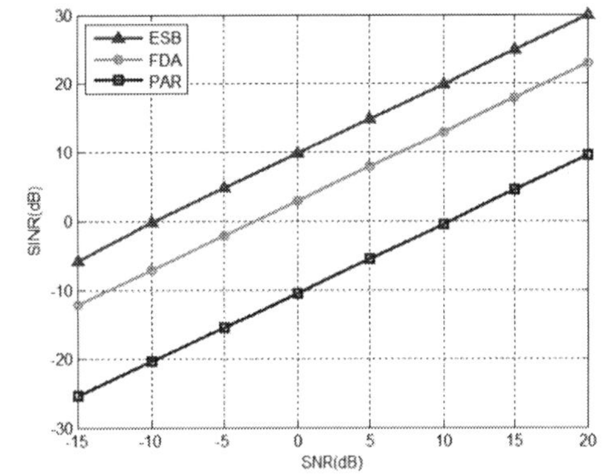

Figure 9. The change diagram of SINR with SNR.

The green circle represents the target location, and the blue square represents the interference location. From the graph, we can see that using the ESB algorithm can form high gain at the target position and form zero trapping in the interference location, so as to realize the function of the beamforming algorithm.

In order to observe the convenience, the gain of the angel dimension and range dimension of the target position are simulated respectively. It is clear from the Figure 7-8 that the ESB algorithm is used to maximize the gain value at the target location.

Finally, the performance of signal-to-interference-plus-noise ratio after using the algorithm is simulated, and compared with the basic FDA and PAR, as shown in Figure 9. As can be seen from Figure 9, after using the ESB algorithm, the SINR performance of the system is the best, with higher noise and robustness.

6. Conclusion

FDA radar can produce range-angle-time correlation beam. It is the innovation of the radar system, which has brought more potential for radar application. Based on the analysis of FDA beam characteristics, we introduce a signal receiving mechanism which accepts the frequency of the corresponding transmitting array element only. With the application of ESB algorithm, we simulate the transmitting beampattern, receiving beampattern, dual path beampattern, and the results show the effectiveness of the application of ESB algorithm in FDA. The next step will be further studied in the improved beamforming algorithm of FDA.

References

[1] McManamon P. F., Bos P. J., Escuti M. J., et al 2009 *Proc. IEEE* vol 6 pp 1078–1096
[2] Li J., Stoica, P. 2010 *IEEE Signal Process Mag.* vol 2 pp 143–144
[3] Jin X. L., Min G. Y., Velentzas S., and Jiang J. M. 2012 *IEEE Trans. Wireless Commun.* vol 11 pp 562–570
[4] Wu J. X., Wang T., Zhang L. F., and Bao Z. 2012 *IEEE Trans. Aerosp. Electron. Syst.* vol 48 pp 3647–3654
[5] Yang X. S., Poor H. V., and Petropulu A. 2004 *IEEE Trans. Signal Process.* vol 52 pp 1607–1619
[6] Antonik, P., Wicks, M. C., Griffiths, H. D., et al 2006 *Proc. of the IEEE Radar Conf.* Verona NY pp 215 - 217
[7] Wicks M. C., Antonik P. 2008 U.S.A Patent 7,319,427
[8] Huang J. and Tong K. F. 2008 *2008 IEEE Antennas and Propagation Society International Symposium* pp 1–4
[9] Zhuang L. and Liu X. Z. 2009 *Proc. Int. Radar Conf. Dig.*, Bordeaux, France pp 1-4
[10] Wang W. Q. 2015 *IEEE Antennas and Propagation Magazine* vol 57 pp 145-152
[11] Wang W. Q. 2016 *IET Radar, Sonar & Navigation* vol 6 pp 1001-1012
[12] Wang W. Q., Shao H. Z., Chen H. 2016 *Journal of Electronics & Information Technology* vol 4 pp 1000-1011
[13] Basit A., Qureshi I. M., Khan W., Malik A. N. and Shoaib B. 2016 *Proceedings. of 2016 International Bhurban Conferen on Applied Sciences & Technology* Islamabad Pakistan pp 682-688
[14] Khan W., Qureshi I. M., and Saeed S.2015 *IEEE Antennas And Wireless Propagation Letters* vol 14 pp 499-502
[15] Deng X., Liao G. S., Liu H. Q. 2007 *Systems Engineering and Electronics* vol 29 pp 449-452
[16] Liu X. J., Liu C. F., Liao G. S. 2010 *Systems Engineering and Electronics* vol 32 pp 669-673

IOP Conf. Series: Journal of Physics: Conf. Series **1060** (2018) 012083 doi:10.1088/1742-6596/1060/1/012083

Topology Optimization Design of Launcher Bracket Based on Multi-body Dynamics

Min Li, Xuezhi Wang, Ming Liu and Chao Li

Air Force Engineering University, Xi'an, Shaanxi, China

Min Li: 17691037413m@sina.cn, Xuezhi Wang: xuezhi@126.com

Abstract: According to the topology optimization theory and engineering application,a launcher bracket topology optimization finite element model was built ,which was based on the multi-body dynamics simulation.The optimal material distribution and the force transmission path of the bracket under the constraint conditions were obtained.The stiffness evaluation standard for bracket structure is established by introducing the specific stiffness structural effectiveness. After comparative and analysis ,the simulation results show that the topology optimization design of launcher bracket based on multi-body dynamics can achieve the lightweight design of launcher bracket, strengthen structure stiffness, increase the bracket modal frequency , which are helpful to reduce the initial disturbance of missile launching.The method provides reference for the design of similar products.

1. Introduction

Traditionally, the parameters of bracket design is gotten by adopting trial and error approach which depends on relevant experience or direct judgement. This design process requires enough raw materials and lengthen the product development cycle as well [1]. Topology optimization technology, as a promising and innovative technology, has been widely used in the development of aviation and automobile [2]. For the optimization of continuum variable structure, topology optimization mainly includes homogenization method, SIMP method, variable thickness method and topological function description method [3].

2. The principle of topology optimization

The topology optimization method of continuum structure based on SIMP method is mainly to establish the function relation between the material parameters and the structure density, through the change of the design variable of the discrete element to control the change of the stiffness of the element. Thus the change of the overall stiffness matrix of the structure is adjusted. Such iterative optimization will enable structural stiffness to be optimized and the material distribution to be ideal. [4] The mathematical model of topology optimization is based on elastic mechanics, variational principle and energy quasi rigid establishment [5]. In the view of the mathematical, it is the problem that refer to solving the boundary value under the effect of face force, body force and concentrated force [6]. The interpolation function of the SIMP model is an exponential function, and its mathematical model is [7]

$$\begin{cases} E_{ijkl}(x) = \rho(x)^p E^0{}_{ijkl}, p > 1 \\ \int_{\Omega} \rho(x) d\Omega \le V \\ 0 \le \rho(x) \le 1, x \in \Omega \end{cases} \tag{1}$$

IOP Conf. Series: Journal of Physics: Conf. Series **1060** (2018) 012083 doi:10.1088/1742-6596/1060/1/012083

Where E_{ijkl} is the elasticity modulus of the selected material, Ω is the design area, V is the volume of design area, and p is penalty factor.

The density value is between the material characteristic values:

$$E_{ijkl}(\rho=0)=0, E_{ijkl}(\rho=1)=E_{ijkl}{}^{0} \tag{2}$$

For the three dimensional model penalty factor ρ , the rule of value is as follows:

$$\rho \geq \max\left\{15\frac{1-\upsilon_0^2}{7-5\upsilon_0^2}, 1.5\frac{1-\upsilon_0^2}{1-2\upsilon_0^2}\right\} \tag{3}$$

It is assumed that the pseudo elastic modulus are all of the same character and Poisson's ratio is a constant. The establishment of mathematical equation of parameter by SIMP method is:

$$\begin{cases} E(\lambda)/E_0 = \rho(\lambda)^{\rho} \\ \int_{\Omega} \rho(\lambda) d\Omega \leq V \end{cases} \tag{4}$$

Where $E(\lambda)$ is Pseudo elastic modulus, E_0 is the real elastic modulus of material, $\rho(\lambda)$ is pseudo density.

Under volume constraint, and on the basis of SIMP method, the mathematical model of topology optimization for continuous structures is described as

$$\underset{u\in U,\rho}{Min}: l(u): s.t.: \begin{cases} a_E(u,v) = l(v), v \in U \\ E_{ijkl}(x) = \rho(x)^p E_{ijkl}^0, p > 1 \\ \int_{\Omega} \rho(\lambda) d\Omega \leq V \\ 0 \leq \rho_{\min} \leq 1, x \in \Omega \end{cases} \tag{5}$$

This model can be applied to many occasions, such as the minimization of flexibility, and the problem of quality minimization.

3. The topology optimization of bracket

3.1. The establishment of the optimization model

In order to get the best optimization result of the bracket, the bracket was reasonably simplified. Using the outline entity of the bracket as the initial model and preserving the necessary feature structures. The necessary structures mainly are electric cabinet, support body, ear shaft and bearing race. The ear shaft is connected with the hydraulic rod and the landing gear to provide support and restrict the bracket. The bearing race is fixed on the frame frame by bolt, so that the bracket can realize gyrate azimuth. The hexahedral grids are used to mesh the brackets with each grid size of 50mm,and the division is shown in Figure 1. As the picture shows, the blue grid is the design area and the gray grid is the non-design area.

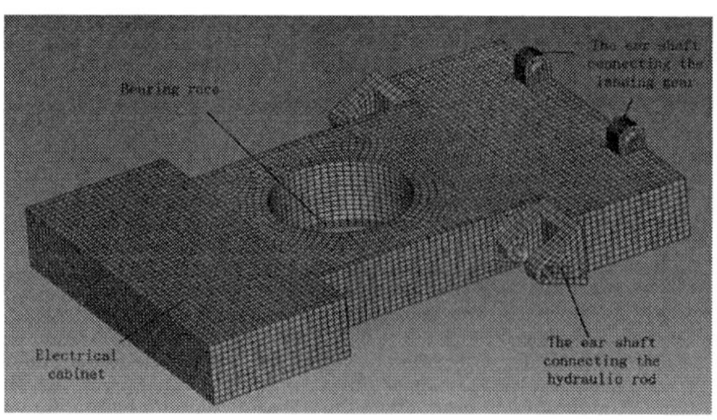

Figure 1. Bracket grid and schematic diagram.

The calculation precision of structural optimization is related to the constraint's quantity and loads of the model. rigid flexible coupling technology is adopt to dynamically analyze the launching process of missile in three typical working conditions: the march condition of launch vehicle in 3g gravitational acceleration caused by the road unevenness, the launching condition of launch vehicle at azimuth zero, pitch 60 degrees and at azimuth 45, pitch 60 degrees . From the dynamic analysis and calculation, the loading conditions of each sports deputy node and the amplitude-frequency characteristics of missile vibration were obtained [8]. Frequency response vibration diagram is shown in Figure 2.

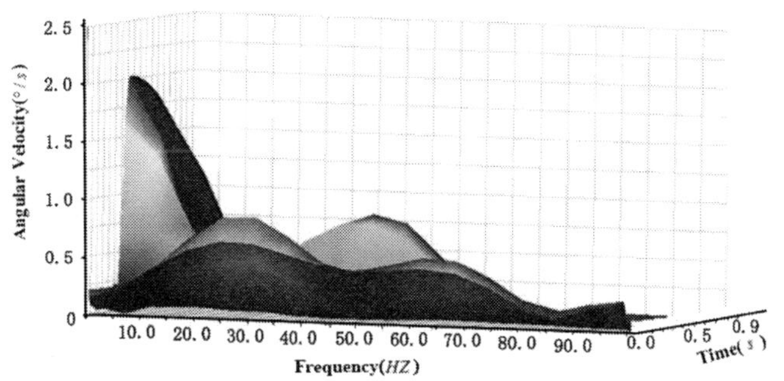

Figure 2. Frequency response vibration diagram.

From Figure 2, we can see that the vibration of missile is the most intense during 10 to 20 *HZ*, and ,in the frequency of 20-90 *HZ*, pitching angular velocity has slowed down, but the maximum pitching angular velocity is up to 1.5deg/sec,which can't be ignored. Therefore, in order to avoid the resonance phenomenon of the launcher, the first order frequency of the bracket structure should be more than 90 *HZ*, so that the impact of the bracket on the initial disturbance and vibration of the missile will be reduced to a minimum [9].

By the non-linear finite element analysis, the bracket has the maximum deformation under the third working condition, and the deformation is located at the end surface of the ear shaft, and the vertical displacement reaches 0.45mm.But under the limit load, the deformation displacement of the bracket are smaller than the design requirements, the maximum stress value is far less than the yield limit of the steel material, so there is a lot of structural optimization margin. When the key position of the bracket is topologically superior, the displacement and deformation of key parts should be within the design range, the modal natural frequency should reach the requirement of device vibration frequency, the volume of design area should be reduced as much as possible, and the manufacturing process conforms to the actual production conditions so that to achieve Optimization purposes successfully.

3.2. Optimization model parameter setting

- Objective Function:Volume minimization function of bracket design area
  - Constraints: the maximum vertical displacement of the joint at the ear axis is less than 1.5mm, and the natural frequency of the bracket is greater than 90HZ.
  - Design Variables: the relative unit density of the force region of the bracket body is the design variable.

3.3. Topology optimization results

The bracket volume optimization iterative curve and the first order modal inherent frequency iteration curve are shown in Figures 3 and Figure 4 and the result of the bracket topology optimization is shown in Figure 5. From Figure 4, the volume of the bracket in the first 6 times of iteration process is greatly reduced, indicating that the initial structure of the bracket has a large number of materials that don't bear load, and can be deleted. After that, the curve is gradually convergent after a small amplitude oscillation. The whole calculation process is calculated by 27 iterations.

IOP Conf. Series: Journal of Physics: Conf. Series **1060** (2018) 012083 doi:10.1088/1742-6596/1060/1/012083

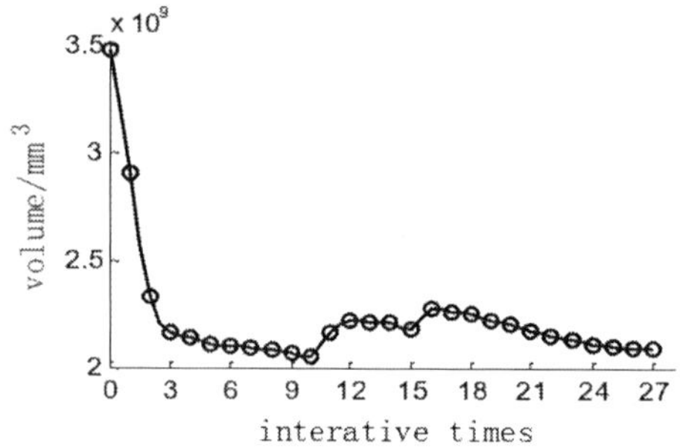

Figure 3. Volume iterative curve.

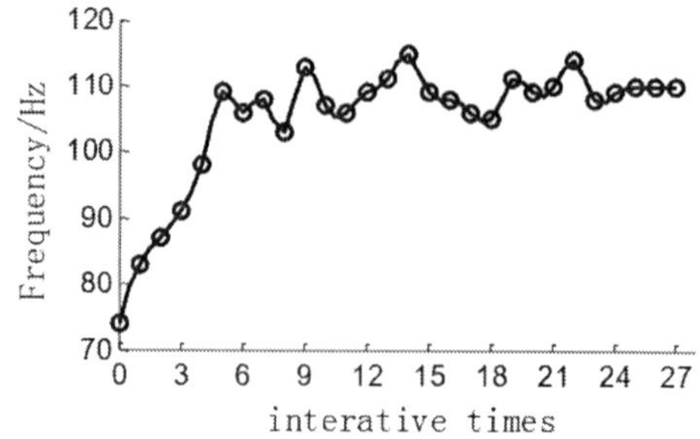

Figure 4. First order modal frequency iterative curve.

After optimization, if the bracket element density is 1, then the material at that unit is very important and needs to be retained. If it is 0 or near 0, then the unit materail is is unimportant and can be removed. Setting the element density threshold as 0.3, so the element whose density threshold is more than 0.3 is retained. Finally, the preliminary optimization structure model is obtained. Compared with the original bracket, the new bracket design mainly adopts the box structure welded by steel plate and proximate matter. By adjusting the structure, the main load bearing material is concentrated on the bottom surface, which is in line with the actual manufacturing and processing requirements. The optimization results are exported through the OSSmooth tool, and the irregular parts of the optimized bracket are regularized, and the final geometry model is obtained considering the actual engineering structure of the bracket.The final geometry model is shown in figure 6.

Figure 5. The result diagram of the topologic optimization of the bracket

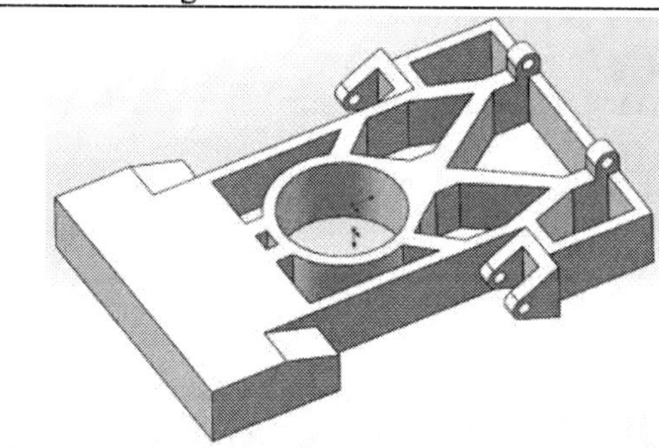

Figure 6. The three-dimensional model of the bracket after optimized.

4. Model verification analysis

4.1. Comparison and analysis of parameter performance

In order to analyze the overall stiffness change of the bracket structure, the specific stiffness effectiveness evaluation was adopted . The specific stiffness formula is shown in formula 4. The greater the value of specific stiffness, the higher the weight stiffness of the bracket unit. The finite element analysis is carried out for the optimized model, and the optimization results are compared with the initial model.

$$\beta = \frac{E / s_d}{m_s} \tag{6}$$

Where β is Specific stiffness efficiency, E is Modulus of elasticity, s_d is Structural displacement, m_s is Structure weight.

Table 1. Comparison and analysis of performance parameters of optimal mode.

	$s\,/\,mm$	$\sigma\,/\,MPa$	$m\,/\,t$	$f\,/\,Hz$	β
Initial model	0.49	95.4	5.4	74	78.9
Opt. model	0.55	143.8	3.8	110	100.5

The results of the analysis are shown in Table 1, where s and σ stand for maximum deformation and maximum displacement of the bracket, m and f stand for the quality of the bracket and the natural frequency of the first order mode. From table 1, the optimization model has a slight increase in deformation, but still within the range of $1.5\,mm$; the maximum stress value increased by $58\,MPa$, mainly due to the stress concentration at the connection of profiles. The rigidity of profile connection can be increased by welding mode, making it smooth, which can effectively reduce the maximum stress value. The specific stiffness effectiveness of the new bracket is increased by 27.4%, which indicates that the stiffness margin of the bracket unit is improved effectively by the optimization model. At the same time, the modal analysis shows that the first order natural frequency of the optimized model is raised by 36, which effectively avoids the resonance frequency range of the launcher.

4.2. Analysis of influence of initial disturbance

The new rigid flexible coupling model is established through the ADAMS, and the dynamic simulation calculation is carried out [10]. In this paper, the launching was analyzed while the launcher is full of missile. The simulation time is 1s, and the step size is 200. Through the dynamic response analysis, the effect of the optimization model on the initial disturbance of the missile is further verified [11]. Figure 7 and Figure 8 is the comparison curve of the missile pitch-angle velocity and the azimuth-angle velocity. Figure 9 is the vibration displacement curve of the launcher barrel.

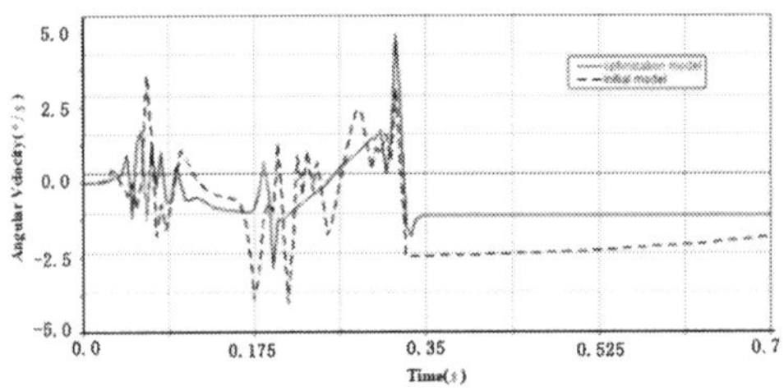

Figure 7. Missile Pitch-angle velocity correlation curve.

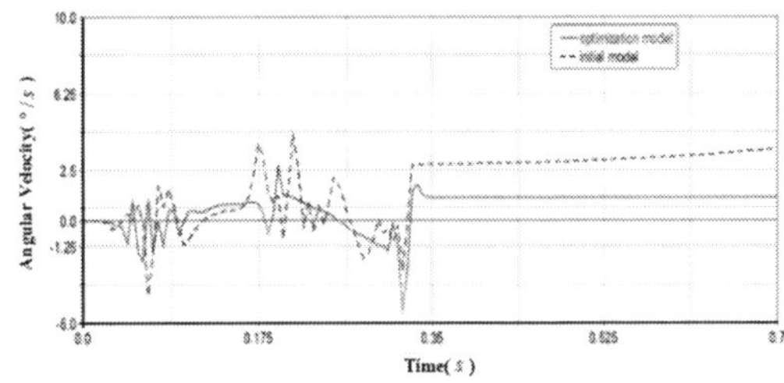

Figure 8. Missile azimuthal-velocity correlation curve.

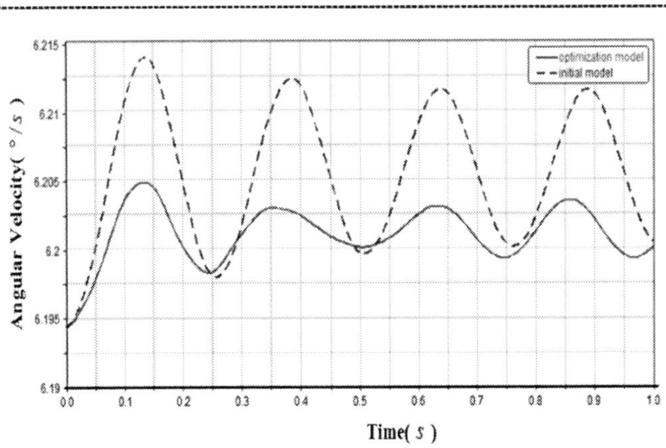

Figure 9. Vibration displacement curve of launch canister.

In Figure 7, Figure 8 and Figure 9, the blue dotted line is the initial model simulation curve, and the red real line is the optimization model simulation curve. The locking force of the missile is unlocked at $0.065\,s$, the front hanger is separated from the guideway at $0.175\,s$, and the rear hanger is separated from the guideway at $0.325\,s$ .When the missile hangs on the guideway only by rear hanger, it forms a bending moment under the gravity and the vibration of launcher.As a result, it leads to the sinking of the missile. Before 0.325S , as seen in Figure 7 and Figure 8, the red curve's peak is always lower than the optimal model. Although there is a moment of red peak is greater than the blue curve during the moving of rear hanger on guideway, the pitch-angular velocity and azimuth-velocity is still lower than the initial model after the rear hanger leaves. For the vibration of launcher's canister, the vibration amplitude of the initial model reaches $0.016\,mm$ and the optimization model is only $0.007\,mm$. To sum up , compared with the initial model, the optimization model can effectively improve the performance of the launcher and reduce the influence of the resonance effect of bracket and launcher on the initial disturbance of the missile.

5. Summary

In this paper, the combination of multi-body dynamics and topology optimization theory is adopted to optimize the launcher bracket with heavy weight, low structural rigidity and low natural frequency, which can cause the resonance effect of the device and increase the initial disturbance of the missile.

1) Compared with the initial structure, the weight of the bracket is obviously reduced, the stiffness and natural frequency are improved, so that the resonant frequency range of the launcher is effectively avoided. It is verified by multi-body dynamics simulation that the optimized model plays a good role in reducing the initial disturbance of the missile and the vibration of canister.

2) After optimization, the displacement deformation of the model increases slightly under the limit load , but the deformation is within the range of design, which proves that the method is feasible.

3) The combination of multi-body dynamics and topology optimization is also instructive and referential for other structural optimization of launchers, and is also suitable for other similar products' initial design and manufacture.

References

[1] FAN Can 2016 *Technology Innovation and Application* vol 18 pp 137
[2] LIN Guowen, MA Dawei, ZHU Sunke 2012 *Journal of Ballistics* vol 24 pp 101-105
[3] ZHANG Guoxian 2012 *Qinghuangdao: Yanshan University*, pp 4-10
[4] HU Ya 2016 *Nan Jing: Nanjing University of Science and Technology* pp 25- 30
[5] Park G J 2011 *Structural & Multidisciplinary Optimization* vol 43 pp 319-337
[6] Choi W S, Park G J. 2012 *Computer Methods in Applied Mechanics and Engineering* vol 191 pp 2105-2122
[7] CEN Haitang. 2004 *Bionic theory of structure, bionic design of light part structure and verification of RP process* Beijing: Beijing University of Aeronautics and Astronautics
[8] CUI Longfei, ZHANG Long, LUO Yun 2013 *Journal of Gun Launch &Control* Vol 4 pp 24-29
[9] YU Hailong, RUI Youting, HE Bin, et al. 2006 Journal of Ballistics vol 18 pp 62-64
[10] ZHANG Shiwei 2010 *Science Technology and Engineering* vol 10 pp 5456-5461
[11] CHEN Zhen, BI Shihua, LIU Guangpu, et al. 2010 *Journal of Solid Rocket Technology* vol 33 pp 112-114.

Research on Evolutionary Game of Information Sharing with Online Health Community

Shen Jiang[1], Zhu Panpan[2] and Xu Man[3]

[1,2] College of Management and Economics, Tianjin University, 300072, China
[3] Business School of NanKai University, Nankai University, 300071, China

Corresponding author's e-mail: ppzhu@tju.edu.cn

Abstract. Online health community has become an important way for patients to share information and has received more and more attention. Based on the theory of evolutionary game theory, this paper establishes a game model of information sharing of online health community members, and studies the factors that affect the information sharing among patients. Results show that the information sharing behavior of the online health community can evolve into a good state and can also be "locked" in a bad state. Online health community managers need to improve the information sharing revenue coefficient, reduce information sharing risk cost, and promote the sustainable information sharing of online community patients.

1. Introduction

With the development of science and technology, patients actively use online platforms to participate in their own health management[1], discuss the problems encountered during the diagnosis and treatment, integrate various resources[2], carry out information sharing, create a unique online health community value creation model [3]. The development of Internet technology and mobile communication technology has enabled the communication between members of online healthy communities not to be affected by time, space and social status[4], providing necessary support for information exchange between patients[5]. Patients join online health communities to seek information, emotional support and opportunities to communicate with others[6]. Experiential knowledge helps all members of the community get the information they want to better fight their own disease[7]. However, because information sharing has certain risks[8], and information provided by patients with community may be wrong or misleading[9], not all patients will share information. Combined with the evolutionary game theory, this paper builds online health information sharing community evolution game model, analyzes the influencing factors of online health community information sharing, to improve the value of online health community, promote the information sharing level and sustainable development of community patients.

The rest of this paper is organized as follows. Section 2 builds an online health community information sharing base model and analyzes its evolution stabilization strategy. The differences of online health community patients will be considered in Section 3. Finally, the results of the two models are compared and suggestions for information sharing are proposed.

2. Basic model of online health community information sharing behavior

2.1. Benefit Matrix

Online health community information sharing is a complex and dynamic system[10].There are two strategies of information sharing and information monopoly. When information is shared, the patient gains the benefit of self-efficacy. If both parties share information, they will benefit from the complementary and superposition of information. At the same time, information sharing needs to pay a certain cost. Suppose the parties involved in information sharing are patient A and patient B. The probability of patient A actively participating in information sharing is x, and the probability of information monopoly is 1-x. Similarly, the probability that patient B actively participates in information sharing is y, and the probability of information monopoly is 1-y.The information sharing benefit matrix is shown in table 1.

Table 1. Information sharing strategy and benefit matrix of online health community

		Patient B	
		Information sharing(y)	Information monopoly(1-y)
PatientA	Information sharing(x)	a - c,a - c	b -c ,p
	Information monopoly(1-x)	P,b -c	0,0

c is the cost of information sharing. when the two parties share information, the benefit of information synergy is a. When one patient shares information, another monopolizes information, the benefits obtained by the information sharer is b, and the benefits obtained by the information monopolist is p.

The expected benefits U_a^{11} of information sharing, expected benefits U_a^{12} of information monopoly and the average benefits $\overline{U_a^{\ 1}}$ of patient A are respectively:

$$U_a^{11} = y(a-c)+(1-y)(b-c) \tag{1}$$

$$U_a^{12} = y*p \tag{2}$$

$$\overline{U_a^{\ 1}} = xU_a^1 + (1-x)U_a^2 \tag{3}$$

The expected benefits U_b^{11} of information sharing, expected benefits U_b^{12} of information monopoly and the average benefits $\overline{U_b^{\ 1}}$ of patient B are respectively:

$$U_b^{11} = x(a-c)+(1-x)(b-c) \tag{4}$$

$$U_b^{12} = xp \tag{5}$$

$$\overline{U_b^{\ 1}} = yU_b^1 + (1-y)U_b^2 \tag{6}$$

According to the dynamic equation of Malthusian, the replication dynamic equation of information sharing of patient A and patient B is respectively:

$$\left\{\begin{aligned} \dot{x} &= \frac{dx}{dt} = x(U_a^{11} - \overline{U_a^{\ 1}}) = x(1-x)[y(a-p-b)+b-c)] \\ \dot{y} &= \frac{dy}{dt} = y(U_b^{11} - \overline{U_b^{\ 1}}) = y(1-y)[x(a-b-p)+b-c] \end{aligned}\right\} \tag{7}$$

2.2. Equilibrium Points and Stability Analysis

When $\dot{x}=0$, $\dot{y}=0$, the dynamic equation of replication is solved, and 5 local equilibrium points of the system(0,0),(0,1),(1,0),(1,1),(G,Q) are obtained. $G=\frac{c-b}{a-b-p}$, $Q=\frac{c-b}{a-b-p}$. The equilibrium points of dynamic equation is not necessarily the evolutionary stability strategy of the system. According to Friedman's method, the evolution stabilization strategy of differential equation system can be derived from the local stable analysis of the Jacobson matrix. The Jacobson matrix

$$J=\begin{pmatrix}\frac{\partial \dot{x}}{\partial x} & \frac{\partial \dot{x}}{\partial y}\\ \frac{\partial \dot{y}}{\partial x} & \frac{\partial \dot{y}}{\partial y}\end{pmatrix}=\begin{pmatrix}a_{11} & a_{12}\\ a_{21} & a_{22}\end{pmatrix} \tag{8}$$

Among them,

$$\begin{Bmatrix}a_{11}=(1-2x)[y(a-b-p)+b-c]\\ a_{12}=x(1-x)(a-b-p)\\ a_{21}=y(1-y)(a-b-p)\\ a_{22}=(1-2y)[x(a-b-p)+b-c]\end{Bmatrix} \tag{9}$$

When the following two conditions are met, the local equilibrium points of the system will become the evolutionary stability strategy (ESS).

Condition 1:

$$\det J=\begin{vmatrix}a_{11} & a_{12}\\ a_{21} & a_{22}\end{vmatrix}=a_{11}a_{22}-a_{12}a_{21}>0 \tag{10}$$

Condition 2:

$$trJ=a_{11}+a_{22}<0 \tag{11}$$

The values of a_{11}, a_{12}, a_{21} and a_{22} of 5 local equilibrium points of the system are shown in table 2.

Table 2. Values of a_{11}, a_{12}, a_{21} and a_{22} at the local equilibrium point

Equilibrium	a_{11}	a_{12}	a_{21}	a_{22}
(0,0)	b-c	0	0	b-c
(0,1)	a-p-c	0	0	c-b
(1,0)	c-b	0	0	a-p-c
(1,1)	p+c-a	0	0	p+c-a
(G,Q)	0	D	W	0

Note: $D=\frac{(a-p-c)(c-b)}{a-b-p}$, $W=\frac{(a-p-c)(c-b)}{a-b-p}$

It is clear that $a_{11}+a_{22}$=0 at the point (G,Q), which does not meet the condition 2, so the local equilibrium points is certainly not an evolutionary stabilization strategy.

2.3. *Results Discussion*

(1) When c>b,a<p+b, the local stability analysis of the system equilibrium points is shown in table 3. At this point, (0,0)is the system stability point. When the cost of information sharing is higher than benefits of information sharing, all members of the community will not share information.

Table 3. Local stability analysis of system equilibrium point

Equilibrium	trJ	det J	Stability
(0,0)	-	+	ESS

(0,1)	-	-	Instability point
(1,0)	-	-	Instability point
(1,1)	+	+	Instability point
(G,Q)	+	0	Instability point

(2) When $a-p<c<b$, local stability analysis of the system equilibrium points is shown in table 4. At this point, (0, 1)and(1, 0)is the system stability point. When the cost of information sharing is lower than the benefits of information sharing, higher than the difference of the collaborative benefits of information sharers minus benefits of information monopolists, information sharing members in the community will always keep information sharing, and information monopolists will keep the information monopoly.

Table 4. Local stability analysis of system equilibrium point

Equilibrium	trJ	det J	Stability
(0, 0)	+	+	Instability point
(0,1)	-	+	ESS
(1,0)	-	+	ESS
(1,1)	+	+	Instability point
(G,Q)	+	0	Instability point

(3) When $b<c<a-p$, local stability analysis of the system equilibrium points is shown in table 5. At this point, (1,1) is the system stability point. When the cost of information sharing is lower than the difference of the collaborative benefits of information sharers minus benefits of information monopolists, eventually members of the community will share information.

Table 5. Local stability analysis of system equilibrium point

Equilibrium	trJ	det J	Stability
(0,0)	+	+	Instability point
(1,0)	+	+	Instability point
(1,1)	-	+	ESS
(G,Q)	+	0	Instability point

3. Online health community information sharing model based on individual differences

3.1. Benefit Matrix

Information sharing in online health communities varies with the subject of information sharing. Each information sharer's information reserve, information expression ability and information absorbing ability are different, and the benefits and costs of information sharing are different. Considering the individual differences of information sharing among online health communities, information sharing game model in online health community based on individual differences is constructed. I is information levels of patients. K is the information absorptive capacity coefficients. F is information transfer capacity coefficients. R is the benefit of information sharing. C is the cost coefficient of information sharing. The benefit matrix of information sharing is shown in table 6.

Table 6. Information sharing strategy and benefit matrix of online health community

		Patient B	
		Information sharing(y)	Information monopoly (1-y)
PatientA	Information sharing (x)	$I_aR_a+I_bF_bK_a-I_aC_a$, $I_bR_b+I_aF_aK_b-I_bC_b$	$I_aR_a-I_aC_a$, $I_aF_aK_b$
	Information	$I_bF_bK_a$, $I_bR_b-I_bC_b$	0, 0

IOP Conf. Series: Journal of Physics: Conf. Series **1060** (2018) 012084 doi:10.1088/1742-6596/1060/1/012084

monopoly(1-x)

The expected benefits of information sharing U_a^{21}, expected benefits of information monopoly U_a^{22} and the average benefits $\overline{U_a^{\ 2}}$ of patient A are respectively:

$$U_a^{21} = y(\mathrm{I_aR_a} + \mathrm{I_bF_bK_a} - \mathrm{I_aC_a}) + (1-y)(\mathrm{I_aR_a} - \mathrm{I_aC_a}) \tag{12}$$

$$U_a^{22} = y * \mathrm{I_bF_bK_a} \tag{13}$$

$$\overline{U_a^{\ 2}} = xU_a^{21} + (1-x)U_a^{22} \tag{14}$$

The expected benefits of information sharing U_b^{21}, expected benefits of and information monopoly U_b^{22} and the average benefits $\overline{U_b^{\ 2}}$ of patient B are respectively:

$$U_b^{21} = x(\mathrm{I_bR_b} + \mathrm{I_aF_aK_b} - \mathrm{I_bC_b}) + (1-x)(\mathrm{I_bR_b} - \mathrm{I_bC_b}) \tag{15}$$

$$U_b^{22} = x\mathrm{I_aF_aK_b} \tag{16}$$

$$\overline{U_b^{\ 2}} = yU_b^1 + (1-y)U_b^2 \tag{17}$$

According to the dynamic equation of Malthusian, the replication dynamic equation of information sharing of patient A and patient B is respectively:

$$\begin{Bmatrix} \dot{x} = \dfrac{dx}{dt} = x(U_a^{21} - \overline{U_a^{\ 2}}) = x(1-x)(U_a^{21} - U_a^{22}) = x(1-x)(I_a\mathrm{R}_a - I_aC_a) \\ \dot{y} = \dfrac{dy}{dt} = y(U_b^{21} - \overline{U_b^{\ 2}}) = y(1-y)(U_b^{21} - U_b^{22}) = y(1-y)(I_aR_b - I_bC_b) \end{Bmatrix} \tag{18}$$

3.2. Equilibrium Points and Stability Analysis

The evolution stabilization strategy of differential equation system can be derived from the local stable analysis of the Jacobson matrix. The Jacobson matrix

$$J = \begin{pmatrix} \dfrac{\partial \dot{x}}{\partial x} & \dfrac{\partial \dot{x}}{\partial y} \\ \dfrac{\partial \dot{y}}{\partial x} & \dfrac{\partial \dot{y}}{\partial y} \end{pmatrix} = \begin{pmatrix} a_{11} & a_{12} \\ a_{21} & a_{22} \end{pmatrix} = \begin{pmatrix} (1-2x)(I_aR_a - I_aC_a) & 0 \\ 0 & (1-2y)(I_bR_b - I_bC_b) \end{pmatrix} \tag{19}$$

The values of a_{11}, a_{12}, a_{21} and a_{22} of 4 local equilibrium points of the system are shown in table 7.

Table 7. Values of a_{11}, a_{12}, a_{21} and a_{22} at the local equilibrium point

Equilibrium	a_{11}	a_{12}	a_{21}	a_{22}
(0,0)	I_aR_a-I_aC_a	0	0	I_bR_b-I_bC_b
(0,1)	I_aR_a-I_aC_a	0	0	I_bC_b-I_bR_b
(1,0)	I_aC_a-I_aR_a	0	0	I_bR_b-I_bC_b
(1,1)	I_aC_a-I_aR_a	0	0	I_bC_b-I_bR_b

When $I_aR_a<I_aC_a$, $I_bR_b<I_bC_b$, local stability analysis of the system equilibrium points is shown in table 8. When $I_aR_a<I_aC_a$, $I_bR_b>I_bC_b$, local stability analysis of the system equilibrium points is shown in table 9. When $I_aR_a>I_aC_a$, $I_bR_b<I_bC_b$, local stability analysis of system equilibrium points is shown in table 10. When $I_aR_a>I_aC_a$, $I_bR_b>I_bC_b$, local stability analysis of system equilibrium points is shown in table 11.

Table 8. Local stability analysis of the system equilibrium point

Equilibrium	trJ	det J	Stability
(0,0)	-	+	ESS
(0,1)	Indeterminate	-	Instability point
(1,0)	Indeterminate	-	Instability point
(1,1)	+	+	Instability point

Table 9. Local stability analysis of the system equilibrium point

Equilibrium	trJ	det J	Stability
(0,0)	Indeterminate	-	Instability point
(0,1)	-	+	ESS
(1,0)	+	+	Instability point
(1,1)	Indeterminate	-	Instability point

Table 10. Local stability analysis of the system equilibrium point

Equilibrium	trJ	det J	Stability
(0,0)	Indeterminate	-	Instability point
(0,1)	+	+	Instability point
(1,0)	-	+	ESS
(1,1)	Indeterminate	-	Instability point

Table 11. Local stability analysis of the system equilibrium point

Equilibrium	trJ	det J	stability
(0,0)	+	+	Instability point
(0,1)	Indeterminate	-	Instability point
(1,0)	Indeterminate	-	Instability point
(1,1)	-	+	ESS

Table 8-11 show that when the income coefficient is less than the risk coefficient, the community ends up with no information sharing;when the income coefficient is greater than the risk factor, community patients will eventually share information. If income coefficient of some patients in community is less than the risk coefficient, income coefficient of other patients is greater than the risk coefficient, the former will not share information and the latter will always share information.

4. Conclusion

Compare the above two evolutionary game model of information sharing, found that the differences of patients in online health community, namely information conversion coefficient, absorption coefficient and patient information reserve, have no direct impact on the development of community information sharing, and the income coefficient and risk coefficient of information sharing determine the development of information sharing. Therefore, to protect patients' privacy, reduce the cost of information sharing, and give information sharing incentives, it is the key to enhance the value of online health community and promote the continuous information sharing of community patients.

References

[1] A. F.Payne, K.Storbacka, P. Frow, "Managing the co-creation of value", Journal of the Academy of Marketing Science, vol.36, 2008, pp. 83-96.
[2] S. L.Vargo, R.F.Lusch, "It's all B2B and beyond: toward a systems perspective of the market", Industrial Marketing Management, vol.40, 2011, pp.181-187.
[3] J. R.McColl-Kennedy, S.L.Vargo, T.S.Dagger, et al. "Health care customer value cocreation practice styles", Journal of Service Research,vol.15, 2012, pp.370-389.

[4] S. A.King, D.Moreggi, "Internet Self-Help and Support Groups"[M] Psychology and the Internet. 2007.
[5] J.Preece, "Etiquette online: from nice to necessary", Communications of the ACM,vol.47, 2004, pp. 56-61.
[6] U.Josefsson, "Coping with illness online: the case of patients' online communities "The Information Society: An International Journal, vol. 21, 2005, pp. 133-141.
[7] J.Zhao, K.Abrahamson, J. G.Anderson, S.Ha, R.Widdows, "Trust, empathy, social identity, and contribution of knowledge within patient online communities", Behaviour & Information Technology, vol. 32, 2013,pp.1041-1048.
[8] M.Etgar, "A descriptive model of the consumer co-production process", Journal of the Academy of Marketing Science, vol. 36, 2008, pp. 97-108.
[9] E. J.Madara, "The mutual-aid self-help online revolu-tion", Social Policy, vol. 27, 1997, pp.20-26.
[10] Chaohua Deng , Jiang Meng. Research on knowledge sharing behavior of online medical health community, National conference on computer simulation and information technology. 2015.

IOP Conf. Series: Journal of Physics: Conf. Series **1060** (2018) 012085 doi:10.1088/1742-6596/1060/1/012085

The Study on Mathematical Model and Simulation of Asynchronous Motor Considering Iron Loss

Zhen GUO [1, 2] and Qing-wei ZHANG[2]

[1]College of Information and Control Engineering, China University of Petroleum(East China), Qingdao, 266580, China
[2]College of Automation Engineering, Qingdao University of Technology, Qingdao, 266520, China

qs2004b@163.com

Abstract. When the dynamic mathematical model of asynchronous motor is established, it usually ignores the motor iron loss. Actually, the motor iron loss exists and it has a great influence on the performance of the motor, especially on the efficiency. A motor model considering iron loss is built in this paper and compared with the motor model ignoring iron loss by Simulink. The paper provides a practical model that has high-performance control and it is closer to the actual physical behaviour of asynchronous motors. The simulation results verify the correctness, validity and accuracy of the model and the conclusions provide a theoretical reference for the research of asynchronous motor and the efficiency optimal design of its control system.

1. Introduction

With the development of AC motor control theory and power electronics technology, motor control system is becoming more and more complex [1,2]. Computer simulation has increasingly become an indispensable means of control system research. The simulation model of the control object is very important to the simulation research of the control system, and the accuracy of the model directly affects the reliability of the simulation results. MATLAB software is a popular powerful simulation tool at present. It also provides great convenience for modeling and simulation of electrical system [3]. However, the simulation model of asynchronous motor provided by the module library ignores iron loss of the motor. When it is used to study the performance of the motor related to efficiency, it is bound to reduce the simulation precision and it can not be used for the accurate simulation of asynchronous motor vector control [4-6]. Therefore, it is necessary to establish a more accurate model of asynchronous motor. The structure of the model should be simple, easy to implement and has strong stability for simulation research. In the text, the dynamic mathematical model of asynchronous motor considering iron loss in two phase stationary alpha and beta coordinate system is derived based on mathematical analysis. Finally, two models are compared with the simulation examples to verify the feasibility of the asynchronous motor model considering the iron loss.

2. Mathematical model of asynchronous motor ignoring iron loss

In the two-phase stationary α, β coordinate system, the voltage equations of asynchronous motor are written as

$$\begin{cases} U_{s\alpha} = R_s i_{s\alpha} + p\psi_{s\alpha} \\ U_{s\beta} = R_s i_{s\beta} + p\psi_{s\beta} \\ 0 = R_r i_{r\alpha} + \omega_r \psi_{r\beta} + p\psi_{r\alpha} \\ 0 = R_r i_{r\beta} - \omega_r \psi_{r\alpha} + p\psi_{r\beta} \end{cases} \tag{1}$$

The flux equations are given as

$$\begin{cases} \psi_{s\alpha} = L_s i_{s\alpha} + L_m i_{r\alpha} \\ \psi_{s\beta} = L_s i_{s\beta} + L_m i_{r\beta} \\ \psi_{r\alpha} = L_r i_{r\alpha} + L_m i_{s\alpha} \\ \psi_{r\beta} = L_r i_{r\beta} + L_m i_{s\beta} \end{cases} \tag{2}$$

The torque equation is given as

$$T_e = n_p L_m (i_{s\beta} i_{r\alpha} - i_{s\alpha} i_{r\beta}) = n_p (\psi_{s\alpha} i_{s\beta} - \psi_{s\beta} i_{s\alpha}) \tag{3}$$

The equation of motion is given as

$$p\omega_r = \frac{n_p}{J}(T_e - T_L) \tag{4}$$

Where the $U_{s\alpha}$ and $U_{s\beta}$ are the stator α-axes and β-axes input voltages; $i_{s\alpha}$ and $i_{s\beta}$ are the α-axes and β-axes stator current; $i_{r\alpha}$ and $i_{r\beta}$ are the α-axes and β-axes rotor current; $\psi_{s\alpha}$ and $\psi_{s\beta}$ are the α-axes and β-axes stator fluxes; $\psi_{r\alpha}$ and $\psi_{r\beta}$ are the α-axes and β-axes rotor fluxes; R_s is the stator resistance; R_r is the rotor resistance; L_m is the magnetizing inductance, L_s is the stator inductance, and L_r is the rotor inductance; ω_r is the rotor angular frequency; n_p is the number of pole-pairs; J is the moment of inertia; T_L is the load torque and T_e is the electromagnetic torque; p is the differential operator. The dynamic mathematical model of asynchronous motor ignoring iron loss in $\alpha\beta$ stationary coordinate can be composed from equation (1) to equation (4).

3. Mathematical model and equivalent circuit of asynchronous motor considering iron loss

The iron loss of motor is related to core structure parameters, voltage and frequency and flux density. In the dynamic model, if the iron loss is to be considered, the loss of the motor iron loss can be expressed by the loss of an equivalent pure resistance winding. Usually, the stator iron loss is the main factor in the asynchronous motor, so the equivalent winding of the stator is assumed only on the stator. In the α β stationary coordinate system, two equivalent windings of iron loss are added to the stator side except the four windings on the original stator and rotor axis. The voltage equations are written as equation (5).

$$\begin{cases} U_{s\alpha} = R_s i_{s\alpha} + p\psi_{s\alpha} \\ U_{s\beta} = R_s i_{s\beta} + p\psi_{s\beta} \\ 0 = R_r i_{r\alpha} + \omega_r \psi_{r\beta} + p\psi_{r\alpha} \\ 0 = R_r i_{r\beta} - \omega_r \psi_{r\alpha} + p\psi_{r\beta} \\ 0 = R_m i_{R_{m\alpha}} - p\psi_{m\alpha} \\ 0 = R_m i_{R_{m\beta}} - p\psi_{m\beta} \end{cases} \tag{5}$$

The motor excitation current equations are given as

$$\begin{cases} i_{L_{m\alpha}} = i_{s\alpha} + i_{r\alpha} - i_{R_{m\alpha}} \\ i_{L_{m\beta}} = i_{s\beta} + i_{r\beta} - i_{R_{m\beta}} \end{cases} \tag{6}$$

The relationships between magnetic flux and current are as follows

IOP Conf. Series: Journal of Physics: Conf. Series **1060** (2018) 012085 doi:10.1088/1742-6596/1060/1/012085

$$\begin{cases} \psi_{s\alpha} = L_{1s} i_{s\alpha} + \psi_{m\alpha} \\ \psi_{s\beta} = L_{1s} i_{s\beta} + \psi_{m\beta} \\ \psi_{r\alpha} = L_{1r} i_{r\alpha} + \psi_{m\alpha} \\ \psi_{r\beta} = L_{1r} i_{r\beta} + \psi_{m\beta} \\ \psi_{m\alpha} = L_m i_{L_{m\alpha}} \\ \psi_{m\beta} = L_m i_{L_{m\beta}} \end{cases} \quad (7)$$

The torque equation is given as

$$T_e = \frac{L_m}{L_r}\left[\psi_{r\alpha}\left(i_{s\beta} - i_{R_{m\beta}}\right) - \psi_{r\beta}\left(i_{s\alpha} - i_{R_{m\alpha}}\right)\right] \quad (8)$$

The equation of motion is given as

$$p\omega_r = \frac{n_p}{J}(T_e - T_L) \quad (9)$$

Where $i_{R_{m\alpha}}$ and $i_{R_{m\beta}}$ are the α-axes and β-axes equivalent iron loss resistance current; $i_{L_{m\alpha}}$ and $i_{L_{m\beta}}$ are the α-axes and β-axes excitation current; R_m is the equivalent iron loss resistance; $\psi_{m\alpha}$ and $\psi_{m\beta}$ are the α-axes and β-axes main flux.

The dynamic mathematical model of asynchronous motor considering iron loss in αβ stationary coordinate can be composed from equation(5) to equation (9) and the equivalent circuit as shown in Figure 1.

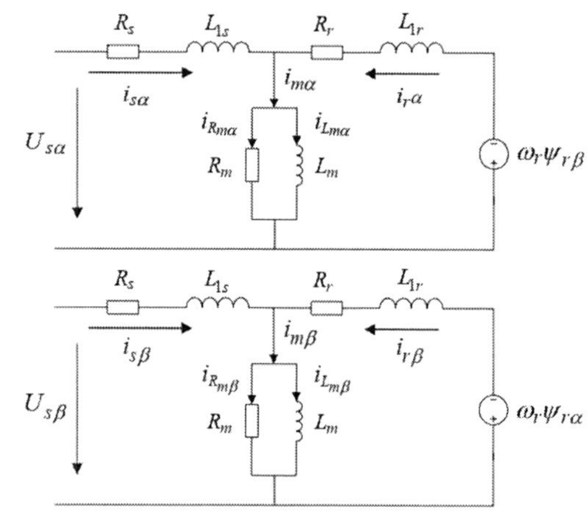

Figure 1. The equivalent circuit of asynchronous motor considering iron loss in $\alpha\beta$ stationary coordinate system

4. Simulation Results and Analysis

The simulation model of asynchronous motor considering iron loss according the equation(5) to equation (9) is established and it is shown in Figure 2.

Simulation experiment 1: set asynchronous motor to start directly under rated voltage and rated load. The electromagnetic torque waveform, the angular velocity waveform of the rotor and the α component waveform of the stator current are shown in Figure 3.

Simulation experiment 2: set asynchronous motor to start under rated voltage and rated load at 0.2s. The electromagnetic torque waveform of the starting process, the rotor angular velocity waveform and the α component waveform of the stator current are shown in Figure 4.

It can be seen that the motor model considering iron loss is more stable and less oscillation than the motor model without considering iron loss. The inertia of the system is larger and the starting speed is slower. The stator current amplitude is larger in the steady state than that without considering iron loss.

IOP Conf. Series: Journal of Physics: Conf. Series **1060** (2018) 012085 doi:10.1088/1742-6596/1060/1/012085

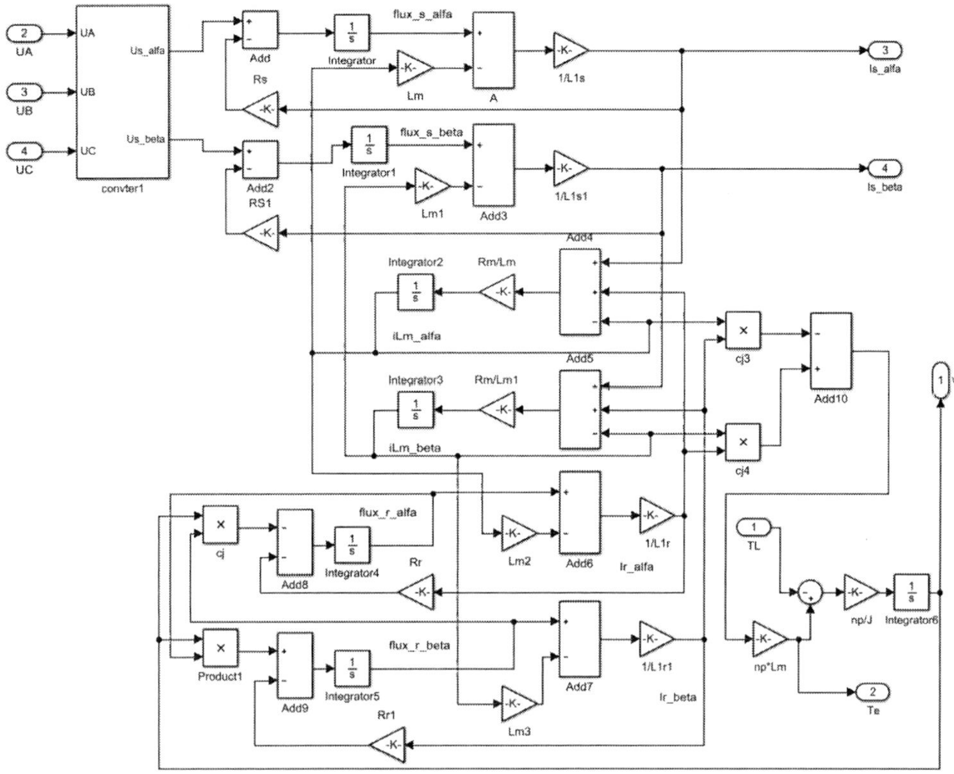

Figure 2. The simulation model of asynchronous motor considering iron loss

A three-phase squirrel cage asynchronous motor with a triangular connection is used in simulation. The parameters of the motor are shown in Table 1.

Table 1. Test motor parameters

parameters	values
Rated voltage	220V
Rated shaft power	80W
Rated frequency	50Hz
Stator resistance	72Ω
Stator leakage inductance	0.26H
Magnetizing inductance	1.64H
Stator inductance	1.9H
Rated load	0.5N · m
Rated current	0.78A
Rated speed	1400r/min
Equivalent iron loss resistance	2400Ω
Rotor resistance	70Ω
Rotor leakage inductance	0.26H
Number of the pole pairs	2
Rotor inductance	1.9H
Rotor inertia	0.0005kg · m^2

IOP Conf. Series: Journal of Physics: Conf. Series **1060** (2018) 012085 doi:10.1088/1742-6596/1060/1/012085

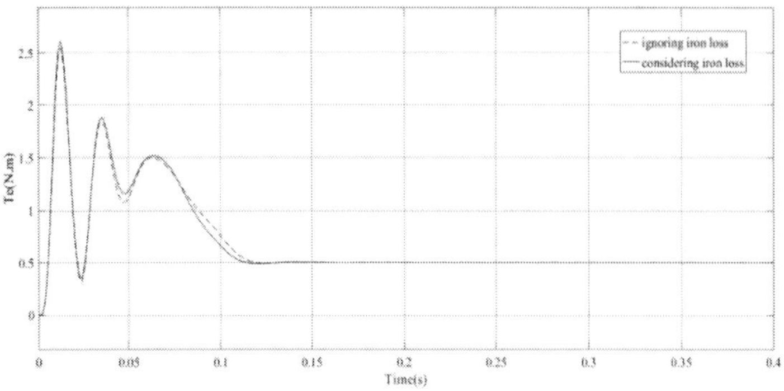

(a) The electromagnetic torque waveform

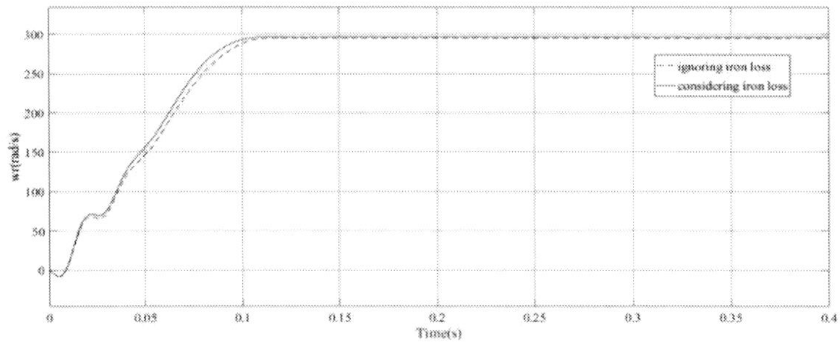

(b) The rotor angular velocity waveform

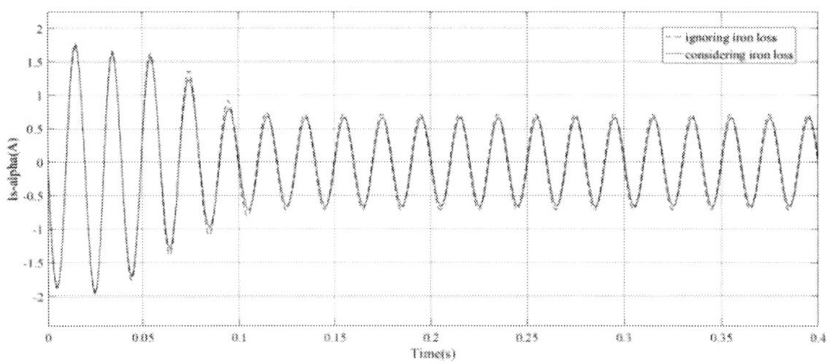

(c) The α component waveform of the stator current

Figure 3. Simulation waves of setting asynchronous motor start directly under rated condition

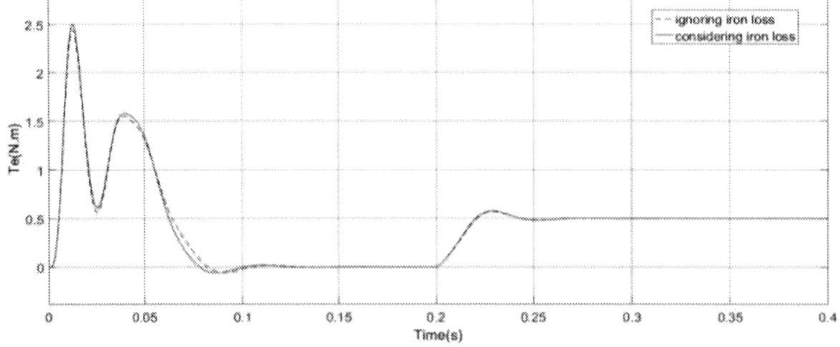

(a) The electromagnetic torque waveform

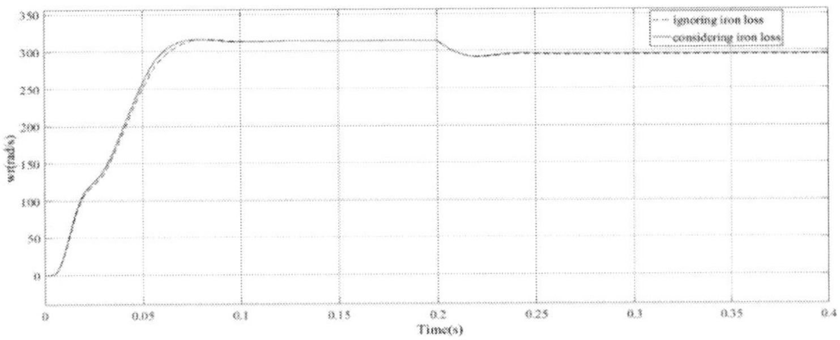

(b)The rotor angular velocity waveform

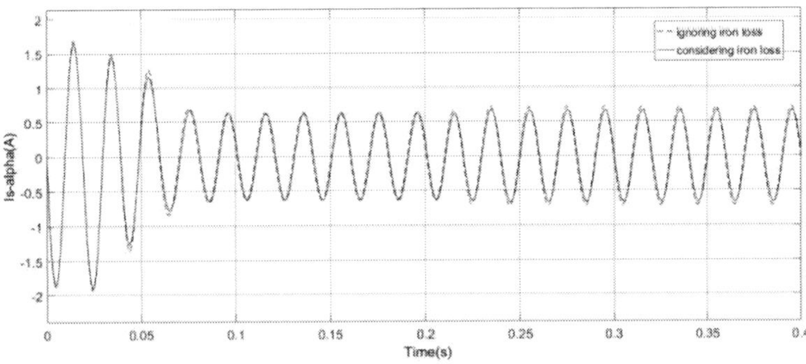

(c) The α component waveform of the stator current

Figure 4. Simulation waves of asynchronous motors load applied at 0.2s

5. Conclusion

The rated operation efficiency is relatively low and the iron loss is great for such small power asynchronous motor. The traditional model which ignores the iron loss will bring more inaccuracy. The difference between the two models is compared by the simulation of the MATLAB/Simulink and simulation results provide that the mathematical model of the motor considering iron loss is more accurate and valid than the mathematical model ignoring the iron loss.

References

[1] ShiyiShao, Abdi. E, and McMahon. R., Vector Control of The BDFM for Wind Power Generation, IEEE International Conference on Sustainable Energy Technologies,2008, pp. 322-327.

[2] Christian Kral, and Anton Haumer, Consistent equivalent circuit parameters of induction motors for the calculation of partial load efficiencies, IEEE Industrial Electronics, 2008, pp. 698-705.

[3] Yao Jun, and Ma Songhui, Simulink modeling and simulation, Xi'an Electronic and Science University press, Xi'an, 2002.

[4] Mohan P.R.,Reddy T.B.,Kumar M.V. A Simple Generalized Pulse Width Modulation Algorithm for Vector Control based Voltage Source Inverter fed Induction Motor Drives [C]//Emerging Research Areas: Magnetics, Machines and Drives, 2014 Annual International Conference on, IEEE, 2014,pp 1-5.

[5] Mark A.P.,Irudayaraj G.C.R.,Vairamani R., et al. Dynamic Performance Analysis for Different Vector-Controlled CSI- Fed Induction Motor Drives, Journal of Power Electronics, 2014, 14(5),pp1-11.

[6] Baby B.,Shajilal A.S. An Improved Indirect Vector Controlled Current Source Inverter Fed Induction Motor Drive With Rotor Resistance Adaptation, Emerging Research Areas: Magnetics, Machines and Drives, 2014 Annual International Conference on, IEEE, 2014,pp 1-6.

Evaluation of the Sustainable Development Ability of the Urban Ecosystem in Jiangsu Province Based on the Information Entropy

Fan Song [1], Xiao-hua Yang [1*] and Tong Liu[1]

[1] State Key Laboratory of Water Environment Simulation, School of Environment, Beijing Normal University, Beijing 100875, China

*Corresponding author: Xiao-hua Yang, Professor, guicai0014@mail.bnu.edu.cn.

Abstract. Urban ecosystem (UE) is a typical "natural-social-economic system". In the process of a new round of urbanization, urban ecosystems are facing challenges on ecological land reduction and environmental degradation. Better understanding and optimizing the evolutionary directions of UE are of great significance for both urban sustainable development and eco-civilization construction. In this study, based on dissipative structure theory and urban nature-economy-society complex ecosystem theory, the indicator system of evaluating UE developing direction was set up through information entropy method. This indicator system was made up by four types of information entropy including backup inflow entropy, imposed output entropy, destructive metabolism entropy, and regenerative metabolism entropy. Then, Jiangsu province were selected as the study area. Year-based information entropy model together with UE evolution index system were applied to analyse the order and complexity of UE during 2005-2015; The weights of indicators evaluating UE were calculated. Furthermore, the index scores of UE were calculated by combing with standardization values of indicators. And urban ecosystem sustainable development level was evaluated based on integrated development degree and coordinated development degree models. This study aims to provide a theoretical reference for the sustainable development and optimization of UE health in the Jiangsu province.

1. Introduction

We are living in an era of rapid economic development, people are enjoying great achievements of economy development as well as facing with many ecological environment problems. Such as environmental pollution, lack of energy, loss of biodiversity and so on, which are destroying the health of UE [1]. Compared with other ecosystems, UE has high vulnerability and human dependence, how to correctly evaluate the sustainable development of urban ecosystem process has been a hotspot in the research of the sustainable development of ecosystem. UE is a typical social - economic - natural compound ecosystem and a people-centric ecosystem. This system generally includes three parts of economic, social and natural systems, they have an impact on the health of urban ecosystems together [2].

A healthy urban ecosystem should be stable and sustainable. It is the integration of society, economy and natural system [3]. In recent years, some scholars imported the concept of entropy into the evaluation of the sustainable development of ecosystem. They studied the evolution of UE from the perspective of "information entropy". For example, Wu et al. [4] analyzed the evolution of UE of

Wuhan based on information entropy and found the driving factors. Wang et al. [5] analyzed the evolution of ecological system in Shanghai from 2003 to 2013 by using the information entropy theory and made a prediction of the sustainable development ability of 2014-2020 with grey model. These studies enrich the methods of urban ecological environmental assessment and prove the feasibility of information entropy method in UE evolution and sustainable development.

2. Study area

Jiangsu province is located in the east coast of China, and the level of economic and social development is at the forefront of the country. However, behind the rapid development of economic is the challenge of ecological environment. With high resource consumption and serious environmental damage, the environmental situation is increasingly serious. Therefore, it is of great significance to analyze the evolution trend of human settlements ecosystem of Jiangsu province so as to ensure the sustainable development of social and economic development in the region and to advise on regional development patterns in other areas.

3. Method

3.1 Information entropy theory of urban ecosystem

"Entropy" was originally a state parameter elicited by the second law of thermodynamics reflecting the irreversibility of the spontaneous process. The larger the entropy is, the more chaotic the system is, and if the entropy of the system is neutralized by some means, the system can be re-formed into a new ordered structure. The entropy change of UE is divided into two parts: first is the "input-output entropy" of material exchange between natural ecological environment and social and economic system, which reflects the support of natural system to urban system. The second is the entropy produced in human habitation, which is referred to as "metabolism entropy production", which reflects the ability to recover and metabolism of UE [6]. The "input-output entropy" and "metabolism entropy production" can be further divided into four parts, namely backup inflow entropy ΔS_1, imposed output entropyΔS_2, destructive metabolism entropyΔS_3, and regenerative metabolism entropyΔS_4. According to "input-output entropy" and "metabolism entropy production", the "total entropy" of the UE can be obtained to quantify the orderness of the system. The specific index system is shown in table 1.

Table 1. Evaluation index system of sustainable development ability

Target hierarchy	Rule hierarchy		Index hierarchy	Units
UE	Input-output entropy	Backup inflow entropy	Per capita GDP	Yuan
			Local financial revenue	Yuan
			Total volume of foreign trade	$\times 10^8$US $
			Grain output	$\times 10^4$t
			Total vegetable production	$\times 10^4$t
			Total output of meat	$\times 10^4$t
			Total fruit production	$\times 10^4$t
			Total aquatic product	$\times 10^4$t
			Steel product quantity	$\times 10^4$t
			Cement output	$\times 10^4$t
			Electric energy production	$\times 10^8$kw·h
			Urban residents' education cultural entertainment expenses	Yuan per person

	Imposed output entropy	Population density	Person/km^2
		Population growth	‰
		Residential electricity consumption	$\times 10^8$kw·h
		Daily water consumption per capita	Litre
		The per capita consumption expenditure of urban residents	Yuan
		The per capita living consumption expenditure of rural residents	Yuan
		Energy consumption	$\times 10^4$t standard coal
		Industrial energy consumption	$\times 10^4$t standard coal
		Consumption of chemical fertilizers	$\times 10^4$t
		Car ownership per 100 households	-
Metabolism entropy production	Destructive metabolism entropy	Total industrial wastewater discharge	$\times 10^8$t
		Urban sewage discharge	$\times 10^8$t
		Total industrial emissions	$\times 10^8$m^3
		Total dust emission of smoke	$\times 10^4$t
		Total sulfur dioxide emissions	$\times 10^4$t
		PM concentrations	μg/m^3
		SO_2 concentration	μg/m^3
		NO_2 concentration	μg/m^3
		Industrial solid waste production	$\times 10^4$t
		Life garbage clearance volume	$\times 10^4$t
	Regenerative metabolism entropy	Treatment rate of domestic sewage	%
		Life garbage treatment rate	%
		The percentage of industrial solid wastes that are comprehensively utilized	%
		Water grades better than III	%
		Green coverage of built-up area	%
		Nature reserve area ratio	%
		Per capita park green area	m^2
		Water-saving irrigation area	10^3hm^2
		Length of drainage pipes	km
		The proportion of environmental protection input	%

3.2. Quantitative model of information entropy

According to the information entropy theory of Shannon C. E., for an uncertain system such as the UE, if the random variable X is used to represent its state characteristics: for discrete random variables, assume the value of X is $X = \{x_1, x_2, \ldots, x_n\}$ $(n \geq 2)$,each value corresponds to probability $P = \{p_1, p_2, \ldots, p_n\}$ $(0 \leq pi \leq 1, i = 1,2, \ldots, n)$, and $\sum_{i=1}^{n} P_i = 1$,then the information entropy of the system can be calculated by following formula[7]:

$$S = -\sum_{i=1}^{n} P_i \cdot \ln(P_i) \tag{1}$$

Where S is the information entropy of the uncertain system, and P_i represents the probability of each discrete random variable X.

Based on the above theories, this model is divided into two parts: the year information entropy and the index information entropy. The evolution direction and complexity of the system are analyzed by year information entropy. The index information entropy is combined with the analytic hierarchy process to determine the index weight, which can calculate the comprehensive score of the system and the comprehensive development degree and the coordinated development degree. Thus, the health and coordination degree of UE are judged.

3.2.1. Year information entropy. Year information entropy Δ S system can be expressed as follows:

$$\Delta S = -\frac{1}{lnm}\sum_{i=1}^{n}\frac{q_{ij}}{q_j}ln\frac{q_{ij}}{q_j}, q_j = \sum_{i=1}^{n} q_{ij}\ (i = 1,2, \ldots, n; j = 1,2, \ldots, m) \tag{2}$$

Where m is the year, n is the index, and the ΔS is the year information entropy of the UE. q_{ij} is the standardized value of each index and q_j is the sum of the normalized values of the j year.

3.2.2. Index information entropy. The multi-dimensional information in the system is comprehensively quantified, and the score of UE in this study adopted the comprehensive weighted index model:

$$G = \sum W_i \cdot X_{ij} \tag{3}$$

Where G is the total score of the sustainable development ability of UE in a certain year, and X_{ij} is the standardized value of each index. W_i is the weight of each index, which can be calculated by the following formulas:

$$W_i = (w_{i1} + w_{i2})/2, w_{i1} = {(1 - e_i)}\Big/{(n - \sum_{i=1}^{n} e_i)} \tag{4}$$

$$e_i = -\frac{1}{lnm}\sum_{j=1}^{m}\frac{X_{ij}}{X_i}ln\frac{X_{ij}}{X_i}, X_i = \sum_{i=1}^{n} X_{ij}\ (i = 1,2, \ldots, n; j = 1,2, \ldots, m) \tag{5}$$

Where w_{i1} is the weight of entropy weight calculation, and w_{i2} is the weight of analytic hierarchy process, e_i is the index information entropy.

The relationship between urban ecological environment and socio-economic development can be analyzed by "comprehensive development degree" and "coordinated development degree" model [8]. Among them, the overall benefit evaluation of ecological environment and economic development adopts "comprehensive development degree T":

$$T = \alpha_1 G_1 + \alpha_2 G_2 + \alpha_3 G_3 + \alpha_4 G_4 \tag{6}$$

Where the α_i (i = 1,2,3,4) is the coefficient of bias, and $\alpha_1 + \alpha_2 + \alpha_3 + \alpha_4 = 1$.G1, G2, G3 and G4 are the information entropy scores of various indicators. The higher the score of T, the higher the comprehensive development of UE. The level of coordinated development of environment and economy adopts the " coordinated development degree D":

$$D = \sqrt{\{\frac{G_1 \cdot G_2 \cdot G_3 \cdot G_4}{\left[\frac{(G_1+G_2+G_3+G_4)}{4}\right]^4}\}^k \cdot T} \tag{7}$$

Where T is the comprehensive development degree; K is a constant which is 8 in this study. Similarly, the higher the score of D, the better coordinated development of UE.

4. Results and analysis

4.1. Entropy movement analysis of UE evolution

It can be seen from Figure.1 that the backup inflow entropy of Jiangsu province generally showed a parabolic upstate during 2005 to 2015. With the rapid development of social economy, the demand of UE for natural resource system is increasing, but this state is slowing down and the system is stabilizing. The imposed output entropy showed a decline trend, indicating that the pressure of population and society on the UE is getting smaller. The destructive metabolism entropy showed a decreasing state of oscillation, indicating that the emission of various pollutants in human production and life has decreased. However, in individual years it has been a significant rise, the governance effect has been repeated, and needed to be intensified to prevent deterioration. The line of regenerative metabolism entropy was a parabola, and in 2010 it exceeded the destructive metabolism entropy, which indicating that the pollution control capacity of urban environment is increasing, and the municipal engineering is also being improved. At the same time, it is found that there was no significant change in entropy in recent years. In order to prevent the occurrence of deterioration, the governance should not be relaxed.

From 2005 to 2015, the input-output entropy, metabolism entropy production and total entropy of the UE in Jiangsu province showed a wavelike decrease change. Although there were slight fluctuations in the year, the overall downward trend was obvious. The input-output entropy is negative, and the decline trend has slowed down in recent years, indicating that the capacity of natural resource system to UE is increasing. Entropy production has experienced the change from positive to negative and the overall trend in declining, which showing that the metabolism capacity of UE in Jiangsu province is increasing, but the entropy production is the most volatile of all three kinds of entropy. How to stabilize the entropy will be the focus of future work. The total entropy value is negative and decreasing continuously, the UE is changing in a healthy and orderly direction, and its stability is continuously enhanced.

4.2. Index movement analysis of UE evolution

As can be seen from Figure.2, the comprehensive development degree of Jiangsu province was almost consistent with the comprehensive score, and there was an obvious process of going up at first and then going down. The development trend of coordinated development D was to go up and then down, just the opposite of T. During the period of 2005-2008, the coordinated development degree D was on the rise, and the development type was environmental lag type. At this time, the metabolic and reduction ability of the UE was still at a low level, its carrying capacity to population is still weak. Therefore, although the coordinated development D was on the rise, the comprehensive development degree T and the comprehensive score have declined. In 2008-2015, it showed a trend of rapid decline, the development type was the economic lag type. At this time, the focus of development is to develop economy and improve the coordinated development of environment and economy. Due to the economic level and environmental control level has reached a higher level, so comprehensive development degree T and comprehensive score began to rise, especially after 2011.

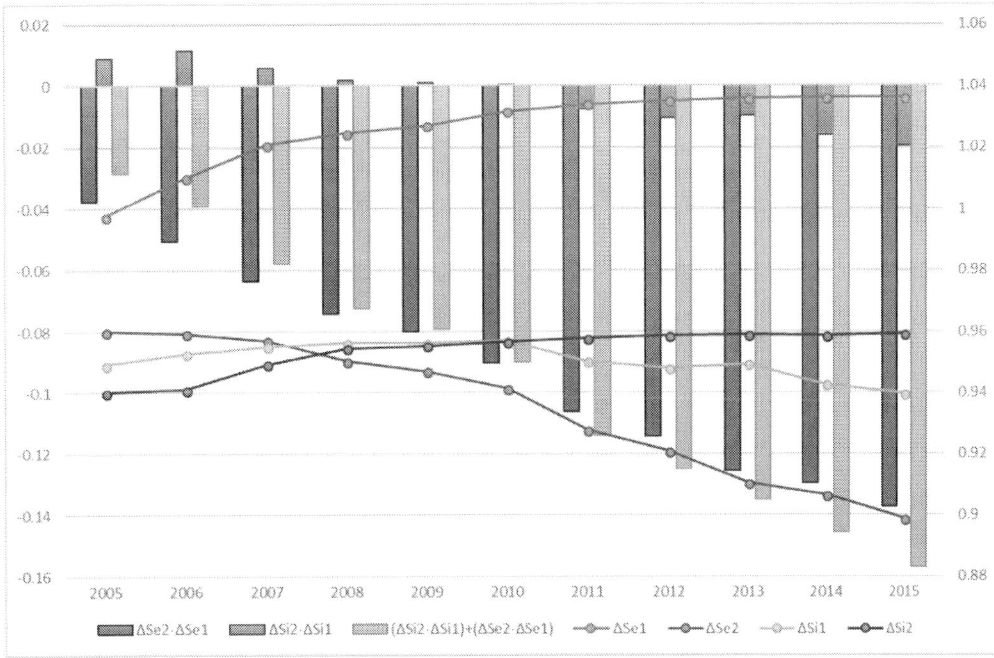

Figure 1. Entropy movement of UE in Jiangsu province

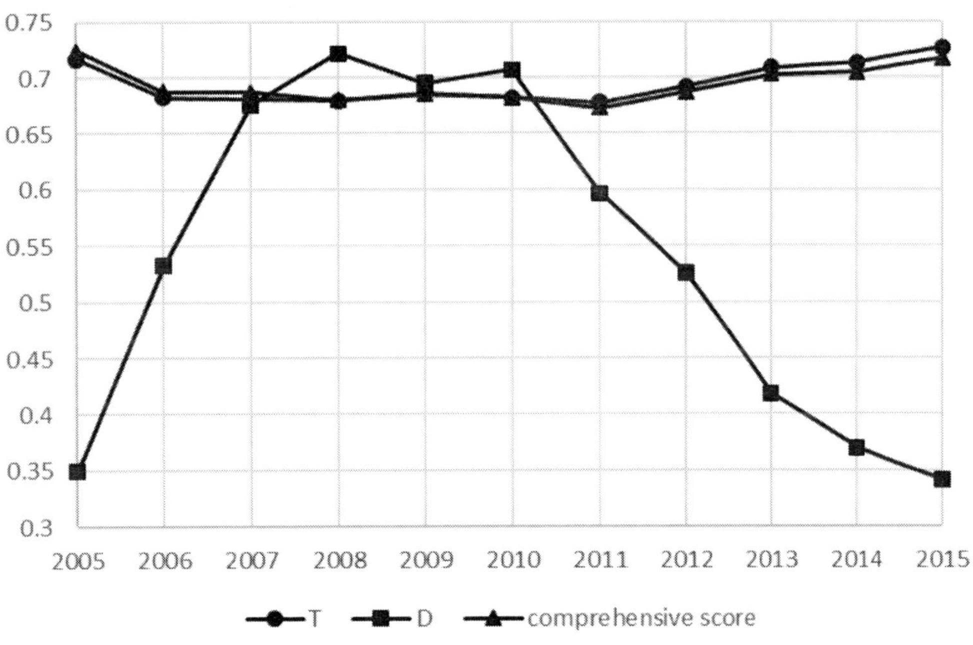

Figure 2. Index movement of UE in Jiangsu province

5. Conclusion

The results of the study preliminarily reveal the evolution of urban ecological system and the trend of sustainable development in the past period of Jiangsu province, which can provide some scientific reference for the sustainable development. However, constrained by data acquisition and modeling theory, the algorithm and conclusion have some room for improvement. The study of dynamic evolution of UE with higher "spatial resolution" is an important direction for further research.

Acknowledgements
This work was supported by the National Key Research Program of China (No. 2017YFC0506603, 2016YFC0401305), the Project of National Natural Foundation of China (No. 51379013, 51679007), and the State Key Program of National Natural Science of China (No. 41530635).

References
[1] Shi H C, Liu W, He J, Liu L, Shi X J, Wan II Y 2012 An urban ecosystem assessment method and its application *Acta Ecologica Sinica* 32(17):5542-5549.
[2] Yang J, Li Z, Song Z J, Wang J 2016 Study on evaluation of ecological system state of inland key cities of Yangtze River Economic Belt: case of cities in midstream of Yangtze River *Yangtze River* 47 (2):11-14.
[3] Wang P, Deng X Z, Zhou H M, Qi W 2018 Responses of urban ecosystem health to precipitation extreme: A case study in Beijing and Tianjin *Journal of Cleaner Production* 177:124-133.
[4] Wu Y J, Liao Le, Yuan X Y 2013 Evolution of Wuhan urban complex ecosystem based on information entropy *Resources and Environment in the Yangtze Basin* 22(1):21-26.
[5] Wang L, Xu G, Liu M 2016 Analysis and forecasting of shanghai urban ecosystem evolution based on information entropy and GM (1,1) *Acta Scientiae Circumstantiae* 36(6): 2262-2271.
[6] Wang L J, Dong X F, Liu X G, Yin H 2016 Evaluation of the sustainable development ability of the ecosysem in Lanzhou based on the information entropy *Journal of Lanzhou University: Natural Sciences* 52(6): 764-770.
[7] Lin Z M, Xia B 2013 Sustainability analysis of the urban ecosystem in Guangzhou City based on information entropy between 2004 and 2010 *Journal of Geographical Sciences* 23(3): 417-435.
[8] Liao C B 1999 Quantitaitve judgement and classification system for coordinated development of environment amd economy-A case study of the city group in the pearl river delta. *Tropical Geography* 19(2):171-177.

IOP Conf. Series: Journal of Physics: Conf. Series **1060** (2018) 012087 doi:10.1088/1742-6596/1060/1/012087

Time Reversal MIMO Radar for Multi-targets DOA Estimation

Mengbo Liu, Guoping Hu, Bingbing Fan, Jinghao Gu and Zhi Li

Air and Missile Defense College, Air Force Engineering University, Xi'an 710051, China

Email:619736846@qq.com

Abstract. For direction finding problem of multi-target in the multi-input multi-output (MIMO) radar, a multi-target direction of arrival (DOA) estimation method with time reversal (TR) MIMO radar is proposed. Firstly, according to the echo signal matrix of MIMO radar, we use TR theory to establish the echo signal model of TR MIMO radar. Then, the echo signal of TR MIMO is performed with matched filter and vectorized processing on the basis of the waveform diversity of MIMO radar, which extends the virtual aperture of MIMO radar. Finally, the Multiple Signal Classification (MUSIC) algorithm is used to estimate the DOAs, improving the estimation performance greatly. Compared with the conventional MIMO, TR MIMO can effectively improve the echo signal gain and increase the virtual aperture. The simulation results show that TR MIMO can accurately estimate the angles even when the (signal noise ratio) SNR is -20dB, which demonstrates the effectiveness of proposed configuration.

1. Introduction

Since the advent of the MIMO radar, it has been widely concerned and made great progress due to its unique performance advantages in targets detection, direction finding accuracy, anti-interference ability, anti stealth [1]-[2] and so on. It is useful to improve the estimation accuracy and angle resolution of radar system that MIMO radar transmits mutually orthogonal waveforms from transmitters and the signal is then performed with matched filter and vectorized processing in receivers, obtaining the extended virtual matrix. Several kinds of DOA algorithms in MIMO radar are introduced in [3].

The origin of time reversal (TR) is the study of phase conjugation method for optics, also called time inversion. In complex and multi scattering environment, TR technology [4]-[5] based on time symmetry and spatial reciprocity of wave equation in static medium, uses multipath components to coherently stack and performs incoherent superposition of other clutter signals, so as to achieve great space-time aggregation performance. At present, TR technology is widely used in the field of MIMO communication.

In recent years, more and more foreign scholars have combined TR technology with radar detection. The TR technology has been applied to radar in [6]-[8], showing unique performance advantages. TR MIMO radar is firstly extended to target location in [7], deducing the expressions of the deterministic Cramer-Rao bounds (CRB), having higher estimation accuracy in strong clutter environment proved by computer simulation. Compared with the conventional MIMO [9] radar, TR MIMO radars adapt the transmission waveforms to the multipath environment and thus provide a relatively straightforward procedure for adaptive waveform coding [10] and also have higher signal-to-noise ratio (SNR) and potential gain for the DOA and velocity estimation, resulting in the improved CBR and DOA estimation performance [11]. However, the internal research on TR technology used for DOA estimation in MIMO radar is relatively small. In [4], TR technology is applied to Ultra Wideband (UWB) MIMO radar, performing DOA and DOD estimation. According to the time domain expression, the paper uses MUSIC algorithm to estimate the DOAs, improving the estimation accuracy under the condition of low SNR. However, the paper does not give a complete signal model of TR MIMO radar based multi-target DOA estimation, which makes the establishment of TR MIMO radar target positioning system unsystematic.

In this letter, according to the echo signal model of MIMO radar and DOA estimation algorithm, combined with the characteristic of TR, a multi-target DOA estimation method with time reversal MIMO radar is proposed. The method can extend the virtual aperture of MIMO system by the echo signal of TR MIMO radar performed with matched filter and vectorized processing. The simulation results verify the authenticity and validity of the method.

2. MIMO Radar Array Signal Model

To establish the echo signal model of TR MIMO radar, we take monostatic MIMO radar system as an example in this paper. The system contains colocated transmit and receive arrays both equipped with uniform linear array, having M antennaes and N antennaes separated by d (To ensure the uniqueness of the angle, $d=\lambda/2$), respectively. The aperture and heights of uniform linear arrays are assumed far less than the target distance. The DOA for each target concerning the colocated receiver and transmitter is denoted by θ_k ($k=1,2,\cdots,K$). Assume there are K far-field targets, which belong to incoherent targets, probed by narrowband orthogonal waveforms. The received signal can be expressed as

$$X(t)=\sum_{k=1}^{K}a_r(\theta_k)\beta_k e^{j2\pi f_k t}a_t^{\mathrm{T}}(\theta_k)s(t)+W(t) \tag{1}$$

Where β_k is the radar cross section fading coefficient, f_k is Doppler frequency of the kth target. $\mathbf{a}_t(\theta_k)=[1,\cdots,e^{-j\pi(M-1)\sin\theta_k}]^T$, $\mathbf{a}_r(\theta_k)=[1,\cdots,e^{-j\pi(N-1)\sin\theta_k}]^T$, $(\cdot)^{\mathrm{T}}$ denotes transpose, and $\mathbf{s}(t)=[s_1(l)\,s_2(l)\cdots s_K(l)]^T, l=0,\cdots,L$, L represent s the number of snapshots. $\mathbf{W}(t)$ is an additive Gaussian white noise vector whose mean and covariance matrix is zero and σ^2, respectively. After matched filtering, we can have

$$D=E\left[X(t)s^{\mathrm{H}}(t)\right]=B\Lambda A^{\mathrm{T}}+W \tag{2}$$

Where $B=[a_r(\theta_1),\cdots,a_r(\theta_K)]$, $A=[a_t(\theta_1),\cdots,a_t(\theta_K)]$, $\Lambda=diag(\beta_1 e^{j2\pi f_1 t},\cdots\beta_K e^{j2\pi f_K t})$, $(\cdot)^{\mathrm{H}}$ represents conjugate transpose. W denotes the noise matrix after matched filtering. Thus, after the vectorization operator of the matrix D, a composite data vector can be expressed as

$$\overline{D}=vec(D)=A\circ B\psi+vec(W) \tag{3}$$

$$A\circ B=\left[a_t(\theta_1)\otimes a_r(\theta_1),\cdots,a_t(\theta_K)\otimes a_r(\theta_K)\right] \tag{4}$$

Where $vec(\cdot)$ is the vectorization operator of matrix, $\psi=[\beta_1 e^{j2\pi f_1 t},\cdots,\beta_k e^{j2\pi f_K t}]^{\mathrm{T}}$, " $\circ$ " , " $\otimes$ " is Khatri-Rao product and Kronecker product, respectively.

3. TR MIMO Radar Array Signal Model

In the TR MIMO radar, the output vector $X(t)$ is energy normalized by factor ε, phase conjugated, time reversed, and retransmitted into the medium. Then the TR transmitted signal is $\varepsilon X^*(-t)$, and the TR output can be given as

$$\begin{aligned}\boldsymbol{Y}(t) &= \varepsilon\sum_{k=1}^{K}\boldsymbol{a}_t(\theta_k)\mathrm{e}^{\mathrm{j}4\pi f_k t}\beta_k\boldsymbol{a}_r^{\mathrm{T}}(\theta_k)\boldsymbol{X}^*(-t)+\boldsymbol{V}(t)\\ &=\varepsilon\sum_{k=1}^{K}|\beta_k|^2\mathrm{e}^{\mathrm{j}4\pi f_k t}\boldsymbol{a}_t(\theta_k)\boldsymbol{a}_r^{\mathrm{T}}(\theta_k)\boldsymbol{a}_r^*(\theta_k)\boldsymbol{a}_t^{\mathrm{H}}(\theta_k)\boldsymbol{s}^*(-t)+\tilde{\boldsymbol{V}}(t)\end{aligned} \tag{5}$$

Where $\varepsilon=\sqrt{\int\|s(t)\|^2dt\Big/\int\|\boldsymbol{X}(t)\|^2dt}$, $(\cdot)^*$ represents conjugation; $\boldsymbol{V}(t)$ and $\tilde{\boldsymbol{V}}(t)$ are the observation noise for the TR stage and the accumulated noise from both the conventional and TR stages, respectively. Similar to the noises $\boldsymbol{W}(t)$, the noises $\boldsymbol{V}(t)$ are also spatially and temporally complex Gaussian white processes with zero-mean and variance σ^2. As described in [12] , the accumulated noises can be proved as approximately white noise. Thus, after matched filtering, we can obtain

$$\begin{aligned}\boldsymbol{Z} &= E\left[\boldsymbol{Y}(t)\boldsymbol{s}^{\mathrm{T}}(-t)\right]=\sum_{k=1}^{K}|\beta_k|^2\mathrm{e}^{\mathrm{j}4\pi f_k t}\boldsymbol{a}_t(\theta_k)\boldsymbol{a}_r^{\mathrm{T}}(\theta_k)\boldsymbol{a}_r^*(\theta_k)\boldsymbol{a}_t^{\mathrm{H}}(\theta_k)E\left[\mathbf{s}^*(-t)\boldsymbol{s}^{\mathrm{T}}(-t)\right]+E\left[\tilde{\boldsymbol{V}}(t)\boldsymbol{s}^{\mathrm{T}}(-t)\right]\\ &= N\sum_{k=1}^{K}|\beta_k|^2\mathrm{e}^{\mathrm{j}4\pi f_k t}\boldsymbol{a}_t(\theta_k)\boldsymbol{a}_t^{\mathrm{H}}(\theta_k)+\tilde{\boldsymbol{V}}\\ &= N\cdot\boldsymbol{A}\boldsymbol{\Delta}\boldsymbol{A}^{\mathrm{H}}+\tilde{\boldsymbol{V}}\end{aligned} \tag{6}$$

Where N are the number of MIMO radar receiving array elements, $\boldsymbol{A}=[\boldsymbol{a}_t(\theta_1),\cdots,\boldsymbol{a}_t(\theta_K)]$, $\boldsymbol{\Delta}=diag(|\beta_1|^2\mathrm{e}^{\mathrm{j}4\pi f_1 t},\cdots,|\beta_k|^2\mathrm{e}^{\mathrm{j}4\pi f_K t})$, $\tilde{\boldsymbol{V}}$ is an additive Gaussian white noise vector whose mean and covariance matrix is zero and σ^2. Similar to the vectorization operator of the matrix $\boldsymbol{D}$ for MIMO radar, the vector can be expressed as

$$\tilde{\boldsymbol{Z}}=vec(\boldsymbol{Z})=N\cdot vec(\boldsymbol{A}\boldsymbol{\Delta}\boldsymbol{A}^{\mathrm{H}})+vec(\tilde{\boldsymbol{V}})=N\cdot\boldsymbol{A}^*\circ\boldsymbol{A}\boldsymbol{\eta}+\boldsymbol{v} \tag{7}$$

$$\boldsymbol{A}^*\circ\boldsymbol{A}=\left[\mathbf{a}_t^*(\theta_1)\otimes\mathbf{a}_t(\theta_1),\cdots,\mathbf{a}_t^*(\theta_K)\otimes\mathbf{a}_t(\theta_K)\right] \tag{8}$$

Where $\boldsymbol{\eta}=[|\beta_1|^2\mathrm{e}^{\mathrm{j}4\pi f_1 t},\cdots,|\beta_K|^2\mathrm{e}^{\mathrm{j}4\pi f_K t}]^{\mathrm{T}}$, $\boldsymbol{v}=vec(\tilde{\boldsymbol{V}})$.

4. DOA Estimation Method based TR MIMO Radar

4.1 DOA Estimation Method

Firstly, calculate the covariance matrix of $\tilde{\boldsymbol{Z}}$

$$\boldsymbol{R}_{\tilde{Z}\tilde{Z}}=E\left[\tilde{\boldsymbol{Z}}\tilde{\boldsymbol{Z}}^{\mathrm{H}}\right] \tag{9}$$

In fact, the calculation formula of $\boldsymbol{R}_{\tilde{Z}\tilde{Z}}$ is

$$\hat{\boldsymbol{R}}_{\tilde{Z}\tilde{Z}}=\frac{1}{L}\sum_{l=1}^{L}\tilde{\boldsymbol{Z}}(l)\tilde{\boldsymbol{Z}}(l)^{\mathrm{H}} \tag{10}$$

Where L are the sampling times for the virtual matrix, also called the number of snapshots.

Then, Perform eigenvalue decomposition of (10), and we can get

$$\hat{\boldsymbol{R}}_{\tilde{Z}\tilde{Z}}=\boldsymbol{U}_s\boldsymbol{\Sigma}_s\boldsymbol{U}_s^{\mathrm{H}}+\boldsymbol{U}_n\boldsymbol{\Sigma}_n\boldsymbol{U}_n^{\mathrm{H}} \tag{11}$$

Where $\boldsymbol{\Sigma}_s$ is a diagonal matrix consisting of the largest K eigenvalues; $\boldsymbol{\Sigma}_n$ stands for a diagonal matrix consisting of the smallest $MM-K$ eigenvalues; $\boldsymbol{U}_s$ and $\boldsymbol{U}_n$ are the signal subspace and the noise subspace, respectively. It is well known that the signal subspace is orthogonal to the noise subspace.

Finally, Assume $\boldsymbol{a}(\theta)=\mathbf{a}_t^*(\theta)\otimes\mathbf{a}_t(\theta)$, using MUSIC algorithm, we can obtain

$$\boldsymbol{P}_{\mathrm{MUSIC}}(\theta)=\frac{1}{\boldsymbol{a}^{\mathrm{H}}(\theta)\boldsymbol{U}_n\boldsymbol{U}_n^{\mathrm{H}}\boldsymbol{a}(\theta)} \tag{12}$$

According to (12), seeking the spectrum peak, we can get the DOAs.

4.2 Performance Analysis of the Proposed Method

(1) Estimation accuracy and resolution

Comparing (3) and (7), we can see that after the vectorization operator , TR MIMO radar can produce N times gain. Thus, In the condition that the number of MIMO radar receiver and transmitter elements is properly set up ($M>N$), the DOA estimation accuracy can be improved in TR MIMO radar system when the SNR is low. Where $\boldsymbol{A}\circ\boldsymbol{B}=[\boldsymbol{a}_t(\theta_1)\otimes\boldsymbol{a}_r(\theta_1),\cdots,\boldsymbol{a}_t(\theta_K)\otimes\boldsymbol{a}_r(\theta_K)]$ is the direction vector matrix of MIMO radar, $\boldsymbol{A}^*\circ\boldsymbol{A}=[\mathbf{a}_t^*(\theta_1)\otimes\mathbf{a}_t(\theta_1),\cdots,\mathbf{a}_t^*(\theta_K)\otimes\mathbf{a}_t(\theta_K)]$ is the direction vector matrix of TR MIMO radar. The dimension of $\overline{\boldsymbol{D}}$ is $MN\times 1$, The dimension of $\tilde{\boldsymbol{Z}}$ is $M^2\times 1$, showing that the free degree of MIMO radar is M+N-1, the free degree of TR MIMO radar is 2M-1. So in the condition of $M>N$, TR MIMO radar extends the virtual aperture of the conventional MIMO, improving the angular resolution and estimation accuracy greatly.

(2) Computational complexity

The complexity of the proposed method mainly contains calculating the covariance matrix, eigenvalue decomposition and seeking the spectrum peak, so we can analyse from these three aspects. Where $\tilde{\boldsymbol{Z}}$ is $M^2\times 1$ dimensional matrix, $\overline{\boldsymbol{D}}$ is $MN\times 1$ dimensional matrix. Thus, the computational complexity of TR MIMO radar is about $O\{(M^4L+M^6+n[M^2(M^2-K)+M^2-K]\}$, however, the computational complexity of the conventional MIMO is about $O\{(M^2N^2L+M^3N^3+n[MN(MN-K)+MN-K]\}$, n is the total search times. Assume $M>N$, the algorithm in this paper has a higher computational complexity.

5. Simulation Results

In the simulation experiment, we construct a MIMO radar system consisting of colocated receive and transmit arrays both equipped with uniform linear array (ULA), having 6 antennaes and 8 antennaes separated by half a wavelength, respectively.And the array elements transmit mutually orthogonal signals, the number of snapshots is L=200. Three incoherent narrowband signal sources coming from $\theta_1=-10^\circ,\theta_2=20^\circ,\theta_3=40^\circ$ are considered. We test the estimation performance by 300 Monte Carlo trials.

Example 1: Compare the estimation of TR MIMO and MIMO radar in the condition of low SNR. We compare the spatial spectrum of TR MIMO and MIMO when the SNR is 10dB and -20dB in Figure.1 and Figure.2, respectively. From Figure.1, in case of SNR=10dB, both TR MIMO and MIMO radar can estimate the DOAs, but the spectrum peaks of TR MIMO are higher. From Figure.2, in case of SNR=-20dB, the MUSIC algorithm of MIMO radar is invalidated and can not estimate the DOAs. However, TR MIMO radar is still able to detect the angle of targets, showing the better performance advantages.

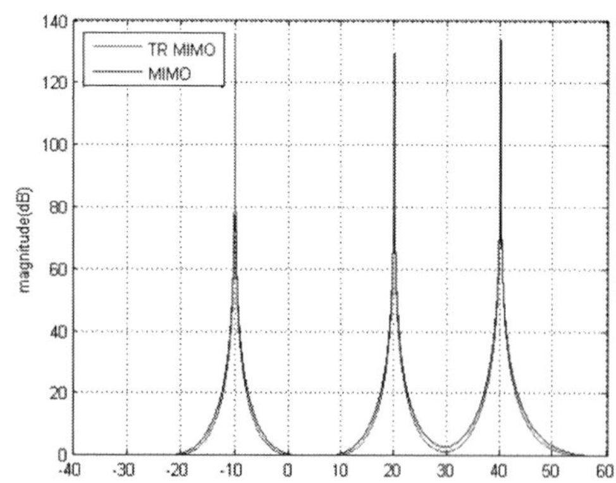

Figure.1 The spatial spectrum of TR MIMO and MIMO

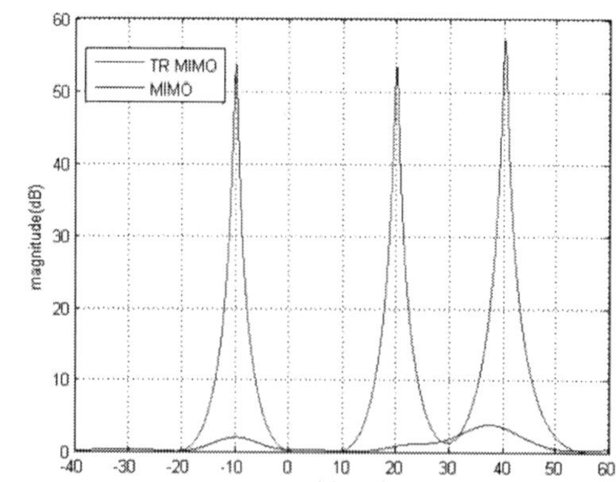

Figure.2 The spatial spectrum of TR MIMO and MIMO

Example 2: In this example, we test the estimation performance by the RMSE versus SNR. As we can see from Figure.3, the RMSE of TR MIMO radar is smaller than MIMO radar, and TR MIMO still has a low error when the SNR is -20dB.

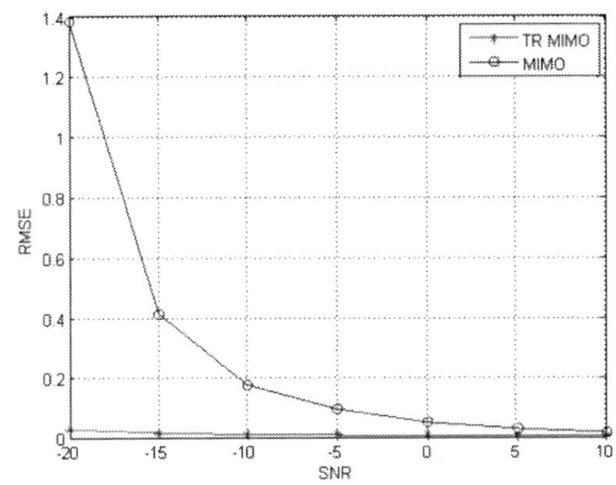

Figure.3 The RMSE of TR MIMO and MIMO versus the SNR

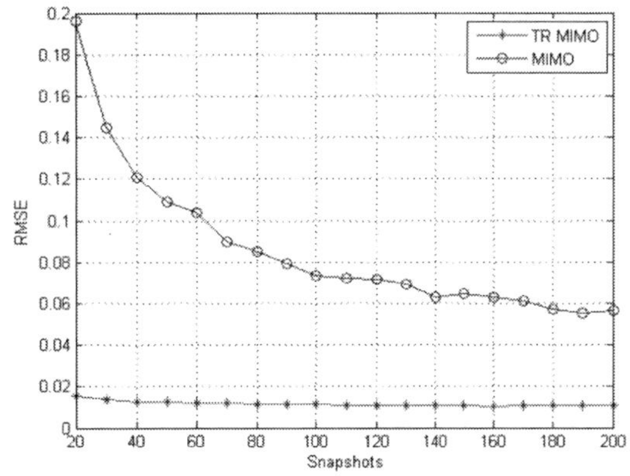

Figure.4 The RMSE of TR MIMO and MIMO versus the number of snapshots

Example 3: In this example, we examine the estimation performance by the RMSE versus the number of snapshots. In Figure.4, we assume the SNR is 0dB. In the case of the different number of snapshots, the performance of TR MIMO is better than the conventional MIMO radar. Especially, when the number of snapshots is smaller than 50, the estimation performance of TR MIMO is not affected, showing that the proposed method is still superior performance under relatively low snapshots.

Example 4: In this example, we examine the estimation performance by the RMSE versus the number of arrays. Figure.5 shows the DOA estimation performance with different number of transmitter and receiver antennas. The more the number of transmitter or receiver antennas has, the better the performance of estimation will be gotten for TR MIMO radar.

Example 5: In this example, we examine the angle resolution by the spatial spectrum of TR MIMO and MIMO radar. Three close non-coherent narrowband signal sources coming from $\theta_1=22^\circ,\theta_2=25^\circ,\theta_3=28^\circ$ are considered. In the case of SNR=10dB, compared with the conventional MIMO radar, TR MIMO can still distinguish three targets in Figure.6, showing the better angle resolution.

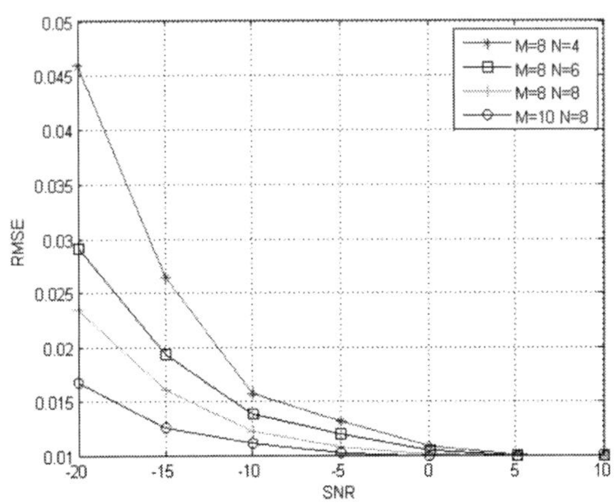

Figure.5 DOA estimation performance with different number of sensors

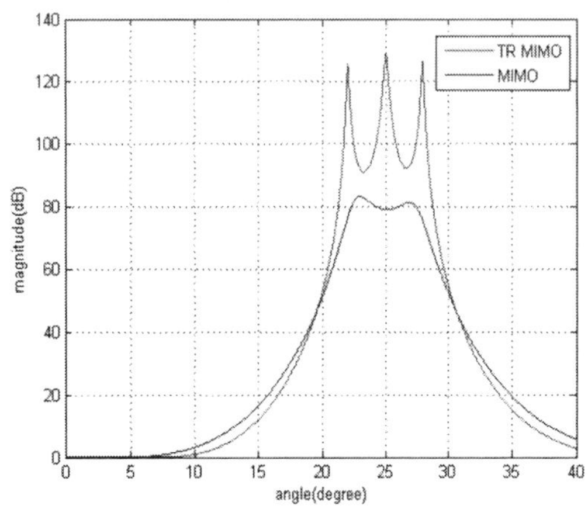

Figure.6 The spatial spectrum of TR MIMO

6. Conclusion

A new TR MIMO radar for multi-target DOA estimation is presented, which is optimized from MIMO radar and TR technology. As evident from the Monte Carlo simulations, in the condition of $M > N$, TR MIMO radar extends the virtual aperture of the conventional MIMO, improving the angle resolution and estimation accuracy greatly, increasing computational complexity.

References

[1] Aittomaki T and Koivunen V. Perfermance of MIMO radar with angular diversity under swerling scattering models [J]. IEEE Journal on Selected Topics in Signal Processing, 2010, 4(1): 101-114.
[2] XIE Rong. Research on MIMO radar angle estimation algorithm [D]. Xi'an:Xi'an Electronic and Science University, 2011. (in Chinese)
[3] WANG Lei,SUI Kai-kai,LI Hong. DOA estimation of multi-target in the MIMO radar [J]. Electronic Design Engineering, 2012, 20(3): 44-47. (in Chinese)
[4] JIANG Yan-ying,OU Yang-shan,JIN Liang-nian,et al. Time Reversal Detection in UWB MIMO Radar[J]. Journal of Guilin University of Electronic Technology, 2013, 33(3): 173-176. (in Chinese)
[5] LIU Ze-long,YANG Wei,MIAO Nan-nan,et al. Application of Time Reversal Technology in Target Detection of Linear Frequency Modulation Continuous Wave Radar[J]. Journal of Terahertz Science and Electronic Information Technology, 2015, 13(3): 409-414. (in Chinese)
[6] Y. Jin, J. Moura, and N. O'Donoughue. Time Reversal Adaptive Waveform in MIMO Radar[C]. In Proceedings of 2010 International Conference on Electromagnetics in Advanced Applications, 2010: 741-744.
[7] F. Foroozan, A. Asif, Y. Jin, and J. Moura. Direction Finding Algorithms for Time Reversal MIMO Radars[C]. In Proceedings of IEEE Statistical Signal Processing Workshop, 2011: 433-436.
[8] F. Foroozan, A. Asif, and Y. Jin. Cramér-Rao Bounds for Time Reversal MIMO Radars with Multipath[J]. IEEE Trans. Aerospace and Electronic System, 2016, 52(1): 137-154.
[9] YANG Ming-lei,GUO Wei-na,YANG Qian,et al. Estimation Method of Low Complexity DOA for MIMO Radar[J]. Journal of Air Force Early Warning Academy, 2013, 27(3): 163-168. (in Chinese)
[10] H. T. Nguyen, J. Andersen, and G. Pedersen. The Potential Use of Time Reversal Techniques in Multiple Element Antenna Systems[J]. IEEE Communications Letters, 2005, 9(1): 40–42.
[11] H. Mohammad. S. Sajjadieh, and A Asif. Compressive Sensing Time Reversal MIMO Radar: Joint Direction and Doppler Frequency Estimation[J]. IEEE Signal Processing letters, 2015, 22(9): 1283-1287.
[12] Y. Jin, J. Moura, N. O'Donoughue, and J. Harley, "Single antenna time reversal detection of moving target," in IEEE ICASSP, March 2010, pp. 3558-3561.

IOP Conf. Series: Journal of Physics: Conf. Series **1060** (2018) 012088 doi:10.1088/1742-6596/1060/1/012088

Research on State Estimation of Hypersonic Glide Vehicle

Z C Huang, Y S Zhang and Yao Liu

Space Engineering University, Beijing, 101416, China

zichen_huang_sc@163.com

Abstract:In order to estimate the motion state of the hypersonic glide vehicle, a motion model based on the aerodynamic parameters of the prototype CAV-H and an observation model of GEO infrared early-warning satellite were established. The exponential weighted recursive least square method with forgetting factor was used to measure the early-warning satellite observation data with measuring error in angle to obtain a fitting parameter describing the polynomial of the target state, The method can focus on the local motion of the target, dilute the influence of the early observation data on the fitting parameters, reduce the estimation error caused by the target manoeuvring. The simulation result shows that the estimation of both position and velocity of the hypersonic glide vehicle is highly accurate.

1 Introduction

As a new concept weapon, Hypersonic Glide Vehicle(HGV)brings new challenges to existing air defense anti-missile systems with the character of long range, low ballistic, fast speed, strong mobility [1].When studying its state estimation problem, Most scholars choose to use Interacting Multiple Model(IMM)filtering algorithm [2][3][4][5], but IMM needs parallel computing which is not conducive to rapid response in wartime. And the accuracy depends on the accuracy of the motion model. This paper describes the spatial motion of HGV with high order polynomials, fit discrete observations with errors from GEO infrared warning satellite with Exponential Weighted Recursive Least Squares (EW-RLS), Using interpolation to obtain the position and velocity of the observation time and its neighborhood, Provides foundation for rapid ballistic and tracking observations of HGV targets.

2 Simulation of typical trajectories

After HGV re-enters the atmosphere, it can use aerodynamic force to continuously transform kinetic energy and potential energy, Jump in vertical plane to get out of the warning system to track. Horizontal manoeuvres can also be used to circumvent the impact of defensive weapons [6], In this section, a three-dimensional motion model of HGV reentry segment with control conditions is established based on the published aerodynamic parameters of the US Army's CAV-H prototype, Typical reentry trajectories are simulated and analyzed, which will lay the foundation for subsequent research on reentry interception methods.

2.1 Reentry model of HGV

HGV has a flat lift body structure, using Bank-to-Turn (BTT) turning method [7], If we ignore the effect of Earth's rotation, think of the Earth as a uniform sphere, The reentry kinetic equation of HGV can be expressed as:

$$m\ddot{r} = \boldsymbol{R} + m\boldsymbol{g} \tag{1}$$

Where, m is quality, $\boldsymbol{r}$ is position vector, $\boldsymbol{R}$ is aerodynamic vector, $\boldsymbol{g}$ is gravitational acceleration vector.

Projects items in the above equation into the semi-velocity coordinate system [8], we have,

$$\begin{aligned}
&\dot{v} = \frac{-\rho v^2 C_D S_r}{2\cdot m} - g\sin\theta && \dot{r} = v\sin\theta \\
&\dot{\theta} = \frac{\rho v^2 C_L S_r \cos\gamma}{2\cdot mv} + (\frac{v}{r} - \frac{g}{v})\cos\theta && \dot{\lambda} = \frac{v\sin\sigma\cos\theta}{r\cos\psi} \\
&\dot{\sigma} = -\frac{\rho v^2 C_L S_r \sin\gamma}{2\cdot mv\cos\theta} + \frac{v\tan\varphi\cos\theta\sin\sigma}{r} && \dot{\psi} = \frac{v\cos\sigma\cos\theta}{r}
\end{aligned} \tag{2}$$

Where, v is velocity, θ is ballistic inclination, σ is Ballistic declination, r is Geocentric distance, λ is longitude, ψ is latitude, ρ is atmospheric density, S_r is Reference pneumatic area, C_D is resistance coefficient, C_L is lift coefficient, γ is tilt angle.

According to the 1976 American Standard Atmosphere Model USSA76 [9],

$$\rho = \rho_0 \exp(-\beta h) \tag{3}$$

Where, ρ_0=1.225kg/m^3, β=1.389×10^{-4}.

CAV-H's drag coefficient and lift coefficient can be expressed as a function of attack angle and speed. [10]:

$$\begin{cases} C_D = 0.000724\cdot\alpha^2 + 0.406\exp(-0.00095\cdot v) + 0.024 \\ C_L = 0.0513\cdot\alpha + 0.2945\exp(-0.000302\cdot v) - 0.2317 \end{cases} \tag{4}$$

Where, α is attack angle.

Simultaneous Formula (2) and Formula (4), Given initial position, initial speed, total mass, and control variables $\alpha(t)$ and $\gamma(t)$, The position (spherical coordinates) in the ECEF coordinate system and the speed (spherical coordinates) in the local coordinate system can be obtained.

Convert spherical coordinates to rectangular coordinates,

$$\boldsymbol{r} = \begin{bmatrix} r\cos\psi\cos\lambda \\ r\cos\psi\sin\lambda \\ r\sin\psi \end{bmatrix} \qquad \boldsymbol{v}^d = \begin{bmatrix} v\cos\theta\cos\sigma \\ v\cos\theta\sin\sigma \\ v\sin\theta \end{bmatrix} \tag{5}$$

Further, the speed in the ECEF coordinate system,

$$\boldsymbol{v} = \boldsymbol{\Omega}_d^e \cdot \boldsymbol{v}^d = \begin{bmatrix} -\sin\psi\cos\lambda & -\sin\psi\sin\lambda & \cos\psi \\ \cos\psi\cos\lambda & \cos\psi\sin\lambda & \sin\psi \\ -\sin\lambda & \cos\lambda & 0 \end{bmatrix} \begin{bmatrix} v\cos\theta\cos\sigma \\ v\cos\theta\sin\sigma \\ v\sin\theta \end{bmatrix} \tag{6}$$

2.2 Simulation Example

Guidance control and trajectory optimization of HGV reentry section are not the focus of this paper, we assume constant control conditions.

- Initial conditions: as shown in Table 1.

Table 1. HGV initial reentry conditions

parameter	trajectory 1	trajectory 2	trajectory 3
longitude	0°E	0°E	0°E
latitude	0°N	0°N	0°N
height	80km	80km	80km

velocity	6000m/s	6000m/s	5000m/s
inclination	-3°	-10°	-3°
declination	30°	45°	60°

- Control conditions: $\alpha(t) \equiv 8^\circ$, $\gamma(t) \equiv 0^\circ$.
- Termination conditions: $v(t)$=1000m/s.
- Simulation results: as shown in Figure 1.

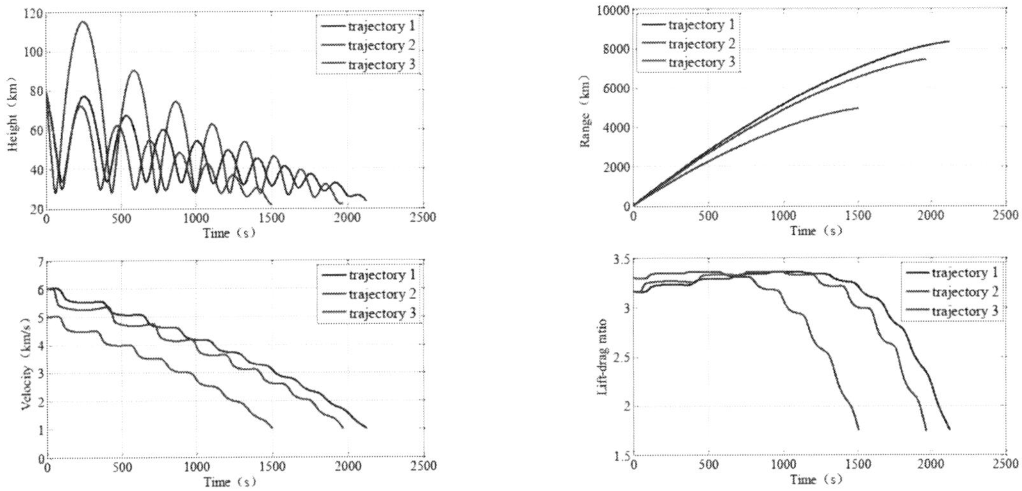

Figure 1 Trajectory Simulation curve

We can see from Figure 1 that affected by aerodynamic drag, the speed of the HGV is declining overall. Especially when the HGV moves up to the position which is the minimum value of the ballistic altitude, the speed will be obviously reduced.

In addition, the lift-drag ratio of HVG was maintained at more than 3 in the early stage of reentry, which represented good aerodynamic efficiency and was consistent with relevant documents. Due to the decrease of HGV's own speed in the later period, lift-drag ratio is getting lower.

Import ballistic data into STK software:

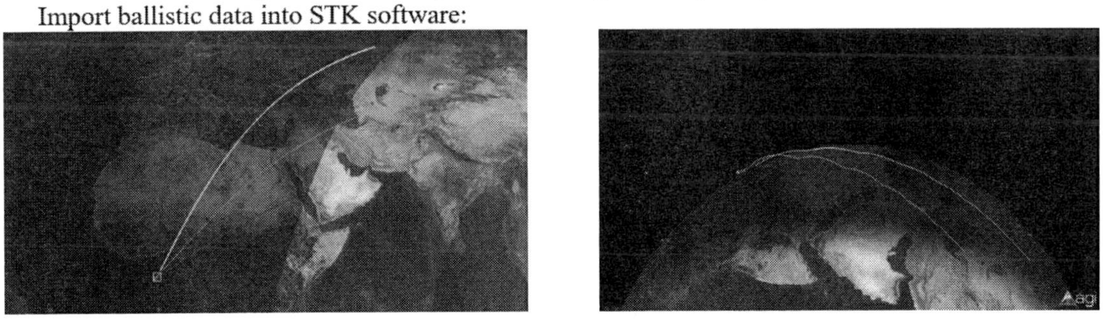

Figure 2 Bullet point and Space trajectory

In Figure 2, the yellow trajectory represents the trajectory 1, the green trajectory represents the trajectory 2, and the red trajectory represents the trajectory 3.

3 Binary observation error

In the ballistic missile defense process, early warning missions and tracking missions are generally completed by space-based infrared early warning satellites and ground-based active phased array radars. Ground-based active phased array radar is relatively high in terms of positioning and tracking accuracy. However, a layer of plasma [11-12] that absorbs and scatters radar waves will appear on the surface when HGV performs hypersonic gliding in the near space, As a result, ground-based radars are difficult to accurately locate HGV. On the other hand, when the ground-based radar detects a flying target of near space, the effective distance is only about 700km due to the curvature of the earth.

In response to this problem, many scholars have proposed the idea of using space-based early warning satellites to complete the tracking mission and studied Infrared detectability of HGV reentry. The conclusion seems to be more consistent, that is, The infrared characteristics of HGV reentry section are obvious and stable [13-14] and can be accurately recognized by space-based infrared early warning satellites [15-16]. References to U.S. Space Based Infra-Red System-high, (SBIRS-high), two GEO infrared warning satellite are used to observe HGV in this section.

3.1 Geometric positioning error

Figure 3 shows the definition of the line of sight angle in the satellite measurement coordinate system.

Observing azimuth q_a is the angle between the projection of line-of-sight vector in $x_c O_S y_c$ plane and x_c axis, $q_a \in [0°, 360°)$, when we see along z_c axis, q_a increases in clockwise. Observing pitch angle q_b is the angle between line-of-sight vector and $x_c O_S y_c$ plane, $q_b \in [-90°, 90°]$, q_b is negative when the line-of-sight vector is on the same side of the $x_c O_S y_c$ plane as z_c axis.

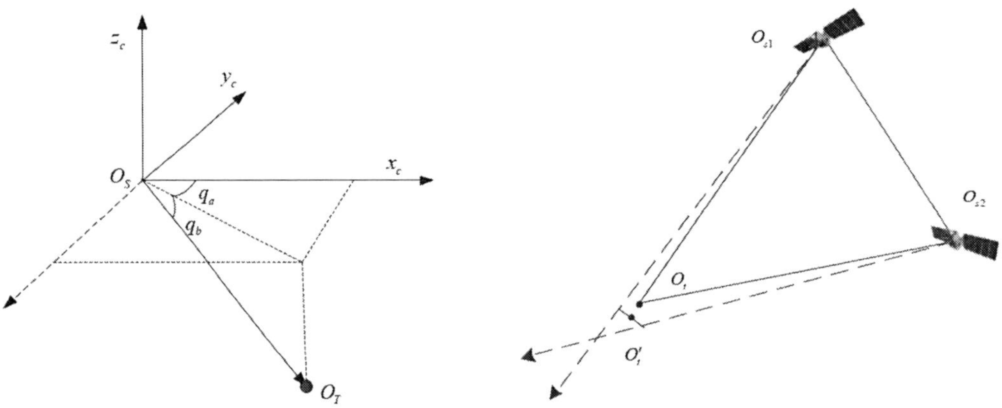

Figure 3 Double Star Positioning Principle

The unit vector of the line of sight can be expressed in the ECEF coordinate system:

$$\boldsymbol{L}^e = \boldsymbol{\Omega}_c^e \boldsymbol{L}^c = \begin{bmatrix} -\sin\lambda & 0 & -\cos\lambda \\ \cos\lambda & 0 & -\sin\lambda \\ 0 & -1 & 0 \end{bmatrix} \begin{bmatrix} \cos q_b \cos q_a \\ \cos q_b \sin q_a \\ \sin q_b \end{bmatrix} \tag{7}$$

Suppose the two line-of-sight unit vectors are $\boldsymbol{L}_1^e$ and $\boldsymbol{L}_2^e$, In the absence of error in the measured value, the intersection of the two lines of sight is the target centroid O_t ,However, there is measurement noise in actual observations, line-of-sights may not intersect in space, The midpoint O_t' of the vertical line of the two line-of-sights can replace the target position approximately, as shown in Figure 3.

Geostationary satellites are stationary with respect to the Earth, Assume the location of two infrared early warning satellites in the ECEF coordinate system are $\boldsymbol{O}_{s1} = [x_1, y_1, z_1]^T$ and $\boldsymbol{O}_{s2} = [x_2, y_2, z_2]^T$,two line-of-sight vectors are $\boldsymbol{L}_1^e = [a_1, b_1, c_1]^T$ and $\boldsymbol{L}_2^e = [a_2, b_2, c_2]^T$, than the O_t' point can be Calculated according to the following procedure [17]:

$$\begin{bmatrix} a_1 & -a_2 & b_1c_2 - c_1b_2 \\ b_1 & -b_2 & a_2c_1 - a_1c_2 \\ c_1 & -c_2 & a_1b_2 - b_1a_2 \end{bmatrix} \begin{bmatrix} \lambda_1 \\ \lambda_2 \\ \lambda_3 \end{bmatrix} = \begin{bmatrix} x_2 - x_1 \\ y_2 - y_1 \\ z_2 - z_1 \end{bmatrix} \tag{8}$$

$$\overline{\boldsymbol{r}} = \begin{bmatrix} \dfrac{x_1 + a_1\lambda_1 + x_2 + a_2\lambda_2}{2} & \dfrac{y_1 + b_1\lambda_1 + y_2 + b_2\lambda_2}{2} & \dfrac{z_1 + c_1\lambda_1 + z_2 + c_2\lambda_2}{2} \end{bmatrix}^T \tag{9}$$

3.2 Simulation Example

- HGV trajectory: trajectory 1 in section 2.2
- Longitude of GEO satellite: λ_{s1}=10° E, λ_{s2}=40° E
- Early warning satellite detection frequency:1Hz
- Early warning satellite measurement angle error: Gaussian distribution from a mean of 0° and a standard deviation of 0.001°

The simulation scenario set up in STK software is shown in Figure 4.

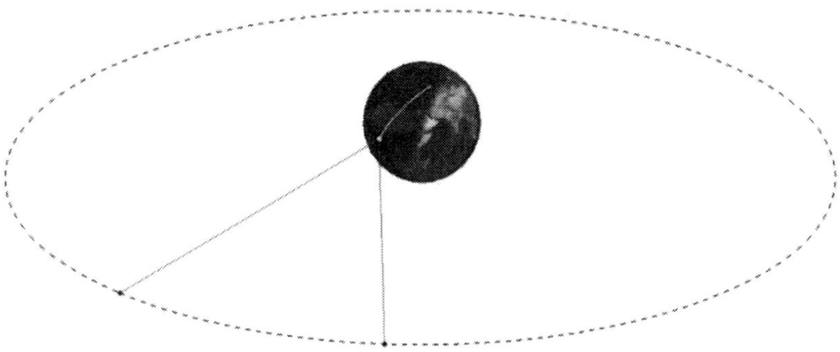

Figure 4. 3D view of STK scene

- GEO satellite measurement data

Using the ACESS Toolbox in STK Software to Generate the True Value of Observation Angle of Two Early Warning Satellites to HGV, Add Gaussian white noise with a mean of 0 and a standard deviation of 0.001° as the measured value, as shown in Figure 5:

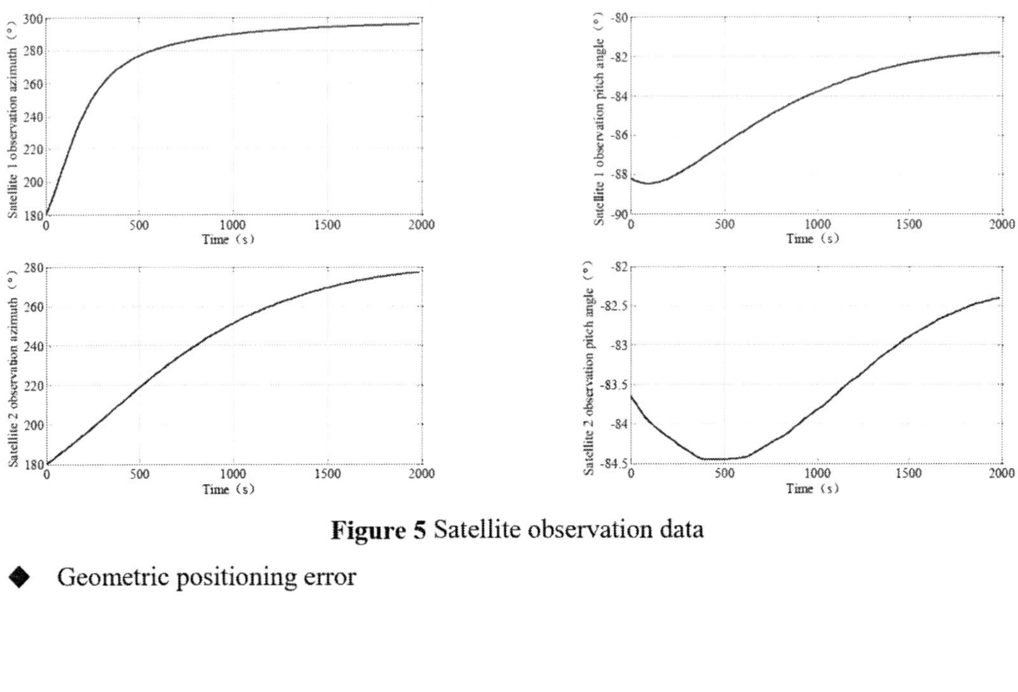

Figure 5 Satellite observation data

- Geometric positioning error

By the formula (9), observation position $\bar{\boldsymbol{R}}$ of HGV can be expressed in the ECEF coordinate system as:

$$\bar{\boldsymbol{R}} = \left[\bar{\boldsymbol{r}}(1) \quad \bar{\boldsymbol{r}}(2) \quad \bar{\boldsymbol{r}}(3) \quad \cdots \quad \bar{\boldsymbol{r}}(n)\right] \tag{10}$$

By the formula (5), real position $\boldsymbol{R}$ of HGV can be known, so the geometric positioning error can be expressed in the ECEF coordinate system as:

$$\bar{\boldsymbol{E}} = \bar{\boldsymbol{R}} - \boldsymbol{R} = \left[\bar{E}_x \quad \bar{E}_y \quad \bar{E}_z\right]^{\mathrm{T}} \tag{11}$$

Geometric positioning error in the ECEF coordinate system is shown in Figure 6.

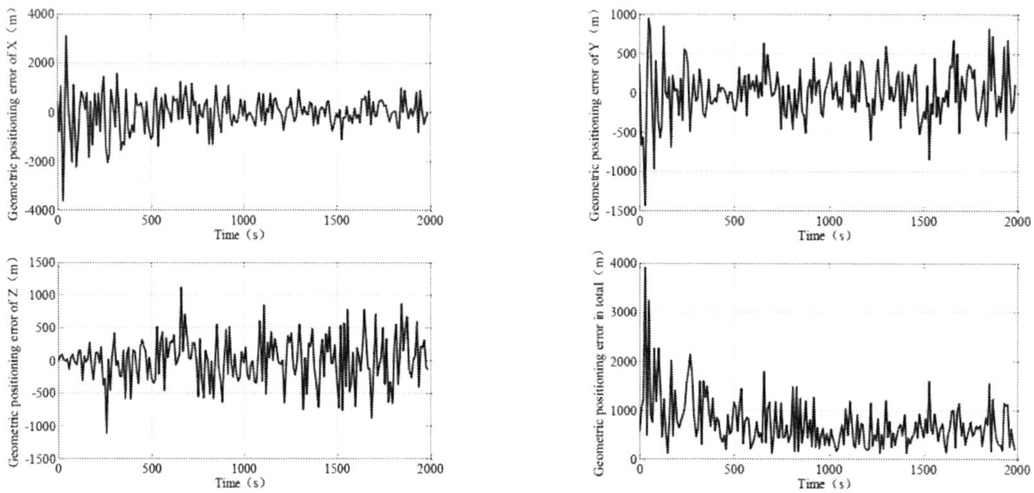

Figure 6. Geometric positioning error

4 State Estimation of HGV

Interpolation polynomial method is a common method for ballistic missile target state estimation. However, the reentry trajectory of HGV is highly nonlinear, if the conventional least squares method is used to process observation data, the resulting state estimation error will be very large. This paper use Exponential Weighted Recursive Least Square [18](EW-RLS)with forgetting factor to estimate the polynomial coefficient, EW-RLS is able to focus on the local fit of the target motion, inhibit the influence of HGV frequent maneuvers, obtain more accurate position and velocity estimates.

4.1 EW-RLS fitting method

Assuming the observation position with the error sequence

$$\boldsymbol{Z}_x = \left[z_x(1) \quad z_x(2) \quad \cdots \quad z_x(n)\right]^T \tag{12}$$

$\boldsymbol{w}_x$ is a certain estimate for the vector to be sought

$$\boldsymbol{w}_x = \left[\hat{a}_0 \quad \hat{a}_1 \quad \hat{a}_2 \quad \hat{a}_3 \quad \hat{a}_4\right]^T \tag{13}$$

Residual square sum due to fitting can be expressed as

$$S_x = \left(\boldsymbol{Z}_x - \boldsymbol{U}\boldsymbol{w}_x\right)^T \left(\boldsymbol{Z}_x - \boldsymbol{U}\boldsymbol{w}_x\right) \tag{14}$$

Where,

$$\boldsymbol{U} = \begin{bmatrix} 1 & (t_1 - t_0) & (t_1 - t_0)^2 & (t_1 - t_0)^3 & (t_1 - t_0)^4 \\ 1 & (t_2 - t_0) & (t_2 - t_0)^2 & (t_2 - t_0)^3 & (t_2 - t_0)^4 \\ \cdots & \cdots & \cdots & \cdots & \cdots \\ 1 & (t_n - t_0) & (t_n - t_0)^2 & (t_n - t_0)^3 & (t_n - t_0)^4 \end{bmatrix} = \begin{bmatrix} \boldsymbol{u}(1)^T \\ \boldsymbol{u}(2)^T \\ \cdots \\ \boldsymbol{u}(n)^T \end{bmatrix} \tag{15}$$

According to Least Squares Theory, the w_x that makes S_x minimal is the Optimal estimate, therefore, S_x is called cost function. Make improvements based on least squares, take the cost function as

$$\tilde{S}_x = (Z_x - Uw_x)^T \begin{bmatrix} \xi(n,1) & 0 & 0 & 0 \\ 0 & \xi(n,2) & 0 & 0 \\ 0 & 0 & \cdots & 0 \\ 0 & 0 & 0 & \xi(n,n) \end{bmatrix} (Z_x - Uw_x) \tag{16}$$

$$= (Z_x - Uw_x)^T B (Z_x - Uw_x)$$

Where, $\xi(n,i)$ is weighting factor, meets $0 < \xi(n,i) < 1$, Weighting factors enable parameter estimates to "forget" old observations and focus on new observations, let:

$$\xi(n,i) = \vartheta^{n-i}, \quad i = 1,2,...n \tag{17}$$

Where, ϑ is Forgetting factor, ϑ equals 1 in classical least squares method.

By the formula (16), optimal estimation based on new cost function $\tilde{S}_x$ is:

$$\hat{w}_x = (U^T BU)^{-1} U^T BZ_x \tag{18}$$

EW-RLS uses the new observations to update the old estimates. The calculation steps are as follows

(1) Given initial value, include $w_x(0)$ and Inverse correlation matrix $p(0)$.

(2) Recursive according to the following formula

$$\hat{w}_x(i) = \hat{w}_x(i-1) + k(i)\left[z_x(i) - u^T(i)\hat{w}(i-1)\right] \tag{19}$$

$$k(i) = \frac{p(i-1)u(i)}{\vartheta + u^T(i)p(i-1)u(i)} \tag{20}$$

$$p(i) = \vartheta^{-1}p(i-1) - \vartheta^{-1}k(i)u^T(i)p(i-1) \tag{21}$$

Where, k is Gain vector.

4.2 Simulation Example

Single estimate position of HGV can be expressed in the ECEF coordinate system as:

$$\hat{r}(t) = \left[\hat{r}_x(t) \quad \hat{r}_y(t) \quad \hat{r}_z(t)\right]^{\mathrm{T}} \qquad \hat{v}(t) = \left[\hat{v}_x(t) \quad \hat{v}_y(t) \quad \hat{v}_z(t)\right]^{\mathrm{T}} \tag{22}$$

Where,

$$\begin{cases} \hat{r}_x(t) = a_0 + a_1 t + a_2 t^2 + a_2 t^3 + a_4 t^4 \\ \hat{r}_y(t) = b_0 + b_1 t + b_2 t^2 + b_2 t^3 + b_4 t^4 \\ \hat{r}_z(t) = c_0 + c_1 t + c_2 t^2 + c_2 t^3 + c_4 t^4 \end{cases} \qquad \begin{cases} \hat{v}_x(t) = a_1 t + 2a_2 t + 3a_2 t^2 + 4a_4 t^3 \\ \hat{v}_y(t) = b_1 t + 2b_2 t + 3b_2 t^2 + 4b_4 t^3 \\ \hat{v}_z(t) = c_1 t + 2c_2 t + 3c_2 t^2 + 4c_4 t^3 \end{cases} \tag{23}$$

Where, a_i, b_i and c_i are coefficient to be estimated.

Let:

$$w_x(0) = (U^{\mathrm{T}}U)^{-1}U^{\mathrm{T}}Z \qquad p(0) = (\sum_{i=1}^{n} u(i)u^{\mathrm{T}}(i))^{-1} \tag{24}$$

By the formula (23), estimate position of HGV and estimate velocity of HGV can be expressed in the ECEF coordinate system as:

$$\hat{\boldsymbol{R}} = \left[\hat{\boldsymbol{r}}(1) \quad \hat{\boldsymbol{r}}(2) \quad \hat{\boldsymbol{r}}(3) \quad \cdots \quad \hat{\boldsymbol{r}}(n)\right] \qquad \hat{\boldsymbol{V}} = \left[\hat{\boldsymbol{v}}(1) \quad \hat{\boldsymbol{v}}(2) \quad \hat{\boldsymbol{v}}(3) \quad \cdots \quad \hat{\boldsymbol{v}}(n)\right] \tag{25}$$

By the formula (5) and (6), real position $\boldsymbol{R}$ and real velocity $\boldsymbol{V}$ of HGV can be known, so the estimated error of position and velocity can be expressed in the ECEF coordinate system as:

$$\hat{\boldsymbol{E}}_r = \hat{\boldsymbol{R}}_t - \boldsymbol{R}_t = \left[\hat{E}_{rx} \quad \hat{E}_{ry} \quad \hat{E}_{rz}\right]^{\mathrm{T}} \qquad \hat{\boldsymbol{E}}_v = \hat{\boldsymbol{V}}_t - \boldsymbol{V}_t = \left[\hat{E}_{vx} \quad \hat{E}_{vy} \quad \hat{E}_{vz}\right]^{\mathrm{T}} \tag{26}$$

Estimated error of position and velocity in the ECEF coordinate system is shown in Figure 7 and Figure 8.

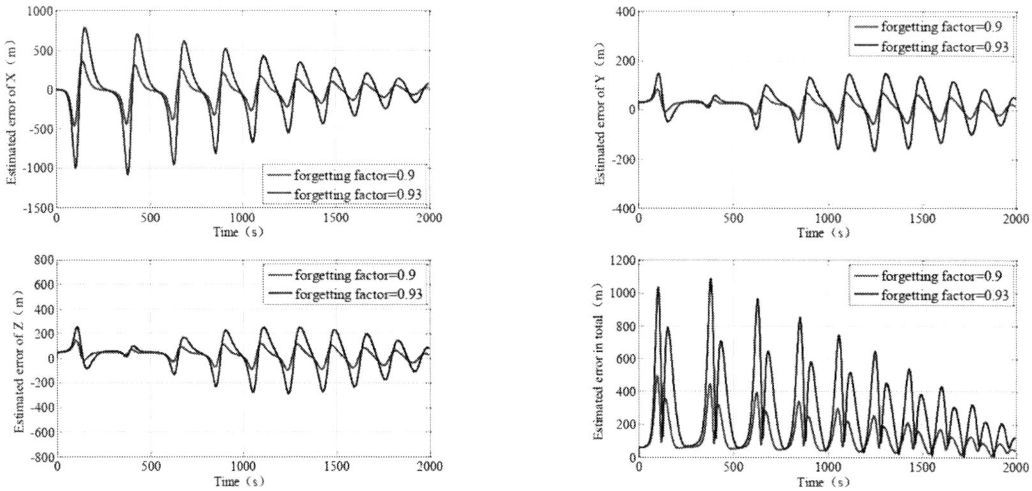

Figure 7. Estimated error of position

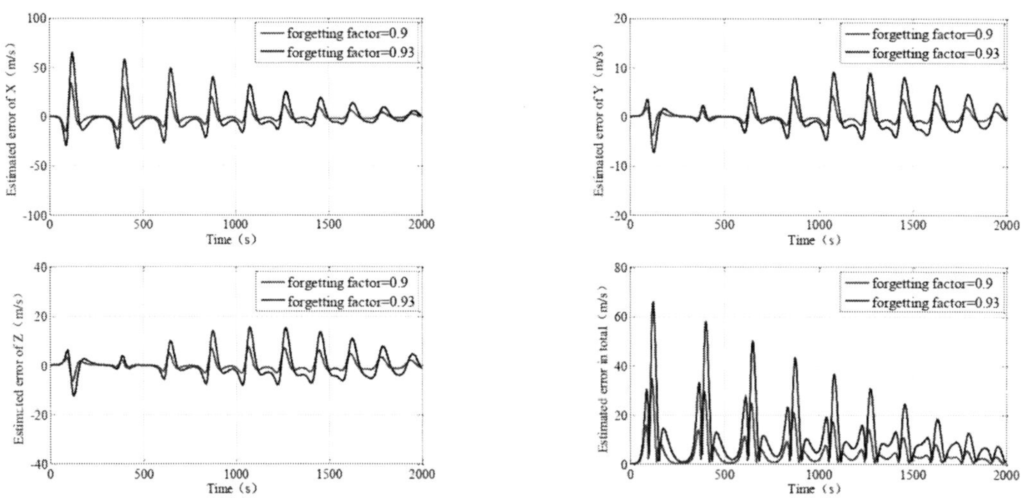

Figure 8. Estimated error of velocity

5 Analysis and discussion

Under the initial conditions and the control conditions, HGV has made several jumps in the near space with a range of up to 8,000 km, which is close to its published design specifications.

When the standard deviation of satellite measurement angle is 0.001°, the observation error of GEO infrared early-warning satellite for HGV reentry section is within 5km.

Whenever the HGV moves near to the position which is a minimum altitude or a maximum altitude, the absolute value of the estimated error in position and velocity will increase rapidly. This is due to the drastic changes in the stress conditions of the HGV. The smaller the forgetting factor, the smaller the fluctuation of the absolute value of the error.

When the forgetting factor is taken as 0.9, the position estimation error for HGV is within 1km, and the speed estimation error is within 80m/s. When the forgetting factor is taken as 0.93, the position estimation error for HGV is within 400m, and the speed estimation error is 30m/s. Within, the accuracy is more ideal.

Reference

[1] H L Xu, J X Li. Study on Defensive Techniques of Hypersonic Vehicles in Proximity Space. Air Navigation Missile, 2014(5):43-47.

[2] W Miao, C B Li, C Wu. Interactive Multiple Model Tracking Algorithm Based on Modified Turning Model. Modern Defense Technology, 2015, 43(3).

[3] X Y Zhang, G H Wang, J J Li, et al. Hypersonic Trajectory Target Tracking Technology in Near Space. Acta Aeronautica et Astronautica Sinica, 2015, 36(6):1983-1994.

[4] J J Li, G H Wang, X Y Zhang, et al. IMM algorithm for hypersonic maneuvering target tracking in near space. Electro-optics & Control, 2015(9):15-19.

[5] L Qin, J L Li. Tracking near-space targets based on multi-station interactive multi-model algorithm. Journal of System Simulation, 2014, 26(10): 2486-2491.

[6] E M Yong. Research on trajectory optimization and guidance method for hypersonic gliding reentry vehicle. National University of Defense Technology, 2008.

[7] G H Li, H B Zhang, G J Tang. Analysis of Typical Trajectory Characteristics of Hypersonic Glide Vehicles. Journal of Astronautics, 2015, 36(4):397-403.

[8] H Y Zhao. Aircraft Reentry Dynamics and Guidance. National University of Defense Technology Press, 1997.

[9] Y Zhang. Ballistic Missile Ballistics. National University of Defense Technology Press, 1999.

[10] G R Duan, Y Sun, M R Zhang, Z Zhang, X Y Gao. Aerodynamic Coefficients Models of Hypersonic Vehicle Based on Aero Database.2010 First International Conference on Pervasive Computing Signal Processing and Applications (PCSPA), 2010,09: 1001-1004.

[11] K Minkwan. Electrostatic Manipulation of a Hypersonic Plasma Images of the Two-Dimensional Sheath. IEEE transactions on Layer plasma science, 2008, 36,(4): 1198-1199.

[12] J T Li, L X Guo, Q J Fang, et al. Electromagnetic wave propagation in plasma sheath of hypersonic vehicle. Systems Engineering and Electronics, 2011, 33(5): 969-973.

[13] Z J Zhang, Z G Cao, W W Wang. Effects of hypersonic vehicle's optical dome on infrared imaging. Optical Engineering, 2011, 50(9):119-124.

[14] M D Clemente, G Rufolo, A Ianiro. Hypersonic test analysis by means of aerothermal coupling methodology and infrared thermography. AIAA Journal, 2013, 51(7): 1755-1769.

[15] H A Quevedo, P F Singer. Synthetic generation of IRST observations for tracker performance studies. Proceedings of SPIE - The International Society for Optical Engineering, 1998, 3373:78-86.

[16] S T Zhang, F Chen, H Liu. Numerical Simulation of Aerothermal Infrared Characteristics of Hyperspace Spacecraft near Space. Computer Simulation, 2010, 27(1):114-118.

[17] H Chai. Analysis of uncertainty warfare of the out-atmosphere missile defense system based on reachable set. National University of Defense Technology, 2015.

[18] S Haykin. Adaptive filter theory. Word publishing corporation, 1986.

The Study of Simulation Architecture in Meta-Synthetic Hall

X J Zhao[1,*], L J Kong[1] and J Y Zhao[1]

[1]Science and Technology on Complex Land Systems Simulation Laboratory Beijing China

[*]zhaoxjmail@126.com

Abstract. The meta-synthetic methodology is the scientific and effective way to solve the problems of the complicated large system. Through the meta-synthetic Hall, we combined the expert, repository, and simulation test together, building an environment to realize the Meta-Synthesis methodology. In this article, we put forward simulation architecture for the Meta-Synthesis Hall. With the architecture, we can build the simulation system and do experiment for the Hall, supporting expert discussion and repository accumulation. We can solve the system experiment and data collection problems for complicated large system analysis, enforces system improve and perfect.

1. Requirement

The complicated large system consists of many functional subsystems, which hinge and merge each other, is a highly synthesis and integration system. It is a huge and difficult work to construct and research Complex land system; there are a lot of uncertain and complicated factors. To construct the system is facing great difficulties and risks.

In order to solve the problems in the process of constructing complicated large system, and reduce developing risk, it is become more important to built a suitable demonstrate method in the developing complicated large system. The Meta-Synthesis method [1] based on simulation is an effective demonstrate method to carry out complicated large system.

Through the simulate architecture, we can build the parallel simulation system which can replace real complicated large system for some degree to analysis the real system. With the simulation and other systems, we can build the Meta-Synthesis Hall which can put forward Meta-Synthesis method conveniently and support the research on the complicated large system efficiently.

2. The Meta-Synthetic Hall based on simulation

The Meta-Synthetic Hall is a concrete application of meta-synthetic methodology. The Meta-Synthetic Hall consists of three parts: expert system, machine system and knowledge system, the expert system and machine system is the carrier of knowledge. The three system combine high intelligently to form man-machine system, not only has the knowledge and information collection, storage, transmission, analysis and synthesis function, more important is has the function to generate new knowledge and intelligence.

2.1. The Meta-Synthetic Hall architecture

The Meta-Synthetic Hall based on simulation is a special application of the meta-synthetic hall using system simulation [2] technology. The Hall was build based on the simulation test, the expert demonstration and information resources. The Hall organizes simulation system, expert group and the information repository together with the distributed interactive network, is shown in Figure 1. The Hall form a kind of high level matrix system, vertical hierarchy, horizontal distribution and interaction each other. The Meta-Synthetic Hall Based on Simulation provides a standardized, structured form to solve the problem of the complicated large system.

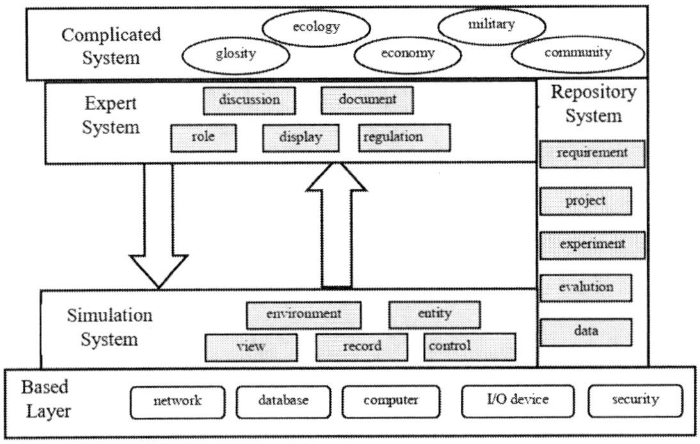

Figure 1. Meta-Synthetic Hall based on simulation

The Expert Group is core of the Hall, was constituted with the persons from all kinds of related field. They not only give the question that we met, but also put forward the project to solving the question. Of course, the project is the hypothesis first, but in the end it will become the true mostly. The Expert Group is the intelligent generate place, it cannot be replace by any machine. With the help of simulation test and repository information, The Expert Group can adjust and optimize the project to meet requirement of the question.

The Simulation System is the parallel system of complicated large system (it may be exist or may be not) with simulation technology. The simulation system has same characters with the complicated huge system and can replace it to some extent. It is serial and has different accuracy level; it may be simple and may be complex to satisfy the requirement of the research. The most important function of the Simulation System is the test on it. The Simulation System can run all kinds of tests that were give out to verify the project of the Expert Group, and generate the data to be analysis by experts.

The Information Repository is a huge database, it store and manage all the data of the complicated large system. The expert and simulation test can get all the information they need from the Information Repository and put the new information into it also. How to organize the huge information with high efficiency is the big problem the Information Repository will meet.

2.2. The procession of Meta-Synthetic Hall

In order to study the complicated large system, we need to do a series science and experience activities. The Meta-Synthetic Hall Based on Simulation follows the scientific and experiment integration means, combining intelligence and knowledge. The Hall has the ability to study and solve the problem of the complicated large system.

The Meta-Synthetic Hall Based on Simulation is an autonomous system and operates with the definite process, is shown in Figure 2, generally the process have five phases. First, we should give the problem related with the complicated large system to The Hall, experts of the Hall need explicit the problems in order to analysis it. Then, the Expert Group will study the complicated large system to get the project that should solve the problems. Third, the Hall should build the simulation system of the complicated large system and put the information related with the system into the repository. Fourth, the Hall carries out the tests according with the requirement that study need and record the test data into repository. Fifth, the experts analysis the data in repository and evaluate the efficiency of project. If through the experiment we can prove the project can solve the problems, the Hall will get the conclusion and give out the project to solve the problem of the complicated large system, otherwise, the Hall will began the other circle at second phase. Generally, the Hall should circle the process many times to get the conclusion, and in the process the Hall need to interchange information with other system frequently. So, the Meta-Synthetic Hall itself is an open, dynamic system, continuous development and evolution.

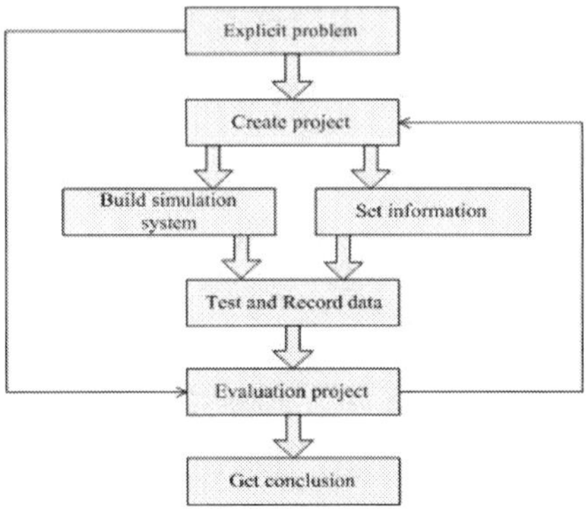

Figure 2. The procession of Meta-Synthetic Hall

3. The simulation test system

In order to support the complicated large system research and analysis, we need to build a large-scale simulation system. The simulation system is the parallel system of the actual system, they are similarity, but different actually. The simulation system can be construct, virtual, or live system [3] to be built with the simulation technology. It have the same component, logistic, characteristic with the actual one, but it is simply and can be destroy and rebuilt repeatedly many time easily. Because we cannot test with the actual complicated large system frequently, for example the globe, the society, the war, the environment, so we need to build the simulation system to study the actual complicated system.

For the different research purpose, we can build different simulation system to satisfy the different requirement. One complicated large system can have many simulation systems, some are complex, and the other may be simple. Simple or complex is not important point; the important is that the simulation system can replace the actual one to some extent and at some field, so the test at the simulation system should be same with the actual system. Generally speaking, the simulation of complicated large system is a huge and difficult work, and the simulation system is also complicated. We need to spend more time and money on it to build a parallel simulation system; it is comprehensive and accuracy enough with the research on the system.

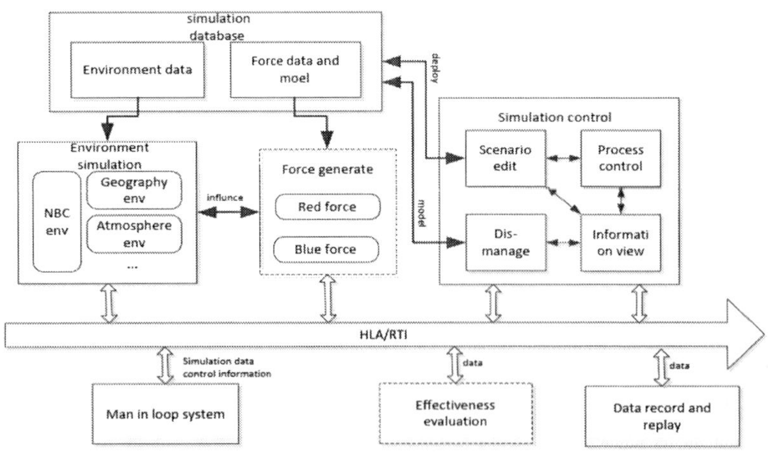

Figure 3. The simulation system architecture

For example, with the efficiency research of the complex land system, we build a simulation system to launch test. The simulation system of the complex land system include: environment simulation system, computer generate force system, man in the loop control system, information display system, HLA/RTI [4], system control and other related systems, is shown in Figure 3.

3.1. The force generate

Computer Generate Force System is the subsystem of simulation system to generating simulation entities of combat force. The simulation entities have the ability to decision and activity, simulate all kinds of force unites, including fighting unit, command unit, communication unit, and detection unit, ensure unit, etc. they can simulate weapon ability, operation activity, information transmission, equipment and logistics insurances, etc. The Computer Generate Force System reflected combat effectiveness of the complex land system.

In order to improve the unity and efficiency of develop CGF system, we put forward a common model structure [5] for all kinds of weapon system entity as shown in Figure 3. The common model structure have nine kinds of common modules, the module has a serials fixed interfaces which can ensure the entity run smoothly. With the structure, we can build every kind of weapon system entity through assemble different kinds of modules in the CGF system. With the same model structure, entities can interact each other easily by the unity interfaces which have been defined in the common structure.

The modules of common structure must could cover as large field as possible, to meet the different requirement of all kinds of weapon system entity. So we divided eight kinds of modules in the common model structure, including command and control module, communication module, detector module, motility module, protection module, strike module, service module. Every kind of model has special structure and interface based on the real function characters in real weapon system. Every simulation entity in the CGF system can be assembling by some kind of modules according with the concrete condition of the real weapon system. In the different kind of entity, the module maybe has different achievement method, but they must have unity structure and interface.

The interface of structure is other key in the common structure, which connected the different modules together to form the simulation entity of weapon system. The CGF system has an interface list which is defined in the common structure, every interface in the list should be defined concretely in the system. Every module fixed some interfaces from the list, but they can't change them. Through the interfaces, the different module can interact each other in the simulation entity or beyond it.

3.2. The man in loop simulation

Man in the Loop system [6] is the subsystem of simulation system to give the platform which put the human being into the simulation system, lead in the perception, analysis, judgment, decision, and executive ability of the person. The person in the simulation not only input signals to the system, but also sensing variety output signal of the system. Man can make decisions according to the variation of system output signal, and feedback at the same time.

Man in the Loop system is mainly used in the field of user training, operator interface design, teamwork training, the relation analysis of man and weapon. In these fields, the most accurate artificial intelligent system still can't accomplish the task, the man in loop is the only way to realize the precisely simulation. The forms of the man in loop system include virtual reality system, training simulator, and network drilling system.

The virtual reality system is a major man in loop simulation system which was used in the early stage. With the virtual reality technology, the simulation system can built the virtual object and environment that is realistic with really object you want. The virtual reality system is also suitable to build the object which is a concept one, so it is superiority with the study in the concept stage. The virtual reality system uses the three dimension graphic and multi-sensor input technologies to build the virtual object. Generally the virtual reality system comprises three parts, including function deployment platform, tracking and positioning module, and human manipulating module. The system can simulate the visual, auditory and tactile conditions for the user and at the same time it can realize interact between the virtual object and real person.

3.3. The environment simulation

The Environment Simulation System is responsible for generating land, sea, and air environment for the simulation test. The system was able to provide dynamic simulation of battlefield environment, produce the real-time data of environment, and meet the requirements of complex land system simulation for the dynamic environment data change. The forces simulated by the Compute Generated Force System have to run under the constraint of simulation environment. The system was able to objectively and accurately reflect the environment effects on the force unities operation.The Environment Simulation System has six modules to achieve different function, including simulation control, scenario compile, dynamic simulation, information display, data publish, and database compile.

The environment simulation was control by the control federate in the environment simulation federation, the other federates will accept the instruction from the control federate and execute it to ensure the environment simulation running normally. The scenario compile module can compile the environment scenario file to instruct how to build and run environment with simulation part, and the user can edit the parameter of the environment easily, especially for the atmosphere and ocean parameters.

Dynamic simulation module is the core part of environment simulation, it can simulate land, atmosphere, ocean, and space change continually. At the same time, it can support CGF system through providing various services to CGF, including road topology, obstacle interdiction, and other information. The information display system can provide vary kinds of view for environment information and support the information search with specific time and position. Database compile module is deal with kinds of original data, the original data must be convert, cut, clean, and mix to generate the product data for CGF and other users.

3.4. The data record and replay

The Multi-Info Display System will visualize the information of simulation test in the process. The simulation process can produce all kinds of data, and the amount of the data is huge. How to see and realize the huge data is a big problem, the graphical of the data is the efficiency method to solve it. Through different kinds of information display way, the Information Display System cans convenience the scientific person to analysis and study of simulation test data.

The Multi-Info Display System can display simulation data distinctly in three dimensions, the user can view the kinds of data which was produced by simulation in real time and in late time by replay. Through the display system, the user can cruise in battle field and acquire each side force deployment. The display system comprise four parts, including status control, three dimensions generate, synchronize handle, and edge fuse. The display process is driven by data from the CGF.

3.5. The other systems

The simulation system has other kinds of subsystem to maintain the simulation run smoothly, including simulate Control system, record system, playback system. Simulation control system is responsible to control the simulation process, record system can record simulation data completely, and playback system can run simulation process once more with the simulation data.

4. The simulation experiment for the Meta-Synthetic Hall

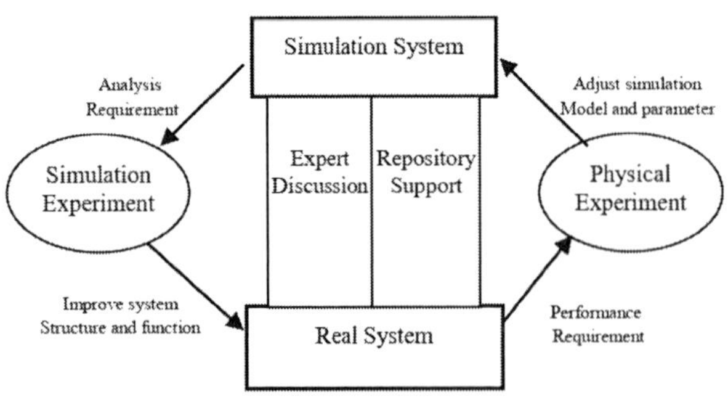

Figure 4. The simulation and physical experiment

In the Meta-Synthetic Hall, the simulation system is the major role to complete the experiment which is used to test and evaluate the object. The object and problem in the MSH is very complicate and cannot get conclusion through the simple discussion. Usually, we need to do lots of experiments to get useful data, and through analysis the data to support expert discuss. The experiments are verity and include logic, physic, simulation experiment, and so on. With the development of the computer, network, simulation technology, the simulation experiment has become a major tool to support expert discussion.

4.1. The superiority of simulation experiment

The simulation system can implement experiment for the concept system which only stay on the paper no physical. When we face a new system which wasn't be built ago, the big problem is how to do the experiment with it, otherwise we can not to find the mistake in the design of the system, especially for the concept model. If we cannot solve the mistake in the initial stage, we would spend more time and resources in the fellow stage. The simulation system put forward the solve solution to this situation, we can build a simulation system, it have the same property with the physical system. The simulation system is a parallel system of the physical one, through the experiment of the simulation system, we can analysis and evaluation the physical system of the future, improving design and helping build a reasonable sterol system, as shown in figure 4.

Simulation experiment has superiority of high efficiency and low cost. In the process of simulation experiment, we needn't to build lots of physical model, just need to develop the digital models and to assemble the simulation system. So, we can increase the experiment efficiency greatly, building the simulation system quickly and put forward lots of experiment, and need not to do much more preparing work for every experiment. At the same time, the simulation experiment cut down cost hardly, not only save lots of money to build the physical model, but also reduce the cost of the experiment itself.

4.2. The shortage of simulation

Although the simulation experiment has above mentioned advantages, it still cannot replace the physical experiment. Because there are essential referents between simulate and physical experiment, the simulation system cannot reflect all character of the real system forever. So the simulation system can help solve lots of mistakes for us, but cannot replace the experiment of the real system. The real system experiment is the final stage to evaluate the efficiency of the real system. In the Meta-Synthetic Hall, there are simulate and physical system to be both exist usually, the compare between the simulation and real system experiment can improve and perfect the function of system. For the study of complicated system, the simulation system has more useful value, for the application of complicated system; the real system is the final target.

5. The conclusion

The complex land system is an open complicated giant system, so the system research and development need more scientific systems engineering methodology. The meta-synthetic is the scientific methodology to solve giant complex system problem. Through combining expert panel, simulation system and repository, we can build Meta-Synthetic Hall. The Meta-Synthetic Hall based on simulation is the scientific and effective way to solve the problems of the complex land system demonstration and construction.

References

[1] D. C. Sun, F. Y. Lin, Systems engineering methodology, In Chinese, Tinghua university press, 2005.08
[2] C.L.Kang, National Defense System Analysis Methods, In Chinese, National Defense Industry press 2003.2.
[3] Ralph Liebert, Michael J. O'Connor, Richard Wyman, Technical Aspects of the Multi-Service Distributed Environment to Support Joint Test and Evaluation, Simulation Interoperability Workshop Sping Conference, 2006 Spring.
[4] Distributed Simulation Systems Specification, Version 2.0, OMG, Inc, 2002.11.
[5] Dirk Brade, Conceptual Modeling Meets Formal Specification, Simulation Interoperability Workshop Spring Conference, 2003.
[6] Craig T. Doescher, Randy Kubik, Michael Wilmer, Is It Live or Is It Memorex: The Power of Live/ Sim Integration, Simulation Interoperability Workshop Spring Conference, 2003.

IOP Conf. Series: Journal of Physics: Conf. Series **1060** (2018) 012090 doi:10.1088/1742-6596/1060/1/012090

Fault-tolerant Control of a Class of Uncertain Systems with Actuator Failure

Han-Wen SHEN[1], Jing-Bo ZHAO [1] and Peng-Hao Liao[1]

[1]College of Information and Control Engineering, Qingdao University of Technology, Qingdao, 266520, China

644479770@qq.com

Abstract. The fault-tolerant control of uncertain networked control systems with actuator failure and packet loss is studied. In the event of a failure of the actuator component, there is a problem of packet loss. The uncertain control system with actuator failure and packet loss is modeled as a Markov jump linear system based on the different locations of packet loss during execution. Then, analysing the stability of the fault system by using Lyapunov stability theory and linear matrix inequalities technology ,the adequacy of the random stability of the closed-loop fault system is proved. At the same time, the mature cone repair linearization method is adopted for the controller design. Finally, the feasibility of the design method was verified by using the example of a double rotor helicopter simulation model.

1. Introduction

With the rapid development of computer technology, control technology and communication technology, network control systems have received great attention in modern industry due to their advantages of resource sharing, less cabling, low cost. The network control system is a closed-loop control system composed of sensors, controllers and actuators as well as mutual networks. The network control system inevitably causes delay, packet loss, and timing inconsistency. Fault-tolerant control has become a research hotspot in the control field and has achieved great results. [1-3]

Packet loss is a common problem after the control system is introduced into the network. The upper and lower limits of the signal or control action scale factor [4]. An active fault-tolerant control system based on sliding mode control theory is studied in [5]. But the previous researches were mostly based on the time-delay network control system. The system components were faulted and the delay was considered for the component failure.

Loss of different situations for analysis, in the event of failure of the actuator parts based on the different locations of packet loss will be divided into four subsystems of uncertain control system, the four faulty control system is modeled,as four Markov jumps Linear system, then use Lyapunov stability theory and matrix inequality technology to analyze the stability of the system, use Schur's complement lemma to design a suitable controller. Finally, simulation to prove the design method are given in this paper.

2. Problem Description

Figure 1 is a block diagram of the network system studied in this paper.

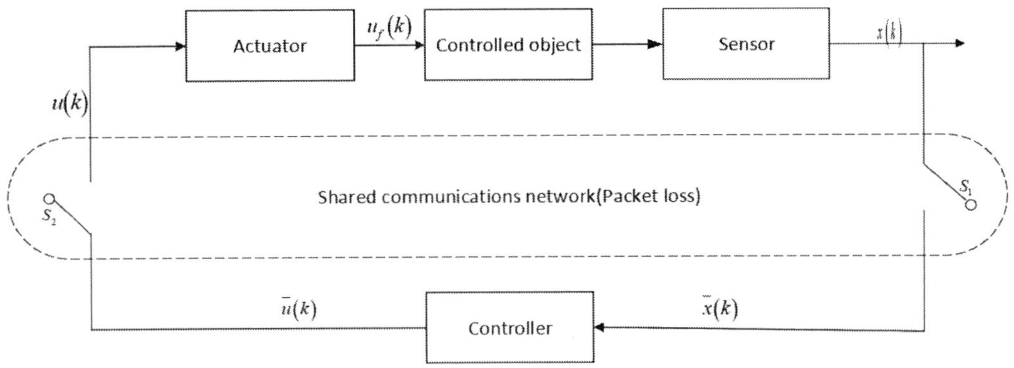

Figure 1. Network Control System Structure

Network control system controlled objects:

$$x(k+1) = Ax(k) + Bu(k) \tag{1}$$

In the actual control system, due to the disturbance of the external environment, changes in system parameters, and disturbances of unknown factors, the uncertainty in the control system makes it difficult to obtain the ideal system model. Considering the uncertainty, the controlled object model is:

$$x(k+1) = (A + \Delta A)x(k) + (B + \Delta B)u(k) \tag{2}$$

$\Delta A, \Delta B$ are unknown matrices and there are uncertainties in the system. Assume that ΔA and ΔB satisfy the following equations:

$$\Delta A(k) = D_a F_a(k) E_a, \Delta B(k) = D_b F_b(k) E_b \tag{3}$$

Where D_a, D_b, E_a, E_b are known constant matrices, $F_a(k), F_b(k)$ are unknown matrices, and $F_a(k)$ $F_b(k)$ satisfy:

$$F_a^T(k)F_a(k) \leq I, \quad F_b^T(k)F_b(k) \leq I \tag{4}$$

The uncertainty parameters $\Delta A(k), \Delta B(k))$ existing in the system are related to the unknown matrices $F_a(k), F_b(k)$, which are different from the previously used exact matching uncertainties $[\Delta A(k), \Delta B(k)] = DF(k)[E_a, E_b]$[6]. For possible actuator failures, it must be studied to define the actuator form and actuator failure model:

$$u(k) = Kx(k), \; u^F(x) = Mu(k) \tag{5}$$

When $m_i = 0$, the i-th channel of the actuator completely fails; when $m_i = 1$, the i-th channel can work normally; when $0 \leq m_{il} \leq m_i \leq m_{iu}, 1 > m_{il}, m_{iu} \geq 1$ *and* $m_i \neq 1$ *In*dicates that the i th channel has not completely failed. In order to represent the actuator fault matrix, a partial matrix is now introduced:

$$M_0 = diag[m_{01}, m_{02}, \dots, m_{0p}], L = diag[l_1, l_2, \dots, l_p], H = diag[h_1, h_2, \dots, h_p], |L| = diag[|l_1|, |l_2|, \dots, |l_p|], m_{0i} = \frac{m_{il}+m_{iu}}{2}, l_i = \frac{m_i - m_{0i}}{m_{0i}}, h_i = \frac{m_{iu}-m_{il}}{m_{iu}+m_{il}}, i = 1,2,\dots,p \tag{6}$$

From m_{0i}, l_i, h_i you can get:

$$\mathrm{M} = \mathrm{M_0(I + L)}, \; |\mathrm{L}| \leq \mathrm{H} \leq \mathrm{I} \tag{7}$$

It can be concluded that the controlled equation with an uncertain system of actuator failure is

$$x(k+1) = \bar{A}x(k) + \bar{B}Mu(k) \tag{8}$$

$\bar{A}, \bar{B}$ represent the status matrix parameters and control input parameters, respectively:

$$\bar{A} = A + D_a F_a(k) E_a, \bar{B} = B + D_b F_b(k) E_b \tag{9}$$

When the switch $S_1$is closed, the sensor data can be directly trans-mitted to the controller, the sensor output $x(k)$. The data that the controller needs to process is the control signa $l\bar{x}(k)$ after the switch. If the signal received by the controller is the switch off state, then the data packet is lost. The signal that the controller can receive is actually a switch. Keep a signal before the output, that is $\bar{x}(k-1)$, at this time we represent the model of the switch S_1:

$$\bar{x}(k) = \begin{cases} x(k) & S_1\ closed \\ \bar{x}(k-1) & S_1\ opened \end{cases} \tag{10}$$

In the same way, calculate the model of S_2:

$$u(k) = \begin{cases} \bar{u}(k) & S_2\ closed \\ u(k-1) & S_2\ opened \end{cases} \tag{11}$$

When analyzing the position of the packet loss, let the dimension vector be

$$\xi(k) = [x^T(k)\bar{x}^T(k-1), u^T(k-1)]^T \tag{12}$$

(1) No packet loss occurred
(2) Switch S_1 opened, packet loss is only between sensor and controller.
(3) Switch S_2 opened, packet loss is only between controller and actuator.
(4) Switch S_1, S_2 opened, packet loss occurs at the same time between sensor and controller and controller and actuator.

The packet loss situation is divided into four categories. Now the four situations are handled in a unified manner. The closed-loop fault network control system model:

$$\xi(k+1) = \Phi_{\sigma(k)} \xi(k) \tag{13}$$

Definition 1:For the initial condition(ξ_0, σ_0), that has been given, if the constant $\Psi(\xi_0, \sigma_0)$ exists and the solution of the control system (13) satisfies $lim_{T\to\infty} E\{\sum_{k=0}^{T} \|\xi(k)\|^2 \xi_0, \sigma_0\} \le \Psi(\xi_0, \sigma_0)$, the uncertain network control system, which can indicate actuator failure, is stable.

3. Stability analysis

If there is an actuator failure, there is a known matrix $P_i > 0$, controller parameter K, $\alpha_{ij}, \beta_{ij}, \gamma_{ij}, \kappa_{ij}$. scalar for the control system (13), such that if equation (14) holds at any $i, j \in S$, the control system (13) is stable.

$$\begin{bmatrix} -P_i & * & * & * & * & * & * & * & * \\ \bar{\Xi}_{i1} & \bar{Z}_{i1} & * & * & * & * & * & * & * \\ \bar{\Xi}_{i2} & 0 & \bar{Z}_{i2} & * & * & * & * & * & * \\ \bar{\Xi}_{i3} & 0 & 0 & \bar{Z}_{i3} & * & * & * & * & * \\ \bar{\Xi}_{i4} & 0 & 0 & 0 & \bar{Z}_{i4} & * & * & * & * \\ \bar{\Psi}_{i1} & 0 & 0 & 0 & 0 & \bar{\Sigma}_{i1} & * & * & * \\ \bar{\Psi}_{i2} & 0 & 0 & 0 & 0 & 0 & \bar{\Sigma}_{i2} & * & * \\ \bar{\Psi}_{i3} & 0 & 0 & 0 & 0 & 0 & 0 & \bar{\Sigma}_{i3} & * \\ \bar{\Psi}_{i4} & 0 & 0 & 0 & 0 & 0 & 0 & 0 & \bar{\Sigma}_{i4} \end{bmatrix} < 0 \tag{14}$$

Lemma 1: Known symmetric matrix M, D, E matrix of a certain dimension, when the matrix $F(k)$ is satisfied $M + DF(k)E + E^T F^T(k) D^T < 0$, And if and only positive scalar ε exists, $M + \varepsilon DD^T + \varepsilon^{-1} E^T E < 0$ is established.

Lemma 2: If R_1, R_2, Y is a symmetric real matrix of a certain dimension, L is a diagonal matrix of a certain dimension, H is a positive definite diagonal matrix of a certain dimension ,and $|L| \le H$, If there is a normal number $\beta > 0$ such that $Y + \beta R_1 H R_1^T + \beta^{-1} R_2^T H R_2 < 0$ holds, then $R_1 L R_2 +$

IOP Conf. Series: Journal of Physics: Conf. Series **1060** (2018) 012090 doi:10.1088/1742-6596/1060/1/012090

$R_2^T L^T R_1^T + Y < 0$ is established.

Lemma 3(Schur Theorem): For symmetric matrix A, matrix B and symmetrical positive definite matrix C, then $A + B^T CB < 0$ can be written as $\begin{bmatrix} A & B^T \\ B & -C^{-1} \end{bmatrix} < 0$.

Proof: Construct a Lyapunov function,

$$V(k) = \xi^T(k) P_{\sigma(k)} \xi(k) \tag{15}$$

$P_{\sigma(k)}$ is a positive definite matrix.

Assumptions

$$\beta = min_{i\in S} \lambda_{min}(-\Omega_{1i}),\ \lim_{T\to\infty} E\{\textstyle\sum_{k=0}^{T} \|\xi(k)\|^2\} \le \frac{1}{\beta} E\{V(0)\} = \frac{1}{\beta}\xi^T(k0) P_{\sigma 0} \xi(0) \tag{16}$$

If there is a constant $\Psi(\xi_0, \sigma_0)$ such that $\Psi(\xi_0, \sigma_0) = \frac{1}{\beta}\xi^T(k0) P_{\sigma 0}\xi(0)$, then according to definition 1 it can be concluded that the control system (13) is stable.

From Lemma 3, $\Omega_{1i} \triangleq \sum_{j=1}^{4} \pi_{ij} \Phi_j^T P_j \Phi_j - P_i < 0$ can be expressed as

$$\begin{bmatrix} -P_i & * & * & * & * \\ \sqrt{\pi_{i1}}\Phi_1 & -P_1^{-1} & * & * & * \\ \sqrt{\pi_{i2}}\Phi_2 & 0 & -P_2^{-1} & * & * \\ \sqrt{\pi_{i3}}\Phi_3 & 0 & 0 & -P_3^{-1} & * \\ \sqrt{\pi_{i4}}\Phi_4 & 0 & 0 & 0 & -P_4^{-1} \end{bmatrix} < 0 \tag{17}$$

Applying Lemma 1, equation (17) can be expressed as:

$$\begin{bmatrix} -P_i + Y_{i1} + Y_{i2} & * & * & * & * \\ \Xi_{i1} & Z_{i1} & * & * & * \\ \Xi_{i2} & 0 & Z_{i2} & * & * \\ \Xi_{i3} & 0 & 0 & Z_{i3} & * \\ \Xi_{i4} & 0 & 0 & 0 & Z_{i4} \end{bmatrix} < 0 \tag{18}$$

Equation (18) is equivalent to Lemma 3:

$$\begin{bmatrix} -P_i & * & * & * & * & * & * \\ \Xi_{i1} & Z_{i1} & * & * & * & * & * \\ \Xi_{i2} & 0 & Z_{i2} & * & * & * & * \\ \Xi_{i3} & 0 & 0 & Z_{i3} & * & * & * \\ \Xi_{i4} & 0 & 0 & 0 & Z_{i4} & * & * \\ \Psi_{i1} & 0 & 0 & 0 & 0 & \Sigma_{i1} & * \\ \Psi_{i2} & 0 & 0 & 0 & 0 & 0 & \Sigma_{i2} \end{bmatrix} < 0 \tag{19}$$

Using Lemma 3 again, we can conclude that Eq. (19) is equivalent to Eq. (14). $-P_i$ is nonlinear Items, cannot be directly solved, so new variables Q_i need to be introduced. $Q_1 = P_1^{-1}, Q_2 = P_2^{-1}, Q_3 = P_3^{-1}, Q_4 = P_4^{-1}$, Substitute (14), available

$$\begin{bmatrix} -P_i & * & * & * & * & * & * & * & * \\ \bar{\Xi}_{i1} & X_{i1} & * & * & * & * & * & * & * \\ \bar{\Xi}_{i2} & 0 & X_{i2} & * & * & * & * & * & * \\ \bar{\Xi}_{i3} & 0 & 0 & X_{i3} & * & * & * & * & * \\ \bar{\Xi}_{i4} & 0 & 0 & 0 & X_{i4} & * & * & * & * \\ \bar{\Psi}_{i1} & 0 & 0 & 0 & 0 & \bar{\Sigma}_{i1} & * & * & * \\ \bar{\Psi}_{i2} & 0 & 0 & 0 & 0 & 0 & \bar{\Sigma}_{i2} & * & * \\ \bar{\Psi}_{i3} & 0 & 0 & 0 & 0 & 0 & 0 & \bar{\Sigma}_{i3} & * \\ \bar{\Psi}_{i4} & 0 & 0 & 0 & 0 & 0 & 0 & 0 & \bar{\Sigma}_{i4} \end{bmatrix} < 0 \tag{20}$$

$X_{ij} = \alpha_{ij} D_{0a} D_{0a}^T + \beta_{ij} D_{0b} D_{0b}^T + \gamma_{ij} V_1 M_0 H M_0^T V_1^T - Q_j$, Using the cone complementarity linearization method proposed in Article 7, the nonlinear matrix inequality of Theorem 1 can be solved by solving the nonlinear minimization problem (13):

$$\min tr\left(\sum_{i=1}^{4} P_i Q_i\right), \quad s.t.(20) \text{ and } (22) \tag{21}$$

$$\begin{bmatrix} P_i & I \\ I & Q_i \end{bmatrix} \geq 0, \quad i \in S \tag{22}$$

From equation (22), $P_i Q_i \geq I$ can be obtained, andmin $tr(\sum_{i=1}^{4} P_i Q_i) \geq 4(2n+p)$ can be deduced. $\mathrm{tr}(\sum_{i=1}^{4} P_i Q_i) = 4(2n+p)$, then $P_i Q_i = I$, the matrix inequality (10) is established. The iterative algorithm of the fault-tolerant controller is now designed to solve the minimization problem.

① Solve the feasible solutions $\left(P_i^0, Q_i^0, K^0, \alpha_{ij}^0, \beta_{ij}^0, \lambda_{ij}^0, \kappa_{ij}^0\right)$ and $\mathrm{k} = 0$ for (20) and (22).

② Solve the optimization problem with $\left(P_i, Q_i, K, \alpha_{ij}, \beta_{ij}, \gamma_{ij}, \kappa_{ij}\right)$ as variables.
$\min tr\left(\sum_{i=1}^{4} P_i^k Q_i + Q_i^k P_i\right)$, $s.t.(20)$ and (22), $P_i^{k+1} = P_i^k$, $Q_i^{k+1} = Q_i^k$.

③ Substituting (P_i, K) into the formula (13), if the formula (13) is true, then the algorithm terminates and is no longer executed. If it is not established and $\mathrm{k} < \mathrm{n}_1$, then $\mathrm{k}+1$ returns to step 2 to continue execution.

4. Simulation example

Consider the system model of a horizontal-rotation double rotor helicopter [7]:

$$\begin{cases} \dot{x} = Ax + Bu \\ y = Cx \end{cases} \tag{23}$$

And $A = \begin{bmatrix} -0.02 & 0.005 & 2.4 & -32 \\ -0.14 & 0.44 & -1.3 & -30 \\ 0 & 0.018 & -1.6 & 1.2 \\ 0 & 0 & 1 & 0 \end{bmatrix}, B = \begin{bmatrix} 0.14 & -0.12 \\ 0.36 & -8.6 \\ 0.35 & 0.009 \\ 0 & 0 \end{bmatrix}, C = \begin{bmatrix} 0 & 1 & 0 & 0 \\ 0 & 0 & 0 & 57.3 \end{bmatrix}$

The input is the combined propeller thrust and the differential collecting propeller thrust, which are denoted by u_1, u_2, respectively. The output is the vertical speed and the inclination angle, which are respectively expressed by y_1, y_2.With a radius of 2, the center of the circle $(-3,0)$ represents a stable area,the fluctuation range of the fault value are $0.6 \leq m_1 \leq 1.25, 0.4 \leq m_1 \leq 1.45$, then the pole $(-2.23, 0.07, 0.49 + 0.42i + 0.49 - 0.42i)$ of the dual-rotor helicopter system (23) is not within the stable area.

$$K_{nomal} = \begin{bmatrix} 2.1161 & -0.0981 & -21.059 & -81.282 \\ 0.076 & 0.3958 & -1.0262 & -6.9184 \end{bmatrix} \tag{24}$$

$$K_{reliable} = \begin{bmatrix} 0.7861 & -0.0822 & -15.2475 & -46.3392 \\ -0.1947 & 0.428 & -1.0237 & -3.7203 \end{bmatrix} \tag{25}$$

In the absence of actuator failure, the pole sets of the normal controller and the fault-tolerant controller system are $\{-1.17, -2.23, -3.46, -3.24\}$,$\{-2.52, -3.09 + 0.18i, -3.09 - 0.18i, -2.99\}$.It can be concluded that the system is stable because the collection is in a stable circular area.

From Figure 2,we can see that in the event of system failure while considering about the phenomenon of packet loss, the system under the control of the original normal controller can't stabilize the system, not in the stable area, the designed fault-tolerant controller pole is still within the stable area, indicating that the designed controller has a role.

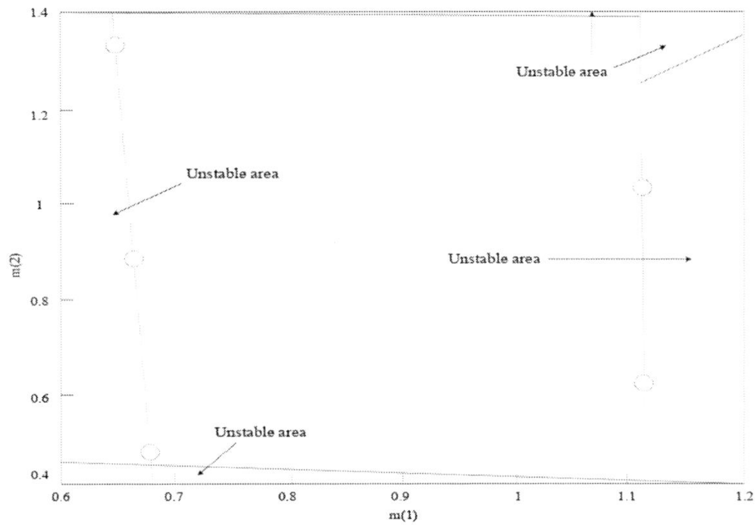

Figure 2. Fault distribution of unstable system

5. Conclusion

The uncertain control system is divided into four sub-systems based on the different positions of the lost parts when the actuator parts fail, and these four faulty control systems are modeled as four Markov jump linear systems. Then use Lyapunov stability theory and matrix inequality technology to analyze the stability of the system, prove the stability of the system, and finally use Schur's complement lemma to design a suitable controller. The example of a dual-rotor aircraft is used to prove the effectiveness of the designed controller.

References

[1] Zhai D, Xi C, Dong J, et al. Adaptive Fuzzy Fault-Tolerant Tracking Control of Uncertain Nonlinear Time-Varying Delay Systems[J]. IEEE Transactions on Systems Man & Cybernetics Systems, 2018, PP(99):1-10.

[2] Lin Z, Guan X, et al. Observer - based robust control for uncertain systems with time - varying delay[J]. Ima Journal of Mathematical Control & Information, 2018, 18(3):439-450.

[3] Tao H, Paszke W, et al. Iterative learning fault-tolerant control for differential time-delay batch processes in finite frequency domains [J]. Journal of Process Control, 2017.

[4] Yang G H, Wang J L, Soh Y C. Reliable H∞ controller design for linear systems.[J]. Automatica, 2001, 37(5):717-725.

[5] ZH MAO, Bin JIANG. Fault Estimation and Accommodation for Networked Control Systems with Transfer Delay[J]. Acta Automatica,Sinica, 2007, 33(7):738-743.

[6] Gao H J, Chen T W, Lam J. A new delay system approach to network-based control[J]. Automatica,2008, 44(1): 39-52.

[7] El Ghaoui L, Oustry F, Aitrami M. A cone complementarity linearization algorithm for static output-feedback and related problems[J]. IEEE Trans.autom.control, 1997, 42(8):1171-1176.

Modeling a FCC Riser with the RBF Neural Network

Qian Liu[a], Jizheng Chu[b] and Minghui Chu[c]

Academy of Information Science and Technology, Beijing University of Chemical Technology, Beijing, 100020, China

[a]LQbuct@163.com,[b]chujz@mail.buct.edu.cn, [c]chumh0130@163.com

Abstract. A radial basis function(RBF) neural network model has been proposed for predicting production yields and reducing simulation times of a modified special pseudo-components(SPCs) fluid catalytic cracking riser. Subsequently, a modified special pseudo-components riser, served as the source of reliable data for artificial neural network(ANN) model training and testing. The ANN model experimental results show a substantial reduction of computation time and good match compared to special pseudo-components fluid catalytic cracking riser mechanism model. Its potential use for modeling optimize and control of the riser is enormous.

1. Introduction

Fluid catalytic cracking unit (FCCU) is a core technology which used in modern refineries to modify heavy gas oil and residuum. FCCU is a process of transforming heavy hydrocarbon molecules into light hydrocarbons, any enhancement of this process leads to important economic reward [1-2].

Developing a high quality and robust physical model is very hard due to the complexity of FCC process [3-4]. Therefore, many scientists have proposed a statistical model [5-6]. Where, artificial neural networks could be a good alternative to the whole FCC process [7-8].

Ali, Rohani et al. [9] used artificial neural network in identification and control of a riser-type FCC unit, Malay et al. modified and integrated dynamic model by neural adaptive control, discussed the robust of feedforward neural network in steady state. While the accuracy of predicting oil production remains to be improved. Maria Mihet et al. based on the data of five-lumped catalytic cracking reaction models, established a feedforward neural network. LI Quan-shan et al. continue to train the catalytic cracking reaction model with an improved RBF neural networks. However, petroleum oils are complicated mixtures of numerous hydrocarbons and non-hydrocarbons, cracking reaction schemes using a few lumps suffer from the disadvantage that a change in product specifications or in the number of products requires reformulating the model and refitting the data.

According to Zhang [10] and Li [11] et al. proposed and improved a special pseudo-components model, assuming all pseudo components are composed of a group of characteristic factors ($K_{W,L}$, $K_{W,H}$), when the properties of crude oil changed, keep temperature constant, it can be seen that mixing ratio of light and heavy fraction ($K_{W,L}$, $K_{W,H}$) changes. As a mechanism model proposed by our group have same weakness with traditional mechanism model that more partial differential equations, algebraic equations need to be solved and long simulation time caused by over parameters. There are 9 tune parameters which took several months for parameter identification. In this paper, a modified special pseudo-component riser was used as the first principle model. Which served as reliable data for artificial neural network (ANN) model. Based on it, a RBF neural network model was built and compared the fitting degree of BP (Back-Propagation) neural network model.

2. The physical model of riser

Cracking reactions take place in riser as shown in figure 1. The length of a riser is generally 25-40 meters, fresh raw oil enters from the bottom of it. After steam atomization at the bottom entrance, catalyst vaporized and then moves upward along the riser with catalytic cracking reaction. This process generally lasts for 3-5s.

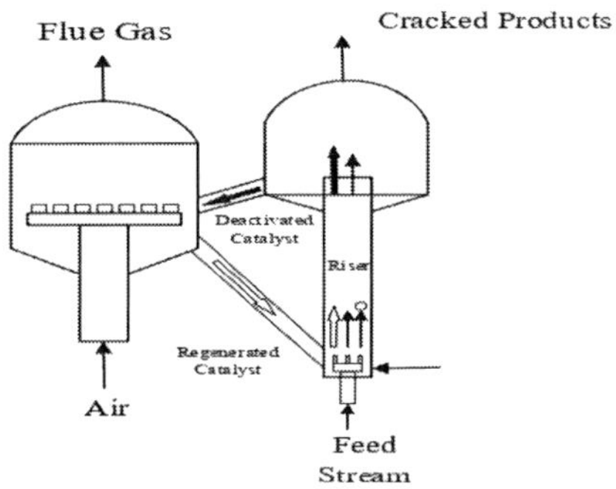

Figure 1. FCCU system structure

Gupta [12] proposed a pseudo component model of FCC riser, used to divide crude oil. While, in industrial practice, when the property of crude oil changed and pseudo component model also needs to be rebuilt. Special pseudo components (SPCs) are used to express the feedstock and product oils, as suggested by Zhang [10] et al. Competition among reactant PCs is estimated from the transition state theory, as suggested by Li et al. [11].

This paper modified the formula of catalyst concentration C_{CAT}, component concentration C_i, the calculation formula of entropy correction reaction based on a steady state model of riser proposed Li et al. [11]:

$$c_i = \frac{F_i}{(v_g\delta_g + v_c\delta_c)\Omega_{RS}} \tag{1}$$

$$c_{\mathrm{CAT}} = \frac{F_{\mathrm{CAT}}}{(v_g\delta_g + v_c\delta_c)\Omega_{RS}} \tag{2}$$

The formula for calculating the concentration in Li et al. as follows:

$$c_i = \frac{F_i}{v_g\delta_g\Omega_{RS}} \tag{3}$$

$$c_{\mathrm{CAT}} = \frac{F_{\mathrm{CAT}}}{v_c\Omega_{RS}} \tag{4}$$

Where υ_g and υ_c are respectively gas and cluster phase velocities in m/s, δ_g is volume fraction of the gas phase, Ω_{RS} is cross section area of the riser in m^2, and $F_C = F_{CK} + F_{CAT}$ the cluster phase flowrate in kg/s. In comparison with (1-2) and (3-4), cracking reaction take place in the whole riser. To get a better fitting, solid phase and gas phase mixed with each other as shown in (1-2).

According to transition state theory [13-14], If feedstock and product oils are mixture of pseudo components, PC_i, i = 1, 2, ..., c, cracking of PC_i can be generally expressed as:

$$PC_i \rightleftharpoons [PC_i]^* \rightarrow PC_m + \text{surplusage} + \lambda_{i,m}CK \qquad (5)$$

PC_m is produced from cracking PC_i, surplusage in the above reaction is not given a name, CK the coke, $\lambda_{i,m}$ the amount of the coke in kg/kmol-PC_i, and $\Delta H_{i,m}$ the heat consumed in kJ/ kmol-PC_i. Detailed equation of modeling a riser is referred to Li et al. [11].

Predictive production as follows:

Table 1. Fitting error of the riser model

		case1	case2
	Prod.	Fitting error	
RO, wt%	10.13	-0.86	0.18
LCO, wt%	22.15	2.79	3.65
GSL, wt%	40.85	0.65	-0.32
LPG, wt%	11.96	-0.38	-0.46
GS, wt%	4.37	-0.36	-0.78
CK, wt%	7.63	-0.14	-0.5
ROT, °C	515.00	1.4	2.4

Case 1 respectively the correct first principle model in this paper, case 2 respectively listed in Li. et al. The first principle model simulation results show that oil production and temperature are closer to industrial value than case 2, which provides a more stable and accurate data for RBF simplified model training.

Fluid catalytic cracking riser is a complex unit due to its multivariable nature, with nonlinear, large delay, strong coupling [15]. Li et al. makes parameter identification extremely tedious. There are 9 tune parameters which took several months for parameter identification. The operating conditions of fluid catalytic cracking riser are as shown in Table 2.

Table 2. Range of operating variables

Process variable	Range
Crude oil flow, kg/s	3.9667-4.2667
Recycle oil flow, kg/s	33.8260-36.8260
Temperature of crude oil, K	473.15-503.15
Temperature of recycle oil, K	553.15-583.15
Flow rate of raw catalyst, kg/s	296-300
Temperature of spent catalyst, K	776-782
Catalyst to oil ratio	7.79-8.09

3. Artifical neural network model

Feedforward neural network has simple structure and easy to achieve, researchers have proved that 3-layer feedforward neural network model has ability to approximate any nonlinear function [16]. To solve the challenge of long computational time and complex modeling mechanism of the first principle model by Li et al. Back Propagation(BP) algorithm and Radial Basis Function(RBF) algorithm as classical feedforward neural networks are simulated in simplified riser modeling.

Radial Basis Function (RBF) network is a feedforward neural network, which is used to establish the nonlinear mapping relationship between process parameters [17]. Mapping from input layer to hidden layer is nonlinear, while mapping from hidden layer to output layer is linear [18]. Radial basis function which from hidden layer adopts the Gauss function, can be applied to multi output nonlinear system [19].

$X = (x_1, x_2, \ldots x_d)^T$ and $\hat{y} = (\hat{y}_1, \hat{y}_2, \ldots \hat{y}_p)^T$ represent inputs and outputs. $\Phi = (\varphi_1, \varphi_2, \ldots, \varphi_m)^T$ as hidden layer neurons, m represents the number of hidden layer neural, w_{jk} represents the weight of the hidden and output layers.

Radial basis Gauss function of hidden layer can be applied to multi output nonlinear system, which is defined as follows:

$$\hat{y}_k = w_0 + \sum_{j=1}^{m} w_{jk}\varphi(x, c_j, \delta_j) \tag{6}$$

$$\varphi_j = \exp(\| x - c_j \|^2 / \delta_j^2) \tag{7}$$

C_j and δ_j RBF as center and width, $\|\cdot\|$ the euclidean distance of x and C_j, w_0 Output bias. Some result represents that center and width of RBF affect the performance of the neural network.

The regression and prediction capabilities of RBF model are characterized by the following:

$$\text{RMSE} = ([\sum_{p}(Y_p / Y_p^* - 1)^2 + (t_{RO} / t_{RO}^* - 1)^2]/N_P)^{0.5} \tag{8}$$

$$E_{\max} = \max\{|Y_p / Y_p^* - 1|_{\substack{p=GS,LPG,GSL,\\LCO,RO,CK}}, |t_{RO} / t_{RO}^* - 1|\} \tag{9}$$

Where RMSE is the root mean square relative error, E_{max} the maximum absolute relative error, and N_p the number of products.

BP [11] network uses gradient descent method to train the network weights, which has a large amount of computation. While RBF network solves the linear equation, using the recursive least square method to solve the network weights. It's learning speed is faster and avoid falling into a local minimum.

According to table 3, 3100 sets. The operating conditions of riser reactor can be expressed as $X = (x_1, x_2, \ldots x_7)^T$. In this work, 3050 points were allocated to train set and remaining 50 sets were utilized as the test set. The accuracy of Y is as follows:

Table 3 proved that RBF neural network model has a better predictability than BP model, worth mentioning that training a RBF neural network riser model took only 31.59 seconds. Figure2-9 compared prediction output error in RBF&BP model, images verified that the error of RBF is smaller than BP algorithm. Figure 9 explained a good fitting degree with predictive temperature and expected temperature.

Table 3. Prediction results

	RBF Testing error	BP Testing error
MSE(max), %	$1.18e^{-5}$	$7.68e^{-4}$
RMSE(average), %	$3.95e^{-4}$	$8.71e^{-3}$
Time, s	31.59	44.23

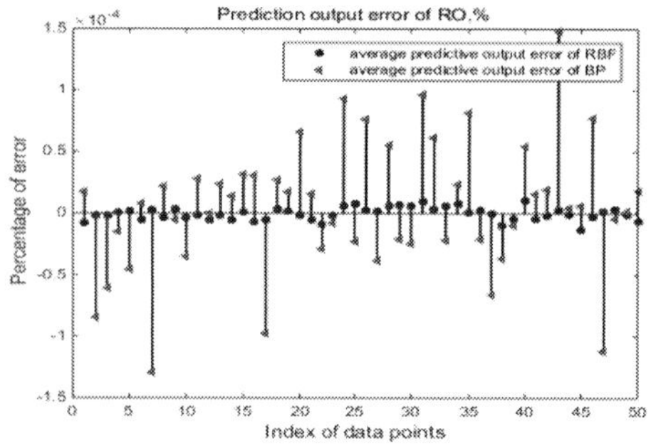

Figure 2. The prediction output error of RO with RBF&BP caption

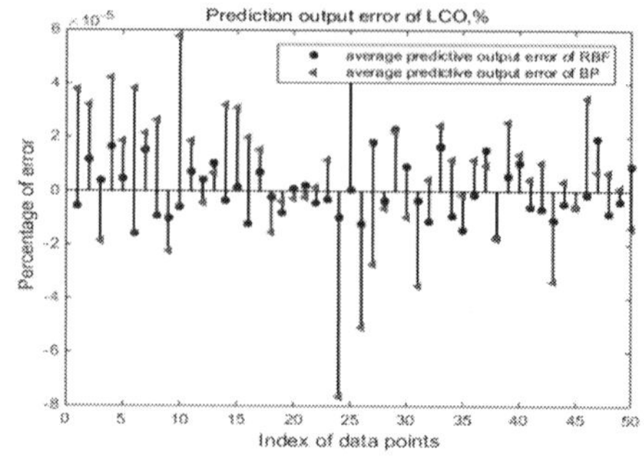

Figure 3. The prediction output error of LCO with RBF&BP caption

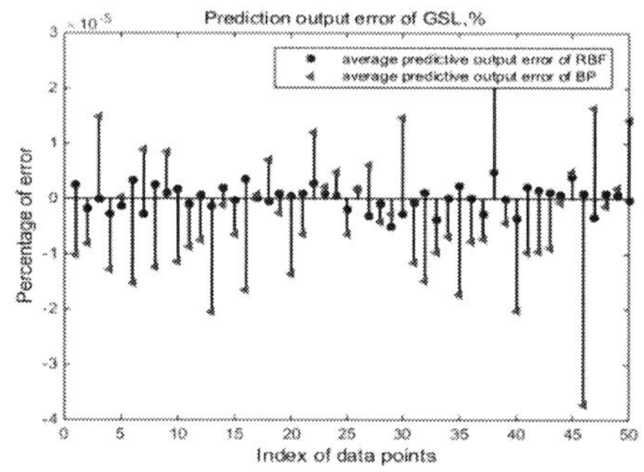

Figure 4. The prediction output error of GSL with RBF&BP caption

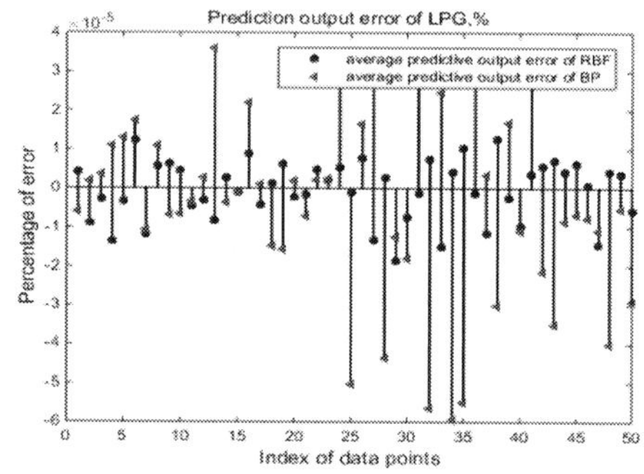

Figure 5. The prediction output error of LPG with RBF&BP caption

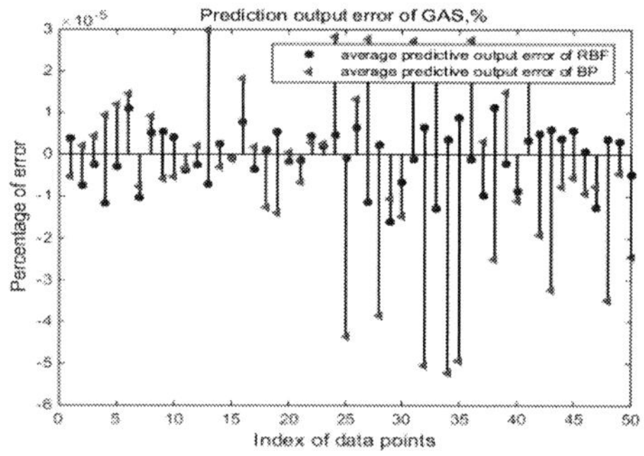

Figure 6. The prediction output error of GAS with RBF&BP caption

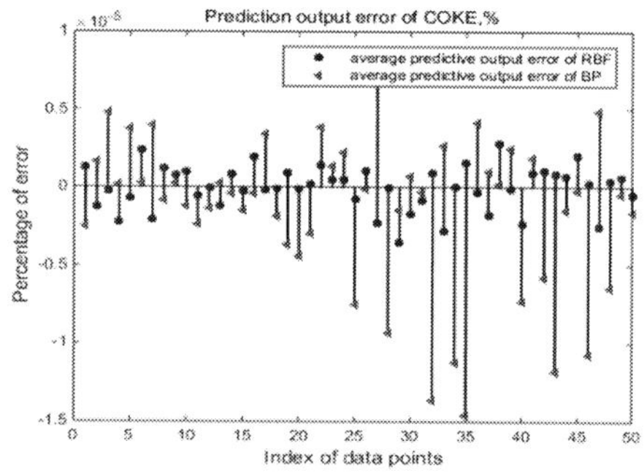

Figure 7. The prediction output error of COKE with RBF&BP caption

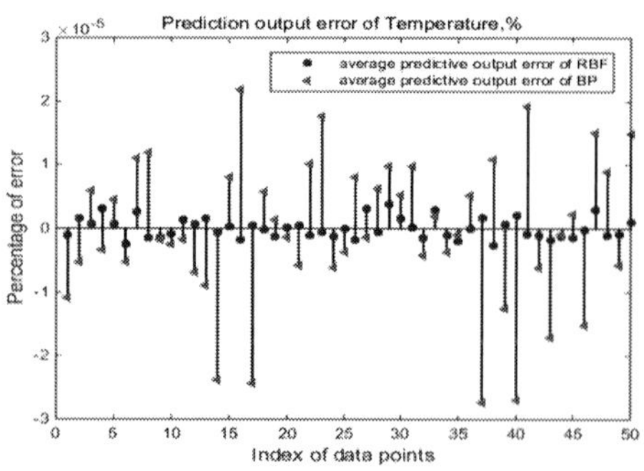

Figure 8. The prediction output error of temperature with RBF&BP caption

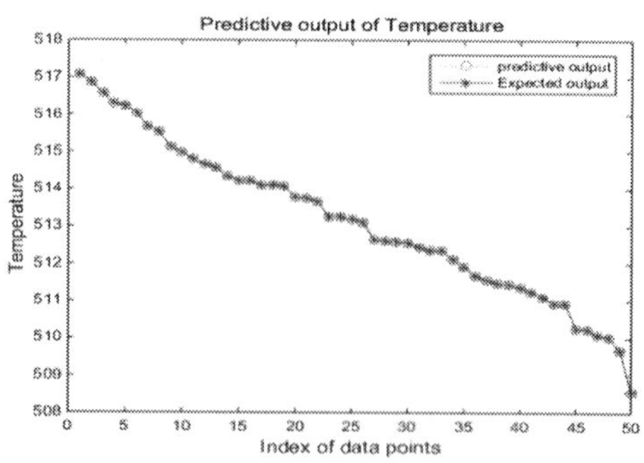

Figure 9. Temperature with predictive and expected output in RBF.

4. Conclusion

When modified the concentration of components(C_i) and catalyst(C_{CAT}), results as shown in table 1 that oil production and temperature are closer to industrial value than Li.

A modified special pseudo-components fluid catalytic cracking riser was used to provide a random varying inputs and outputs data for the ANN model. Results as shown in table3 and Fig2-8, RBF neural network model shown a very good regression and generalization performance with first principle model than BP neural network model.

Set up a RBF neural network model needs only 31.59 seconds, while the optimization process of first principle model takes several months. Compared with complex mechanism modeling, RBF neural network model can better excavate latent knowledge behind process data.

References

[1] Zeydan M 2008 *J. International Journal of Industrial Engineering Theory Applications & Practice,* **15** 1
[2] Miheţ M, Cristea V M and Agachi P Ş 2010 J. Asia-Pacific Journal of Chemical Engineering, 4 878
[3] Ławryńczuk M 2011 J.Chemical Engineering Journal 166 269
[4] Nair A and Daryapurkar R 2016 C. International Conference on Advances in Computing, Communications and Informatics (India: IEEE) PP1891-5
[5] Bispo V D D S and Meleiro L A D C 2013 Modeling, optimization and control of a fcc unit using neural networks and evolutionary methods J. Engevista. E(2013)
[6] Yeh T M, Huang M C and Huang C T 2003 J. Computers & Chemical Engineering 27 55
[7] Weekman V W and Nace D M 1970 J. AIChE Journal 16 397
[8] Cristina P 2015 J. Procedia Engineering 100 602
[9] Alaradi A A and Rohani S 2002 J. Computers & Chemical Engineering 26 401
[10] Zhang J, Wang Z, Jiang H, Chu J, Zhou J and Shao S 2013 J. Chemical Engineering Science 102 87
[11] Yong Li, Jizheng Chu and Jiarui Zhang 2017 J. Chinese Journal of Chemical Engineering 25 1449
[12] Gupta R K, Kumar V and Srivastava V K 2007 J. Chemical Engineering Science 62 4510
[13] H Eyring 1935 J. Journal of Chemical Physics 3 107
[14] H Eyring 1935 J. Chemical Reviews 1 65
[15] Yakubu M John, Raj Patel and Iqbal M Mujtaba 2017 J. Energy & Fuels 31 5645
[16] Sanger T D 1989 J. Neural Networks 2 459
[17] Zhou J, Ren J and Yao C 2017 J. Measurement 102 271
[18] Barlow M T and Chen X 2016 J. Mathematische Annalen 366 1
[19] LI Quan-shan, Zhang Yi-shan and Cao Liu-lin, Lin Xiao-lin and CUI Jia 2015 J. Journal of Central South University 22 1342

A Simplified Simulation Model of Steady Temperature for Tubular Furnace

Tao Zhang[1, a] and Jizheng Chu[1, b]

[1]Academy of Mathematics and Systems Science, Beijing University of Chemical Technology, Beijing, 100020, China

[a]ztaobuct@163.com; [b]chujz@mail.buct.cn,

Abstract. To solve the problems of complex radiative, convective heat transfer and difficult parameter correction in tubular furnaces, a simplified model of a new tubular heating furnace was proposed, which coupling each pipe heat transfer to a whole pipe and simplified the radiation heat transfer formula parameters of each pipe to a whole single heat transfer parameter. In addition, the simplified method takes the flue gas generated by combustion as the sole heat source, each section of the pipe wall acts as the intermediate medium to receive the heat of the burning flue gas and heat the fluid. The results shows that the new model was effective and accurate.

1. Introduction

As the most important heating equipment, the reheating furnace plays an important role in the process of industrial production. As the second largest energy consuming equipment in China, it is next to thermal power generation, but its average combustion efficiency is about 30%, which is far lower than that of 50% in other developed countries [1]. As the important heating equipment in oil refineries and petrochemical plants, tubular reheating furnace plays an indispensable role in oil production and catalytic refining. In a refinery, the tubular furnace with multi pipes and multi burners is often used to heat the crude oil. It can improve the heat transfer efficiency and reduce the possibility of coking [2]. Therefore, simulating its combustion and heat transfer process can provide a reliable basis for its production and processing.

The structure of tube heating furnace started as a Heap-type furnace, which based on the principle of kettle type steamer. The tube bundles were used as an endothermic surface and the elbow of each tube was in the furnace. The burner was installed under the tube bundle, each tube was heated unevenly, it was easy to burn and fire at the elbow [3]. After that, the structure of tube furnace was adjusted to be a pure convective furnace, but it still had uneven heating phenomenon. To solve this problem, people designed a tubular heating furnace with a radiation chamber and a convection chamber, such as box furnace.

In the 1950s, researchers began to study furnace simulation models. SiddallR G and Selcuk [4] simplified the whole heat transfer model as a gas, a heating surface and a reflector by Lobo-Evens method, but it was too simple to get the surface temperature of radiation chamber and furnace tube temperature. Hottel [5] proposed a regional method to calculate heat transfer in the furnace, but it has a large amount of calculation and need a computer for numerical calculation. Howell [6] proposed a Monte-Carlo method to divide the radiation energy into several parts, calculated the radiative heat transfer of different regions by statistical method and got temperature and heat flux distribution.

Costik [7] based on statistical method, studied a two-dimensional radiation model. Fitzgerald et al. [8] studied a three-dimensional radiative heat transfer model. Spalding et al. [9] developed a turbulent combustion model, and then the model of reheating furnace began to take into turbulence and combustion. Yuan J et al.Wei-jun Zhang, Zhi Yi et al[10] proposed an zonal method to simplified the two-dimensional heat transfer model of a multi-zone reheating furnace. NC Mehta, DD Shukla, P K and oliya [11] had done a heat transfer analysis of induction furnace wall made of zirconia using explicit finite difference method. Gupta A K et al. [12] studied the relationship between flame volume, air preheat temperature and oxygen concentration by PDF combustion model and regional radiation model. Prins et al. [13] simulated flow field and pressure field in cracking furnace by complete unstructured grid partition scheme. Stefanidis et al. [14] used detailed combustion models in cracking furnace simulation to get a suitable combustion model. Habibi et al. [15] studied the effect of different radiation model included DOM, P-1, and Rosseland on flow field, temperature, and flame shape. The simulation results provides a basis for the simulation of the heating furnace model.

Due to complexity of the heat and energy transfer process, the calculation speed will become very slow [10]. In this paper, according to the actual production needs, a simplified model was developed by using single heat transfer parameter to simplify radiation and convection formula. This model can get the flue gas, the wall of each pipe and the heating oil temperature quickly and effectively, which are the important parameters of oil production.

2. Simplified model of reheating furnace and methodology

The tubular furnace investigated in this paper is from Shengli Oil Refinery of Qilu Petrochemical Corp. It consists of four parts, which are radiation chamber, convection chamber, chimney and residual heat recovery system. The bottom of the radiation chamber is equipped with burners, convection chamber and radiation chamber have heating pipes. The chimney is equipped with a flue baffle, and the flue gas pipeline is connected with residual heat recovery system.

The fuel oil or fuel gas is mixed with the incoming air at the bottom of radiation chamber through nozzles. The flue gas generated by combustion is used as a radiation source to heat tubes in the form of thermal radiation. Then it enters convection chamber and heat the tubes, after that enters the air preheater to heat air, finally enters the atmosphere through chimney. The radiation chamber tubes absorbs most of heat in the whole heat transfer process about 70% [16]. Figure 1 is the simplified schematic diagram of tubular heating furnace.

In this paper, the flue gas produced by fuel combustion are simplified as a process that heat source to heat the tube wall. So the energy equation [17]:

$$F_g h_g + F_a h_a + \sum_i Y_{s,i} F_g \mathrm{dH_i} = F_s h_s \tag{1}$$

Where F_g, F_a, F_s are the fuel gas, air and flue gas flow. h_g, h_a and h_s are the enthalpy of fuel, gas and air, dH_i is the combustion heat of each component in fuel gas.

$$h_k = \sum_i Y_{k,i} h_{k,i} \quad (\mathrm{k = g,a,s}) \tag{2}$$

$Y_{k,i}$ is the mass fraction of species i and h_s is defined as:

$$h_{k,i} = \int_{298.15k}^{T} C_{p,i} dT \quad (\mathrm{k = g,a,s}) \tag{3}$$

Where $C_{c,j}$ is the constant pressure specific heat of species i. The fuel gas composition and combustion heat as shown in table 1.

The flue gas transfers heat to the tube wall in radiation chamber and convection chamber, then each tube wall transfers heat to heated oil. This process follows the radiative and convective heat transfer formula. In this paper, multichannel tubes are heated at the same time. Figure 2 is the radiative heat transfer model.

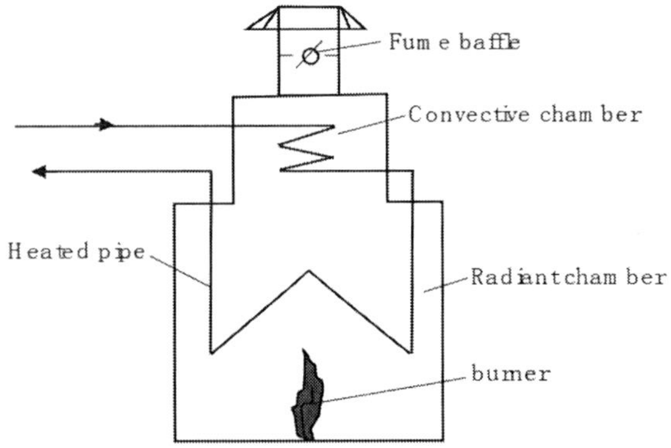

Figure 1. Simplified model of tubular reheating furnace

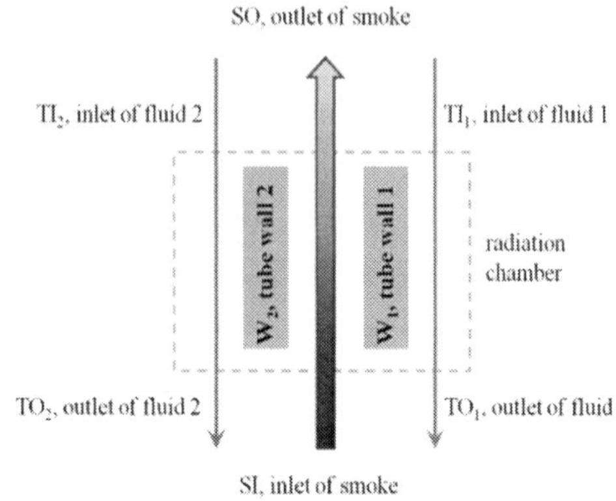

Figure 2. Radiation chamber model

Table 1. Fuel gas composition and combustion heat

Component	Volume content %	Combustion heat kJ/Nm3
CO	1.84	12636
CO_2	0.38	0
H_2	13.34	11096
O_2 and N_2	9.19	0
CH_4	49.82	35902
C_2H_6	16.17	64397
C_2H_4	4.14	59477
C_3H_8	1.94	93240
C_3H_6	1.86	87667
iso-C_4H_{10}	0.02	122853
n-C_4H_{10}	0.42	123412
n-C_4H_8	0.46	123649
t-C_4H_8	0.07	112964
c-C_4H_8	0.05	112964
i-C_5H_{12}	0.07	134821
n-C_5H_{12}	0	156467
neo-C_5H_{12}	0	140375
C_6	0	176273
H_2S	0.23	23368

For simplicity, analysis is carried out under the following assumptions:

1) The flue gas is mixed uniformly in radiation chamber and radiation chamber.

2) There is no heat transfer between the tube walls.

SI is the flue gas of entering radiation chamber, which is generated by the combustion of fuel and air in adiabatic condition. SO is the flue gas of leaving radiation chamber. TI_l and TO_l are the heated fluid of L road entering and leaving radiation chamber. W_l is the tube wall of L road.

By the radiative transfer equation, the heat release of flue gas to wall as:

$$Q_{s,l} = \alpha_{s,l} A_{s,l} (\bar{T}_{SO}^4 - \hat{T}_{Wl}^4) \quad (4)$$

$Q_{s,l}$ is the radiation quantity that flue gas to the wall of L road. $\alpha_{s,l}$ is the heat transfer coefficient, $A_{s,l}$ is the heated area of tube wall. $\overline{T}_{SO}$ and $\hat{T}_{Wl}$ are the temperature of flue gas and tube wall. Heat balance equation of heat transfer between flue gas and tube wall as:

$$Q_s = \sum_{l=1}^{L} Q_{s,l} = \overline{F}_{SOl}\overline{c}_{p,SOl}(\overline{T}_{SI} - \overline{T}_{SO}) \tag{5}$$

$\overline{F}_{SOl}$ is the flow of the flue gas, $\overline{c}_{p,SOl}$ is the average specific heat of flue gas, $\overline{T}_{SI}$ and $\overline{T}_{SO}$ are the temperature at the inlet and outlet of radiation chamber. L is the number of total heated pipes. To simplify the whole heat transfer process, $\lambda_{s,l}$, λ_s, $T_{W,av}$ can be as:

$$\lambda_{s,l} = \frac{\overline{F}_{SO}\overline{c}_{p,SO}}{\alpha_{s,l}A_{s,l}} \tag{6}$$

$$\lambda_s = \frac{1}{\sum_{l=1}^{L} 1/\lambda_{s,l}} \tag{7}$$

$$T_{W,av} = \left(\frac{\sum_{l=1}^{L} \hat{T}_{Wl}^4/\lambda_{s,l}}{\sum_{l=1}^{L} 1/\lambda_{s,l}} \right)^{1/4} \tag{8}$$

From Eq. (4), Eq. (5) can be obtained:

$$\overline{T}_{SO}^4 + \lambda_s\overline{T}_{SO} = \lambda_s\overline{T}_{SI} + T_{W,av}^4 \tag{9}$$

Where $T_{W,av}$ is the average temperature of all the tube walls.

Heat transfer equation and heat balance equation between tube and wall as

$$Q_{o,l} = \alpha_{o,l}A_{o,l}(\hat{T}_{Wl} - \frac{\overline{T}_{TOl} + \overline{T}_{TIl}}{2}) = \overline{F}_{TOl}\overline{c}_{p,TOl}(\overline{T}_{TOl} - \overline{T}_{TIl}) \tag{10}$$

$Q_{o,l}$ is the contact heat transfer quantity of the tube wall to the heated oil. $\overline{T}_{TIL}$, $\overline{T}_{TOl}$ are the heated oil temperature at the inlet and outlet of radiation chamber. $\overline{F}_{TOl}$ is the flow of heated oil. $\overline{C}_{p,SOl}$ is the average specific heat of heated oil. So, $\overline{T}_{TOl}$ can be obtained:

$$\overline{T}_{TOl} = \frac{\hat{T}_{Wl}/\overline{T}_{TIl} - 0.5 + \lambda_{o,l}}{0.5 + \lambda_{o,l}}\overline{T}_{TIl} \tag{11}$$

$$\lambda_{o,l} = \frac{\overline{F}_{TOl}\overline{c}_{p,TOl}}{\alpha_{o,l}A_{o,l}} \tag{12}$$

Energy balance equation of gas and oil as:

$$Q_o = \sum_{l=1}^{L} Q_{o,l} = \overline{F}_{TOl}\overline{c}_{p,TOl}(\overline{T}_{TOl} - \overline{T}_{TIl}) = \lambda Q_s = \lambda\sum_{l=1}^{L} Q_{s,l} = \lambda\overline{F}_{SOl}\overline{c}_{p,SOl}(\overline{T}_{SI} - \overline{T}_{SO}) \tag{13}$$

λ is the thermal efficiency with the range of 0 to 1.From Eq. (9), Eq. (10), Eq. (11) can be obtained

$$\lambda \hat{T}_{W1}^{4} + \frac{2\lambda\lambda_{s,1}}{1+2\lambda_{o,1}}\hat{T}_{W1} = \lambda \overline{T}_{SO}^{4} + \frac{2\lambda\lambda_{s,1}}{1+2\lambda_{o,1}}\overline{T}_{TI1} \tag{14}$$

Thus, the heat transfer formula of every tube can be simplified. $\lambda_{s,1}$ and $\lambda_{o,1}$ above can be obtained through the steady state data of the industrial field. To solving the equation of higher degree, one way is using Newton method and Secant method.

In convective model, the convective heat transfer is the first square function about flue gas temperature and tube wall temperature. The other heat transfer equations of convection model are the same as radiation model that need not be detailed here. It is worth noting that the outlet temperature of flue gas obtained by radiation model was used as the input condition for convection model, and the outlet temperature of oil obtained by convection model was used as the input condition of radiation model. Therefore, in the process of coupling radiation model and convection model, the oil temperature at the inlet of radiation chamber need to be assumed in advance, and then solved the real oil temperature by Secant method. The main parameters about the heating furnace are in table 2.

In this paper, the parameters were corrected by a set of industrial data. Then using the parameters, this model predicts three sets of outlet temperature of oil and flue gas. Table 3 is prediction error of this model. In figure 3-4, 15 sets of data was obtained by changing the feed volume of fuel gas, and then draw the relationship between the combustion calorific value, flue gas temperature, oil temperature and wall temperature. In figure 5-6, when changing the feed amount of the heated oil, get the relationship between gas, tube temperature and oil flow.

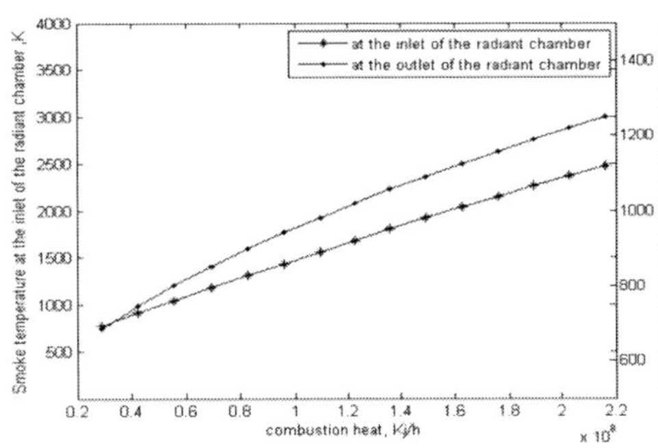

Figure 3. Relation diagram of combustion heat value and flue gas in the radiation chamber

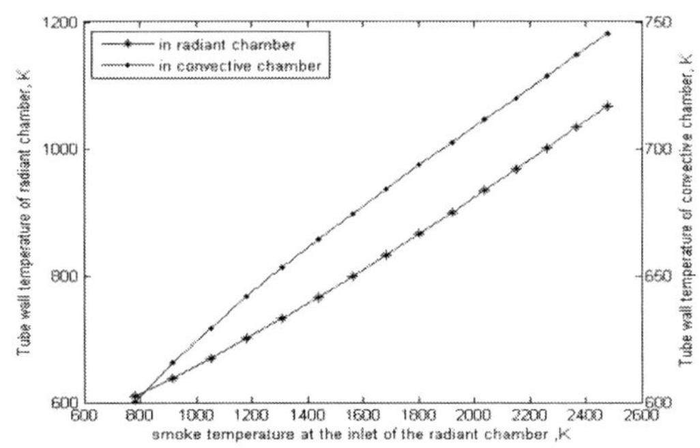

Figure 4. Relation diagram of tube wall temperature and oil flow temperature

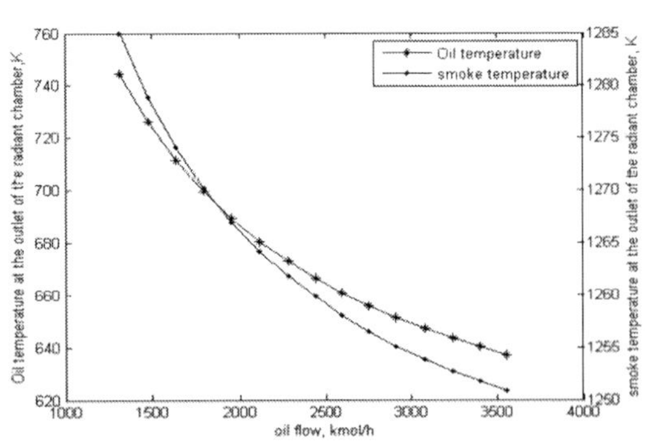

Figure 5. Relation diagram of oil flow and smoke temperature in radiation chamber

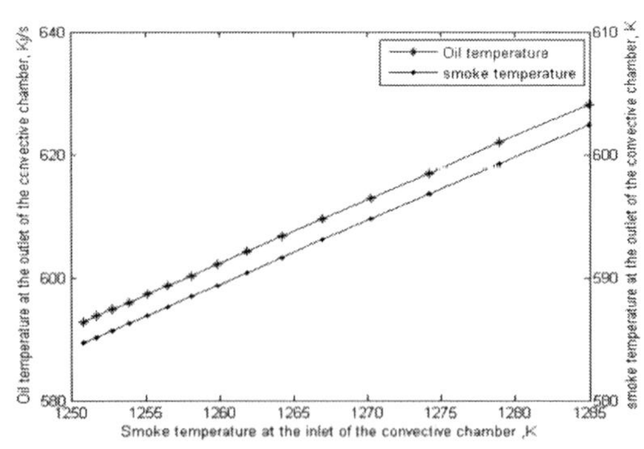

Figure 6. Relation diagram of smoke temperature and oil temperature in the convection chamber

Table 2. Parameter value of reheating furnace.

Parameter	Calibrated value
Fuel flow,Nm^3/h	5960.19
combustion heat of fuel gas, kJ/Nm3	37309.25
Thermal load of Radiation chamber, GJ/h	139.15
thermal efficiency, %	93.5
surface area of convection tubes, m^2	1509
Thermal strength of convection tubes, W/m^2	8916.87
Surface area of radiation tubes, m^2	1577
Thermal strength of convection tubes, W/ m^2	24510.32
Mass flow of radiation tubes, kg/h	892256
Internal diameter of pipe, mm	177.3
Number of tube passes	6
excess air factor	1.08

Table 3. Prediction error of the simplified model.

	1	2	3
Prod. oil Tem,°C	365.03	369.95	367.05
Prediction oil Tem,°C	369.8648	362.5119	363.6381
error	-4.8348	7.4381	3.4119
Prod. Smoke Tem,°C	309.5	314.9	312.6
Prediction.Smoke Tem,°C	315.8064	310.6808	311.3591
error	-6.3064	4.2192	1.2409

3. Conclusions

By the research on simplified model of tubular furnace, the simulation results proved that using flue gas as the heat source is feasible and the single heat transfer parameter can simplify the transfer heat formula. Compared the results with the actual production data, It is obvious that the simplified model is effective and accurate. Therefore, it can simulate and predict the actual tubular furnace for production and research.

References

[1] Wang J, Liu Y and Sundén B 2017 *J. Energy Conversion & Management* JECM(2017)
[2] Wochele J and Stucki S 2015 J. Chemical Engineering & Technology 22(3) 209
[3] Garg A. 1999 J. Hydrocarbon Processing 3 161
[4] Siddall R G and Selcuk N 1976 J. International Journal of Heat and Mass Transfer 49 10
[5] Hottel H C and Sarofim A F 1967 J. Landolt-Börnstein - Group VI Astronomy and Astrophysics 142 42
[6] Howell J R 1969 J. Advances in Heat Transfer 5 1
[7] Hottel H C and Cohen E S 1958 J. Aiche Journal 4 3
[8] F M Satler and J A Costik 1974 J. Ins. Fuel 47 3
[9] Patankar S V and Spalding D B 1972 J. International Journal of Heat & Mass Transfer 15 1787
[10] Wei-jun, Zhang and Zhi Yi 2014 J. Journal of Iron and Steel Research 21 419
[11] NC Mehta, DD Shukla, P K and oliya 2017 J. International Journal of Advance Research & Innovative Ideas in Education 3 796
[12] Gupta A K, Bolz S and Hasegawa T 1999 J. Journal of Energy Resources Technology 121 209
[13] Oprins A J M and Heynderickx G J 2003 J. Chemical Engineering Science 58 4883
[14] Stefanidis G D, Merci B and Heynderickx G J 2006 J. Computers & Chemical Engineering 30 635
[15] Habibi A, Merci B and Heynderickx G J 2007 J. Computers & Chemical Engineering 31 1389
[16] Demidenko N D, Kulagina L V 2017 J. Chemical & Petroleum Engineering 53 1.
[17] Tang G, Wu B and Bai D 2017 J. International Journal of Heat & Mass Transfer 113 1142.

Design of Pump Fault Diagnosis System Based on T-FMEA

Wang Chaowei*, Li Peng, Chen Boren and Zhou Xingpan

China Ship Develop and Design Center, Wuhan 430064, China

*wangcwbu@163.com

Abstract. Water pump is a widely used machine, which is applied to vehicles, vessels, aircraft and other equipment. Once pump develops faults, critical accidents may occur. So the fault diagnosis research and fault diagnosis system design of pump has great realistic significance. Aiming at the centrifugal pump in common use, the study discusses the diagnosis requirement based on testability-orient failure mode and effect analysis (T-FMEA). Then design of the hardware of fault diagnosis system is presented. The software framework is also given in the paper. BP-nerve network is selected to diagnose the failures of pump. The validity and valid of the system is verified through experiment. It reveals that T-FMEA has a certain practical significance for the water pump fault diagnosis research, which can guide the design of fault diagnosis system.

1. Introduction

As the development of science and technology, the military equipment such as vessels and aircraft is becoming more complex, which consists of many mechanical and electronic components. Pump is the typical one with the features and characteristics of mechanical and electronic components. The improvement of fault diagnosis capability would enhance the optional availability and reliability of the whole equipment [1-2].

In the paper, the failure mechanisms and modes of the centrifugal pump are studied based on T-FMEA. Then the failure signals are presented. The fault diagnosis system is built according to centralized framework. Aiming at the signals and modes, the corresponding sensors are selected to monitor the condition of pump. The axis track sensor is newly developed to meet the monitor requirement. The software of the are developed based on labview and BP-nerve network algorithm. The validity and valid of the system is verified through experiment with failures injected. The consequence shows that the system can effectively diagnosis faults of pump and the valid can meet the engineering practical need.

2. Pump failure mechanisms, modes and signals anabasis based on T-FMECA

2.1. failure mechanism, mode and signal

2.1.1. Failure and fault. Failure refers to the event or inoperable state, in which any item or part of an item does not, or would not, perform as previously specified. Fault refers to immediate cause of failure, such as maladjustment, misalignment defect and etc. In this paper, failure equals to fault.

2.1.2. Failure mechanism. Failure mechanism refers to the fundamental factors/reasons that lead to the failures. The physical, chemical, electrical, thermal or other process results in failures in other words. The influence on failure mechanism can be heat, chemistry, motion and so on.

2.1.3 Failure mode. Failure mode refers to the consequence of the mechanism through which the failure occurs, such as short, open, fracture, excessive. When the pump is the objection for diagnosis, the failure mode should be focus on the next level.

2.1.4. Failure signal. Failure signal refers to the signals which caused by failure. They can be optics, heat, vibration or other forms. Not all of the signal can be observed by operator directly. Some would be measure by sensors. It is important for fault diagnosis to get the failure signals.

2.2. Testability-orient failure mode and effect analysis

2.2.1. Failure analysis. Failure analysis refers to subsequent to a failure, the logical systematic examination of an item, its construction, application, and documentation to identify the failure made and determine the failure mechanism and its basic course.

2.2.2. Failure mode and effect analysis. Failure mode and effect analysis refers to a procedure by which each potential failure mode in a system is .analyzed, to determine the results or effects thereof on the system and to classify each potential failure mode according to its severity. The formats of FMEA could vary according to the different targets of analysis, such as testability-orient FMEA, maintainability-orient FMEA.

2.2.3 Testability. Testability refers to the capability of detecting and locating faults for equipment. It usually used in electronic components. For mechanical and electronic components, testability design is usually equals to the progress of faults diagnosis. Testability is realized by built-in-test (BIT) and other support resources. However, faults diagnosis is realized by a system which consists of hardware and software.

2.2.4. T-FMEA. Testability-orient failure mode and effect analysis is used to determine the basic information and incomes for testability design and faults diagnosis. This method consists of six parts. 1) make out the breakdown of the objection, 2) determine the failure mode, mechanism and other basic information of each component, 3) determine the failure evolution feature, 4) determine the failure signal feature, 5) determine the test point or sensors for diagnosis, 6) determine the diagnosis method or technique of each failure mode. When implements T-FMEA, it should be noted that one failure mode may cause one more signals. So the sensors and diagnosis method or technique should be determined for every mode.

2.3. Pump T-FMECA

2.3.1. Pump introduction. The pump studied in the paper is a centrifugal pump. It consists of electrmot -or, axis, frame, substrate and pump, as shown in Figure 1.

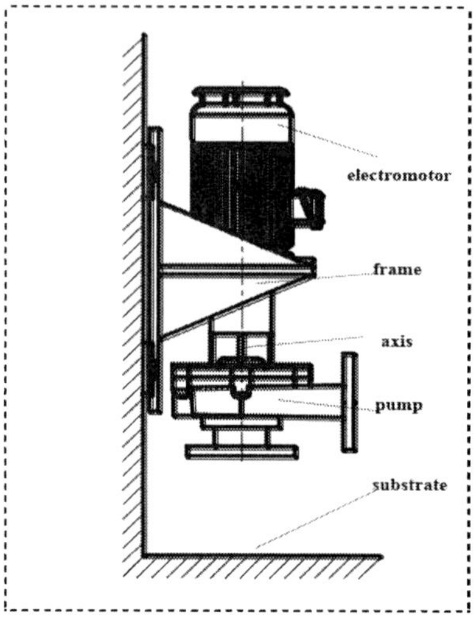

Figure 1. Centrifugal pump.

2.3.2 Pump T-FMECA. According to the steps mentioned in 2.2.4, the analysis of the centrifugal pump T-FMECA is conducted. The four failure modes of four component are studied. The frame and substrate is kinds of structure component, which is not complicated. Their capability can be ensued by lifetime design. The fault diagnosis analysis would be based on the four kinds of failure signals. All of the signals should be monitored by corresponding sensors. Some signal can determine the condition of the pump directly without further analysis. Whereas some other signal, such as vibration, have to be processed by algorithm and software to fault diagnosis [3-4]. Hardware design of pump fault diagnosis system

2.4. Framework of pump fault diagnosis system hardware

The common fault diagnosis system can be dived into three kinds, centralized framework, distributed framework. Centralized framework is simple and of high efficiency. The information flow among the framework is clear. However, it cannot handle a huge number of complex system monitor signals. So it applied to small and simple equipment. Distributed framework is usually used in giant equipment fault diagnosis [5]. The signal mentoring is realised on lower lever in this framework. The framework of pump fault diagnosis system is shown in figure 2.

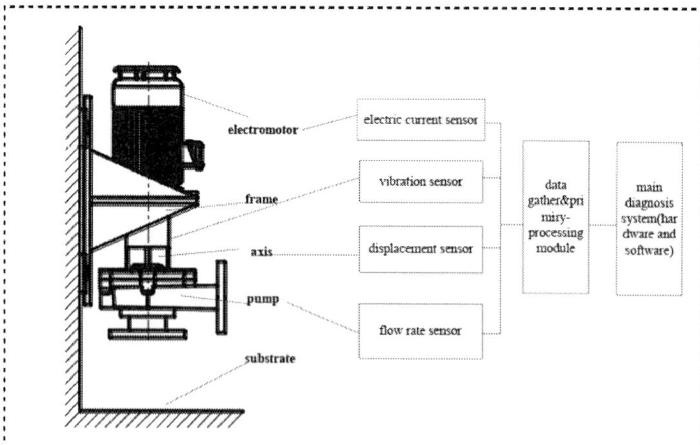

Figure 2. framework of pump fault diagnosis system.

2.5. Sensors of pump fault diagnosis system

2.5.1. Electric current sensor. The electric current se-nsor in the system is the Hall sensor. The working theory of the sensor is based on Hall effect. It can be fixed in the wires of electromotor, so the electric current can be measured directly.

2.5.2. Flow rate sensor. The flow rate sensor in the system is one kind of low rate sensor with type of LWGY, which is a kind of volute flow rate sensors. Volute flow rate sensors are delicacy and easily to maintenance. It would be fixed on the outcome pipe of the pump with the flanges.

2.5.3. Vibration sensor. The vibration sensor in the system is electric charge vibration sensor with Type PE. It can prevent the interference of circumstance. The sensor can be fixed on the shell of the pump with nuts. Three vibration sensors are needed to collect the vibration signals from X-Y-Z axises.

2.5.4. Displacement sensor. The displacement sensor in the system is permanent magnet perturbation sensor, which is newly developed. The traditional displacement sensor is based on the theory of eddy current, whose size is too big to fix in the pump. Whereas the permanent magnet perturbation sensor is much smaller and the measure accuracy is higher than the latter one.

2.6. Data gather and primary-processing module of pump fault diagnosis system

Due to the various failure signals and kinds of sensors in the pump fault diagnosis system, the signal types are not unique. So it is necessary to primary-process and digitalize for the data gather. The primary-processing consists of magnifying and filtering. The data gather and primary-processing module contain the four outcomings and inputs of four kinds of sensors. The inputs are linked by plugs. The signals enter the module and the main system through primary-processing.

2.7. Main fault diagnosis system

The failure signal would upload into the main system. They would be shown to the operator after analysed. The display pages contains original data module, time domain and frequency domain analysis module, feature value extract module and fault diagnosis module. All the functions are realised by software.

3. Software design of pump fault diagnosis system

The framework of pump fault diagnosis system software is shown in figure 3. The software would firstly recognize and distinguish the pump's condition based on the failure signals. If the signals are abnormal, the software should confirm and prompt remarks for repairing, maintenance or checking. If they are regular, then continue. The signals are not the original ones but the information after feature value extraction. The software in the study is developed by labview, as shown in figure 4.

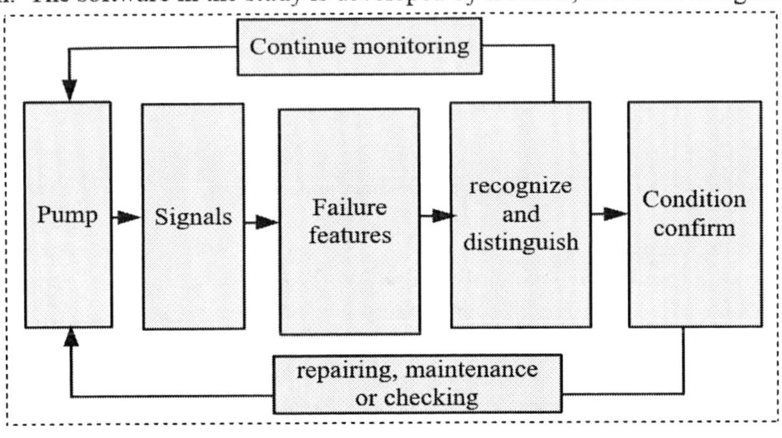

Figure 3. Framework of pump fault diagnosis system software.

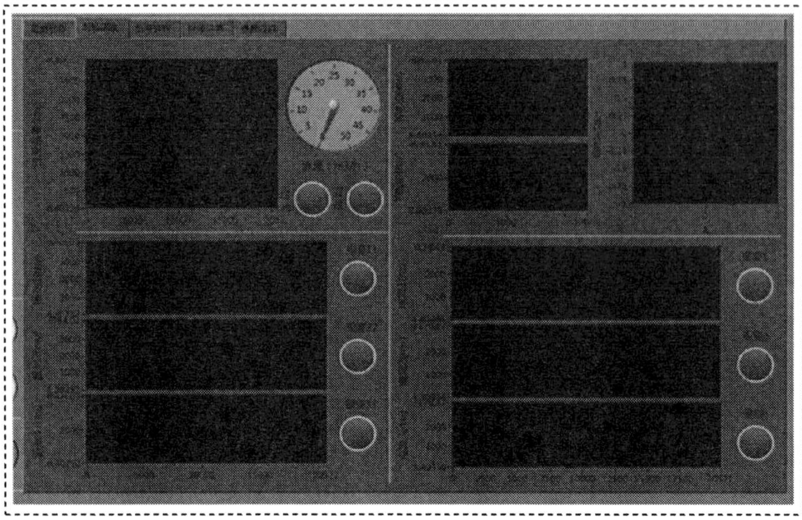

Figure 4. Display pages of system .

4. Verify of the pump fault diagnosis system

4.1. Experiment system
The experiment system is established as shown in figure.5. The 380V AC is linked with the electromotor though the wire. Then the electromotor controls the centrifugal pump in the system to realise the function of uptaking and draining away water. The entrance and exit of the pump is linked with water tank by pipes and flanges. The flow rate sensor is fixed on the exit pipe. Other sensors are fixed on the pump. The pump is fixed on the iron frame with bolts.

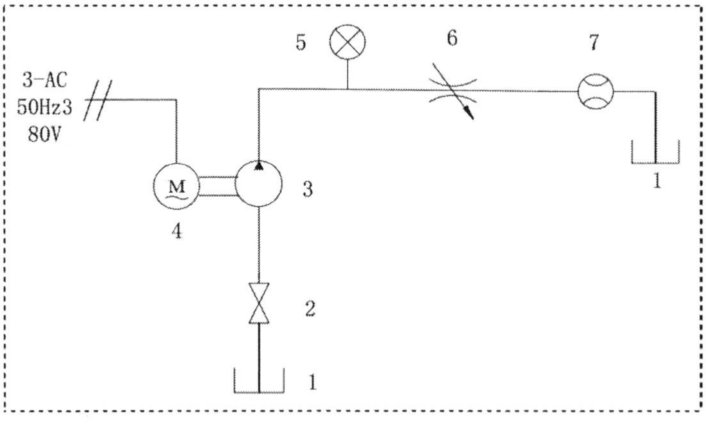

Figure 5. Experiment system.

4.2. Experiment data analysis
Due to the limit of the length of the paper, the vibration data collected in the experiment is analysed as an example. The following six valve after processing is used to train the BP-nerve network to conduct the bearing fault diagnosis, which is maximum(mm, P1), minimum(mm, P2), mean value(mm, P3), center frequency(Hz, P4), number of feature frequency(integer, P5) and various mean value(mm, P6).The modes in bearing fault diagnosis contains inner ring fault, outer ring fault and rolling fault.

The failure data will be processed by the well-trained BP-never network. It shows that the fault detecting rate can reach the goal of 0.95, which reveals that the pump fault diagnosis system can accurately detect and locate the fault of pump.

5. Conclusion

This study presented a method of failure mechanisms, modes and signal analysis based on T-FMEA. The fault diagnosis techniques are also given by T-FMEA aiming at the different modes. The fault diagnosis system is built according to the consequence of T-FMEA. The corresponding sensors are selected to monitor the condition of pump. Then the fault diagnosis system is built. The validity and valid of the system is verified through experiment. It reveals that T-FMEA has a certain practical significance for the water pump fault diagnosis research, which can guide the design of fault diagnosis system.

References

[1] L. T. Dvornikov, V. A. Turov.Estimating the reliability level of hydraulic systems of drilling installations at the time of design and development [J].Soviet Mining Science. 1987.6 : 472-477.
[2] P. Athanasatos, D. Koulocheris. Experimental Verification of Fault Predictions in High Pressure Hydraulic Systems.Modern Mechanical Engineering [J]. 2014.04:67-83.
[3] Irem Y. Tumer, Kemper Lewis. A robust system reliability analysis using partitioning and parallel processing of Markov chain[J]. Intelligence for Engineering Design, Analysis and Manufacturing.2014.04:311-322.
[4] NouraRezikaBellahseneHatem, Mohamed Mostefai. New Approach to the Diagnosis of Control Hydraulic Systems[J]. Computational Water , Energy and Environmental Engineering.2013.04: 97-102.
[5] Ante Bukša, IvicaŠegulja. Ship system reliability[J]. Scientific Journal of Maritime Research 2008.8 .

IOP Conf. Series: Journal of Physics: Conf. Series **1060** (2018) 012094 doi:10.1088/1742-6596/1060/1/012094

Research of Anti-destroy Optimization in Combat System Complex Network Based on Fractal Theory

Deng Qing[1], Xue Qing[1], Chen Lin[1] and Yu Ping-Gang[2]

[1] Army Armored Force Institute, Beijing 100072, China;
[2] National Defence University of PLA, Beijing 100091, China

154247597@qq.com

Abstract: Studying anti-destroy forecast of combat system complex network, which is of great resisting damage and improving survival ability. Optimizing the network anti-destroy, it may change the system into network composed of nodes and routes, measure network connection with intercommunication and the shortest path length to optimize anti-destroy property. Traditional methods improve network by increasing or decreasing side and get less center node and more collect ratio, but ignore command grade of combat system and affect forecast result. The article proposes a method for optimizing the network anti-destroy based on fractal theory. The network is constructed by case calculation means of fractal dimensionality. According to the connect way of relevance hybrid dimensionality and plus-minus related, the thesis realized fractal optimization of existing system network based on isolating colony node and extending path. By simulation test, the method can effectively improve network anti-destroy property.

1. Introduction

With the information technology continuous development and extensive use in the military field, the composition of combat systems has become increasingly complex, interrelated and influenced each other. The combat effectiveness has increasingly depended on the overall combat capability formed by close cooperation with each other. In actual war, the combat system is under the enemy's random and selective attack. It is of great military value enhancing the combat capability of the combat system to grasp the initiative on the battlefield and win the war. Existing research shows that combat systems reflect the small-world, scale-free features of complex networks [1]. Therefore, how to optimize the anti-destroy of the combat systems complex network and find out the key node or connection has become an important issue to be solved [2, 3].

At present, the main method of research on anti-destroy in combat system complex network is mainly based on the study of complex network invulnerability. Also it has obtained some research findings. For example, in literature [4], the problem of network anti-destroy is converted into seepage of a generalized random graph. Using the more general origin function method, it random gave the critical value of removal ratio under failure condition. There is only one selective attack type in the network failure mode of this method, which is not completely consistent with reality. In literature [5], tabu search algorithm is used to study the invulnerability optimization of complex networks with arbitrary degree distribution. In the process of invulnerability optimization, the network efficiency is taken as the objective function, and the mean value of the network degree distribution is used as cost constraint to construct optimization model with optimal network topological structure. The algorithm can search for the optimal network structure with given cost constraints in the topological network space of the current network and the global coupling network, but there may be a problem of getting into a local optimum.

In view of the existing problems above, a method based on fractal theory is proposed to optimize the invulnerability of the combat system complex network on the basis of summarizing relevant literatures. The experimental results show that the proposed method can effectively improve the destruction resistance.

2. Fractal Characteristic of Combat System Complex Network

In the process of optimizing the invulnerability of combat system complex network, the combat system is transformed into a complex network according to complex network theory. Evaluation index of damage resistance is selected to analyze the invulnerability of different degree distribution networks against different strike modes, which can realize the invulnerability optimization of the combat system complex network. The specific process is as follows:

In the military system, due to the demand of operational command and control, combat system formed a top-down and hierarchical structure[6-7]. This hierarchical network can be seamlessly integrated into a scaleless topology. It has a fixed module structure and the power degree index is as follows:

$$\gamma = 1 + \frac{\ln 4}{\ln 3} = 2.26 \tag{1}$$

The most important feature of above modular level network is agglomeration factor:

$$C(k) \approx k^{-1} \tag{2}$$

The curve of function C(k) is a straight line with a slope of -1 in the double logarithmic coordinate system. The hierarchical network structure means that sparsely connected nodes are part of a highly connected clustered area and communicate connections among highly connected clusters, which are composed of cluster nodes. The self-similar feature presented by the combat system complex network lays the foundation for the application of the invulnerability of the combat system complex network.

Based on above characteristics, a combat system complex network model is constructed. For any combat system G, the nodes of system are composed of targets, reconnaissance, accusations, and blows. The interconnected edges are consisted of single combat system network information flow and energy flow against enemy, with directed edges. These nodes and edges form a complex network model of combat system. Its formal representation is:

$$G=(P,L) \tag{3}$$

$$P \in [P_t \cup P_o \cup P_c \cup P_f] \tag{4}$$

$$L \in [L_i \cup L_e] \tag{5}$$

In the formula, $P_t\, P_o\, P_c\, P_f$ respectively represents the target, reconnaissance, accusation and attack node. $L_i\, L_e$ represents single communication network information flow and energy flow against enemy.

Anti-destroy of complex network refers to the ability of the network to maintain its functions under the condition that nodes or edges in the network are naturally defeated or deliberately attacked. In the combat system complex network, this ability maintaining its function is mainly reflected in the ability of the topology structure to hold connectivity. Therefore, the anti-destroy of a complex network is mainly due to the redundancy of alternative approaches between nodes, including concepts of efficacy and vulnerability. For the efficiency of the network, the network's average shortest path CPL, diameter D, clustering coefficient C are used to measure effectiveness of the network.

$$CPL = \frac{1}{N(N+1)/2} \sum_{i \geq j} d_{ij} \tag{6}$$

N is the total number of nodes, and d_{ij} is the shortest path between nodes i and j.

$$D = \max_{i,j} d_{ij} \tag{7}$$

$$C = \frac{1}{n}\sum_{i=1}^{n}\frac{E_i}{k_i(k_i - 1)/2} \tag{8}$$

In the formula, n is the total number of nodes, k_i is the neighbor node of node i, and E_i is the number of edges actually existing among ki neighbor nodes.

The vulnerability of network is related to decrease of network performance after some nodes or edges in the network are attacked. The critical value f_c of the node removal ratio of network collapse is used to calculate the network vulnerability.

$$f_c = 1 - \frac{1}{\kappa_0^{\,2}/\kappa_0 - 1} \tag{9}$$

k_0 is the degree distribution of the initial network.

3. Anti-destroy Optimization of Combat System Complexity Network Based on Fractal Theory

3.1 Algorithm Principle

In the process of the anti-destroy optimization of the combat system complex network, the important concept that characterizes the fractal characteristics of the combat system complex network is the fractal dimension dB, and the box counting method[8] is used to calculate the fractal dimension. The specific steps are detailed as follows:

First, all the network nodes are covered with a box of size S_B. The information such as Euclidean distance between network nodes is not considered here, and only the shortest path length between abstract extracted nodes is considered. All nodes covered by a box are connected to each other with a minimum distance smaller than the box scale S_B. Then a node is used to represent a box in the newly constructed network, which can obtain the minimum number of boxes $N(S_B)$ to cover the network..

According to existing research, the actual complex network (in Figure 1) is calculated using the box counting method:

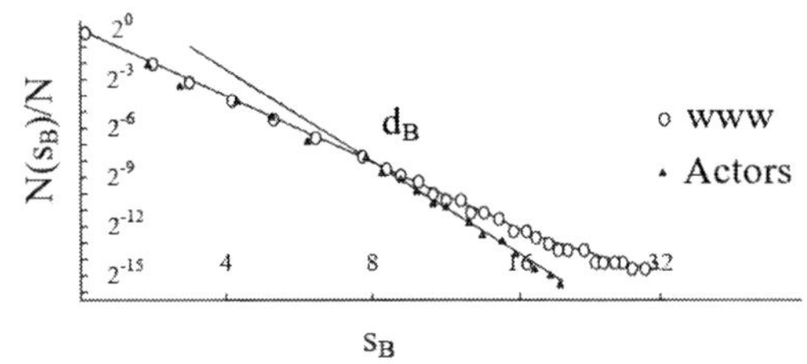

Figure 1. *N*(*SB*) Logarithm of network.

$$\frac{N(s_B)}{N} \approx s_B^{\,-d_B} \tag{10}$$

Transforming different box sizes, renormalize the network and give new connection probability distributions. Obviously, because there is no change in the complex network, only the method of representation is different, the scale of its degree distribution remains invariant.

$$\begin{cases} P(k) \approx k^{-r} \\ P(k') \approx (k')^{-r} \end{cases} \tag{11}$$

From Figure 1, we can see that in the real-world network, there is a relationship between scale factor S_B and d_k:

$$I(s_B) \approx s_B^{\,-d_k} \tag{12}$$

From the renormalization process of www network as shown in Figure 2, there is a relationship between the degree k' of the nodes in the network after renormalization and the maximum number k of connections with each box:

$$k' = I(s_B)k \tag{13}$$

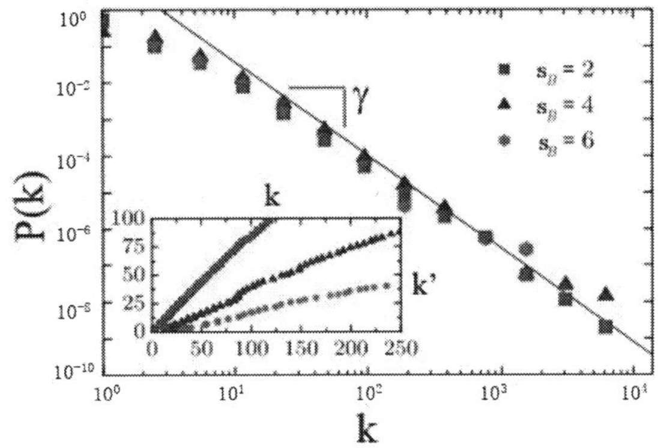

Figure 2. P(k) Logarithm of network.

Suppose n(k) denote the number of nodes with a degree k in the network before renormalization.

$$n(k) = NP(k) \tag{14}$$

$n'(k)$ denotes the number of boxes with moderate k' in the network after renormalization.

$$n'(k) = N'P(k') \tag{15}$$

Since the network was renormalized, its structure did not change.

$$n(k)dk = n'(k')dk' \tag{16}$$

Substitute formula (12)

$$n(k)dk = n'(k')dk' = n'(Ik)d(Ik) \tag{17}$$

Substitute formula (14)

$$NP(k)dk = N'P(Ik)d(Ik) \tag{18}$$

Substitute formula (10)

$$N = N'I^{1-r} \tag{19}$$

Substitute formula (2)

$$d_B = (r-1)d_k \tag{20}$$

From formula (20), if r>1, then $N(S_B)/N$ obeys power-law distribution, and the entire complex network presents fractal features; conversely, if r<1, $N(S_B)/N$ obeys exponential distribution, complex network It presents a non-fractal characteristic. this provides assurance for us to achieve the goal of improving the destruction resistance of combat system complex networks by changing the topological structure of the complex network.

3.2 Optimization Model

In the course of the invulnerability optimization of the complex network of combat systems, based on the fractal dimension, the network growth rules are determined according to the fractal theory and the growth equation is obtained. Then use the degree of node negative correlation to achieve network reconstruction and optimize the network anti-destroy. The detailed steps are described as follows:

The connection relationship between cluster nodes of a complex network determines that complex networks exhibit different characteristics. If the connection relationship between the cluster nodes is positively related, that means the cluster nodes in the network tend to connect with other cluster nodes, as a whole the network presents a small-world feature; conversely, if there is a negatively correlated connection relationship between the cluster nodes, the cluster nodes are more likely to connect to non-cluster nodes with less connectivity, the network exhibits self-similar fractal features. In other words, weak negative correlation or irrelevant growth mechanisms lead to the characteristics of non-fractal topologies[9]. It can be seen that any kind of complex network is determined to exhibit fractal characteristics as connection strength between cluster nodes.

Evolve the growth process by time and build an improved combat system model as follows:

1) Build initial network. Assuming that the initial network is a star network, the total number of node is n(0)=5, and the number of edge k(0)=4. Each initial node becomes a cluster node with time.

2) Network growth rules. Suppose every new time step, add hk(t-1) new nodes to nodes with degree k(t-1). Growth equation of the network can be expressed :

$$n(t) = n(t-1) + 2hk(t-1) \quad (21)$$

Without considering ring structure, then:

$$k(t-1) = n(t-1) - 1 \quad (22)$$

By equations (21), (22)

$$\lim_{t\to\infty} \frac{n(t)}{n(t-1)} = 2h+1 \quad (23)$$

It can be seen that the structure presented by the entire network has nothing to do with the initial structure of the network with the continuous growth of the network. Therefore, the initial structure of network can not determine the characteristics of the final network.

3) Reconnection rules of the network. There are two different network reconnection rules

Rule-1: any two boxes are directly connected to the cluster node by the probability p through the cluster nodes in the box. The network is reconnected according to the degree of positive correlation of the nodes;

Rule-2: Any two boxes are connected through a non-cluster node in the box with a probability of 1-p, that means they are reconnected to the network according to the negative correlation of nodes.

From the different reconnection principles above, it can be seen that Rule-1 will cause the network to exhibit high clustering and short-path characteristics of small-world features; on the other hand, Rule-2 leads to cluster nodes separation in the network. the path becomes longer, the network gradually presents a fractal feature. The simultaneous appearance of small-world features and fractal features in real networks is due to the different combinations of two different growth models. Figure 3 shows the network growth model under two different time evolution mechanisms.

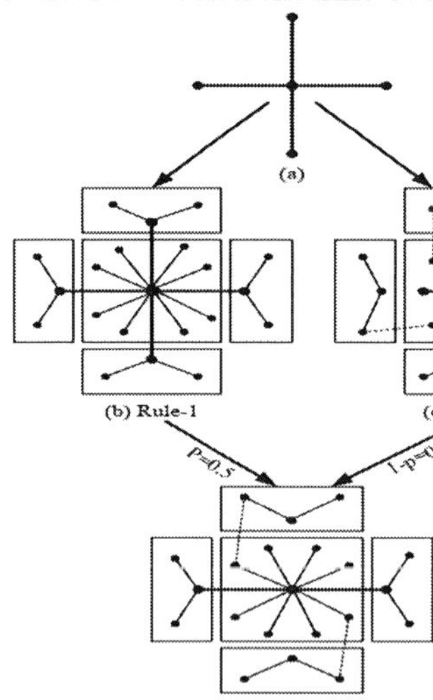

(a) Initial network, (b) Rule-1, (c) Rule-2, (d) Rule-1&Rule-2

Figure 3. Two different network growth models.

4. Simulation Test

In order to prove the effectiveness of the proposed method based on fractal theory, an experiment is needed. The experiment adopts the object-oriented programming method and uses VC++ development environment and Matlab simulation tools to simulate. In the experiment, a complex network of two groups of combat systems is constructed. Assume that the reconnaissance nodes that make up each team are composed of reconnaissance and detection devices that are assigned to them, specifically two battlefield televisions, three-step tactical radars, three light armored reconnaissance vehicles, and two aircrafts. Unmanned reconnaissance aircraft, a total of 10 nodes; command and control unit consists of 4 battalion command vehicles and 1 regiment command post, a total of 5 alleged nodes; the strike unit consists of 4 tank companies, 8 machine companies, and 3 artillery units. Even composed, a total of 15 combat nodes; then the two missions cooperative combat system has a total of 20 reconnaissance nodes, 10 alleged nodes, 30 cracked nodes, and 30 target nodes.

The complex network model of the two groups of combat systems is reconnected based on fractal theory. The changes of anti-destroy index are shown in Table 1:

Table. 1 Comparison of anti-destroy index.

	Efficiency			Vulnerability
Anti-destroy	*CPL*	*D*	*C*	f_c
Before fractal optimization	2.238	6	0.67	0.022
After fractal optimization	2.243	7	0.56	0.082

From table 1, we can analyse the efficiency index of anti-destroy. The length of the average shortest path increases slightly after fractal optimization. The network diameter also increases from 6 to 7, which shows that the response speed of combat system complex networks has slightly decreased. Because the path passing among nodes becomes long after the cluster node is reconnected.

From the vulnerability index of anti-destroy, it is difficult to compare the differences between the two networks because of small scale. So we magnify ten times, using a focused attack method which is based on the degree of the cluster node and remove node one by one. In this case, we can find the rule of the maximum cluster C, average value <c>[10] and removal ratio f. The results are shown in figure 4 and 5.

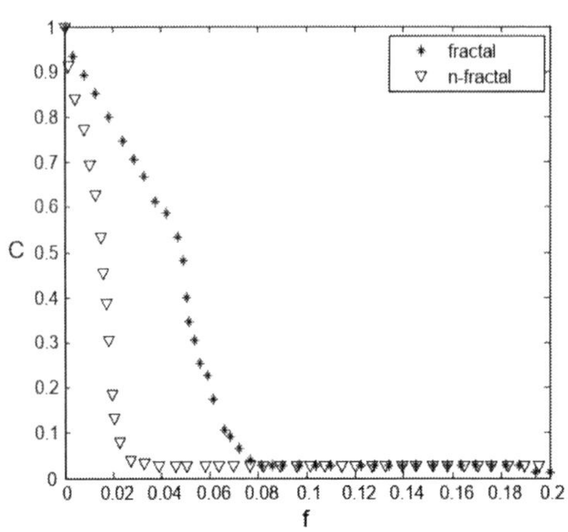

Figure 4. Curve of maximum cluster C with removal ratio f.

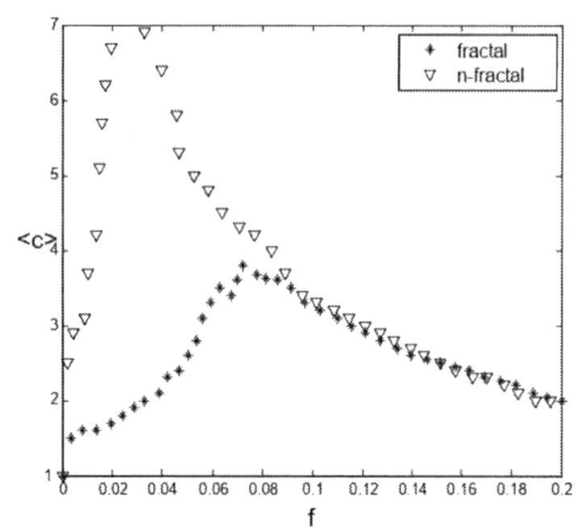

Figure 5. Curve of average value <c> with removal ratio f.

From Figures 4 and 5, it can be seen that when the proportion of cluster nodes are removed from the network, f = 0.022, the current maximum cluster C of the complex network with non-fractal combat systems has fallen to zero, while the complex network with fractal combat systems, f=0.082, it drops to 0; similarly, when the proportion of cluster nodes are removed from the network, f=0.022, average value <c> of the complex network with non-fractal combat systems has risen to near the peak value. However, the fractal battle system complex network rose to its peak value about f=0.082. we can draw a conclusion that the fractal combat system complex network has significantly stronger destruction resistance than other battle system complex networks under the same condition and.

5. Conclusions

Based on the fact of the self-similar fractal feature in the actual complex network and the characteristics of hierarchical networks in the combat system structure, this paper proposes an anti-destroy optimization method for combat system complex network based on fractal theory. Through simulation test and insurability comparison, it is proved that the optimized combat network has a high resistance to destruction as a whole.

References

[1] David.A.Jarvis. a methodology for analyzing complex military command and control(C2) networks[R]. in the 10th International Command and Control Research and Technology Symposium. 2005. U.S.A.
[2] Wang B, T.H.W., Guo C.H, etc, Entropy Optimization of Scale-free Networks Robustness to Random Failures[J]. Physica A, 2005. 363: p. 591-596.
[3] Wang Bing, T.H., etc, Optimization of network structure to random failures[J]. Physica A, 2006. 368: p. 607-614.
[4] Wang Bing, T.H., etc, Optimization of network structure to random failures[J]. Physica A, 2006. 368: p. 607-614.
[5] Callaway D S,Newman M E J,Strogatz S H,et al. Network robustness and fragility:percolation on random graphs[J]. Phys.Rev.Lett.,2000,85(25):5468-5471.
[6] Han Zhu, Jianyang Zhu. Network architecture [J]. Physics, 2003. 32(6): p. 364-369.
[7] Ravasz E, B.A.L., Hierarchical organization in complex networks[J]. Physical.Review.Lett, 2003. 67: p. 026112.
[8] Chaoming Song, S.H., Hernan A, Makse Self-similarity of complex networks[J]. Nature 2005. 433: p. 392-395.
[9] Chaoming Song, S.H., Hernan A, Makse. Origins of fractality in the growth of complex networks[J]. Nature 2008. 433: p. 382-386.
[10] Albert R, J.H.B.A.L., Attack and error tolerance in complex networks[J]. Nature, 2000. 406: p. 378-382.